Using Multivariate Statistics

Using Multivariate Statistics

Third Edition

Barbara G. Tabachnick
Linda S. Fidell

California State University, Northridge

HarperCollins*College*Publishers

Acquisitions Editor: Catherine Woods
Project Coordination: Interactive Composition Corporation
Text and Cover Design: John Massey
Cover Photograph: Rick Strange / The Picture Cube
Art Studio: Interactive Composition Corporation
Electronic Production Manager: Eric Jorgensen
Manufacturing Manager: Hilda Koparanian
Electronic Page Makeup: Interactive Composition Corporation
Printer and Binder: RR Donnelley & Sons Company
Cover Printer: New England Book Components, Inc.

Using Multivariate Statistics, Third Edition

Library of Congress Cataloging in Publication Data

Tabachnick, Barbara G.
    Using multivariate statistics / Barbara G. Tabachnick, Linda S. Fidell.—3rd ed.
        p. cm.
    Includes bibliographical references (p. 851–860) and index.
    ISBN 0-673-99414-7
    1. Multivariate analysis. I. Fidell, Linda S. II. Title.
QA278.T3 1996
519.5'35—dc20   95-16499
                                                                                CIP

96 97 98 9 8 7 6 5 4 3

To Friendship

Brief Contents

Contents

CHAPTER 2   A Guide to Statistical Techniques: Using the Book   19

CHAPTER 6 Canonical Correlation 195

CHAPTER 7   Multiway Frequency Analysis   239

CHAPTER 9   Multivariate Analysis of Variance and Covariance   375

CHAPTER 10 Profile Analysis of Repeated Measures 441

CHAPTER 12   Logistic Regression   575

CHAPTER 13 Principal Components and Factor Analysis 635

CHAPTER 14 Structural Equation Modeling   709
by Jodie B. Ullman

Preface

Every five or six years, or so, we seem to be overcome with a desire to tinker with this book. Part of the goal is to introduce new material. A second goal is to update the output and discussion of the computer packages. Third is to correct problems that have been brought to our attention by our students and colleagues. This edition introduces new chapters on logistic regression and structural equation modeling, upgrades all output to 1995 versions of the programs, and is responsive, we hope, to the many helpful comments of our colleagues.

Logistic regression (Chapter 12) is an alternative to discriminant function analysis that has become quite popular in the medical sciences. Because the analysis provides data on relative risk and is nearly assumption-free, it should also appeal to social scientists. Structural equation modeling (Chapter 14) is an ever more popular technique that allows the researcher to examine statistical linkages among observed and latent variables. With development of more user-friendly computer programs for this analysis, it is now accessible to many of us.

However, structural equation modeling is so slippery that we were reluctant to write the chapter ourselves. We are all fortunate that a former student of ours, Jodie Ullman, was willing to author this chapter. Jodie is currently completing a Ph.D. with Dr. Peter Bentler at UCLA after finishing a Master's degree at CSUN. She has both theoretical and hands-on experience with structural equation modeling, and is familiar with the organization of this book. She wrote the chapter, which was reviewed by Dr. Bentler (Thank you) and then edited by us, and returned to her for final run-through. If there are errors, they no doubt crept in during our editing.

Another goal of this edition is inclusion of some of the snazzy new features of the programs. Developments in understanding the implications of violation of assumptions are often reflected in new tests of assumptions available in SPSS, BMDP, SAS, and SYSTAT. Some of the many new developments in multiple regression (Chapter 5) are included, along with updates in data screening in Chapter 4 and elsewhere. There have been minor wording changes throughout based on confusion we created among our students with the second edition. Some of the examples are now from other disciplines in acknowledgement of the use of this book by colleagues from other areas. Most of our small sample examples are still silly in the belief that one should derive pleasure from mastery of this material whenever possible.

We have found, over the years, that there are two distinct skills involved in statistics: knowledge of the mathematical underpinnings of a statistical technique and practical knowledge of the benefits and limitations of application of the technique to a data set. Although almost everyone understands the importance of the math, we have found that it alone is not enough to insure wisdom in analysis of data. This book is an attempt to foster understanding of what can be learned through correct practical application of statistical techniques to data.

On another front, we are aware of the continuing controversy over the statistical hypothesis test (e.g., Cohen, 1994) and a bit uncertain about the long term outcome. Thus our firm decision was to straddle the fence (and hope to increase sales) by continuing to include both the hypothesis test and the estimation of effect size (we still usually call it strength of association) for every analysis. Those of you who feel that the hypothesis test is misguided can ignore it in favor of reporting effect size, although in a complicated multivariate study this might prove cumbersome.

For those of you who would like to play with the setups and data for the large sample examples in your own computing facilities, we have developed an MS-DOS diskette of ASCII files. It is included in the back of your text.

We are indebted to numerous individuals who have shared their time and expertise to help us prepare this edition. Along with Jodie Ullman from CSUN and UCLA and Peter Bentler from UCLA, we want to thank John Bartko from NIMH, Dave DeYoung from California Department of Motor Vehicles, Joan Foster from Simon Fraser University, Gregg Gold from CSUN and UCLA, Michael Gold from CSUN and UCLA, Samuel Kotz from the University of Maryland, Dave Nichols from SPSS, Belasz Schreil from CSUN, John Vokey from the University of Lethbridge, and, belatedly, Keith Widaman from the University of California at Riverside. We also would like to thank the following reviewers: Dr. L. E. Banderet, Quinsigamond Community College; Dr. Dale Berger, Claremont Graduate School; Dr. Dennis Doverspike, University of Akron; Dr. Jim Fleming, California State University, Northridge; Dr. Lisa Harlow, University of Rhode Island; Dr. Jennifer Rose, Indiana University, Bloomington; and Dr. L. James Tromater, University of Richmond. All of these people have done their level best to help us make this book as clear and accurate as possible. If it isn't, the fault is certainly ours.

Barbara G. Tabachnick
Linda S. Fidell

CHAPTER 1
Introduction

1.1 ■ MULTIVARIATE STATISTICS: WHY?

Multivariate statistics are increasingly popular techniques used for analyzing complicated data sets. They provide analysis when there are many independent variables (IVs) and/or many dependent variables (DVs), all correlated with one another to varying degrees. Because of the difficulty of addressing complicated research questions with univariate analyses and because of the availability of canned software for performing multivariate analyses, multivariate statistics are becoming widely used. Indeed, the day is near when a standard univariate statistics course ill prepares a student to read research literature or a researcher to produce it.

But how much harder are the multivariate techniques? Compared with the multivariate methods, univariate statistics are so straightforward and neatly structured that it is hard to believe they once took so much effort to master. Yet many researchers apply and correctly interpret results of intricate analysis of variance before the grand structure is apparent to them. The same can be true of multivariate statistics. While we are delighted if you gain insights into the full multivariate general linear model,[1] we have accomplished our goal if you feel comfortable selecting and setting up multivariate analyses and interpreting the computer output.

Multivariate methods are more complex than univariate by at least an order of magnitude. But for the most part the greater complexity requires few conceptual leaps. Familiar concepts such as sampling distributions and homogeneity of variance simply become more elaborate.

Multivariate models have not gained popularity by accident—or even by sinister design. Their growing popularity parallels the greater complexity of contemporary research. In psychology, for example, we are less and less enamored of the simple, clean, laboratory study where pliant, first-year college students each provide us with a single behavioral measure on cue.

---

[1]Chapter 15 is an attempt to foster such insights.

1.1.1 ■ The Domain of Multivariate Statistics: Numbers of IVs and DVs

Multivariate statistics are an extension of univariate and bivariate statistics. Multivariate statistics are the "complete" or general case, while univariate and bivariate statistics are special cases of the multivariate model. If your design has many variables, multivariate techniques often let you perform a single analysis instead of a series of univariate or bivariate analyses.

Variables are roughly dichotomized into two major types—independent and dependent. Independent variables (IVs) are the differing conditions (treatment vs. placebo) to which you expose your subjects, or characteristics (tall or short) the subjects themselves bring into the research situation. IVs are usually considered either predictor or causal variables because they predict or cause the DVs—the response or outcome variables. Note that IV and DV are defined within a research context; a DV in one research setting may be an IV in another.

Additional terms for IVs and DVs are predictor-criterion, stimulus-response, task-performance, or simply input-output. We use IV and DV throughout this book to identify variables that belong on one side of an equation or the other, without causal implication. That is, the terms are used for convenience rather than to indicate that one of the variables caused or determined the size of the other.

The term "univariate statistics" refers to analyses in which there is a single DV. There may, however, be more than one IV. For example, the amount of social behavior of graduate students (the DV) is studied as a function of course load (one IV) and type of training in social skills to which students are exposed (another IV). Analysis of variance is a commonly used univariate statistic.

"Bivariate statistics" frequently refers to analysis of two variables where neither is an experimental IV and the desire is simply to study the relationship between the variables (e.g., the relationship between income and amount of education). Bivariate statistics can, of course, be applied in an experimental setting, but usually they are not. Prototypical examples of bivariate statistics are the Pearson product-moment correlation coefficient and chi square analysis. (Ch. 3 reviews univariate and bivariate statistics.)

With multivariate statistics, you simultaneously analyze multiple dependent and multiple independent variables. This capability is important in both nonexperimental (correlational or survey) and experimental research.

1.1.2 ■ Experimental and Nonexperimental Research

A critical distinction between experimental and nonexperimental research is whether the researcher manipulates the levels of the IVs. In an experiment, the researcher has control over the levels (or conditions) of at least one IV to which a subject is exposed. Further, the experimenter randomly assigns subjects to levels of the IV, and controls all other influential factors by holding them constant, counterbalancing, or randomizing their influence. Scores on the DV are expected to be the same, within random variation, except for the influence of the IV (Campbell & Stanley, 1966). If there are systematic differences in the DV associated with levels of the IV, these differences are attributed to the IV.

For example, if groups of undergraduates are randomly assigned to the same material but different types of teaching techniques, and afterward some groups of undergraduates perform better than others, the difference in performance is said, with some degree of confidence, to be caused by the difference in teaching technique. In this type of research the terms independent and dependent have obvious meaning: the value of the DV depends on the manipulated level of the IV. The IV is manipulated by the experimenter and the score on the DV depends on the level of the IV.

In nonexperimental (correlational or survey) research, the levels of the IV(s) are not manipulated by the researcher. The researcher can define the IV, but has no control over the assignment of subjects to levels of it. For example, groups of people may be categorized into geographic area of residence (Northeast, Midwest, etc.), but only the definition of the variable is under researcher control. Except for the military or prison, place of residence is rarely subject to manipulation by a researcher. Nevertheless, a naturally occurring difference like this is often considered an IV and is used to predict some other nonexperimental (dependent) variable such as income. In this type of research the distinction between IVs and DVs is usually arbitrary and many researchers prefer to call IVs predictors and DVs criterion variables.

In nonexperimental research, it is very difficult to attribute causality to an IV. If there is a systematic difference in a DV associated with levels of an IV, the two variables are said (with some confidence) to be related, but the cause of the relationship is unclear. For example, income is related to geographic area, but no causal association is implied.

The distinctions not relevant to choice of *statistical* technique is whether the data are experimental. Statistics "work" whether or not the researcher manipulated the IV. The choice is not crucially affected by the experimental-nonexperimental

Multivariate Statistics in Nonexperimental Research

In nonexperimental research, a common example is the survey. Typically, many people are surveyed, and each provides answers to many questions, producing a large number of variables. These variables are usually interrelated in highly complex ways, univariate or bivariate statistics are not sensitive to this complexity. Bivariate correlations between all pairs of variables, for example, could reveal that the 20 to 25 variables measured really represent...

If differences are sought among subgroups in a sample (e.g., between Catholics and Protestants) on a set of attitudinal variables, we could use several univariate *t* tests (or analyses of variance) to examine group differences on each variable separately. But if the variables are related, which is highly likely, the results of many *t* tests are misleading and statistically suspect.

With the use of multivariate statistical techniques, complex interrelationships among variables are revealed and assessed in statistical inference. Further, it is possible to keep the overall Type I error rate at, say, 5%, no matter how many variables are tested.

1.1.2.2 ■ Multivariate Statistics in Experimental Research

Although multivariate techniques were developed for use in nonexperimental research, they are also useful in experimental research where there may be multiple IVs and multiple DVs. With multiple IVs, the research is usually designed so that the IVs are independent of each other and a straightforward correction for numerous statistical tests is available (see Ch. 3). With multiple DVs, a problem of inflated error rate arises if each DV is tested separately. Further, at least some of the DVs are likely to be correlated with each other, so separate tests of each DV reanalyze some of the same variance. Therefore, multivariate tests are used.

Experimental research designs with multiple DVs were unusual at one time. Now, however, with attempts to make experimental designs more realistic, and with the availability of computer programs, experiments often have several DVs. It is dangerous to run an experiment with only one DV and risk missing the impact of the IV because the most sensitive DV is not measured. Multivariate statistics help the experimenter design more efficient and more realistic experiments by allowing measurement of multiple DVs without violation of acceptable levels of Type I error.

1.1.3 ■ Computers and Multivariate Statistics

One answer to the question "Why multivariate statistics?" is that the techniques are now accessible by computer. Only the most dedicated number cruncher would consider doing real-life sized problems in multivariate statistics without a computer. Fortunately, excellent multivariate programs are available in a number of computer packages.

In this book, four packages are demonstrated. Examples are based on programs in SPSS (Statistical Package for the Social Sciences) (Norušis, 1990; SPSS Inc., 1992), BMDP Statistical Software (Dixon, 1992), SAS (SAS Institute Inc., 1990), and SYSTAT (Wilkinson & Hill, 1994; Steinberg & Colla, 1991). Each package contains a variety of specialized programs for both univariate and multivariate statistics. The manuals are excellent guides to all these packages and additional documentation appears in various newsletters, pocket guides, technical reports, and other materials that are printed as the programs are continuously corrected, revised, and upgraded.

These four packages are selected because of their popularity and ready availability at most large- and medium-sized computing facilities. If you are affiliated with a university or with a research organization that has access to a computer network or to well-equipped microcomputers, it is likely that several packages are available to you.

If you have access to all four packages, you are indeed fortunate. Programs within the packages do not completely overlap, and some problems are better handled through one package than another. For example, doing several versions of the same basic analysis on the same set of data is particularly easy with SPSS. BMDP, on the other hand, is well designed to do preliminary analysis of data and to test assumptions. SAS has the most extensive capabilities for saving derived scores from data screening or from intermediate analyses. And SYSTAT is good for fast, interactive work.

These four packages are available for mainframe computers and IBM-compatible microcomputers. SYSTAT and SPSS are available in full for Macintosh computers. Microcomputer versions tend to have more "user-friendly" procedures for entering data and instructions but setup and output are very similar, if not identical, to those of the mainframe versions of the programs. All of the MS-DOS versions offer both batch processing and at least some level of menu structure to supplement or replace batch commands. Although the examples in this book are based on batch processing from a command file, the output is the same if the same instructions are entered through a menu.

Recent advances include IBM-compatible Windows versions of all the packages. Some are based on DOS versions of the packages (e.g., SYSTAT) while others are complete rewrites (e.g., BMDP). Companion workbooks to this book translate examples to Windows versions, where applicable.

Documentation for the packages has changed considerably over the years. All started out with a single manual and perhaps a supplement or two. Now most of the manuals have several volumes and there are extensive publications lists, with a choice among several partially overlapping reference guides, user's manuals, and the like.

The packages differ in ease of use, with SPSS among the easiest. In SPSS, all programs use the same basic setup language. Once you learn the language, it is easy to describe the data set, and you need to refer to the manual only to specify the desired statistical analysis. The original SPSS manual (Nie et al., 1975) was better than the more recent one because the statistical techniques associated with the various programs were fully discussed. The current base manual (SPSS Inc., 1992) is simply a reference guide to the most fundamental programs, and does not include the multivariate techniques. However, the advanced and professional statistics guides (e.g., Norušis, 1990) present useful discussion and examples for several multivariate procedures. Several different forms of each type of analysis are illustrated, showing setup, language and annotated segments of computer output. Discussion of the output in the guide tends to be minimal but the output itself is beautifully formatted and easy to follow; everything is nicely labeled and can be found without much difficulty. However, the manuals do not provide details of the statistical algorithms; if a technique has several alternative algorithms, the reader must refer to the supplementary manual (SPSS Inc., 1991) to find out which one is used.

SPSS has had several major releases. The original package (Nie et al., 1975) was designed primarily for survey-type research and did not include programs for multivariate analysis of variance or repeated-measures analysis of variance. This lack was remedied in the SPSS 7–9 update (Hull & Nie, 1981), which included both SPSS MANOVA and an expanded program for multiple regression (SPSS NEW REGRESSION). These new programs and upgraded versions of many old programs were consolidated into SPSS[x] and also are now available in SPSS (SPSS, Inc., 1992; Norušis, 1990) which also features simplified yet flexible control language. Most computing facilities have implemented SPSS Version 5 (or SPSS/PC+) but there are a few computers on which only the older programs such as SPSS[x] can be used.[2]

In BMDP, some, but not all, conventions for setup are standard for all the programs. The two-volume manual includes helpful discussion and many examples of setup and annotated output; the index improves with each update. The output itself is well formatted and labeled, and there are numerous explanatory comments for statistics that are uncommon. For regular users, each chapter concludes with a checklist containing an abbreviated form of information critical for setup; this material is also available in the BMDP User's Digest (BMDP Statistical Software, Inc., 1992). The BMDP programs are statistically quite sophisticated but present the information at a level that most researchers can handle. The package once was more comprehensive than SPSS, but with additional programs regularly being added to the SPSS package, the gap is diminishing.

SAS is really two sets of programs in one. The basic programs provide for extensive data manipulation and some simple statistics, mostly descriptive. The statistics programs include a comprehensive, sophisticated assortment of techniques. The two-volume manual has examples of input and annotated output for many variations on the basic statistical techniques gathered together at the end of each chapter. Output, however, is more difficult to interpret than that of the other packages because labeling is often esoteric and unfamiliar unless your statistical training was in a mathematics/statistics department. SAS output lacks the helpful comments of BMDP and the pleasing visual format of both SPSS and BMDP.

SYSTAT is less comprehensive than the three other packages but is the easiest to use on microcomputers. The program now comes with three manuals: a quick reference guide, advanced applications, and the basic user's manual. The manuals are written in an informal style and are full of examples of input and output. The authors of these manuals have strong opinions on many of the controversial issues in statistics, and reading the manuals is pleasurable as well as educational. Among the four packages, these are the only manuals you would consider curling up with in front of a warm fire.

---

[2]Setup differs somewhat between SPSS[x] and SPSS/PC+. Version 5 of SPSS recognizes both setups. The setups in this book are compatible with SPSS Version 5 and SPSS/PC+, but not SPSS[x].

Chapters 5 through 14 of this book (the chapters that cover the specialized multivariate techniques) offer explanations and illustrations of a variety of programs within each package and a comparison of the features of the programs. We hope that once you understand the techniques, you will be able to generalize to virtually any multivariate program, and will not be limited to one program or only these four packages.

1.1.3.1 ■ Program Updates

With commercial computer packages, you need to know which version of the package you are using. Programs are continually being changed, and not all changes are immediately implemented at each facility. Therefore, many versions of the various programs are simultaneously in use at different institutions; even at one institution, more than one version of a package is sometimes available.

Program updates are often corrections of errors discovered in earlier versions. Occasionally, though, there are major revisions in one or more programs or a new program is added to the package. Sometimes defaults change with updates, so that output looks different although setups are the same. Check with your computing facility to find out which version of each package is in use. Then be sure that the manual you are using is consistent with the version in use at your facility. Also check updates for error correction in previous releases that may be relevant to some of your previous runs.

Except where noted, this book reviews SPSS Version 5, the 1992 update of BMDP (Release 7.0), SAS Version 6, and SYSTAT 6.

1.1.3.2 ■ Garbage In, Roses Out?

The trick in multivariate statistics is not in computation; that is easily done by computer. The trick is to select reliable and valid measurements, choose the appropriate program, use it correctly, and know how to interpret the output. Output from commercial computer programs, with their beautifully formatted tables, graphs, and matrices, can make garbage look like roses. Throughout this book we try to suggest clues that reveal when the true message in the output more closely resembles the fertilizer than the flowers.

When you use multivariate statistics, you rarely get as close to the raw data as you do when you apply univariate statistics to a relatively few cases. Errors and anomalies in the data that would be obvious if the data were processed by hand are less easy to spot when processing is entirely by computer. But the computer packages have programs to graph and describe your data in the simplest univariate terms and to display bivariate relationships among your variables. As discussed in Ch. 4, these programs provide preliminary analyses that are absolutely necessary if the results of multivariate programs are to be believed.

1.1.4 ■ Why Not?

There is, as usual, "no free lunch." Certain costs are associated with the benefits of using multivariate procedures. Benefits of increased flexibility in research design, for instance, are sometimes paralleled by increased ambiguity in interpretation of results. In addition, multivariate results can be quite sensitive to which analytic strategy is chosen (cf. Section 1.2.5) and do not always provide better protection against statistical errors than their univariate counterparts. Add to this the fact that occasionally you still can't get a firm statistical answer to your research questions, and you may wonder if the increase in complexity and difficulty is warranted.

Frankly, we think it is. Slippery as some of the concepts and procedures are, these statistics provide insights into relationships among variables that may more closely resemble the complexity of the "real" world. And sometimes you get at least partial answers to questions that couldn't be asked at all in the univariate framework. For a complete analysis, making sense of your data usually requires a judicious mix of multivariate and univariate statistics.

And the addition of multivariate statistics to your repertoire makes data analysis a lot more fun. If you liked univariate statistics, you'll love multivariate statistics![3]

1.2 ■ SOME USEFUL DEFINITIONS

In order to describe multivariate statistics easily, it is useful to review some common terms in research design and basic statistics. Distinctions were made in preceding sections between IVs and DVs and between experimental and nonexperimental research. Additional terms are described in this section.

1.2.1 ■ Continuous, Discrete, and Dichotomous Data

In applying statistics of any sort, it is important to consider the type of measurement and the nature of the correspondence between numbers and the events that they represent. The distinction made here is between continuous, discrete, and dichotomous variables; you may prefer to substitute the terms "interval" or "quantitative" for continuous and "nominal," "categorical," or "qualitative" for dichotomous and discrete.

Continuous variables are measured on a scale that changes values smoothly rather than in steps. Continuous variables take on any value within the range of the scale, and the size of the number reflects the amount of the variable. Precision is limited by the measuring instrument, not by the nature of the scale itself. Some examples of continuous variables are time as measured on an old-fashioned analog clock face, annual income, age, temperature, distance, and grade point average (GPA).

Discrete variables take on a finite and usually small number of values and there is no smooth transition from one value or category to the next. Examples include time as displayed by a digital clock, continents, categories of religious affiliation, and type of community (rural or urban).

Sometimes discrete variables are used in multivariate analyses in place of continuous ones if there are numerous categories and the categories represent a quantitative attribute. For instance, a variable that represents age categories (where, say, 1 stands for 0 to 4 years, 2 stands for 5 to 9 years, 3 stands for 10 to 14 years, and so on up through the normal age span) can be used because there are a lot of categories and the numbers designate a quantitative attribute (increasing age). But if the same numbers are used to designate categories of religious affiliation, they are not in appropriate form for analysis with many of the techniques[4] because religions do not fall along a quantitative continuum.

Discrete variables composed of qualitatively different categories are sometimes analyzed after being changed into a number of dichotomous or two-level variables (e.g., Catholic vs. non-Catholic,

---

[3]Don't even think about it.

[4]See, however, the chapters on multiway frequency analysis and logistic regression where discrete and dichotomous variables are welcome.

Protestant vs. non-Protestant, Jewish vs. non-Jewish, and so on until the degrees of freedom are used). Recategorization of a discrete variable into a series of dichotomous ones is called dummy variable coding. The conversion of a discrete variable into a series of dichotomous ones is done to limit the relationship between the dichotomous variables and others to linear relationships. A discrete variable with more than two categories can have a relationship of any shape with another variable, and the relationship is changed arbitrarily if assignment of numbers to categories is changed. Dichotomous variables, however, with only two points, can have only linear relationships with other variables; they are, therefore, appropriately analyzed by methods using correlation where only linear relationships are analyzed.

The distinction between continuous and discrete variables is not always clear. If you add enough digits to the digital clock, for instance, it becomes for all practical purposes a continuous measuring device, while time as measured by the analog device can also be read in discrete categories such as hours or half hours. In fact, any continuous measurement may be rendered discrete (or dichotomous) by specifying cutoffs on the continuous scale.

The property of variables that is crucial to application of multivariate procedures is not type of measurement so much as shape of distribution, as discussed in Ch. 4 and elsewhere. Nonnormally distributed continuous variables and dichotomous variables with very uneven splits between the categories present problems to several of the multivariate analyses. This issue and its resolution are discussed at some length in Ch. 4.

Another type of measurement that is used sometimes produces a rank order (ordinal) scale. This scale assigns a number to each subject to indicate the subject's position vis-à-vis other subjects along some dimension. For instance, ranks are assigned to contestants (first place, second place, third place, etc.) to provide an indication of who was best—but not by how much. A problem with ordinal measures is that their distributions are rectangular (one frequency per number) instead of normal, unless tied ranks are permitted and they pile up in the middle of the distribution. However, as with discrete scales representing a quantitative dimension, rank order data can be analyzed through multivariate techniques if a reasonably large number of ranks is assigned (say 20 or more for each subsample to be analyzed) and if the variable has a linear relationship with other variables.

1.2.2 ■ Samples and Populations

Samples are measured in order to make generalizations about populations. Ideally, samples are selected, usually by some random process, so that they represent the population of interest. In real life, however, populations are frequently best defined in terms of samples, rather than vice versa; the population is the group from which you were able to randomly sample.

Sampling has somewhat different connotations in nonexperimental and experimental research. In nonexperimental research, you investigate relationships among variables in some predefined population. Typically you take elaborate precautions to ensure that you have achieved a representative sample of that population; you define your population, then do your best to randomly sample from it.[5]

In experimental research, you attempt to create different populations by treating subgroups from an originally homogeneous group differently. The sampling objective here is to assure that all subjects come from the same population before you treat them differently. Random sampling consists

---

[5]Strategies for random sampling are discussed in detail in such texts as Kish (1965) and Moser and Kalton (1972).

of randomly assigning subjects to treatment groups (levels of the IV) to ensure that, before differ-ential treatment, all subsamples come from the same population. Statistical tests provide evidence as to whether, after treatment, all samples still come from the same population. Generalizations about treatment effectiveness are made to the type of subjects who participated in the experiment.

1.2.3 ■ Descriptive and Inferential Statistics

Descriptive statistics describe samples of subjects in terms of variables or combinations of vari-ables. Inferential statistics test hypotheses about differences in populations on the basis of mea-surements made on samples of subjects. If reliable differences are found, descriptive statistics are then used to provide estimations of central tendency, and the like, in the population. Descriptive sta-tistics used in this way are called parameter estimates.

Use of inferential and descriptive statistics is rarely an either-or proposition. We are usually interested in both describing and making inferences about a data set. We describe the data, find reli-able differences or relationships, and estimate population values for the reliable findings. However, there are more restrictions on inference than there are on description. Many assumptions of multi-variate statistics are necessary only for inference. If simple description of the sample is the major goal, many assumptions are relaxed, as discussed in Chapters 5 through 14.

1.2.4 ■ Orthogonality

Orthogonality is perfect nonassociation between variables. If two variables are orthogonal, know-ing the value of one variable gives no clue as to the value of the other; the correlation between them is zero.

Orthogonality is often desirable in statistical applications. For instance, factorial designs for experiments are orthogonal when two or more IVs are completely crossed with equal sample sizes in each combination of levels. Except for use of a common error term, tests of hypotheses about main effects and interactions are independent of each other; the outcome of each test gives no hint as to the outcome of the others. In orthogonal experimental designs with random assignment of sub-jects, manipulation of the levels of the IV, and good controls, changes in value of the DV can be unambiguously attributed to various main effects and interactions.

Similarly, in multivariate analyses, there are advantages if sets of IVs or DVs are orthogonal. If all pairs of IVs in a set are orthogonal, each IV adds, in a simple fashion, to prediction of the DV. Consider income as a DV with education and occupational prestige as IVs. If education and occu-pational prestige are orthogonal, and if 35% of the variability in income may be predicted from edu-cation while a different 45% is predicted from occupational prestige, then 80% of the variance in income is predicted from education and occupational prestige together.

Orthogonality can easily be illustrated in Venn diagrams as in Figure 1.1. Venn diagrams repre-sent shared variance (or correlation) as overlapping areas between two (or more) circles. The total variance for income is one circle. The section with horizontal stripes represents the part of income predictable from education and the section with vertical stripes represents the part predictable from occupational prestige; the circle for education overlaps the circle for income 35% while the circle for occupational prestige overlaps 45%. Together, they account for 80% of the variability in income because education and occupational prestige are orthogonal and do not themselves overlap.

There are similar advantages if a set of DVs is orthogonal. The overall effect of an IV can be partitioned into effects on each DV in an additive fashion.

1.2.5 ■ Standard and Sequential Analyses

Usually, however, the variables are correlated with each other (nonorthogonal). IVs in nonexperimental designs are often correlated naturally; in experimental designs IVs become correlated when unequal numbers of subjects are measured in different cells of the design. DVs are usually correlated because individual differences among subjects tend to be consistent over many attributes.

When variables are correlated, they have shared or overlapping variance. In the example of Figure 1.2, education and occupational prestige correlate with each other. Although the independent contribution made by education is still 35% while that by occupational prestige is 45%, their joint contribution to prediction of income is not 80% but rather something smaller due to the overlapping area shown by the arrow in Figure 1.2(a). A major decision for the multivariate analyst is how to handle the variance that is predictable from more than one variable. Many multivariate techniques have at least two strategies for handling it; some have more.

In standard analysis, the overlapping variance contributes to the size of summary statistics of the overall relationship but is not assigned to either variable. Overlapping variance is disregarded in assessing the contribution of each variable to the solution. Figure 1.2(a) is a Venn diagram of a standard analysis where overlapping variance is shown as overlapping areas in circles; the unique contributions of X_1 and X_2 to prediction of Y are shown as horizontal and vertical areas, respectively, and the total relationship between Y and the combination of X_1 and X_2 is those two areas plus the area with the arrow. If X_1 is education and X_2 is occupational prestige, then in standard analysis, X_1 is "credited with" the area marked by the horizontal lines and X_2 by the area marked by vertical lines. Neither of the IVs is assigned the area designated with the arrow. When X_1 and X_2 substantially overlap each other, very little horizontal or vertical area may be left for either of them despite the fact that they are both related to Y. They have essentially knocked each other out of the solution.

Sequential analyses differ in that the researcher assigns priority for entry of variables into equations and the first one to enter is assigned both unique variance and any overlapping variance it has with other variables. Lower-priority variables then are assigned on entry their unique and any remaining overlapping variance. Figure 1.2(b) shows a sequential analysis for the same case as Figure 1.2(a) where X_1 (education) is given priority over X_2 (occupational prestige). The total variance explained is the same as in Figure 1.2(a) but the relative contributions of X_1 and X_2 have changed; education now shows a stronger relationship with income than in the standard analysis, while the relation between occupational prestige and income remains the same.

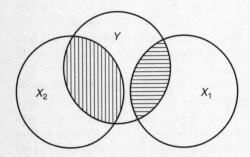

FIGURE 1.1   VENN DIAGRAM FOR Y
(INCOME), X_1 (EDUCATION), AND
X_2 (OCCUPATIONAL PRESTIGE).

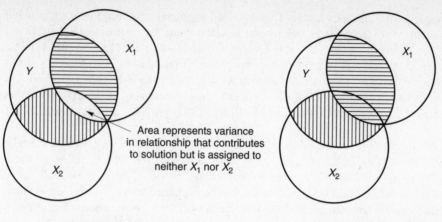

(a) Standard analysis

(b) Sequential analysis in which X_1 is given priority over X_2

FIGURE 1.2 STANDARD (a) AND SEQUENTIAL (b) ANALYSES OF THE RELATIONSHIP BETWEEN Y, X_1, AND X_2. HORIZONTAL SHADING DEPICTS VARIANCE ASSIGNED TO X_1. VERTICAL SHADING DEPICTS VARIANCE ASSIGNED TO X_2.

The choice of strategy for dealing with overlapping variance is not trivial. If variables are correlated, the overall relationship remains the same but the apparent importance of variables to the solution changes depending on whether a standard or a sequential strategy is used. If the multivariate procedures have a reputation for unreliability it is because solutions change, sometimes dramatically, when different strategies for entry of variables are chosen. However, the strategies also ask different questions of the data and it is incumbent on the researcher to determine exactly which question to ask. We try to make the choices clear in the chapters that follow.

1.3 ■ COMBINING VARIABLES

Multivariate analyses combine variables to do useful work such as predict scores or predict group membership. The combination that is formed depends on the relationships among the variables and the goals of analysis, but in most cases the combination that is formed is a linear combination.[6] A linear combination is one in which each variable is assigned a weight (e.g. W_1), and then products of weights and variable scores are summed to produce a score on a combined variable. In equation 1.1, Y' (the predicted DV) is predicted by a linear combination of X_1 and X_2 (the IVs).

$$Y' = W_1X_1 + W_2X_2 \tag{1.1}$$

If, for example, Y' is predicted income, X_1 is education, and X_2 is occupational prestige, the best prediction of income is obtained by weighting education (X_1) by W_1, and occupational prestige (X_2) by W_2 before summing. No other values of W_1 and W_2 produce as good a prediction of income.

---

[6]Except for Multiway Frequency Analysis and Logistic Regression.

Notice that Equation 1.1 includes neither X_1 or X_2 raised to powers (exponents) nor a product of X_1 and X_2. This seems to severely restrict multivariate solutions until one realizes that X_1 could itself be a product of two different variables or a single variable raised to a power. For example, X_1 might be education squared. A multivariate solution does not produce exponents or cross products of IVs to improve a solution, but the researcher can include Xs that are cross products of IVs or are IVs raised to powers. Inclusion of variables raised to powers or cross products of variables has both theoretical and practical implications for the solution. Berry (1993) provides a useful discussion of many of the issues.

The size of the W values (or some function of them) often reveals a great deal about the relationship between DV and IVs. If, for instance, the W value for some IV is zero, the IV is not needed in the best DV-IV relationship. Or if some IV has a large W value, then the IV tends to be important to the relationship. Although complications (to be explained later) prevent interpretation of the multivariate solution from the sizes of the W values alone, they are nonetheless important in most multivariate procedures.

The combination of variables can be considered a supervariable, not directly measured but worthy of interpretation. The supervariable may represent an underlying dimension that predicts something or optimizes some relationship. Therefore the attempt to understand the meaning of the combination of IVs is worthwhile in many multivariate analyses.

In the search for the best weights to apply in combining variables, computers do not try out all possible sets of weights. Various algorithms have been developed to compute the weights. Most algorithms involve manipulation of a correlation matrix, a variance-covariance matrix, or a sum-of-squares and cross-products matrix. Section 1.5 describes these matrices in very simple terms and shows their development from a very small data set. Appendix A describes some terms and manipulations appropriate to matrices. In Chapters 5 through 14, a small hypothetical sample of data is analyzed by hand to show how the weights are derived for each analysis. Though this information is useful for a basic understanding of multivariate statistics, it is not necessary for applying multivariate techniques fruitfully to your research questions.

1.4 ■ NUMBER AND NATURE OF VARIABLES TO INCLUDE

Attention to the number of variables included in analysis is important. A general rule is to get the best solution with the fewest variables. As more and more variables are included, the solution usually improves, but only slightly. Sometimes the improvement does not compensate for the cost in degrees of freedom of including more variables, so the power of the analyses diminishes. If there are too many variables relative to sample size, the solution provides a wonderful fit to the sample that may not generalize to the population, a condition known as overfitting. To avoid overfitting, include only a limited number of variables in each analysis.[7]

Considerations for inclusion of variables in a multivariate analysis include cost, availability, meaning, and theoretical relationships among the variables. Except in analysis of structure, one usually wants a small number of valid, cheaply obtained, easily available, uncorrelated variables that assess all the theoretically important dimensions of a research area. Another important consideration is reliability. How stable is the position of a given score in a distribution of scores when mea-

---

[7] The exceptions are analysis of structure, such as factor analysis, where numerous variables are measured.

sured at different times or in different ways? Unreliable variables degrade an analysis while reliable ones enhance it. A few reliable variables give a more meaningful solution than a large number of unreliable variables. Indeed, if variables are sufficiently unreliable, the entire solution may reflect only measurement error.

Further considerations for variable selection are mentioned as they apply to each analysis.

1.5 ■ DATA APPROPRIATE FOR MULTIVARIATE STATISTICS

For multivariate statistics, an appropriate data set consists of values on a number of variables for each of several subjects or cases. For continuous variables, the values are scores on variables. For example, if the continuous variable is the GRE, the values for the various subjects are scores such as 500, 650, 420, and so on. For discrete variables, values are number codes for group membership or treatment. For example, if there are three teaching techniques, students who receive one technique are arbitrarily assigned a "1," those receiving another technique are assigned a "2," and so on.

1.5.1 ■ The Data Matrix

The data matrix is an organization of scores in which rows (lines) represent subjects and columns represent variables. An example of a data matrix with six subjects[8] and four variables is in Table 1.1. For example, X_1 might be type of teaching technique, X_2 score on the GRE, X_3 GPA, and X_4 gender, with women coded 1 and men coded 2.

Data are entered into a data file with long-term storage accessible by computer in order to apply computer techniques to them. Each subject starts with a new row (line). Information identifying the subject is typically entered first, followed by the value of each variable for that subject. Scores for each variable are entered in the same order for each subject. If there are more data for each subject than can be accommodated on a single line, the data are continued on additional lines, but all of the data for each subject are kept together. All of the computer package manuals provide information on setting up a data matrix.

In this example, there are values for every variable for each subject. This is not always the case with research in the real world. With large numbers of subjects and variables, scores frequently are missing on some variables for some subjects. For instance, respondents may refuse to answer some kinds of questions, or some students may be absent the day that one of the tests is given, and so forth. This creates missing values in the data matrix. To deal with missing values, first build a data

TABLE 1.1 A DATA MATRIX OF HYPOTHETICAL SCORES

Student	X_1	X_2	X_3	X_4
1	1	500	3.20	1
2	1	420	2.50	2
3	2	650	3.90	1
4	2	550	3.50	2
5	3	480	3.30	1
6	3	600	3.25	2

[8]Normally, of course, there are many more than six subjects.

file where some symbol is used to indicate that a value on a variable is missing in data for a subject. The various programs have standard symbols, such as *, for this purpose. You can use other symbols, but it is often just as convenient to use one of the default symbols. Once the data set is available, consult Ch. 4 for various options to deal with this messy (but often unavoidable) problem.

1.5.2 ■ The Correlation Matrix

Most readers are familiar with **R,** a correlation matrix. **R** is a square, symmetrical matrix. Each row (and each column) represents a different variable, and the value at the intersection of each row and column is the correlation between the two variables. For instance, the value at the intersection of the second row, third column, is the correlation between the second and the third variables. The same correlation also appears at the intersection of the third row, second column. Thus correlation matrices are said to be symmetrical about the main diagonal, which means they are mirror images of themselves above and below the diagonal going from top left to bottom right. For this reason it is common practice to show only the bottom half or the top half of an **R** matrix. The entries in the main diagonal are often omitted as well, since they are all ones—correlations of variables with themselves.[9]

Table 1.2 shows the correlation matrix for X_2, X_3 and X_4 of Table 1.1. The .85 is the correlation between X_2 and X_3 and it appears twice in the matrix (as do others). Other correlations are as indicated in the table.

Many programs allow the researcher a choice between analysis of a correlation matrix and analysis of a variance-covariance matrix. If the correlation matrix is analyzed, a unit-free result is produced. That is, the solution reflects the relationships among the variables but not in the metric in which they are measured. If the metric of the scores is somewhat arbitrary, analysis of R is appropriate.

1.5.3 ■ The Variance-Covariance Matrix

If, on the other hand, scores are measured along a meaningful scale, it is sometimes appropriate to analyze a variance-covariance matrix. A variance-covariance matrix, Σ, is also square and symmetrical, but the elements in the main diagonal are the variances of each variable, while the off-diagonal elements are covariances between pairs of different variables.

Variances, as you recall, are averaged squared deviations of each score from the mean of the scores. Because the deviations are averaged, the number of scores included in computation of a

TABLE 1.2		CORRELATION MATRIX FOR PART OF HYPOTHETICAL DATA FOR TABLE 1.1		
		X_2	X_3	X_4
	X_2	1.00	.85	−.13
R =	X_3	.85	1.00	−.46
	X_4	−.13	−.46	1.00

[9] Alternatively, other information such as standard deviations are inserted.

variance is not relevant, but the metric in which the scores are measured is. Scores measured in large numbers tend to have largish numbers as variances; scores measured in small numbers tend to have small variances.

Covariances are averaged cross products (the deviation between one variable and its mean times the deviation between a second variable and its mean). Covariances are similar to correlations except that they, like variances, retain information concerning the scales in which the variables are measured. The variance-covariance matrix for the continuous data of Table 1.1 appears in Table 1.3

1.5.4 ■ The Sum-of-Squares and Cross-Products Matrix

This matrix, **S**, is a precursor to the variance-covariance matrix in which deviations are not yet averaged. Thus, the size of the entries depends on the number of cases as well as on the metric in which the elements were measured. The sum-of-squares and cross-products matrix for X_2, X_3, and X_4 in Table 1.1 appears in Table 1.4.

The entry in the major diagonal of the **S** matrix is the sum of squared deviations of scores from the mean for that variable, hence "sum of squares", or SS. That is, for each variable, the value in the major diagonal is

$$SS(X_j) = \sum_{i=1}^{N} (X_{ij} - \overline{X}_j)^2 \tag{1.2}$$

where $i = 1, 2, \ldots, N$. N is the number of subjects. j identifies the variable. X_{ij} is the score on variable j by subject i, and $\overline{X}_j$ is the mean of all scores on the jth variable.

For example, for X_4, the mean is 1.5. The sum of squared deviations around the mean and the diagonal value for the variable is

$$\sum_{i=1}^{6}(X_{i4} - \overline{X}_4) = (1 - 1.5)^2 + (2 - 1.5)^2 + (1 - 1.5)^2$$
$$+ (2 - 1.5)^2 + (1 - 1.5)^2 + (2 - 1.5)^2$$
$$= 1.50$$

The off-diagonal elements of the sum-of-squares and cross-products matrix are the cross products—the sum of products (SP)—of the variables. For each pair of variables, represented by row

TABLE 1.3 VARIANCE-COVARIANCE MATRIX FOR PART OF HYPOTHETICAL DATA OF TABLE 1.1

		X_2	X_3	X_4
	X_2	7026.66	32.80	-6.00
$\Sigma =$	X_3	32.80	.21	-.12
	X_4	-6.00	-.12	.30

TABLE 1.4 SUM-OF-SQUARES AND CROSS-PRODUCTS MATRIX FOR PART OF HYPOTHETICAL DATA OF TABLE 1.1

		X_2	X_3	X_4
	X_2	35133.33	164.00	-30.00
$S =$	X_3	164.00	1.05	-0.59
	X_4	-30.00	-0.59	1.50

and column labels in Table 1.4, the entry is the sum of the product of the deviation of one variable around its mean times the deviation of the other variable around its mean.

$$SP(X_j X_k) = \sum_{i=1}^{N} (X_{ij} - \overline{X}_j)(X_{ik} - \overline{X}_k) \tag{1.3}$$

j identifies the first variable, k identifies the second variable and all other terms are as defined in Equation 1.1. (Note that if $j = k$, Equation 1.3 becomes identical to Equation 1.2.)

For example, the cross-product term for variables X_2 and X_3 is

$$\sum_{i=1}^{6}(X_{i2} - \overline{X}_2)(X_{i3} - \overline{X}_3) = (500 - 533.33)(3.20 - 3.275)$$

$$+ (420 - 533.33)(2.50 - 3.275)$$

$$+ \cdots + (600 - 533.33)(3.25 - 3.275)$$

$$= 164.00$$

Most computations start with $\mathbf{S}$ and proceed to $\sum$ or $\mathbf{R}$. The progression from a sum-of-squares and cross-products matrix to a variance-covariance matrix is simple.

$$\sum = \frac{1}{N-1}\,\mathbf{S} \tag{1.4}$$

The variance-covariance matrix is produced by dividing every element in the sum-of-squares and cross-products matrix by $N - 1$, where N is the number of cases.

The correlation matrix is derived from an $\mathbf{S}$ matrix by dividing each sum of squares by itself (to produce the 1s in the main diagonal of $\mathbf{R}$) and each cross product of the $\mathbf{S}$ matrix by the square root of the product of the sum of squared deviations around the mean for each of the variables in the pair. That is, each cross product is divided by

$$Denominator(X_j X_k) = \sqrt{\sum_{i=1}^{N} (X_{ij} - \overline{X}_j)^2 \sum_{i=1}^{N} (X_{ik} - \overline{X}_k)^2} \tag{1.5}$$

where terms are defined as in Equation 1.3.

For some multivariate operations, it is not necessary to feed the data matrix to a computer program. Instead, an $\mathbf{S}$ or an $\mathbf{R}$ matrix is entered, with each row (representing a variable) starting a new line. Often considerable computing time and expense are saved by entering one or the other of these matrices rather than raw data.

1.5.5 ■ Residuals

Often a goal of analysis, or test of its efficiency, is its ability to reproduce the values of a DV or the correlation matrix of a set of variables. For example, we might want to predict scores on the GRE (X_2 of Table 1.1) from knowledge of GPA (X_3) and gender (X_4). After applying the proper statistical operations, a multiple regression in this case, a predicted GRE score for each student is computed by applying the proper weights for GPA and gender to the GPA and gender scores for each student. But because we already obtained GRE scores for our sample of students, we are able to compare the predicted with the obtained GRE score. The difference between the predicted and obtained values is known as the residual and is a measure of error of prediction.

In most analyses the residuals for the entire sample sum to zero. That is, sometimes the prediction is too large and sometimes it is too small, but the average of all the errors is zero. The squared value of the residuals, however, provides a measure of how good the prediction is. When the predictions are close to the obtained values, the squared errors are small. The way that the residuals are distributed is of further interest in evaluating the degree to which the data meet the assumptions of multivariate analyses, as discussed in Ch. 4 and elsewhere.

1.6 ■ ORGANIZATION OF THE BOOK

Chapter 2 gives a guide to the multivariate techniques that are covered in this book and places them in context with the more familiar univariate and bivariate statistics where possible. Included in Chapter 2 is a flow chart that organizes statistical techniques on the basis of the major research questions asked. Chapter 3 provides a brief review of univariate and bivariate statistics for those who are interested.

Chapter 4 deals with the assumptions and limitations of multivariate statistics. Assessment and violation of assumptions are discussed, along with alternatives for dealing with violations when they occur. Chapter 4 is meant to be referred to often, and the reader is guided back to it frequently in Chapters 5 through 14.

Chapters 5 through 14 cover specific multivariate techniques. They include descriptive, conceptual sections as well as a guided tour through a real-world data set for which the analysis is appropriate. The tour includes an example of a Results section describing the outcome of the statistical analysis appropriate for submission to a professional journal. Each technique chapter includes a comparison of computer programs available in the four packages described in Section 1.1.3. You may want to vary the order in which you cover these chapters.

Chapter 15 is an attempt to integrate univariate, bivariate, and multivariate statistics through the multivariate general linear model. The common elements underlying all the techniques are emphasized, rather than the differences among them. Chapter 15 is meant to pull together the material in the remainder of the book with a conceptual rather than pragmatic emphasis. Some may wish to consider this material earlier, for instance, immediately after Chapter 2.

CHAPTER 2

A Guide to Statistical Techniques: Using the Book

2.1 ■ RESEARCH QUESTIONS AND ASSOCIATED TECHNIQUES

This chapter organizes the statistical techniques in this book by major research question. A decision tree at the end of this chapter leads you to an appropriate analysis for your data. On the basis of your major research question and a few characteristics of your data set, you determine which statistical technique(s) is appropriate. The first, most important criterion for choosing a technique is the major research question to be answered by the statistical analysis. Research questions are categorized here into degree of relationship among variables, significance of group differences, prediction of group membership, and structure. This chapter emphasizes differences in research questions answered by the different techniques, while Chapter 15, the last chapter, provides an integrated overview of the techniques.[1]

2.1.1 ■ Degree of Relationship among Variables

If the major purpose of analysis is to assess the associations among two or more variables, some form of correlation/regression or chi square is appropriate. The choice among five different statistical techniques is made by determining the number of independent and dependent variables, the nature of the variables (continuous or discrete), and whether any of the IVs are best conceptualized as covariates.[2]

---

[1] You may find it helpful to read Chapter 15 soon instead of waiting for the end.

[2] If the effects of some IVs are assessed after the effects of other IVs are statistically removed, the latter are called covariates.

2.1.1.1 ■ Bivariate *r*

Bivariate correlation and regression, as reviewed in Chapter 3, assess the degree of relationship between two continuous variables such as belly dancing skill and years of musical training. Bivariate correlation measures the association between two variables with no distinction necessary between IV and DV. Bivariate regression, on the other hand, predicts a score on one variable from knowledge of the score on another variable (e.g., predicts skill in belly dancing from years of musical training). The predicted variable is considered the DV, while the predictor is considered the IV. Bivariate correlation and regression are not multivariate techniques but they are integrated into the general linear model in Chapter 15.

2.1.1.2 ■ Multiple *R*

Multiple correlation assesses the degree to which one continuous variable (the DV) is related to a set of other (usually) continuous variables (the IVs) that have been combined to create a new, composite variable. Multiple correlation is a bivariate correlation between the original DV and the composite variable created from the IVs. For example, how large is the association between belly dancing skill and a number of IVs such as years of musical training, body flexibility, and age?

Multiple regression is used to predict the score on the DV from scores on several IVs. Other examples are prediction of success in an educational program from scores on a number of aptitude tests, prediction of the sizes of earthquakes from a variety of geological and electromagnetic variables, or stock market behavior from a variety of political and economic variables.

In multiple correlation and regression the IVs may or may not be correlated with each other. With some ambiguity, the techniques also allow assessment of the relative contribution of each of the IVs toward predicting the DV, as discussed in Chapter 5.

2.1.1.3 ■ Sequential *R*

In sequential (sometimes called hierarchical) multiple regression, IVs are given priorities by the researcher before their contributions to prediction of the DV are assessed. For example, the researcher might first assess the effects of age and flexibility on belly dancing skill before looking at the contribution that years of musical training makes to that skill. Differences among dancers in age and flexibility are statistically "removed" before assessment of the effects of years of musical training.

In the example of an educational program, success of outcome might first be predicted from variables such as age and IQ. Then scores on various aptitude tests are added to see if prediction of outcome is enhanced after adjustment for age and IQ.

In general, then, the effects of IVs that enter first are assessed and removed before the effects of IVs that enter later are assessed. For each IV in a sequential multiple regression, higher-priority IVs act as covariates for lower-priority IVs.

The degree of relationship between the DV and the IVs is reassessed at each step of the hierarchy. That is, multiple correlation is recomputed as each new IV (or set of IVs) is added. Sequential multiple regression, then, is also useful for developing a reduced set of IVs (if that is desired) by determining when IVs no longer add to predictability. Sequential multiple regression is discussed in Chapter 5.

2.1.1.4 ■ Canonical *R*

In canonical correlation, there are several continuous DVs as well as several continuous IVs and the goal is to assess the relationship between the two sets of variables. For example, we might study the relationship between a number of indices of belly dancing skill (the DVs, such as knowledge of steps, ability to play finger cymbals, responsiveness to the music) and the IVs (flexibility, musical training, and age). Or we might ask whether there is a relationship between achievements in arithmetic, reading, and spelling as measured in elementary school and a set of variables reflecting early childhood development (e.g., age at first speech, walking, toilet training). Such research questions are answered by canonical correlation, the subject of Chapter 6.

2.1.1.5 ■ Multiway Frequency Analysis

A goal of multiway frequency analysis is to assess relationships among discrete variables where none is considered a DV. For example, you might be interested in the relationships among gender, occupational category, and preferred type of reading material. Or the research question might involve relationships among gender, categories of religious affiliation, and attitude toward abortion. Chapter 7 deals with multiway frequency analysis.

When one of the variables is considered a DV with the rest serving as IVs, multiway frequency analysis is called logit analysis, as described in Section 2.1.3.3.

2.1.2 ■ Significance of Group Differences

When subjects are randomly assigned to groups (treatments), the major research question usually is the extent to which reliable mean differences on DVs are associated with group membership. Once reliable differences are found, the researcher often assesses the degree of relationship (strength of association) between IVs and DVs.

The choice among techniques hinges on the number of IVs and DVs, and whether some variables are conceptualized as covariates. Further distinctions are made as to whether all DVs are measured on the same scale, and how within-subjects IVs are to be treated.

2.1.2.1 ■ One-Way ANOVA and *t* Test

These two statistics, reviewed in Chapter 3, are strictly univariate in nature and are adequately covered in most standard statistical texts.

2.1.2.2 ■ One-Way ANCOVA

One-way analysis of covariance is designed to assess group differences on a single DV after the effects of one or more covariates are statistically removed. Covariates are chosen because of their known association with the DV; otherwise there is no point to their use. For example, age and degree of reading disability are usually related to outcome of a program of educational therapy (the DV). If groups are formed by randomly assigning children to different types of educational therapy (the IV), it is useful to remove differences in age and degree of reading disability before examining the relationship between outcome and type of therapy. Prior differences among children in age and reading disability are used as covariates. The ANCOVA question is: Are there mean differences in outcome associated with type of educational therapy after adjusting for differences in age and degree of reading disability?

ANCOVA gives a more powerful look at the IV-DV relationship by minimizing error variance (cf. Chapter 3). The stronger the relationship between the DV and the covariate(s), the greater the power of ANCOVA over ANOVA. ANCOVA is discussed in Chapter 8.

ANCOVA is also used to adjust for differences among groups when groups are naturally occurring and random assignment to them is not possible. For example, one might ask if attitude toward abortion (the DV) varies as a function of religious affiliation. However, it is not possible to randomly assign people to religious affiliation. In this situation, there could easily be other systematic differences between groups, such as level of education, that are also related to attitude toward abortion. Apparent differences between religious groups might well be due to differences in education rather than differences in religious affiliation. To get a "purer" measure of the relationship between attitude and religious affiliation, attitude scores are first adjusted for educational differences, that is, education is used as a covariate. Chapter 8 also discusses this somewhat problematical use of ANCOVA.

When there are more than two groups, planned or post hoc comparisons are available in ANCOVA just as in ANOVA. With ANCOVA, selected and/or pooled group means are adjusted for differences on covariates before differences in means on the DV are assessed.

2.1.2.3 ■ Factorial ANOVA

Factorial ANOVA, reviewed in Chapter 3, is the subject of numerous statistics texts (e.g., Keppel, 1991; Myers & Well, 1991; Winer, 1971) and is introduced in most elementary texts. Although there is only one DV in factorial ANOVA, its place within the general linear model is discussed in Chapter 15.

2.1.2.4 ■ Factorial ANCOVA

Factorial ANCOVA differs from one-way ANCOVA only in that there is more than one IV. The desirability and use of covariates are the same. For instance, in the educational therapy example of Section 2.1.2.2, another interesting IV might be gender of the child. The effects of gender, type of educational therapy and their interaction on outcome are assessed after adjusting for age and prior degree of reading disability. The interaction of gender with type of therapy asks if boys and girls differ as to which type of educational therapy is more effective after adjustment for covariates.

2.1.2.5 ■ Hotelling's T^2

Hotelling's T^2 is used when the IV has only two groups and there are several DVs. For example, there might be two DVs, such as score on an academic achievement test and attention span in the classroom, and two levels of type of educational therapy, emphasis on perceptual training vs. emphasis on academic training. It is not legitimate to use separate t tests for each DV to look for differences between groups because that inflates Type I error due to unnecessary multiple significance tests with (likely) correlated DVs. Instead, Hotelling's T^2 is used to see if groups differ on the two DVs combined. The researcher asks if there are reliable differences in the centroids (average on the combined DVs) for the two groups.

Hotelling's T^2 is a special case of multivariate analysis of variance, just as the t test is a special case of univariate analysis of variance, when the IV has only two groups. Multivariate analysis of variance is discussed in Chapter 9.

2.1.2.6 ■ One-Way MANOVA

Multivariate analysis of variance evaluates differences among centroids for a set of DVs when there are two or more levels of an IV (groups). MANOVA is useful for the educational therapy example in the preceding section with two groups and also when there are more than two groups (e.g., if a nontreatment control group is added).

With more than two groups, planned and post hoc comparisons are available. For example, if a main effect of treatment is found in MANOVA, it might be interesting to ask post hoc if there are differences in the centroids of the two groups given different types of educational therapy, ignoring the control group, and, possibly, if the centroid of the control group differs from the centroid of the two educational therapy groups combined.

Any number of DVs may be used; the procedure deals with correlations among them and the entire analysis is accomplished within the preset level for Type I error. Once reliable differences are found, techniques are available to assess which DVs are influenced by which IV. For example, assignment to treatment group might affect the academic DV but not attention span.

MANOVA is also available when there are within-subject IVs. For example, children might be measured on both DVs three times: 3, 6, and 9 months after therapy begins.

MANOVA is discussed in Chapter 9 and a special case of it (profile analysis) in Chapter 10. Profile analysis is an alternative to one-way between-subjects MANOVA when the DVs are all measured on the same scale. Discriminant function analysis is also available for one-way between-subjects designs, as described in Section 2.1.3.1 and Chapter 11.

2.1.2.7 ■ One-Way MANCOVA

In addition to dealing with multiple DVs, multivariate analysis of variance can be applied to problems where there are one or more covariates. In this case, MANOVA becomes multivariate analysis of covariance—MANCOVA. In the educational therapy example of Section 2.1.2.6, it might be worthwhile to adjust the DV scores for pretreatment differences in academic achievement and attention span. Here the covariates are pretests of the DVs, a classic use of covariance analysis. After adjustment for pretreatment scores, differences in posttest scores (DVs) can more clearly be attributed to treatment (the two types of educational therapy plus control group that make up the IV).

In the one-way ANCOVA example of religious groups in Section 2.1.2.2, it might be interesting to test political liberalism vs. conservatism, and attitude toward ecology, as well as attitude toward abortion, to create three DVs. Here again, differences in attitudes might be associated with both differences in religion and differences in education (which, in turn, varies with religious affiliation). In the context of MANCOVA, education is the covariate, religious affiliation the IV, and attitudes the DVs. Differences in attitudes among groups with different religious affiliation are assessed after adjustment for differences in education.

If the IV has more than two levels, planned and post hoc comparisons are useful, with adjustment for covariates. MANCOVA (Chapter 9) is available for both the main analysis and comparisons.

2.1.2.8 ■ Factorial MANOVA

Factorial MANOVA is the extension of MANOVA to designs with more than one IV and multiple DVs. For example, gender (a between-subjects IV) might be added to type of educational therapy (another between-subjects IV) with both academic achievement and attention span used as DVs. In this case the analysis is a two-way between-subjects factorial MANOVA that provides tests of the main effects of gender and type of educational therapy and their interaction on the centroids of the DVs.

Duration of therapy (3, 6, and 9 months) might be added to the design as a within-subjects IV with type of educational therapy a between-subjects IV to examine the effects of duration, type of educational therapy, and their interaction on the DVs. In this case, the analysis is a factorial MANOVA with one between- and one within-subjects IV.

Comparisons can be made among margins or cells in the design, and the influence of various effects on combined or individual DVs can be assessed. For instance, the researcher might plan (or decide post hoc) to look for linear trends in scores associated with duration of therapy for each type of therapy separately (the cells) or across all types of therapy (the margins). The search for linear trend could be conducted among the combined DVs or separately for each DV with appropriate adjustments for Type I error rate.

Virtually any complex ANOVA design (cf. Chapter 3) with multiple DVs can be analyzed through MANOVA, given access to appropriate computer programs. Factorial MANOVA is covered in Chapter 9.

2.1.2.9 ■ Factorial MANCOVA

It is sometimes desirable to incorporate one or more covariates into a factorial MANOVA design to produce factorial MANCOVA. For example, pretest scores on academic achievement and attention span could serve as covariates for the two-way between-subjects design with gender and type of educational therapy serving as IVs and posttest scores on academic achievement and attention span serving as DVs. The two-way between-subjects MANCOVA provides tests of gender, type of educational therapy and their interaction on adjusted, combined centroids for the DVs.

Here again procedures are available for comparisons among groups or cells and for evaluating the influences of IVs and their interactions on the various DVs. Factorial MANCOVA is discussed in Chapter 9.

2.1.2.10 ■ Profile Analysis

A special form of MANOVA is available when all of the DVs are measured on the same scale (or on scales with the same psychometric properties) and you want to know if groups differ on the scales. For example, you might use the subscales of the Profile of Mood States as DVs to assess whether mood profiles differ between a group of belly dancers and a group of ballet dancers.

There are two ways to conceptualize this design. The first is as a one-way between-subjects design where the IV is the type of dancer and the DVs are the Mood States subscales; one-way MANOVA provides a test of the main effect of type of dancer on the combined DVs. The second way is as a profile study with one grouping variable (type of dancer) and the several subscales; profile analysis provides tests of the main effects of type of dancer and of subscales as well as their interaction (frequently the effect of greatest interest to the researcher).

If there is a grouping variable and a repeated measure such as trials where the same DV is measured several times, there are three ways to conceptualize the design. The first is as a one-way between-subjects design with several DVs (the score on each trial); MANOVA provides a test of the main effect of the grouping variable. The second is as a two-waybetween- and within-subjects design; ANOVA provides tests of groups, trials, and their interaction, but with some very restrictive assumptions that are likely to be violated. Third is as a profile study where profile analysis provides tests of the main effects of groups and trials and their interaction, but without the restrictive assumptions. This is sometimes called the "multivariate approach to repeated measures ANOVA."

Finally, you might have a between- and within-subjects design (groups and trials) where several DVs are measured on each trial. For example, you might assess groups of belly and ballet dancers on the Mood State subscales at various points in their training. This application of profile analysis is frequently referred to as doubly multivariate. Chapter 10 deals with all these forms of profile analysis.

2.1.3 ■ Prediction of Group Membership

In research where groups are identified, the emphasis is frequently on predicting group membership from a set of variables. Discriminant function analysis, logit analysis, and logistic regression are designed to accomplish this prediction. Discriminant function analysis tends to be used when all IVs are continuous and nicely distributed, logit analysis when IVs are all discrete, and logistic regression when IVs are a mix of continuous and discrete and/or poorly distributed.

2.1.3.1 ■ One-Way Discriminant Function

In one-way discriminant function analysis, the goal is to predict membership in groups (the DV) from a set of IVs. For example, the researcher might want to predict category of religious affiliation from attitude toward abortion, liberalism vs. conservatism, and attitude toward ecological issues. The analysis tells us if group membership is predicted reliably. Or, the researcher might try to discriminate belly dancers from ballet dancers from scores on Mood State subscales.

These are the same questions as those addressed by MANOVA, but turned around. Group membership serves as the IV in MANOVA and the DV in discriminant function analysis. If groups differ significantly on a set of variables in MANOVA, the set of variables reliably predicts group membership in discriminant function analysis. One-way between-subjects designs can be fruitfully analyzed through either procedure and are often best analyzed with a combination of both procedures.

As in MANOVA, there are techniques for assessing the contribution of various IVs to prediction of group membership. For example, the major source of discrimination among religious groups might be abortion attitude, with little predictability contributed by political and ecological attitudes.

In addition, discriminant function analysis offers classification procedures to evaluate how well individual cases are classified into their appropriate groups on the basis of their scores on the IVs. One-way discriminant function analysis is covered in Chapter 11.

2.1.3.2 ■ Sequential One-Way Discriminant Function

Sometimes IVs are assigned priorities by the researcher so their effectiveness as predictors of group membership is evaluated in the established order in sequential discriminant function analysis. For example, when attitudinal variables are predictors of religious affiliation, variables might be prioritized according to their expected contribution to prediction, with abortion attitude given highest priority, political liberalism vs. conservatism second priority, and ecological attitude lowest priority.

Sequential discriminant function analysis first assesses the degree to which religious affiliation is reliably predicted from abortion attitude. Gain in prediction is, then, assessed with addition of political attitude, and then with addition of ecological attitude.

Sequential analysis provides two types of useful information. First, it is helpful in eliminating predictors that do not contribute more than predictors already in the analysis. For example, if political and ecological attitudes do not add appreciably to abortion attitude in predicting religious affiliation, they can be dropped from further analysis. Second, sequential discriminant function analysis is a covariance analysis. At each step of the hierarchy, higher priority predictors are covariates for lower priority predictors. Thus, the analysis permits you to assess the contribution of a predictor with the influence of other predictors removed.

Sequential discriminant function analysis is also useful for evaluating sets of predictors. For example, if a set of continuous demographic variables is given higher priority than an attitudinal set in prediction of group membership, one can see if attitudes reliably add to prediction after adjustment for demographic differences. Sequential discriminant function analysis is discussed in Chapter 11. However, it is usually more efficient to answer such questions through sequential logistic regression, particularly when some of the predictor variables are continuous and others discrete (see Section 2.1.3.5).

2.1.3.3 ■ Multiway Frequency Analysis (Logit)

The logit form of multiway frequency analysis may be used to predict group membership when all of the predictors are discrete. For example, you might want to predict whether or not someone is a belly dancer (the DV) from knowledge of gender, occupational category, and preferred type of reading material (science fiction, romance, history, statistics).

This technique allows evaluation of the odds that a case is in one group (e.g., belly dancer), based on membership in various categories of predictors (e.g., female professors who read science fiction.) This form of multiway frequency analysis is discussed in Chapter 7.

2.1.3.4 ■ Logistic Regression

Logistic regression allows prediction of group membership when predictors are continuous, discrete, or a combination of the two. Thus, it is an alternative to both discriminant function analysis and logit analysis. For example, prediction of whether someone is a belly dancer may be based on gender, occupational category, preferred type of reading material, and age.

Logistic regression allows one to evaluate the odds (or probability) of membership in one of the groups (e.g., belly dancer) based on the combination of values of the predictor variables (e.g., 35-year-old female professors who read science fiction). Chapter 12 covers logistic regression analysis.

2.1.3.5 ■ Sequential Logistic Regression

As in sequential discriminant function analysis, sometimes predictors are assigned priorities and then assessed in terms of their contribution to prediction of group membership given their priority. For example, one can assess how well preferred type of reading material predicts whether someone is a belly dancer after adjusting for differences associated with age, gender, and occupational category. Sequential logistic regression is also covered in Chapter 12.

2.1.3.6 ■ Factorial Discriminant Function

If groups are formed on the basis of more than one attribute, prediction of group membership from a set of IVs can be performed through factorial discriminant function analysis. For example, respondents might be classified on the basis of both gender and religious affiliation. One could use attitudes toward abortion, politics, and ecology to predict gender (ignoring religion), or religion (ignoring gender), or both gender and religion. But this is the same problem as addressed by factorial MANOVA. For a number of reasons, programs designed for discriminant function analysis do not readily extend to factorial arrangements of groups. Unless some special conditions are met (cf. Chapter 11), it is usually better to rephrase the research question so that factorial MANOVA can be used.

2.1.3.7 ■ Sequential Factorial Discriminant Function

Difficulties inherent in factorial discriminant function analysis extend to sequential arrangements of predictors. Usually, however, questions of interest can readily be rephrased in terms of factorial MANCOVA.

2.1.4 ■ Structure

A final set of questions is concerned with the latent structure underlying a set of variables. Depending on whether the search for structure is empirical or theoretical, the choice is principal components, factor analysis, or structural equation modeling. Principal components is an empirical approach while factor analysis and structural equation modeling tend to be theoretical approaches.

2.1.4.1 ■ Principal Components

If scores on numerous variables are available from a group of subjects, the researcher might ask if and how the variables group together. Can the variables be combined into a smaller number of supervariables on which the subjects differ? For example, suppose people are asked to rate the effectiveness of numerous behaviors for coping with stress (e.g., "talking to a friend," "going to a movie," "jogging," "making lists of ways to solve the problem"). The numerous behaviors may represent just a few basic coping mechanisms, such as increasing or decreasing social contact, engaging in physical activity, and instrumental manipulation of stress producers.

Principal components analysis uses the correlations among the variables to develop a small set of components which empirically summarize the correlations among the variables. This analysis is discussed in Chapter 13.

2.1.4.2 ■ Factor Analysis

When there are hypotheses about underlying structure, or when the researcher wants to understand underlying structure, factor analysis is often used. In this case the researcher believes that responses to many different questions are driven by just a few underlying structures called factors. In the example of mechanisms for coping with stress, one might hypothesize ahead of time that there are two major factors: general approach to problems (escape vs. direct confrontation) and use of social supports (withdrawing from people vs. seeking them out).

It is sometimes useful to explore differences between groups in terms of latent structure. For example, young college students might use the two coping mechanisms hypothesized above, whereas older adults may have a substantially different factor structure for coping styles.

As implied in this discussion, factor analysis is useful in developing and testing theories. What is the structure of personality? Are there some basic dimensions of personality on which people differ? By collecting scores from many people on numerous variables that may reflect different aspects of personality, researchers address questions about underlying structure through factor analysis, as discussed in Chapter 13.

2.1.4.3 ■ Structural Equation Modeling

Structural equation modeling combines factor analysis, canonical correlation, and multiple regression. Like factor analysis, some of the variables can be latent, while others are directly observed. Like canonical correlation, there can be many IVs and many DVs. And like multiple regression, the goal may be prediction.

For example, one may want to predict birth outcome (the DVs) from several demographic, personality, and attitudinal measures (the IVs). The DVs are a mix of several observed variables such as birth weight, a latent assessment of mother's acceptance of the child based on several measured attitudes, and a latent assessment of infant responsiveness; the IVs are several demographic variables such as socioeconomic status, race, and income, several latent IVs based on personality measures, and prebirth attitudes toward parenting.

The technique evaluates whether the model provides a reasonable fit to the data, and the contribution of each of the IVs to the DVs. Comparisons among alternative models are also possible, as well as evaluation of differences between groups. Chapter 14 covers structural equation modeling.

2.2 ■ A DECISION TREE

A decision tree starting with major research questions appears in Table 2.1. For each question, choice among techniques depends on number of IVs and DVs (sometimes an arbitrary distinction) and whether some variables are usefully viewed as covariates. The table also briefly describes analytic goals associated with some techniques.

The paths in Table 2.1 are only recommendations concerning an analytic strategy. Researchers frequently discover that they need two or more of these procedures or, even more frequently, a judicious mix of univariate and multivariate procedures to answer fully their research questions. We recommend a flexible approach to data analysis where both univariate and multivariate procedures are used to clarify the results.

2.3 ■ TECHNIQUE CHAPTERS

Chapters 5 through 14, the basic technique chapters, follow a common format. First, the technique is described and the general purpose briefly discussed. Then the specific kinds of questions that can be answered through application of that technique are listed. Next, both the theoretical and practical limitations of the technique are discussed; this section lists assumptions particularly associated with the technique, describes methods for checking the assumptions for your data set, and gives suggestions for dealing with violations. Then a small hypothetical data set is used to illustrate the statistical development of the procedure. Simple analyses by programs from four computer packages follow.

The next section describes the major types of the technique, where appropriate. Then some of the most important issues to be considered when using the technique are covered, including special statistical tests, data snooping, and the like. A detailed comparison of features available in the SPSS, BMDP, SAS, and SYSTAT programs is then made.

The next section shows a step-by-step application of the technique to actual data gathered as described in Appendix B. Assumptions are tested and violations dealt with, where necessary. Major hypotheses are evaluated and follow-up analyses are performed as indicated. Then a Results section is developed, as might be appropriate for submission to a professional journal. When more than one major type of technique is available, there are additional complete examples using real data.

Finally, each technique chapter ends with a description of articles from the literature that use the technique. The only criteria for selection of examples were that they were recent (at the time of search) and covered a variety of fields. They are not meant to be taken as models, but rather as examples of the way techniques are currently being used.

In working with these technique chapters, it is suggested that the student/researcher apply the various analyses to some interesting large data set. Many data banks are readily accessible through computer installations.

Further, although we recommend methods of reporting multivariate results, it may be inappropriate to report them fully in all publications. Certainly one would at least want to mention that univariate results were supported and guided by multivariate inference. But the details associated with a full disclosure of multivariate results at a colloquium, for instance, might require more attention than one could reasonably expect from an audience. Likewise, a full multivariate analysis may be more than some journals are willing to print.

2.4 ■ PRELIMINARY CHECK OF THE DATA

Before applying any technique, or sometimes even before choosing a technique, you should determine the fit between your data and some very basic assumptions underlying most of the multivariate statistics. Though each technique has specific assumptions, as well, most require consideration of material in Chapter 4.

TABLE 2.1 CHOOSING AMONG STATISTICAL TECHNIQUES

Major research question	Number (kind) of dependent variables	Number (kind) of independent variables	Covariates	Analytic strategy	Goal of analysis
Degree of relationship among variables	One (continuous)	One (continuous)		Bivariate r	Create a linear combination of IVs to optimally predict DV
		Multiple (continuous)	None	Multiple R	
			Some	Sequential multiple R	
	Multiple (continuous)	Multiple (continuous)		Canonical R	Maximally correlate a linear combination of DVs with a linear combination of IVs
	None	Multiple (discrete)		Multiway frequency analysis	Create a log-linear combination of IVs to optimally predict category frequencies

TABLE 2.1 Cont.

Major research question	Number (kind) of dependent variables	Number (kind) of independent variables	Covariates	Analytic strategy	Goal of analysis
Significance of group differences	One (continuous)	One (discrete)	None	One-way ANOVA or t test	Determine reliability of mean group differences
			Some	One-way ANCOVA	
		Multiple (discrete)	None	Factorial ANOVA	
			Some	Factorial ANCOVA	
	Multiple (continuous)	One (discrete)	None	One-way MANOVA or Hotelling's T^2	Create a linear combination of DVs to maximize mean group differences
			Some	One-way MANCOVA	
		Multiple (discrete)	None	Factorial MANOVA	
			Some	Factorial MANCOVA	
	One (continuous)	Multiple (one discrete within-S)		Profile analysis of repeated measures	Create linear combinations of DVs to maximize mean group differences and differences between levels of within-subjects IVs
	Multiple (continuous/ commensurate)	One (discrete)		Profile analysis	
	Multiple (continuous)	Multiple (one discrete within-S)		Doubly multivariate profile analysis	

TABLE 2.1 Cont.

Major research question	Number and nature of DVs	Number and nature of IVs	Covariates	Analytic strategy	Goal of analysis
Prediction of group membership	One (discrete)	Multiple (continuous)	None	One-way discriminant function	Create a linear combination of IVs to maximize group differences
			Some	Sequential one-way discriminant function	
		Multiple (discrete)		Multiway frequency analysis (logit)	Create a loglinear combination of IVs to optimally predict DV
		Multiple (continuous and/or discrete)	None	Logistic regression	Create a linear combination of the log of the odds of being in one group
			Some	Sequential logistic regression	
	Multiple (discrete)	Multiple (continuous)	None	Factorial discriminant function	Create a linear combination of IVs to maximize group differences (DVs)
			Some	Sequential factorial discriminant function	
Structure	Multiple (continuous observed)	Multiple (latent)		Factor analysis (theoretical)	Create linear combinations of observed variables to represent latent variables
	Multiple (latent)	Multiple (continuous observed)		Principal components (empirical)	
	Multiple (continuous observed and/or latent)	Multiple (continuous observed and/or latent)		Structural equation modeling	Create linear combinations of observed and latent IVs to predict linear combinations of observed and latent DVs

CHAPTER 3
Review of Univariate and Bivariate Statistics

This chapter provides a brief review of univariate and bivariate statistics. While it is probably too "dense" to be a good source from which to learn, it will hopefully serve as a useful reminder of material already mastered and a help in establishing a common vocabulary. Section 3.1 goes over the logic of the statistical hypothesis test, while Sections 3.2, 3.3, and 3.4 skim many topics in analysis of variance and are background for Chapters 8 to 11. Section 3.5 summarizes correlation and regression, which are background for Chapters 5 and 6 while Section 3.6 summarizes chi square, background for Chapter 7.

3.1 ■ HYPOTHESIS TESTING

Statistics are used to make rational decisions under conditions of uncertainty. Inferences (decisions) are made about populations based on data from samples that contain incomplete information. Different samples taken from the same population probably differ from one another and from the population. Therefore, inferences regarding the population are always a little risky.

The traditional solution to this problem is statistical decision theory. Two hypothetical states of reality are set up, each represented by a probability distribution. Each distribution represents an alternative hypothesis about the true nature of events. Given sample results, a best guess is made as to which distribution the sample was taken from using formalized statistical rules to define "best."

3.1.1 ■ One-Sample z Test as Prototype

Statistical decision theory is most easily illustrated through a one-sample z test, using the standard normal distribution as the model for two hypothetical states of reality. Suppose there is a sample of 25 IQ scores and a need to decide whether this sample of scores is a random sample of a "normal" population with $\mu = 100$ and $\sigma = 15$, or a random sample from a population with $\mu = 108$ and $\sigma = 15$.

First, note that hypotheses are tested about means, not individual scores. Therefore the distributions representing hypothetical states of reality are distributions of means rather than distributions of individual scores. Distributions of means produce "sampling distributions of means" that differ systematically from distributions of individual scores; the mean of a population distribution, μ, is equal to the mean of a sampling distribution, μ, but the standard deviation of a population of individual scores, σ, is not equal to the standard deviation of a sampling distribution, $\sigma_{\bar{Y}}$. Sampling distributions have smaller standard deviations than distributions of scores, and the decrease is related to N, the sample size.

$$\sigma_{\bar{Y}} = \frac{\sigma}{\sqrt{N}} \tag{3.1}$$

For the sample, then,

$$\sigma_{\bar{Y}} = \frac{15}{\sqrt{25}} = 3$$

The question being asked, then, is, "Does our mean, taken from a sample of size 25, come from a sampling distribution with $\mu_{\bar{Y}} = 100$ and $\sigma_{\bar{Y}} = 3$ or does it come from a sampling distribution with $\mu_{\bar{Y}} = 108$ and $\sigma_{\bar{Y}} = 3$?" Figure 3.1(a) shows the first sampling distribution, defined as the null hypothesis, H_0, that is the sampling distribution of means calculated from all possible samples of size 25 taken from a population where $\mu = 100$ and $\sigma = 15$.

The sampling distribution for the null hypothesis has a special, fond place in statistical decision theory because it alone is used to define "best guess." A decision axis for retaining or rejecting H_0 cuts through the distribution so that the probability of rejecting H_0 by mistake is small. "Small" is defined probabilistically as α; an error in rejecting the null hypothesis is referred to as an α, or Type I, error. There is little choice in picking α. Tradition and journal editors decree that it is .05 or smaller, meaning that the null hypothesis is rejected no more than 5% of the time when it is true.

With a table of areas under the standard normal distribution (the table of z scores or standard normal deviates), the decision axis is placed so that the probability of obtaining a sample mean above that point is 5% or less. Looking up 5% in Table C.1, the z corresponding to a 5% cutoff is 1.645 (between 1.64 and 1.65). Notice that the z scale is one of two abscissas in Figure 3.1(a). If the decision axis is placed where $z = 1.645$, one can translate from the z scale to the Y scale to locate the decision axis along Y. The transformation equation is

$$\bar{Y} = \mu + z\sigma_{\bar{Y}} \tag{3.2}$$

Equation 3.2 is a rearrangement of terms from the z test for a single sample:[1]

$$z = \frac{\bar{Y} - \mu}{\sigma_{\bar{Y}}} \tag{3.3}$$

Applying Equation 3.2 to the example,

$$\bar{Y} = 100 + (1.645)(3) = 104.935$$

---

[1] The more usual procedure for testing a hypothesis about a single mean is to solve for z on the basis of the sample mean and standard deviation to see if the sample mean is sufficiently far away from the mean of the sampling distribution under the null hypothesis. If z is 1.645 or larger, the null hypothesis is rejected.

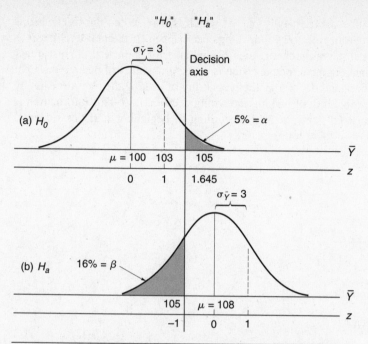

FIGURE 3.1 SAMPLING DISTRIBUTION FOR MEANS WITH $N = 25$ AND $\sigma = 15$ UNDER TWO HYPOTHESES: (a) H_0: $\mu = 100$ AND (b) H_a: $\mu = 108$.

The null hypothesis that the mean IQ of the sampling distribution is 100 is rejected if the mean IQ of the sample is equal to or greater than 104.935; call it 105.

Frequently this is as far as the model is taken—the null hypothesis is either retained or rejected. However, if the null hypothesis is rejected, it is rejected in favor of an alternative hypothesis, H_a. The alternative hypothesis is not always stated explicitly,[2] but when it is one can evaluate the probability of retaining the null hypothesis when it should be rejected because H_a is true.

This second type of error is called a β, or Type II, error. Because in the example the μ for H_a is 108, the sampling distribution of means for H_a can be graphed, as in Figure 3.1(b). The decision axis is placed with regard to H_0, so we need to find the probability associated with the place it crosses H_a. The first step is to find z corresponding to an IQ score of 105 in a distribution with $\mu_{\bar{Y}} = 108$ and $\sigma_{\bar{Y}} = 3$. Applying Equation 3.3, we find that

$$z = \frac{105 - 108}{3} = -1.00$$

By looking up $z = -1.00$, about 16% of the time sample means are equal to or less than 105 when the population $\mu = 108$ and the alternative hypothesis is true. Therefore $\beta = .16$.

---

[2] Often the alternative hypothesis is simply that the sample is taken from a population that is not equal to the population represented by the null hypothesis. There is no attempt to specify "not equal to."

H_0 and H_a represent alternative realities, only one of which is true. When the researcher is forced to decide whether to retain or reject H_0, four things can happen. If the null hypothesis is true, a correct decision is made if the researcher retains H_0 and an error is made if the researcher rejects it. If the probability of making the wrong decision is α, the probability of making the right decision is $1 - \alpha$. If, on the other hand, H_a is true, the probability of making the right decision by rejecting H_0 is $1 - \beta$ and the probability of making the wrong decision is β. This information is summarized in a "confusion matrix" (aptly named, according to beginning statistics students) showing the probabilities of each of these four outcomes:

		Reality	
		H_0	H_a
	"H_0"	$1 - \alpha$	β
Statistical decision	"H_a"	α	$1 - \beta$
		1.00	1.00

For the example, the probabilities are

		Reality	
		H_0	H_a
	"H_0"	.95	.16
Statistical decision	"H_a"	.05	.84
		1.00	1.00

3.1.2 ■ Power

The lower right-hand cell of the confusion matrix represents the most desirable outcome and the power of the research. Usually the researcher believes that H_a is true and hopes that the sample data lead to rejection of H_0. Power is the probability of rejecting H_0 when H_a is true. In Figure 3.1(b), power is the portion of the H_a distribution that falls above the decision axis. Many of the choices in designing research are made with an eye toward increasing power because research with low statistical power usually is not worth the effort.

Figure 3.1 and Equations 3.1 and 3.2 suggest some ways to enhance power. One obvious way to increase power is to move the decision axis to the left. However, it can't be moved far or Type I error rates reach an unacceptable level. Given the choice between .05 and .01 for α error, though, a decision in favor of .05 increases power. A second strategy is to move the curves farther apart by applying a stronger treatment. Other strategies involve decreasing the standard deviation of the sampling distributions either by decreasing variability in scores (e.g., exerting greater experimental control) or by increasing sample size, N.

This model for statistical decisions and these strategies for increasing power generalize to other sampling distributions and to tests of hypotheses other than a single sample mean against a hypothesized population mean.

There is occasionally the danger of *too much power*. The null hypothesis is probably never exactly true and any sample is likely to be slightly different from the population value. With a large enough sample, rejection of H_0 is virtually certain. For that reason, a "minimal meaningful difference" should guide the selection of sample size. The sample size should be large enough to be likely to reveal a minimal meaningful difference. Rejection of the null hypothesis may be trivial if the sample is large enough to reveal any difference whatever. If so, rejection of H_0 is nontrivial. This issue is considered further in Section 3.4.

3.1.3 ■ Extensions of the Model

The z test for the difference between a sample mean and a population mean readily extends to a z test of the difference between two sample means. A sampling distribution is generated for the difference between means under the null hypothesis that $\mu_1 = \mu_2$ and is used to position the decision axis. The power of an alternative hypothesis is calculated with reference to the decision axis, just as before.

When population variances are unknown, it is desirable to evaluate the probabilities using Student's t rather than z, even for large samples. Numerical examples of use of t to test differences between two means are available in most univariate statistics books and are not presented here. The logic of the process, however, is identical to that described in Section 3.1.1.

3.2 ■ ANALYSIS OF VARIANCE

Analysis of variance is used to compare two or more means to see if there are any reliable differences among them. Distributions of scores for three hypothetical samples are provided in Figure 3.2. Analysis of variance evaluates the differences among means relative to the dispersion in the sampling distributions. The null hypothesis is that $\mu_1 = \mu_2 = \cdots = \mu_k$ as estimated from $\overline{Y}_1, \overline{Y}_2, \ldots \overline{Y}_k$, with k equal to the number of means being compared.

Analysis of variance (ANOVA) is really a set of analytic procedures based on a comparison of two estimates of variance. One estimate comes from differences among scores within each group; this estimate is considered random or error variance. The second estimate comes from differences

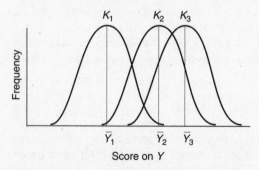

FIGURE 3.2 IDEALIZED FREQUENCY DISTRIBU-
TION OF THREE SAMPLES AND
THEIR MEANS.

in group means and is considered a reflection of group differences or treatment effects plus error. If these two estimates of variance do not differ appreciably, one concludes that all of the group means come from the same sampling distribution of means, and that the slight differences among them are due to random error. If, on the other hand, the group means differ more than expected, it is concluded that they were drawn from different sampling distributions of means and the null hypothesis that the means are the same is rejected.

Differences among variances are evaluated as ratios, where the variance associated with differences among sample means is in the numerator, and the variance associated with error is in the denominator. The ratio between these two variances forms an F distribution. F distributions change shape depending on degrees of freedom in both numerator and denominator of the F ratio. Thus tables of critical F, for testing the null hypothesis, depend on two degree-of-freedom parameters (cf. Appendix C, Table C.3).

The many varieties of analysis of variance are conveniently summarized in terms of the partition of *sums of squares*, that is, sums of squared differences between scores and their means. A sum of squares (SS) is simply the numerator of a variance, S^2.

$$S^2 = \frac{\Sigma(Y - \overline{Y})^2}{N - 1} \tag{3.4}$$

$$SS = \Sigma(Y - \overline{Y})^2 \tag{3.5}$$

The square root of variance is standard deviation, S, the measure of variability that is in the metric of the original scores.

$$S = \sqrt{S^2} \tag{3.6}$$

3.2.1 ■ One-Way Between-Subjects ANOVA

DV scores appropriate to one-way between-subjects ANOVA with equal n are presented in a table, with k columns representing groups (levels of the IV) and n scores within each group.[3] Table 3.1 shows how subjects are assigned to groups within this design.

Each column has a mean, $\overline{Y}_j$ where $j = 1, 2, \ldots, k$. Each score is designated Y_{ij}, where $i = 1, 2, \ldots, n$. The symbol GM represents the grand mean of all scores over all groups.

The difference between each score and the grand mean $(Y_{ij} - GM)$ is considered the sum of two component differences, the difference between the score and its own group mean and the difference between that mean and the overall mean.

$$Y_{ij} - GM = (Y_{ij} - \overline{Y}_j) + (\overline{Y}_j - GM) \tag{3.7}$$

This result is achieved by first subtracting and then adding the group mean to the equation. Each term is then squared and summed separately to produce the sum of squares for error and the sum

---

[3]Throughout the book, n is used for sample size within a single group or cell and N is used for total sample size.

TABLE 3.1 ASSIGNMENT OF SUBJECTS IN A ONE-WAY BETWEEN-SUBJECTS ANOVA

Treatment		
K_1	K_2	K_3
S_1	S_4	S_7
S_2	S_5	S_8
S_3	S_6	S_9

of squares for treatment, respectively. The basic partition holds because, conveniently, the cross-product terms produced by squaring and summing cancel each other out. Across all scores, the partition is

$$\sum_i \sum_j (Y_{ij} - \text{GM})^2 = \sum_i \sum_j (Y_{ij} - \overline{Y}_j)^2 + n \sum_j (\overline{Y}_j - \text{GM})^2 \tag{3.8}$$

Each of these terms is a sum of squares (SS)—a sum of squared differences between scores (with means sometimes treated as scores) and their associated means. That is, each term is a special case of Equation 3.5.

The term on the left of the equation is the total sum of squared differences between scores and the grand mean, ignoring groups with which scores are associated, designated SS_{total}. The first term on the right is the sum of squared deviations between each score and its group mean. When summed over all groups, it becomes sum of squares within groups, SS_{wg}. The last term is the sum of squared deviations between each group mean and the grand mean, the sum of squares between groups, SS_{bg}. Equation 3.8 is also symbolized as

$$\text{SS}_{\text{total}} = \text{SS}_{wg} + \text{SS}_{bg} \tag{3.9}$$

Degrees of freedom in ANOVA partition the same way as sums of squares:

$$\text{df}_{\text{total}} = \text{df}_{wg} + \text{df}_{bg} \tag{3.10}$$

Total degrees of freedom is the number of scores minus 1. The 1 df is lost when the grand mean is estimated. Therefore

$$\text{df}_{\text{total}} = N - 1 \tag{3.11}$$

Within-groups degrees of freedom are the number of scores minus k, lost when the means for each of the k groups are estimated. Therefore

$$\text{df}_{wg} = N - k \tag{3.12}$$

Between-groups degrees of freedom are k "scores" (each group mean treated as a score) minus 1, lost when the grand mean is estimated, so that

$$\text{df}_{bg} = k - 1 \tag{3.13}$$

Verifying the equality proposed in Equation 3.10, we get

$$N - 1 = N - k + k - 1$$

As in the partition of sums of squares, the term associated with group means is subtracted out of the equation and then added back in.

Another common notation for the partition of Equation 3.7 is

$$\text{SS}_{\text{total}} = \text{SS}_K + \text{SS}_{S(K)} \tag{3.14}$$

as shown in Table 3.2(a). In this notation, the total sum of squares is partitioned into a sum of squares due to the k groups, SS_K, and a sum of squares due to subjects within the groups, $\text{SS}_{S(K)}$. (Notice that the order of terms on the right side of the equation is the reverse of that in Equation 3.9.)

The division of a sum of squares by degrees of freedom produces variance, called mean square, MS, in ANOVA. Variance, then, is an average sum of squares. ANOVA produces three variances: one associated with total variability among scores, MS_{total}; one associated with variability within groups, MS_{wg} or $\text{MS}_{S(K)}$; and one with variability between groups, MS_{bg} or MS_K.

MS_K and $\text{MS}_{S(K)}$ provide the variances for an F ratio to test the null hypothesis that $\mu_1 = \mu_2 = \cdots = \mu_k$.

$$F = \frac{\text{MS}_K}{\text{MS}_{S(K)}} \qquad \text{df} = (k - 1), N - k \tag{3.15}$$

Once F is computed, it is tested against critical F obtained from a table, such as Table C.3, with numerator df $= k - 1$ and denominator df $= N - k$ at desired alpha level. If obtained F exceeds critical F, the null hypothesis is rejected in favor of the hypothesis that there is a difference among the means in the k groups.

Anything that increases obtained F increases power. Power is increased by decreasing the error variability or increasing the sample size in the denominator ($\text{MS}_{S(K)}$) or by increasing differences among means in the numerator (MS_K).

3.2.2 ■ Factorial Between-Subjects ANOVA

If groups are formed along more than one dimension, differences among means are attributable to more than one source. Consider an example with six groups, three of women and three of men, where the DV is scores on a final examination in statistics. One source of variation in means is due to gender, SS_G. If the three groups within each gender are exposed to three different methods of teaching statistics, a second source of differences among means is teaching method, SS_T. The final

TABLE 3.2   PARTITION OF SUMS OF SQUARES AND DEGREES OF FREEDOM FOR SEVERAL ANOVA DESIGNS

(a) One-way between-subjects ANOVA

$$SS_{total}{}^{a}$$

$$SS_K \qquad SS_{S(K)}$$

$$df = k - 1, \qquad N - k$$

(b) Factorial between-subjects ANOVA

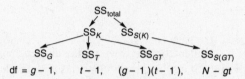

$$SS_{total}$$

$$SS_K \qquad SS_{S(K)}$$

$$SS_G \qquad SS_T \qquad SS_{GT} \qquad SS_{S(GT)}$$

$$df = g - 1, \qquad t - 1, \qquad (g-1)(t-1), \qquad N - gt$$

(c) One-way within-subjects ANOVA

$$SS_{total}$$

$$SS_K \qquad SS_{S(K)}$$

$$SS_K \qquad SS_S \qquad SS_{SK}$$

$$df = k - 1, \qquad s - 1, \qquad (s-1)(k-1)$$

(d) One-way matched-randomized ANOVA

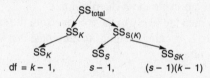

$$SS_{total}$$

$$SS_K \qquad SS_{S(K)}$$

$$SS_K \qquad SS_B \qquad SS_{BK}$$

$$df = k - 1, \qquad b - 1, \qquad (b-1)(k-1)$$

(e) Factorial within-subjects ANOVA

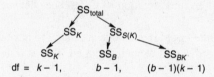

$$SS_{total}$$

$$SS_K \qquad SS_{S(K)}$$

$$SS_A \quad SS_B \quad SS_{AB} \quad SS_S \quad SS_{SA} \quad SS_{SB} \quad SS_{SAB}$$

$$df = a - 1, \, b - 1, \, (a-1)(b-1), \, s - 1, \, (s-1)(a-1), \, (s-1)(b-1), \, (s-1)(a-1)(b-1)$$

(f) Mixed within-between-subjects ANOVA

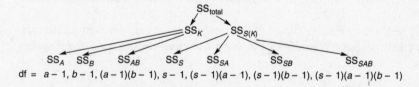

$$SS_{total}$$

$$SS_{bs} \qquad SS_{ws}$$

$$SS_G \qquad SS_{S(G)} \qquad SS_T \qquad SS_{GT} \qquad SS_{TS(G)}$$

$$df = g - 1, \qquad s - g, \qquad t - 1, \qquad (g-1)(t-1), \qquad (t-1)(s-g)$$

[a] For all SS_{total}, $df = N - 1$.

TABLE 3.3   ASSIGNMENT OF SUBJECTS IN A
              FACTORIAL BETWEEN-SUBJECTS
              DESIGN

		Teaching techniques		
		T_1	T_2	T_3
	G_1	S_1 S_2	S_5 S_6	S_9 S_{10}
Gender	G_2	S_3 S_4	S_7 S_8	S_{11} S_{12}

source of known differences among means is the interaction between gender and teaching methods, SS_{GT}. The interaction tests whether effectiveness of teaching methods varies with gender.

Allocation of subjects in this design is shown in Table 3.3. Sums of squares and degrees of freedom are partitioned as in Table 3.2(b). Error is estimated by variation in scores within each of the six cells, $SS_{S(GT)}$. Three null hypotheses are tested using the F distribution.

The first test asks if means for men and women are likely to have come from the same sampling distribution of means. Scores are averaged across teaching methods to eliminate that source of variability. Gender differences are tested in the F ratio:

$$F = \frac{MS_G}{MS_{S(GT)}} \qquad df = (g - 1), N - gt \qquad (3.16)$$

Rejection of the null hypothesis supports an interpretation of differences between women and men in performance on the final exam.

The second test asks if means from the three teaching methods are likely to have come from the same sampling distribution of means, averaged across women and men. This is tested as:

$$F = \frac{MS_T}{MS_{S(GT)}} \qquad df = (t - 1), N - gt \qquad (3.17)$$

Rejection of the null hypothesis supports an interpretation of differences in effectiveness of the three teaching methods.

The third test asks if the cell means, the means for women and the means for men within each teaching method, are likely to have come from the same sampling distribution of *differences* between means.

$$F = \frac{MS_{GT}}{MS_{S(GT)}} \qquad df = (g - 1)(t - 1), N - gt \qquad (3.18)$$

Rejection of the null hypothesis supports an interpretation that men and women differ regarding the most effective teaching methods.

In each case, the estimate of normally occurring variability in test scores, error, is $MS_{S(GT)}$, or within-cell variance. In each case, critical F is read from a table with appropriate degrees of freedom and desired alpha, and if obtained F (Equations 3.16, 3.17, or 3.18) is greater than critical F, the null hypothesis is rejected. When there is an equal number of scores in each cell, the three tests are independent (except for use of a common error term): the test of each main effect (gender and teaching method) is not related to the test of the other main effect or the test of the interaction.

In a one-way between-subjects design (Section 3.2.1), $k - 1$ degrees of freedom are used to test the null hypothesis of differences among groups. If k is equal to the number of cells in a two-way design, tests of G, T, and GT use up the $k - 1$ degrees of freedom. With proper partitioning, then, of a two-way factorial design, you get three tests for the price of one.

With higher-order between-subjects factorial designs, variation due to differences among groups is partitioned into main effects for each IV, two-way interactions between each pair of IVs, three-way interactions among each trio of IVs, and so on. In any between-subjects factorial design, error sum of squares is the sum of squared differences within each cell of the design.

3.2.3 ■ Within-Subjects ANOVA

In some designs, the means that are tested are derived from the same subjects measured on different occasions, as shown in Table 3.4, rather than from different groups of subjects.[4]

In these designs, computation of sum of squares and mean square for the effect of the IV is the same as for the between-subject designs. However, the error term is further partitioned into individual differences due to subjects, SS_S, and interaction of individual differences with treatment, SS_{SK}. Because subjects are measured repeatedly, their effect as a source of variability in scores is estimated and subtracted from $SS_{S(K)}$, the error term in a corresponding between-subjects design. The interaction of individual differences with treatment, MS_{SK}, is used as the error term;

$$F = \frac{MS_K}{MS_{SK}} \qquad df = (k - 1), (k - 1)(s - 1) \tag{3.19}$$

The partition of sums of squares for a one-way within-subjects design with k levels is shown in Table 3.2(c), where s is the number of subjects.

MS_{SK} is used as the error term because, once SS_S is subtracted, no variation is left within cells of the design—there is, in fact, only one score per cell. The interaction of individual differences with treatment is all that remains to serve as an estimate of error variance. If there are individual differences among subjects in scores, and if individuals react similarly to the IV, the interaction is a good estimate of error. Once individual differences are subtracted, the error term is usually smaller than the error term in a corresponding between-subjects design, so the within-subjects design is more sensitive than the between-subjects design.

TABLE 3.4   ASSIGNMENT OF SUBJECTS IN A ONE-WAY WITHIN-SUBJECTS DESIGN

		Treatment		
		K_1	K_2	K_3
Subjects	S_1	S_1	S_1	S_1
	S_2	S_2	S_2	S_2
	S_3	S_3	S_3	S_3

[4]This design is also called repeated measures, one-score-per-cell, randomized blocks, matched-randomized changeover, or crossover

But if there are no consistent individual differences in scores,[5] or if there is an interaction between subjects and treatment, the error term may be larger than that of a between-subjects design. The statistical test is then conservative, it is more difficult to reject the null hypothesis of no difference between means, and the power of the test is reduced. Because Type I error is unaffected, the statistical test is not in disrepute, but, in this case, a within-subjects design is a poor choice of research design.

A within-subjects analysis is also used with a matched-randomized blocks design, as shown in Table 3.5 and Table 3.2(d). Subjects are first matched on the basis of variables thought to be highly related to the DV. Subjects are then divided into blocks, with as many subjects within each block as there are levels of the IV. Finally, members of each block are randomly assigned to levels of the IV. Although the subjects in each block are actually different people, they are treated statistically as if they were the same person. In the analysis, there is a test of the IV, and a test of blocks (the same as the test of subjects in the within-subjects design), with the interaction of blocks and treatment used as the error term. Because matching is used to produce consistency in performance within blocks and the effect of blocks is subtracted from the error term, this design should also be more sensitive than the between-subjects design. It will not be, however, if the matching fails.

Factorial within-subjects designs, as shown in Table 3.6, are an extension of the one-way within-subjects design. The partition of a two-way within-subjects design is in Table 3.2(e). In the analysis, the error sum of squares is partitioned into a number of "subjects by effects" inter-

TABLE 3.5 ASSIGNMENT OF SUBJECTS IN A MATCHED-RANDOMIZED DESIGN[a]

		Treatment		
		A_1	A_2	A_3
	B_1	S_1	S_2	S_3
Blocks	B_2	S_4	S_5	S_6
	B_3	S_7	S_8	S_9

[a] Where subjects in the same block have been matched on some relevant variable.

TABLE 3.6 ASSIGNMENT OF SUBJECTS IN A FACTORIAL WITHIN-SUBJECTS DESIGN

		Treatment A		
		A_1	A_2	A_3
	B_1	S_1 S_2	S_1 S_2	S_1 S_2
Treatment B				
	B_2	S_1 S_2	S_1 S_2	S_1 S_2

[5]Notice that the degrees of freedom for error, $(k - 1)(s - 1)$, are fewer than in the between-subjects design. Unless the reduction in error variance due to subtraction of SS_S is substantial, the loss of degrees of freedom may overcome the gain due to smaller SS when MS_{SK} is computed.

actions just as the sum of squares for effects is partitioned into numerous sources. It is common (though not universal) to develop a separate error term for each F test; for instance, the test of the main effect of A is

$$F = \frac{MS_A}{MS_{SA}} \qquad df = (a - 1), (a - 1)(s - 1) \qquad\qquad (3.20)$$

For the main effect of B, the test is

$$F = \frac{MS_B}{MS_{SB}} \qquad df = (b - 1), (b - 1)(s - 1) \qquad\qquad (3.21)$$

and for the interaction,

$$\frac{MS_{AB}}{MS_{SAB}} \qquad df = (a - 1)(b - 1), (a - 1)(b - 1)(s - 1) \qquad\qquad (3.22)$$

For higher-order factorial designs, the partition into sources of variability grows prodigiously, with an error term developed for each main effect and interaction tested.

There is controversy in within-subject analyses concerning conservatism of the F tests and whether or not separate error terms should be used. In addition, if the repeated measurements are on single subjects, there are often carry-over effects that limit generalizability to situations where subjects are tested repeatedly. Finally, when there are more than two levels of the IV, the analysis has the assumption of homogeneity of covariance. Homogeneity of covariance is, roughly speaking, the assumption that subjects "line up" in scores the same for all pairs of levels of the IV. If some pairs of levels are close in time (e.g., trial 2 and trial 3), while other pairs are distant in time (e.g., trial 1 and trial 10), the assumption is often violated. Such violation is serious because Type I error rate is affected. Homogeneity of covariance is discussed in greater detail in Chapters 8, 9, and, especially, 10, and in Frane (1980).

For these reasons, within-subjects ANOVA is sometimes replaced by profile analysis, where repetitions of DVs are transformed into separate DVs (Chapter 10) and a multivariate statistical test is used.

3.2.4 ■ Mixed Between-Within-Subjects ANOVA[6]

Often in factorial designs, one or more IVs are measured between subjects while other IVs are measured within subjects.[7] The simplest example of the mixed between-within-subjects design involves one between-subjects and one within-subjects IV, as seen in Table 3.7.[8]

To show the partition, the total SS is divided into a source attributable to the between-subjects part of the design (Groups), and a source attributable to the within-subjects part (Trials), as shown

---

[6] This design is also called split plot, repeated measures, and randomized block factorial design.

[7] Mixed designs can also have "blocks" rather than repeated measures on individual subjects as the within-subjects segment of the design.

[8] When the between-subjects variables are based on naturally occurring differences among subjects (e.g., age, sex), the design is said to be "blocked" on the subject variables. This is a different use of the term blocks from that of the preceding section. In a mixed design, both kinds of blocking can occur.

TABLE 3.7   ASSIGNMENT OF SUBJECTS IN
              A BETWEEN-WITHIN-
              SUBJECTS DESIGN

		Trials		
		T_1	T_2	T_3
Groups	G_1	S_1 S_2	S_1 S_2	S_1 S_2
	G_2	S_3 S_4	S_3 S_4	S_3 S_4

in Table 3.2(f). Each source is then further partitioned into effects and error components: between-subjects into groups and subjects-within-groups error term; and within-subjects into trials, the group-by-trials interaction; and the trial-by-subjects-within-groups error term.

As more between-subjects IVs are added, between-subjects main effects and interactions expand the between-subjects part of the partition. For all the between-subjects effects, there is a single error term consisting of variance among subjects confined to each combination of the between-subjects IVs. As more within-subjects IVs are added, the within-subjects portion of the design expands. Sources of variability include main effects and interactions of within-subjects IVs and interactions of between- and within-subjects IVs. Separate error terms are developed for each source of variability in the within-subjects segment of the design.[9]

Problems associated with within-subjects designs (e.g., homogeneity of covariance) carry over to mixed designs, and profile analysis is sometimes used to circumvent some of these problems.

3.2.5 ■ Design Complexity

Discussion of analysis of variance has so far been limited to factorial designs where there are equal numbers of scores in each cell and levels of each IV are purposely chosen by the researcher. Several deviations from these straightforward designs are possible. A few of the more common types of design complexity are mentioned below, but the reader actually faced with use of these designs is referred to one of the more complete analysis of variance texts such as Winer (1971), Keppel (1991), or Myers and Well (1991).

3.2.5.1 ■ Nesting

In between-subjects designs, subjects are said to be nested within levels of the IV. That is, each subject is confined to only one level of each IV or combination of IVs. Nesting also occurs with IVs when levels of one IV are confined to only one level of another IV, rather than factorially crossing over the levels of the other IV.

---

[9] "Subjects" is no longer available as a source of variance for analysis. Because subjects are confined to levels of the between-subjects variable(s), differences between subjects in each group are used to estimate error for testing variance associated with between-subjects variables.

Take the example where the IV is various levels of teaching methods. Children within the same classroom cannot be randomly assigned to different methods but whole classrooms can be so assigned. The design is one-way between-subjects where teaching methods is the IV and classrooms serve as subjects. For each classroom, the mean score for all children on the test is obtained, and the means serve as DVs in one-way ANOVA.

If the effect of classroom is also assessed, the design is nested or hierarchical, as seen in Table 3.8(a). Classrooms are randomly assigned to and nested in teaching methods, and children are nested in classrooms. The error term for the test of classroom is subjects within classrooms and teaching method, while the error term for the test of teaching method is classrooms within teaching technique.

3.2.5.2 ■ Latin Square Designs

The order of presentation of levels of an IV often produces differences in the DV. In within-subjects designs, subjects become practiced or fatigued or experiment-wise as they experience more levels of the IV. In between-subjects designs, there are often time of day or experimenter effects that change scores on the DV. To get an uncontaminated look at the effects of the IV, it is important to counterbalance the effects of increasing experience, time of day, and the like, so that they are independent of levels of the IV. If the within-subjects IV is something like trials, counterbalancing is not possible because the order of trials cannot be changed. But when the IV is something like background color of slide used to determine if background color affects memory for material on the slide, a Latin square arrangement is often used to control order effects.

A Latin square design is shown in Table 3.8(b). If A_1 is a yellow background, A_2 a blue background, and A_3 a red background, then subjects are presented the slides in the order specified by the Latin square. The first subject is presented the slide with blue background, then yellow, then red. The second subject is presented yellow, then red, then blue, etc. The yellow slide (A_1) appears once in first position, once in second, and once in third, and so on for the other colors, so that order effects are distributed evenly across the levels of the IV.

The simple design of Table 3.8(b) produces a test of the IV (A), a test of subjects (if desired), and a test of order. The effect of order (like the effect of subjects) is subtracted out of the error term, leaving it smaller than it is when order effects are not analyzed. The error term itself is composed of fragments of interactions that are not available for analysis because effects are not fully crossed in the design. Thus, the design is more sensitive than a comparable between-subjects design when there are order effects and no interactions and less sensitive when there are no order effects but there are interactions. Consult Keppel (1991) or one of the other ANOVA texts for greater detail on this fascinating topic.

TABLE 3.8 SOME COMPLEX ANOVA DESIGNS

(a) Nested designs Teaching techniques					(b) Latin square designs[a] Order		
T_1	T_2	T_3			1	2	3
Classroom 1	Classroom 2	Classroom 3		S_1	A_2	A_1	A_3
Classroom 4	Classroom 5	Classroom 6	Subjects	S_2	A_1	A_3	A_2
Classroom 7	Classroom 8	Classroom 9		S_3	A_3	A_2	A_1

[a] Where the three levels of treatment *A* are experienced by different subjects in different orders, as indicated.

3.2.5.3 ■ Unequal *n* and Nonorthogonality

In a simple one-way between-subjects ANOVA, problems created by unequal group sizes are relatively minor. Computation is slightly more difficult, but that is no real disaster, especially if analysis is by computer. However, as group sizes become more discrepant, the assumption of homogeneity of variance is more important. If the group with the smaller *n* has the larger variance, the *F* test is too liberal, leading to increased Type I error rate and an inflated alpha level.

In factorial designs with more than one between-subjects IV, unequal sample sizes in each cell create difficulty in computation and ambiguity of results. With unequal *n*, a factorial design is nonorthogonal. Hypotheses about main effects and interactions are not independent, and sums of squares are not additive. The various sources of variability contain overlapping variance and the same variance can be attributed to more than one source, as discussed in Chapter 1. If effects are tested without taking the overlap into account, the probability of a Type I error increases because systematic variance contributes to more than one test. A variety of strategies is available to deal with the problem, none of them completely satisfactory.

The simplest strategy is to randomly delete cases from cells with greater *n* until all cells are equal. If unequal *n* is due to random loss of a few subjects in an experimental design originally set up for equal *n*, deletion is often a good choice. An alternative strategy with random loss of subjects in an experimental design is an unweighted-means analysis, described in Chapter 8 and ANOVA textbooks such as Winer (1971) and Keppel (1991). The unweighted-means approach has greater power than random deletion of cases and is the preferred approach as long as computational aids are available.

But in nonexperimental work, unequal *n* often results from the nature of the population. Differences in sample sizes reflect true differences in numbers of various types of subjects. To artificially equalize the *n* is to distort the differences and lose generalizability. In these situations, decisions are made as to how tests of effects are to be adjusted for overlapping variance. Standard methods for adjusting tests of effects with unequal *n* are discussed in Chapter 8.

3.2.5.4 ■ Fixed and Random Effects

In all the ANOVA designs discussed so far, levels of each IV are selected by the researcher on the basis of their interest in testing significance of the IV. This is the usual fixed effects model. Sometimes, however, there is a desire to generalize to a population of levels of an IV. In order to generalize to the population of levels of the IVs, a large number of levels are randomly selected from the population, just as subjects are randomly selected from the population of subjects when the desire is to generalize results to the population of subjects. Consider, for example, an experiment to study effects of word familiarity[10] on recall where the desire is to generalize results to all levels of word familiarity. A finite set of familiarity levels is randomly selected from the population of familiarity levels. Word familiarity is considered a random-effects IV.

The analysis is set up so that results generalize to levels other than those selected for the experiment—generalize to the population of levels from which the sample was selected. During analysis, alternative error terms for evaluating the statistical significance of random effects IVs are used. Although computer programs are available for analysis of random effects IVs, use of them is fairly rare. The interested reader is referred to one of the more sophisticated ANOVA texts, such as Winer (1971), for a full discussion of the random-effects model.

---

[10] Word familiarity is usually operationalized by frequency of usage of words in the English language.

3.2.6 ■ Specific Comparisons

When an IV has more than one degree of freedom (more than two levels) or when there is an inter-action between two or more IVs, the overall test of the effect is ambiguous. The overall test, with $k - 1$ degrees of freedom, is pooled over $k - 1$ single degree of freedom subtests. If the overall test is significant, so usually are one or more of the subtests, but there is no way to tell which one(s). To find out which single degree of freedom subtests are significant, comparisons are performed.

In analysis, degrees of freedom are best thought of as a nonrenewable resource. They are ana-lyzed once with conventional alpha levels, but further analyses require very stringent alpha levels. For this reason, the best strategy is to plan the analysis very carefully so that the most interesting comparisons are tested with conventional alpha levels. Unexpected findings or less interesting effects are tested later with stringent alpha levels. This is the strategy used by the researcher who has been working in an area for a while and knows precisely what to look for.

Regrettably, research is often more tentative; so the researcher "spends" the degrees of freedom on omnibus (routine) ANOVA testing main effects and interactions at conventional alpha levels, and then snoops the single degree of freedom comparisons of significant effects at stringent alpha levels. Snooping through data after results of ANOVA are known is called "conducting post hoc comparisons."

We present here the most flexible method of conducting comparisons, with mention of other methods as they are appropriate. The procedure for conducting comparisons is the same for planned and post hoc comparisons up to the point where an obtained F is evaluated against a critical F.

3.2.6.1 ■ Weighting Coefficients for Comparisons

Comparison of treatment means begins by assigning a weighting factor (w) to each of the cell or marginal means so the weights reflect your null hypotheses. Suppose you have a one-way design with k means and you want to make comparisons. For each comparison, a weight is assigned to each mean. Weights of zero are assigned to means (groups) that are left out of a comparison, although at least two of the means must have nonzero weights. Means that are contrasted with each other are assigned weights with opposite signs (positive or negative) with the constraint that the weights sum to zero, that is,

$$\sum_1^k w_j = 0$$

For example, consider an IV with four levels, producing $\overline{Y}_1$, $\overline{Y}_2$, $\overline{Y}_3$, and $\overline{Y}_4$. If you want to test the hypothesis that $\mu_1 - \mu_3 = 0$, weighting coefficients are $1, 0, -1, 0$ producing $1\overline{Y}_1 + 0\overline{Y}_2 + (-1)\overline{Y}_3 + 0\overline{Y}_4$. $\overline{Y}_2$ and $\overline{Y}_4$ are left out while $\overline{Y}_1$ is compared with $\overline{Y}_3$. Or, if you want to test the null hypothesis that $(\mu_1 + \mu_2)/2 - \mu_3 = 0$ (to compare the average of means from the first two groups with the mean of the third group leaving out the fourth group), weighting coefficients are $\frac{1}{2}, \frac{1}{2}, -1$, 0, (or any multiple of them, such as $1, 1, -2, 0$), respectively. Or, if you want to test the null hypoth-esis that $(\mu_1 + \mu_2)/2 - (\mu_3 + \mu_4)/2 = 0$ (to compare the average mean of the first two groups with the average mean of the last two groups), the weighting coefficients are $\frac{1}{2}, \frac{1}{2}, -\frac{1}{2}, -\frac{1}{2}$, (or 1, $1, -1, -1$).

The idea behind the test is that the sum of the weighted means is equal to zero when the null hypothesis is true. The more the sum diverges from zero, the greater the confidence with which the null hypothesis is rejected.

3.2.6.2 ■ Orthogonality of Weighting Coefficients

In a design with equal number of cases in each group, any pair of comparisons is orthogonal if the sum of the cross products of the weights for the two comparisons is equal to zero. For example, in the three comparisons below,

	w_1	w_2	w_3
Comparison 1	1	−1	0
Comparison 2	½	½	−1
Comparison 3	1	0	−1

the sum of the cross products of weights for comparison 1 and comparison 2 is

$$(1)(\tfrac{1}{2}) + (-1)(\tfrac{1}{2}) + (0)(-1) = 0$$

Therefore the two comparisons are orthogonal.

Comparison 3, however, is orthogonal to neither of the first two comparisons. For instance, checking it against comparison 1,

$$(1)(1) + (-1)(0) + (0)(-1) = 1$$

In general, there are as many orthogonal comparisons as there are degrees of freedom. Because $k = 3$ in the example, df = 2. There are only two orthogonal comparisons when there are three levels of an IV, and only three orthogonal comparisons when there are four levels of an IV.

There are advantages to use of orthogonal comparisons, if they suit the needs of the research. First, there are only as many of them as there are degrees of freedom, so the temptation to "overspend" degrees of freedom is avoided. Second, orthogonal comparisons analyze nonoverlapping variance. If one of them is significant, it has no bearing on the significance of another of them. Last, because they are independent, if all $k - 1$ orthogonal comparisons are performed, the sum of the sum of squares for the comparisons is the same as the sum of squares for the IV in omnibus ANOVA. That is, the sum of squares for the effect has been completely broken down into the $k - 1$ orthogonal comparisons that comprise it.

3.2.6.3 ■ Obtained *F* for Comparisons

Once the weighting coefficients are chosen, the following equation is used to obtain F for the comparison if sample sizes are equal in each group:

$$F = \frac{n_c(\Sigma w_j \overline{Y}_j)^2 / \Sigma w_j^2}{\text{MS}_{\text{error}}} \tag{3.23}$$

where n_c is the number of scores in each of the means to be compared, $(\Sigma w_j \overline{Y}_j)^2$ is the squared sum of the weighted means, Σw_j^2 is the sum of the squared coefficients, and MS_{error} is the mean square for error in the ANOVA.

The numerator of Equation 3.23 is both sum of squares and mean square for the comparison because a comparison has only one degree of freedom.

For factorial designs, comparisons are done on either marginal or cell means, corresponding to comparisons on main effects and interactions respectively. The number of scores per mean and the error term follow from the ANOVA design used. However, if comparisons are made on within-subjects effects, it is customary to develop a separate error term for each comparison, just as separate error terms are developed for omnibus tests of within-subjects IVs.

Chapter 10 has much more information on comparisons of both main effects and interactions, including control language for performing them through some of the more popular computer programs.

Once you have obtained F for a comparison, whether by hand calculation or computer, the obtained F is compared with critical F to see if it is statistically reliable. If obtained F exceeds critical F, the null hypothesis for the comparison is rejected. But which critical F is used depends on whether the comparison is planned or performed post hoc.

3.2.6.4 ■ Critical *F* for Planned Comparisons

If you are in the enviable position of having planned your comparisons prior to data collection, and if you have planned no more of them than you have degrees of freedom for effect, critical F is obtained from the tables just as in routine ANOVA. Each comparison is tested against critical F at routine alpha with one degree of freedom in the numerator and degrees of freedom associated with the MS_{error} in the denominator. If obtained F is larger than critical F, the null hypothesis represented by the weighting coefficients is rejected.

With planned comparisons, omnibus ANOVA is not performed;[11] the researcher moves straight to comparisons. Once the degrees of freedom are spent on the planned comparisons, however, it is perfectly acceptable to snoop the data at more stringent alpha levels (Section 3.2.6.5), including main effects and interactions from omnibus ANOVA if they are appropriate.

Sometimes, however, the researcher can't resist the temptation to plan more comparisons than degrees of freedom for effect. When there are too many tests, even if comparisons are planned, the α level across all tests exceeds the α level for any one test and some adjustment of α for each test is needed. It is common practice to use a Bonferroni-type adjustment where slightly more stringent α levels are used with each test to keep α across all tests at reasonable levels. For instance, when 5 comparisons are planned, if each one of them is tested at $\alpha = .01$, the alpha across all tests is an acceptable .05 (roughly .01 times 5, the number of tests). However, if 5 comparisons are each tested at $\alpha = .05$, the alpha across all tests is approximately .25 (roughly .05 times 5), unacceptable by most standards.

If you want to keep overall α at, say, .10, and you have 5 tests to perform, you can assign each of them $\alpha = .02$, or you can assign two of them $\alpha = .04$ with the other three evaluated at $\alpha = .01$, for an overall Type I error rate of roughly .11. The decision about how to apportion α through the tests is also made prior to data collection.

As an aside, it is important to realize that routine ANOVA designs with numerous main effects and interactions suffer from the same problem of inflated Type I error rate across all tests as planned comparisons where there are too many tests. Some adjustment of alpha for separate tests is needed in big ANOVA problems, as well, if all effects are evaluated even if the tests are planned.

---

[11] You might perform routine ANOVA to compute the error term(s).

3.2.6.5 ■ Critical *F* for Post Hoc Comparisons

If you are unable to plan your comparisons and choose to start with routine ANOVA instead, you want to follow up significant main effects (with more than two levels) and interactions with post hoc comparisons to find the treatments that are different from one another. Because you have already spent your degrees of freedom in routine ANOVA and are likely to capitalize on chance differences among means that you notice in the data, some form of adjustment of α is necessary.

Many procedures for dealing with inflated Type I error rate are available as described in standard ANOVA texts such as Keppel (1991). The tests differ in the number and type of comparisons they permit and the amount of adjustment required of α. The tests that permit more numerous comparisons have correspondingly more stringent adjustments to critical *F*. For instance, the Dunnett test, which compares the mean from a single control group with each of the means of the other groups, in turn, has a less stringent correction than the Tukey test which allows all pairwise comparisons of means. The name of this game is to choose the most liberal test that permits you to perform the comparisons of interest.

The test described here (Scheffé, 1953) is the most conservative and most flexible of the popular methods. Once critical *F* is computed with the Scheffé adjustment, there is no limit to the number and complexity of comparisons that can be performed. You can perform all pairwise comparisons and all combinations of treatment means pooled and contrasted with other treatment means, pooled or not, as desired. Some possibilities for pooling are illustrated in Section 3.2.6.1. Once you pay the "price" in conservatism for this flexibility, you might as well conduct all the comparisons that make sense, given your research design.

The Scheffé method for computing critical *F* for a comparison on marginal means is

$$F_s = (k - 1)F_c \tag{3.24}$$

where F_s is adjusted critical *F*, $(k - 1)$ is degrees of freedom for the effect, and F_c is tabled *F* with $k - 1$ degrees of freedom in the numerator and degrees of freedom associated with the error term in the denominator.

If obtained *F* is larger than critical F_s, the null hypothesis represented by the weighting coefficients for the comparison is rejected. (See Chapter 10 for a more extended discussion of the appropriate correction.)

3.3 ■ PARAMETER ESTIMATION

If a reliable difference among means is found, one is usually interested in reporting the likely population value for each mean. Since sample means are unbiased estimators of population means, the best guess about the size of a population mean (μ) is the mean of the sample randomly selected from that population. In most reports of research, therefore, sample mean values are reported along with statistical results.

Sample means are only estimations of population means. They are unbiased because they are systematically neither large nor small, but they are rarely precisely at the population value—and there is no way to know when they are. Thus the error in estimation, the familiar confidence interval of introductory statistics, is often reported along with the estimated means. The size of the confidence interval depends on sample size, the estimation of population variability, and the degree of confidence one wishes to have in estimating μ. Alternatively, cell standard deviations or standard errors are presented along with sample means so that the reader can compute the confidence interval if it is desired.

3.4 ■ STRENGTH OF ASSOCIATION[12]

Although significance testing, comparisons, and parameter estimation help illuminate the nature of group differences, they do not assess the degree to which the IV(s) and DV are related. It is important to assess the degree of relationship to avoid publicizing trivial results as though they had practical utility. As discussed in Section 3.1.2, overly powerful research sometimes produces results that are statistically significant but realistically meaningless.

Strength of association assesses the proportion of variance in the DV that is associated with levels of an IV. It assesses the amount of total variance in the DV that is predictable from knowledge of the levels of the IV. If the total variances of the DV and the IV are represented by circles as in a Venn diagram, strength of association assesses the degree of overlap of the circles. Statistical significance testing assesses the *reliability* of the association between the IV and DV. Strength of association measures *how much* association there is.

A rough estimate of strength of association is available for any ANOVA through η^2 (eta squared).

$$\eta^2 = \frac{SS_{effect}}{SS_{total}} \tag{3.25}$$

When there are 2 levels of the IV, η^2 is the (squared) point biserial correlation between the continuous variable (the DV) and the dichotomous variable (the two levels of the IV).[13] After finding a significant main effect or interaction, η^2 shows the proportion of variance in the DV (SS_{total}) attributable to the effect (SS_{effect}). In a balanced, equal-n design, η^2's are additive; the sum of η^2 for all significant effects is the proportion of variation in the DV that is predictable from knowledge of the IVs.

This simple, popular measure of strength of association is flawed for two reasons. The first is that η^2 for a particular IV depends on the number and significance of other IVs in the design. η^2 for an IV tested in a one-way design is likely to be larger than η^2 for the same IV in a two-way design where the other IV and the interaction increase the size of the total variance, especially if one or both of the additional effects are large. This is because the denominator of η^2 contains systematic variance for other effects in addition to error variance and systematic variance for the effect of interest.

Therefore, an alternative form of η^2, called partial η^2, is available in which the denominator contains only variance attributable to the effect of interest plus error.

$$\text{partial } \eta^2 = \frac{SS_{effect}}{SS_{effect} + SS_{error}} \tag{3.26}$$

---

[12] This is also called effect size or treatment magnitude.

[13] All strength of association values are associated with the particular levels of the IV used in the research and do not generalize to other levels.

With this alternative, η^2's for all significant effects in the design do not sum to proportion of systematic variance in the DV. Indeed, the sum is sometimes greater than 1.00. It is imperative, therefore, to be clear in your report when this version of η^2 is used.

A second flaw is that η^2 describes proportion of systematic variance in a sample with no attempt to estimate proportion of systematic variance in the population. A statistic developed to estimate strength of association between IV and DV in the population is ω^2 (omega squared).

$$\omega^2 = \frac{SS_{effect} - (df_{effect})(MS_{error})}{SS_{total} + MS_{error}} \tag{3.27}$$

This is the additive form of ω^2 where the denominator represents total variance, not just variance due to effect plus error, and *is limited to between-subjects analysis of variance designs with equal sample sizes in all cells*. Forms of ω^2 are available for designs containing repeated measures (or randomized blocks) as described by Vaughn and Corballis (1969).

A separate measure of strength of association is computed and reported for each main effect and interaction of interest in a design.

3.5 ■ BIVARIATE STATISTICS: CORRELATION AND REGRESSION

Strength of association as described in Section 3.4 is assessed between a continuous DV and discrete levels of an IV. Frequently, however, a researcher wants to measure the strength of association between two continuous variables where the IV–DV distinction is blurred. For instance, the association between years of education and income is of interest even though neither is manipulated and inferences regarding causality are not possible. Correlation is the measure of the size and direction of relationship between the two variables, and squared correlation is the measure of strength of association between them.

Correlation is used to measure the association between variables; regression is used to predict one variable from the other (or many others). However, the equations for correlation and bivariate regression are very similar, as indicated below.

3.5.1 ■ Correlation

The Pearson product-moment correlation coefficient, r, is easily the most frequently used measure of association and the basis of many multivariate calculations. The most interpretable equation for Pearson r is

$$r = \frac{\Sigma Z_X Z_Y}{N - 1} \tag{3.28}$$

where Pearson r is the average cross product of standardized X and Y variable scores.

$$Z_y = \frac{Y - \overline{Y}}{S} \text{ and } Z_X = \frac{X - \overline{X}}{S}$$

and S is as defined in Equations 3.4 and 3.6.

Pearson r is independent of scale of measurement (because both X and Y scores are converted to standard scores) and independent of sample size (because of division by $N - 1$). The value of r ranges between $+1.00$ and -1.00 where values close to .00 represents no relationship or predictability between the X and Y variables. An r value of $+1.00$ or -1.00 indicates perfect predictibility of one score when the other is known. When correlation is perfect, scores for all subjects in the X distribution have the same relative positions as corresponding scores in the Y distribution.[14]

The raw score form of Equation 3.28 also sheds light on the meaning of r:

$$r = \frac{N \Sigma XY - (\Sigma X)(\Sigma Y)}{\sqrt{[N \Sigma X^2 - (\Sigma X)^2][N \Sigma Y^2 - (\Sigma Y)^2]}} \qquad (3.29)$$

Pearson r is the covariance between X and Y relative to (the square root of the product of) the X and Y variances. Only the numerators of variance and covariance equations appear in Equation 3.29 because the denominators cancel each other out.

3.5.2 ■ Regression

Whereas correlation is used to measure the size and direction of the relationship between two variables, regression is used to predict a score on one variable from a score on the other. In bivariate (two-variable) regression where Y is predicted from X, a straight line between the two variables is found. The best-fitting straight line goes through the means of X and Y and minimizes the sum of the squared distances between the data points and the line.

To find the best-fitting straight line, an equation is solved of the form

$$Y' = A + BX \qquad (3.30)$$

where Y' is the predicted score, A is the value of Y when X is 0.00, B is the slope of the line (change in Y divided by change in X), and X is the value from which Y is to be predicted.

The difference between the predicted and the obtained values of Y at each value of X represents errors of prediction or residuals. The best-fitting straight line is the line that minimizes the squared errors of prediction.

To solve Equation 3.30 both B and A are found.

$$B = \frac{N \Sigma XY - (\Sigma X)(\Sigma Y)}{N \Sigma X^2 - (\Sigma X)^2} \qquad (3.31)$$

The bivariate regression coefficient, B, is the ratio of the covariance of the variables and the variance of the one from which predictions are made.

Note the differences and similarities between Equation 3.29 (for correlation), and Equation 3.31 (for the regression coefficient). Both have the covariance between the variables as a numerator but differ in denominator. In correlation, the variances of both are used in the denominator. In regression, the variance of the predictor variable serves as the denominator; if Y is predicted from X, X variance is the denominator while if X is predicted from Y, Y variance is the denominator.

---

[14] When correlation is perfect, $Z_X = Z_Y$ for each pair and the numerator of Equation 3.28 is, in effect, $\Sigma Z_X Z_X$. Because $\Sigma Z_X^2 = N - 1$, Equation 3.28 reduces to $(N - 1)/(N - 1)$, or 1.00.

To complete the solution, the value of the intercept, A, is also calculated.

$$A = \overline{Y} - B\overline{X} \tag{3.32}$$

The intercept is the mean of the predicted variable minus the product of the regression coefficient times the mean of the predictor variable.

3.6 ■ CHI-SQUARE ANALYSIS

Analysis of variance examines the relationship between a discrete variable (the IV) and a continuous variable (the DV), correlation and regression examine the relationship between two continuous variables, and the chi-square (χ^2) test of independence is used to examine the relationship between two discrete variables. If, for instance, one wants to examine a potential relationship between region of the country (Northeast, Southeast, Midwest, South, and West) and approval vs. disapproval of current political leadership, χ^2 is the appropriate analysis.

In χ^2 analysis, the null hypothesis generates expected frequencies against which observed frequencies are tested. If the observed frequencies are similar to the expected frequencies, then the value of χ^2 is small and the null hypothesis is retained; if they are sufficiently different, then the value of χ^2 is large and the null hypothesis is rejected. The relationship between the size of χ^2 and the difference in observed and expected frequencies can readily be seen from the computational equation for χ^2 (Equation 3.33) below.

$$\sum_{ij} (fo - Fe)2/Fe \tag{3.33}$$

where fo represents observed frequencies and Fe represents the expected frequencies in each cell. Summation is over all the cells in a two-way table.

Usually the expected frequencies for a cell are generated from its row sum and its column sum.

$$\text{Cell } Fe = (\text{row sum})(\text{column sum})/N \tag{3.34}$$

When this procedure is used to generate the expected frequencies, the null hypothesis tested is that the variable on the row (say, region of the country) is independent of the variable on the column (attitude toward current political leadership). If the fit to the observed frequencies is good (so that χ^2 is small), then one concludes that the two variables are independent; a poor fit leads to a large χ^2, rejection of the null hypothesis, and the conclusion that the two variables are related.

The techniques discussed in this chapter for making decisions about differences, estimating population means, assessing association between two variables, and predicting a score on one variable from a score on another are important to, and widely used in, the social and behavioral sciences. They form the basis for most undergraduate—and some graduate—statistics courses. It is hoped that this brief review reminds you of material already mastered so that, with common background and language, we begin in earnest the study of multivariate statistics.

CHAPTER 4
Cleaning Up Your Act: Screening Data Prior to Analysis

This chapter deals with a set of issues that are resolved before the main data analysis is run. Careful consideration of these issues is time-consuming and sometimes tedious; it is common, for instance, to spend many days in careful examination of data prior to running the main analysis. But consideration and resolution of these issues before the main analysis is fundamental to an honest analysis of the data.

The first issues concern the accuracy with which data have been entered into the data file and consideration of factors that could produce distorted correlations. Next, missing data, the bane of (almost) every researcher, are assessed and dealt with. Next, many multivariate procedures are based on assumptions; the fit between your data set and the assumptions is assessed before the procedure is applied. Transformations of variables to bring them into compliance with requirements of analysis are considered. Outliers, cases that are extreme, create other headaches because solutions are unduly influenced and sometimes distorted by them. Finally, perfect or near-perfect correlations among variables can threaten a multivariate analysis.

This chapter deals with issues that are relevant to most analyses. However, the issues are not all applicable to all analyses all the time; for instance, multiway frequency analysis (Chapter 7) and logistic regression (Chapter 12), two procedures that use loglinear techniques, have far fewer assumptions than the other techniques. Other analyses have additional assumptions that are not covered in this chapter. For these reasons, assumptions and limitations specific to each analysis are reviewed in the third section of the chapter describing the analysis.

There are differences in data screening for grouped and ungrouped data. If you are performing multiple regression, canonical correlation, factor analysis, or structural equation modeling where subjects are not subdivided into groups, there is one procedure for screening data. If you are performing analysis of covariance, multivariate analysis of variance or covariance, profile analysis, or discriminant function analysis, where subjects are in groups, there is another procedure for screening data. Differences in these procedures are illustrated by example in Section 4.2.

You may find the material in this chapter difficult from time to time. Sometimes it is necessary to refer to material covered in subsequent chapters to explain some issue, material that is more understandable after those chapters are studied. Therefore, you may want to read this chapter now to get an overview of the tasks to be accomplished prior to the main data analysis and then read it again after mastering the remaining chapters.

4.1 ■ IMPORTANT ISSUES IN DATA SCREENING

4.1.1 ■ Accuracy of Data File

The best way to insure the accuracy of a data file is to proofread the original data against a computerized listing of it, using, for instance, SPSS LIST or BMDP PRINT. With a small data file, proofreading is highly recommended, but with a large data file, it may not be possible. In this case, screening for accuracy involves examination of descriptive statistics and graphic representations of the variables.

The first step with a large data set is to examine univariate descriptive statistics through one of the descriptive programs such as SPSS FREQUENCIES, BMDP1D or 2D, SYSTAT STATS, or SAS MEANS or UNIVARIATE. For continuous variables, are all the values within range? Are means and standard deviations plausible? If you have discrete variables (such as categories of religious affiliation), are there any out-of-range numbers? Have you accurately programmed your codes for missing values?

4.1.2 ■ Honest Correlations

Most multivariate procedures analyze patterns of correlation (or covariance) among variables. It is important that the correlations, whether between two continuous variables or between a discrete and continuous variable, be as accurate as possible. Under some rather common research conditions, correlations are inflated, deflated, or simply inaccurately computed.

4.1.2.1 ■ Inflated Correlation

When composite variables are constructed from several individual items by pooling responses to individual items, correlations are inflated if some items are reused. Scales on personality inventories, measures of socioeconomic status, health indices, and many other variables in social and behavioral studies are composites of several items. If composite variables are used and they contain, in part, the same items, correlations are inflated. Don't overinterpret a high correlation between two measures composed, in part, of the same items. If there is enough overlap, consider using only one of the composite variables in the analysis.

4.1.2.2 ■ Deflated Correlation

Sample correlations may be lower than population correlations when there is restricted range in sampling of cases, very uneven splits in the categories of discrete variables, and computational inaccuracy with small coefficients of variation. Problems with distributions that lead to lowered correlations are discussed in Section 4.1.5.

A falsely small correlation between two continuous variables is obtained if the range of responses to one of the variables is restricted in the sample. Correlation is a measure of the extent to which scores on two variables go up together (positive correlation), or one goes up while the other goes down (negative correlation). If the range of scores on one of the variables is narrow because of restricted sampling, then it is effectively a constant and cannot correlate highly with another variable. In a study of success in graduate school, for instance, quantitative ability could not emerge as highly correlated with other variables if all students had about the same high scores in quantitative skills.

When a correlation is too small because of restricted range in sampling, you can estimate its magnitude in a nonrestricted sample by using Equation 4.1 if you can estimate the standard deviation in the nonrestricted sample. The standard deviation in the nonrestricted sample is estimated from prior data or from knowledge of the population distribution.

$$\tilde{r}_{xy} = \frac{r_{t(xy)} \left[\dfrac{S_x}{S_{t(x)}} \right]}{\sqrt{1 + r^2_{t(xy)} \left[\dfrac{S^2_x}{S^2_{t(x)}} \right] - r^2_{t(xy)}}} \tag{4.1}$$

where $\tilde{r}_{xy}$ is adjusted correlation, $r_{t(xy)}$ is the correlation between X and Y with the range of X truncated, S_x is the unrestricted standard deviation of X, and $S_{t(x)}$ is the truncated standard deviation of X.

Many programs allow input of a correlation matrix instead of raw data. The estimated correlation is inserted in place of the truncated correlation prior to analysis of the correlation matrix. (However, insertion of estimated correlations may create internal inconsistencies in the correlation matrix, as discussed in Section 4.1.3.3.)

The correlation between a continuous variable and a dichotomous variable, or between two dichotomous variables (unless they have the same peculiar splits), is also too low if most (say over 90%) responses to the dichotomous variable fall into one category. Even if the continuous and dichotomous variables are strongly related in the population, the highest correlation that could be obtained is well below 1. Some recommend dividing the obtained (but deflated) correlation by the maximum it could achieve given the split between the categories and then using the resulting value in subsequent analyses. This procedure is attractive, but not without hazard, as discussed by Comrey and Lee (1992).

When means are very large numbers and standard deviations are very small, the values in the correlation matrix are also sometimes too small. The programs encode the first several digits of a very large number and then round off the rest. The number of digits rounded off depends on the precision with which numbers are being handled by the computer. If the variability is in the digits that are dropped, then correlations between the variable and others are falsely low. The coefficient of variation (the standard deviation divided by the mean) is an indicator of this problem that is available in many descriptive statistics programs. When the coefficient of variation is very small (0.0001 or less), computational inaccuracy may occur because roundoff errors are too influential to the size of correlation. The solution is to subtract a large constant from every score of a variable with very large numbers before calculating correlation. Subtracting a constant from every score does not affect the size of a variable's correlation with other variables.

4.1.3 ■ Missing Data

Missing data is one of the most pervasive problems in data analysis. The problem occurs when rats die, equipment malfunctions, respondents become recalcitrant, or somebody goofs. Its seriousness depends on the pattern of missing data, how much is missing, and why it is missing.

The pattern of missing data is more important than the amount missing. Missing values scattered randomly through a data matrix pose less serious problems. Nonrandomly missing values, on the other hand, are serious no matter how few of them there are because they affect the generalizability of results. Suppose that in a questionnaire with both attitudinal and demographic questions several respondents refuse to answer questions about income. It is likely that refusal to answer questions about income is related to attitude. If respondents with missing data on income are deleted, the sample values on the attitude variables are distorted. Some method of estimating income is needed to retain the cases for analysis of attitude.

If only a few data points are missing in a random pattern from a large data set, the problems are less serious and almost any procedure for handling missing values yields similar results. If, however, a lot of data are missing from a small to moderately sized data set, the problems can be very serious. Unfortunately, there are as yet no firm guidelines for how much missing data can be tolerated for a sample of a given size.

Although the temptation to assume that data are missing randomly is nearly overwhelming, the safest thing to do is to test it. Use the information you have to test for patterns in missing data. For instance, construct a dummy variable with two groups, cases with missing and nonmissing values on income, and perform a test of mean differences in attitude between the groups. If there are no differences, decisions about how to handle missing data are not so critical (except, of course, for inferences about income). If there are reliable differences and η^2 is substantial (cf. Equation 3.25), care is needed to preserve the cases with missing values for other analyses, as discussed in Section 4.1.3.2. BMDP8D is useful for this analysis because it performs t tests to help determine whether a variable's values are missing at random and provides several options for generating a complete correlation matrix despite missing data.

BMDPAM is a program designed specifically to investigate the relationships among variables with missing values in those cases that have missing values. Table 4.1 shows selected AM output for a data set with missing values on ATTHOUSE ("1" is missing) and INCOME ("99" is missing). Case label 67 (case number 52), among others, is missing INCOME; Case label 338 (case number 253) is missing ATTHOUSE. Although the program does not reveal the relationships between variables with missing values and variables with complete values, it does reveal the relationships among variables with missing values. For this example, the correlation between a $0 - 1$ ($0 =$ case has missing value, $1 =$ case complete) distribution for ATTHOUSE and a $0 - 1$ distribution for INCOME is $-.011$; there is no association between failure to respond regarding INCOME and ATTHOUSE.

The decision about how to handle missing data is important. At best, the decision is among several bad alternatives, four of which are discussed in the subsections that follow. The alternatives are listed with the most frequently used alternatives first. For greater detail on these alternatives and others, consult Rummel (1970) and Cohen and Cohen (1983).

TABLE 4.1   BMDPAM SETUP AND SELECTED OUTPUT FOR MISSING DATA

```
    /INPUT          VARIABLES ARE 8.  FORMAT IS FREE.
                        FILE='SCREEN.DAT'.
                    TITLE IS 'BMDPAM RUN FOR MISSING DATA'.
    /VARIABLE       NAMES ARE SUBNO,TIMEDRS,ATTDRUG,ATTHOUSE,
                    INCOME,EMPLMNT,MSTATUS,RACE.
                    LABEL = SUBNO.
                    MISSING = (4)1, (5)99.
    /END
```

NUMBER OF CASES READ.   465

   NUMBER OF CASES WITH NO DATA MISSING AND WITH
     POSITIVE CASE WEIGHT.   438

PERCENTAGES OF MISSING DATA FOR EACH VARIABLE IN EACH GROUP

THESE PERCENTAGES ARE BASED ON SAMPLE SIZES AND GROUP SIZES
REPORTED WITH THE UNIVARIATE SUMMARY STATISTICS BELOW.
VARIABLES WITHOUT MISSING DATA ARE NOT INCLUDED.

```
                     1

ATTHOUSE    4      0.2
INCOME      5      5.6
```

SAMPLE SIZES FOR EACH PAIR OF VARIABLES

(NUMBER OF TIMES BOTH VARIABLES ARE AVAILABLE)
IN ORDER TO SAVE SPACE, VARIABLES WITH NO MISSING
DATA OR THOSE THAT HAVE NO DATA ARE NOT INCLUDED.

```
              ATTHOUSE INCOME

                  4        5
ATTHOUSE    4    464
INCOME      5    438      439
```

PAIRWISE PERCENTAGES OF MISSING DATA

DIAGONAL ELEMENTS ARE THE PERCENTAGES THAT EACH VARIABLE
IS MISSING. OFF-DIAGONAL ELEMENTS ARE THE PERCENTAGES
EITHER VARIABLE IS MISSING. THESE PERCENTAGES DO NOT INCLUDE
CASES WITH MISSING GROUP OR WEIGHT VARIABLES, CASES WITH
ZERO WEIGHTS, CASES EXCLUDED BY SETTING USE EQUAL TO A
NON-POSITIVE VALUE BY TRANSFORMATIONS, OR CASES WITH GROUPING
VALUES NOT USED. VARIABLES WITH NO MISSING DATA OR THAT
HAVE NO DATA ARE NOT INCLUDED HERE.

```
              ATTHOUSE INCOME
                  4        5

ATTHOUSE    4     0.2
INCOME      5     5.8      5.6
```

CORRELATIONS OF THE DICHOTOMIZED VARIABLES

WHERE, FOR EACH VARIABLE, ZERO INDICATES THAT THE VALUE WAS
MISSING AND ONE INDICATES THAT THE VALUE WAS PRESENT.
VARIABLES WITH NO MISSING DATA OR THAT ARE COMPLETELY
MISSING ARE NOT INCLUDED.

TABLE 4.1   *(CONTINUED)*

```
              ATTHOUSE INCOME
                    4       5

ATTHOUSE   4    1.000
INCOME     5   -0.011    1.000
```

```
PATTERN OF MISSING DATA AND DATA BEYOND LIMITS
------------------------------------------------
COUNT OF MISSING VARIABLES INCLUDES CASES WITH TOO MANY MISSING
VALUES OR DATA BEYOND LIMITS.   THE COLUMN LABELED WT. IS FOR
THE CASE WEIGHT, IF ANY.   M REPRESENTS A MISSING VALUE.
B REPRESENTS A VALUE GREATER THAN THE MAXIMUM LIMIT.
S REPRESENTS A VALUE LESS THAN THE MINIMUM LIMIT.
```

```
                                T A A I E M R
                                I T T N M S A
                                M T T C P T C
                                E D H O L A E
                                D R O M M T
                    NO OF       W R U U E N U
        CASE     CASE MISS.     T S G S   T S
        LABEL     NO. VARS.  GROUP  .     E

        67         52    1                 M
        79         64    1                 M
        84         69    1                 M
        95         77    1                 M
        138       118    1                 M
        156       135    1                 M
        228       161    1                 M
        239       172    1                 M
        240       173    1                 M
        241       174    1                 M
        248       181    1                 M
        265       196    1                 M
        272       203    1                 M
        317       236    1                 M
        321       240    1                 M
        338       253    1             M
        343       258    1                 M
        420       304    1                 M
        447       321    1                 M
        453       325    1                 M
        486       352    1                 M
        512       378    1                 M
        513       379    1                 M
        552       409    1                 M
        568       419    1                 M
        570       421    1                 M
        584       435    1                 M
```

4.1.3.1 ■ Deleting Cases or Variables

One procedure for handling missing values is simply to drop any cases with them. If only a few cases have missing data and they seem to be a random subsample of the whole sample, deletion is a good alternative. Deletion of cases with missing values is the default option for most programs in the BMDP, SPSS, SAS, and SYSTAT packages.[1]

---

[1] Because this is the default option, numerous cases can be deleted without the researcher's knowledge. For this reason, it is important to check the number of cases in your analyses to make sure that all of the desired cases are used.

Be especially aware that some of the programs in the BMDP series delete cases with missing values, even if the variables on which the scores are missing are not currently in use. For example, if the income variable has numerous missing values, some programs will delete all of the cases with missing value on income even from an analysis that does not include income. The solution to this problem is to insert a USE statement into the control language for the variables so that only those variables currently undergoing analysis are considered when cases are deleted for missing values.

If missing values are concentrated in a few variables and those variables are not critical to the analysis, or are highly correlated with other, complete variables, the variable(s) with missing values are profitably dropped.

But if missing values are scattered throughout cases and variables, deletion of cases can mean substantial loss of subjects. This is particularly serious when data are grouped in an experimental design because loss of even one case requires adjustment for unequal n (see Chapter 8). Further, the researcher who has expended considerable time and energy collecting data is not likely to be eager to toss some out. And as previously noted, if cases with missing values are not randomly distributed through the data, distortions of the sample occur if they are deleted.

4.1.3.2 ■ Estimating Missing Data

A second option is to estimate (impute) missing values and then use the estimates during data analysis. There are at least three popular schemes for doing so; using prior knowledge, inserting mean values, and using regression.

Prior knowledge is used when a researcher replaces a missing value with a value from a well-educated guess. If the researcher has been working in an area for a while, and if the sample is large and the number of missing values small, this is often a reasonable procedure. The researcher is often confident that the value would have been about at the median, or whatever, for a particular case. Alternatively, the researcher can downgrade a continuous variable to a dichotomous variable (e.g., "high" vs. "low") to predict with confidence into which category a case with a missing value falls. The discrete variable replaces the continuous variable in the analysis, but it has less information than the continuous variable.

Or, *means* are calculated from available data and used to replace missing values prior to analysis. In the absence of all other information, the mean is the best guess about the value of a variable. Part of the attraction of this procedure is that it is conservative; the mean for the distribution as a whole does not change and the researcher is not required to guess at missing values. On the other hand, the variance of a variable is reduced because the mean is probably closer to itself than to the missing value it replaces, and the correlation the variable has with other variables is reduced because of the reduction in variance. The extent of loss in variance depends on the amount of missing data and on the actual values that are missing.

A compromise is to insert a group mean for the missing value. If, for instance, the case with a missing value is a Republican, the mean value for Republicans is computed and inserted in place of the missing value. This procedure is not as conservative as inserting overall mean values and not as liberal as using prior knowledge.

The BMDPAM program has provision for inserting the overall mean value or the mean value for a group if the cases are classified into groups through use of METHOD IS MEAN in the ESTIMATE paragraph. The same goal is accomplished through the MEANSUB option in some of the SPSS programs or with variable modification (e.g., RECODE, IF . . . THEN) language in the other packages. SAS STANDARD allows a data set to be created with missing values replaced by the variable mean.

A sophisticated method for estimating missing values uses *regression* (see Chapter 5). Other variables are used as IVs to write a regression equation for the variable with missing data serving as DV. Cases with complete data generate the regression equation; the equation is then used to predict missing values for incomplete cases. Sometimes the predicted values from a first round of regression are inserted for missing values and then all the cases are used in a second regression. The predicted values for the variable with missing data from round two are used to develop a third equation, and so forth until the predicted values from one step to the next are similar. The predictions from the last round are the ones used to replace missing values.

Advantages to regression are that it is more objective than the researcher's guess but not as blind as simply inserting the grand mean. One disadvantage to use of regression is that the scores fit together better than they should: because the missing value is predicted from other variables, it is likely to be more consistent with them than a real score is. A second problem is reduced variance because the estimate is probably too close to the mean. A third problem is the requirement that good IVs be available in the data set; if none of the other variables is a good predictor of the one with missing data, the estimate from regression is about the same as simply inserting the mean. Finally, estimates from regression are used only if the estimated value falls within the range of values for complete cases; out of range estimates are not acceptable.

Using regression to estimate missing values is convenient in BMDPAM through the METHOD IS sentence of the ESTIMATE paragraph. If SINGLE is specified, the missing value is predicted from bivariate regression with the best complete variable. TWOSTEP uses the best two complete variables, STEP only uses variables that correlate with the DV, and REGR uses all variables available for the case. The data file with estimates inserted is created via the SAVE paragraph and used in subsequent analyses. PRELIS (Chapter 14) estimates missing data by inserting a value from another case that has a similar response pattern across the variables.

4.1.3.3 ■ Using a Missing Data Correlation Matrix

Another option involves analysis of a missing data correlation matrix. In this option, all available pairs of values are used to calculate each of the correlations in **R**. A variable with 10 missing values has all its correlations with other variables based on 10 fewer pairs of numbers. If some of the other variables also have missing values, but in different cases, the number of complete pairs of variables is further reduced. Thus each correlation in **R** can be based on a different number and a different subset of cases, depending on the pattern of missing values. Because the standard error of the sampling distribution for r is based on N, some correlations are less stable than others in the same correlation matrix.

But that is not the only problem. In a correlation matrix based on complete data, the sizes of some correlations place constraints on the sizes of others. In particular,

$$r_{13}r_{23} - \sqrt{(1 - r_{13}^2)(1 - r_{23}^2)} \le r_{12} \le r_{13}r_{23} + \sqrt{(1 - r_{13}^2)(1 - r_{23}^2)} \tag{4.2}$$

The correlation between variables 1 and 2, r_{12}, cannot be smaller than the value on the left or larger than the value on the right in a three-variable correlation matrix. If $r_{13} = .60$ and $r_{23} = .40$, then r_{12} cannot be smaller than $-.49$ or larger than $.97$. If, however, r_{12}, r_{23}, and r_{13} are all based on different subsets of cases because of missing data, the value for r_{12} may go out of range.

Most multivariate statistics involve calculation of eigenvalues and eigenvectors from a correlation matrix (see Appendix A). With loosened constraints on size of correlations in a missing data correlation matrix, eigenvalues can become negative. Because eigenvalues represent variance, negative eigenvalues represent something akin to negative variance. Moreover, because the total variance that is partitioned among eigenvalues is a constant (usually equal to the number of variables), positive eigenvalues are inflated by the size of negative eigenvalues, resulting in inflation of variance. The statistics derived under these conditions are very likely distorted.

However, with a large sample and only a few missing values, eigenvalues are often all positive even if some correlations are based on slightly different pairs of cases. Under these conditions, a missing data correlation matrix provides a reasonable multivariate solution and has the advantage of using all available data. Use of this option for the missing data problem should not be rejected out of hand but should be used cautiously with a wary eye to negative eigenvalues.

A missing value correlation matrix is prepared through the TYPE = ALLVALUE option in the ESTIMATE paragraph of BMDPAM, through the CORPAIR (default) option in BMDP8D, through the PEARSON/PAIRWISE option of the SYSTAT CORR program, or through the PAIRWISE deletion option in some of the SPSS programs. It is the default option for SAS CORR. If this is not an option of the program you want to run then generate a missing data correlation matrix through another program for input to the one you are using.

4.1.3.4 ■ Treating Missing Data as Data

It is possible that the fact that a value is missing is itself a very good predictor of the variable of interest in your research. If a dummy variable is created where cases with complete data are assigned 0 and cases with missing data 1, the liability of missing data could become an asset. The mean is inserted for missing values so that all cases are analyzed, and the dummy variable is used as simply another variable in analysis, as discussed by Cohen and Cohen (1983, pp. 281–292).

4.1.3.5 ■ Repeating Analyses with and without Missing Data

If you use some method of estimating missing values or a missing data correlation matrix, consider repeating your analyses using only complete cases. This is particularly important if the data set is small, the proportion of missing values high, or data are missing in a nonrandom pattern. If the results are similar, you can have confidence in them. If they are different, however, you need to investigate the reasons for the difference, and either evaluate which result more nearly approximates "reality" or report both sets of results.

4.1.4 ■ Outliers

Outliers are cases with such extreme values on one variable or a combination of variables that they distort statistics. For example, consider a regression coefficient; outliers, more than other cases, determine which one of a number of possible regression lines is chosen. In the bivariate scatterplot of Figure 4.1, several regression lines, all at slightly different tilts, provide a good fit to the data points inside the swarm. But when the data point labeled A in the upper right-hand portion of the scatterplot is also considered, the regression coefficient that is computed is the one from among the several good alternatives that provides the best fit to the extreme case. The case is an outlier because it has much more impact on the value of the regression coefficient than any of those inside the swarm.

Outliers are found in both univariate and multivariate situations, among both dichotomous and continuous variables, among both IVs and DVs, and in both data and results of analyses. They lead

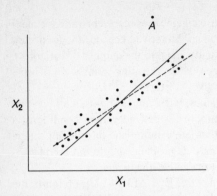

FIGURE 4.1 BIVARIATE SCATTERPLOT
FOR SHOWING IMPACT OF
AN OUTLIER.

to both Type I and Type II errors, frequently with no clue as to which effect they have in a particular analysis. And they lead to results that do not generalize because the results are overly determined by the outlier(s). Outliers are a pervasive problem in data analyses.

There are four reasons for the presence of an outlier. First is incorrect data entry. Cases that are extreme should be checked carefully to see that data are correctly entered. Second is failure to specify missing value codes in computer control language so that missing value indicators are read as real data. Third is that the outlier is not a member of the population from which you intended to sample. If the case should not have been sampled, it is deleted once it is detected. Fourth is that the case is from the intended population but the distribution for the variable in the population has more extreme values than a normal distribution. In this event, the researcher retains the case but considers changing the value on the variable(s) so that the case no longer has as much impact. Although errors in data entry and missing values specification are easily found and remedied, deciding between alternatives three and four, between deletion and retention with alteration, is difficult.

4.1.4.1 ■ Detecting Univariate and Multivariate Outliers

Univariate outliers are cases with an extreme value on one variable; multivariate outliers are cases with an unusual combination of scores on two or more variables. For example, a 15-year-old is perfectly within bounds regarding age, and someone who earns $45,000 a year is in bounds regarding income, but a 15-year-old who earns $45,000 a year is very unusual and is likely to be a multivariate outlier.

Among dichotomous variables, those with very uneven splits between two categories produce outliers. Rummel (1970) suggests deleting dichotomous variables with 90–10 splits between categories both because the correlation coefficients between these variables and others are truncated and because the scores in the category with 10% of the cases are more influential than those in the category with 90% of the cases. Dichotomous variables with extreme splits are easily found in the programs for frequency distributions (SPSS FREQUENCIES, BMDP2D, SYSTAT TABLES, or SAS UNIVARIATE) used during routine preliminary data screening.

Among continuous variables, the method of searching for outliers depends on whether or not data are grouped. If you are going to perform one of the analyses with ungrouped data (regression, canonical correlation, factor analysis, or structural equation modeling), univariate and multivariate outliers are sought among all cases at once, as illustrated in Sections 4.2.1.1 (univariate) and 4.2.1.4 (multivariate). If you are going to perform one of the analyses with grouped data (ANCOVA, MANOVA or MANCOVA, profile analysis, discriminant function analysis, or logistic regression) outliers are sought separately within each group, as illustrated in Sections 4.2.2.1 and 4.2.2.3.

Among continuous variables, *univariate outliers* are cases with very large standardized scores, z scores, on one or more variables. Cases with standardized scores in excess of 3.29 ($p < .001$, two-tailed test) are potential outliers. However, the extremeness of a standardized score depends on the size of the sample; with a very large N, a few standardized scores in excess of 3.29 are expected. Z scores are available through BMDP1D, SPSS DESCRIPTIVES (plus Option 3 in some versions), SYSTAT (in any procedure), and SAS STANDARD (with MEAN = 0 and STD = 1). In addition, many of the BMDP programs give the z score associated with the highest and lowest raw scores as part of routine output.

As an alternative or in addition to inspection of z scores, there are graphical methods for finding univariate outliers. Helpful plots are histograms, box plots, normal probability plots, or detrended normal probability plots. Histograms of variables are readily understood and available and may reveal one or more univariate outliers. There is usually a pileup of cases near the mean with cases trailing away in either direction. An outlier is a case (or a very few cases) that seems to be unattached to the rest of the distribution. Histograms for continuous variables are produced by BMDP2D (ungrouped) and BMDP5D or 7D (grouped data), SPSS FREQUENCIES (plus SORT and SPLIT for grouped data), SYSTAT DENSITY (HIST and BOX), and SAS UNIVARIATE or CHART (with BY for grouped data).

Box plots are simpler and literally box in observations that are around the median; cases that fall far away from the box are extreme. Normal probability plots and detrended normal probability plots are also very useful for assessing normality of distributions of variables and are discussed in that context in Section 4.1.5.1. However, points that lie a considerable distance from others on these plots are also potential univariate outliers.

Once potential univariate outliers are located, the search for multivariate outliers begins. It is often a good idea to see if univariate outliers are also multivariate outliers before deciding what to do with them. Often the same cases show up in both analyses; sometimes they do not. It is usually better to decide what to do with outliers when the total extent of the problem is known.

Multivariate outliers are cases that have an unusual pattern of scores. The z score on each variable may be within the expected range, but when two or more variables are considered together, the unusual combination of scores becomes apparent. Cases with such unusual combinations are sometimes called discrepant. It is almost impossible to screen for multivariate outliers by hand, but both statistical methods and graphical methods are available, particularly among the BMDP programs. Of the two methods, the statistical procedure is preferred.

One statistical procedure is computation of Mahalanobis distance for each case. Mahalanobis distance is the distance of a case from the centroid of the remaining cases where the centroid is the point created by the means of all the variables. Mahalanobis distance is also a discriminant function analysis (see Chapter 11) where an equation is computed that best separates one case from the rest of the cases. If a case has an unusual combination of scores, then those scores are weighted heavily in the equation (the discriminant function) and the Mahalanobis distance of the case from the rest of the cases is statistically significant.

Mahalanobis distance of each case, in turn, from all other cases, is available in several of the BMDP programs (4M, 7M, AM, and 9R) and in SYSTAT GLM and ANOVA (where the value produced is the square root of the Mahalanobis distance produced by the other programs). Mahalanobis distances for all cases are also found through SPSS REGRESSION by using the CASEWISE subcommand; the 10 cases with the largest Mahalanobis distances are identified through the RESIDUALS subcommand with the OUTLIERS(MAHAL) specification.

Mahalanobis distances are requested and interpreted in Sections 4.2.1.4 and 4.2.2.3, and numerous other places throughout the book. A very conservative probability estimate for a case being an outlier, say $p < .001$, is appropriate with Mahalanobis distance.

A number of regression programs, including those in SAS, provide a value of leverage, h_{ii}, which converts easily to Mahalanobis distance[2]:

$$Mahalanobis\ distance = (N - 1)(h_{ii} - 1/N) \tag{4.3}$$

When multivariate outliers are sought in grouped data, they are sought within each group. Mahalanobis distance as a test for within-group multivariate outliers is provided in BMDPAM, BMDP7M, and SYSTAT GLM and ANOVA. Use of other programs, including SPSS REGRESSION, requires separate runs for each group. In these runs you must specify a dummy DV, such as the case number, to find outliers among the set of IVs. When separate runs for each group are used instead of a single run for within-group outliers, different error terms are developed and different cases may be identified as outliers.

Frequently some outliers hide behind other outliers. If a few cases identified as outliers are deleted, the data become more consistent and then other cases become extreme. Sometimes it is a good idea to screen for multivariate outliers several times, each time dealing with cases identified as outliers on the last run, until finally no new outliers are identified. But if the process of identifying ever more outliers seems to stretch into infinity, do a trial run with and without outliers to see if ones identified later are truly influencing results. If not, do not delete the later-identified outliers.

4.1.4.2 ■ Describing Outliers

Once multivariate outliers are identified, you need to discover why the cases are extreme. (You already know why univariate outliers are extreme.) It is important to identify the variables on which the cases are deviant for three reasons. First, this procedure helps you decide whether the case is properly part of your sample, or not. Second, if you are going to modify scores instead of delete cases, you have to know which scores to modify. Third, it provides an indication of the kinds of cases to which your results do not generalize.

If there are only a few multivariate outliers, it is reasonable to examine them individually. If there are several, you may choose to examine them as a group to see if there are any variables that separate the group of outliers from the rest of the cases.

Whether you are examining one or a group of outliers, the trick is to create a dummy grouping variable where the outlier(s) have one value and the rest of the cases another value. The dummy variable is then used as the grouping DV in discriminant function analysis (Chapter 11) or logistic regression (Chapter 12) or as the DV in regression (Chapter 5). If it is available to you, BMDP9R, the program that performs all possible subsets regression, is the program of choice for investigat-

---

[2] Leverage is discussed more fully in Section 5.3.2.2.

ing outliers. Variables on which the outlier(s) differ from the rest of the cases show up again and again in the various subsets. Once those variables are identified, means on those variables for outlying and nonoutlying cases are found through any of the routine descriptive programs. Description of outliers is illustrated in Sections 4.2.1.4 and 4.2.2.3.

4.1.4.3 ■ Reducing the Influence of Outliers

Once outliers have been identified, there are several strategies for reducing their influence. But before you use one of them, *check the data for the case* to make sure that they are accurately entered into the data file. If the data are accurate, consider the possibility that one variable is responsible for most of the outliers. If so, *elimination of the variable* would reduce the number of outliers. If the variable is highly correlated with others or is not critical to the analysis, deletion of it is a good alternative.

If neither of these simple alternatives is reasonable, you must decide whether the cases that are outliers are properly part of the population from which you intended to sample. *If the cases are not part of the population, they are deleted* with no loss of generalizability of results to your intended population. The description of the outliers is a description of the kinds of cases to which the results do not apply.

If you decide that the *outliers are sampled from your target population*, they remain in the analysis, but steps are taken to reduce their impact—*variables are transformed or scores changed.*

For univariate outliers, a first option for reducing impact is variable transformation, undertaken to change the shape of the distribution to more nearly normal. In this case, outliers are considered part of a nonnormal distribution with tails that are too heavy so that too many cases fall at extreme values of the distribution. Cases that were outliers in the untransformed distribution are still on the tails of the transformed distribution, but their impact is reduced. Transformation of variables has other salutary effects, as described in Section 4.1.6.

For univariate outliers, a second option is to change the score(s) on the variable(s) for the outlying case(s) so that they are deviant, but not as deviant as they were. For instance, assign the outlying case(s) a raw score on the offending variable that is one unit larger (or smaller) than the next most extreme score in the distribution. Because measurement of variables is sometimes rather arbitrary anyway, this is often an attractive alternative to reduce the impact of a univariate outlier.

Transformation or score alteration may not work for a truly multivariate outlier because the problem is with the combination of scores on two or more variables, not with the score on any one variable. The case is discrepant from the rest in its combinations of scores. Although the number of apparent outliers is often substantially reduced after transformation or alteration, there are sometimes a few cases that still are distant from the rest. These cases are either deleted or allowed to remain with the knowledge that the results may be distorted in almost any direction by them.

Any deletions, changes of scores, and transformations are reported in the Results section together with the rationale.

4.1.4.4 ■ Outliers in a Solution

Some cases may not fit well within a solution; for example, their predicted scores may differ markedly from their actual scores in multiple regression. Such cases are identified *after* an analysis is completed, not as part of the screening process. To identify and eliminate or change scores for such cases before the major analysis could produce overfitting, and in any event would make the solution look better than it actually is. Therefore, it is a procedure best limited to exploratory analysis. Chapters describing techniques for ungrouped data deal with outliers in the solution in sections devoted to limitations of the technique.

4.1.5 ■ Normality, Linearity, and Homoscedasticity

Underlying some multivariate procedures and most statistical tests of their outcomes is the assumption of multivariate normality. Multivariate normality is the assumption that each variable and all linear combinations of the variables are normally distributed. When the assumption is met, the residuals[3] of analysis are also normally distributed and independent. The assumption of multivariate normality is not readily tested because it is impractical to test an infinite number of linear combinations of variables for normality. Those tests that are available are overly sensitive for most analyses.

Although the assumption of multivariate normality is made in derivation of many significance tests, its importance in analysis of a data set is not, at present, known. It is tempting to conclude that most inferential statistics are robust[4] to violations of the assumption, but that conclusion may not be warranted.[5] Multivariate statistics probably are sometimes robust to violation of their assumptions but, unfortunately, the literature on robustness is far from conclusive regarding when assumptions can be violated. The safest strategy is to use transformations of variables to improve their normality unless there is some compelling reason not to.

Evaluation of the normality of distributions is almost always relevant when statistical inference is planned. Bradley (1982) reports that statistical inference becomes less and less robust as distributions depart from normality, rapidly so under many conditions. And even when the statistics are used purely descriptively, normality, linearity, and homoscedasticity of variables enhance the analysis.

The assumption of multivariate normality applies differently to different multivariate statistics. For analyses where subjects are not grouped, the assumption applies to the distributions of the variables themselves or to the residuals of the analyses; for analyses where subjects are grouped, the assumption applies to the sampling distributions[6] of means of variables.

For ungrouped data, if there is multivariate normality, each variable is itself normally distributed and the relationships between pairs of variables, if present, are linear and homoscedastic (i.e., the variance of one variable is the same at all values of the other variable). The assumption of multivariate normality can be checked partially through normality, linearity, and homoscedasticity of

---

[3] Residuals are left-overs. They are the segments of scores not accounted for by the multivariate analysis. They are also called "errors" between predicted and obtained scores where the analysis provides the predicted scores. Note that the practice of using a dummy DV such as case number to investigate multivariate outliers will *not* produce meaningful residuals plots.

[4] Robust means that the researcher is led to correctly reject the null hypothesis at a given alpha level the right number of times even if the distributions do not meet the assumptions of analysis. Often Monte Carlo procedures are used where a distribution with some known properties is put into a computer, sampled from repeatedly, and repeatedly analyzed; the researcher studies the rates of retention and rejection of the null hypothesis against the known properties of the distribution in the computer.

[5] The univariate *F* test of mean differences, for example, is frequently said to be robust to violation of assumptions of normality and homogeneity of variance with large and equal samples, but Bradley (1984) questions this generalization.

[6] A sampling distribution is a distribution of statistics (not of raw scores) computed from random samples of a given size taken repeatedly from a population. For example, in univariate ANOVA, hypotheses are tested with respect to the sampling distribution of means (Chapter 3).

variables (or through examination of residuals in analyses involving prediction).[7] The assumption is certainly violated, at least to some extent, if the variables (or the residuals) are not normally distributed, and do not have pairwise linearity and homoscedasticity.

For grouped data, it is the sampling distributions of the means of variables that are to be normally distributed. The central limit theorem reassures us that, with large sample sizes, sampling distributions of means are normally distributed regardless of the distributions of variables. For example, if there are at least 20 degrees of freedom for error in a univariate ANOVA, the F test is said to be robust to violations of normality of variables (provided that there are no outliers). The degree to which robustness extends to multivariate analyses is not yet clear, but the larger the sample size the smaller the effect nonnormality is likely to have. If you have grouped data with large samples, transformation of variables is less imperative.

These issues are discussed again in the third sections of Chapters 5 through 14 as they apply directly to one or another of the multivariate procedures. For nonparametric procedures such as multiway frequency analysis (Chapter 10) or nonlinear analyses such as logistic regression (Chapter 12), there are no distributional assumptions. Instead, distributions of scores typically are hypothesized and observed distributions are tested against hypothesized distributions.

4.1.5.1 ■ Normality

Screening continuous variables for normality is an important early step in almost every multivariate analysis, particularly when inference is a goal. Although normality of the variables is not always required for analysis, the solution is usually quite a bit better if the variables are all normally distributed. If the variables are not normally distributed, and particularly if they are nonnormal in very different ways (e.g., some positively and some negatively skewed), the solution is degraded.

Normality of variables is assessed by either statistical or graphical methods. Two components of normality are skewness and kurtosis. Skewness has to do with the symmetry of the distribution; a skewed variable is a variable whose mean is not in the center of the distribution. Kurtosis has to do with the peakedness of a distribution; a distribution is either too peaked (with long, thin tails) or too flat (with short, heavy tails).[8] Figure 4.2 shows a normal distribution, distributions with skewness, and distributions with nonnormal kurtosis. A variable can have significant skewness, kurtosis, or both.

When a distribution is normal, the values of skewness and kurtosis are zero. If there is positive skewness, there is a pileup of cases to the left and the right tail is too long; with negative skewness, there is a pileup of cases to the right and the left tail is too long. Kurtosis values above zero indicate a distribution that is too peaked with long tails, while kurtosis values below zero indicate a distribution that is too flat (also with too many cases in the tails).[9] Nonnormal kurtosis produces an underestimate of the variance of a variable.

---

[7] Analysis of residuals to screen for normality, linearity and homoscedasticity in multiple regression is discussed in Section 5.3.2.4.

[8] If you decide that outliers are sampled from the intended population but that there are too many cases in the tails, you are saying that the distribution from which the outliers are sampled has kurtosis that departs from normal.

[9] The equation for kurtosis gives a value of 3 when the distribution is normal, but all of the statistical packages subtract 3 before printing kurtosis so that the expected value is zero.

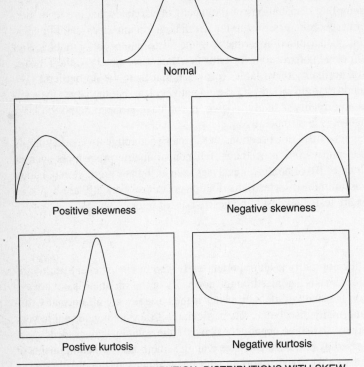

FIGURE 4.2 NORMAL DISTRIBUTION, DISTRIBUTIONS WITH SKEW-
NESS, AND DISTRIBUTIONS WITH KURTOSES.

There are significance tests for both skewness and kurtosis that test the obtained value against null hypotheses of zero. For instance, the standard error for skewness is approximately

$$s_s = \sqrt{\frac{6}{N}} \tag{4.4}$$

where N is the number of cases. The obtained skewness value is then compared with zero using the z distribution, where

$$z = \frac{S - 0}{s_S} \tag{4.5}$$

and S is the value reported for skewness. The standard error for kurtosis is approximately

$$s_k = \sqrt{\frac{24}{N}} \tag{4.6}$$

and the obtained kurtosis value is compared with zero using the z distribution, where

$$z = \sqrt{\frac{K - 0}{S_k}} \tag{4.7}$$

and K is the value reported for kurtosis.

Conventional but conservative (.01 or .001) alpha levels are used to evaluate the significance of skewness and kurtosis with small to moderate samples, but if the sample is large, it is a good idea to look at the shape of the distribution instead of using formal inference tests. Because the standard errors for both skewness and kurtosis decrease with larger N, with large samples the null hypothesis is likely to be rejected when there are only minor deviations from normality.

In a large sample, a variable with statistically significant skewness often does not deviate enough from normality to make a substantive difference in the analysis. In other words, with large samples the significance level of skewness is not as important as its actual size (worse the farther from zero) and the visual appearance of the distribution. In a large sample, the impact of departure from zero kurtosis also diminishes. For example, underestimates of variance associated with negative kurtosis (flat distributions) disappear with samples of 100 or more cases; with positive kurtosis underestimation of variance disappears with samples of 200 or more (Waternaux, 1976).

Values for skewness and kurtosis are available in several programs. SPSS FREQUENCIES, for instance, prints as options skewness, kurtosis, and their standard errors and, in addition, superimposes a normal distribution over a frequency histogram for a variable if HISTOGRAM = NORMAL is specified. BMDP2D gives both skewness and kurtosis values and their significance levels, along with a histogram of the variable. Several other BMDP programs routinely print out values for skewness and kurtosis, but without significance levels. SYSTAT STATS produces skewness and kurtosis as options; the distribution is inspected through DENSITY. SAS MEANS and UNIVARIATE provide skewness and kurtosis values. A histogram or stem and leaf plot is also available in SAS UNIVARIATE. In structural equation modeling (Chapter 14), skewness and kurtosis for each variable are available in EQS and Mardia's Coefficient (the multivariate kurtosis measure) is available in EQS, PRELIS, and CALIS. In addition, PRELIS can be used to deal with nonnormality through alternative correlation coefficients, such as polyserial or polychoric (cf. Section 14.5.6).

Frequency histograms are an important graphical device for assessing normality, especially with the normal distribution as an overlay, but even more helpful than frequency histograms are expected normal probability plots and detrended expected normal probability plots. In these plots, the scores are ranked and sorted; then an expected normal value is computed and compared with the actual normal value for each case. The expected normal value is the z score that a case with that rank holds in a normal distribution; the normal value is the z score it has in the actual distribution. If the actual distribution is normal, then the points for the cases fall along the diagonal running from lower left to upper right, with some minor deviations due to random processes. Deviations from normality shift the points away from the diagonal.

Consider the expected normal probability plots for ATTDRUG and TIMEDRS through BMDP5D in Figure 4.3. As reported in Section 4.2.1.1, ATTDRUG is reasonably normally distributed (kurtosis = -0.447, skewness = -0.123) while TIMEDRS is too peaked and positively skewed (kurtosis = 13.101, skewness = 3.248, both significantly different from 0). The cases for ATTDRUG line up along the diagonal while those for TIMEDRS do not. At low values of TIMEDRS, there are too many cases above the diagonal and at high values there are too many cases below the diagonal, reflecting the patterns of skewness and kurtosis.

```
/INPUT              VARIABLES ARE 8.  FORMAT IS FREE
                    FILE='SCREEN.DAT'.
                    TITLE IS 'NORMALITY AS ASSESSED THROUGH EXPECTED NORMAL
                        PROBABILITY PLOTS'.
/VARIABLE           NAMES ARE SUBNO, TIMEDRS, ATTDRUG, ATTHOUSE,
                        INCOME, EMPLMNT, MSTATUS, RACE.
                    LABEL = SUBNO.
                    MISSING = (4)1, (5) 99.
                    USE = ATTDRUG, TIMEDRS.
/PLOT               TYPE = NORM, DNORM.
/END
```

NORMAL PLOT OF VARIABLE   3 ATTDRUG

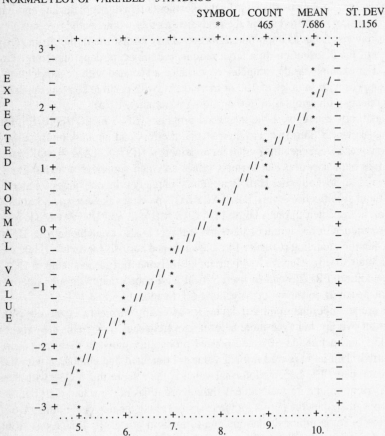

ATTDRUG
VALUES FROM NORMAL DISTRIBUTION WOULD LIE
ON THE LINE INDICATED BY THE SYMBOL "/"

FIGURE 4.3 EXPECTED NORMAL PROBABILITY PLOT AND DETRENDED NORMAL PROB-
ABILITY PLOT FOR ATTDRUG AND TIMEDRS CREATED THROUGH BMDP5D.

DEVIATIONS FROM NORMAL PLOT OF VARIABLE   3 ATTDRUG

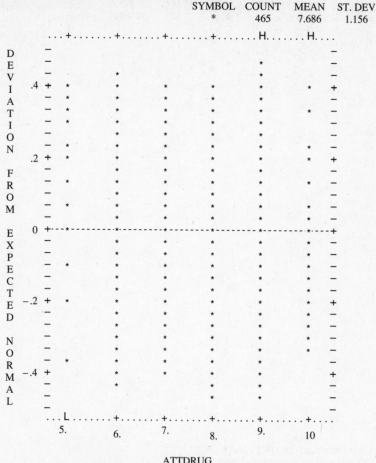

ATTDRUG
VALUES FROM NORMAL DISTRIBUTION WOULD LIE
ON THE LINE INDICATED BY THE SYMBOL "–".

FIGURE 4.3 *(CONTINUED)*

Detrended normal probability plots for TIMEDRS and ATTDRUG are also in Figure 4.3. These plots are similar to expected normal probability plots except that deviations from the diagonal are plotted instead of values along the diagonal. In other words, the linear trend from lower left to upper right is removed. If the distribution of a variable is normal, as is ATTDRUG, the cases distribute themselves evenly above and below the horizontal line that intersects the Y axis at 0.0, the line of zero deviation from expected normal values. The skewness and kurtosis of TIMEDRS are again apparent from the cluster of points above the line at low values of TIMEDRS and below the line at

NORMAL PLOT OF VARIABLE   2 TIMEDRS

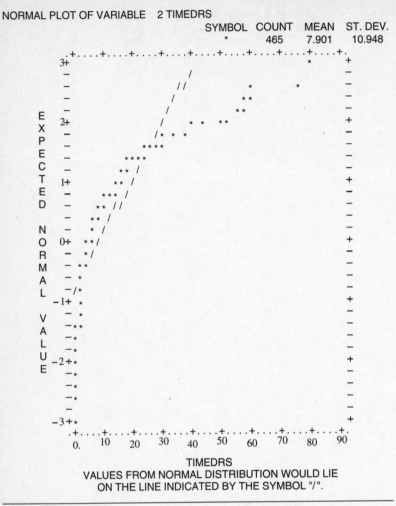

FIGURE 4.3 *(CONTINUED)*

high values of TIMEDRS. Normal probability plots for variables are also available in SAS UNI-VARIATE, SPSS MANOVA, and SYSTAT PPLOT. Many of these programs also produce detrended normal plots.

Univariate outliers show up on these plots as cases that are far away from the other cases. Note, for instance, the two cases in the upper right-hand corner of the expected normal probability plot for TIMEDRS. The cases are a considerable distance from the others and do not appear to be connected to them. These cases are most likely to show up as univariate outliers when z scores are considered.

If you are going to perform multiple regression, an alternative to screening variables prior to analysis is conducting the analysis and then screening through the residuals (the differences between the predicted and obtained DV values). If normality is present, the residuals are normally and independently distributed. That is, the differences between predicted and obtained scores in

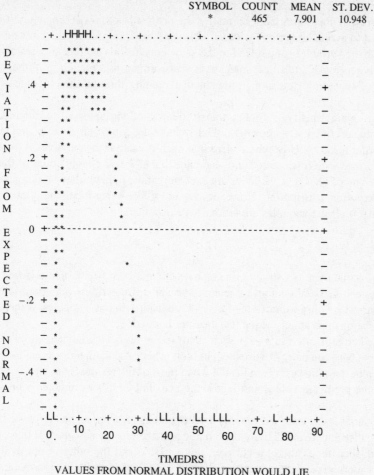

DEVIATIONS FROM NORMAL PLOT OF VARIABLE 2 TIMEDRS

	SYMBOL	COUNT	MEAN	ST. DEV.
	*	465	7.901	10.948

FIGURE 4.3 (CONTINUED)

multiple regression—the errors—are symmetrically distributed around a mean value of zero and there are no contingencies among the errors. In multiple regression, residuals are also screened for normality through the expected normal probability plot and the detrended normal probability plot.[10] All the BMDP regression programs and SPSS REGRESSION provide this diagnostic technique (and others, as discussed in Chapter 5). If the residuals are normally distributed, the expected normal probability plot and the detrended normal probability plot look just the same as they do if a

---

[10] For grouped data, residuals have the same shape as within-group distributions because the predicted value is the mean, and subtracting a constant does not change the shape of the distribution. Many of the programs for grouped data plot the within-group distribution as an option, as discussed in the next few chapters when relevant.

variable is normally distributed. In regression, if the residuals plot looks normal, there is no reason to screen the individual variables for normality.

Although residuals will reveal departures from normality, the analyst has to resist temptation to look at the rest of the output to avoid "tinkering" with variables and cases to produce an anticipated result. Because screening the variables should lead to the same conclusions as screening residuals, it may be more objective to make one's decisions about transformations, deletion of outliers, and the like, on the basis of screening runs alone rather than screening through the outcome of analysis.[11]

With ungrouped data, if nonnormality is found, transformation of variables is considered. Common transformations are described in Section 4.1.6. Unless there are compelling reasons not to transform, it is probably better to do so. However, realize that even if each of the variables is normally distributed, or transformed to normal, there is no guarantee that all linear combinations of the variables are normally distributed. That is, if variables are each univariate normal, they do not necessarily have a multivariate normal distribution. However, it is more likely that the assumption of multivariate normality is met if all the variables are normally distributed.

4.1.5.2 ■ Linearity

The assumption of linearity is that there is a straight line relationship between two variables (where one or both of the variables can be combinations of several variables). Linearity is important in a practical sense because Pearson's r only captures the linear relationships among variables; if there are substantial nonlinear relationships among variables, they are ignored.

Nonlinearity is diagnosed either from residuals plots in analyses involving a predicted variable, or from bivariate scatterplots between pairs of variables. In plots where standardized residuals are plotted against predicted values, nonlinearity is indicated when most of the residuals are above the zero line on the plot at some predicted values and below the zero line at other predicted values (see Chapter 5).

Linearity between two variables is assessed roughly by inspection of bivariate scatterplots. If both variables are normally distributed and linearly related, the scatterplot is oval-shaped. If one of the variables is nonnormal, then the scatterplot between this variable and the other is not oval. Examination of bivariate scatterplots is demonstrated in Section 4.2.1.2, along with transformation of a variable to enhance linearity.

However, sometimes the relationship between variables is simply not linear. Consider, for instance, number of symptoms and dosage of drug, as shown in Figure 4.4(a). It seems likely that there are lots of symptoms when the dosage is low, only a few symptoms when the dosage is moderate, and lots of symptoms again when the dosage is high. Number of symptoms and drug dosage are curvilinearly related. One alternative in this case is to use the square of number of symptoms instead of number of symptoms in the analysis. Another alternative is to do a three-way equal-n on number dosage—code it into high, medium and low on a couple of dummy variables (high vs. low on one dummy variable and a combination of high and low vs. medium on another dummy variable)—and then use the dummy variables in place of dosage in analysis.[12] The dichotomous dummy variables can only have a linear relationships with other variables, if, indeed, there is any relationship at all after recoding.

---

[11] We realize that others (e.g., Berry, 1993; Fox 1991) have very different views about the wisdom of screening from residuals.

[12] A nonlinear analytic strategy is most appropriate here, such as nonlinear regression through BMDP3R or SYSTAT NONLIN, but such strategies are beyond the scope of this book.

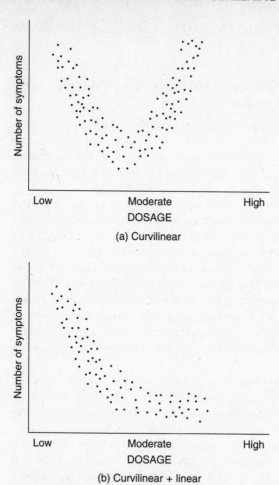

FIGURE 4.4 CURVILINEAR RELATIONSHIP AND
CURVILINEAR AND MIXED LINEAR
RELATIONSHIP.

Often two variables have a mix of linear and curvilinear relationships, as shown in Figure 4.4(b). One variable generally gets smaller (or larger) as the other gets larger (or smaller) but there is also a curve to the relationship. For instance, symptoms might drop off with increasing dosage, but only to a point; increasing dosage beyond the point does not result in further reduction or increase in symptoms. In this case, the linear component may be strong enough that not much is lost by ignoring the curvilinear component unless it has important theoretical implications.

Assessing linearity through bivariate scatterplots is reminiscent of reading tea leaves, especially with small samples. And there are many cups of tea if there are several variables and all possible pairs are examined, especially when subjects are grouped and the analysis is done separately within each group. If there are only a few variables, screening all possible pairs is not burdensome; if there are numerous variables, you may want to use statistics on skewness to screen only pairs that are likely to depart from linearity. Think, also, about pairs of variables that might have true nonlinearity and examine them through bivariate scatterplots.

Bivariate scatterplots are produced by BMDP6D, SPSS PLOT (or SPSS SCATTERGRAM), SYSTAT PLOT, and SAS PLOT, among other programs.

4.1.5.3 ■ Homoscedasticity, Homogeneity of Variance, and Homogeneity of Variance-Covariance Matrices

For ungrouped data, the assumption of homoscedasticity is that the variability in scores for one continuous variable is roughly the same at all values of another continuous variable. For grouped data, this is the same as the assumption of homogeneity of variance when one of the variables is discrete (the grouping variable) the other is continuous (the DV): the variability in the DV is expected to be about the same at all levels of the grouping variable.

Homoscedasticity is related to the assumption of normality because when the assumption of multivariate normality is met, the relationships between variables are homoscedastic. The bivariate scatterplots between two variables are of roughly the same width all over with some bulging toward the middle. Homoscedasticity for a bivariate plot is illustrated in Figure 4.5(a).

Heteroscedasticity, the failure of homoscedasticity, is caused either by nonnormality of one of the variables or by the fact that one variable is related to some transformation of the other. Consider, for example, the relationship between age (X_1) and income (X_2), as depicted in Figure 4.5(b). People start out making about the same salaries, but with increasing age, people spread farther apart on income. The relationship is perfectly lawful, but it is not homoscedastic. In this example, income is likely to be positively skewed and transformation of income is likely to improve the homoscedasticity of the relationship.

Another source of heteroscedasticity is greater error of measurement at some levels of an IV. For example, people in the age range 25 to 45 might be more concerned about their weight than people who are younger or older. Older and younger people would, as a result, give less reliable estimates of their weight, increasing the variance of weight scores at those ages.

It should be noted that heteroscedasticity is not fatal to an analysis. The linear relationship between variables is captured by the analysis but there is even more predictability if the heteroscedasticity is accounted for. If it is not, the analysis is weakened, but not invalidated. Homoscedasticity is evaluated through bivariate scatterplots and corrected, if possible, by transformation of variables.

When data are grouped, homoscedasticity is known as homogeneity of variance. A great deal of research has assessed the robustness (or lack thereof) of ANOVA to violation of homogeneity of variance. Recent guidelines have become more stringent than earlier, more cavalier ones. However, it still is generally agreed that most formal tests of homogeneity of variance are too strict because they are too highly influenced by nonnormality. Instead F_{max} in conjunction with sample size ratios is recommended.

F_{max} is the ratio of the largest cell variance to the smallest. If sample sizes are relatively equal (within a ratio of 4 to 1 or less for largest to smallest cell size), an F_{max} as great as 10 is probably acceptable. As the cell size discrepancy increases (e.g., goes to 9 to 1), an F_{max} as small as 3 has been found to be associated with inflated Type I error if the larger variance is associated with the smaller cell size (Milligan, Wong, & Thompson, 1987). Violation of homogeneity of variance is dealt with by variable transformation or use of a more stringent criterion for α, for example .025 or .01 rather than .05.

The multivariate analog of homogeneity of variance is homogeneity of variance-covariance matrices. As for univariate homogeneity of variance, inflated Type I error rate occurs when the

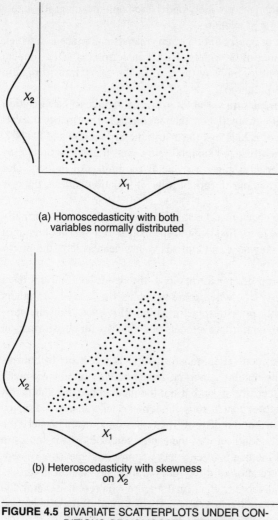

(a) Homoscedasticity with both
    variables normally distributed

(b) Heteroscedasticity with skewness
    on X_2

FIGURE 4.5 BIVARIATE SCATTERPLOTS UNDER CON-
DITIONS OF HOMOSCEDASTICITY AND
HETEROSCEDASTICITY.

greatest dispersion is associated with the smallest sample size. The formal test, Box's *M*, is too strict
with the large sample sizes usually necessary for multivariate applications of ANOVA. Chapter 11
demonstrates a graphical assessment of homogeneity of variance-covariance matrices.

4.1.6 ■ Common Data Transformations

Although data transformations are recommended as a remedy for outliers and for failures of normality, linearity, and homoscedasticity, they are not universally recommended. The reason is that
an analysis is interpreted from the variables that are in it and transformed variables are sometimes

harder to interpret. For instance, although IQ scores are widely understood and meaningfully interpreted, the logarithm of IQ scores may be harder to explain.

Whether transformation increases difficulty of interpretation often depends on the scale in which the variable is measured. If the scale is meaningful or widely used, transformation often hinders interpretation but if the scale is somewhat arbitrary anyway (as is often the case), transformation does not notably increase the difficulty of interpretation.

With ungrouped data, it is probably best to transform variables to normality unless interpretation is not feasible with the transformed scores. With grouped data, the assumption of normality is evaluated with respect to the sampling distribution of means (not the distribution of scores) and the central limit theorem predicts normality with decently sized samples. However, transformations may improve the analysis, and may have the further advantage of reducing the impact of outliers. Our recommendation, then, is to consider transformation of variables in all situations unless there is some reason not to.

If you decide to transform, it is important to check that the variable is normally or near-normally distributed after transformation. Often you need to try first one transformation and then another until you find the transformation that produces the skewness and kurtosis values nearest zero, the prettiest picture, and/or the fewest outliers.

With almost every data set where we have used transformations, the results of analysis have been substantially improved. This is particularly true when some variables are skewed and others are not, or variables are skewed very differently prior to transformation. However, if all the variables are skewed to about the same moderate extent, improvements of analysis with transformation are often marginal.

With grouped data, the test of mean differences after transformation is a test of differences between medians in the original data. After a distribution is normalized by transformation, the mean is equal to the median. The transformation affects the mean but not the median because the median depends only on rank order of cases. Therefore, conclusions about means of transformed distributions apply to medians of untransformed distributions. Transformation is undertaken because the distribution is skewed and the mean is not a good indicator of the central tendency of the scores in the distribution. For skewed distributions, the median is often a more appropriate measure of central tendency than the mean, anyway, so interpretation of differences in medians is appropriate.

Variables differ in the extent to which they diverge from normal. Figure 4.6 presents several distributions together with the transformations that are likely to render them normal. If the distribution differs moderately from normal, a square root transformation is tried first. If the distribution differs substantially, a log transformation is tried. If the distribution differs severely, the inverse is tried. According to Bradley (1982) the inverse is the best of several alternatives for J-shaped distributions, but even it may not render the distribution normal. Finally, if the departure from normality is severe and no transformation seems to help, you may want to try dichotomizing the variable.

The direction of the deviation is also considered. When distributions have positive skewness, as discussed above, the long tail is to the right. When they have negative skewness, the long tail is to the left. If there is negative skewness, the best strategy is to "reflect" the variable and then apply the appropriate transformation for positive skewness.[13] To reflect a variable, find the largest score

---

[13] Remember, however, that the interpretation of a reflected variable is just the opposite of what it was; if big numbers meant good things prior to reflecting the variable, big numbers mean bad things afterwards.

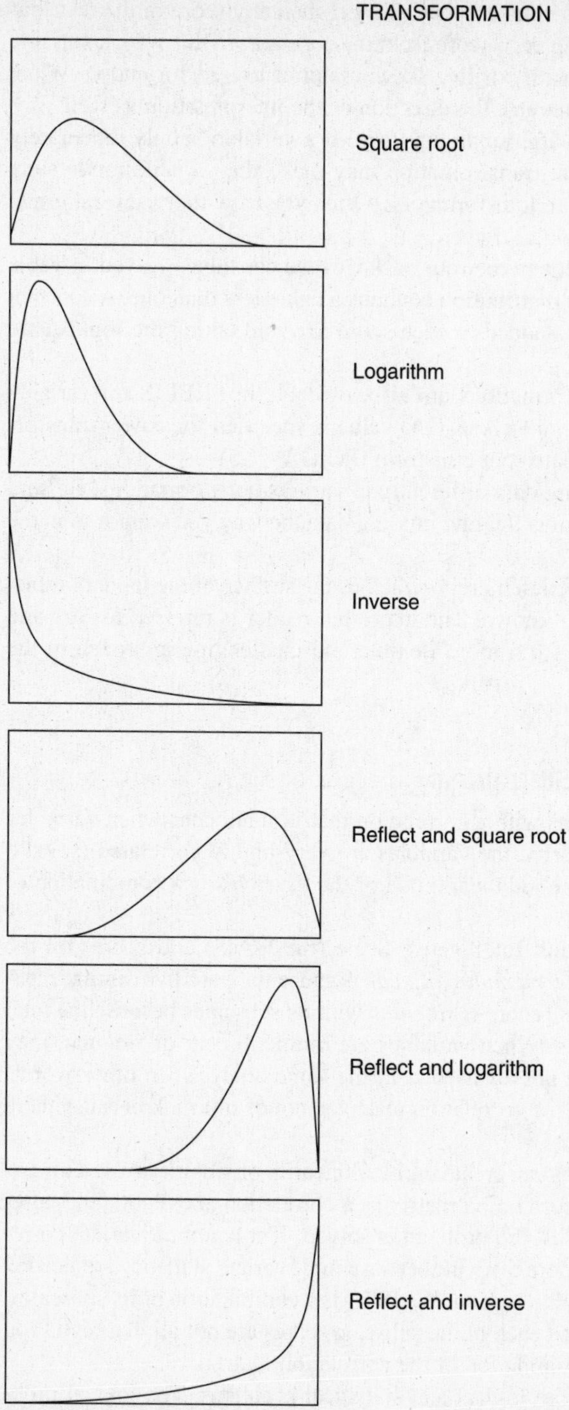

TRANSFORMATION

Square root

Logarithm

Inverse

Reflect and square root

Reflect and logarithm

Reflect and inverse

FIGURE 4.6 ORIGINAL DISTRIBUTIONS AND COMMON
TRANSFORMATIONS TO PRODUCE NORMALITY.

in the distribution and add one to it to form a constant that is larger than any score in the distribution. Then create a new variable by subtracting each score from the constant. In this way, a variable with negative skewness is converted to one with positive skewness prior to transformation. When you interpret a reflected variable, be sure to reverse the direction of the interpretation as well.

Remember to check your transformations after applying them. If a variable is only moderately positively skewed, for instance, a square root transformation may make the variable moderately negatively skewed, and there is no advantage to transformation. Often you have to try several transformations before you find the most helpful one.

Control language for transforming variables in the four packages we consider is given in Table 4.2. Notice that a constant is also added if the distribution contains a value less than one. A constant (to bring the smallest value to at least one) is added to each score to avoid taking the log, square root, or inverse of zero.

Logarithmic (LO) and power (PO) transformations are also available in PRELIS for variable used in structural equation modeling (Chapter 14). A γ (GA) value is specified for power transformations; for example $\gamma = 1/2$ provides a square root transform (PO GA = .5)

Different software packages handle missing data differently in various transformations. Be sure to check the manual to insure that the program is treating missing data the way you want it to in the transformation.

It should be clearly understood that this section merely scratches the surface of the topic of transformations, about which a great deal more is known. The interested reader is referred to Box and Cox (1964) or Mosteller and Tukey (1977) for a more flexible and challenging approach to the problem of transformation.

4.1.7 ■ Multicollinearity and Singularity

Multicollinearity and singularity are problems with a correlation matrix that occur when variables are too highly correlated. With multicollinearity, the variables are very highly correlated (say .90 and above); with singularity, the variables are redundant; one of the variables is a combination of two or more of the other variables.

For example, scores on the Wechsler Adult Intelligence Scale (the WAIS) and scores on the Stanford-Binet Intelligence Scale are likely to be *multicollinear* because they are two similar measures of the same thing. But the total WAIS IQ score is *singular* with its subscales because the total score is found by combining subscale scores. When variables are multicollinear or singular, they contain redundant information and they are not all needed in the same analysis. In other words, there are fewer variables than it appears and the correlation matrix is not of full rank because there are not really as many variables as columns.

Either bivariate or multivariate correlations can create multicollinearity or singularity. If a bivariate correlation is too high, it shows up in a correlation matrix as a correlation above .90, and, after deletion of one of the two redundant variables, the problem is solved. If it is a multivariate correlation that is too high, diagnosis is slightly more difficult because multivariate statistics are needed to find the offending variable. For example, although the WAIS IQ is a combination of its subscales, the bivariate correlations between total IQ and each of the subscale scores are not all that high. You would not know there was singularity by examination of the correlation matrix.

Multicollinearity and singularity cause both logical and statistical problems. The logical problem is that, unless you are doing analysis of structure (factor analysis, principal components analysis, and structural equation modeling), it is not a good idea to include redundant variables in the

TABLE 4.2 CONTROL LANGUAGE FOR COMMON DATA TRANSFORMATIONS

	BMDP /TRANSFORM	SPSS COMPUTE	SAS DATA Procedure	SYSTAT Data Module
Moderate positive skewness	NEWX=SQRT(X).	NEWX=SQRT(X)	NEWX=SQRT(X)	LET NEWX=SQR(X)
Substantial positive skewness	NEWX=LOG(X).	NEWX=LG10(X)	NEWX=LOG10(X)	LET NEWX=LOG(X)/LOG(10)
With zero	NEWX=LOG(X + C).	NEWX=LG10(X + C)	NEWX=LOG10(X + C)	LET NEWX=LOG(X + C)/LOG(10)
Severe positive skewness, L-shaped	NEWX=1/X.	NEWX=1/X	NEWX=1/X	LET NEWX=1/X
With zero	NEWX=1/(X + C).	NEWX=1/(X + C)	NEWX=1/(X + C)	LET NEWX=1/(X + C)
Moderate negative skewness	NEWX=SQRT(K − X).	NEWX=SQRT(K − X)	NEWX=SQRT(K − X)	LET NEWX=SQR(K − X)
Substantial negative skewness	NEWX=LOG(K − X).	NEWX=LG10(K − X)	NEWX=LOG10(K − X)	LET NEWX=LOG(K − X)/LOG(10)
Severe negative skewness, J-Shaped	NEWX=1/(K − X).	NEWX=1/(K − X)	NEWX=1/(K − X)	NEWX=1/(K − X)

C = a constant added to each score so that the smallest score is 1.

K = a constant from which each score is subtracted so that the smallest score is 1; usually equal to the largest score + 1.

same analysis. They are not needed and, because they inflate the size of error terms, they actually weaken an analysis. Unless you are doing analysis of structure, think carefully before including two variables with a bivariate correlation of, say, .70 or more in the same analysis. You might omit one of the variables, or you might create a composite score from the redundant variables.

The statistical problems created by singularity and multicollinearity occur at much higher correlations (.90 and higher). The problem is that singularity prohibits, and multicollinearity renders unstable, matrix inversion. Matrix inversion is the logical equivalent of division; calculations requiring division (and there are many of them—see the fourth sections of Chapters 5 through 14) cannot be performed on singular matrices because they produce determinants equal to zero that cannot be used as divisors (see Appendix A).

With multicollinearity, the determinant is not exactly zero, but it is zero to several decimal places. Division by a near-zero determinant produces very large and unstable numbers in the inverted matrix. The sizes of numbers in the inverted matrix fluctuate wildly with only minor changes (say, in the second or third decimal place) in the sizes of the correlations in **R**. The portions of the multivariate solution that flow from an inverted matrix that is unstable are also unstable. In regression, for instance, error terms get so large that none of the coefficients is significant (Berry, 1993). Further, when r is .9, the precision of estimation of weighting coefficients is halved (Fox, 1991).

Most programs protect against multicollinearity and singularity by computing SMCs for the variables. SMC is the squared multiple correlation of a variable where it serves as DV with the rest as IVs in multiple correlation (see Chapter 5). If the SMC is high, the variable is highly related to the others in the set and you have multicollinearity. If the SMC is 1, the variable is perfectly related to others in the set and you have singularity. Many programs convert the SMC values for each variable to tolerance $(1 - SMC)$ and deal with tolerance instead of SMC.

If you use BMDPAM during preliminary data screening for missing values, multicollinearity shows up as very high SMC values given as routine printout. If there is singularity, it is so stated and the offending variables are identified.

If you do not use BMDPAM, screening for singularity takes the form of running your main analysis to see if the computer balks. Singularity aborts most runs except those for principal components analysis (see Chapter 13) where matrix inversion is not required. If the run aborts, you need to identify and delete the offending variable. A first step is to think about the variables. Did you create any of them from others of them; for instance, did you create one of them by adding together two others? If so, deletion of one removes singularity. If not, then use BMDPAM to identify the offending variable.

Screening for multicollinearity that causes statistical instability is also routine with most programs because they have tolerance criteria for inclusion of variables. If the tolerance $(1 - SMC)$ is too low, the variable does not enter the analysis. Default tolerance levels range between .01 and .0001, so SMCs are .99 to .9999 before variables are excluded. You may wish to take control of this process, however, by adjusting the tolerance level (an option with many programs) or deciding yourself which variable(s) to delete instead of letting the program make the decision on purely statistical grounds. For this you need SMCs for each variable. SMCs are given as routine output by many BMDP programs, and are also available through factor analysis programs in all packages. PRELIS provides SMCs for structural equation modeling.

SAS, SYSTAT, and SPSS have incorporated "collinearity diagnostics" proposed by Belsely, Kuh, and Welsch (1980) in which a conditioning index is produced, as well as variance proportions associated with each variable, after standardization, for each root (see Chapter 15 and Appendix A for a discussion of roots). Variables with large variance proportions are those with problems.

Criteria for multicollinearity are a conditioning index > 30 and at least two variance prop
$> .50$ for a given root number. Collinearity diagnostics are demonstrated in Section 4.2.1.6.

4.1.8 ■ A Checklist and Some Practical Recommendations

A checklist for screening is in Table 4.3. It is important to consider all the issues prior to the fundamental analysis, lest you be tempted to make some of your decisions on the basis of how they influence the analysis. If you choose to screen through residuals, you can't avoid doing an analysis at the same time; however, in these cases you concentrate on the residuals and not on the other features of the analysis while making your screening decisions.

The order in which screening takes place is important because the decisions that you make at one step influence the outcomes of later steps. In a situation where you have both nonnormal variables and potential outliers, a fundamental decision is whether you would prefer to transform variables or to delete cases or change scores on cases. If you transform variables first, you are likely to find fewer outliers. If you delete or modify the outliers first, you are likely to find fewer variables with nonnormality.

Of the two choices, transformation of variables is usually preferable. It typically reduces the number of outliers. It is likely to produce normality, linearity, and homoscedasticity among the variables. It increases the likelihood of multivariate normality to bring the data into conformity with one of the fundamental assumptions of most inferential tests. And on a very practical level, it usually enhances the analysis even if inference is not a goal. On the other hand, transformation may threaten interpretation, in which case all the statistical niceties are of no avail.

Or, if the impact of outliers is reduced first, you are less likely to find variables that are skewed because significant skewness is sometimes caused by extreme cases on the tails of the distributions. If you have cases that are outliers because they are not part of the population from which you intended to sample, by all means delete them before checking distributions.

Lastly, as will become obvious in the next two sections, although the issues are different, the runs on which they are screened are not necessarily different. That is, the same run often provides you with information regarding two or more issues.

TABLE 4.3 CHECKLIST FOR SCREENING DATA

1. Inspect univariate descriptive statistics for accuracy of input
   a. Out-of-range values
   b. Plausible means and standard deviations
   c. Coefficient of variation
   d. Univariate outliers
2. Evaluate amount and distribution of missing data: deal with problem
3. Check pairwise plots for nonlinearity and heteroscedasticity
4. Identify and deal with nonnormal variables
   a. Check skewness and kurtosis, probability plots
   b. Transform variables (if desirable)
   c. Check results of transformation
5. Identify and deal with multivariate outliers
   a. Variables causing multivariate outliers
   b. Description of multivariate outliers
6. Evaluate variables for multicollinearity and singularity

4.2 ■ COMPLETE EXAMPLES OF DATA SCREENING

Evaluation of assumptions is somewhat different for ungrouped and grouped data. That is, if you are going to perform multiple regression, canonical correlation, factor analysis or structural equation modeling on ungrouped data, there is one approach to screening. If you are going to perform univariate or multivariate analysis of variance (including profile analysis) or discriminant function analysis on grouped data, there is another approach to screening.[14]

Therefore, two complete examples are presented that use the same set of variables taken from the research described in Appendix B; number of visits to health professionals (TIMEDRS), attitudes toward drug use (ATTDRUG), attitudes toward housework (ATTHOUSE), INCOME, marital status (MSTATUS),and RACE. The grouping variable used in the analysis of grouped data is current employment status (EMPLMNT).[15]

Where possible in these examples, and for illustrative purposes, screening for ungrouped data is performed using SPSS, and screening for grouped data is performed using BMDP programs. BMDP programs are generally richer in screening procedures, especially when the data are grouped. SAS has better screening procedures for ungrouped than grouped data while SYSTAT has better procedures for grouped data. However, SAS is generally less useful than BMDP, SPSS for ungrouped data, or SYSTAT for grouped data because there is no significance test for identifying multivariate outliers through Mahalanobis distance. Instead, leverage (Equation 4.3) must be used to calculate Mahalanobis distance if a significance test is desired.

4.2.1 ■ Screening Ungrouped Data

A flow diagram for screening ungrouped data appears as Figure 4.7. The direction of flow assumes that data transformation is undertaken, as necessary. If transformation is not acceptable, then other procedures for handling outliers are used.

4.2.1.1 ■ Accuracy of Input, Missing Data, Distributions, and Univariate Outliers

A check on accuracy of data entry, missing data, skewness, and kurtosis for the data set is done through SPSS FREQUENCIES, as seen in Table 4.4.

The minimum and maximum values, means, and standard deviations of each of the variables are inspected for plausibility. For instance, the MINIMUM number of visits to health professionals (TIMEDRS) is 0 and the MAXIMUM is 81, higher than expected but found to be accurate upon checking the data sheets.[16] The MEAN for the variable is 7.901, higher than the national average but not extremely so, and the standard deviation (STD DEV) is 10.948. These values are all reasonable, as are the values on the other variables. For instance, the ATTDRUG variable is constructed with a range of 5 to 10 so it is reassuring to find these values as MINIMUM and MAXIMUM.

TIMEDRS shows no MISSING CASES but has strong positive SKEWNESS (3.248). The significance of SKEWNESS is evaluated by dividing it by S E Skew, as in Equation 4.5,

---

[14] If you are using multiway frequency analysis or logistic regression there are far fewer assumptions than with these analyses.

[15] This is a motley collection of variables chosen primarily for their statistical properties.

[16] The woman with this number of visits was terminally ill when she was interviewed.

Searching for:

Programs:

Plausible range
Missing Values
Normality
Univariate Outliers

SPSS FREQUENCIES
BMDP2D
SYSTAT STATISTICS
SAS MEANS

Pairwise Linearity

SPSS PLOT
BMDP6D
SYSTAT PLOT
SAS PLOT

Or Check Residuals:

SPSS REGRESSION
BMDP2R
SYSTAT REG + PLOT
    GLM + PLOT
SAS REG + PLOT
    GLM + PLOT

Multivariate Outliers

SPSS REGRESSION
BMDPAM
    4M

(Describe Outliers)

SPSS DISCRIMINANT
    LIST CASES
BMDP9R
    1D
SYSTAT STATISTICS
    BASIC
SAS MEANS
    DATA

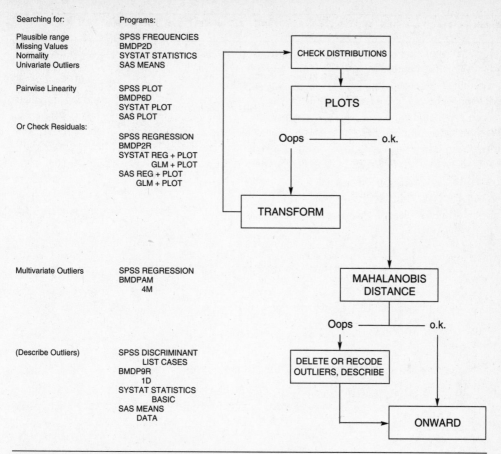

FIGURE 4.7 FLOW DIAGRAM FOR SCREENING UNGROUPED DATA.

$$z = \frac{3.248}{.113} = 28.74$$

to reveal a clear departure from symmetry. The distribution also has significant KURTOSIS as evaluated by Equation 4.7,

$$z = \frac{13.101}{.226} = 57.97$$

The departures from normality are also obvious from inspection of the difference between frequencies expected under the normal distribution (the dots) and obtained frequencies. Because this variable is a candidate for transformation, evaluation of univariate outliers is deferred.

ATTDRUG, on the other hand, is well behaved. There are no MISSING CASES and SKEWNESS and KURTOSIS are well within expected values. ATTHOUSE has a single missing value but is otherwise well distributed except for the two extremely low scores. A score of 2 is 4.8 standard deviations below the mean of ATTHOUSE. Because $z = -4.8$ is well beyond the $p = .001$ criterion of 3.29 (2-tailed), the decision is made to delete from further analysis the data from the two

```
    TITLE                DESCRIPTION OF UNGROUPED DATA.
    DATA LIST            FILE = 'SCREEN.DAT'      FREE
                            /SUBNO,TIMEDRS,ATTDRUG,ATTHOUSE,
                         INCOME,EMPLMNT,MSTATUS,RACE.
    MISSING VALUES       ATTHOUSE(1), INCOME(99).
    VAR LABELS           TIMEDRS 'VISITS TO HEALTH PROFESSIONALS'/
                         MSTATUS 'MARITAL STATUS'.
    VALUE LABELS         EMPLMNT 0 'PAIDWORK' 1 'HOUSEWIFE'/
                         MSTATUS 1 'SINGLE' 2 'MARRIED'/
                         RACE 1 'WHITE' 2 'NONWHITE'.
    FREQUENCIES          VARIABLES = TIMEDRS TO INCOME,
                                MSTATUS, RACE/
                    HISTOGRAM = NORMAL/
                    FORMAT = NOTABLE/ STATISTICS = ALL.

TIMEDRS    VISITS TO HEALTH PROFESSIONALS

   Count    Midpoint    One symbol equals approximately  4.00 occurrences

      86           0  |************:*********
     187           4  |***************:********************************
      74           8  |*****************:**
      44          12  |***********      .
      27          16  |*******          .
      13          20  |***          .
       7          24  |**        .
       6          28  |**.
       3          32  |*.
       4          36  |:
       2          40  |*
       1          44  |
       1          48  |
       2          52  |*
       3          56  |*
       3          60  |*
       0          64  |
       0          68  |
       0          72  |
       1          76  |
       1          80  |
                      +----+----+----+----+----+----+----+----+----+----+
                      0        40        80       120       160       200
                              Histogram frequency

Mean          7.901    Std err       .508    Median        4.000
Mode          2.000    Std dev     10.948    Variance    119.870
Kurtosis     13.101    S E Kurt      .226    Skewness      3.248
S E Skew       .113    Range       81.000    Minimum        .000
Maximum      81.000    Sum       3674.000

Valid cases    465    Missing cases       0
- - - - - - - - - - - - - - - - - - - - - - - - - - - - - - - - - - - -
ATTDRUG

   Count     Value    One symbol equals approximately  4.00  occurrences

      13      5.00  |**:
      60      6.00  |**************:*
     126      7.00  |*******************************.
     149      8.00  |*************************************.
      95      9.00  |*********************:***
      22     10.00  |*****:
                    +----+----+----+----+----+----+----+----+----+----+
                    0        40        80       120       160       200
                            Histogram frequency
```

TABLE 4.4 *(CONTINUED)*

```
Mean            7.686      Std err        .054      Median      8.000
Mode            8.000      Std dev       1.156      Variance    1.337
Kurtosis        -.447      S E Kurt       .226      Skewness    -.123
S E Skew         .113      Range         5.000      Minimum     5.000
Maximum        10.000      Sum        3574.000

Valid cases      465      Missing cases       0
```

ATTHOUSE

```
    Count   Midpoint    One symbol equals approximately  2.00 occurrences

        0       -2  |
        0        0  |
        2        2  |*
        0        4  |
        0        6  |
        0        8  |
        0       10  |
        2       12  |*.
        5       14  |***.
       15       16  |********  .
       33       18  |****************  .
       53       20  |***************************  .
       73       22  |**************************************  .
       83       24  |*****************************************.*
       80       26  |*************************************.*****
       54       28  |**********************.**
       41       30  |***************.******
       15       32  |******.*
        6       34  |**:
        2       36  |:
        0       38  |
                    +----+----+----+----+----+----+----+----+----+----+
                    0    20        40        60        80        100
                            Histogram frequency
```

```
Mean           23.541      Std err        .208      Median     24.000
Mode           23.000      Std dev       4.484      Variance   20.102
Kurtosis        1.556      S E Kurt       .226      Skewness    -.457
S E Skew         .113      Range        33.000      Minimum     2.000
Maximum        35.000      Sum       10923.000

Valid cases      464      Missing cases       1
```

INCOME

```
    Count   Value      One symbol equals approximately  2.00 occurrences

       71     1.00   |**************:*********************
       39     2.00   |********************  .
       79     3.00   |********************************.********
       84     4.00   |***********************************:******
       46     5.00   |***********************
       36     6.00   |******************
       36     7.00   |******************.
       19     8.00   |**********.
       14     9.00   |****:**
       15    10.00   |*:******
                     +----+----+----+----+----+----+----+----+----+----+
                     0    20        40        60        80        100
                             Histogram frequency
```

TABLE 4.4   *(CONTINUED)*

Mean	4.210	Std err	.115	Median	4.000
Mode	4.000	Std dev	2.419	Variance	5.851
Kurtosis	-.359	S E Kurt	.233	Skewness	.582
S E Skew	.117	Range	9.000	Minimum	1.000
Maximum	10.000	Sum	1848.000		

Valid cases   439   Missing cases   26

MSTATUS   MARITAL STATUS

```
   Count     Value    One symbol equals approximately  8.00 occurrences

    103       1.00 |************* .
    362       2.00 |*************************************:****
                   +----+----+----+----+----+----+----+----+----+----+
                   0        80       160      240      320      400
                             Histogram frequency
```

Mean	1.778	Std err	.019	Median	2.000
Mode	2.000	Std dev	.416	Variance	.173
Kurtosis	-.190	S E Kurt	.226	Skewness	-1.346
S E Skew	.113	Range	1.000	Minimum	1.000
Maximum	2.000	Sum	827.000		

Valid cases   465   Missing cases   0

RACE

```
   Count     Value    One symbol equals approximately 10.00 occurrences

    424       1.00 |*****************************************:
     41       2.00 |**:*
                   +----+----+----+----+----+----+----+----+----+----+
                   0       100      200      300      400      500
                             Histogram frequency
```

Mean	1.088	Std err	.013	Median	1.000
Mode	1.000	Std dev	.284	Variance	.081
Kurtosis	6.521	S E Kurt	.226	Skewness	2.914
S E Skew	.113	Range	1.000	Minimum	1.000
Maximum	2.000	Sum	506.000		

Valid cases   465   Missing cases   0

women with extremely favorable attitudes toward housework. Information about these deletions is included in the report of results. The single missing value is replaced with the mean.

On INCOME, however, there are 26 cases with missing values—more than 5% of the sample. If INCOME is not critical to the hypotheses, we delete it in subsequent analyses. If INCOME is important to the hypotheses, we could either replace the missing values or delete the cases that failed to provide scores (if we determined that they are a random subsample).

The two remaining variables are dichotomous and not evenly split. MSTATUS has a 362 to 103 split, roughly a 3.5 to 1 ratio, that is not particularly disturbing. But RACE, with a split greater than 10 to 1 is marginal. For this analysis, we choose to retain the variable, realizing that its association with other variables is deflated because of the uneven split.

Table 4.5 shows the distribution of ATTHOUSE with elimination of the univariate outliers. The mean for ATTHOUSE changes to 23.634, the value used to replace the missing ATTHOUSE score

TABLE 4.5   INPUT AND SPSS OUTPUT SHOWING DESCRIPTIVE STATISTICS
AND HISTOGRAMS FOR ATTHOUSE WITH UNIVARIATE OUTLIERS DELETED

```
TITLE           DESCRIPTION OF ATTHOUSE WITHOUT OUTLIERS.
DATA LIST       FILE = 'SCREEN.DAT'    FREE
                /SUBNO,TIMEDRS,ATTDRUG,ATTHOUSE,
                INCOME,EMPLMNT,MSTATUS,RACE.
MISSING VALUE   ATTHOUSE(1), INCOME(99).
VAR LABELS      TIMEDRS 'VISITS TO HEALTH PROFESSIONALS'/
                MSTATUS 'MARITAL STATUS'.
VALUE LABELS    EMPLMNT 0 'PAIDWORK' 1 'HOUSEWIFE'/
                MSTATUS 1 'SINGLE' 2 'MARRIED'/
                RACE 1 'WHITE' 2 'NONWHITE'.
SELECT IF       (ATTHOUSE NE 2).
FREQUENCIES     VARIABLES = ATTHOUSE/
                HISTOGRAM = NORMAL/
                FORMAT = NOTABLE/ STATISTICS = ALL.
```

ATTHOUSE

```
   Count   Midpoint    One symbol equals approximately  2.00 occurrences

      0       8.0  |
      0       9.5  |
      1      11.0  |*
      2      12.5  |:
      4      14.0  |**.
     15      15.5  |****:***
     13      17.0  |*******   .
     50      18.5  |***************:********
     23      20.0  |************               .
     73      21.5  |****************************:********
     46      23.0  |***********************            .
     80      24.5  |*********************************:********
     37      26.0  |*******************       .
     54      27.5  |********************:******
     27      29.0  |**************.
     25      30.5  |********:****
      4      32.0  |**   .
      6      33.5  |*:*
      2      35.0  |:
      0      36.5  |
      0      38.0  |
                   +----+----+----+----+----+----+----+----+----+----+
                   0        20        40        60        80       100
                              Histogram frequency
```

Mean	23.634	Std err	.198	Median	24.000
Mode	23.000	Std dev	4.262	Variance	18.167
Kurtosis	-.260	S E Kurt	.227	Skewness	-.038
S E Skew	.114	Range	24.000	Minimum	11.000
Maximum	35.000	Sum	10919.000		

Valid cases	462	Missing cases	0

in subsequent analyses. The case with a missing value on ATTHOUSE becomes complete and available for use in all computations.

At this point we have investigated the accuracy of data entry and the distributions of all variables, determined the number of missing values, found the mean for replacement of missing data, and found two univariate outliers which, when deleted, results in $N = 463$.

4.2.1.2 ■ Linearity and Homoscedasticity

Because of nonnormality on at least one variable, SPSS PLOT is run to check the bivariate plots for departures from linearity and homoscedasticity, as reproduced in Figure 4.8. The variables picked as "worst case" are those with the most discrepant distributions: TIMEDRS which has the greatest departure from normality, and ATTDRUG, which is nicely distributed. (The SELECT IF instruction eliminates the univariate outliers on ATTHOUSE.)

In Figure 4.8 ATTDRUG is along the Y axis; turn the page so that the Y axis becomes the X axis and you can see the symmetry of the ATTDRUG distribution. TIMEDRS is along the X axis. The asymmetry of the distribution is apparent from the pileup of scores at low values of the variable. The overall shape of the scatterplot is not oval; the variables are not linearly related. Heteroscedasticity is evident in the greater variability in ATTDRUG scores for low than high values of TIMEDRS.

4.2.1.3 ■ Transformation

In this case, the decision is made to transform variables prior to searching for multivariate outliers. With strong skewness for TIMEDRS, a logarithmic transformation is applied to TIMEDRS. Because the smallest value on the variable is zero, one is added to each score as the transformation is performed, as indicated in the COMPUTE statement. Table 4.6 shows the distribution of TIMEDRS as transformed to LTIMEDRS.

SKEWNESS is reduced from 3.248 to 0.221 and KURTOSIS reduced from 13.101 to -0.183 by the transformation. The frequency plot is not exactly pleasing (the frequencies are still too high at small scores) but the statistical evaluation of the distribution is much improved.

Figure 4.9 is a bivariate scatterplot between ATTDRUG and LTIMEDRS. Although still not perfect, the overall shape of the scatterplot is more nearly oval. The nonlinearity associated with non-normality of one of the variables is "fixed" by transformation of the variable.

4.2.1.4 ■ Detecting Multivariate Outliers

The 463 cases, with transformation applied to LTIMEDRS, are screened for multivariate outliers through SPSS REGRESSION (Table 4.7) using the RESIDUALS=OUTLIERS(MAHAL) control language. Case labels are used as the dummy DV, since multivariate outliers among IVs are unaffected by the DV.[17] The remaining VARIABLES are considered independent variables.

The criterion for multivariate outliers is Mahalanobis distance at $p < .001$. Mahalanobis distance is evaluated as χ^2 with degrees of freedom equal to the number of variables, in this case five: LTIMEDRS, ATTDRUG, ATTHOUSE, MSTATUS, and RACE. Any case with a Mahalanobis Distance in Table 4.7 greater than $\chi^2(5) = 20.515$ (cf. Appendix C, Table C.4), then, is a multivariate outlier. As seen in Table 4.7, cases 117 and 193 are outliers among these variables in this data set.

There are 461 cases remaining if the two multivariate outliers are deleted. Little is lost by deleting the additional two outliers from the sample. However, it is also necessary to determine why the two cases are multivariate outliers, to know how their deletion limits generalizability, and to include that information in the Results section.

---

[17] For a multiple regression analysis, the actual DV would be used here rather than SUBNO as a dummy DV.

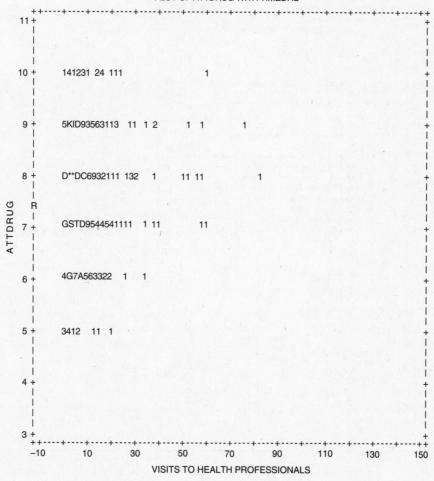

```
TITLE           SCATTERPLOT FOR UNGROUPED DATA.
DATA LIST       FILE = 'SCREEN.DAT'      FREE
                /SUBNO, TIMEDRS, ATTDRUG, ATTHOUSE,
                    INCOME, EMPLMNT, MSTATUS, RACE.
VAR LABELS      TIMEDRS 'VISITS TO HEALTH PROFESSIONALS'/
                MSTATUS 'MARITAL STATUS'.
VALUE LABEL     EMPLMNT 0 'PAIDWORK' 1 'MARRIED'/
                RACE 1 'WHITE' 2 'NONWHITE'.
SELECT IF       ATTHOUSE NE 2.
RECODE          ATTHOUSE (1 = 23.634).
PLOT            HORIZONTAL = MIN (-10)/
                FORMAT = REGRESSION/
                PLOT = ATTDRUG WITH TIMEDRS.
```

PLOT OF ATTDRUG WITH TIMEDRS

```
        ++----+----+----+----+----+----+----+----+----+----+----+----+----+----+----++
    11 +                                                                            +
       |                                                                            |
       |                                                                            |
       |                                                                            |
    10 +   141231 24 111              1                                             +
       |                                                                            |
       |                                                                            |
       |                                                                            |
     9 +   5KID93563113  11 1 2      1 1        1                                    +
       |                                                                            |
       |                                                                            |
       |                                                                            |
     8 +   D**DC6932111 132    1     11 11           1                              +
       |                                                                            |
   R   |                                                                            |
     7 +   GSTD9544541111  1 11        11                                           +
A      |                                                                            |
T      |                                                                            |
T      |                                                                            |
D    6 +   4G7A563322   1  1                                                        +
R      |                                                                            |
U      |                                                                            |
G      |                                                                            |
     5 +   3412   11  1                                                             +
       |                                                                            |
       |                                                                            |
       |                                                                            |
     4 +                                                                            +
       |                                                                            |
       |                                                                            |
       |                                                                            |
     3 +                                                                            +
        ++----+----+----+----+----+----+----+----+----+----+----+----+----+----+----++
       -10      10      30      50      70      90     110     130     150
```

VISITS TO HEALTH PROFESSIONALS

463 cases plotted. Regression statistics of ATTDRUG on TIMEDRS:
Correlation   .10502 R Squared   .01103   S.E. of Est   1.15325 2-tailed Sig.   .0238
Intercept (S.E.)   7.59673(.06616) Slope (S.E.)   .01109(.00489)

FIGURE 4.8 ASSESSMENT OF LINEARITY THROUGH BIVARIATE SCATTERPLOTS, AS
PRODUCED BY SPSS PLOT. THIS INDICATES ATTDRUG IS NORMAL; TIME-
DRS IS NONNORMAL.

TABLE 4.6   INPUT AND SPSS OUTPUT SHOWING DESCRIPTIVE STATISTICS AND
HISTOGRAMS FOR TIMEDRS AFTER LOGARITHMIC TRANSFORM

```
TITLE            DESCRIPTION OF TRANSFORMED UNGROUPED DATA.
DATA LIST        FILE = 'SCREEN.DAT'    FREE
                 /SUBNO,TIMEDRS,ATTDRUG,ATTHOUSE,
                  INCOME,EMPLMNT,MSTATUS,RACE.
VAR LABELS       TIMEDRS 'VISITS TO HEALTH PROFESSIONALS'/
                 MSTATUS 'MARITAL STATUS'.
VALUE LABELS     EMPLMNT 0 'PAIDWORK' 1 'HOUSEWIFE'/
                 MSTATUS 1 'SINGLE' 2 'MARRIED'/
                 RACE 1 'WHITE' 2 'NONWHITE'.
COMPUTE          LTIMEDRS = LG10(TIMEDRS + 1).
SELECT IF        (ATTHOUSE NE 2).
RECODE           ATTHOUSE(1 = 23.634).
FREQUENCIES      VARIABLES = LTIMEDRS/
                 HISTOGRAM = NORMAL/
                 FORMAT = NOTABLE/ STATISTICS = ALL.


LTIMEDRS

   Count    Midpoint    One symbol equals approximately  1.50 occurrences

     42       .0   |*****:********************
      0       .1   |            .
      0       .2   |                    .
     44       .3   |*****************:************
      0       .4   |                          .
     65       .5   |***********************:******************
     52       .6   |*****************************:*******
     41       .7   |***************************  .
     53       .8   |*****************************:******
     25       .9   |*****************         .
     34      1.0   |***********************.
     33      1.1   |*******************:**
     24      1.2   |***************:
     16      1.3   |**********.
     11      1.4   |*******.
      6      1.5   |**** .
      6      1.6   |***:
      3      1.7   |*:
      6      1.8   |:***
      2      1.9   |:
      0      2.0   |
                   +----+----+----+----+----+----+----+----+----+----+
                   0        15        30        45        60        75
                              Histogram frequency
```

Mean	.742	Std err	.019	Median	.699
Mode	.477	Std dev	.416	Variance	.173
Kurtosis	-.183	S E Kurt	.226	Skewness	.221
S E Skew	.113	Range	1.914	Minimum	.000
Maximum	1.914	Sum	343.743		

Valid cases	463	Missing cases	0

4.2.1.5 ■ Variables Causing Cases to be Outliers

SPSS does not currently have a program designed to test regressions with all subsets of IVs. Therefore BMDP9R, which does all possible subsets regression, is used to identify the combination of variables on which case 117 and case 193 deviate from the remaining 462 cases. Each outlying case is evaluated in a separate BMDP9R run where a dummy variable is created to separate the outlying case from

```
TITLE           SCATTERPLOT FOR TRANSFORMED UNGROUPED DATA.
DATA LIST       FILE = 'SCREEN.DAT'        FREE
                /SUBNO, TIMEDRS, ATTDRUG, ATTHOUSE
                    INCOME, EMPLMNT, MSTATUS, RACE.
VAR LABELS      TIMDRS 'VISITS TO HEALTH PROFESSIONALS'/
                MSTATUS 'MARITAL STATUS'.
VALUE LABEL     EMPLMNT 0 'PAIDWORK' 2 'HOUSEWIFE'/
                MSTATUS 1 'SINGLE' 2 'MARRIED'/
                RACE 1 'WHITE' 2 'NONWHITE'.
COMPUTE         LTIMEDRS = LG10 (TIMEDRS + ).
SELECT IF       ATTHOUSE NE 2.
RECODE          ATTHOUSE (1 = 23.634).
PLOT            HORIZONTAL = MIN (0)/
                FORMAT = REGRESSION/
                PLOT = ATTDRUG WITH LTIMEDRS.
```

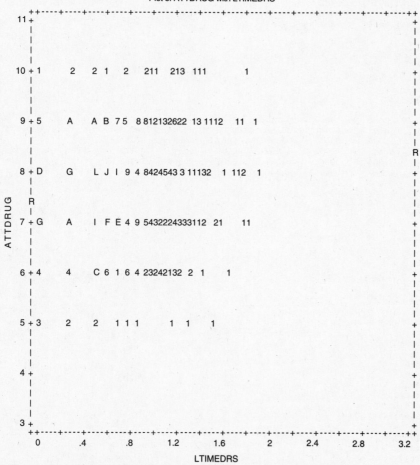

Plot of ATTDRUG with LTIMEDRS

463 cases plotted. Regression statistics of ATTDRUG on LTIMEDRS:
Correlation   .11099 R Squared .01232   S.E. of Est   1.15250 2-tailed Sig. .0169
Intercept (S.E.)   7.45509(.10971)   Slope (S.E.)   .30923(.12896)

FIGURE 4.9 ASSESSMENT OF LINEARITY AFTER LOG TRANSFORMATION OF TIME-
DRS, AS PRODUCED BY SPSS PLOT.

TABLE 4.7   SETUP AND SELECTED SPSS REGRESSION OUTPUT FOR MULTIVARIATE
                     OUTLIERS AND MULTICOLLINEARITY

```
DATA LIST              FILE = 'SCREEN.DAT'  FREE
                       /SUBNO,TIMEDRS,ATTDRUG,ATTHOUSE,
                          INCOME,EMPLMNT,MSTATUS,RACE.
VAR LABELS             TIMEDRS 'VISITS TO HEALTH PROFESSIONALS'
                       /MSTATUS 'MARITAL STATUS'.
VALUE LABELS           EMPLMNT 0 'PAIDWORK' 1 'HOUSEWIFE'
                       /MSTATUS 1 'SINGLE' 2 'MARRIED'
                       /RACE 1 'WHITE' 2 'NONWHITE'.
COMPUTE                LTIMEDRS = LG10(TIMEDRS + 1).
SELECT IF              (ATTHOUSE NE 2).
RECODE                 ATTHOUSE(1 = 23.634).
REGRESSION             VARIABLES=SUBNO LTIMEDRS ATTDRUG ATTHOUSE
                          MSTATUS RACE/
                       STATISTICS COLLIN/
                       DEPENDENT = SUBNO/ENTER/
                       RESIDUALS=OUTLIERS(MAHAL).
```

```
        * * * *   M U L T I P L E   R E G R E S S I O N   * * * *

Total Cases =           463

Collinearity Diagnostics
```

Number	Eigenval	Cond Index	Constant	LTIMEDRS	ATTDRUG	ATTHOUSE	MSTATUS
1	5.65579	1.000	.00027	.00590	.00067	.00093	.00150
2	.20969	5.193	.00092	.91802	.00167	.00150	.01114
3	.06026	9.688	.00061	.01011	.00208	.00589	.28898
4	.04271	11.508	.00400	.06346	.03493	.28983	.45696
5	.02476	15.113	.00416	.00212	.53466	.41480	.06256
6	.00679	28.872	.99003	.00039	.42600	.28707	.17886

Number	RACE
1	.00186
2	.01980
3	.66215
4	.16056
5	.03940
6	.11622

```
Outliers - Mahalanobis' Distance
```

Case #	*MAHAL
117	21.83684
193	20.64977
433	19.96803
99	18.49884
333	18.46936
291	17.51804
58	17.37338
71	17.17153
102	16.94198
196	16.72324

the remaining cases. In Table 4.8, the dummy variable for case 117 is created in the TRANSFORM paragraph with DUMMY = 0 and IF (KASE EQ 117) THEN DUMMY = 1.0.[18] With the dummy

---

[18] KASE is a variable that BMDP uses to identify the sequential order of a case in the data file.

```
/INPUT          VARIABLES ARE 8.  FORMAT IS FREE.
                FILE='SCREEN.DAT'.
                TITLE IS 'BMDP9R RUN FOR CASE 117'.
/VARIABLE       NAMES ARE SUBNO,TIMEDRS,ATTDRUG,ATTHOUSE,
                    INCOME,EMPLMNT,MSTATUS,RACE.
                LABEL = SUBNO.
/TRANSFORM        USE = ATTHOUSE NE 2.
                  IF (ATTHOUSE EQ 1) THEN ATTHOUSE = 23.634.
                  LTIMEDRS = LOG(TIMEDRS + 1).
                  DUMMY = 0.
                  IF (KASE EQ 117) THEN DUMMY = 1.0.
/REGRESS        DEPENDENT IS DUMMY.
                INDEPENDENT ARE LTIMEDRS, ATTDRUG, ATTHOUSE, MSTATUS,
                    RACE.
/END
```

SUBSETS WITH   1 VARIABLES

R-SQUARED	ADJUSTED R-SQUARED	CP	
0.022279	0.020158	10.04	RACE
0.011651	0.009507	15.13	ATTDRUG
0.007038	0.004884	17.35	LTIMEDRS
0.000604	-0.001564	20.43	MSTATUS
0.000016	-0.002153	20.72	ATTHOUSE

SUBSETS WITH   2 VARIABLES

R-SQUARED	ADJUSTED R-SQUARED	CP		
0.034562	0.030365	6.14	VARIABLE	COEFFICIENT T-STATISTIC
			3 ATTDRUG	-0.00444724 -2.42
			8 RACE	0.0247388 3.30
			INTERCEPT	0.00940593
0.031190	0.026977	7.76	LTIMEDRS RACE	
0.023193	0.018946	11.60	MSTATUS RACE	
0.022395	0.018144	11.98	ATTHOUSE RACE	
0.020957	0.016700	12.67	LTIMEDRS ATTDRUG	
0.012234	0.007940	16.85	ATTDRUG MSTATUS	
0.011703	0.007406	17.11	ATTDRUG ATTHOUSE	
0.007846	0.003532	18.96	LTIMEDRS MSTATUS	
0.007107	0.002790	19.31	LTIMEDRS ATTHOUSE	
0.000638	-0.003707	22.42	ATTHOUSE MSTATUS	

TABLE 4.8 *(CONTINUED)*

SUBSETS WITH   3 VARIABLES

R-SQUARED	ADJUSTED R-SQUARED	CP			
0.046132	0.039897	2.59	VARIABLE	COEFFICIENT	T-STATISTIC
			9 LTIMEDRS	0.0121284	2.36
			3 ATTDRUG	-0.00493636	-2.68
			8 RACE	0.0259985	3.48
			INTERCEPT	0.00278895	
0.035456	0.029152	7.71	ATTDRUG MSTATUS RACE		
0.034763	0.028454	8.05	ATTDRUG ATTHOUSE RACE		
0.032404	0.026080	9.18	LTIMEDRS MSTATUS RACE		
0.031197	0.024865	9.76	LTIMEDRS ATTHOUSE RACE		
0.023366	0.016982	13.51	ATTHOUSE MSTATUS RACE		
0.021773	0.015380	14.28	LTIMEDRS ATTDRUG MSTATUS		
0.021002	0.014604	14.65	LTIMEDRS ATTDRUG ATTHOUSE		
0.012316	0.005860	18.82	ATTDRUG ATTHOUSE MSTATUS		
0.007887	0.001402	20.94	LTIMEDRS ATTHOUSE MSTATUS		

SUBSETS WITH   4 VARIABLES

R-SQUARED	ADJUSTED R-SQUARED	CP			
0.047366	0.039046	4.00	VARIABLE	COEFFICIENT	T-STATISTIC
			9 LTIMEDRS	0.0123197	2.39
			3 ATTDRUG	-0.00493969	-2.68
			7 MSTATUS	0.00395668	0.77
			8 RACE	0.0262357	3.51
			INTERCEPT	-0.00463605	
0.046133	0.037802	4.59	VARIABLE	COEFFICIENT	T-STATISTIC
			9 LTIMEDRS	0.0121444	2.34
			3 ATTDRUG	-0.00493578	-2.68
			4 ATTHOUSE	-0.0000110362	-0.02
			8 RACE	0.0259926	3.47
			INTERCEPT	0.00303981	
0.035728	0.027307	9.58	ATTDRUG ATTHOUSE MSTATUS RACE		
0.032404	0.023953	11.18	LTIMEDRS ATTHOUSE MSTATUS RACE		
0.021796	0.013253	16.27	LTIMEDRS ATTDRUG ATTHOUSE MSTATUS		

SUBSETS WITH   5 VARIABLES

R-SQUARED	ADJUSTED R-SQUARED	CP			
0.047368	0.036946	6.00	VARIABLE	COEFFICIENT	T-STATISTIC
			9 LTIMEDRS	0.0122966	2.36
			3 ATTDRUG	-0.00494057	-2.68
			4 ATTHOUSE	0.0000163019	0.03
			7 MSTATUS	0.00396837	0.77
			8 RACE	0.0262452	3.50
			INTERCEPT	-0.00502853	

TABLE 4.8   *(CONTINUED)*

```
STATISTICS FOR 'BEST' SUBSET
----------------------------
MALLOWS' CP                          2.59
SQUARED MULTIPLE CORRELATION      0.04613
MULTIPLE CORRELATION              0.21478
ADJUSTED SQUARED MULT. CORR.      0.03990
RESIDUAL MEAN SQUARE             0.002074
STANDARD ERROR OF EST.           0.045537
F-STATISTIC                          7.40
NUMERATOR DEGREES OF FREEDOM            3
DENOMINATOR DEGREES OF FREEDOM       459
SIGNIFICANCE (TAIL PROB.)         0.0001

*** N O T E *** THE ABOVE F-STATISTIC AND ASSOCIATED SIGNIFICANCE TEND TO BE
                LIBERAL WHENEVER A SUBSET OF VARIABLES IS SELECTED BY THE CP
OR
                ADJUSTED R-SQUARED CRITERIA.
```

VARIABLE NO.	NAME	REGRESSION COEFFICIENT	STANDARD ERROR	STAND. COEF.	T-STAT.	2TAIL SIG.	TOL-ERANCE	TO CONTRIBUTION R-SQ
	INTERCEPT	0.00278895	0.0164722	0.060	0.17	0.866		
9	LTIMEDRS	0.0121284	0.00514020	0.109	2.36	0.019	0.982635	0.01157
3	ATTDRUG	-0.00493636	0.00184093	-0.123	-2.68	0.008	0.986955	0.01494
8	RACE	0.0259985	0.00746970	0.159	3.48	0.001	0.994523	0.02517

```
THE CONTRIBUTION TO R-SQUARED FOR EACH VARIABLE IS THE AMOUNT
BY WHICH R-SQUARED WOULD BE REDUCED IF THAT VARIABLE WERE
REMOVED FROM THE REGRESSION EQUATION.
```

variable as the DV and the remaining variables as IVs, you can find the variables that distinguish the outlier from the other cases.

For the 117th case, LTIMEDRS, ATTDRUG, and RACE show up consistently in the best subsets of any size. The biggest R-SQUARED for regressions with one variable (RACE) is .022279. With two variables, the best combination is ATTDRUG and RACE, R-SQUARED = .034562. With three variables, the best combination is LTIMEDRS, ATTDRUG, and RACE, R-SQUARED = .046132. With four variables, the best combination is LTIMEDRS, ATTDRUG, MSTATUS, and RACE, but R-SQUARED is only .047366, not substantially better than the three variable solution. The three variable solution is picked as the 'BEST' SUBSET for distinguishing this outlier. For the 193rd case (Table 4.9), the 'BEST' SUBSET is LTIMEDRS, ATTHOUSE, and RACE.

The final step in evaluating outlying cases is to determine how their scores on the variables that cause them to be outliers differ from the scores of the remaining sample. The SPSS LIST and DESCRIPTIVES procedures are used, as seen in Table 4.10. The LIST procedure is run for each outlying case to show its values on all the variables of interest (including a newly created variable called SEQ to verify that the appropriate case is chosen from the data file). Then DESCRIPTIVES is used to show the average values for the remaining sample against which the outlying cases are compared.[19]

---

[19] These values are equal to those shown in the earlier FREQUENCIES runs but for deletion of the two univariate outliers.

TABLE 4.9 SETUP AND PARTIAL BMDP9R OUTPUT SHOWING VARIABLES CAUSING THE 193RD CASE TO BE AN OUTLIER

```
/INPUT       VARIABLES ARE 8.  FORMAT IS FREE.
             FILE='SCREEN.DAT'.
             TITLE IS 'BMDP9R RUN FOR CASE 193'.
/VARIABLE    NAMES ARE SUBNO,TIMEDRS,ATTDRUG,ATTHOUSE,
                   INCOME,EMPLMNT,MSTATUS,RACE.
             LABEL = SUBNO.
/TRANSFORM       USE = ATTHOUSE NE 2.
                 IF (ATTHOUSE EQ 1) THEN ATTHOUSE = 23.634.
                 LTIMEDRS = LOG(TIMEDRS + 1).
                 DUMMY = 0.
                 IF (KASE EQ 193) THEN DUMMY = 1.0.
/REGRESS     DEPENDENT IS DUMMY.
             INDEPENDENT ARE LTIMEDRS, ATTDRUG, ATTHOUSE, MSTATUS,
                       RACE.
/END
```

                                        SUBSETS WITH   1 VARIABLES

| | ADJUSTED | | |
R-SQUARED	R-SQUARED	CP	
0.022279	0.020158	8.77	RACE
0.012096	0.009953	13.64	LTIMEDRS
0.006492	0.004337	16.32	ATTHOUSE
0.002797	0.000634	18.09	ATTDRUG
0.000604	-0.001564	19.14	MSTATUS

                                        SUBSETS WITH   2 VARIABLES

| | ADJUSTED | | | | |
R-SQUARED	R-SQUARED	CP	VARIABLE	COEFFICIENT	T-STATISTIC
0.036811	0.032623	3.82	9 LTIMEDRS	0.0135064	2.63
			8 RACE	0.0257504	3.44
			INTERCEPT	-0.0358984	
0.029923	0.025705	7.11	ATTHOUSE RACE		
0.024781	0.020541	9.57	ATTDRUG RACE		
0.023193	0.018946	10.33	MSTATUS RACE		
0.016349	0.012073	13.61	LTIMEDRS ATTHOUSE		
0.013771	0.009483	14.84	LTIMEDRS ATTDRUG		
0.012973	0.008682	15.22	LTIMEDRS MSTATUS		
0.009045	0.004736	17.10	ATTDRUG ATTHOUSE		
0.007432	0.003117	17.87	ATTHOUSE MSTATUS		
0.003411	-0.000922	19.80	ATTDRUG MSTATUS		

TABLE 4.9   *(CONTINUED)*

SUBSETS WITH   3 VARIABLES

R-SQUARED	ADJUSTED R-SQUARED	CP			
0.041835	0.035573	3.42	VARIABLE	COEFFICIENT	T-STATISTIC
			9 LTIMEDRS	0.0123561	2.39
			4 ATTHOUSE	0.000782596	1.55
			8 RACE	0.0261658	3.49
			INTERCEPT	-0.0539926	
0.038156	0.031870	5.18	VARIABLE	COEFFICIENT	T-STATISTIC
			9 LTIMEDRS	0.0130407	2.53
			3 ATTDRUG	0.00148125	0.80
			8 RACE	0.0255874	3.41
			INTERCEPT	-0.0467581	
0.038115	0.031828	5.20	LTIMEDRS MSTATUS RACE		
0.032168	0.025842	8.04	ATTDRUG ATTHOUSE RACE		
0.031294	0.024963	8.46	ATTHOUSE MSTATUS RACE		
0.025705	0.019337	11.13	ATTDRUG MSTATUS RACE		
0.017953	0.011535	14.84	LTIMEDRS ATTDRUG ATTHOUSE		
0.017517	0.011096	15.05	LTIMEDRS ATTHOUSE MSTATUS		
0.014645	0.008205	16.42	LTIMEDRS ATTDRUG MSTATUS		
0.009990	0.003520	18.65	ATTDRUG ATTHOUSE MSTATUS		

SUBSETS WITH   4 VARIABLES

R-SQUARED	ADJUSTED R-SQUARED	CP			
0.043532	0.035179	4.60	VARIABLE	COEFFICIENT	T-STATISTIC
			9 LTIMEDRS	0.0125328	2.42
			4 ATTHOUSE	0.000814610	1.61
			7 MSTATUS	0.00465033	0.90
			8 RACE	0.0264612	3.53
			INTERCEPT	-0.0634882	
0.043107	0.034749	4.81	VARIABLE	COEFFICIENT	T-STATISTIC
			9 LTIMEDRS	0.0119119	2.29
			3 ATTDRUG	0.00143989	0.78
			4 ATTHOUSE	0.000776866	1.54
			8 RACE	0.0260044	3.47
			INTERCEPT	-0.0644166	
0.039454	0.031065	6.55	LTIMEDRS ATTDRUG MSTATUS RACE		
0.033542	0.025101	9.38	ATTDRUG ATTHOUSE MSTATUS RACE		
0.019115	0.010548	16.29	LTIMEDRS ATTDRUG ATTHOUSE MSTATUS		

SUBSETS WITH   5 VARIABLES

R-SQUARED	ADJUSTED R-SQUARED	CP			
0.044793	0.034342	6.00	LTIMEDRS ATTDRUG ATTHOUSE MSTATUS RACE		

TABLE 4.9   *(CONTINUED)*

```
STATISTICS FOR 'BEST' SUBSET
---------------
MALLOWS' CP                         3.42
SQUARED MULTIPLE CORRELATION      0.04184
MULTIPLE CORRELATION              0.20454
ADJUSTED SQUARED MULT. CORR.      0.03557
RESIDUAL MEAN SQUARE             0.002083
STANDARD ERROR OF EST.           0.045640
F-STATISTIC                         6.68
NUMERATOR DEGREES OF FREEDOM           3
DENOMINATOR DEGREES OF FREEDOM       459
SIGNIFICANCE (TAIL PROB.)         0.0002

*** N O T E *** THE ABOVE F-STATISTIC AND ASSOCIATED SIGNIFICANCE TEND TO BE
LIBERAL WHENEVER A SUBSET OF VARIABLES IS SELECTED BY
                  THE CP OR ADJUSTED R-SQUARED CRITERIA.

-----------------------------------------
```

VARIABLE NO. NAME	REGRESSION COEFFICIENT	STANDARD ERROR	STAND. COEF.	T- STAT.	2TAIL SIG.	TOL- ERANCE	TO CONTIRBUTION R-SQ
INTERCEPT	-0.0539926	0.0150198	-1.162	-3.59	0.000		
9 LTIMEDRS	0.0123561	0.00517241	0.111	2.39	0.017	0.974807	0.01191
4 ATTHOUSE	0.000782596	0.000504431	0.072	1.55	0.121	0.977440	0.00502
8 RACE	0.0261658	0.00748854	0.160	3.49	0.001	0.993982	0.02549

```
THE CONTRIBUTION TO R-SQUARED FOR EACH VARIABLE IS THE AMOUNT
BY WHICH R-SQUARED WOULD BE REDUCED IF THAT VARIABLE WERE
REMOVED FROM THE REGRESSION EQUATION.
```

The 117th case is nonwhite on RACE, has very unfavorable attitudes regarding use of drugs (the lowest possible score on ATTDRUG), and a high score on LTIMEDRS. The 193rd case is also non-white on RACE, has very unfavorable attitudes toward housework (a high score on ATTHOUSE), and a very high score on LTIMEDRS. There is some question, then, about the generalizability of subsequent findings to nonwhite women who make numerous visits to physicians, especially in combination with unfavorable attitude toward either use of drugs or housework.

4.2.1.6 ■ Multicollinearity

The test for multicollinearity is produced in SPSS through the STATISTICS COLLIN instruction. As seen by the Collinearity Diagnostics output of Table 4.7, no multicollinearity is evident. Although the last root has a Cond(itioning) Index that approaches 30, only one of the Variance Proportions is greater than .5.

Screening information as it might be described in a Results section of a journal article appears next.

4.2.2 ■ Screening Grouped Data

For this example, the cases are divided into two groups according to the EMPLMNT (employment) variable; there are 246 cases who have PAIDWORK, and 219 cases who are HOUSEWFEs. For illustrative purposes, variable transformation is considered inappropriate for this example, to be undertaken only if proved necessary. A flow diagram for screening grouped data appears in Figure 4.10.

TABLE 4.10 SETUP AND SPSS OUTPUT SHOWING VARIABLE SCORES FOR MULTIVARIATE
OUTLIERS AND DESCRIPTIVE STATISTICS FOR ALL CASES

```
TITLE                DESCRIPTION OF MULTIVARIATE OUTLIERS.
DATA LIST            FILE = 'SCREEN.DAT'     FREE
                      /SUBNO,TIMEDRS,ATTDRUG,ATTHOUSE,
                       INCOME,EMPLMNT,MSTATUS,RACE.
VAR LABELS           TIMEDRS 'VISITS TO HEALTH PROFESSIONALS'/
                     MSTATUS 'MARITAL STATUS'.
VALUE LABELS         EMPLMNT 0 'PAIDWORK' 1 'HOUSEWIFE'/
                     MSTATUS 1 'SINGLE' 2 'MARRIED'/
                     RACE 1 'WHITE' 2 'NONWHITE'.
COMPUTE          LTIMEDRS = LG10(TIMEDRS + 1).
SELECT IF        ATTHOUSE NE 2.
COMPUTE          SEQ = $CASENUM.
RECODE                ATTHOUSE(1 = 23.634).
LIST             CASES FROM 117 TO 117.
LIST             CASES FROM 193 TO 193
DESCRIPTIVES         ATTDRUG, ATTHOUSE, MSTATUS, RACE, LTIMEDRS, SEQ.
```

SUBNO	TIMEDRS	ATTDRUG	ATTHOUSE	INCOME	EMPLMNT	MSTATUS	RACE	LTIMEDRS	SEQ
137.00	30.00	5.00	24.00	10.00	.00	2.00	2.00	1.49	117.00

Number of cases read: 117   Number of cases listed: 1

SUBNO	TIMEDRS	ATTDRUG	ATTHOUSE	INCOME	EMPLMNT	MSTATUS	RACE	LTIMEDRS	SEQ
262.00	52.00	9.00	31.00	4.00	1.00	2.00	2.00	1.72	193.00

Number of cases read: 193   Number of cases listed: 1

Number of valid observations (listwise) =   463.00

Variable	Mean	Std Dev	Minimum	Maximum	Valid N	Label
ATTDRUG	7.68	1.16	5.00	10.00	463	
ATTHOUSE	23.63	4.26	11.00	35.00	463	
MSTATUS	1.78	.41	1.00	2.00	463	MARITAL STATUS
RACE	1.09	.28	1.00	2.00	463	
LTIMEDRS	.74	.42	.00	1.91	463	
SEQ	232.00	133.80	1.00	463.00	463	

Results

Prior to analysis, number of visits to health professionals, attitude
toward drug use, attitude toward housework, income, marital status, and
race were examined through various SPSS programs for accuracy of data
entry, missing values, and fit between their distributions and the
assumptions of multivariate analysis. The single missing value on attitude toward housework was replaced by the mean for all cases, while
income, with missing values on more than 5% of the cases, was deleted.
The poor split on race (424 to 41) truncates its correlations with

```
other variables, but it was retained for analysis. To improve pairwise
linearity and to reduce the extreme skewness and kurtosis, visits to
health professionals was logarithmically transformed.
    Two cases with extremely low z scores on attitude toward housework
were found to be univariate outliers; two other cases were identified
through Mahalanobis distance as multivariateoutliers with p < .001.²⁰
All four outliers were deleted, leaving 461 cases for analysis.
```

4.2.2.1 ■ Accuracy of Input, Missing Data, Distributions, Homogeneity of Variance, and Univariate Outliers

BMDP7D provides histograms and descriptive statistics for each group separately, as seen in Table 4.11. As with ungrouped data, accuracy of input is judged by plausible MEANs and STD.DEV.s and reasonable MAXIMUM and MINIMUM values. The distributions are judged by their overall shapes within each group. TIMEDRS is just as badly skewed when grouped as when ungrouped, but this is of less concern when dealing with sampling distributions based on over 200 cases unless the skewness causes nonlinearity among variables or there are outliers. ATTDRUG remains well distributed within each group.

As seen in Table 4.10, the ATTHOUSE variable is nicely distributed, as well, but the two cases in the PAIDWORK group with very low scores are outliers. With scores of 2, each case is 4.48 standard deviations below the mean for her group—beyond the $\alpha = .001$ criterion of 3.29 for a two-tailed test. Because there are more cases in the PAIDWORK group, it is decided to delete these two women with extremely favorable attitudes toward housework from further analysis, and to report the deletion in the Results section. There is also a score missing within the group of women with PAIDWORK. ATTDRUG and most of the other variables have 246 cases in this group, but ATTHOUSE has only 245 cases. Because the case with the missing value is from the larger group, it is decided to delete the case from subsequent analyses.

On INCOME, however, it is the smaller group, HOUSEWFEs, with the greater number of missing values; within that group almost 7% of the cases do not have INCOME scores. INCOME, then, is a good candidate for variable deletion, although other remedies are available should deletion seriously interfere with hypothesis testing.

The splits in the two dichotomous variables, MSTATUS and RACE, are about the same for grouped as for ungrouped data, but the appearance of the graphs is misleading because lengths of the lines do not reflect the actual numbers in the groups. For instance, on the MSTATUS variable among HOUSEWFEs there are 193 cases with value 2 (married) and only 26 cases with value 1 (unmarried) despite the fact that the frequency bars look about the same length for values 1 and 2. Similarly for RACE; there are only 27 cases with value 2 (nonwhite) among women with PAID-WORK and 14 cases with value 2 among HOUSEWFEs. The splits for both MSTATUS (for the HOUSEWFE group) and for RACE (both groups) are disturbing, but we choose to retain them here.

---

[20] Case 117 was nonwhite with very unfavorable attitudes regarding use of drugs but numerous visits to physicians. Case 193 was also nonwhite with very unfavorable attitudes toward housework and numerous visits to physicians. Results of analyses may not generalize to nonwhite women with numerous visits to physicians, if they have either very unfavorable attitude toward use of drugs or toward housework.

Searching for:

Plausible range
Missing Values
Univariate Outliers
Homogeneity of Variance

Threats to Pairwise Linearity:
    Opposite Skewness
    Differential Skewness

Pairwise Linearity

Multivariate Outliers

(Describe Outliers)

Programs:

SPSS FREQUENCIES
    (SORT, SPLIT)
BMDP7D
SYSTAT STATS
SAS MEANS

SPSS PLOT
BMDP1V
    6D
SYSTAT PLOT
SAS PLOT

SPSS REGRESSION
BMDP7M
    AM
SYSTAT GLM

SPSS DISCRIMINANT
    LIST CASES
BMDP9R
    1D
SYSTAT STATISTICS
SAS MEANS
DATA

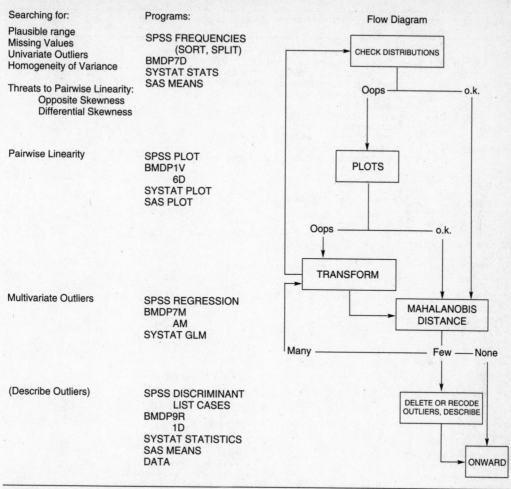

FIGURE 4.10 FLOW DIAGRAM FOR SCREENING GROUPED DATA.

For the remaining analyses, INCOME is deleted as a variable, and the case with the missing value as well as the two univariate outliers on ATTHOUSE are deleted, leaving a sample size of 462: 243 cases in the PAIDWORK group and 219 cases in the HOUSEWFE group.

Because cell sample sizes are not very discrepant, variance ratios as great as 10 can be tolerated. All F_{max} are well below this criterion. For example, squaring the standard deviations for the two groups for TIMEDRS, $F_{max} = (1.197)^2/(1.113)^2 = 1.16$.

4.2.2.2 ■ Linearity

Because of the poor distribution on TIMEDRS, a check of scatterplots is warranted to see if TIMEDRS has a linear relationship with other variables. There is no need to check for linearity with MSTATUS and RACE because variables with two levels have only linear relationships with other

TABLE 4.11   SETUP AND SELECTED BMDP7D OUTPUT SHOWING DESCRIPTIVE STATISTICS AND HISTOGRAMS FOR GROUPED DATA

```
/INPUT              VARIABLES ARE 8.  FORMAT IS FREE.
                    FILE='SCREEN.DAT'.
                    TITLE IS 'UNIVARIATE CHECK FOR GROUPED DATA'.
/VARIABLE           NAMES ARE SUBNO,TIMEDRS,ATTDRUG,ATTHOUSE,
                       INCOME,EMPLMNT,MSTATUS,RACE.
                    LABEL = SUBNO.
                    MISSING = (4)1, (5)99.
/GROUP              CODES(EMPLMNT) = 0, 1.
                    NAMES(EMPLMNT) = PAIDWORK, HOUSEWFE.
/HISTOGRAM          GROUPING = EMPLMNT.
                    VAR = TIMEDRS TO INCOME, MSTATUS, RACE.
/END


                 ************                     ************
HISTOGRAM OF   * TIMEDRS  * (   2)   GROUPED  BY  * EMPLMNT  * (   6)
                 ************                     ************
          PAIDWORK                          HOUSEWFE
MIDPOINTS.................................+..................................+
 87.5000)
 84.0000)
 80.5000)*
 77.0000)
 73.5000)*
 70.0000)
 66.5000)
 63.0000)
 59.5000)**                                    *
 56.0000)                                       ***
 52.5000)*                                      *
 49.0000)*
 45.5000)
 42.0000)                                       *
 38.5000)***                                    **
 35.0000)                                       *
 31.5000)*                                      **
 28.0000)**                                     ****
 24.5000)***                                    **
 21.0000)****                                   ****
 17.5000)*******                                ***********
 14.0000)********                               ******************
 10.5000)****************                       ***********************
  7.0000)M**********************************40 M**********************
  3.5000)*******************************************108 *************************************79
  0.0000)************************************47 ****************************************
          LEGEND FOR GROUP MEANS:     M - MEAN COINCIDES WITH AN ASTERISK
                                      N - MEAN DOES NOT COINCIDE WITH ANY ASTERISK
MEAN         7.293                          8.584
STD.DEV.    11.066                         10.800
S. E. M.     0.706                          0.730
MAXIMUM     81.000                         60.000
MINIMUM      0.000                          0.000
CASES INCL.  246                            219
```

variables. Of the two remaining variables, ATTHOUSE and ATTDRUG, the distribution of ATTDRUG differs most from that of TIMEDRS after univariate outliers are deleted.

Most appropriately checked first, then, are within-group scatterplots of ATTHOUSE versus TIMEDRS. In the within-group scatterplots of Figure 4.11, there is ample evidence of skewness in the bunching up of scores at low values of TIMEDRS, but no suggestion of nonlinearity for these

TABLE 4.11   *(CONTINUED)*

```
              ************                    ************
HISTOGRAM OF * ATTDRUG *  (   3)   GROUPED  BY   * EMPLMNT  * (   6)
              ************                    ************

           PAIDWORK                        HOUSEWFE
MIDPOINTS...........................................+.................................................+
 10.0000)******                    ****************
  9.8000)
  9.6000)
  9.4000)
  9.2000)
  9.0000)********************************************49 ***********************************************46
  8.8000)
  8.6000)
  8.4000)
  8.2000)
  8.0000)************************************************78 ***************************************71
  7.8000)                                  N
  7.6000)N
  7.4000)
  7.2000)
  7.0000)*******************************************72 *****************************************54
  6.8000)
  6.6000)
  6.4000)
  6.2000)
  6.0000)********************************* ****************************
  5.8000)
  5.6000)
  5.4000)
  5.2000)
  5.0000)*******                    ******
       LEGEND FOR GROUP MEANS:    M - MEAN COINCIDES WITH AN ASTERISK
                                  N - MEAN DOES NOT COINCIDE WITH ANY ASTERISK

MEAN         7.593                        7.790
STD.DEV.     1.113                        1.197
S. E. M.     0.071                        0.081
MAXIMUM     10.000                       10.000
MINIMUM      5.000                        5.000
CASES INCL.    246                          219
```

variables in the group of PAIDWORKers ("P" on the scatterplot) or HOUSEWFEs ("H" on the scatterplot). Because the plots are acceptable, there is no evidence that the extreme skewness of TIMEDRS produces a harmful departure from linearity. Nor is there any reason to expect nonlinearity with the symmetrically distributed ATTDRUG.

4.2.2.3 ■ Multivariate Outliers

Multivariate outliers within the groups are sought using BMDPAM, one of two convenient BMDP programs (the other is BMDP7M) for this purpose. Table 4.12 shows univariate summary statistics for both groups and a selected portion of the section which provides Mahalanobis distance for each case from its group centroid. In the TRANSFORM paragraph, only cases with ATTHOUSE GT 2 are selected to eliminate the case with the missing value (coded 1) and the two univariate outliers (value 2). In BMDPAM, if missing values are declared, cases with missing data are included during the

TABLE 4.11   *(CONTINUED)*

```
                     ************                    ************
HISTOGRAM OF * ATTHOUSE * (   4)   GROUPED  BY   * EMPLMNT  * (   6)
                     ************                    ************

               PAIDWORK                          HOUSEWFE
MIDPOINTS.....................................+.................................+
  37.5000)
  36.0000)
  34.5000)**                                    **
  33.0000)****
  31.5000)********                              ******
  30.0000)*********                             ****
  28.5000)*************************              *************************
  27.0000)***************                       *************
  25.5000)**********************************44  ****************************************
  24.0000)M**************                       M********************
  22.5000)*********************************42   ***************************************43
  21.0000)*****************                     *****************
  19.5000)****************************           ****************************
  18.0000)************                          ********
  16.5000)************                          ************
  15.0000)**                                    **
  13.5000)***                                   **
  12.0000)*
  10.5000)                                      *
   9.0000)
   7.5000)
   6.0000)
   4.5000)
   3.0000)
   1.5000)**
   0.0000)
            LEGEND FOR GROUP MEANS:    M - MEAN COINCIDES WITH AN ASTERISK
                                       N - MEAN DOES NOT COINCIDE WITH ANY ASTERISK


MEAN        23.641                       23.429
STD.DEV.     4.831                        4.068
S. E. M.     0.309                        0.275
MAXIMUM     34.000                       35.000
MINIMUM      2.000                       11.000
CASES EXCL.  (   1)                      (   0)
CASES INCL.    245                          219
```

search for multivariate outliers. Because we have decided to eliminate those cases in any event, TRANSFORM is used to omit them.

Because $\alpha = .001$ is the criterion chosen for identifying multivariate outliers, they are readily found through BMDPAM since the program places an asterisk after each case that is an outlier at that alpha level. Table 4.12 shows that cases 193 and 206 are outliers in the HOUSEWFE group (SUBNO 262 and 276, respectively).

Altogether, nine cases (about 2%), almost evenly distributed in the two groups, are identified as multivariate outliers. While this is not an exceptionally large number of cases to delete, it is worth investigating alternative strategies for dealing with outliers. The univariate summary statistics for TIMEDRS in Table 4.12 show a LARGEST STANDARD SCORE of 6.62 among those with PAIDWORK and 4.76 among HOUSEWFEs; the poorly distributed variable produces univariate

TABLE 4.11 *(CONTINUED)*

```
                    ***********                    ***********
HISTOGRAM OF  * INCOME  *  (   5)    GROUPED  BY   * EMPLMNT  *  (   6)
                    ***********                    ***********

              PAIDWORK                             HOUSEWFE
MIDPOINTS...................................+..........................................+
 10.4000)
 10.0000)*******                             *******
  9.6000)
  9.2000)
  8.8000)**********                          ****
  8.4000)
  8.0000)***********                         *******
  7.6000)
  7.2000)
  6.8000)****************                     *********************
  6.4000)
  6.0000)******************                   *****************
  5.6000)
  5.2000)
  4.8000)****************************         *****************
  4.4000)N
  4.0000)****************************************  M************************************46
  3.6000)
  3.2000)
  2.8000)*******************************46 ***********************************
  2.4000)
  2.0000)*********************             ******************
  1.6000)
  1.2000)
  0.8000)***************************       *********************************
  0.4000)
        LEGEND FOR GROUP MEANS:    M - MEAN COINCIDES WITH AN ASTERISK
                                   N - MEAN DOES NOT COINCIDE WITH ANY ASTERISK

MEAN          4.238                        4.176
STD.DEV.      2.440                        2.400
S. E. M.      0.159                        0.168
MAXIMUM      10.000                       10.000
MINIMUM       1.000                        1.000
CASES EXCL.   ( 11)                        ( 15)
CASES INCL.    235                          204
```

outliers in both groups. The skewed histograms of Table 4.10 suggest a logarithmic transformation of TIMEDRS.

Table 4.13 shows output from a second run of BMDPAM identical to the run in Table 4.12 except that TIMEDRS is replaced by LTIMEDRS, its logarithmic transform. The 193rd case remains extreme, but case 206 is no longer an outlier. With the transformed variable, the entire data set contains only two multivariate outliers (the same two identified in ungrouped data).

4.2.2.4 ■ Variables Causing Cases to be Outliers

Identification of the variables causing outliers to be extreme proceeds in the same manner as for ungrouped data except that the values for the case are compared with the means for the group the case comes from. For case 193, a housewife, the BMDP9R all-subsets-regression run is limited to

TABLE 4.11   *(CONTINUED)*

```
                 * * * * * * * * * * * *              * * * * * * * * * * * *
HISTOGRAM OF  *  MSTATUS  *  (   7)   GROUPED  BY   *  EMPLMNT  *  (    6)
                 * * * * * * * * * * * *              * * * * * * * * * * * *

           PAIDWORK                             HOUSEWFE
MIDPOINTS.....................................+.................................+
  2.00000)*********************************169 ***************************************193
  1.96000)
  1.92000)
  1.88000)                                        N
  1.84000)
  1.80000)
  1.76000)
  1.72000)
  1.68000)N
  1.64000)
  1.60000)
  1.56000)
  1.52000)
  1.48000)
  1.44000)
  1.40000)
  1.36000)
  1.32000)
  1.28000)
  1.24000)
  1.20000)
  1.16000)
  1.12000)
  1.08000)
  1.04000)
  1.00000)*****************************************77 *************************
          LEGEND FOR GROUP MEANS:      M - MEAN COINCIDES WITH AN ASTERISK
                                       N - MEAN DOES NOT COINCIDE WITH ANY ASTERISK


MEAN         1.687                          1.881
STD.DEV.     0.465                          0.324
S. E. M.     0.030                          0.022
MAXIMUM      2.000                          2.000
MINIMUM      1.000                          1.000
CASES INCL.   246                            219
```

housewives, as seen in the TRANSFORM paragraph USE = EMPLMNT EQ 1 sentence of Table 4.14. As seen in the table, the same variables cause this woman to be an outlier from her group as from the entire sample: she differs on the combination of RACE, ATTHOUSE, and LTIMEDRS. Similarly, case 117 differs from her group on the same variables that make her extreme with respect to the entire sample (output not shown).

As with ungrouped data, identification of variables on which cases are outliers is followed by an analysis of the scores on the variables for those cases. This analysis is done through BMDP1D where outlying cases are declared to have a missing value on a dummy variable (Table 4.15) and /PRINT MISS is used to provide scores on all variables for the outliers. BMDP1D now also lists

TABLE 4.11   *(CONTINUED)*

```
                * * * * * * * * * * * *                  * * * * * * * * * * * *
HISTOGRAM OF  *  RACE      *  (   8)   GROUPED  BY   *  EMPLMNT  *  (   6)
                * * * * * * * * * * * *                  * * * * * * * * * * * *

           PAIDWORK                             HOUSEWFE
MIDPOINTS...............................................+.....................................................+
 2.00000)*************************               **************
 1.96000)
 1.92000)
 1.88000)
 1.84000)
 1.80000)
 1.76000)
 1.72000)
 1.68000)
 1.64000)
 1.60000)
 1.56000)
 1.52000)
 1.48000)
 1.44000)
 1.40000)
 1.36000)
 1.32000)
 1.28000)
 1.24000)
 1.20000)
 1.16000)
 1.12000)N
 1.08000)                                      N
 1.04000)
 1.00000)*******************************219 ***********************************205

          LEGEND FOR GROUP MEANS:    M - MEAN COINCIDES WITH AN ASTERISK
                                     N - MEAN DOES NOT COINCIDE WITH ANY ASTERISK

MEAN          1.110                          1.064
STD.DEV.      0.313                          0.245
S. E. M.      0.020                          0.017
MAXIMUM       2.000                          2.000
MINIMUM       1.000                          1.000
CASES INCL.    246                            219
```

scores for the first 10 cases by default. The two outliers are nonwhite with frequent visits to health professionals. Case 117, in addition, has very unfavorable attitudes toward drug use, while case 193 has favorable attitudes toward housework.

4.2.2.5 ■ Multicollinearity

In the BMDPAM run of Table 4.13, SMCs are shown for each variable as predicted by all other variables. Since all R^2 are well below .7, there is no indication of even logical multicollinearity.

Screening information as it might be described in a Results section of a journal article appears next.

```
/INPUT          VARIABLES ARE 8. FORMAT IS FREE.
                FILE='SCREEN.DAT'.
                TITLE IS 'BMDP9R RUN FOR CASE 193'.
/VARIABLE       NAMES ARE SUBNO,TIMEDRS,ATTDRUG,ATTHOUSE,
                    INCOME,EMPLMNT,MSTATUS,RACE.
                LABEL = SUBNO.
                MISSING = (4)1, (5)99.
/GROUP          VAR = EMPLMNT.
                CODES(EMPLMNT) = 0, 1.
                NAMES (EMPLMNT) = PAIDWORK, HOUSEWFE.
/PLOT           XVAR = TIMEDRS. YVAR = ATTDRUG.
                GROUP = EACH.
/END
```

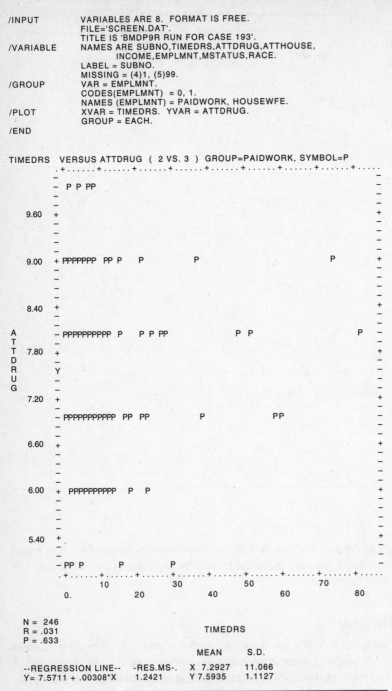

```
TIMEDRS  VERSUS ATTDRUG ( 2 VS. 3 ) GROUP=PAIDWORK, SYMBOL=P
       .+......+......+......+......+......+......+......+......+......+....
       -                                                                 -
       -  P P PP                                                         -
       -                                                                 -
       -                                                                 -
9.60   +                                                                 +
       -                                                                 -
       -                                                                 -
       -                                                                 -
9.00   + PPPPPPP PP P     P        P                           P         +
       -                                                                 -
       -                                                                 -
       -                                                                 -
8.40   +                                                                 +
       -                                                                 -
A      - PPPPPPPPPP P    P P PP        P P               P              -
T      -                                                                 -
T 7.80 +                                                                 +
D      -                                                                 -
R      Y                                                                 -
U      -                                                                 -
G 7.20 +                                                                 +
       -                                                                 -
       - PPPPPPPPPP PP PP      P        PP                              -
       -                                                                 -
6.60   +                                                                 +
       -                                                                 -
       -                                                                 -
       -                                                                 -
6.00   + PPPPPPPPP  P   P                                                +
       -                                                                 -
       -                                                                 -
       -                                                                 -
5.40   +                                                                 +
       -                                                                 -
       - PP P      P        P                                           -
       .+......+......+......+......+......+......+......+......+......+....
           10        30        50        70
       0.        20        40        60        80
```

```
N = 246
R = .031                          TIMEDRS
P = .633
                                  MEAN    S.D.
--REGRESSION LINE--  -RES.MS-.   X 7.2927  11.066
Y= 7.5711 + .00308*X    1.2421   Y 7.5935   1.1127
```

FIGURE 4.11 SETUP AND PARTIAL BMDP6D OUTPUT SHOWING WITHIN-
GROUP SCATTERPLOT OF ATTHOUSE VS. TIMEDRS.

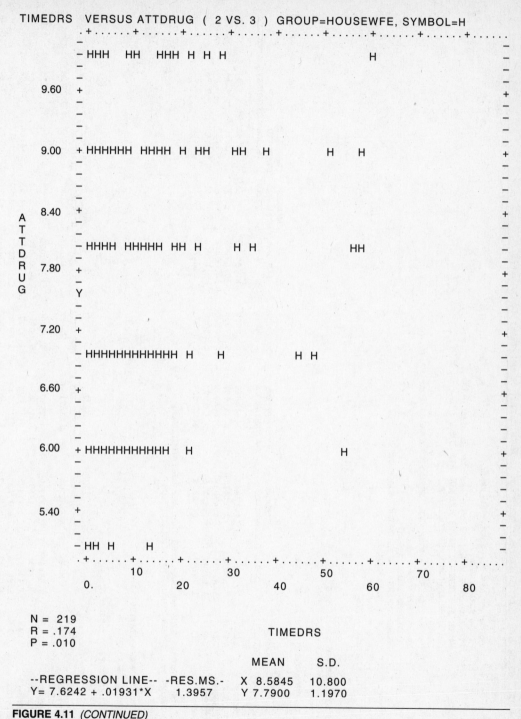

TIMEDRS VERSUS ATTDRUG (2 VS. 3) GROUP=HOUSEWFE, SYMBOL=H

```
        .+......+......+......+......+......+......+......+......+......
      – HHH   HH   HHH H H H                       H          –
                                                              –
9.60  +                                                       +
      –                                                       –
      –                                                       –
      –                                                       –
9.00  + HHHHHH  HHHH  H HH    HH   H          H     H          +
      –                                                       –
      –                                                       –
      –                                                       –
8.40  +                                                       +
A     –                                                       –
T     –                                                       –
T     – HHHH  HHHHH  HH H     H H            HH               –
D     –                                                       –
R 7.80 +                                                      +
U     –                                                       –
G     Y                                                       –
      –                                                       –
7.20  +                                                       +
      –                                                       –
      – HHHHHHHHHHHH  H     H         H H                     –
      –                                                       –
6.60  +                                                       +
      –                                                       –
      –                                                       –
      –                                                       –
6.00  + HHHHHHHHHHH   H                    H                  +
      –                                                       –
      –                                                       –
      –                                                       –
5.40  +                                                       +
      –                                                       –
      – HH  H     H                                          –
        .+......+......+......+......+......+......+......+......+......
          10        30        50        70
      0.        20        40        60        80
```

N = 219
R = .174   TIMEDRS
P = .010

                          MEAN   S.D.

--REGRESSION LINE-- -RES.MS.-   X 8.5845   10.800
Y= 7.6242 + .01931*X   1.3957   Y 7.7900   1.1970

FIGURE 4.11 *(CONTINUED)*

TABLE 4.12 SETUP AND SELECTED BMDPAM OUTPUT FOR IDENTIFICATION OF MULTIVARIATE OUTLIERS

```
/INPUT          VARIABLES ARE 8.  FORMAT IS FREE.
                FILE='SCREEN.DAT'.
                TITLE IS 'MULTIVARIATE OUTLIERS FOR GROUPED DATA'.
/VARIABLE       NAMES ARE SUBNO,TIMEDRS,ATTDRUG,ATTHOUSE,
                INCOME, EMPLMNT,MSTATUS,RACE.
                LABEL = SUBNO.
                USE = TIMEDRS TO ATTHOUSE, EMPLMNT TO RACE.
/GROUP          VAR = EMPLMNT.
                CODES(EMPLMNT) = 0, 1.
                NAMES(EMPLMNT) = PAIDWORK, HOUSEWFE.
/TRANSFORM      USE = ATTHOUSE GT 2.
/EST            METHOD = REGR.
/PRINT          MATRIX = DIS.
/END
```

UNIVARIATE SUMMARY STATISTICS

GROUP IS PAIDWORK   SIZE IS   243

VARIABLE	SAMPLE SIZE	MEAN	STANDARD DEVIATION	COEFFICIENT OF VARIATION	SMALLEST VALUE	LARGEST VALUE	SMALLEST STANDARD SCORE	LARGEST STANDARD SCORE	SKEWNESS	KURTOSIS
2 TIMEDRS	243	7.35802	11.11842	1.511060	0.00000	81.00000	-0.66	6.62	3.80	17.29
3 ATTDRUG	243	7.59671	1.11429	0.146681	5.00000	10.00000	-2.33	2.16	-0.14	-0.51
4 ATTHOUSE	243	23.81893	4.43038	0.186002	12.00000	34.00000	-2.67	2.30	-0.04	-0.49
7 MSTATUS	243	1.69547	0.46116	0.271993	1.00000	2.00000	-1.51	0.66	-0.84	-1.29
8 RACE	243	1.10700	0.30975	0.279808	1.00000	2.00000	-0.35	2.88	2.53	4.40

VALUES FOR KURTOSIS GREATER THAN ZERO INDICATE DISTRIBUTIONS
WITH HEAVIER TAILS THAN THE NORMAL DISTRIBUTION.

UNIVARIATE SUMMARY STATISTICS

GROUP IS HOUSEWFE   SIZE IS   219

VARIABLE	SAMPLE SIZE	MEAN	STANDARD DEVIATION	COEFFICIENT OF VARIATION	SMALLEST VALUE	LARGEST VALUE	SMALLEST STANDARD SCORE	LARGEST STANDARD SCORE	SKEWNESS	KURTOSIS
2 TIMEDRS	219	8.58447	10.79950	1.258027	0.00000	60.00000	-0.79	4.76	2.53	7.56
3 ATTDRUG	219	7.78995	1.19697	0.153656	5.00000	10.00000	-2.33	1.85	-0.14	-0.48
4 ATTHOUSE	219	23.42922	4.06803	0.173631	11.00000	35.00000	-3.06	2.84	-0.06	-0.04
7 MSTATUS	219	1.88128	0.32420	0.172330	1.00000	3.00000	-2.72	0.37	-2.34	3.50
8 RACE	219	1.06393	0.24518	0.230451	1.00000	2.00000	-0.26	3.82	3.54	10.59

VALUES FOR KURTOSIS GREATER THAN ZERO INDICATE DISTRIBUTIONS
WITH HEAVIER TAILS THAN THE NORMAL DISTRIBUTION.

TABLE 4.12   *(CONTINUED)*

SQUARED MULTIPLE CORRELATIONS OF EACH VARIABLE WITH ALL OTHER VARIABLES

(MEASURES OF MULTICOLLINEARITY OF VARIABLES)
AND TESTS OF SIGNIFICANCE OF MULTIPLE REGRESSION
DEGREES OF FREEDOM FOR F-STATISTICS ARE   4 AND   456

VARIABLE NO.	NAME	R-SQUARED	F-STATISTIC	SIGNIFICANCE (P LESS THAN)
2	TIMEDRS	0.031708	3.73	0.00531
3	ATTDRUG	0.012315	1.42	0.22577
4	ATTHOUSE	0.021730	2.53	0.03976
7	MSTATUS	0.011355	1.31	0.26558
8	RACE	0.004709	0.54	0.70693

NOTE THAT WHEN YOU USE THE ML OR ALLVALUE OPTIONS, THE
DEGREES OF FREEDOM USED IN COMPUTING THE ABOVE F STATISTICS
INCLUDE CASES WITH MISSING VALUES. IF THE AMOUNT OF
MISSING DATA IS VERY LARGE, THE SIGNIFICANCE OF THE
F STATISTICS IS EXAGGERATED.

MAHALANOBIS DISTANCES ARE COMPUTED FROM EACH CASE TO THE
CENTROID OF ITS GROUP. ONLY THOSE VARIABLES WHICH WERE
ORIGINALLY AVAILABLE ARE USED-ESTIMATED VALUES ARE NOT USED.
FOR LARGE MULTIVARIATE NORMAL SAMPLES, THE MAHALANOBIS
DISTANCES HAVE AN APPROXIMATELY CHI-SQUARE DISTRIBUTION WITH
THE NUMBER OF DEGREES OF FREEDOM EQUAL TO THE NUMBER OF
NONMISSING VARIABLES. SIGNIFICANCE LEVELS REPORTED BELOW
THAT ARE LESS THAN .001 ARE FLAGGED WITH AN ASTERISK (*).

ESTIMATES OF MISSING DATA, MAHALANOBIS D-SQUARED (CHI-SQUARED)
AND SQUARED MULTIPLE CORRELATIONS WITH AVAILABLE VARIABLES

CASE LABEL	CASE NUMBER	MISSING VARIABLE	ESTIMATE	R-SQUARED	GROUP	CHI-SQ	CHISQ/DF	D.F.	SIGNIFICANCE
261	192				PAIDWORK	5.325	1.065	5	0.3775
262	193				HOUSEWFE	30.453	6.091	5	0.0000 *
263	194				PAIDWORK	1.269	0.254	5	0.9381
264	195				HOUSEWFE	1.525	0.305	5	0.9102
265	196				HOUSEWFE	18.503	3.701	5	0.0024
266	197				HOUSEWFE	1.131	0.226	5	0.9514
267	198				PAIDWORK	13.920	2.784	5	0.0161
268	199				PAIDWORK	5.202	1.040	5	0.3918
269	200				HOUSEWFE	3.917	0.783	5	0.5615
270	201				PAIDWORK	1.295	0.259	5	0.9355
271	202				HOUSEWFE	0.974	0.195	5	0.9646
272	203				HOUSEWFE	5.460	1.092	5	0.3623
273	204				PAIDWORK	6.912	1.382	5	0.2273
274	205				HOUSEWFE	3.796	0.759	5	0.5792
276	206				HOUSEWFE	20.530	4.106	5	0.0010 *

TABLE 4.13 SETUP AND SELECTED BMDPAM OUTPUT AFTER TRANSFORMATION OF TIMEDRS

```
/INPUT           VARIABLES ARE 8.  FORMAT IS FREE.
                 FILE='SCREEN.DAT'.
                 TITLE IS 'MULTIVARIATE OUTLIERS FOR GROUPED DATA'.
/VARIABLE        NAMES ARE SUBNO,TIMEDRS,ATTDRUG,ATTHOUSE,
                 INCOME,EMPLMNT,MSTATUS,RACE.
                 LABEL = SUBNO.
                 USE = ATTDRUG, ATTHOUSE, EMPLMNT TO RACE, LTIMEDRS.
/GROUP           VAR = EMPLMNT.
                 CODES(EMPLMNT) = 0, 1.
                 NAMES(EMPLMNT) = PAIDWORK, HOUSEWFE.
/TRANSFORM       USE = ATTHOUSE GT 2.
                 LTIMEDRS = LOG(TIMEDRS + 1).
/EST             METHOD = REGR.
/PRINT           MATRIX = DIS.
/END
```

UNIVARIATE SUMMARY STATISTICS

GROUP IS PAIDWORK   SIZE IS   243

VARIABLE	SAMPLE SIZE	MEAN	STANDARD DEVIATION	COEFFICIENT OF VARIATION	SMALLEST VALUE	LARGEST VALUE	SMALLEST STANDARD SCORE	LARGEST STANDARD SCORE	SKEWNESS	KURTOSIS
3 ATTDRUG	243	7.59671	1.11429	0.146681	5.00000	10.00000	-2.33	2.16	-0.14	-0.51
4 ATTHOUSE	243	23.81893	4.43038	0.186002	12.00000	34.00000	-2.67	2.30	-0.04	-0.49
7 MSTATUS	243	1.69547	0.46116	0.271993	1.00000	2.00000	-1.51	0.66	-0.84	-1.29
8 RACE	243	1.10700	0.30975	0.279808	1.00000	2.00000	-0.35	2.88	2.53	4.40
9 LTIMEDRS	243	0.72247	0.39158	0.542001	0.00000	1.91381	-1.85	3.04	0.41	0.31

VALUES FOR KURTOSIS GREATER THAN ZERO INDICATE DISTRIBUTIONS
WITH HEAVIER TAILS THAN THE NORMAL DISTRIBUTION.

UNIVARIATE SUMMARY STATISTICS

GROUP IS HOUSEWFE   SIZE IS   219

VARIABLE	SAMPLE SIZE	MEAN	STANDARD DEVIATION	COEFFICIENT OF VARIATION	SMALLEST VALUE	LARGEST VALUE	SMALLEST STANDARD SCORE	LARGEST STANDARD SCORE	SKEWNESS	KURTOSIS
3 ATTDRUG	219	7.78995	1.19697	0.153656	5.00000	10.00000	-2.33	1.85	-0.14	-0.48
4 ATTHOUSE	219	23.42922	4.06803	0.173631	11.00000	35.00000	-3.06	2.84	-0.06	-0.04
7 MSTATUS	219	1.88128	0.32420	0.172330	1.00000	2.00000	-2.72	0.37	-2.34	3.50
8 RACE	219	1.06393	0.24518	0.230451	1.00000	2.00000	-0.26	3.82	3.54	10.59
9 LTIMEDRS	219	0.76578	0.44141	0.576423	0.00000	1.78533	-1.73	2.31	0.03	-0.60

VALUES FOR KURTOSIS GREATER THAN ZERO INDICATE DISTRIBUTIONS
WITH HEAVIER TAILS THAN THE NORMAL DISTRIBUTION.

TABLE 4.13   (CONTINUED)

SQUARED MULTIPLE CORRELATIONS OF EACH VARIABLE WITH ALL OTHER VARIABLES
───

(MEASURES OF MULTICOLLINEARITY OF VARIABLES)
AND TESTS OF SIGNIFICANCE OF MULTIPLE REGRESSION
DEGREES OF FREEDOM FOR F-STATISTICS ARE   4 AND   456

VARIABLE NO.	NAME	R-SQUARED	F-STATISTIC	SIGNIFICANCE (P LESS THAN)
3	ATTDRUG	0.013816	1.60	0.17398
4	ATTHOUSE	0.027438	3.22	0.01276
7	MSTATUS	0.007811	0.90	0.46531
8	RACE	0.007550	0.87	0.48344
9	LTIMEDRS	0.038094	4.51	0.00138

NOTE THAT WHEN YOU USE THE ML OR ALLVALUE OPTIONS, THE
DEGREES OF FREEDOM USED IN COMPUTING THE ABOVE F STATISTICS
INCLUDE CASES WITH MISSING VALUES. IF THE AMOUNT OF
MISSING DATA IS VERY LARGE, THE SIGNIFICANCE OF THE
F STATISTICS IS EXAGGERATED.

MAHALANOBIS DISTANCES ARE COMPUTED FROM EACH CASE TO THE
CENTROID OF ITS GROUP. ONLY THOSE VARIABLES WHICH WERE
ORIGINALLY AVAILABLE ARE USED-ESTIMATED VALUES ARE NOT USED.
FOR LARGE MULTIVARIATE NORMAL SAMPLES, THE MAHALANOBIS
DISTANCES HAVE AN APPROXIMATELY CHI-SQUARE DISTRIBUTION WITH
THE NUMBER OF DEGREES OF FREEDOM EQUAL TO THE NUMBER OF
NONMISSING VARIABLES. SIGNIFICANCE LEVELS REPORTED BELOW
THAT ARE LESS THAN .001 ARE FLAGGED WITH AN ASTERISK (*).

ESTIMATES OF MISSING DATA, MAHALANOBIS D-SQUARED (CHI-SQUARED)
AND SQUARED MULTIPLE CORRELATIONS WITH AVAILABLE VARIABLES

CASE LABEL	CASE NUMBER	MISSING VARIABLE	R-SQUARED ESTIMATE	GROUP	CHI-SQ	CHISQ/DF	D.F.	SIGNIFICANCE
261	192			PAIDWORK	5.886	1.177	5	0.3175
262	193			HOUSEWFE	20.695	4.139	5	0.0009 *
263	194			PAIDWORK	1.349	0.270	5	0.9298
264	195			HOUSEWFE	2.396	0.479	5	0.7921
265	196			HOUSEWFE	19.353	3.871	5	0.0017
266	197			HOUSEWFE	3.766	0.753	5	0.5835
267	198			PAIDWORK	14.166	2.833	5	0.0146
268	199			HOUSEWFE	5.967	1.193	5	0.3094
269	200			HOUSEWFE	4.013	0.803	5	0.5475
270	201			PAIDWORK	1.897	0.379	5	0.8632
271	202			HOUSEWFE	0.891	0.178	5	0.9709
272	203			HOUSEWFE	5.667	1.133	5	0.3399
273	204			PAIDWORK	9.668	1.934	5	0.0852
274	205			HOUSEWFE	5.966	1.193	5	0.3095
276	206			HOUSEWFE	6.667	1.333	5	0.2466

```
/INPUT              VARIABLES ARE 8.  FORMAT IS FREE.
                    FILE='SCREEN.DAT'.
                    TITLE IS 'DESCRIBE CASE 193 FOR GROUPED DATA'.
/VARIABLE           NAMES ARE SUBNO,TIMEDRS,ATTDRUG,ATTHOUSE,
                       INCOME,EMPLMNT,MSTATUS,RACE.
                    LABEL = SUBNO.
                    USE = ATTDRUG, ATTHOUSE, EMPLMNT TO RACE, LTIMEDRS,
                       DUMMY.
/GROUP              CODES(EMPLMNT) = 0, 1.
                    NAMES(EMPLMNT) = PAIDWORK, HOUSEWFE.
/TRANSFORM          USE = EMPLMNT EQ 1.
                    LTIMEDRS = LOG(TIMEDRS + 1).
                    DUMMY = 0.
                    IF (KASE EQ 193) THEN DUMMY = 1.
/REGRESS            DEPENDENT IS DUMMY.
                    INDEPENDENT ARE LTIMEDRS, ATTDRUG, ATTHOUSE,
                                    MSTATUS, RACE.
/END
```

```
NUMBER OF CASES READ. . . . . . . . . . . . .    465
    CASES WITH USE SET TO ZERO . . . . . . . .    246
        REMAINING NUMBER OF CASES . . . . . . .    219
```

                                    SUBSETS WITH   1 VARIABLES

	ADJUSTED		
R-SQUARED	R-SQUARED	CP	
0.067169	0.062870	8.88	RACE
0.021728	0.017220	19.79	LTIMEDRS
0.015960	0.011426	21.17	ATTHOUSE
0.004709	0.000123	23.87	ATTDRUG
0.000618	-0.003987	24.85	MSTATUS

                                    SUBSETS WITH   2 VARIABLES

	ADJUSTED		
R-SQUARED	R-SQUARED	CP	

0.097723	0.089369	3.55	VARIABLE	COEFFICIENT	T-STATISTIC
			9 LTIMEDRS	0.0268997	2.70
			8 RACE	0.0763766	4.27
			INTERCEPT	-0.0972922	

0.083742	0.075258	6.90	ATTHOUSE RACE
0.070865	0.062262	9.99	ATTDRUG RACE
0.068064	0.059435	10.66	MSTATUS RACE
0.031735	0.022769	19.38	LTIMEDRS ATTHOUSE
0.023764	0.014725	21.30	LTIMEDRS MSTATUS
0.023478	0.014436	21.37	LTIMEDRS ATTDRUG
0.020641	0.011573	22.05	ATTDRUG ATTHOUSE
0.017830	0.008736	22.72	ATTHOUSE MSTATUS
0.005346	-0.003864	25.72	ATTDRUG MSTATUS

TABLE 4.14   (CONTINUED)

SUBSETS WITH   3 VARIABLES

R-SQUARED	ADJUSTED R-SQUARED	CP			
0.107160	0.094702	3.28	VARIABLE	COEFFICIENT	T-STATISTIC
			9 LTIMEDRS	0.0239913	2.37
			4 ATTHOUSE	0.00164393	1.51
			8 RACE	0.0760936	4.26
			INTERCEPT	-0.133280	
0.100689	0.088141	4.83	VARIABLE	COEFFICIENT	T-STATISTIC
			9 LTIMEDRS	0.0280625	2.79
			7 MSTATUS	0.0114618	0.84
			8 RACE	0.0768860	4.29
			INTERCEPT	-0.120288	
0.098504	0.085925	5.36	LTIMEDRS ATTDRUG RACE		
0.087407	0.074673	8.02	ATTDRUG ATTHOUSE RACE		
0.086116	0.073364	8.33	ATTHOUSE MSTATUS RACE		
0.071778	0.058826	11.77	ATTDRUG MSTATUS RACE		
0.035041	0.021577	20.59	LTIMEDRS ATTHOUSE MSTATUS		
0.033795	0.020313	20.89	LTIMEDRS ATTDRUG ATTHOUSE		
0.025439	0.011841	22.89	LTIMEDRS ATTDRUG MSTATUS		
0.022543	0.008904	23.59	ATTDRUG ATTHOUSE MSTATUS		

SUBSETS WITH   4 VARIABLES

R-SQUARED	ADJUSTED R-SQUARED	CP			
0.111586	0.094980	4.22	VARIABLE	COEFFICIENT	T-STATISTIC
			9 LTIMEDRS	0.0251827	2.48
			4 ATTHOUSE	0.00177923	1.62
			7 MSTATUS	0.0141022	1.03
			8 RACE	0.0766971	4.29
			INTERCEPT	-0.164535	
0.108148	0.091478	5.04	VARIABLE	COEFFICIENT	T-STATISTIC
			9 LTIMEDRS	0.0230161	2.23
			3 ATTDRUG	0.00180983	0.49
			4 ATTHOUSE	0.00166301	1.52
			8 RACE	0.0756492	4.22
			INTERCEPT	-0.146606	
0.101404	0.084608	6.66	LTIMEDRS ATTDRUG MSTATUS RACE		
0.089809	0.072796	9.45	ATTDRUG ATTHOUSE MSTATUS RACE		
0.037019	0.019019	22.12	LTIMEDRS ATTDRUG ATTHOUSE MSTATUS		

SUBSETS WITH   5 VARIABLES

R-SQUARED	ADJUSTED R-SQUARED	CP	
0.112501	0.091667	6.00	LTIMEDRS ATTDRUG ATTHOUSE MSTATUS RACE

TABLE 4.14 *(CONTINUED)*

```
STATISTICS FOR 'BEST' SUBSET
---------------
MALLOWS' CP                            3.28
SQUARED MULTIPLE CORRELATION       0.10716
MULTIPLE CORRELATION               0.32735
ADJUSTED SQUARED MULT. CORR.       0.09470
RESIDUAL MEAN SQUARE              0.004134
STANDARD ERROR OF EST.           0.064294
F-STATISTIC                            8.60
NUMERATOR DEGREES OF FREEDOM              3
DENOMINATOR DEGREES OF FREEDOM         215
SIGNIFICANCE (TAIL PROB.)         0.0000
```

*** N O T E *** THE ABOVE F-STATISTIC AND ASSOCIATED SIGNIFICANCE TEND TO BE LIBERAL WHENEVER A SUBSET OF VARIABLES IS SELECTED BY
                    THE CP OR ADJUSTED R-SQUARED CRITERIA.

VARIABLE NO. NAME	REGRESSION COEFFICIENT	STANDARD ERROR	STAND. COEF.	T- STAT.	2TAIL SIG.	TOL- ERANCE	CONTRI- BUTION TO R-SQ
INTERCEPT	-0.133280	0.0321994	-1.972	-4.14	0.000		
9 LTIMEDRS	0.0239913	0.0101029	0.157	2.37	0.018	0.953473	0.02342
4 ATTHOUSE	0.00164393	0.00109055	0.099	1.51	0.133	0.963449	0.00944
8 RACE	0.0760936	0.0178549	0.276	4.26	0.000	0.989452	0.07543

THE CONTRIBUTION TO R-SQUARED FOR EACH VARIABLE IS THE AMOUNT
BY WHICH R-SQUARED WOULD BE REDUCED IF THAT VARIABLE WERE
REMOVED FROM THE REGRESSION EQUATION.

TABLE 4.15 SETUP AND BMDP1D OUTPUT SHOWING VARIABLE SCORES FOR MULTIVARIATE OUTLIERS AND DESCRIPTIVE STATISTICS FOR ALL CASES

```
/INPUT        VARIABLES ARE 8.  FORMAT IS FREE.
              FILE-'SCREEN.DAT'.
              TITLE IS 'DESCRIBE CASE 193 FOR GROUPED DATA'.
/VARIABLE     NAMES ARE SUBNO,TIMEDRS,ATTDRUG,ATTHOUSE,
                 INCOME,EMPLMNT,MSTATUS,RACE.
              LABEL = SUBNO.
              USE = ATTDRUG, ATTHOUSE, EMPLMNT TO RACE, LTIMEDRS,
                 DUMMY.
/GROUP        VAR = EMPLMNT.
                 CODES(EMPLMNT) = 0, 1.
              NAMES(EMPLMNT) = PAIDWORK, HOUSEWFE.
/TRANSFORM    USE = ATTHOUSE NE 2.
              LTIMEDRS = LOG(TIMEDRS + 1).
              DUMMY = 0.
              IF (KASE EQ 193 OR KASE EQ 117) THEN DUMMY = XMIS.
/PRINT        MISS.
/END
```

CASE NO.	LABEL	3 ATTDRUG	4 ATTHOUSE	6 EMPLMNT	7 MSTATUS	8 RACE	9 LTIMEDRS	10 DUMMY
1	1	8.00	27.00	HOUSEWFE	2.00	1.00	0.30	0.00
2	2	7.00	20.00	PAIDWORK	2.00	1.00	0.60	0.00
3	3	8.00	23.00	PAIDWORK	2.00	1.00	0.00	0.00
4	4	9.00	28.00	HOUSEWFE	2.00	1.00	1.15	0.00
5	5	7.00	24.00	HOUSEWFE	2.00	1.00	1.20	0.00
6	6	8.00	25.00	PAIDWORK	2.00	1.00	0.60	0.00
7	7	7.00	30.00	HOUSEWFE	2.00	1.00	0.48	0.00
8	8	7.00	24.00	HOUSEWFE	2.00	1.00	0.00	0.00
9	9	7.00	20.00	HOUSEWFE	2.00	1.00	0.90	0.00
10	10	8.00	30.00	PAIDWORK	1.00	1.00	0.70	0.00
117	137	5.00	24.00	PAIDWORK	2.00	2.00	1.49	MISSING
193	262	9.00	31.00	HOUSEWFE	2.00	2.00	1.72	MISSING

```
NUMBER OF CASES READ. . . . . . . .        465
CASES WITH USE SET TO ZERO . . . .           2
   REMAINING NUMBER OF CASES              463
```

TABLE 4.15 *(CONTINUED)*

VARIABLE NO. NAME	GROUPING VARIABLE LEVEL	TOTAL FREQUENCY	MEAN	STANDARD DEVIATION	ST.ERR OF MEAN	COEFF. OF VARIATION	SMALLEST VALUE	SMALLEST Z-SCORE	LARGEST VALUE	LARGEST Z-SCORE	RANGE
3 ATTDRUG		463	7.6847	1.1584	.05384	.15074	5.0000	-2.32	10.000	2.00	5.0000
	EMPLMNT PAIDWORK	244	7.5902	1.1167	.07149	.14712	5.0000	-2.32	10.000	2.16	5.0000
	HOUSEWFE	219	7.7900	1.1970	.08088	.15366	5.0000	-2.33	10.000	1.85	5.0000
4 ATTHOUSE		463	23.585	4.3857	.20382	.18595	1.0000	-5.15	35.000	2.60	34.000
	EMPLMNT PAIDWORK	244	23.725	4.6563	.29809	.19626	1.0000	-4.88	34.000	2.21	33.000
	HOUSEWFE	219	23.429	4.0680	.27489	.17363	11.000	-3.06	35.000	2.84	24.000
7 MSTATUS		463	1.7819	.41343	.01921	.23202	1.0000	-1.89	2.0000	0.53	1.0000
	EMPLMNT PAIDWORK	244	1.6926	.46236	.02960	.27316	1.0000	-1.50	2.0000	0.66	1.0000
	HOUSEWFE	219	1.8813	.32420	.02191	.17233	1.0000	-2.72	2.0000	0.37	1.0000
8 RACE		463	1.0886	.28440	.01322	.26127	1.0000	-0.31	2.0000	3.20	1.0000
	EMPLMNT PAIDWORK	244	1.1107	.31435	.02012	.28303	1.0000	-0.35	2.0000	2.83	1.0000
	HOUSEWFE	219	1.0639	.24518	.01657	.23045	1.0000	-0.26	2.0000	3.82	1.0000
9 LTIMEDRS		463	.74243	.41579	.01932	.56004	0.0000	-1.79	1.9138	2.82	1.9138
	EMPLMNT PAIDWORK	244	.72146	.39109	.02504	.54208	0.0000	-1.84	1.9138	3.05	1.9138
	HOUSEWFE	219	.76578	.44141	.02983	.57642	0.0000	-1.73	1.7853	2.31	1.7853
10 DUMMY		461	0.0000	0.0000	0.0000		0.0000		0.0000		0.0000
	EMPLMNT PAIDWORK	243	0.0000	0.0000	0.0000		0.0000		0.0000		0.0000
	HOUSEWFE	218	0.0000	0.0000	0.0000		0.0000		0.0000		0.0000

Results

Prior to analysis, number of visits to health professionals, attitude toward drug use, attitude toward housework, income, marital status, and race were examined through various BMDP programs for accuracy of data entry, missing values, and fit between their distributions and the assumptions of multivariate analysis. The variables were examined separately for the 246 employed women and the 219 housewives.

A case with a single missing value on attitude toward housework was deleted from the group of employed women, leaving 245 cases in that group. Income, with missing values on more than 5% of the cases, was deleted. Pairwise linearity was checked using within-group scatterplots and found to be satisfactory.

Two cases in the employed group were univariate outliers because of their extremely low z scores on attitude toward housework; these cases were deleted. Using Mahalanobis distance with $p < .001$, 9 cases (about 2%) were identified as multivariate outliers in their own groups. Because several of these cases had extreme z scores on visits to health professionals and because that variable was severely skewed, a logarithmic transformation was applied. With the transformed variable in the variable set, only two cases were identified as multivariate outliers. One multivariate outlier was from the employed group and the other from the housewife group.[21] With all four outliers and the case with missing values deleted, 242 cases remained in the employed group and 218 in the group of housewives.

---

[21] Case 117, an employed woman, was nonwhite with very unfavorable attitudes regarding use of drugs but numerous visits to physicians. Case 193, a housewife, was also nonwhite with very favorable attitudes toward housework and numerous visits to physicians. Results of analyses may not generalize to nonwhite women with numerous visits to physicians, if they have either very unfavorable attitudes toward use of drugs or very favorable attitudes toward housework.

CHAPTER 5
Multiple Regression

5.1 ■ GENERAL PURPOSE AND DESCRIPTION

Regression analyses are a set of statistical techniques that allow one to assess the relationship between one DV and several IVs. For example, is reading ability in primary grades (the DV) related to several IVs such as perceptual development, motor development, and age? The terms regression and correlation are used more or less interchangeably to label these procedures although the term regression is often used when the intent of the analysis is prediction, and the term correlation is used when the intent is simply to assess the relationship between the DV and the IVs.

Regression techniques can be applied to a data set in which the IVs are correlated with one another and with the DV to varying degrees. One can, for instance, assess the relationship between a set of IVs such as education, income, and socioeconomic status with a DV such as occupational prestige. Because regression techniques can be used when the IVs are correlated, they are helpful both in experimental research when, for instance, correlation among IVs is created by unequal numbers of cases in cells, and in observational or survey research where nature has "manipulated" correlated variables. The flexibility of regression techniques is, then, especially useful to the researcher who is interested in real-world or very complicated problems that cannot be meaningfully reduced to orthogonal designs in a laboratory setting.

Multiple regression is an extension of bivariate regression (see Chapter 3) in which several IVs instead of just one are combined to predict a value on a DV for each subject. The result of regression is an equation that represents the best prediction of a DV from several continuous (or dichotomous) IVs. The regression equation takes the following form:

$$Y' = A + B_1X_1 + B_2X_2 + \ldots + B_kX_k$$

where Y' is the predicted value on the DV, A is the Y intercept (the value of Y when all the X values are zero), the Xs represent the various IVs (of which there are k), and the Bs are the coefficients assigned to each of the IVs during regression. Although the same intercept and coefficients are used to predict the values on the DV for all cases in the sample, a different Y' value is predicted for each subject as a result of inserting the subject's own X values into the equation.

The goal of regression is to arrive at the set of B values, called regression coefficients, for the IVs that bring the Y values predicted from the equation as close as possible to the Y values obtained by measurement. The regression coefficients that are computed accomplish two intuitively appealing and highly desirable goals: they minimize (the sum of the squared) deviations between predicted and obtained Y values and they optimize the correlation between the predicted and obtained Y values for the data set. In fact, one of the important statistics derived from a regression analysis is the multiple correlation coefficient, the Pearson product moment correlation coefficient between the obtained and predicted Y values: $R = r_{yy'}$ (see Section 5.4.1).

Regression techniques consist of standard multiple regression, sequential (hierarchical) regression, and statistical (stepwise or setwise) regression. Differences between these techniques involve the way variables enter the equation: what happens to variance shared by variables and who determines the order in which variables enter the equation?

5.2 ■ KINDS OF RESEARCH QUESTIONS

The primary goal of regression analysis is usually to investigate the relationship between a DV and several IVs. As a preliminary step, one determines how strong the relationship is between DV and IVs; then, with some ambiguity, one assesses the importance of each of the IVs to the relationship.

A more complicated goal might be to investigate the relationship between a DV and some IVs with the effect of other IVs statistically eliminated. Researchers often use regression to perform what is essentially a covariates analysis in which they ask if some critical variable (or variables) adds anything to a prediction equation for a DV after other IVs—the covariates—have already entered the equation. For example, does gender add to prediction of mathematical performance after statistical adjustment for extent and difficulty of mathematical training?

Another strategy is to compare the ability of several competing sets of IVs to predict a DV. Is use of Valium better predicted by a set of health variables or by a set of attitudinal variables?

All too often, regression is used to find the best prediction equation for some phenomenon regardless of the meaning of the variables in the equation, a goal met by statistical (stepwise) regression. In the several varieties of statistical regression, statistical criteria alone, computed from a single sample, determine which IVs enter the equation and the order in which they enter.

Regression analyses can be used with either continuous or dichotomous IVs. A variable that is initially discrete can be used if it is first converted into a set of dichotomous variables (numbering one fewer than the number of discrete categories) by dummy variable coding with 1s and 0s. For example, consider an initially discrete variable assessing religious affiliation on which 1 stands for Protestant, 2 for Catholic, 3 for Jewish, and 4 for none or other. The variable may be converted into a set of three new variables (Protestant = 1 vs. non-Protestant = 0, Catholic = 1 vs. non-Catholic = 0, Jewish = 1 vs. non-Jewish = 0), one variable for each degree of freedom. When the new variables are entered into regression as a group (as recommended by Fox, 1991), the variance due to the original discrete IV is analyzed, and, in addition, one can examine effects of the newly created dichotomous components. Dummy variable coding is covered in glorious detail by Cohen and Cohen (1983, pp. 183-220).

ANOVA (Chapter 3) is a special case of regression where main effects and interactions have been dummy variable coded. ANOVA problems can be handled through multiple regression, but multiple regression problems often cannot readily be converted into ANOVA because of correlations among IVs and the presence of continuous IVs. If analyzed through ANOVA, continuous IVs have to be rendered discrete (e.g., high, medium, and low), a process that often results in loss of information and unequal cell sizes. In regression, the full range of continuous IVs is maintained.

As a statistical tool, regression is very helpful in answering a number of practical questions, as discussed in Sections 5.2.1 through 5.2.8.

5.2.1 ■ Degree of Relationship

How good is the regression equation? Does the regression equation really provide better-than-chance prediction? Is the multiple correlation really any different from zero when allowances for naturally occurring fluctuations in such correlations are made? For example, can one reliably predict reading ability given knowledge of perceptual development, motor development, and age? The statistical procedures described in Section 5.6.2.1 allow you to determine if your multiple correlation is reliably different from zero.

5.2.2 ■ Importance of IVs

If the multiple correlation is different from zero, you may want to ask which IVs are important in the equation and which IVs are not. For example, is knowledge of motor development helpful in predicting reading ability, or can we do just as well with knowledge of only age and perceptual development? The methods described in Section 5.6.1 help you to evaluate the relative importance of various IVs to a regression solution.

5.2.3 ■ Adding IVs

Suppose that you have just computed a regression equation and you want to know whether or not you can improve your prediction of the DV by adding one or more IVs to the equation. For example, is prediction of a child's reading ability enhanced by adding a variable reflecting parental interest in reading to the three IVs already included in the equation? A test for improvement of the multiple correlation after addition of one new variable is given in Section 5.6.1.2, and for improvement after addition of several new variables in Section 5.6.2.3.

5.2.4 ■ Changing IVs

Although the regression equation is a linear equation (that is, it does not contain squared values, cubed values, cross products of variables, and the like), the researcher may include nonlinear relationships in the analysis by redefining IVs. Curvilinear relationships, for example, can be made available for analysis by squaring or raising to a higher power the original IVs. Interaction can be made available for analysis by creating a new IV that is a cross product of two or more original IVs and including it with the originals in the analysis. Aiken and West (1991) discuss interpretation of these interactions.

For an example of a curvilinear relationship, suppose a child's reading ability increases with increasing parental interest up to a point, and then levels off. Greater parental interest does not result in greater reading ability. If the square of parental interest is added as an IV, better prediction of a child's reading ability could be achieved.

Inspection of a scatterplot between predicted and obtained Y values (known as residuals analysis—see Section 5.3.2.4) may reveal that the relationship between the DV and the IVs also has more complicated components such as curvilinearity and interaction. To improve prediction or because of theoretical considerations one may want to include some of these more complicated IVs. Procedures for using regression for nonlinear curve fitting are discussed in Cohen and Cohen (1983) and McNeil, Kelly, and McNeil (1975). There is danger, however, in too liberal use of powers or cross products of IVs; the sample data may be overfit to the extent that results no longer generalize to a population.

5.2.5 ■ Contingencies among IVs

You may be interested in the way that one IV behaves in the context of one, or a set, of other IVs. Sequential regression can be used to statistically adjust for the effects of some IVs while examining the relationship between an especially interesting IV and the DV. For example, after adjustment for differences in perceptual development and age, does motor development predict reading ability? This procedure is described in Section 5.5.2.

5.2.6 ■ Comparing Sets of IVs

Is prediction of a DV from one set of IVs better than prediction from another set of IVs? For example, is prediction of reading ability based on perceptual and motor development and age as good as prediction from family income and parental educational attainments? Section 5.6.2.5 demonstrates a method for comparing the solutions given by two sets of predictors.

5.2.7 ■ Predicting DV Scores for Members of a New Sample

One of the more important applications of regression involves predicting scores on a DV for subjects for whom only data on IVs are available. This application is fairly frequent in personnel selection for employment, selection of students for graduate training, and the like. Over a fairly long period, a researcher collects data on a DV, say, success in graduate school, and on several IVs, say, undergraduate GPA, GRE verbal scores, and GRE math scores. Regression analysis is performed and the regression equation obtained. If the IVs are strongly related to the DV, then for a new sample of applicants to graduate school, regression coefficients are applied to IV scores to predict success in graduate school ahead of time. Admission to the graduate school may, in fact, be based on prediction of success through regression.

The generalizability of a regression solution to a new sample is checked within a single large sample by a procedure called cross validation. A regression equation is developed from a portion of a sample and then applied to the other portion of the sample. If the solution generalizes, the regression equation predicts DV scores better than chance for the new cases, as well.

5.2.8 ■ Parameter Estimates

Parameter estimates in multiple regression are the unstandardized regression coefficients (B weights). A B weight for a particular IV represents the change in the DV associated with a one-unit change in that IV, all other IVs held constant. Suppose, for example, we want to predict graduate record exam scores (GRE) from grade point averages (GPA) and our analysis produces the following equation:

$$GRE' = 200 + 100(GPA)$$

The $B = 100$ tells us that for each one unit increase in GPA (e.g., from a GPA of 2.0 to one of 3.0), we expect a 100 point increase in GRE scores. Sometimes this is usefully expressed in terms of percentage of gain in the DV. For example, assuming the mean GRE is 500, an increase of one grade point represents a 20% (100/500) average increase in GRE.

Accuracy of parameter estimates depends on agreement with the assumptions of multiple regression analysis (cf. Section 5.3.2.4), including the assumption that IVs are measured without error. Therefore, interpretation has to be tempered by knowledge of the reliability of the IVs. You need to be cautious when interpreting regression coefficients with transformed variables, because the coefficients and interpretations of them apply only to the variable after transformation.

5.3 ■ LIMITATIONS TO REGRESSION ANALYSES

Attention to issues surrounding assumptions of regression analysis has become a growth industry, partly because of the relative simplicity of regression compared to the multivariate techniques and partly because of the extensive use of multiple regression in all facets of science and commerce. A glance at the myriad of diagnostic tests available in regression programs confirms this view. However, it should also be noted that many of the popular diagnostic tests are concerned with poor fit of regression models to some cases—outliers in the solution—rather than tests conducted as part of screening.

This discussion barely skims the surface of the goodies available for screening your data and assessing the fit of cases to your solution, but should adequately cover most gross violations of assumptions. Recent Sage publications by Berry (1993) and Fox (1991) offer some other interesting insights into regression assumptions and diagnostics.

5.3.1 ■ Theoretical Issues

Regression analyses reveal relationships among variables but do not imply that the relationships are causal. Demonstration of causality is a logical and experimental, rather than statistical, problem. An apparently strong relationship between variables could stem from many sources, including the influence of other, currently unmeasured variables. One can make an airtight case for causal relationship among variables only by showing that manipulation of some of them is followed inexorably by change in others when all other variables are controlled.

Another problem for theory rather than statistics is inclusion of variables. Which DV should be used and how is it to be measured? Which IVs should be examined and how are they to be measured? If one already has some IVs in an equation, which IVs should be added to the equation for the most improvement in prediction? The answers to these questions can be provided by theory, astute observation, good hunches, or sometimes by careful examination of the distribution of residuals.

There are, however, some general considerations for choosing IVs. Regression will be best when each IV is strongly correlated with the DV but uncorrelated with other IVs. A general goal of regression, then, is to identify the fewest IVs necessary to predict a DV where each IV predicts a substantial and independent segment of the variability in the DV.

There are other considerations to selection of variables. If the goal of research is manipulation of some DV (say body weight), it is strategic to include as IVs variables that can be manipulated (e.g., caloric intake, physical activity) as well as those that cannot (e.g., genetic predisposition). Or, if one is interested in predicting a variable such as annoyance caused by noise for a neighborhood, it is strategic to include cheaply obtained sets of IVs (e.g., neighborhood characteristics published by the Census Bureau) rather than expensively obtained ones (e.g., attitudes from in-depth interviews) if both sets of variables predict equally well.

It should be clearly understood that a regression solution is extremely sensitive to the combination of variables that is included in it. Whether or not an IV appears particularly important in a solution depends on the other IVs in the set. If the IV of interest is the only one that assesses some important facet of the DV, the IV will appear important; if the IV of interest is only one of several that assess the same important facet of the DV, it will appear less important. An optimal set of IVs is the smallest reliable, uncorrelated set that "covers the waterfront" with respect to the DV.

Regression analysis assumes that IVs are measured without error, a clear impossibility in most social and behavioral science research. The best we can do is choose the most reliable IVs possible. More subtly, it is assumed that important unmeasured IVs, which contribute to error, are not correlated with any of the measured IVs. If, as Berry (1993) points out, an unmeasured IV is correlated with a measured IV, then the components of error are correlated with the measured IV, a violation of the assumption of independence of errors. Worse, the relationship between an unmeasured IV and a measured IV can change the estimates of the regression coefficients; if the relationship is positive, the coefficient for the measured IV is overestimated, if negative, underestimated. If the regression equation is to accurately reflect the contribution of each IV to prediction of the DV, then, all of the relevant IVs have to be included.

5.3.2 ■ Practical Issues

In addition to theoretical considerations, use of multiple regression requires that several practical matters be attended, as described in Sections 5.3.2.1 through 5.3.2.4.

5.3.2.1 ■ Ratio of Cases to IVs

The cases-to-IVs ratio has to be substantial or the solution will be perfect—and meaningless. With more IVs than cases, one can find a regression solution that completely predicts the DV for each case, but only as an artifact of the cases-to-IV ratio.

Required sample size depends on a number of issues, including the desired power, alpha level, number of predictors, and expected effect sizes. Green (1991) provides a thorough discussion of these issues and some procedures to help decide how many cases are necessary. The simplest rules of thumb are $N \geq 50 + 8m$ (m is the number of IVs) for testing the multiple correlation and $N \geq 104 + m$ for testing individual predictors. These rules of thumb assume a medium-size relationship between the IVs and the DV, $\alpha = .05$ and $\beta = .20$. For example, if you plan six predictors, you need $50 + (8)(6) = 98$ cases to test regression and $104 + 6 = 110$ cases for testing individual predictors. If you are interested in both the overall correlation and the individual IVs, *calculate N both ways and choose the larger number of cases.*

A higher cases-to-IV ratio is needed when the DV is skewed, a small effect size is anticipated, or substantial measurement error is expected from unreliable variables. That is, if the DV is not normally distributed and transformations are not undertaken, more cases are required. The size of anticipated effect is also relevant because more cases are needed to demonstrate a small effect than a large one. Green (1991) offers the following, more complex, rule of thumb that takes into account effect size: $N \geq (8/f^2) + (m - 1)$, where $f^2 = .01, .15$, and $.35$ for small, medium, and large effects, respectively. For more precisely estimated effect sizes, note that $f^2 = R^2/(1 - R^2)$, where R^2 is the expected squared multiple correlation. Finally, if variables are unreliable, measurement error is larger and more cases are needed.

It is also possible to have too many cases. As the number of cases becomes quite large, almost any multiple correlation will depart significantly from zero, even one that predicts negligible variance in the DV. For both statistical and practical reasons, then, one wants to measure the smallest number of cases that has a decent chance of revealing a relationship of a specified size.

If stepwise regression is to be used, yet more cases are needed. A cases-to-IV ratio of 40 to 1 is reasonable because stepwise regression can produce a solution that does not generalize beyond the sample unless the sample is large. An even larger sample is needed in stepwise regression if cross validation (deriving the solution with some of the cases and testing it on the others) is used to test the generalizability of the solution.

If you cannot measure as many cases as you would like, there are some strategies that may help. You can delete some IVs or create one (or more than one) IV that is a composite of several others. The new, composite IV is used in the analysis in place of the original IVs. Lastly, you can employ a setwise regression strategy, as available in BMDP9R or SAS REG and described in Section 5.5.3. In this frankly exploratory strategy, regressions are computed for all possible subsets of IVs and the researcher decides which regression is best.

Be sure to verify that the analysis included as many cases as you think it should have. By default, regression programs delete cases for which there are missing values on any of the variables which can result in substantial loss of cases. Consult Chapter 4 if you have missing values and wish to estimate them rather than delete the cases.

5.3.2.2 ■ Outliers Among the IVs and on the DV

Extreme cases have too much impact on the regression solution and should be deleted, rescored, or the variable transformed. Consult Chapter 4 for a summary of general procedures for detecting and dealing with univariate and multivariate outliers using both statistical tests and graphical methods.

In regression, cases are evaluated for univariate extremeness with respect to the DV and each IV. Univariate outliers show up in initial screening runs (e.g., with BMDP2D or SPSS FREQUENCIES) as cases far from the mean and unconnected with other cases on either plots or z scores. Multivariate outliers among the IVs are sought using either statistical methods such as Mahalanobis distance (through BMDPAM, SPSS REGRESSION, or SYSTAT GLM as described in Chapter 4) or by using graphical methods.

Screening for outliers can be performed either prior to a regression run (as recommended in Chapter 4) or through a residuals analysis after an initial regression run. The problems with an initial regression run are, first, the temptation to make screening decisions based on desired outcome, and second, the overfitting that may occur if outliers in the solution are deleted along with outliers among the variables. It seems safer to deal with outliers among the variables in initial screening runs, and then determine the fit of the solution to the cases.

Regression programs offer more specialized tests for identifying outliers than most programs for the other techniques. The goal is that all of the cases contribute equally to the regression solution. However, cases that are far away from the others have more impact than the others on the size of regression coefficients. The idea, then, is to assess the size of the impact of each case on the solution. Since potential outliers have substantially more impact than cases that are close to other cases, outliers are identified by assessing their impact.

Three of the most important measures of impact are leverage, discrepancy, and influence. *Leverage* assesses outliers in the set of IVs. Leverage statistics are related to Mahalanobis distance or variations of the diagonal elements of a "hat" matrix, called HATDIAG, RHAT, and h_{ii}. The diagonal elements of the hat matrix are similar to Mahalanobis distance, but on a different scale so that significance tests based on a χ^2 distribution do not apply.[1] Equation 4.3 shows the relationship between leverage and Mahalanobis distance.

Cases with high leverage are far from the others, but they can be far out on basically the same line as the other cases, or far away and off the line. *Discrepancy* measures the extent to which a case is in line with the others. Figure 5.1(a) shows a case with high leverage and low discrepancy; Figure 5.1(b) shows a case with high leverage and high discrepancy. In Figure 5.1(c) is a case with low leverage and high discrepancy.

Influence is a product of leverage and discrepancy (Fox 1991). It assesses change in regression coefficients when a case is deleted; cases with influence scores larger than 1.00 are suspected of being outliers. Measures of influence are variations of Cook's distance and are identified in output as Cook's distance, modified Cook's distance, DFFITS, and DBETAS. For the interested reader, Fox (1991, p. 29–30) describes these terms in more detail. Fox (1991) indicates that outliers influence the precision of estimation of the regression weights. With high leverage and low discrepancy, the standard errors of the regression coefficients are too small; with low leverage and high discrepancy, the standard errors of the regression coefficients are too large. Neither situation generalizes well to population values.

Several varieties of leverage and influence are available in the four statistical packages. Tables 5.11 through 5.13 show which programs in the four computer package produce these measures; consult the program manuals for a more comprehensive list.

5.3.2.3 ■ Multicollinearity and Singularity

Calculation of regression coefficients requires inversion of the matrix of correlations among the IVs (Equation 5.6), an inversion that is impossible if IVs are singular and unstable if they are multicollinear, as discussed in Chapter 4. Singularity and multicollinearity can be identified in screening runs through perfect or very high squared multiple correlations (SMC) among IVs, where each IV in turn serves as DV while the others are IVs, or very low tolerances $(1 - SMC)$.

In regression, these conditions are also signaled by very large (relative to the scale of the variables) standard errors for regression coefficients. Berry (1993) reports that when r is 0.9, the standard errors of the regression coefficients are doubled; when multicollinearity is present, none of the regression coefficients may be significant because of the large size of standard errors. For an extended discussion of the complicated relationship between outliers and collinearity, see Fox (1991).

Most multiple regression programs have default values for tolerance $(1 - SMC)$ that protect the user against inclusion of multicollinear IVs. If the default values for the programs are in place, IVs that are very highly correlated with IVs already in the equation are not entered. This makes sense both statistically and logically because the IVs threaten the analysis due to inflation of regression coefficients and because they are not needed due to their correlations with other IVs.

---

[1] It is suggested (e.g., Lunneborg, 1994) that outliers be defined as cases with $h_{ii} \geq 2(k > N)$.

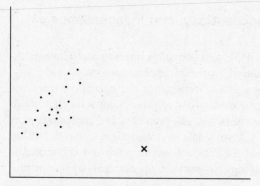

(a) High leverage, low discrepancy, moderate influence

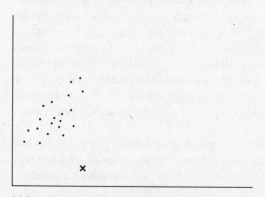

(b) High leverage, high discrepancy, high influence

(c) Low leverage, high discrepancy, moderate influence

FIGURE 5.1 THE RELATIONSHIPS AMONG LEVER-
AGE, DISCREPANCY, AND INFLUENCE.

However, you may want to make your own choice about which IV to delete on logical rather than statistical grounds by considering issues such as the reliability of the variables or the cost of measuring the variables. You may want to delete the least reliable variable, for instance, rather than the variable identified by the program with very low tolerance. With a less reliable IV deleted from the set of IVs, the tolerance for the IV in question may be sufficient for entry.

If multicollinearity is detected but you want to maintain your set of IVs anyway, ridge regression might be considered. Ridge regression is a controversial procedure that attempts to stabilize estimates of regression coefficients by inflating the variance that is analyzed. For a more thorough description of ridge regression, see Chapter 7 of Dillon and Goldstein (1984). Although originally greeted with enthusiasm (cf. Price, 1977), serious questions about the procedure have been raised by Rozeboom (1979), Fox (1991) and others. If, after consulting this literature, you still want to employ ridge regression, it is available through BMDP4R.

5.3.2.4 ■ Normality, Linearity, Homoscedasticity, and Independence of Residuals

Routine pre-analysis screening procedures can be used to assess normality, linearity, and homoscedasticity, or they can be assessed through residuals produced by one of the regression programs.

Examination of residuals scatterplots provides a test of assumptions of normality, linearity, and homoscedasticity between predicted DV scores and errors of prediction. Assumptions of analysis are that the residuals (differences between obtained and predicted DV scores) are normally distributed about the predicted DV scores, that residuals have a straight line relationship with predicted DV scores, and that the variance of the residuals about predicted DV scores is the same for all predicted scores.[2] When these assumptions are met, the residuals appear as plot (a) in Figure 5.2.

Residuals scatterplots may be examined in lieu of or after initial screening runs. If residuals scatterplots are examined in lieu of initial screening, and the assumptions of analysis are deemed met, further screening of variables and cases is unnecessary. That is, if the residuals show normality, linearity, and homoscedasticity, if no outliers are evident, if the number of cases is sufficient, and if there is no evidence of multicollinearity or singularity, then regression requires only one run. (Parenthetically, we might note that we have never, in many years of multivariate analyses with many data sets, found this to be the case.) If, on the other hand, the residuals scatterplot from an initial run looks yucky, further screening via the procedures in Chapter 4 is warranted.

Residuals scatterplots are provided by all the statistical programs discussed in this chapter. All provide a scatterplot in which one axis is predicted scores and the other axis is errors of prediction. Which axis is which, however, and whether or not the predicted scores and residuals are standardized differ from program to program.

SPSS, SAS, and BMDP provide the plots directly in their regression programs. In SPSS, both predicted scores and errors of prediction are standardized; in SAS and BMDP they are not. For SYSTAT, you have to save the predicted values and residuals, and then plot them using SYSTAT PLOT, in either standardized or unstandardized form. In any event, it is the overall shape of the

---

[2] Note that there are no distributional assumptions about the IVs, other than their relationship with the DV. However, a prediction equation often is enhanced if IVs are normally distributed, primarily because linearity between the IV and DV is enhanced.

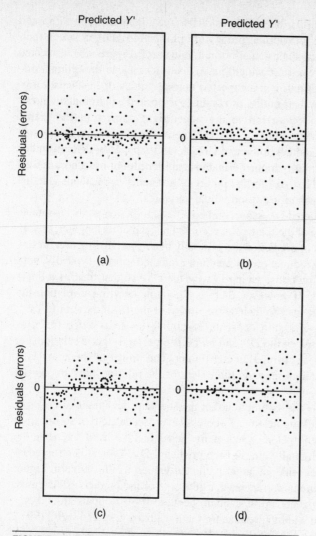

FIGURE 5.2 PLOTS OF PREDICTED VALUES OF THE DV
(Y') AGAINST RESIDUALS, SHOWING (a) ASSUMPTIONS
MET, (b) FAILURE OF NORMALITY, (c) NONLINEARITY,
AND (d) HETEROSCEDASTICITY.

scatterplot that is of interest. If all assumptions are met, the residuals will be nearly rectangularly distributed with a concentration of scores along the center. As mentioned above, Figure 5.2(a) illustrates a distribution in which all assumptions are met.

The assumption of normality is that errors of prediction are normally distributed around each and every predicted DV score. The residuals scatterplot should reveal a pileup of residuals in the center of the plot at each value of predicted score, and a normal distribution of residuals trailing off symmetrically from the center. Figure 5.2(b) illustrates a failure of normality, with a skewed distribution of residuals.

A second test for the normality of the distribution of residuals is available through SPSS and BMDP programs for regression. The test is a normal probability plot of residuals in which their expected normal values are plotted against their actual normal values (cf. Figure 4.3). Expected normal values are estimates of the z score a score should have, given its rank in the original distribution if the original distribution is normal. If the expected normal values of residuals correspond to actual normal values (i.e., if the distribution of residuals is normal), the points will fall along a straight line running from the bottom left to the upper right corners of the graph. Distributions that are not normal will deviate from the straight line by curving above or below it in specific ways, depending on how the residuals are skewed. Bock (1975, pp. 156–160) illustrates the effects of several deviations from normality in the normal probability plot of residuals, as well as other details for the interested reader. The "fix" for the failure of normality of residuals is transformation of variables, after viewing distributions of individual variables.

Linearity of relationship between predicted DV scores and errors of prediction is also assumed. If nonlinearity is present, the overall shape of the scatterplot is curved instead of rectangular, as seen in Figure 5.2(c). In this illustration, errors of prediction are generally in a negative direction for low and high predicted scores and in a positive direction for medium predicted scores. Typically, non-linearity of residuals can be made linear by transforming IVs (or the DV) so that there is a linear relationship between each IV and the DV. If, however, there is a genuine curvilinear relationship between an IV and the DV, it may be necessary to include the square of the IV in the set of IVs.

Failure of linearity of residuals in regression does not invalidate an analysis so much as weaken it. A curvilinear relationship between the DV and an IV is a perfectly good relationship that is not completely captured by a linear correlation coefficient. The power of the analysis is reduced to the extent that the analysis cannot map the full extent of the relationships among the IVs and the DV.

The assumption of homoscedasticity is the assumption that the standard deviations of errors of prediction are approximately equal for all predicted DV scores. Heteroscedasticity also does not invalidate the analysis so much as weaken it. Homoscedasticity means that the band enclosing the residuals is approximately equal in width at all values of the predicted DV. Typical heteroscedasticity is a case in which the band becomes wider at larger predicted values, as illustrated in Figure 5.2(d). In this illustration, the errors of prediction increase as the size of the prediction increases. Serious heteroscedasticity occurs when the spread in standard deviations of residuals around predicted values is three times higher for the widest spread as for the most narrow spread (Fox, 1991). Heteroscedasticity may occur when some of the variables are skewed and others are not. Transformation of the variables may reduce or eliminate heteroscedasticity.

Heteroscedasticity can also result from interaction of an IV with another variable that is not part of the regression equation. For example, it may be that increasing variability in income with age is associated with education; for those with higher education, there is greater growth in income with age. Including education as well as age as predictors of income will strengthen the model as well as eliminate heteroscedasticity.

Another remedy is to use weighted (generalized) least squares regression, available as an option in all major regression programs. In this procedure, you weight the regression by the variance of the variable which produces the heteroscedasticity. For example, if you know that variance in the DV (e.g., income) increases with increasing values of an IV (e.g., age), you weight the regression by age. This latter remedy is less appealing than inclusion of the "interacting" variable (education), but may be more practical if you cannot identify or measure the interacting variable, or if the heteroscedasticity is a result of measurement error.

Under special and somewhat rare conditions, significance tests are available for linearity and homoscedasticity. Fox (1991, pp. 64–66) summarizes some of these significance tests for failure of linearity and heteroscedasticity, useful when some of the IVs are discrete with only a few categories.

Another assumption of regression is that errors of prediction are independent of one another. In some instances, this assumption is violated as a function of something associated with the order of cases. Often "something" is time or distance. For example, time produces nonindependence of errors when subjects who are interviewed early in a survey exhibit more variability of response because of interviewer inexperience with a questionnaire. Distance produces nonindependence of errors when subjects who are farther away from a toxic source exhibit more variable reactions. Nonindependence of errors is, then, either a nuisance factor to be eliminated or of considerable research interest.

Nonindependence of errors associated with order of cases can be checked in SPSS REGRESSION, BMDP2R and 9R, or SYSTAT GLM or REGRESS by entering cases in order and requesting a plot of residuals against sequence of cases. The associated Durbin-Watson statistic is a measure of autocorrelation of errors over the sequence of cases, and, if significant, indicates nonindependence of errors. Positive autocorrelation makes estimates of error variance too small, and results in inflation of the Type I error rate. Negative autocorrelation makes the estimates too large, and results in loss of power. Details on the use of this statistic and a test for its significance are given by Wesolowsky (1976). If nonindependence is found, consult Dillon and Goldstein (1984) for the options available to you.

5.3.2.5 ■ Outliers in the Solution

Some cases may be poorly fit by the regression equation. These cases lower the multiple correlation. Examination of these cases is informative since it shows which kinds of cases cannot be predicted by your solution.

Residuals, the difference between predicted and obtained Y values, identify outliers in the solution and are available in raw or standardized form—with or without the outlying case deleted. A graphical method based on residuals uses leverage on the X axis and residuals on the Y axis. As in routine residuals plots, outlying cases in the solution fall outside the swarm of points produced by the remainder of the cases.

Examine the residuals plot. If outlying cases are evident, identify them by examining the list of standardized residuals for individual cases. The statistical criterion for identifying an outlier in the solution depends on the sample size; the larger the sample, the more likely that one or more residuals will be discrepant. A criterion of $p = .001$ is appropriate for $N < 1000$, defining outlying cases as those with standardized residuals in excess of about ± 3.3.

5.4 ■ FUNDAMENTAL EQUATIONS FOR MULTIPLE REGRESSION

A data set appropriate for multiple regression consists of a sample of research units (e.g., graduate students) for whom scores are available on a number of IVs and on one DV. A small sample of hypothetical data with three IVs and one DV is illustrated in Table 5.1.

Table 5.1 contains six students, with scores on three IVs: a measure of professional motivation (MOTIV), a composite rating of qualifications for admissions to graduate training (QUAL), and a composite rating of performance in graduate courses (GRADE). The DV is a rating of performance on graduate comprehensive exams (COMPR). We ask how well we can predict COMPR from scores on MOTIV, QUAL, and GRADE.

In actually addressing that question, a sample of six cases is highly inadequate, but the sample is sufficient to illustrate calculation of multiple correlation and to demonstrate some analyses by canned computer programs. The reader is encouraged to work problems involving these data by hand as well as by available computer programs. Setup and selected output for this example through SPSS REGRESSION, BMDP1R, SAS REG, and SYSTAT REGRESS appear in Section 5.4.3.

A variety of ways is available to develop the "basic" equation for multiple correlation.

5.4.1 ■ General Linear Equations

One way of developing multiple correlation is to obtain the prediction equation for Y' in order to compare the predicted value of the DV with obtained Y.

$$Y' = A + B_1X_1 + B_2X_2 + \ldots + B_kX_k \qquad (5.1)$$

where Y' is the predicted value of Y, A is the value of Y' when all Xs are zero, B_1 to B_k represent regression coefficients, and X_1 to X_k represent the IVs.

The best-fitting regression coefficients produce a prediction equation for which squared differences between Y and Y' are at a minimum. Because squared errors of prediction—$(Y - Y')^2$—are minimized, this solution is called a least-squares solution.

In the sample problem, $k = 3$. That is, there are three IVs available to predict the DV, COMPR.

$$(\text{COMPR})' = A + B_M(\text{MOTIV}) + B_Q(\text{QUAL}) + B_G(\text{GRADE})$$

To predict a student's COMPR score, the available IV scores (MOTIV, QUAL, and GRADE) are multiplied by their respective regression coefficients. The coefficient-by-score products are summed and added to the intercept, or baseline, value (A).

Differences among the observed values of the DV (Y), the mean of Y ($\overline{Y}$), and the predicted values of Y (Y') are summed and squared, yielding estimates of variation attributable to different

---

TABLE 5.1 SMALL SAMPLE OF HYPOTHETICAL DATA FOR ILLUSTRATION OF MULTIPLE CORELATION

Case No.	IVs			DV
	MOTIV(X_1)	QUAL (X_2)	GRADE (X_3)	COMPR (Y)
1	14	19	19	18
2	11	11	8	9
3	8	10	14	8
4	13	5	10	8
5	10	9	8	5
6	10	7	9	12
Mean	11.00	10.17	11.33	10.00
Standard deviation	2.191	4.834	4.367	4.517

sources. Total sum of squares for Y is partitioned into a sum of squares due to regression and a sum of squares left over or residual.

$$SS_Y = SS_{reg} + SS_{res} \tag{5.2}$$

Total sum of squares of Y

$$SS_y = \Sigma(Y - \bar{Y})^2$$

is, as usual, the sum of squared differences between each individual's observed Y score and the mean of Y over all N cases. The sum of squares for regression

$$SS_{reg} = \Sigma(Y' - \bar{Y})^2$$

is the portion of the variation in Y that is explained by use of the IVs as predictors. That is, it is the sum of squared differences between predicted Y' and the mean of Y because the mean of Y is the best prediction for the value of Y in the absence of any useful IVs. Sum of squares residual

$$SS_{res} = \Sigma(Y - Y')^2$$

is the sum of squared differences between observed Y and the predicted scores, Y', and represents errors in prediction.

The squared multiple correlation is

$$R^2 = \frac{SS_{reg}}{SS_y} \tag{5.3}$$

The squared multiple correlation, R^2, is the proportion of sum of squares for regression in the total sum of squares for Y.

The squared multiple correlation is, then, the proportion of variation in the DV that is predictable from the best linear combination of the IVs. The multiple correlation itself is the correlation between the obtained and predicted Y values; that is, $R = r_{yy'}$.

Total sum of squares (SS_y) is calculated directly from the observed values of the DV. For example, in the sample problem, where the mean on the comprehensive examination is 10,

$$SS_C = (18 - 10)^2 + (9 - 10)^2 + (8 - 10)^2 + (8 - 10)^2 + (5 - 10)^2 + (12 - 10)^2 = 102$$

To find the remaining sources of variation, it is necessary to solve the prediction equation (Equation 5.1) for Y', which means finding the best-fitting A and B_i. The most direct method of deriving the equation involves thinking of multiple correlation in terms of individual correlations.

5.4.2 ■ Matrix Equations

Another way of looking at R^2 is in terms of the correlations between each of the IVs and the DV. The squared multiple correlation is the sum across all IVs of the product of the correlation between the DV and IV and the (standardized) regression coefficient for the IV.

$$R^2 = \sum_{i=1}^{k} r_{yi} \beta_i \tag{5.4}$$

where each r_{yi} is the correlation between the DV and the ith IV, and β_i is the standardized regression coefficient, or beta weight[3].

The standardized regression coefficient is the regression coefficient that would be applied to the standardized X_i value—the z score of the X_i value—to predict standardized Y'.

Because r_{yi} are calculated directly from the data, computation of R^2 involves finding the standardized regression coefficients (β_i) for the k IVs. Derivation of the k equations in k unknowns is beyond the scope of this book. However, solution of these equations is easily illustrated using matrix algebra. For those who are not familiar with matrix algebra, the rudiments of it are available in Appendix A. Sections A.4 (matrix multiplication) and A.5 (matrix inversion or division) are the only portions of matrix algebra necessary to follow the next few steps.

In matrix form:

$$R^2 = \mathbf{R}_{yi} \mathbf{B}_i \tag{5.5}$$

where $\mathbf{R}_{yi}$ is the row matrix of correlations between the DV and the k IVs, and $\mathbf{B}_i$ is a column matrix of standardized regression coefficients for the same k IVs.

The standardized regression coefficients can be found by inverting the matrix of correlations among IVs and multiplying that inverse by the matrix of correlations between the DV and the IVs.

$$\mathbf{B}_i = \mathbf{R}_{ii}^{-1} \mathbf{R}_{iy} \tag{5.6}$$

$\mathbf{B}_i$ is the column matrix of standardized regression coefficients, $\mathbf{R}_{ii}^{-1}$ is the inverse of the matrix of correlations among the IVs, and $\mathbf{R}_{iy}$ is the column matrix of correlations between the DV and the IVs.

Because multiplication by an inverse is the same as division, the column matrix of correlations between the IVs and the DV is divided by the correlation matrix of IVs.

These equations,[4] then, are used to calculate R^2 for the sample COMPR data from Table 5.1. All the required correlations are in Table 5.2.

Procedures for inverting a matrix are amply demonstrated elsewhere (e.g., Cooley & Lohnes, 1971; Harris, 1975) and are typically available in computer installations and spreadsheet programs. Because the procedure is extremely tedious by hand, and becomes increasingly so as the matrix becomes larger, the inverted matrix for the sample data is presented without calculation in Table 5.3.

Using Equation 5.6, the $\mathbf{B}_i$ matrix is found by postmultiplying the $\mathbf{R}_{ii}^{-1}$ matrix by the $\mathbf{R}_{iy}$ matrix.

$$\mathbf{B}_i = \begin{bmatrix} 1.20255 & -0.31684 & -0.20435 \\ -0.31684 & 2.67113 & -1.97305 \\ -0.20435 & -1.97305 & 2.62238 \end{bmatrix} \begin{bmatrix} .58613 \\ .73284 \\ .75043 \end{bmatrix} = \begin{bmatrix} 0.31931 \\ 0.29117 \\ 0.40221 \end{bmatrix}$$

---

[3] β is used to indicate a sample standardized regression coefficient, rather than a population estimate of the unstandardized coefficient, consistent with usage in software packages.

[4] Similar equations can be solved in terms of Σ (variance-covariance) matrices or $\mathbf{S}$ (sum-of-squares and cross-product) matrices as well as correlation matrices. If Σ or $\mathbf{S}$ matrices are used, the regression coefficients are nonstandardized coefficients, as in Equation 5.1.

TABLE 5.2 CORRELATIONS AMONG IVS AND THE DV FOR SAMPLE
DATA IN TABLE 5.1

		R_{ii}			R_{iy}
		MOTIV	QUAL	GRADE	COMPR
	MOTIV	1.00000	.39658	.37631	.58613
	QUAL	.39658	1.00000	.78329	.73284
	GRADE	.37631	.78329	1.00000	.75043
R_{yi}	COMPR	.58613	.73284	.75043	1.00000

TABLE 5.3 INVERSE OF MATRIX OF
INTERCORRELATIONS AMONG IVS FOR
SAMPLE DATA IN TABLE 5.1

	MOTIVE	QUAL	GRADE
MOTIVE	1.20255	−0.31684	−0.20435
QUAL	−0.31684	2.67113	−1.97305
GRADE	−0.20435	−1.97305	2.62238

so that $\beta_M = 0.319$, $\beta_Q = 0.291$, and $\beta_G = 0.402$. Then, using Equation 5.5, we obtain

$$R^2 = [.58613 \quad .73284 \quad .75043] \begin{bmatrix} 0.31931 \\ 0.29117 \\ 0.40221 \end{bmatrix} = .70237$$

In this example, 70% of the variance in graduate comprehensive exam scores is predictable from knowledge of motivation, admission qualifications, and graduate course performance.

Once the standardized regression coefficients are available, they are used to write the equation for the predicted values of COMPR (Y'). If z scores are used throughout, the beta weights (β_i) are used to set up the prediction equation. The equation is similar to Equation 5.1 except that there is no A (intercept) and both the IVs and predicted DV are in standardized form.

If instead the equation is needed in raw score form, the coefficients must first be transformed to unstandardized B_i coefficients.

$$B_i = \beta_i \left(\frac{S_y}{S_i} \right) \tag{5.7}$$

Unstandardized coefficients (B_i) are found by multiplying standardized coefficients (beta weights—β_i) by the ratio of standard deviations of the DV and IV, where S_i is the standard deviation of the ith IV and S_y is the standard deviation of the DV and

$$A = \overline{Y} - \sum_{i=1}^{k} (B_i \overline{X_i}) \tag{5.8}$$

The intercept is the mean of the DV less the sum of the means of the IVs multiplied by their respective unstandardized coefficients.

For the sample problem of Table 5.1:

$$B_M = 0.319 \left(\frac{4.517}{2.191} \right) = 0.658$$

$$B_Q = 0.291 \left(\frac{4.517}{4.834} \right) = 0.272$$

$$B_G = 0.402 \left(\frac{4.517}{4.366} \right) = 0.416$$

$$A = 10 - [(0.658)(11.00) + (0.272)(10.17) + (0.416)(11.33) = -4.72$$

The prediction equation for raw COMPR scores, once scores on MOTIV, QUAL, and GRADE are known, is

$$(COMPR)' = -4.72 + 0.658(MOTIV) + 0.272(QUAL) + 0.416(GRADE)$$

If a graduate student has ratings of 12, 14, and 15, respectively, on MOTIV, QUAL, and GRADE, the predicted rating on COMPR is

$$(COMPR)' = -4.72 + 0.658(12) + 0.272(14) + 0.416(15) = 13.22$$

The prediction equation also shows that for every one unit change in GRADE there is change of about 0.4 points on COMPR if the values on the other IVs are held constant.

5.4.3 ■ Computer Analyses of Small Sample Example

Tables 5.4 through 5.7 demonstrate setup and selected output for computer analyses of the data in Table 5.1, using default values. Table 5.4 illustrates SPSS REGRESSION. The simplest BMDP program, BMDP1R, is illustrated in Table 5.5. Tables 5.6 and 5.7 show runs through SAS REG and SYSTAT REGRESS programs, respectively.

In SPSS REGRESSION all of the variables, IVs and DV, are listed in the VARS= sentence of the REGRESSION instruction. Then the DV is specified. METHOD=ENTER is the instruction that specifies standard multiple regression. In standard multiple regression, results are given for only one step. At the left are R, R^2, adjusted R^2 (see Section 5.6.3) and Standard Error, the standard error of the predicted score, Y'. To the right is the ANOVA table, showing details of the F test of the hypothesis that multiple regression is zero (see Section 5.6.2.1). Below are the regression coefficients and their significance tests, including B weights, the standard error of B (SE B), β weights (Beta), t tests for the coefficients (see Section 5.6.2.2), and their significance levels (Sig T). The term (Constant) refers to the intercept (A).

BMDP1R is limited to standard multiple regression so no special instruction is necessary to specify that procedure. The REGRESS instruction identifies the DV and the IVs. BMDP1R provides the same information as SPSS but in slightly different format and with some different labels. For example, β is referred to as STD. REG COEFF and B weights are in the column labeled COEFFICIENT. Other differences are seen by comparing values in Tables 5.4 and 5.5. An addi-

TABLE 5.4	SETUP AND SELECTED SPSS REGRESSION OUTPUT FOR STANDARD MULTIPLE REGRESSION ON SAMPLE DATA IN TABLE 5.1

```
TITLE          SMALL SAMPLE MULTIPLE REGRESSION.
DATA LIST      FILE='TAPE50.DAT'   FREE
               /SUBJNO, MOTIV, QUAL, GRADE, COMPR.
REGRESSION     VARS = MOTIV TO COMPR
               /DEP = COMPR / METHOD=ENTER.


                 * * * *   M U L T I P L E   R E G R E S S I O N   * * * *


Listwise Deletion of Missing Data

Equation Number 1    Dependent Variable..   COMPR

Block Number  1.  Method:  Enter


Variable(s) Entered on Step Number   1..    GRADE
                                     2..    MOTIV
                                     3..    QUAL


Multiple R          .83807       Analysis of Variance
R Square            .70235                     DF     Sum of Squares    Mean Square
Adjusted R Square   .25588       Regression     3          71.64007       23.88002
Standard Error     3.89615       Residual       2          30.35993       15.17997

                       F =      1.57313        Signif F =  .4114


-------------------Variables in the Equation-------------------

Variable          B          SE B        Beta        T    Sig T

GRADE          .416034     .646191     .402209     .644   .5857
MOTIV          .658266     .872131     .319306     .755   .5292
QUAL           .272048     .589114     .291158     .462   .6896
(Constant)   -4.721798    9.065646                -.521   .6544


End Block Number   1   All requested variables entered.
```

tional statistic is TOLERANCE (cf. Chapter 4), but adjusted R^2 is not reported. BMDP1R also provides univariate statistics for each of the variables (see Chapter 4 for definition of the coefficient of variation).

In using SAS REG for standard multiple regression, the variables for the regression equation are specified in the MODEL statement, with the DV on the left side of the equation and the IVs on the right. In the ANOVA table the sum of squares for regression is called Model and residual is called Error. Total SS and df are in the row labeled C Total. Below the ANOVA is the standard error of the estimate, shown as the square root of the error term, MS_{res} (Root MSE). Also printed are the mean of the DV (Dep Mean), R^2, adjusted R^2, and the coefficient of variation (C.V.) — defined here as 100 times the standard error of the estimate divided by the mean of the DV. The section labeled Parameter Estimates includes the usual B coefficients in the Parameter Estimate column, their standard errors, t tests for those coefficients (T For HO: PARAMETER=0) and significance levels (Prob > T). Standardized regression coefficients are not printed unless requested.

TABLE 5.5   SETUP AND SELECTED BMDP1R OUTPUT FOR STANDARD MULTIPLE REGRESSION
                    ON SAMPLE DATA IN TABLE 5.1

```
/INPUT      VARIABLES=5.  FORMAT IS FREE.  FILE = 'TAPE50.DAT'.
      TITLE IS 'SMALL SAMPLE STANDARD MULTIPLE REGRESSION'.
/VARIABLE   NAMES ARE SUBJNO, MOTIV, QUAL, GRADE, COMPR.
/REGRESS    DEP = COMPR.
            IND = MOTIV TO GRADE.
/END
```

VARIABLE	MEAN	STANDARD DEVIATION	COEFFICIENT OF VARIATION	MINIMUM	MAXIMUM
1 SUBJNO	3.5000	1.8708	0.53452	1.00000	6.00000
2 MOTIV	11.0000	2.1909	0.19917	8.00000	14.00000
3 QUAL	10.1667	4.8339	0.47547	5.00000	19.00000
4 GRADE	11.3333	4.3665	0.38528	8.00000	19.00000
5 COMPR	10.0000	4.5166	0.45166	5.00000	18.00000

ALL DATA CONSIDERED AS A SINGLE GROUP

MULTIPLE R	0.8381	STD. ERROR OF EST.	3.8961
MULTIPLE RSQUARE	0.7024		

ANALYSIS OF VARIANCE

	SUM OF SQUARES	DF	MEAN SQUARE	F RATIO	P(TAIL)
REGRESSION	71.6401	3	23.8800	1.573	0.4114
RESIDUAL	30.3599	2	15.1800		

VARIABLE		COEFFICIENT	STD. ERROR	STD. REG COEFF	T	P(2 TAIL)	TOLERANCE
INTERCEPT		-4.72180					
MOTIV	2	0.6583	0.8721	0.32	0.75	0.53	0.8316
QUAL	3	0.2720	0.5891	0.29	0.46	0.69	0.3744
GRADE	4	0.4160	0.6462	0.40	0.64	0.59	0.3813

Setup for SYSTAT REGRESS (or GLM) is similar to SAS REG, except that the IV side of the equation includes CONSTANT (specifying the intercept) and terms are connected by + signs. The listing of information about R is fully labeled, followed by information about the regression coefficients—B (Coefficient), standard error of B (Std Error), β (Std Coef), Tolerance, F test and significance level of F. The ANOVA table is printed last.

Additional features of these programs are discussed in Section 5.7.

5.5 ■ MAJOR TYPES OF MULTIPLE REGRESSION

There are three major analytic strategies in multiple regression: standard multiple regression, sequential (hierarchical) regression, and statistical (stepwise and setwise) regression. Differences among the strategies involve what happens to overlapping variability due to correlated IVs and who determines the order of entry of IVs into the equation.

Consider the Venn diagram in Figure 5.3(a) in which there are three IVs and one DV. IV_1 and IV_2 both correlate substantially with the DV and with each other. IV_3 correlates to a lesser extent

TABLE 5.6   SETUP AND SAS REG OUTPUT FOR STANDARD MULTIPLE REGRESSION ON SAMPLE DATA OF TABLE 5.1

```
DATA SSAMPLE;
INFILE 'TAPE50.DAT';
INPUT SUBJNO MOTIV QUAL GRADE COMPR;
PROC REG;
     MODEL COMPR = MOTIV QUAL GRADE;
```

Model: MODEL1
Dependent Variable: COMPR

Analysis of Variance

Source	DF	Sum of Squares	Mean Square	F Value	Prob>F
Model	3	71.64007	23.88002	1.573	0.4114
Error	2	30.35993	15.17997		
C Total	5	102.00000			

Root MSE	3.89615	Rsquare	0.7024	
Dep Mean	10.00000	Adj Rsq	0.2559	
C.V.	38.96148			

Parameter Estimates

Variable	DF	Parameter Estimate	Standard Error	T for H0: Parameter=0	Prob > |T|
INTERCEP	1	-4.721798	9.06564626	-0.521	0.6544
MOTIV	1	0.658266	0.87213087	0.755	0.5292
QUAL	1	0.272048	0.58911415	0.462	0.6896
GRADE	1	0.416034	0.64619051	0.644	0.5857

TABLE 5.7   SETUP AND SYSTAT REGRESS OUTPUT FOR STANDARD MULTIPLE REGRESSION ON SAMPLE DATA IN TABLE 5.1

```
USE TAPE50
MODEL COMPR = CONSTANT + MOTIV + QUAL + GRADE
ESTIMATE
```

Dep Var: COMPR   N: 6   Multiple R: 0.838   Squared multiple R: 0.702

Adjusted squared multiple R: 0.256   Standard error of estimate: 3.896

Effect	Coefficient	Std Error	Std Coef	Tolerance	F	P(2 Tail)
CONSTANT	-4.722	9.066	0.0	.	0.271	0.654
MOTIV	0.658	0.872	0.319	0.832	0.570	0.529
QUAL	0.272	0.589	0.291	0.374	0.213	0.690
GRADE	0.416	0.646	0.402	0.381	0.415	0.586

Analysis of Variance

Source	Sum-of-Squares	DF	Mean-Square	F-Ratio	P
Regression	71.640	3	23.880	1.573	0.411
Residual	30.360	2	15.180		

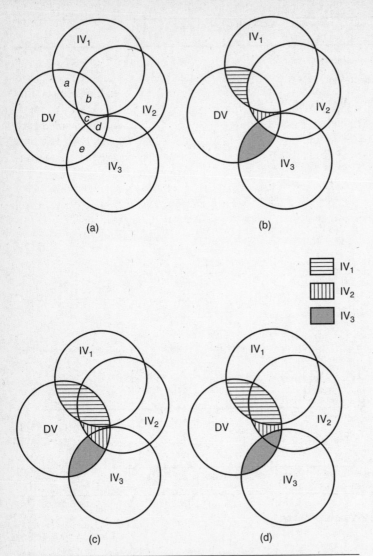

FIGURE 5.3 VENN DIAGRAMS ILLUSTRATING (a) OVERLAPPING VARIANCE SECTIONS; AND ALLOCATION OF OVERLAPPING VARIANCE IN (b) STANDARD MULTIPLE REGRESSION, (c) SEQUENTIAL REGRESSION, AND (d) STEPWISE REGRESSION.

with the DV and to a negligible extent with IV_2. R^2 for this situation is the area $a + b + c + d + e$. Area a comes unequivocally from IV_1; area c unequivocally from IV_2; area e from IV_3. However, there is ambiguity regarding areas b and d. Both areas could be predicted from either of two IVs; area b from either IV_1 or IV_2, area d from either IV_2 or IV_3. To which IV should the contested area be assigned? The interpretation of analysis can be drastically affected by choice of strategy because the apparent importance of the various IVs to the solution changes.

5.5.1 ■ Standard Multiple Regression

The standard model is the one used in the solution for the small sample graduate student data in Table 5.1. In the standard, or simultaneous, model all IVs enter into the regression equation at once; each one is assessed as if it had entered the regression after all other IVs had entered. Each IV is evaluated in terms of what it adds to prediction of the DV that is different from the predictability afforded by all the other IVs.

Consider Figure 5.3(b). The darkened areas of the figure indicate the variability accorded each of the IVs when the procedure of Section 5.6.1.1 is used. IV_1 "gets credit" for area a, IV_2 for area c, and IV_3 for area e. That is, each IV is assigned only the area of its unique contribution. The overlapping areas, b and d, contribute to R^2, but are not assigned to any of the individual IVs.

In standard multiple regression, it is possible for a variable like IV_2 to appear unimportant in the solution when it actually is highly correlated with the DV. If the area of that correlation is whittled away by other IVs, the unique contribution of the IV is often very small despite a substantial correlation with the DV. For this reason, both the full correlation and the unique contribution of the IV need to be considered in interpretation.

Standard multiple regression is handled in the SPSS package by the REGRESSION program, as are all other types of multiple regression. A selected part of output is given in Table 5.4 for the sample problem of Table 5.1. (Full interpretation of program output is available in substantive examples presented later in this chapter.)

Within the BMDP series, standard multiple regression is routinely handled by BMDP1R. A selected sample of output from that program is given in Table 5.5. Standard multiple regression is also available through the more elaborate stepwise or setwise programs—BMDP2R or BMDP9R. SAS REG and SYSTAT REGRESS or GLM are used for standard analyses, as illustrated in Tables 5.6 and 5.7.

5.5.2 ■ Sequential Multiple Regression

In sequential regression (sometimes called hierarchical regression), IVs enter the equation in an order specified by the researcher. Each IV is assessed in terms of what it adds to the equation at its own point of entry. Consider the example in Figure 5.3(c). Assume that the researcher assigns IV_1 first entry, IV_2 second entry, and IV_3 third entry. In assessing importance of variables by the procedure of Section 5.6.1.2, IV_1 "gets credit" for areas a and b, IV_2 for areas c and d, and IV_3 for area e. Each IV is assigned the variability, unique and overlapping, left to it at its own point of entry. Notice that the apparent importance of IV_2 would increase dramatically if it were assigned first entry and, therefore, "got credit" for b, c, and d.

The researcher normally assigns order of entry of variables according to logical or theoretical considerations. For example, IVs that are presumed (or manipulated) to be causally prior are given higher priority of entry. For instance, height might be considered prior to amount of training in assessing success as a basketball player and accorded a higher priority of entry. Variables with greater theoretical importance could also be given early entry.

Or the opposite tack could be taken. The research could enter manipulated or other variables of major importance on later steps, with "nuisance" variables given higher priority for entry. The lesser, or nuisance, set is entered first; then the major set is evaluated for what it adds to the prediction over and above the lesser set. For example, we might want to see how well we can predict reading speed (the DV) from intensity and length of a speed reading course (the major IVs) while holding constant initial differences in reading speed (the nuisance IV). This is the basic analysis of covariance problem in regression format.

IVs can be entered one at a time or in blocks. The analysis proceeds in steps, with information about variables both in and out of the equation given in computer output at each step. Finally, after all variables are entered, summary statistics are provided along with the information available at the last step.

In the SPSS package, sequential regression is performed by the REGRESSION program. In Table 5.8, selected output is shown for the sample problem of Table 5.1, with higher priority given to admission qualifications and course performance, and lower priority given to motivation. In the BMDP package, sequential regression is run with BMDP2R. Output from BMDP2R is illustrated in the next section. SAS REG, SYSTAT GLM and REGRESS, and BMDP2R have interactive modes, in which individual IVs can be entered sequentially.

5.5.3 ■ Statistical (Stepwise) and Setwise Regression

Statistical regression (sometimes generically called stepwise regression) is a rather controversial procedure, in which order of entry of variables is based solely on statistical criteria. The meaning or interpretation of the variables is not relevant. Decisions about which variables are included and which omitted from the equation are based solely on statistics computed from the particular sample drawn; minor differences in these statistics can have profound effect on the apparent importance of an IV.

Consider the example in Figure 5.3(d). IV_1 and IV_2 both correlate substantially with the DV; IV_3 correlates less strongly. The choice between IV_1 and IV_2 for first entry is based on which of the two IVs has the higher full correlation with the DV, even if the higher correlation shows up in the second or third decimal place. Let's say that IV_1 has the higher correlation with the DV and enters first. It "gets credit" for areas a and b. At the second step, IV_2 and IV_3 are compared, where IV_2 has areas c and d available to add to prediction, and IV_3 has areas e and d. At this point IV_3 contributes more strongly to R^2 and enters the equation. IV_2 is now assessed for whether or not its remaining area, c, contributes significantly to R^2. If it does, IV_2 enters the equation; otherwise it does not despite the fact that it is almost as highly correlated with the DV as the variable that entered first. For this reason, interpretation of a statistical regression equation is hazardous unless the researcher takes special care to remember the message of the initial DV–IV correlations.

There are actually three versions of statistical regression, forward selection, backward deletion, and stepwise regression. In forward selection, the equation starts out empty and IVs are added one at a time provided they meet the statistical criteria for entry. Once in the equation, an IV stays in. Bendel and Afifi (1977) recommend a liberal criterion for entry of predictors in forward regression. Important variables are less likely to be excluded from the model with a probability level for entry in the range of .15 to .20 rather than .05. In backward deletion, the equation starts out with all IVs entered and they are deleted one at a time if they do not contribute significantly to regression. Stepwise regression is a compromise between the two other procedures in which the equation starts out empty and IVs are added one at a time if they meet statistical criteria, but they may also be deleted at any step where they no longer contribute significantly to regression. Of the three procedures, stepwise regression is considered the surest path to the best prediction equation.

TABLE 5.8 SETUP AND SELECTED SPSS REGRESSION OUTPUT FOR SEQUENTIAL
MULTIPLE REGRESSION ON SAMPLE DATA IN TABLE 5.1

```
TITLE        SMALL SAMPLE MULTIPLE REGRESSION.
DATA LIST    FILE='TAPE50.DAT'   FREE
             /SUBJNO, MOTIV, QUAL, GRADE, COMPR.
REGRESSION   VARS = MOTIV TO COMPR
             /DEP = COMPR
             /METHOD=ENTER QUAL GRADE  /METHOD=ENTER MOTIV.
```

**** MULTIPLE REGRESSION ****

Listwise Deletion of Missing Data

Equation Number 1   Dependent Variable..   COMPR

Block Number 1. Method: Enter   QUAL   GRADE

Variable(s) Entered on Step Number   1..   GRADE
                                     2..   QUAL

Multiple R	.78586	Analysis of Variance

		DF	Sum of Squares	Mean Square
Multiple R	.78586			
R Square	.61757	Regression 2	62.99218	31.49609
Adjusted R Square	.36262	Residual 3	39.00782	13.00261
Standard Error	3.60591			

F =   2.42229   Signif F = .2365

------------ Variables in the Equation ------------

Variable	B	SE B	Beta	T	Sig T
GRADE	.472160	.594081	.456469	.795	.4848
QUAL	.350654	.536642	.375286	.653	.5601
(Constant)	1.083369	4.440596		.244	.8229

------------ Variables not in the Equation ------------

Variable	Beta In	Partial	Min Toler	T	Sig T
MOTIV	.319306	.470846	.374374	.755	.5292

End Block Number   1   All requested variables entered.

TABLE 5.8 *(CONTINUED)*

* * * * MULTIPLE REGRESSION * * * *

Equation Number 1   Dependent Variable..   COMPR

Block Number 2.   Method: Enter   MOTIV

Variable(s) Entered on Step Number   3..   MOTIV

Multiple R	.83807
R Square	.70235
Adjusted R Square	.25588
Standard Error	3.89615

Analysis of Variance

	DF	Sum of Squares	Mean Square
Regression	3	71.64007	23.88002
Residual	2	30.35993	15.17997

F =   1.57313   Signif F = .4114

------------ Variables in the Equation ------------

Variable	B	SE B	Beta	T	Sig T
GRADE	.416034	.646191	.402209	.644	.5857
QUAL	.272048	.589114	.291158	.462	.6896
MOTIV	.658266	.872131	.319306	.755	.5292
(Constant)	-4.721798	9.065646		-.521	.6544

End Block Number   2   All requested variables entered.

Statistical regression is typically used to develop a subset of IVs that is useful in predicting the DV, and to eliminate those IVs that do not provide additional prediction to the IVs already in the equation. For this reason, statistical regression may have some utility if the only aim of the researcher is a prediction equation. Even so, the sample from which the equation is drawn should be large and representative because of the following considerations.

The procedure's controversy lies primarily in capitalization on chance and overfitting of data. Decisions about which variables are included and which omitted from the equation are dependent on potentially minor differences in statistics computed from a single sample, where some variability in the statistics from sample to sample is expected. In this way, the technique capitalizes on chance differences within a single sample. For similar reasons, the equation derived from a sample is too close to the sample and may not generalize well to the population. In this way, statistical regression may overfit the data. For statistical regression, cross validation with a second sample is highly recommended. At the very least, separate analyses of two halves of an available sample should be conducted, with conclusions limited to results that are consistent for both analyses. (This caution applies to setwise regression as well.)

Further, a statistical analysis may not lead to the optimum solution in terms of R^2. Several IVs considered together may increase R^2, while any one of them considered alone does not. In simple statistical regression, none of the IVs enters. By specifying that IVs enter in blocks, one can set up combinations of sequential and statistical regression. A block of high-priority IVs is set up to compete among themselves stepwise for order of entry; then a second block of IVs compete among themselves for order of entry. The regression is sequential over blocks, but statistical within blocks.

The SPSS REGRESSION program handles stepwise regression in a manner similar to that of sequential regression. Both SPSS REGRESSION and BMDP2R provide a variety of statistical criteria and stepping options for statistical regression. For the data in Table 5.1, output from a BMDP2R run with default values is shown in Table 5.9.

SAS REG is the procedure provided by the SAS system for statistical regression (incorporating the former SAS STEPWISE), while SYSTAT GLM and REGRESS allow statistical regression as well as standard multiple regression.

An alternative form of statistical regression for choosing an optimal subset of variables is setwise regression. In setwise regression, separate regressions are computed for all IVs singly, all possible pairs of IVs, all possible trios of IVs and so forth until the best subset of IVs is identified according to some criterion (such as maximum R^2) from among all possible subsets. One computer program in the BMDP series, BMDP9R, is designed to do this. In the SAS package, SAS RSQUARE was formerly used for this process. In current versions of SAS, RSQUARE is an option available in SAS REG, along with other options for setwise regression. No provision for setwise regression is available in SPSS or SYSTAT.

5.5.4 ■ Choosing among Regression Strategies

To simply assess relationships among variables, and answer the basic question of multiple correlation, the method of choice is standard multiple regression. As a matter of fact, unless there is good reason to use some other technique, standard multiple regression is recommended. Reasons for using sequential regression are theoretical or for testing explicit hypotheses.

Sequential regression allows the researcher to control the advancement of the regression process. Importance of IVs in the prediction equation is determined by the researcher according to logic or

TABLE 5.9   SETUP AND SELECTED BMDP2R OUTPUT FOR STEPWISE REGRESSION ON SAMPLE DATA IN TABLE 5.1

```
/INPUT     VARIABLES=5.  FORMAT IS FREE.  FILE = 'TAPE50.DAT'.
           TITLE IS 'SMALL SAMPLE STEPWISE REGRESSION'.
/VARIABLE  NAMES ARE SUBJNO, MOTIV, QUAL, GRADE, COMPR.
           USE = MOTIV TO COMPR.
/REGRESS   DEP = COMPR.
/END
```

STEP NO.   0

STD. ERROR OF EST.   4.5166

ANALYSIS OF VARIANCE

	SUM OF SQUARES	DF	MEAN SQUARE
RESIDUAL	102.00000	5	20.40000

VARIABLES IN EQUATION FOR COMPR

VARIABLE	COEFFICIENT	STD. ERROR OF COEFF	STD REG COEFF	TOLERANCE	F TO REMOVE	LEVEL
(Y-INTERCEPT	10.00000)					

VARIABLES NOT IN EQUATION

VARIABLE		PARTIAL CORR.	TOLERANCE	F TO ENTER	LEVEL
MOTIV	2	0.58613	1.00000	2.09	1
QUAL	3	0.73284	1.00000	4.64	1
GRADE	4	0.75043	1.00000	5.16	1

STEP NO.   1

VARIABLE ENTERED   4 GRADE

MULTIPLE R	0.7504
MULTIPLE R-SQUARE	0.5631
ADJUSTED R-SQUARE	0.4539

STD. ERROR OF EST.   3.3376

ANALYSIS OF VARIANCE

	SUM OF SQUARES	DF	MEAN SQUARE	F RATIO
REGRESSION	57.440563	1	57.44056	5.16
RESIDUAL	44.559437	4	11.13986	

VARIABLES IN EQUATION FOR COMPR

VARIABLE	COEFFICIENT	STD. ERROR OF COEFF	STD REG COEFF	TOLERANCE	F TO REMOVE	LEVEL
(Y-INTERCEPT	1.20280)					
GRADE 4	0.77622	0.3418	0.750	1.00000	5.16	1

VARIABLES NOT IN EQUATION

VARIABLE		PARTIAL CORR.	TOLERANCE	F TO ENTER	LEVEL
MOTIV	2	0.49600	0.85839	0.98	1
QUAL	3	0.35297	0.38645	0.43	1

***** F LEVELS(4.000, 3.900) OR TOLERANCE INSUFFICIENT FOR FURTHER STEPPING

TABLE 5.9 *(CONTINUED)*

STEPWISE REGRESSION COEFFICIENTS

VARIABLES	0 Y-INTCPT	2 MOTIV	3 QUAL	4 GRADE
STEP				
0	10.0000*	1.2083	0.6847	0.7762
1	1.2028*	0.7295	0.3507	0.7762*

*** N O T E *** 1) REGRESSION COEFFICIENTS FOR VARIABLES IN THE EQUATION ARE
                    INDICATED BY AN ASTERISK.
               2) THE REMAINING COEFFICIENTS ARE THOSE WHICH WOULD BE
                  OBTAINED IF THAT VARIABLE WERE TO ENTER IN THE NEXT STEP.

SUMMARY TABLE

STEP NO.	VARIABLE ENTERED	REMOVED	MULTIPLE R	RSQ	CHANGE IN RSQ	F TO ENTER	F TO REMOVE	NO. OF VAR. INCLUDED
1	4 GRADE		0.7504	0.5631	0.5631	5.16		1

theory. Explicit hypotheses are tested about proportion of variance attributable to some IVs after variance due to IVs already in the equation is accounted for.

Although there are similarities in programs used and output produced for sequential and statistical regression, there are fundamental differences in the way that IVs enter the prediction equation and in the interpretations that can be made from the results. In sequential regression the researcher controls entry of variables, while in statistical and setwise regression statistics computed from sample data control order of entry. Statistical and setwise regression are, therefore, model-building rather than model-testing procedures. As exploratory techniques, they may be useful for such purposes as eliminating variables that are clearly superfluous in order to tighten up future research. However, clearly superfluous IVs will show up in any of the procedures.

When multicollinearity or singularity are present, setwise regression can be indispensable in identifying multicollinear variables, as indicated in Chapter 4.

For the example of Section 5.4, in which performance on graduate comprehensive exam (COMPR) is predicted from professional motivation (MOTIV), qualifications for graduate training (QUAL), and performance in graduate courses (GRADE), the differences among regression strategies might be phrased as follows. If standard multiple regression is used, two fundamental questions are asked: (1) What is the size of the overall relationship between COMPR and MOTIV, QUAL, and GRADE? and (2) How much of the relationship is contributed uniquely by each IV? If sequential regression is used, with QUAL and GRADE entered before MOTIV, the question is, Does MOTIV significantly add to prediction of COMPR after differences among students in QUAL and GRADE have been statistically eliminated? If statistical regression is used, one asks, What is the best linear combination of IVs to predict the DV in this sample? And use of setwise regression leads to the query, What is the size of R^2 from each one of the IVs, from all possible combinations of two IVs, and from all three IVs for this sample?

5.6 ■ SOME IMPORTANT ISSUES

5.6.1 ■ Importance of IVs

If the IVs are uncorrelated with each other, assessment of the contribution of each of them to multiple regression is straightforward. IVs with bigger correlations or higher standardized regression coefficients are more important to the solution than those with lower (absolute) values. (Because unstandardized regression coefficients are in a metric that depends on the metric of the original variables, their sizes are harder to interpret. A large regression coefficient for an IV with a low correlation with the DV can also be misleading because the IV predicts the DV well only after another IV suppresses irrelevant variance, as seen in Section 5.6.4.)

If the IVs are correlated with each other, assessment of the importance of each of them to regression is more ambiguous. The correlation between an IV and the DV reflects variance shared with the DV, but some of that variance may be predictable from other IVs.

To get the most straightforward answer regarding the importance of an IV to regression, one needs to consider the type of regression it is, and both the full and unique relationship between the IV and the DV. This section reviews several of the issues to be considered when assessing the importance of an IV to standard multiple, sequential, or statistical regression. In all cases, one needs to compare the total relationship of the IV with the DV, the unique relationship of the IV with the DV, and the correlations of the IVs with each other in order to get a complete picture of the function of an IV in regression. The total relationship of the IV with the DV (correlation) and the correlations of the IVs with each other are given in the correlation matrix. The unique contribution of an IV to predicting a DV is generally assessed by either partial or semipartial correlation.

For standard multiple and sequential regression, the relationships between correlation, partial correlation, and semipartial correlation are given in Figure 5.4 for a simple case of one DV and two IVs. In the figure, (squared) correlation, partial correlation, and semipartial correlation coefficients are defined as areas created by overlapping circles. Area $a + b + c + d$ is the total area of the DV and reduces to a value of 1 in many equations. Area b is the segment of the variability of the DV that can be explained by either IV_1 or IV_2, and is the segment that creates the ambiguity. Notice that it is the denominators that change between squared semipartial and partial correlation.

In partial correlation, the contribution of the other IVs is taken out of both the IV and the DV. In semipartial correlation, the contribution of other IVs is taken out of only the IV. Thus (squared) semipartial correlation expresses the unique contribution of the IV to the total variance of the DV. Squared semipartial correlation (sr_i^2) is the more useful measure of importance of an IV.[5] The interpretation of sr_i^2 differs, however, depending on the type of multiple regression employed.

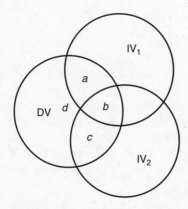

	Standard Multiple	Sequential
r_i^2	IV_1 $(a + b)/(a + b + c + d)$ IV_2 $(c + b)/(a + b + c + d)$	$(a + b)/(a + b + c + d)$ $(c + b)/(a + b + c + d)$
sr_i^2	IV_1 $a/(a + b + c + d)$ IV_2 $c/(a + b + c + d)$	$(a + b)/(a + b + c + d)$ $c/(a + b + c + d)$
pr_i^2	IV_1 $a/(a + d)$ IV_2 $c/(c + d)$	$(a + b)/(a + b + d)$ $c/(c + d)$

FIGURE 5.4 AREAS REPRESENTING SQUARED CORRELATION, SQUARED SEMIPARTIAL CORRELATION, AND SQUARED PARTIAL CORRELATION IN STANDARD MULTIPLE AND SEQUENTIAL REGRESSION (WHERE IV_1 IS GIVEN PRIORITY OVER IV_2).

---

[5] Procedures for obtaining partial correlations are available in earlier editions of this book but are omitted in this edition.

5.6.1.1 ■ Standard Multiple and Setwise Regression

In standard multiple regression and setwise regression, sr_i^2 for an IV is the amount by which R^2 is reduced if that IV is deleted from the regression equation. That is, sr_i^2 represents the unique contribution of the IV to R^2 in that set of IVs.

When the IVs are correlated, squared semipartial correlations do not necessarily sum to multiple R^2. The sum of sr_i^2 is usually smaller than R^2 (although under some rather extreme circumstances, the sum can be larger than R^2). When the sum is smaller, the difference between R^2 and the sum of sr_i^2 for all IVs represents shared variance, variance that is contributed to R^2 by two or more IVs. It is rather common to find substantial R^2, with sr_i^2 for all IVs quite small.

Table 5.10 summarizes procedures for finding sr_i^2 (and pr_i^2) for both standard multiple and sequential regression through SPSS, BMDP, SAS, and SYSTAT.

SPSS and SAS provide versions of sr_i^2 as part of their output. If you use SPSS REGRESSION, sr_i is optionally available as Part Cor by requesting STATISTICS = ZPP. SAS REG provides two squared semipartial correlations with different error terms. For standard multiple regression, the appropriate form is SCORR2, which uses the Type II error term.

For BMDP1R, BMDP9R, or SYSTAT REGRESS or GLM, sr_i^2 is most easily calculated using Equation 5.9.

$$sr_i^2 = \frac{T_i^2}{\mathrm{df}_{\mathrm{res}}}(1 - R^2) \tag{5.9}$$

The squared semipartial correlation (sr_i^2) for the ith IV is calculated from the T_i^2 (or F_i) ratio for that IV, the residual degrees of freedom ($\mathrm{df}_{\mathrm{res}}$), and R^2.

In BMDP1R, R^2 is labeled MULTIPLE R-SQUARE and $\mathrm{df}_{\mathrm{res}}$ is found in the analysis of variance table (cf. Table 5.5). With BMDP9R, information for calculating sr_i^2 is available for the best subset, with $\mathrm{df}_{\mathrm{res}}$ shown as DENOMINATOR DEGREES OF FREEDOM. R^2 is shown as SQUARED MULTIPLE CORRELATION. SYSTAT REGRESS and GLM offer R^2 as Squared multiple R (cf. Table 5.7) and show $\mathrm{df}_{\mathrm{res}}$ in the analysis of variance table. F is shown instead of t, so squaring is unnecessary.

TABLE 5.10   PROCEDURES FOR FINDING sr_i^2 AND pr_i^2 THROUGH SPSS, SAS, BMDP, AND SYSTAT FOR STANDARD MULTIPLE AND SEQUENTIAL REGRESSION

	sr_i^2	pr_i^2
Standard Multiple Regression		
SPSS REGRESSION	STATISTICS=ZPP Part Cor	STATISTICS=ZPP Partial
SAS REG	SCORR2	PCORR2
BMDP1R, 9R	Use Eq. 5.9	
SYSTAT GLM	Use Eq. 5.9	
Sequential Regression		
SPSS REGRESSION	RSQCH in SUMMARY TABLE	
BMDP2R	CHANGE IN RSQ	
SAS REG	Get by subtraction of R^2 in sequential steps	
SYSTAT GLM	Get by subtraction of R^2 in sequential steps	

From the output shown in Table 5.5 the squared semipartial correlation for GRADE is

$$sr_i^2 = \frac{(0.64)^2}{2} (1 - .7024) = .06 \qquad (5.10)$$

In this set of IVs, then, R^2 would be reduced by .06 if GRADE were deleted from the equation.

In all standard multiple regression programs, T_i (or F) is the significance test for sr_i^2, B_i, and β_i, as discussed in Section 5.6.2.

5.6.1.2 ■ Sequential or Statistical Regression

In these two forms of regression, sr_i^2 is interpreted as the amount of variance added to R^2 by each IV at the point that it enters the equation. The research question is, How much does this IV add to multiple R^2 after IVs with higher priority have contributed their share to prediction of the DV? Thus, the apparent importance of an IV is very likely to depend on its point of entry into the equation. In sequential and statistical regression, the sr_i^2 do, indeed, sum to R^2; consult Figure 5.3 if you want to review this point.

As reviewed in Table 5.10, SPSS REGRESSION and BMDP2R provide squared semipartial correlations as part of routine output for sequential and statistical regression.[6] For SPSS, sr_i^2 is RsqCh for each IV in the Summary table (see Table 5.8). For BMDP2R, sr_i^2 is CHANGE IN RSQ for each IV (see Table 5.9).

SYSTAT GLM and REGRESS and SAS REG (for sequential regression), provide R^2 for each step. You can calculate sr_i^2 by subtraction between subsequent steps.

5.6.2 ■ Statistical Inference

This section covers significance tests for multiple regression and for regression coefficients for individual IVs. A test, F_{inc}, is also described for evaluating the statistical significance of adding two or more IVs to a prediction equation in sequential or statistical analysis. Calculations of confidence limits for unstandardized regression coefficients and procedures for comparing the predictive capacity of two different sets of IVs conclude the section.

When the researcher is using either statistical or setwise regression as an exploratory tool, inferential procedures of any kind may be inappropriate. Inferential procedures require that the researcher have a hypothesis to test. When statistical or setwise regression is used to snoop data, there may be no hypothesis, even though the statistics themselves are available.

5.6.2.1 ■ Test for Multiple R

The overall inferential test in multiple regression is whether the sample of scores is drawn from a population in which multiple R is zero. This is equivalent to the null hypothesis that all correlations between DV and IVs and all regression coefficients are zero. With large N, the test of this hypothesis becomes trivial because it is almost certain to be rejected.

For standard multiple and sequential regression, the test of this hypothesis is presented in all computer outputs as analysis of variance. For sequential regression (and for standard multiple

---

[6] If you request one of the statistical regression options in PROC REG, squared semipartial correlations will be printed out in a summary table as Partial R**2.

regression performed through stepwise programs), the analysis of variance table at the last step gives the relevant information. The F ratio for mean square regression over mean square residual tests the significance of multiple R. Mean square regression is the sum of squares for regression in Equation 5.2 divided by k degrees of freedom; mean square residual is the sum of squares for residual in the same equation divided by $(N - k - 1)$ degrees of freedom.

If you insist on inference in statistical or setwise regression, adjustments are necessary because all potential IVs do not enter the equation and the test for R^2 is not distributed as F. Therefore the analysis of variance table at the last step (or for the "best" equation) is misleading; the reported F is biased so that the F ratio actually reflects a Type I error rate in excess of α.

Wilkinson and Dallal (1981) have developed tables for critical R^2 when forward selection procedures are used for statistical addition of variables and the selection stops when the F-to-enter for the next variable falls below some preset value. Appendix C contains Table C.5 that shows how large multiple R^2 must be to be statistically significant at .05 or .01 levels, given N, k, and F, where N is sample size, k is the number of potential IVs, and F is the minimum F-to-enter that is specified. F-to-enter values that can be chosen are 2, 3, or 4.

For example, for a statistical regression in which there are 100 subjects, 20 potential IVs, and an F-to-enter value of 3 chosen for the solution, a multiple R^2 of approximately .19 is required to be considered significantly different from zero at $\alpha = .05$ (and approximately .26 at $\alpha = .01$). Wilkinson and Dallal report that linear interpolation on N and k works well. However, they caution against extensive extrapolation for values of N and k beyond those given in the table. In sequential regression this table is also used to find critical R^2 values if a post hoc decision is made to terminate regression as soon as R^2 reaches statistical significance, if the appropriate F-to-enter is substituted for R^2 probability value as the stopping rule.

Wilkinson and Dallal recommend forward selection procedures in favor of other selection methods (e.g., stepwise selection, subset regression). They argue that, in practice, results using different procedures are not likely to be substantially different. Further, forward selection is computationally simple and efficient, and allows straightforward specification of stopping rules. If you are able to specify in advance the number of variables you wish to select, an alternative set of tables is provided by Wilkinson (1979) to evaluate significance of multiple R^2 with forward selection.

After looking at the data, you may wish to test the significance of some subsets of IVs in predicting the DV where a subset may even consist of a single IV. If several post hoc tests like this are desired, Type I errors become increasingly likely. Larzelere and Mulaik (1977) recommend the following conservative F test to keep Type I error rate below alpha for all combinations of IVs:

$$F = \frac{R_S^2/k}{(1 - R_S^2)/(N - k - 1)} \tag{5.11}$$

where R_i^2 is the squared multiple (or bivariate) correlation to be tested for significance, and k is the total number of IVs. Obtained F is compared with tabled F, with k and $(N - k - 1)$ degrees of freedom. That is, the critical value of F for each subset is the same as the critical value for the overall multiple R.

In the sample problem of Section 5.4, the bivariate correlation between MOTIV and COMPR (from Table 5.2) is tested post hoc as follows:

$$F = \frac{.58613^2/3}{(1 - .58613^2)/2} = 0.349, \qquad \text{with df} = 3, 2$$

which is obviously not significant.

5.6.2.2 ■ Test of Regression Components

In standard multiple regression, for each IV the same significance test evaluates B_i, β_i, and sr_i. The test is straightforward, and results are given in computer output. In SPSS, F_i tests the unique contribution of each IV and appears in the output section labeled Variables in the Equation (see Table 5.4). Degrees of freedom are 1 and df_{res}, which appears in the accompanying analysis of variance table. In BMDP1R, SYSTAT GLM and REGRESS, and SAS REG, T_i or F_i values are given for each IV, tested with df_{res} from the analysis of variance table (see Tables 5.5 to 5.7). If BMDP2R is used for standard multiple regression, F_i values are available at the last step of the analysis as F-TO-REMOVE; they are interpreted as for SPSS.

Recall the limitations of these significance tests. The significance tests are sensitive only to the unique variance an IV adds to R^2. A very important IV that shares variance with another IV in the analysis may be nonsignificant although the two IVs in combination are responsible in large part for the size of R^2. An IV that is highly correlated with the DV but has a nonsignificant regression coefficient may have suffered just such a fate. For this reason, it is important to report and interpret r_{iy} in addition to F_i for each IV, as shown in Table 5.18, summarizing the results of a large sample example.

For setwise regression through BMDP9R the individual IVs are best interpreted as if for standard multiple regression where each variable enters last. The T_i values given for each variable in the "best" subset are interpreted the same as those for BMDP1R noted earlier.

For statistical and sequential regression, assessment of contribution of variables is more complex, and appropriate significance tests may not appear in the computer output. First, there is inherent ambiguity in the testing of each variable. In statistical and sequential regression, tests of sr_i^2 are not the same as tests of the regression coefficients (B_i and β_i). Regression coefficients are independent of order of entry of the IVs, whereas sr_i^2 depend directly on order of entry. Because sr_i^2 reflects "importance" as typically of interest in sequential or statistical regression, tests based on sr_i^2 are discussed here.[7]

SPSS, BMDP, and SAS provide significance tests for sr_i^2 in summary tables. For SPSS, the test is FCh—F ratio for change in R^2—that is accompanied by a significance value, SigCh. In BMDP2R, the F ratio for sr_i^2 is reported in the summary table as F TO ENTER. No significance value is given, but you can evaluate this F with numerator df = 1 and denominator df_{res} for the last step in the equation.

If you use SYSTAT REGRESS or GLM or SAS REG, you need to calculate F for sr_i^2 as found by subtraction (cf. Section 5.6.1.2) using the following equation:

$$F_i = \frac{sr_i^2}{(1 - R^2)/df_{res}} \tag{5.12}$$

The F_i for each IV is based on sr_i^2 (the squared semipartial correlation), multiple R^2 at the final step; and residual degrees of freedom from the analysis of variance table for the final step.

5.6.2.3 ■ Test of Added Subset of IVs

For sequential and statistical regression, one can test whether a block of two or more variables significantly increases R^2 above the R^2 predicted by a set of variables already in the equation.

---

[7] For combined standard-sequential regression, it might be desirable to use the "standard" method for all IVs simply to maintain consistency. If so, be sure to report that the F test is for regression coefficients.

$$F_{inc} = \frac{(R^2_{wi} - R^2_{wo})/M}{(1 - R^2_{wi})/df_{res}} \qquad (5.13)$$

where F_{inc} is the incremental F ratio; R^2_{wi} is the multiple R^2 achieved with the new block of IVs in the equation; R^2_{wo} is the multiple R^2 without the new block of IVs in the equation; M is the number of IVs in the block; and $df_{res} = (N - k - 1)$ is residual degrees of freedom in the final analysis of variance table.

Both R^2_{wi} and R^2_{wo} are found in the summary table of any program that produces one. If you are using SPSS REGRESSION, F_{inc} is available through the TEST command.

The null hypothesis of no increase in R^2 is tested as F with M and df_{res} degrees of freedom. If the null hypothesis is rejected, the new block of IVs does significantly increase the explained variance.

Although this is a poor example because only one variable is in the new block, we can use the sequential example in Table 5.8 to test whether MOTIV adds significantly to the variance contributed by the first two variables to enter the equation, QUAL and GRADE.

$$F_{inc} = \frac{(.70235 - .61757)/1}{(1 - .70235)/2} = 0.570 \qquad \text{with df} = 1, 2$$

(Because only one variable was entered, this test is the same as the test provided for MOTIV in Table 5.8 under the heading Variables not in the Equation. When the T value of .755 is squared, it is the same as F_{inc} above. Thus, F_{inc} can be used when there is only one variable in the block, but the information is already provided in the output.)

5.6.2.4 ■ Confidence Limits around B

To estimate population values, confidence limits for unstandardized regression coefficients (B_i) are calculated. Standard errors of unstandardized regression coefficients, unstandardized regression coefficients, and the critical two-tailed value of t for the desired confidence level (based on $N - 2$ degrees of freedom, where N is the sample size) are used in Equation 5.14.

$$CL_{Bi} = B_i \pm SE_{Bi}(t_{\alpha/2}) \qquad (5.14)$$

The $(1 - \alpha)$ confidence limits for the unstandardized regression coefficient for the ith IV (CL_{Bi}) is the regression coefficient (B_i) plus or minus the standard error of the regression coefficient (SE_{Bi}) times the critical value of t, with $(N - 2)$ degrees of freedom at the desired level of α.

If 95% confidence limits are desired, they are given in SPSS REGRESSION output as 95 Confdnce Intrvl B under the segment of output titled Variables in the Equation. With other output or when 99% confidence limits are desired, Equation 5.14 is used. Unstandardized regression coefficients and the standard errors of those coefficients appear in the sections labeled VARIABLES IN THE EQUATION. Standard errors are called STD ERROR OF COEFF in BMDP2R and are called STD ERROR in SYSTAT MGLH and REGRESS, SAS, BMDP1R, and BMDP9R.

For the example in Table 5.4, the 95% confidence limits for GRADE, with df = 4, are

$$CL_{B_G} = 0.416 \pm 0.646(2.78) = 0.416 \pm 1.796 = -1.380 \leftrightarrow 2.212$$

If the confidence interval contains zero, one retains the null hypothesis that the population regression coefficient is zero.

5.6.2.5 ■ Comparing Two Sets of Predictors

It is sometimes of interest to know whether one set of IVs predicts a DV better than another set of IVs. For example, can ratings of current belly dancing ability be better predicted by personality tests or by past dance and musical training?

The procedure for finding out is fairly convoluted, but if you have a large sample and you have access to BMDP or are willing to enter a pair of predicted scores for each subject in your sample, a test for the significance of the difference between two "correlated correlations" (both correlations are based on the same sample and share a variable) is available (Steiger, 1980). (If sample size is small, nonindependence among predicted scores for cases can result in serious violation of the assumptions underlying the test.)

As suggested in Section 5.4.1, a multiple correlation can be thought of as a simple correlation between obtained DVs and predicted DVs; that is, $R = r_{yy'}$. If there are two sets of predictors, $Y_{a'}$ and $Y_{b'}$ (where, for example $Y_{a'}$ is prediction from personality scores and $Y_{b'}$ is prediction from past training), a comparison of their relative effectiveness in predicting Y is made by testing for the significance of the difference between $r_{yy'a}$ and $r_{yy'b}$. For simplicity, let's call these r_{ya} and r_{yb}.

To test the difference we need to know the correlation between the predicted scores from set A (personality) and those from set B (training), that is, $r_{ya'yb'}$ or, simplified, r_{ab}. This is where the BMDP file manipulation procedures or hand entering become necessary.

The z test for the difference between r_{ya} and r_{yb} is

$$\overline{Z}^* = (z_{ya} - z_{yb}) \sqrt{\frac{N-3}{2 - 2\overline{s}_{ya,\,yb}}} \tag{5.15}$$

where N is, as usual, the sample size,

$$z_{ya} = \frac{1}{2}\ln\left(\frac{1 + r_{ya}}{1 - r_{ya}}\right) \text{ and } z_{yb} = \frac{1}{2}\ln\left(\frac{1 + r_{yb}}{1 - r_{yb}}\right)$$

and

$$\overline{s}_{ya,\,yb} = \frac{(r_{ab})(1 - 2\overline{r}^2) - 1/2(\overline{r}^2)(1 - 2\overline{r}^2 - r^2_{ab})}{(1 - \overline{r}^2)^2}$$

where $\overline{r} = 1/2(r_{ya} + r_{yb})$.

So, for the example, if the correlation between currently measured ability and ability as predicted from personality scores is .40 ($R_a = r_{ya} = .40$), the correlation between currently measured ability and ability as predicted from past training is .50 ($R_b = r_{yb} = .50$), and the correlation between ability as predicted from personality and ability as predicted from training is .10 ($r_{ab} = .10$), and $N = 103$,

$$\overline{r} = \frac{1}{2}(.40 + .50) = .45$$

$$\overline{s}_{ya,\,yb} = \frac{(.10)(1 - 2(.45)^2) - 1/2(.45)^2(1 - 2(.45)^2 - .10^2)}{(1 - .45^2)} = .0004226$$

$$z_{ya} = \frac{1}{2}\ln\left(\frac{1 + .40}{1 - .40}\right) = .42365$$

$$z_{yb} = \frac{1}{2} \ln \left(\frac{1 + .50}{1 - .50} \right) = .54931$$

and, finally,

$$\overline{Z^*} = (.42365 - .54931) \sqrt{\frac{103 - 3}{2 - .000845}} = -0.88874$$

Because $\overline{Z^*}$ is within the critical values of ± 1.96 for a two-tailed test, there is no statistically significant difference between multiple R when predicting Y from Ya' or Yb'. That is, there is no statistically significant difference in predicting current belly dancing ability from past training vs. personality tests.

Steiger (1980) and Steiger and Browne (1984) present additional significance tests for situations where both the DV and the IVs are different, but from the same sample, and for comparing the difference between any two correlations within a correlation matrix.

5.6.3 ■ Adjustment of R^2

Just as simple r_{xy} from a sample is expected to fluctuate around the value of the correlation in the population, sample R is expected to fluctuate around the population value. But multiple R never takes on a negative value, so all chance fluctuations are in the positive direction and add to the magnitude of R. As in any sampling distribution, the magnitude of chance fluctuations is larger with smaller sample sizes. Therefore, R tends to be overestimated, and the smaller the sample the greater the overestimation. For this reason, in estimating the population value of R, adjustment is made for expected inflation in sample R.

All the programs discussed in Section 5.7 routinely provide adjusted R^2. Wherry (1931) provides a simple equation for this adjustment, which is called $\tilde{R}^2$:

$$\tilde{R}^2 = 1 - (1 - R^2)\left(\frac{N - 1}{N - k - 1} \right) \tag{5.16}$$

where N is the sample size, k is the number of IVs, and R^2 is the squared multiple correlation. For the same sample problem,

$$\tilde{R}^2 = 1 - (1 - .70235)(5/2) = .25588$$

as printed out for SPSS, Table 5.4.

For statistical regression, Cohen and Cohen (1983) recommend k based on the number of IVs considered for inclusion, rather than on the number of IVs selected by the program. They also suggest the convention of reporting $\tilde{R}^2 = 0$ when the value spuriously becomes negative.

When the number of subjects is 60 or fewer and there are numerous IVs (say, more than 20), Equation 5.16 may provide inadequate adjustment for R^2. The adjusted value may be off by as much as .10 (Cattin, 1980). In these situations of small N and numerous IVs,

$$R_s^2 = \frac{(N - k - 3)\tilde{R}^4 + \tilde{R}^2}{(N - 2k - 2)\tilde{R}^2 + k} \tag{5.17}$$

Equation 5.17 (Browne, 1975) provides further adjustment, where

$$R^4 = (\tilde{R}^2)^2 - \left(\frac{2k\,(1 - \tilde{R}^2)^2}{(N - 1)(N - k + 1)} \right) \tag{5.18}$$

The adjusted R^2 for small samples is a function of the number of cases, N, the number of IVs, k, and the $\tilde{R}^2$ value as found from Equation 5.16.

When N is less than 50, Cattin (1980) provides an equation that produces even less bias but requires far more computation.

5.6.4 ■ Suppressor Variables

Occasionally you may find an IV that is useful in predicting the DV and in increasing the multiple R^2 solely by virtue of its correlations with other IVs. This IV is called a suppressor variable because it "suppresses" variance that is irrelevant to prediction of the DV. In a full discussion of suppressor variables, Cohen and Cohen (1983) describe and provide examples of several varieties of suppression.

For instance, one might administer as IVs two paper-and-pencil tests, a test of ability to list dance patterns and a test of test-taking ability. By itself the first test poorly predicts the DV (say, belly dancing ability). However, in the context of the test of test-taking ability, the relationship between ability to list dance patterns and belly dancing ability improves. The second IV serves as a suppressor variable because by removing variance due to ability in taking tests, prediction of the DV by the first IV is enhanced.

In output, the presence of a suppressor variable is identified by the pattern of regression coefficients and correlations of each IV with the DV. Compare the simple correlation between each IV and the DV in the correlation matrix with the standardized regression coefficient (beta weight) for the IV. If the beta weight is significantly different from zero, either one of the following two conditions signals the presence of a suppressor variable: (1) the absolute value of the simple correlation between IV and DV is substantially smaller than the beta weight for the IV, or (2) the simple correlation and beta weight have opposite signs.

If you find that a suppressor variable is present, you need to search for it among the IVs for which regression coefficients and correlations are congruent. It is sometimes difficult to identify which variable is doing the suppression. One strategy is to systematically leave each congruent IV out of the equation and examine the changes in regression coefficients for the IV(s) with discrepant regression coefficients and correlations in the original equation.

Once suppressor variables are identified, they are properly interpreted as variables that enhance importance of other IVs by virtue of suppression of irrelevant variance in other IVs or in the DV. It may be difficult to identify which variable is the suppressor even if the choice is narrowed to two variables. Cohen and Cohen (1983) suggest that either variable could be labeled the suppressor.

5.7 ■ COMPARISON OF PROGRAMS

The popularity of multiple regression is reflected in the abundance of applicable programs. BMDP has three separate programs for standard, statistical/sequential, and setwise regression. SPSS and SAS each have a single, highly flexible program for the three types of multiple regression. In SYS-TAT, multiple regression is handled either through a multivariate general linear model program

(GLM) or a subset of it in a strictly multiple regression program (REGRESS). The packages also have programs for the more exotic forms of multiple regression, such as nonlinear regression, probit regression, logistic regression (cf. Chapter 12), and the like.

Direct comparisons of multiple and statistical/sequential programs are summarized in Tables 5.11 and 5.12, respectively. Table 5.13 compares the two programs which can be used for setwise (subsets) regression. Some of these features are elaborated in Sections 5.7.1 through 5.7.4.

5.7.1 ■ SPSS Package

The distinctive feature of SPSS REGRESSION (SPSS Inc. 1990), summarized in Tables 5.11 and 5.12, is flexibility. SPSS REGRESSION offers four options for treatment of missing data (described in the manuals). Provision for expanded variable labels enhances recall of output that has not been looked at in a while. Data can be input raw or as correlation or covariance matrices. And as in all SPSS procedures, analysis can be limited to subsets of cases.

A special option is available so that correlation matrices are printed only when one or more of the correlations cannot be calculated. Also convenient is the optional printing of semipartial correlations for standard multiple regression and 95% confidence intervals for regression coefficients.

The statistical procedure offers forward, backward, and stepwise selection of variables. User-modifiable statistical criteria for statistical selection are (1) the maximum number of steps for model building, (2) the minimum F ratio for adding or maximum F for deleting a variable in the equation, (3) the minimum probability of F for adding or maximum probability of F for removing a variable in the equation, and (4) tolerance. Tolerance is the proportion of variance of a potential IV that is not explained by IVs already in the equation. The smaller the tolerance value, the less the restriction placed on entering variables. One avoids multicollinearity by maintaining reasonable tolerance levels for entry, and thereby disallowing entry of variables that add virtually nothing to predictability.

A series of METHOD=ENTER subcommands can be used for sequential regression. Each ENTER subcommand is evaluated in turn; the IV or IVs listed after each ENTER subcommand are evaluated in that order. Within a single subcommand, variables are entered in order of decreasing tolerance. If there is more than one IV in the subcommand, they are treated as a block in the summary table where changes in the equation are evaluated.

Extensive analysis of residuals is available. For example, a table of predicted scores and residuals can be requested and accompanied by a plot of standardized residuals against standardized predicted values of the DV (z scores of the Y' values). Plots of standardized residuals against sequenced cases are also available. For a sequenced file, one can request a Durbin-Watson statistic, which is used for a test of autocorrelation between adjacent cases. In addition, you can request Mahalanobis distance for the ten most extreme cases as a convenient way of evaluating outliers. This is the only program within the SPSS packages that offers Mahalanobis distance. Partial residual plots (partialing out all but one of the IVs) are available in SPSS REGRESSION.

The flexibility of input for the SPSS REGRESSION program does not carry over into output. The only difference between standard and statistical or sequential regression is in the printing of a progression of steps, and in a summary table available through the HISTORY instruction. Otherwise the statistics and parameter estimates are identical. These values, however, have different meanings, depending on the type of analysis. For example, you can request semipartial correlations (called PART COR) through the ZPP statistics. But these apply only to standard multiple regression. As pointed out in Section 5.6.1, semipartial correlations for statistical or sequential analysis appear in the summary table (obtained through the HISTORY statistic) as RSQ CHANGE.

TABLE 5.11   COMPARISON OF PROGRAMS FOR STANDARD MULTIPLE REGRESSION

Feature	SPSS REGRESSION	BMDP1R	SYSTAT REGRESS and GLM	SAS REG
Input				
Correlation matrix input	Yes	Yes	Yes	Yes
Covariance matrix input	Yes	Yes	Yes	Yes
Missing data options	Yes	Yes	No	No
Regression through the origin	Yes	No	Yes	Yes
Tolerance option	Yes	Yes	Yes	Yes
Test of equality of subsamples	No	Yes	No	No
Post hoc hypotheses[a]	Yes	No	Yes	Yes
Optional error terms	No	No	Yes	Yes
Weighted least squares	Yes	Yes	Yes	Yes
Multivariate multiple regression	No	No	No	Yes
Regression output				
Analysis of variance for regression	Yes	Yes	Yes	Yes
Multiple R	Yes	Yes	Yes	No
R^2	Yes	Yes	Yes	Yes
Adjusted R^2	Yes	Yes	Yes	Yes
Standard error of Y'	Standard Error	STD. ERROR OF EST.	Standard Error of Estimate	Root MSE
Coefficient of variation	No	No	No	Yes
Correlation matrix	Yes	Yes	No	Yes
Significance levels of correlation matrix	Yes[a]	No	No	No
Sum-of-squares and cross-products (SSCP) matrix	Yes	No	No	Yes
Inverse of SSCP matrix	No	No	No	Yes
Covariance matrix	Yes	Yes	No	No
Means and standard deviations	Yes	Yes	No	Yes
Matrix of correlation coefficients if some not computed	Yes	No	N.A.	N.A.
N for each correlation coefficient	Yes	No	N.A.	N.A.
Minimum and maximum of variables	No	Yes	No	No
Sum of squares for each variable	No	No	No	Yes
Coefficient of variation for each variable	No	Yes	No	No
Unstandarized regression coefficients	B	COEFFICIENT	Coefficient	Parameter Estimate
Standard error of regression coefficient	SE B	STD. ERROR	Std Error	Standard Error
F or t test of regression coefficient	T(F optional)	T	F	T
Intercept (constant)	Yes	Yes	Yes	Yes
Standardized regression coefficient	Beta	STD. REG COEFF	Std Coef	Standardized Estimate
Approx. standard error of β	Yes	No	No	No
Partial correlation	Yes	No	No	Yes[b]
Semipartial correlation or sr_i^2	Part Cor	No	No	Yes[b]
Tolerance	Yes	Yes	Yes	Yes
Variance inflation factors	Yes	No	No	Yes

TABLE 5.11   *(CONTINUED)*

Feature	SPSS REGRESSION	BMDP1R	SYSTAT REGRESS and GLM	SAS REG
Variance-covariance matrix for unstandarized *B* coefficients	Yes	No	No	Yes
Correlation matrix for unstandarized *B* coefficients	No	Yes	Yes	Yes
95% confidence interval for *B*	Yes	No	No	No
Sweep matrix	Yes	No	No	No
Eigenvalues	Yes	No	Yes	Yes
Condition indices	Yes	No	Yes	Yes
Variance proportions	Yes	No	Yes	Yes
Hypothesis matrices	No	No	Yes	Yes
Residuals				
Predicted scores, residuals and standardized residuals	Yes	Yes	Data file	Yes
95% confidence interval for predicted value	No	No	No	Yes
Plot of standardized residuals against predicted scores	Yes	Yes	No	Yes
Normal plot of residuals	Yes	Yes	No	No
Durbin-Watson statistic	Yes	Yes	Yes	Yes
Leverage diagnostics (e.g., Mahalanobis distance)	Yes	No	Data file	No
Influence diagnostics (e.g., Cook's distance)	Yes	No	Data file	Yes
Heteroscedasticity test	No	No	No	Yes
Histograms	Yes	No	No	No
Casewise plots	Yes	No	No	No
Partial plots	Yes	Yes	No	Yes
Other plots available	Yes	Yes	No	Yes
Summary statistics for residuals	Yes	No	No	Yes
Save predicted values/residuals	Yes	Yes	Yes	Yes

Note: BMDP2R (cf. Table 5.12) and SAS GLM (cf. Table 9.11) can also be used for standard multiple regression.

[a] Does *not* use Larzelere and Mulaik (1977) correction.

[b] Use Type II for standard MR.

TABLE 5.12   COMPARISON OF PROGRAMS FOR STEPWISE AND/OR SEQUENTIAL REGRESSION

Feature	SAS REG	SPSS REGRESSION	BMDP2R	SYSTAT REGRESS and GLM
Input				
Correlation matrix input	Yes	Yes	Yes	Yes
Covariance matrix input	Yes	Yes	Yes	Yes
Missing data options	No	Yes	No	No
Specify stepping algorithm	Yes	Yes	Yes	No
Specify tolerance	SINGULAR	Yes	Yes	Yes
Specify *F* to enter and/or remove	No	Yes	Yes	Yes
Specify probability of *F* to enter and/or remove	Yes	Yes	Yes	Yes

TABLE 5.12   *(CONTINUED)*

Feature	SAS REG	SPSS REGRESSION	BMDP2R	SYSTAT REGRESS and GLM
Specify maximum number of steps	No	Yes	Yes	No
Specify maximum number of variables	Yes	No	No	No
Regression through the origin	Yes	Yes	Yes	Yes
Force variables into equation	INCLUDE	ENTER	FORCE	FORCE
Weighted least squares	Yes	Yes	Yes	Yes
Specify order of entry (hierarchy)	No	Yes	Yes	No
IV sets for entry in single step	Yes	Yes	Yes	No
Interactive processing	Yes	No	Yes	Yes
Regression Output				
Expanded variable labels	No	Yes	No	No
Analysis of variance for regression, each step	Yes	Yes	Yes	No[a]
Multiple R, each step	No	Yes	Yes	No[a]
R^2, each step	Yes	Yes	Yes	No[a]
Standard error of Y', each step	Root MSE	Standard error	STD. ERROR OF EST.	No[a]
Adjusted R^2, each step	No	Yes	Yes	No[a]
Mallow's C_p, each step	Yes	No	No	No
Sum-of-squares and cross-products matrix	Yes	Yes	No	No
Correlation matrix	Yes	Yes	Yes	No
Correlation significance levels	No	Yes	No	No
Covariance matrix	No	Yes	Yes	No
Correlation matrix of regression coefficients	Yes	No	Yes	Yes
Covariance matrix of regression coefficients	Yes	Yes	No	No
Means and standard deviations	Yes	Yes	Yes	No
Matrix of correlation coefficients if some not computed	N.A.	Yes	No	N.A.
N for each correlation coeff.	N.A.	Yes	No	N.A.
Coefficients of variation	No	No	Yes	No
Minimums and maximums	No	No	Yes	No
Smallest and largest z	No	No	Yes	No
Case number for minimum and maximum values	No	No	Yes	No
Skewness and kurtosis	No	No	Yes	No
95% confidence interval for B	No	Yes	No	No
Sweep matrix	No	Yes	No	No
Condition number bounds	Yes	Yes	No	No
Eigenvalues	Yes	Yes	No	Yes
Condition indices	Yes	Yes	No	Yes
Variance proportions	Yes	Yes	No	Yes
Variables in equation (each step)				
Unstandardized regression coefficients	Parameter Estimate	B	COEFFICIENT	Coefficient
Standard error of regression coefficient	Standard Error	SE B	STD. ERROR OF COEFF	Std Error
95% confidence interval for B	No	Yes	No	No

TABLE 5.12   *(CONTINUED)*

Feature	SAS REG	SPSS REGRESSION	BMDP2R	SYSTAT REGRESS and GLM
Standard regression coefficient	Standardized Estimate	Beta	STD REG COEF	Std Coef
F (or T) to remove	F	T (F optional)	F TO REMOVE	F
p to remove	Yes	Yes	No	Yes
Intercept	INTERCEP	(Constant)	(Y-INTERCEPT)	CONSTANT
Variables not in equation (each step)				
Standardized regression coefficient for entering	No	Beta In	No	No
Partial correlation coefficient for entering	No	Partial	PARTIAL CORR.	Part. corr.
Tolerance	No	Yes	Yes	Yes
F (or T) to enter	No	T (F optional)	F TO ENTER	F
p to enter	No	Yes	No	'P'
Table of stepwise regression coefficients, F to enter and remove, and partial correlations	No	No	Yes	No
Summary table				
Multiple R	No	MultR	Yes	No
R^2	Yes	Rsq	MULTIPLE RSQ	No
Change in R^2 (squared semipartial correlation)	No	Rsqch	CHANGE IN RSQ	No
Adjusted R^2	No	AdjRsq	No	No
F to enter equation	F	F(Ch)	F TO ENTER	No
F to remove from equation	F	F(Ch)	F TO REMOVE	No
p to enter or remove	Prob.F	SigCh	No	No
Standard regression coefficient	No	BetaIn	No	No
Mallow's C_p	Yes	No	No	No
Bivariate correlation	No	Correl	No	No
Equation F and p at each step	No	F(Eqn)	No	No
Residuals				
Predicted scores, residuals, and standardized residuals	Yes	Yes	Yes	Data file
Plot of standardized residuals against predicted scores	Yes	Yes	Yes	No
Normal plot of residuals	No	Yes	Yes	No
Durbin-Watson statistic	Yes	Yes	Yes	Yes
Standardized error of predicted values	Yes	Yes	No	Data file
Leverage diagnostics (e.g., Mahalanobis distance)	No	Yes	Yes	Data file
Influence diagnostics (e.g., Cook's distance)	Yes	Yes	Yes	Data file
Casewise plots	No	Yes	Yes	No
Partial plots	Yes	Yes	Yes	No
Save predicted values/residuals	Yes	Yes	Yes	Yes
Misc. additional diagnostics	No	Yes	Yes	No
Other plots available	Yes	Yes	Yes	No

[a] Available by running separate standard multiple regressions for each step.

TABLE 5.13 COMPARISON OF PROGRAMS FOR SETWISE
REGRESSION

Feature	BMDP9R	SAS REG
Input		
Correlation matrix input	Yes	Yes
Covariance matrix input	Yes	Yes
SSCP matrix input	No	Yes
Specify tolerance	Yes	No
Specify model criterion	Yes	Yes
Regression through the origin	Yes	Yes
Include variables in all models	No	Yes
Weighted least squares	Yes	Yes
Specify penalty for additional variables	Yes	No
Specify number of subsets	Yes	Yes
Specify maximum subset size	Yes	Yes
Regression Output		
Means and standard deviations	Yes	Yes
Largest and smallest values	Yes	No
Skewness and kurtosis	Yes	No
Correlation matrix	Yes	Yes
Covariance matrix	Yes	No
Correlation of regression coefficients	Yes	Yes
Covariance of regression coefficients	Yes	Yes
Shaded, sorted correlation matrix	Yes	No
For each subset		
Variables in subset	Yes	Yes
R^2	Yes	Yes
Adjusted R^2	Yes	If chosen criterion
Mallow's C_p	Yes	If chosen criterion
Regression coefficients (B)	COEFFICIENT	No
T statistics for each variable	Yes	No
For best subset (additional to above)		
Standard errors of B	Yes	No
Standardized coefficients (β)	Yes	No
Tolerances	Yes	No
Contributions to R^2 (sr^2)	Yes	No
Multiple R	Yes	No
F statistics with df and significance	Yes	No
Residual mean square	Yes	No
Residual		
Histograms of standardized residuals	Yes	No
Plots of residuals against predicted scores	Yes	Yes
Normal plot of residuals	Yes	No
Plots of variables against residuals	Yes	Yes
Various other plots and diagnostics[a]	Yes	Yes

[a] See Tables 5.11 and 5.12.

5.7.2 ■ BMDP Series

The BMDP package (Dixon, 1992) offers separate programs for the various types of regression. All the programs do standard multiple regression, but the simplest applicable program is BMDP1R. For statistical and sequential regression, the program is BMDP2R. BMDP9R is available as a setwise program. All these programs exclude all cases with missing data, unless the values are estimated by, perhaps, some of the procedures available through BMDPAM or BMDP8D.

For standard multiple regression, BMDP1R offers analysis of "case combinations" or subsamples. Reduction of residuals due to grouping serves as a test of the equality of regression among the groups. Covariance and correlation matrices are available as output or can be used as input. In addition to tables of predicted values and residuals, the BMDP1R program allows a number of plots of residuals. Although not available in BMDP1R, Mahalanobis distance as well as other measures of extreme cases can be found using any one of several other BMDP programs, including BMDP2R.

The statistical/sequential regression program in the BMDP series, BMDP2R, combines several features of other statistical programs and adds some that are unavailable elsewhere. The user can specify the stepping algorithm (how the program adds or deletes variables in the equation as it progresses) in addition to specifying statistical criteria, as described in the BMDP manual (Dixon, 1992). Specifiable statistical criteria are tolerance, F to enter, F to remove, and maximum number of steps, as for SPSS.

At the end of the output for all steps, a table of stepwise regression coefficients is available with the full regression equation for each step, including regression coefficients for each IV if it were to enter at the next step.

BMDP2R has two ways of specifying sequential regression. Variables can be identified by their level: 0 = no entry at all, 1 = entry on the first step, etc., with more than one variable allowed to be given the same entry level. If this option is chosen, variables given the same entry level compete in a stepwise manner, and only one variable actually enters on each step. Or, sets can be formed and given names, with sets entered in specified sequence. With this option, more than one variable can enter on a step, and summary statistics are given for each set of entry variables.

In BMDP2R, the same input and output matrices are available as in BMDP1R. Residuals tables and plots, however, are far more extensive than BMDP1R. There are 21 diagnostic variables (including Mahalanobis distance) that can be printed or plotted, and 17 of these can be saved. For each plot, you can specify through options the number of extreme cases for which values are printed.

BMDP9R is the most comprehensive setwise program available. Setwise programs search among all possible subsets of variables for those subsets that are best. "Best" is defined by the user from among criteria such as maximum R^2. The researcher can specify the number of best subsets to be identified. Because of extensive output about the contribution of IVs to various subsets, BMDP9R is especially useful in identifying a combination of variables causing outlying cases to be unusual, as illustrated in Section 4.2.1.4. The extent of residuals analysis in BMDP9R is somewhere between that for BMDP1R and BMDP2R. A more extensive list of residuals and other diagnostics (including Mahalanobis distance) can be saved to a file that is printed through BMDP9R.

Cross validation is simplified in BMDP2R and BMDP9R by a case-weighting procedure, where cases to be omitted from computation of the regression equation are given a weight of zero.

5.7.3 ■ SAS System

Currently SAS REG is the all purpose regression program in the SAS system. (SAS Institute, Inc. 1990). Originally SAS had separate programs for different types of regression analysis. Standard multiple regression was handled through REG, statistical (though not sequential) through STEP-WISE, and setwise through RSQUARE—now all these varieties are handled through SAS REG.[8] In addition, GLM can be used for regression analysis; it is more flexible and powerful than any of the other SAS programs, but also more difficult to use.

SAS REG handles correlation, covariance, or SSCP matrix input but has no options for dealing with missing data. A case is deleted if it contains any missing values.

In SAS REG, two types of semipartial correlations are available. The sr_i^2 appropriate for standard multiple regression is the one that uses TYPE II (partial) sums of squares (which can also be printed). SAS REG also does multivariate multiple regression as a form of canonical correlation analysis where specific hypotheses can be tested.

For statistical regression using SAS REG, the usual forward, backward, and stepwise criteria for selection are available, in addition to two "switching" procedures. The statistical criteria available are the probability of F to enter and F to remove an IV from the equation as well as maximum number of variables. Interactive processing can be used to build sequential models.

Three criteria are available for setwise regression: Mallow's C_p, R^2, and adjusted R^2, as for BMDP9R. However, SAS REG differs in several ways from BMDP9R. No "best" subset is identified. You make the choice of the best subset by comparing values on your chosen criterion or on R^2, which is printed for all criteria. No information is given about individual IVs in the various subsets. As a result, the program is not useful for identifying the combination of IVs on which outlying cases differ. You can reduce output by specifying the number of subsets to be evaluated as well as minimum and maximum subset sizes. Sequential regression is handled interactively, in which an initial model statement is followed by instructions to add one or more variables at each subsequent step. Full output is available at the end of each step, but there is no summary table.

Tables of residuals and other diagnostics (including Cook's but not Mahalanobis distance) are extensive, but are saved to file rather than printed. However, a plotting facility within SAS REG allows extensive plotting of residuals. SAS REG also provides partial residual plots where all IVs except one are partialed out.

5.7.4 ■ SYSTAT System

Multiple regression in SYSTAT version 6 (Wilkinson & Hill, 1994) is done through GLM or REGRESS. In former versions of SYSTAT (e.g., Wilkinson, 1990), multiple regression was done through MGLH, a multivariate general linear program that handled ANOVA, MANOVA, canonical correlation, discriminant function analysis, and several types of regression. GLM is the current equivalent of MGLH, while REGRESS is more limited. However, REGRESS handles all multiple regression problems as well as GLM does.

Statistical regression options include forward (stepwise) and backward stepping. Options are also available to modify α level to enter and remove, change tolerance, and force the first k variables into the equation. By choosing INTERACTIVE mode for stepwise regression, you can specify individual variables to enter the equation at each step, allowing a simple form of sequential regression. Output for STEP regression (the subcommand that produces statistical or sequential

---

[8] In Version 6, a request for PROC STEPWISE or PROC RSQUARE will invoke SAS REG with the appropriate MODEL instructions.

regression) is limited, however. Although the regression equation is shown at each step, there is only one ANOVA table, at the end of the stepping process. Further, there is no summary table to show F-to-enter or remove at each step in a convenient form.

Matrix input is accepted in SYSTAT REGRESS or GLM, but processing of output of matrices or descriptive statistics requires the use of other programs in the SYSTAT package. Residuals and other diagnostics are handled by saving values to a file, which can then be printed out through SYSTAT BASIC or plotted through SYSTAT PLOT. In this way, you can take a look at Cook's value or leverage, from which Mahalanobis distance can be computed (Equation 4.3), for each case. Or, you can find summary values for residuals and diagnostics through SYSTAT STATS, the program for descriptive statistics.

5.8 ■ COMPLETE EXAMPLES OF REGRESSION ANALYSIS

To illustrate applications of regression analysis, variables are chosen from among those measured in the research described in Appendix B, Section B.1. Two analyses are reported here, both with number of visits to health professionals (TIMEDRS) as the DV and both using the SPSS REGRESSION program.

The first example is a standard multiple regression between the DV and number of physical health symptoms (PHYHEAL), number of mental health symptoms (MENHEAL), and stress from acute life changes (STRESS). From this analysis, one can assess the degree of relationship between the DV and IVs, the proportion of variance in the DV predicted by regression, and the relative importance of the various IVs to the solution.

The second example demonstrates sequential regression with the same DV and IVs. The first step of the analysis is entry of PHYHEAL to determine how much variance in number of visits to health professionals can be accounted for by differences in physical health. The second step is entry of STRESS to determine if there is a significant increase in R^2 when differences in stress are added to the equation. The final step is entry of MENHEAL to determine if differences in mental health are related to number of visits to health professionals after differences in physical health and stress are statistically accounted for.

5.8.1 ■ Evaluation of Assumptions

Because both analyses use the same variables, this screening is appropriate for both.

5.8.1.1 ■ Ratio of Cases to IVs

With 465 respondents and 3 IVS, the number of cases is well above the minimum requirement of 107 (104 + 3) for testing individual predictors in standard multiple regression. There are no missing data.

5.8.1.2 ■ Normality, Linearity, Homoscedasticity, and Independence of Residuals

We choose for didactic purposes to conduct preliminary screening through residuals. The initial run through SPSS REGRESSION uses untransformed variables in a standard multiple regression to produce the scatterplot of residuals against predicted DV scores that appears in Figure 5.5.

Notice the execrable overall shape of the scatterplot that indicates violation of many of the assumptions of regression. Comparison of Figure 5.5 with Figure 5.2(a) (in Section 5.3.2.4) suggests further analysis of the distributions of the variables. (It was noticed in passing, although we tried not to look, that R^2 for this analysis was significant, but only .22).

SPSS FREQUENCIES is used to examine the distributions of the variables, as seen in Table 5.14. All the variables have significant positive skewness (see Chapter 4), which explains, at least in part, the problems in the residuals scatterplot. Logarithmic and square root transformations are applied as appropriate, and the transformed distributions checked once again for skewness. Thus TIMEDRS and PHYHEAL (with logarithmic transformations) become LTIMEDRS and LPHYHEAL, while STRESS (with a square root transformation) becomes SSTRESS.[9] In the case of MENHEAL, application of the milder square root transformation makes the variable significantly negatively skewed, so no transformation is undertaken.

TITLE	LARGE SAMPLE REGRESSION-RESIDUALS.
DATA LIST	FILE = 'REGRESS.DAT' FREE
	/SUBJNO, TIMEDRS, PHYHEAL, MENHEAL, STRESS.
VAR LABELS	TIMEDRS 'VISITS TO HEALTH PROFESSIONALS'.
REGRESSION	VARS = TIMEDRS TO STRESS/
	DEPENDENT = TIMEDRS/METHOD=ENTER/
	SCATTERPLOT (*RESID, *PRED).

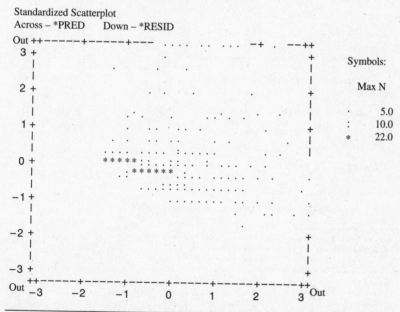

FIGURE 5.5 SPSS REGRESSION SETUP AND RESIDUALS SCATTERPLOT FOR ORIGINAL VARIABLES.

[9] Note the DV (TIMEDRS) is transformed to meet the assumptions of multiple regression. Transformation of the IVs is undertaken to enhance prediction.

TABLE 5.14	SETUP AND OUTPUT FOR EXAMINING DISTRIBUTIONS OF VARIABLES THROUGH SPSS FREQUENCIES

```
TITLE                      LARGE SAMPLE REGRESSIONFREQUENCIES.
DATA LIST                  FILE='REGRESS.DAT'    FREE
                           /SUBJNO,TIMEDRS,PHYHEAL,MENHEAL,STRESS.
VAR LABELS                 TIMEDRS 'VISITS TO HEALTH PROFESSIONALS'
FREQUENCIES                VARS = TIMEDRS TO STRESS/
                           HISTOGRAM/  FORMAT = NOTABLE/
                           STATISTICS = ALL.
```

TIMEDRS   VISITS TO HEALTH PROFESSIONALS

```
   Count   Midpoint   One symbol equals approximately  4.00 occurrences

      86         0  |**********************
     187         4  |*************************************************
      74         8  |*******************
      44        12  |***********
      27        16  |*******
      13        20  |***
       7        24  |**
       6        28  |**
       3        32  |*
       4        36  |*
       2        40  |*
       1        44  |
       1        48  |
       2        52  |*.
       3        56  |*
       3        60  |*
       0        64  |
       0        68  |
       0        72  |
       1        76  |
       1        80  |
                    +----+----+----+----+----+----+----+----+----+----+
                    0        40       80       120      160      200
                              Histogram frequency
```

Mean	7.901	Std err	.508	Median	4.000
Mode	2.000	Std dev	10.948	Variance	119.870
Kurtosis	13.101	S E Kurt	.226	Skewness	3.248
S E Skew	.113	Range	81.000	Minimum	.000
Maximum	81.000	Sum	3674.000		

Valid cases   465   Missing cases   0

PHYHEAL

```
   Count   Value   One symbol equals approximately  2.00 occurrences

      59      2.00  |******************************
      93      3.00  |***********************************************
      78      4.00  |***************************************
      69      5.00  |**********************************
      64      6.00  |********************************
      33      7.00  |*****************
      29      8.00  |***************
      18      9.00  |*********
       8     10.00  |****
       6     11.00  |***
       4     12.00  |**
       2     13.00  |*
       1     14.00  |*
       1     15.00  |*
                    +----+----+----+----+----+----+----+----+----+----+
                    0        20       40       60       80       100
                              Histogram frequency
```

TABLE 5.14   *(CONTINUED)*

Mean	4.972	Std err	.111	Median	5.000
Mode	3.000	Std dev	2.388	Variance	5.704
Kurtosis	1.124	S E Kurt	.226	Skewness	1.031
S E Skew	.113	Range	13.000	Minimum	2.000
Maximum	15.000	Sum	2312.000		

Valid cases   465   Missing cases   0

MENHEAL

```
    Count     Value    One symbol equals approximately  1.00 occurrence

       27      .00 |****************************
       41     1.00 |*****************************************
       35     2.00 |***********************************
       37     3.00 |*************************************
       46     4.00 |**********************************************
       45     5.00 |*********************************************
       48     6.00 |**********************************************
       33     7.00 |*********************************
       31     8.00 |*******************************
       20     9.00 |********************
       24    10.00 |************************
       15    11.00 |***************
       23    12.00 |***********************
       13    13.00 |*************
        6    14.00 |******
        9    15.00 |*********
        7    16.00 |*******
        2    17.00 |**
        3    18.00 |***
                   +----+----+----+----+----+----+----+----+----+----+
                   0        10        20        30        40        50
                             Histogram frequency
```

Mean	6.123	Std err	.194	Median	6.000
Mode	6.000	Std dev	4.194	Variance	17.586
Kurtosis	.292	S E Kurt	.226	Skewness	.602
S E Skew	.113	Range	18.000	Minimum	.000
Maximum	18.000	Sum	2847.000		

Valid cases   465   Missing cases   0

TABLE 5.14   (CONTINUED)

STRESS

Count   Midpoint   One symbol equals approximately 1.50 occurrences

```
   36        20  |************************
   63        64  |*********************************************
   59       108  |******************************************
   65       152  |*********************************************
   53       196  |***********************************
   48       240  |********************************
   41       284  |***************************
   38       328  |*************************
   21       372  |**************
   16       416  |***********
    6       460  |****
    5       504  |***
    8       548  |*****
    3       592  |**
    1       636  |*
    0       680  |
    1       724  |*
    0       768  |
    0       812  |
    0       856  |
    1       900  |*
                 +----+----+----+----+----+----++++++
                 0    15   30       45      60      75
                       Histogram frequency
```

Mean	204.217	Std err	6.297	Median	178.000
Mode	.000	Std dev	135.793	Variance	18439.662
Kurtosis	1.801	S E Kurt	.226	Skewness	1.043
S E Skew	.113	Range	920.000	Minimum	.000
Maximum	920.000	Sum	94961.000		

Valid cases   465   Missing cases   0

Table 5.15 shows output from FREQUENCIES for one of the transformed variables, LPHY-HEAL, the worst prior to transformation. Transformations similarly reduce skewness in the other two transformed variables.

The residuals scatterplot from SPSS REGRESSION following regression with the transformed variables appears as Figure 5.6. Notice that, although the scatterplot is still not perfectly rectangular, its shape is considerably improved over that in Figure 5.5.

5.8.1.3 ■ Outliers

Multivariate outliers are sought using the transformed IVs as part of an SPSS REGRESSION run in which the Mahalanobis distance of each case to the centroid of all cases is computed. The ten cases with the largest distance are printed (see Table 5.16). Mahalanobis distance is distributed as a chi square (χ^2) variable, with degrees of freedom equal to the number of IVs. To determine which cases are multivariate outliers, one looks up critical χ^2 at the desired alpha level (Table C.4). In this case, critical χ^2 at $\alpha = .001$ for 3 df is 16.266. Any case with a value larger than 16.266 in the *MAHAL column of the output is a multivariate outlier among the IVs. None of the cases has a value in excess of 16.266. (If outliers are found, the procedures detailed in Chapter 4 are followed to reduce their influence.)

Note that Figure 5.6 shows no outliers in the solution; none of the standardized residuals exceeds ± 3.29.

TABLE 5.15	SETUP AND OUTPUT FOR EXAMINING DISTRIBUTION OF TRANSFORMED VARIABLE THROUGH SPSS FREQUENCIES

```
TITLE           LARGE SAMPLE REGRESSIONFREQUENCIES (TRANSFORMED).
DATA LIST       FILE='REGRESS.DAT'     FREE
                    /SUBJNO,TIMEDRS,PHYHEAL,MENHEAL,STRESS.
VAR LABELS      TIMEDRS 'VISITS TO HEALTH PROFESSIONALS'.
COMPUTE         LTIMEDRS = LG10(TIMEDRS + 1).
COMPUTE         LPHYHEAL = LG10(PHYHEAL).
COMPUTE         SSTRESS = SQRT(STRESS).
FREQUENCIES     VARS = LTIMEDRS, LPHYHEAL, SSTRESS/
                HISTOGRAM/  FORMAT = NOTABLE/
                STATISTICS = ALL.
```

LTIMEDRS

```
   Count    Midpoint    One symbol equals approximately  1.50 occurrences

     42        .0   |****************************
      0        .1   |
      0        .2   |
     44        .3   |*****************************
      0        .4   |
     67        .5   |*********************************************
     52        .6   |***********************************
     41        .7   |***************************
     53        .8   |***********************************
     25        .9   |****************
     34       1.0   |***********************
     33       1.1   |**********************
     24       1.2   |****************
     16       1.3   |***********
     11       1.4   |*******
      6       1.5   |****
      6       1.6   |****
      3       1.7   |**
      6       1.8   |****
      2       1.9   |*
      0       2.0   |
                    +----+----+----+----+----+----+----+----+----+----+
                    0        15        30        45        60        75
                              Histogram frequency
```

Mean	.741	Std err	.019	Median	.699
Mode	.477	Std dev	.415	Variance	.172
Kurtosis	-.177	S E Kurt	.226	Skewness	.228
S E Skew	.113	Range	1.914	Minimum	.000
Maximum	1.914	Sum	344.698		

Valid cases	465	Missing cases	0

5.8.1.4 ■ Multicollinearity and Singularity

The REGRESSION run of Table 5.17 resolves doubts about possible multicollinearity and singularity among the transformed IVs. All variables enter the equation without violating the default value for tolerance (cf. Chapter 4). Further, the highest correlation among the IVs, between MENHEAL and LPHYHEAL, is .511. (If multicollinearity is indicated, SMCs between each IV serving as DV with the other two as IVs are found through REGRESSION or any other appropriate program and redundant IVs dealt with as discussed in Chapter 4.)

Standardized Scatterplot
Across – *PRED   Down – *RESID

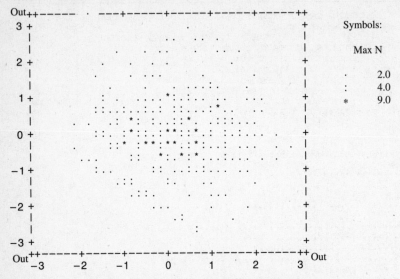

FIGURE 5.6   RESIDUALS SCATTERPLOT FOLLOWING REGRESSION
WITH TRANSFORMED VARIABLES. OUTPUT FROM
SPSS REGRESSION. SEE TABLE 5.16 FOR SETUP.

TABLE 5.16	SETUP AND OUTPUT FROM SPSS REGRESSION SHOWING MULTIVARIATE OUTLIERS
TITLE	REGRESSION-OUTLIERS & RESIDUALS.
DATA LIST	FILE='REGRESS.DAT' FREE
	/SUBJNO,TIMEDRS,PHYHEAL,MENHEAL,STRESS.
VAR LABELS	TIMEDRS 'VISITS TO HEALTH PROFESSIONALS'.
COMPUTE	LTIMEDRS = LG10(TIMEDRS + 1).
COMPUTE	LPHYHEAL = LG10(PHYHEAL).
COMPUTE	SSTRESS = SQRT(STRESS).
REGRESSION	VARS = MENHEAL, LTIMEDRS TO SSTRESS/
	STATISTICS = DEFAULTS,ZPP,CI,F/
	DEPENDENT = LTIMEDRS/METHOD=ENTER/
	RESIDUALS = OUTLIERS(MAHAL)/
	SCATTERPLOT (*RESID,*PRED).

Outliers - Mahalanobis' Distance

Case #	*MAHAL
403	14.13530
125	11.64940
198	10.56870
52	10.54759
446	10.22483
159	9.35066
33	8.62835
280	8.58686
405	8.43122
113	8.35261

TABLE 5.17 STANDARD MULTIPLE REGRESSION ANALYSIS OF LTIMEDRS (THE DV) WITH MENHEAL, SSTRESS, AND LPHYHEAL (THE IVS). SELECTED OUTPUT. SETUP SHOWN IN TABLE 5.16

Listwise Deletion of Missing Data

Equation Number 1   Dependent Variable..   LTIMEDRS

Block Number 1. Method: Enter

Variable(s) Entered on Step Number
1..   SSTRESS
2..   LPHYHEAL
3..   MENHEAL

Multiple R	.61382
R Square	.37678
Adjusted R Square	.37272
Standard Error	.32888

Analysis of Variance

	DF	Sum of Squares	Mean Square
Regression	3	30.14607	10.04869
Residual	461	49.86410	.10817

F =   92.90143   Signif F = .0000

------------------ Variables in the Equation ------------------

Variable	B	SE B	95% Confdnce Intrvl B	Beta	Correl	Part Cor	Partial	F	Sig F
SSTRESS	.015710	.003363	.009100 .022319	.188105	.359048	.171733	.212565	21.815	.0000
LPHYHEAL	1.039970	.087184	.868643 1.211296	.516407	.585748	.438587	.485649	142.289	.0000
MENHEAL	.001884	.004401	-.006766 .010533	.019022	.355130	.015735	.019928	.183	.6689
(Constant)	-.155039	.058256	-.269519 -.040559					7.083	.0081

End Block Number   1   All requested variables entered.

5.8.2 ■ Standard Multiple Regression

SPSS REGRESSION is used to compute a standard multiple regression between LTIMEDRS (the transformed DV), and MENHEAL, LPHYHEAL, and SSTRESS (the IVs).

The output of this REGRESSION run appears as Table 5.17. Included are descriptive statistics, the values of R, R^2, and adjusted R^2, and a summary of the analysis of variance for regression. The significance level for R is found in the source table with $F(3, 461) = 92.90$, $p < .001$. In the portion labeled Variables in the Equation are printed unstandardized and standardized regression coefficients with their significance levels and 95% confidence intervals, and three correlations: total (Correl), semipartial (Part Cor), and partial.

The significance levels for the regression coefficients (see Table 5.17, Variables in the Equation) are assessed through F statistics or through confidence intervals. F statistics are evaluated against 461 df. Only two of the IVs, SSTRESS and LPHYHEAL, contribute significantly to regression with Fs of 26.82 and 142.29, respectively. The significance of SSTRESS and LPHYHEAL is confirmed by their 95% confidence intervals which do not include zero as a possible value.

Semipartial correlations (cf. Equation 5.9) are labeled Part Cor in the Variables in the Equation section. These values, when squared, indicate the amount by which R^2 would be reduced if an IV were omitted from the equation. The sum for the two significant IVs ($.171732^2 + .438592^2 = .2219$) is the amount of R^2 attributable to unique sources. The difference between R^2 and unique variance ($.37678 - .2219 = .1548$) represents variance that SSTRESS, LPHYHEAL, and MENHEAL jointly contribute to R^2.

Information from this analysis is summarized in Table 5.18, in a form that might be appropriate for publication in a professional journal.

It is noted from the correlation matrix in Table 5.17 that MENHEAL correlates with LTIMDRS ($r = .355$) but does not contribute significantly to regression. If Equation 5.11 is used post hoc to evaluate the significance of the correlation coefficient,

$$F = \frac{(.355)^2/3}{(1 - .355^2)/(465 - 3 - 1)} = 22.16$$

the correlation between MENHEAL and LTIMDRS differs reliably from zero; $F(3, 461) = 22.16$, $p < .01$.

Thus, although the bivariate correlation between MENHEAL and LTIMEDRS is reliably different from zero, the relationship seems to be mediated by, or redundant to, the relationship between LTIMDRS and other IVs in the set. Had the researcher measured only MENHEAL and LTIMEDRS, however, the significant correlation might have led to stronger conclusions than are warranted about the relationship between mental health and number of visits to health professionals.

Table 5.19 contains a checklist of analyses performed with standard multiple regression. An example of a Results section in journal format appears next.

TABLE 5.18 STANDARD MULTIPLE REGRESSION OF HEALTH AND STRESS VARIABLES ON NUMBER OF VISITS TO HEALTH PROFESSIONALS

Variables	LTIMEDRS (DV)	LPHYHEAL	SSTRESS	MENHEAL	B	β	sr^2 (unique)
LPHYHEAL	.59				1.040**	0.52	.19
SSTRESS	.36	.32			0.016**	0.19	.03
MENHEAL	.36	.51	.38		0.002	0.02	
				Intercept = −0.155			
Means	0.74	0.65	13.40	6.12			
Standard deviations	0.42	0.21	4.97	4.19		$R^2 = .38^a$	
						Adjusted $R^2 = .37$	
						$R = .61**$	

**$p < .01$.

[a] Unique variability = .22; shared variability = .16.

TABLE 5.19 CHECKLIST FOR STANDARD MULTIPLE REGRESSION

1. Issues

    a. Ratio of cases to IVs
    b. Normality, linearity, and homoscedasticity of residuals
    c. Outliers
    d. Multicollinearity and singularity
    e. Outliers in the solution

2. Major analyses

    a. Multiple R^2, F ratio
    b. Adjusted multiple R^2, overall proportion of variance accounted for
    c. Significance of regression coefficients
    d. Squared semipartial correlations

3. Additional analyses

    a. Post hoc significance of correlations
    b. Unstandardized (B) weights, confidence limits
    c. Standardized (β) weights
    d. Unique versus shared variability
    e. Suppressor variables
    f. Prediction equation

Results

A standard multiple regression was performed between number of visits to health professionals as the dependent variable and physical health, mental health, and stress as independent variables. Analysis was performed using SPSS REGRESSION and SPSS FREQUENCIES for evaluation of assumptions.

Results of evaluation of assumptions led to transformation of the variables to reduce skewness, reduce the number of outliers, and improve the normality, linearity, and homoscedasticity of residuals. A square root transformation was used on the measure of stress. Logarithmic transformations were used on number of visits to health professionals and on physical health. One IV, mental health, was positively skewed without transformation and negatively skewed with it; it was not transformed. With the use of a $p < .001$ criterion for Mahalanobis distance no outliers among the cases were found. No cases had missing data and no suppressor variables were found, $N = 465$.

Table 5.18 displays the correlations between the variables, the unstandardized regression coefficients (B) and intercept, the standardized regression coefficients (β), the semipartial correlations (sr_i^2) and R^2, and adjusted R^2. R for regression was significantly different from zero, $F(3, 461) = 92.90$, $p < .001$. For the two regression coefficients that differed significantly from zero, 95% confidence limits were calculated. The confidence limits for (square root of) stress were 0.0091 to 0.0223, and those for (log of) physical health were 0.8686 to 1.2113.

Only two of the IVs contributed significantly to prediction of number of visits to health professionals as logarithmically transformed, (log of) physical health scores ($sr_i^2 = .19$) and (square root of) acute stress scores ($sr_i^2 = .03$). The three IVs in combination contributed another .15 in shared variability. Altogether, 38% (37% adjusted) of the variability in visits to health professionals was predicted by knowing scores on these three IVs.

Although the correlation between (log of) visits to health professionals and mental health was .36, mental health did not contribute significantly to regression. Post hoc evaluation of the correlation revealed that it was significantly different from zero, $F(3, 461) = 22.16$, $p < .01$. Apparently the relationship between the number of visits to health professionals and mental health is mediated by the relationships between physical health, stress, and visits to health professionals.

5.8.3 ■ Sequential Regression

The second example involves the same three IVs entered one at a time in an order determined by the researcher. LPHYHEAL is the first IV to enter, followed by SSTRESS and then MENHEAL. The main research question is whether information regarding differences in mental health can be used to predict visits to health professionals after differences in physical health and in acute stress are statistically eliminated. In other words, do people go to health professionals for more numerous mental health symptoms if they have physical health and stress similar to other people?

Table 5.20 shows setup and selected portions of the output for sequential analysis using the SPSS REGRESSION program. Notice that a complete regression solution is provided at the end of each step. The significance of the bivariate relationship between LTIMEDRS and LPHYHEAL is assessed at the end of step 1, $F(1, 463) = 241.83$, $p < .001$. The bivariate correlation is .59, accounting for 34% of the variance. After step 2, with both LPHYHEAL and SSTRESS in the equation, $F(2, 462) = 139.51$, $p < .01$, $R = .61$, and $R^2 = .38$. With the addition of MENHEAL, $F(3, 461) = 92.90$, $R = .61$, and $R^2 = .38$. Increments in R^2 at each step are read directly from the RsqCh column of the Summary table. Thus, $sr^2_{LPHYHEAL} = .34$ $sr^2_{SRAHE} = .03$, and $sr^2_{MENHEAL} = .00$

Using procedures of section 5.6.2.3, significance of the addition of SSTRESS to the equation is indicated in the output at Step Number 2 (the step where SSTRESS entered) as F for SSTRESS in the segment labeled Variables in the Equation. Because the F value of 24.772 exceeds critical F with 1 and 461 df (df_{res} at the end of analysis), SSTRESS is making a significant contribution to the equation at this step.

Similarly, significance of the addition of MENHEAL to the equation is indicated at Step Number 3, where the F for MENHEAL is .183. Because this F value does not exceed critical F with 1 and 461 df, MENHEAL is not significantly improving R^2 at its point of entry.

The significance levels of the squared semipartial correlations are also available in the summary table as FCh, with probability value SigCh for evaluating the significance of the added IV.

Thus there is no reliable increase in prediction of LTIMEDRS by addition of MENHEAL to the equation if differences in LPHYHEAL and SSTRESS are already accounted for. Apparently the answer is "no" to the question: Do people with numerous mental health symptoms go to health professionals more often if they have physical health and stress similar to others? A summary of information from this output appears in Table 5.21.

Table 5.22 is a checklist of items to consider in sequential regression. An example of a Results section in journal format appears next.

TABLE 5.20   SETUP AND SELECTED OUTPUT FOR SPSS SEQUENTIAL REGRESSION

```
TITLE         LARGE SAMPLE SEQUENTIAL REGRESSION.
DATA LIST     FILE='REGRESS.DAT'   FREE
              /SUBJNO,TIMEDRS,PHYHEAL,MENHEAL,STRESS.
VAR LABELS    TIMEDRS 'VISITS TO HEALTH PROFESSIONALS'.
COMPUTE       LTIMEDRS = LG10(TIMEDRS + 1).
COMPUTE       LPHYHEAL = LG10(PHYHEAL).
COMPUTE       SSTRESS = SQRT(STRESS).
REGRESSION    VARS = MENHEAL, LTIMEDRS TO SSTRESS/
              STATISTICS = DEFAULTS, HISTORY, CI, F/
              DEPENDENT = LTIMEDRS/METHOD=ENTER LPHYHEAL
                                  /METHOD=ENTER SSTRESS
                                  /METHOD=ENTER MENHEAL.
```

Listwise Deletion of Missing Data

Equation Number 1   Dependent Variable..   LTIMEDRS

Block Number   1.   Method: Enter   LPHYHEAL

Variable(s) Entered on Step Number   1...   LPHYHEAL

		Analysis of Variance			
			DF	Sum of Squares	Mean Square
Multiple R	.58575				
R Square	.34310	Regression	1	27.45151	27.45151
Adjusted R Square	.34168	Residual	463	52.55866	.11352
Standard Error	.33692				
		F = 241.82595 Signif F = .0000			

------------------ Variables in the Equation ------------------

Variable	B	SE B	95% Confdnce Intrvl B		Beta	F	Sig F
LPHYHEAL	1.179612	.075856	1.030548	1.328676	.585748	241.826	.0000
(Constant)	-.023545	.051605	-.124953	.077864		.208	.6484

----------- Variables not in the Equation -----------

Variable	Beta In	Partial	Min Toler	F	Sig F
MENHEAL	.075284	.079823	.738503	2.963	.0859
SSTRESS	.192776	.225590	.899564	24.772	.0000

End Block Number   1   All requested variables entered.

186

TABLE 5.20 *(CONTINUED)*

```
Equation Number 1    Dependent Variable..    LTIMEDRS

Block Number  2.  Method:  Enter      SSTRESS

Variable(s) Entered on Step Number  2..     SSTRESS

Multiple R           .61362        Analysis of Variance
R Square             .37653                          DF   Sum of Squares   Mean Square
Adjusted R Square    .37383        Regression         2         30.12626      15.06313
Standard Error       .32859        Residual         462         49.88391        .10797

                                   F =    139.50724      Signif F =  .0000


------------------ Variables in the Equation ------------------

Variable           B         SE B    95% Confdnce Intrvl B      Beta        F      Sig F

LPHYHEAL      1.055578     .078001     .903297   1.209858     .524654   183.486    .0000
SSTRESS        .016100     .003235     .009743    .022456     .192776    24.772    .0000
(Constant)    -.159502     .057264    -.272033   -.046971               7.758    .0056


---------- Variables not in the Equation ----------

Variable    Beta In    Partial   Min Toler       F     Sig F

MENHEAL     .019022    .019928    .684265      .183    .6689


End Block Number   2   All requested variables entered.


Equation Number 1    Dependent Variable..    LTIMEDRS

Block Number  3.  Method:  Enter      MENHEAL
```

TABLE 5.20 *(CONTINUED)*

Variable(s) Entered on Step Number 3.. MENHEAL

Multiple R	.61382
R Square	.37678
Adjusted R Square	.37272
Standard Error	.32888

Analysis of Variance

	DF	Sum of Squares	Mean Square
Regression	3	30.14607	10.04869
Residual	461	49.86410	.10817

F = 92.90143   Signif F = .0000

------------------ Variables in the Equation ------------------

Variable	B	SE B	95% Confdnce Intrvl B		Beta	F	Sig F
LPHYHEAL	1.033970	.087184	.868643	1.211296	.516407	142.289	.0000
SSTRESS	.015710	.003363	.009100	.022319	.188105	21.815	.0000
MENHEAL	.001884	.004401	-.006766	.010533	.019022	.183	.6689
(Constant)	-.155039	.058256	-.269519	-.040559		7.083	.0081

End Block Number   3   All requested variables entered.

* *

Summary table

Step	MultR	Rsq	AdjRsq	F(Eqn)	SigF	RsqCh	FCh	SigCh		Variable	BetaIn	Correl
1	.5857	.3431	.3417	241.826	.000	.3431	241.826	.000	In:	LPHYHEAL	.5857	.5857
2	.6136	.3765	.3738	139.507	.000	.0334	24.772	.000	In:	SSTRESS	.1928	.3590
3	.6138	.3768	.3727	92.901	.000	.0002	.183	.669	In:	MENHEAL	.0190	.3551

TABLE 5.21 SEQUENTIAL REGRESSION OF HEALTH AND STRESS VARIABLES ON NUMBER OF VISITS TO HEALTH PROFESSIONALS

Variables	LTIMEDRS (DV)	LPHYHEAL	SSTRESS	MENHEAL	B	β	sr^2 (incremental)
LPHYHEAL	.59				1.040	0.52	.34**
SSTRESS	.36	.32			0.016	0.19	.03**
MENHEAL	.36	.51	.38		0.002	0.02	.00
				Intercept = −0.155			
Means	0.74	0.65	13.40	6.12			
Standard deviation	0.42	0.21	4.97	4.19			

$R^2 = .38$

Adjusted $R^2 = .37$

$R = .61$**

**$p < .01$.

189

TABLE 5.22   CHECKLIST FOR SEQUENTIAL REGRESSION ANALYSIS

1. Issues

    a. Ratio of cases to IVs
    b. Normality, linearity, and homoscedasticity of residuals
    c. Outliers
    d. Multicollinearity and singularity
    e. Outliers in the solution

2. Major analyses

    a. Multiple R^2, F ratio
    b. Adjusted R^2, proportion of variance accounted for
    c. Squared semipartial correlations
    d. Significance of regression coefficients
    e. Incremental F

3. Additional analyses

    a. Unstandardized (B) weights, confidence limits
    b. Standardized (β) weights
    c. Prediction equation from stepwise analysis
    d. Post hoc significance of correlations
    e. Suppressor variables
    f. Cross-validation (stepwise and setwise)

Results

Sequential regression was employed to determine if addition of information regarding stress and then mental health symptoms improved prediction of visits to health professionals beyond that afforded by differences in physical health. Analysis was performed using SPSS REGRESSION and SPSS FREQUENCIES for evaluation of assumptions.

These results led to transformation of the variables to reduce skewness, reduce the number of outliers, and improve the normality, linearity, and homoscedasticity of residuals. A square root transformation was used on the measure of stress. Logarithmic transformations were used on the number of visits to health professionals and physical health. One IV, mental health, was positively skewed without transformation and negatively skewed with it; it was not transformed. With the use of a $p < .001$ criterion for Mahalanobis distance no outliers among the cases were identified. No cases had missing data and no suppressor variables were found, $N = 465$.

Table 5.21 displays the correlations between the variables, the unstandardized regression coefficients (B) and intercept, the standardized regression coefficients (β), the semipartial correlations (sr_i^2), and R, R^2, and adjusted R^2 after entry of all three IVs. R was significantly different from zero at the end of each step. After step 3, with all IVs in the equation, $R = .61$, $F(3, 461) = 92.90$, $p < .01$.

After step 1, with log of physical health in the equation, $R^2 = .34$, $F_{inc}(1, 461) = 241.83$, $p < .001$. After step 2, with square root of stress added to prediction of (log of) visits to health professionals by (log of) physical health, $R^2 = .38$, $F_{inc}(1, 461) = 24.77$, $p < .01$. Addition of square root of stress to the equation with physical health results in a significant increment in R^2 After step 3, with mental health added to prediction of visits by (log of) physical health and square root of stress, $R^2 = .38$ (adjusted $R^2 = .37$), $F_{inc}(1, 461) = 0.18$. Addition of mental health to the equation did not reliably improve R^2.

5.9 ■ SOME EXAMPLES FROM THE LITERATURE

The following examples have been included to give you a taste for the ways in which regression can be, and has been, used by researchers.

Multiple regression is the statistical tool used in judgment analysis, a procedure developed to identify important variables underlying policy decisions. In a demonstration involving diagnosis of learning disabilities, Beatty (1977) gave diagnosticians (judges) test profiles for a sample of children. Judges ranked the children in order of severity of learning disability, and these rankings served as the DV.[10] Beatty used as IVs the age of the child and the set of five test scores on which judges had based their rankings.

Judges were analyzed individually, and then in combination, so that judges using similar policies could be grouped. For each group of judges with a distinct policy, it was possible to evaluate how successfully the rankings could be predicted and the relative importance of IVs in influencing the diagnostic rankings. In Beatty's demonstration, the three scores on the Wechsler Intelligence Scale for Children (Verbal, Performance, and Full Scale)[11] contributed the most to multiple correlation for all judges. Some judges seemed to have more predictable policies than others; for one judge, 96% of the variance in rankings was explained by knowledge of the IVs.

In evaluating a negative income tax project in New Jersey, Nicholson and Wright (1977) argue for inclusion of "participant's understanding of the treatment" as one of the IVs in policy experimentation. At each stage of their analysis, separate multiple regression equations were set up for each of six DVs (all related to a family's earnings and hours worked). (These six DVs could have been combined into a single canonical correlation evaluated against the IVs; see Chapter 6.) Separate equations were also run on the basis of ethnicity. IVs for all equations included (1) pretreatment value of the DV; (2) variables related to family age, education, and size; (3) variables related to site and whether or not families received welfare; (4) a dummy variable indicating whether a family was in the treatment or the control group; and (5) variables related to specification of which experimental plan a family was in, with eight combinations of guaranteed income and tax levels.

As the main focus of the evaluation, the regression coefficient for the difference between experimental and control groups was evaluated for statistical significance in each of the regression equations. Then Nicholson and Wright reran the regression equations with 11 binary variables related to participant's knowledge about the tax plan. Finally, a third set of equations was run that included terms related to interactions between knowledge levels and estimated effects of the guarantee and tax levels. It was found that the multiple correlation increased significantly with the addition of variables representing participant knowledge, and increased again when interaction variables were included.

In general, estimates of disincentive effects for families with high understanding were larger than for those who lacked understanding of the system. That is, among families who understood the system, those in the experimental group worked fewer hours and earned less than those in the control group. The difference between experimental and control was less, and in some cases in the opposite direction, for families who showed little understanding of the policy.

---

[10] Opinions differ regarding application of regression to rank order data. However, since rank order data produce rectangular distributions with neither skewness nor outliers, the application may be considered justified.

[11] It is not usually a good idea to include the Full Scale score along with both Verbal and Performance scores in the same analysis, since this could lead to multicollinearity.

S. Fidell (1978) applied stepwise regression procedures to predict annoyance among groups of people. Respondents were interviewed at 24 sites across the United States, stratified by noise exposure and population density. The proportion of respondents at each site who had been highly annoyed by neighborhood noise exposure was predicted from both situational (demographic and physical) variables and attitudinal variables. Since the prevalence of shared attitudes in the community was of greater interest than the intensity of individual beliefs, the unit of analysis was the site rather than the individual respondent. Primary interest in prediction also dictated stepwise entry of variables, many of which were available from census data. Excellent prediction was available from several combinations of variables, with one equation of 3 IVs accounting for 88% of the variance in community annoyance at exposure to noise.

Species density as a function of environmental measures in the United States was studied by Schall and Pianka (1978) in statistical regression analysis. Separate analyses were run with number of birds and number of lizards as DVs. The 11 IVs were the same for both analyses, consisting of such environmental variables as highest and lowest elevation, sunfall, average annual precipitation and several temperature measures. Units of measurement (cases) were 895 quadrants, each 1° of latitude by 1° of longitude. Environmental measures were found to be much better predictors of lizard density (multiple R^2 = .82) than of bird density (multiple R^2 = 0.34). There is also suggestion that different environmental factors predict density for lizards than for birds. Sunfall, entering the regression first for lizards, by itself accounted for .67 of the variance in density. For birds, sunfall appeared to be acted on by a suppressor variable. Simple correlation with bird density was virtually zero, but sunfall entered fifth in the regression and added 9% of predictable variance. Although no cross validation was reported, discrepancies in the two analyses are large enough to be considered important. (Sequential regression would have produced results that could be interpreted with more confidence.)

Maki, Hoffman, and Berk (1978) used standard multiple regression to study the effectiveness of a water conservation campaign. Two parallel analyses were done on sales and production of water (DVs) over 126 months, with months acting as the unit of study. IVs were moratorium (whether the campaign and/or conservation measures were in effect), rainfall, season, population, and dollars spent on public relations (as a measure of intensity of campaign).

Results of the two analyses were similar, with both showing significant effects of the moratorium-conservation efforts. Money spent on the campaign was not a statistically significant predictor in either campaign. Rainfall was a significant predictor of production, but not of sales. Population and season predicted both sales and production.

CHAPTER 6
Canonical Correlation

6.1 ■ GENERAL PURPOSE AND DESCRIPTION

The goal of canonical correlation is to analyze the relationships between two sets of variables. It may be useful to think of one set of variables as IVs and the other set as DVs, or it may not. In either event, canonical correlation provides a statistical analysis for research where each subject is measured on two sets of variables and the researcher wants to know if and how the two sets relate to each other.

Suppose, for instance, a researcher is interested in the relationship between a set of variables measuring medical compliance (willingness to buy drugs, to make return office visits, to use drugs, to restrict activity) and a set of demographic characteristics (educational level, religious affiliation, income, medical insurance). Canonical analysis might reveal that there are two reliable ways that the two sets of variables are related. The first way is between income and insurance on the demographic side and purchase of drugs and willingness to make return office visits on the medical compliance side. Together, these results indicate a relationship between compliance and demography based on ability to pay for medical services. The second way is between willingness to use drugs and restrict activity on the compliance side and religious affiliation and educational level on the demographic side, interpreted, perhaps, as a tendency to accede to authority (or not).

The easiest way to understand canonical correlation is to think of multiple regression. In regression, there are several variables on one side of the equation and a single variable on the other side. The several variables are combined into a predicted value to produce, across all subjects, the highest correlation between the predicted value and the single variable. The combination of variables can be thought of as a dimension among the many variables that predicts the single variable.

In canonical correlation, the same thing happens except that there are several variables on both sides of the equation. Sets of variables on each side are combined to produce, for each side, a predicted value that has the highest correlation with the predicted value on the other side. The combination of variables on each side can be thought of as a dimension that relates the variables on one side to the variables on the other.

There is a complication, however. In multiple regression, there is only one combination of variables because there is only a single variable to predict on the other side of the equation. In canonical correlation there are several variables on both sides and there may be several ways to recombine the variables on both sides to relate them to each other. In the example, the first way of combining the variables had to do with economic issues and the second way had to do with authority. Although there are potentially as many ways to recombine the variables as there are variables in the smaller set, usually only the first two or three combinations are reliable and need to be interpreted.

A good deal of the difficulty with canonical correlation is due to jargon. First there are variables, then there are canonical variates, and finally there are pairs of canonical variates. Variables refers to the variables measured in research (e.g., income). Canonical variates are linear combinations of variables, one combination on the IV side (e.g., income and medical insurance) and a second combination on the DV side (e.g., purchase of drugs and willingness to make return office visits). These two combinations form a pair of canonical variates. However, there may be more than one reliable pair of canonical variates (e.g., a pair associated with economics and a pair associated with authority).

Canonical analysis is one of the most general of the multivariate techniques. In fact, many other procedures—multiple regression, discriminant function analysis, MANOVA—are special cases of it. But it is also the least used and most impoverished of the techniques, for reasons that are discussed below.

6.2 ■ KINDS OF RESEARCH QUESTIONS

Although a large number of research questions are answered by canonical analysis in one of its specialized forms (such as discriminant analysis), relatively few intricate research questions are readily answered through direct application of computer programs currently available for canonical correlation. In part this has to do with the programs themselves and in part it has to do with the kinds of questions researchers consider appropriate in canonical correlation.

In its present stage of development, canonical correlation is best considered a descriptive technique or a screening procedure rather than a hypothesis-testing procedure. The following sections, however, contain questions that can be addressed with the aid of SPSS, BMDP, SAS, and SYSTAT programs.

6.2.1 ■ Number of Canonical Variate Pairs

How many reliable canonical variate pairs are there in the data set? Along how many dimensions are the variables in one set related to the variables in the other? In the example, Is the pair associated with economic issues reliable? And if so, Is the pair associated with authority also reliable? Because canonical variate pairs are computed in descending order of magnitude, the first one or two pairs are often reliable while remaining ones are not. Significance tests for canonical variate pairs are described in Sections 6.4 and 6.5.1.

6.2.2 ■ Interpretation of Canonical Variates

How are the dimensions that relate two sets of variables to be interpreted? What is the meaning in the combination of variables that compose one variate in conjunction with the combination composing the other in the same pair? In the example, all the variables that are important to the first pair of canonical variates have to do with money, so the combination is interpreted as an economic dimension. Interpretation of pairs of canonical variates usually proceeds from matrices of correlations between variables and canonical variates, as described in Sections 6.4 and 6.5.2.

6.2.3 ■ Importance of Canonical Variates

There are several ways to assess the importance of canonical variates. The first is to ask how strongly the variate on one side of the equation relates to the variate on the other side of the equation; that is, how strong is the correlation between variates in a pair? The second is to ask how strongly the variate on one side of the equation relates to the variables on its own side of the equation. The third is to ask how strongly the variate on one side of the equation relates to the variables on the other side of the equation.

For the example, What is the correlation between the economic variate on the compliance side and the economic variate on the demographic side? Then, How much variance does the economic variate on the demographic side extract from the demographic variables? Finally, How much variance does the economic variate on the demographic side extract from the compliance variables? These questions are answered by the procedures described in Sections 6.4 and 6.5.1.

6.2.4 ■ Canonical Variate Scores

Had it been possible to measure directly the canonical variates from both sets of variables, what scores would subjects have received on them? For instance, If directly measurable, what scores would the first subject have received on the economic variate from the compliance side and the economic variate from the demographic side? Examination of canonical variate scores reveals deviant cases, the shape of the relationship between two canonical variates, and the shape of the relationships between canonical variates and the original variables, as discussed briefly in Sections 6.3 and 6.4.

If canonical variates are interpretable, scores on them might be useful as IVs or DVs in other analyses. For instance, the researcher might use scores on the economic variate from the compliance side to examine the effects of publicly supported medical facilities.

6.3 ■ LIMITATIONS

6.3.1 ■ Theoretical Limitations[1]

Canonical correlation has several important theoretical limitations that help explain its scarcity in the literature. Perhaps the most critical limitation is interpretability; procedures that maximize correlation do not necessarily maximize interpretation of pairs of canonical variates. Therefore canonical solutions are often mathematically elegant but uninterpretable. And, although it is common practice in factor analysis and principal components analysis (Chapter 13) to rotate a solution to improve interpretation, rotation of canonical variates is not common practice or even available in some computer programs.

---

[1]The authors are indebted to James Fleming for many of the insights of this section.

The algorithm used for canonical correlation maximizes the linear relationship between two sets of variables. If the relationship is nonlinear the analysis misses some or most of it. If a nonlinear relationship between dimensions in a pair is suspected, use of canonical correlation may be inappropriate unless variables are transformed or combined to capture the nonlinear component.

The algorithm also computes pairs of canonical variates that are independent of all other pairs. In factor analysis (Chapter 13) one has a choice between an orthogonal (uncorrelated) and an oblique (correlated) solution, but in canonical analysis only the orthogonal solution is routinely available. If there was, in the example, a possible relationship between economic issues and authority, canonical correlation might be inappropriate.

An important concern is the sensitivity of the solution in one set of variables to the variables included *in the other set*. In canonical analysis, the solution depends both on correlations among variables in each set and on correlations among variables between sets. Changing the variables in one set may markedly alter the composition of canonical variates in the other set. To some extent, this is expected given the goals of analysis, yet the sensitivity of the procedure to apparently minor changes is a cause for concern.

6.3.2 ■ Practical Issues

6.3.2.1 ■ Ratio of Cases to IVs

The number of cases needed for analysis depends on the reliability of the variables. For variables in the social sciences where reliability is often around .80, about 10 cases are needed for every variable. However, if reliability is very high, as, for instance, in political science where the variables are measures of the economic performance of countries, then a much lower ratio of cases to variables is acceptable.

6.3.2.2 ■ Normality, Linearity, and Homoscedasticity

Although there is no requirement that the variables be normally distributed when canonical correlation is used descriptively, the analysis is enhanced if they are. However, inference regarding number of significant canonical variate pairs proceeds on the assumption of multivariate normality. Multivariate normality is the assumption that all variables and all linear combinations of variables are normally distributed. It is not itself an easily testable hypothesis (most tests available are too strict), but the likelihood of multivariate normality is increased if the variables are all normally distributed.

Linearity is important to canonical analysis in at least two ways. The first is that the analysis is performed on correlation or variance-covariance matrices that reflect only linear relationships. If the relationship between two variables is nonlinear, it is not "captured" by these statistics. The second is that canonical correlation maximizes the linear relationship between a variate from one set of variables and a variate from the other set. Canonical analysis misses potential nonlinear components of relationships between canonical variate pairs.

Finally, canonical analysis is best when relationships among pairs of variables are homoscedastic, that is, when the variance of one variable is about the same at all levels of the other variable.

Normality, linearity, and homoscedasticity can be assessed through normal screening procedures or through the distributions of canonical variate scores produced by a preliminary canonical analysis. If routine screening is undertaken, variables are examined individually for normality through one of the descriptive programs such as BMDP2D or SPSS FREQUENCIES. Pairs of variables, both within sets and across sets, are examined for nonlinearity or heteroscedasticity through programs such as BMDP6D or SPSS PLOT. If one or more of the variables is in violation of the assumptions, transformation is considered, as discussed in Chapter 4 and illustrated in Section 6.7.1.2.

Alternatively, distributions of canonical variate scores produced by a preliminary canonical analysis are examined for normality, linearity, and homoscedasticity and, if found, screening of the original variables is not necessary. Scatterplots where pairs of canonical variates are plotted against each other are available through BMDP6M, and through SAS CANCORR and SYSTAT GLM if canonical variates scores are written to a file for processing through a scatterplot program. If, in the scatterplots, there is evidence of failure of normality, linearity, and/or homoscedasticity, screening of the variables is undertaken. This procedure is illustrated in Section 6.7.1.2.

In the event of persistent heteroscedasticity, you might consider weighting cases based on variables producing unequal variance, or adding a variable which accounts for unequal variance (cf. Section 5.3.2.4).

6.3.2.3 ■ Missing Data

Levine (1977) gives an example of a dramatic change in a canonical solution with a change in procedures for handling missing data. Because canonical correlation is quite sensitive to minor changes in a data set, consider carefully the methods of Chapter 4 for estimating values or eliminating cases with missing data.

6.3.2.4 ■ Outliers

Cases that are unusual often have undue impact on canonical analysis. The search for univariate and multivariate outliers is conducted separately within each set of variables. Consult Chapter 4 and Section 6.7.1.3 for methods of detecting and reducing the effects of both univariate and multivariate outliers.

6.3.2.5 ■ Multicollinearity and Singularity

For both logical and computational reasons, it is important that the variables in each set and across sets are not too highly correlated with each other. This restriction applies to values in $\mathbf{R}_{xx}$, $\mathbf{R}_{yy}$, and $\mathbf{R}_{xy}$ (see Equation 6.1). BMDP6M provides a direct test of multicollinearity and singularity, as illustrated in Section 6.7.1.4. If you are using one of the other programs, consult Chapter 4 for methods of identifying and eliminating multicollinearity and singularity in correlation matrices.

6.4 ■ FUNDAMENTAL EQUATIONS FOR CANONICAL CORRELATION

A data set that is appropriately analyzed through canonical correlation has several subjects, each measured on four or more variables. The variables form two sets with at least two variables in the smaller set. A hypothetical data set, appropriate for canonical correlation, is presented in Table 6.1. Eight intermediate- and advanced-level belly dancers are rated on two sets of variables, the quality of their "top" shimmies (TS), "top" circles (TC), and the quality of their "bottom" shimmies (BS),

TABLE 6.1   SMALL SAMPLE OF HYPOTHETICAL DATA FOR
ILLUSTRATION OF CANONICAL CORRELATION ANALYSIS

ID	TS	TC	BS	BC
1	1.0	1.0	1.0	1.0
2	7.0	1.0	7.0	1.0
3	4.6	5.6	7.0	7.0
4	1.0	6.6	1.0	5.9
5	7.0	4.9	7.0	2.9
6	7.0	7.0	6.4	3.8
7	7.0	1.0	7.0	1.0
8	7.0	1.0	2.4	1.0

and "bottom" circles (BC). Each characteristic of the dance is rated by two judges on a 7-point scale (with larger numbers indicating higher quality) and the ratings averaged. The goal of analysis is to discover patterns, if any, between the quality of the movements on top and the quality of movements on bottom.

You are cordially invited to follow (or dance along with) this example by hand and by computer. Examples of setup and output for this analysis using several popular computer programs appear at the end of this section.

The first step in a canonical analysis is generation of a correlation matrix. In this case, however, the correlation matrix is subdivided into four parts: the correlations between the DVs ($\mathbf{R}_{yy}$), the correlations between the IVs ($\mathbf{R}_{xx}$), and the two matrices of correlations between DVs and IVs ($\mathbf{R}_{xy}$ and $\mathbf{R}_{yx}$).[2] Table 6.2 contains the correlation matrices for the data in the example.

There are several ways to write the fundamental equation for canonical correlation, some more intuitively appealing than others. The equations are all variants on the following equation.

$$\mathbf{R} = \mathbf{R}_{yy}^{-1}\mathbf{R}_{yx}\mathbf{R}_{xx}^{-1}\mathbf{R}_{xy} \tag{6.1}$$

The canonical correlation matrix is a product of four correlation matrices, between DVs (inverted), between IVs (inverted), and between DVs and IVs.

It is helpful conceptually to compare Equation 6.1 with Equation 5.6 for regression. Equation 5.6 indicates that regression coefficients for predicting Y from a set of Xs are a product of (the inverse of) the matrix of correlations among the Xs and the matrix of correlations between the Xs and Y. Equation 6.1 can be thought of as a product of regression coefficients for predicting Xs from Ys ($\mathbf{R}_{yy}^{-1}\mathbf{R}_{yx}$) and regression coefficients for predicting Ys from Xs ($\mathbf{R}_{xx}^{-1}\mathbf{R}_{xy}$).

6.4.1 ■ Eigenvalues and Eigenvectors

Canonical analysis proceeds by solving for the eigenvalues and eigenvectors of the matrix $\mathbf{R}$ of Equation 6.1. Eigenvalues are obtained by analyzing the matrix in Equation 6.1. Eigenvectors are obtained for the Y variables first and then calculated for the Xs using Equation 6.6 in Section 6.4.2. As discussed in Chapter 13 and in Appendix A, solving for the eigenvalues of a matrix is a process

---

[2]Although in this example the sets of variables are neither IVs nor DVs, it is useful to use the terms when explaining the procedure.

TABLE 6.2 CORRELATION MATRICES FOR THE DATA SET IN TABLE 6.1

	R_{xx}	R_{xy}	
	R_{yx}	R_{yy}	

	TS	TC	BS	BC
TS	1.000	−.161	.758	−.341
TC	−.161	1.000	.110	.857
BS	.758	.110	1.000	.051
BC	−.341	.857	.051	1.000

that redistributes the variance in the matrix, consolidating it into a few composite variates rather than many individual variables. The eigenvector that corresponds to each eigenvalue is transformed into the coefficients (e.g., regression coefficients, canonical coefficients) used to combine the original variables into the composite variate.

Calculation of eigenvalues and corresponding eigenvectors is demonstrated in Appendix A but is difficult and not particularly enlightening. For this example, the task is accomplished with assistance from SPSS MANOVA (see Table 6.5 and Section 6.6.1). The goal is to redistribute the variance in the original variables into a very few pairs of canonical variates, each pair capturing a large share of variance and defined by linear combinations of IVs on one side and DVs on the other. Linear combinations are chosen to maximize the canonical correlation for each pair of canonical variates.

Although computing eigenvalues and eigenvectors is best left to the computer, the relationship between a canonical correlation and an eigenvalue[3] is simple, namely,

$$\lambda_i = r_{ci}^2 \tag{6.2}$$

Each eigenvalue, λ_i, is equal to the squared canonical correlation, r_{ci}^2, for the pair of canonical variates.

Once the eigenvalue is calculated for each pair of canonical variates, canonical correlation is found by taking the square root of the eigenvalue. Canonical correlation, r_{ci}^2, is interpreted as an ordinary Pearson product-moment correlation coefficient. When r_{ci}, is squared, it represents, as usual, overlapping variance between two variables, or, in this case, variates. Because $r_{ci}^2 = \lambda_i$, the eigenvalues themselves represent overlapping variance between pairs of canonical variates.

For the data set of Table 6.1, two eigenvalues are calculated, one for each variable in the smaller set (both sets in this case). The first eigenvalue is .83566, which corresponds to a canonical correlation of .91414. The second eigenvalue is .58137, so canonical correlation is .76247. That is, the first pair of canonical variates correlate .91414 and overlap 83.57% in variance, while the second pair correlate .76247 and overlap 58.14% in variance.

Significance tests (Bartlett, 1941) are available to test whether one or a set of r_cs differs from zero.

$$\chi^2 = -\left[N - 1 - \left(\frac{k_x + k_y + 1}{2} \right) \right] \ln \Lambda_m \tag{6.3}$$

---

[3] SPSS and SAS use the terms Sq. Cor and Squared Canonical Correlation, respectively, in place of eigenvalue and use the term "eigenvalue" in a different way.

The significance of one or more canonical correlations is evaluated as a chi square variable where N is the number of cases, k_x is the number of variables in the IV set, k_y is the number in the DV set, and the natural logarithm of lambda, Λ, is defined in Equation 6.4. This chi square has $(k_x)(k_y)$ df.

$$\Lambda_m = \prod_{i=1}^{m} (1 - \lambda_i) \tag{6.4}$$

Lambda, Λ, is the product of differences between eigenvalues and unity, generated across m canonical correlations.

For the example, to test if the canonical correlations as a set differ from zero:

$$\Lambda_2 = (1 - \lambda_1)(1 - \lambda_2) = (1 - .84)(1 - .58) = .07$$

$$\chi^2 = -\left[8 - 1 - \left(\frac{2 + 2 + 1}{2}\right)\right] \ln .07$$

$$= -(4.5)(-2.68)$$

$$= 12.04$$

This χ^2 is evaluated with $(k_x)(k_y) = 4$ df. The two canonical correlations differ from zero: $\chi^2(4) = 12.04$, $p < .02$. The results of this test are interpreted to mean that there is significant overlap in variability between the variables in the IV set and the variables in the DV set, that is, that there is some reliable relationship between quality of top movements and of bottom movements. This result is often taken as evidence that at least the first canonical correlation is significant.

With the first canonical correlate removed, is there still a reliable relationship between the two sets of variables?

$$\Lambda_1 = (1 - .58) = .42$$

$$\chi^2 = -\left[8 - 1 - \left(\frac{2 + 2 + 1}{2}\right)\right] \ln .42$$

$$= -(4.5)(-.87)$$

$$= 3.92$$

This chi square has $(k_x - 1)(k_y - 1) = 1$ df and also differs significantly from zero: $\chi^2(1) = 3.92$, $p < .05$. This result indicates that there is still significant overlap between the two sets of variables after the first pair of canonical variates is removed. It is taken as evidence that the second canonical correlation is also significant.

Significance of canonical correlations is also evaluated using the F distribution as, for example, in SPSS MANOVA.

6.4.2 ■ Matrix Equations

Two sets of canonical coefficients (analogous to regression coefficients) are required for each canonical correlation, one set to combine the DVs and the other to combine the IVs. The canonical coefficients for the DVs are found as follows:

$$B_y = (R_{yy}^{-1/2})' \hat{B}_y \qquad (6.5)$$

Canonical coefficients for the DVs are a product of (the transpose of the inverse of the square root of) the matrix of correlations between DVs and the normalized matrix of eigenvectors, $\hat{B}_y$, for the DVs.

For the example:[4]

$$B_y = \begin{bmatrix} 1.00 & -0.03 \\ -0.03 & 1.00 \end{bmatrix} \begin{bmatrix} -0.45 & 0.89 \\ 0.89 & 0.47 \end{bmatrix} = \begin{bmatrix} -0.48 & 0.88 \\ 0.90 & 0.44 \end{bmatrix}$$

Once the canonical coefficients are computed, coefficients for the IVs are found using the following equation:

$$B_x = R_{xx}^{-1/2} R_{xy} B_y^* \qquad (6.6)$$

Coefficients for the IVs are a product of (the inverse of the square root of) the matrix of correlations between the IVs, the matrix of correlations between the IVs and DVs, and the matrix B_y^* formed by the coefficients for the DVs, each divided by their corresponding canonical correlations.

For the example:

$$B_x = \begin{bmatrix} 1.03 & 0.17 \\ 0.17 & 1.03 \end{bmatrix} \begin{bmatrix} 0.76 & -0.34 \\ 0.11 & 0.86 \end{bmatrix} \begin{bmatrix} -0.48/.91 & 0.89/.76 \\ 0.90/.91 & 0.44/.76 \end{bmatrix} = \begin{bmatrix} -0.63 & 0.80 \\ 0.69 & 0.75 \end{bmatrix}$$

The two matrices of canonical coefficients are used to estimate scores on canonical variates:

$$X = Z_x B_x \qquad (6.7)$$

and

$$Y = Z_y B_y \qquad (6.8)$$

Scores on canonical variates are estimated as the product of the standardized scores on the original variates, Z_x and Z_y, and the canonical coefficients used to weight them, B_x and B_y.

---

[4]These calculations, like others in this section, were carried to several decimal places and then rounded back. The results agree with computer analyses of the same data but the rounded-off figures presented here do not always check out to both decimals.

For the example:

$$
\mathbf{X} = \begin{bmatrix} -1.35 \\ 0.76 \\ 0.76 \\ -1.35 \\ 0.76 \\ 0.55 \\ 0.76 \\ -0.86 \end{bmatrix} \begin{bmatrix} -0.81 \\ -0.81 \\ 1.67 \\ 1.22 \\ -0.02 \\ 0.35 \\ -0.81 \\ -0.81 \end{bmatrix} \begin{bmatrix} -0.63 & 0.80 \\ 0.69 & 0.75 \end{bmatrix} = \begin{bmatrix} 0.34 \\ -1.04 \\ 0.66 \\ 1.74 \\ -0.07 \\ 0.45 \\ -1.04 \\ -1.04 \end{bmatrix} \begin{bmatrix} -1.91 \\ -0.15 \\ 0.39 \\ -0.39 \\ 0.90 \\ 1.47 \\ -0.15 \\ -0.15 \end{bmatrix}
$$

$$
\mathbf{Y} = \begin{bmatrix} -1.54 \\ 0.66 \\ -0.22 \\ -1.54 \\ 0.66 \\ 0.66 \\ 0.66 \\ 0.66 \end{bmatrix} \begin{bmatrix} -0.91 \\ -0.91 \\ 0.76 \\ 1.12 \\ 0.50 \\ 1.26 \\ -0.91 \\ -0.91 \end{bmatrix} \begin{bmatrix} -0.48 & 0.88 \\ 0.90 & 0.44 \end{bmatrix} = \begin{bmatrix} -0.08 \\ -1.14 \\ 0.79 \\ 1.75 \\ 0.13 \\ 0.82 \\ -1.14 \\ -1.14 \end{bmatrix} \begin{bmatrix} -1.76 \\ 0.18 \\ 0.14 \\ -0.86 \\ 0.80 \\ 1.14 \\ 0.18 \\ 0.18 \end{bmatrix}
$$

The first belly dancer, in standardized scores (and appropriate costume), has a z score of -1.35 on TS, -0.81 on TC, -1.54 on BS, and -0.91 on BC. When these z scores are weighted by canonical coefficients, this dancer is estimated to have a score of 0.34 on the first canonical variate and a score of -1.91 on the second canonical variate for the IVs (the Xs), and scores of -0.08 and -1.76 on the first and second canonical variates, respectively, for the DVs (the Ys).

The sum of canonical scores for all belly dancers on each canonical variate is zero, within rounding error. These scores, like factor scores (Chapter 13) are estimates of scores the dancers would receive if they were judged directly on the canonical variates.

Matrices of correlations between the variables and the canonical coefficients, called loading matrices, are used to interpret the canonical variates.

$$
\mathbf{A}_x = \mathbf{R}_{xx}\mathbf{B}_x \tag{6.9}
$$

and

$$
\mathbf{A}_y = \mathbf{R}_{yy}\mathbf{B}_y \tag{6.10}
$$

Correlations between variables and canonical variates are found by multiplying the matrix of correlations between variables by the matrix of canonical coefficients.

For the example:

$$
\mathbf{A}_x = \begin{bmatrix} 1.00 & -0.16 \\ -0.16 & 1.00 \end{bmatrix} \begin{bmatrix} -0.62 & 0.80 \\ 0.69 & 0.75 \end{bmatrix} = \begin{bmatrix} -0.74 & 0.68 \\ 0.79 & 0.63 \end{bmatrix}
$$

$$
\mathbf{A}_y = \begin{bmatrix} 1.00 & 0.05 \\ 0.05 & 1.00 \end{bmatrix} \begin{bmatrix} -0.48 & 0.88 \\ 0.90 & 0.44 \end{bmatrix} = \begin{bmatrix} -0.44 & 0.90 \\ 0.88 & 0.48 \end{bmatrix}
$$

The loading matrices for these data are summarized in Table 6.3. Results are interpreted down columns across sets of variables. For the first canonical variate pair (the first column), TS correlates $-.74$, TC $.79$, BS $-.44$, and BC $.88$. The first pair of canonical variates link low scores on TS and high scores on TC (in the first set of variables) with high scores on BC (in the second set), indicating that poor quality top shimmies and high quality top circles are associated with high quality bottom circles.

For the second canonical variate pair (the second column), TS correlates $.68$, TC $.62$, BS $.90$, and BC $.48$; the second canonical variate pair indicates that high scores on bottom shimmies are associated with high scores on both top circles and top shimmies. Taken together, these results suggest that ability to do bottom circles is related to ability to do top circles but inability to do top shimmies, while ability to do bottom shimmies is associated with ability to do both top movements well.

6.4.3 ■ Proportions of Variance Extracted

How much variance does each of the canonical variates extract from the variables on its own side of the equation? The proportion of variance extracted from the IVs by the canonical variates of the IVs is

$$pv_{xc} = \sum_{i=1}^{k_x} \frac{a_{ixc}^2}{k_x}$$

(6.11)

and

$$pv_{yc} = \sum_{i=1}^{k_y} \frac{a_{iyc}^2}{k_y}$$

(6.12)

The proportion of variance extracted from a set of variables by a canonical variate of the set (called *averaged squared loading* by BMDP6M) is the sum of the squared correlations (loadings from Equations 6.9 and 6.10) divided by the number of variables in the set.

Thus for the first canonical variate in the set of IVs,

$$pv_{x1} = \frac{(-0.74)^2 + 0.79^2}{2} = .58$$

and for the second canonical variate of the IVs,

$$pv_{x2} = \frac{0.68^2 + 0.62^2}{2} = .42$$

TABLE 6.3 LOADING MATRIX FOR THE DATA SET IN TABLE 6.1

		Canonical variate pairs	
	Variable sets	First	Second
First	TS	−0.74	0.68
	TC	0.79	0.62
Second	BS	−0.44	0.90
	BC	0.88	0.48

The first canonical variate extracts 58% of the variance in judgments of top movements while the second canonical variate extracts 42% of the variance in judgments of top movements. In summing for the two variates, almost 100% of the variance in the IVs is extracted by the two canonical variates. As expected in summing the two variables, 100% of the variance in IVs is extracted by the two canonical variates. This happens when the number of variables on one side of the equation is equal to the number of canonical variates. The sum of the pv scores usually is less than 1.00 if there are more variables than canonical variates.

For the DVs and the first canonical variate,

$$pv_{y1} = \frac{(-0.44)^2 + 0.88^2}{2} = .48$$

and for the second canonical variate,

$$pv_{y2} = \frac{0.90^2 + 0.48^2}{2} = .52$$

That is, the first canonical variate extracts 48% of the variance in judgments of bottom movements, while the second canonical variate extracts 52% of variance in judgments of bottom movements. Together the two canonical variates extract almost 100% of the variance in the DVs.

Often, however, one is interested in knowing how much variance the canonical variates from the IVs extract from the DVs, and vice versa. In canonical analysis, this variance is called redundancy.

$$rd = (pv)(r_c^2) \tag{6.13}$$

The redundancy in a canonical variate is the percent of variance it extracts from its own set of variables times the squared canonical correlation for the pair.

Thus for the example:

$$rd_{x1 \to y} = \left(\frac{(-0.44)^2 + 0.88^2}{2}\right)(.84) = .40$$

$$rd_{x2 \to y} = \left(\frac{0.90^2 + .48^2}{2}\right)(.58) = .30$$

$$rd_{y1 \to x} = \left(\frac{(-0.74)^2 + 0.79^2}{2}\right)(.84) = .48$$

and

$$rd_{y2 \to x} = \left(\frac{0.68^2 + 0.62^2}{2}\right)(.58) = .24$$

So, the first canonical variate from the IVs extracts 40% of the variance in judgments of quality of bottom movements. The second canonical variate of the IVs extracts 30% of the variance in judgments of quality of bottom movements. Together the two variates extract 70% of the variance in the DVs.

The first and second canonical variates for the DVs extract 48% and 24% of the variance in judgments of quality of top movements, respectively. Together they extract 72% of the variance in judgments of quality of top movements.

6.4.4 ■ Computer Analyses of Small Sample Example

Tables 6.4 through 6.7 show analyses of this data set by BMDP6M, SPSS MANOVA, SYSTAT GLM, and SAS CANCORR, respectively.

BMDP6M allows specification of the first and second sets of variables in a CANON paragraph (Table 6.4). The output includes numerous univariate descriptive statistics for each variable (cf. Chapter 4), a matrix of correlations among the variables, and SMCs where each variable, in turn, serves as DV while other variables in the same set are IVs.

The table in which the first column is labeled EIGENVALUE allows you to determine the number of significant canonical variate pairs. The first row tests the significance of all pairs considered together as per Equations 6.3 and 6.4, through CHI-SQUARE (= 12.04 in the example). The second row shows the EIGENVALUE (canonical correlation) and the squared CANONICAL CORRELATION for the first canonical variate pair. The "1" refers to the fact that the first canonical variate pair is "peeled off" for the significance test (CHI-SQUARE = 3.92) of the remaining canonical variate(s) (in this case only one). (With more than two variables in both sets, the second test includes all the remaining canonical correlations after the first canonical variate pair is peeled off.) The third row shows the canonical correlation and squared canonical correlation for the second canonical variate pair.

Next are two tables for the first set of variables, and in identical format, two tables for the second set of variables. CANONICAL VARIABLE LOADINGS are the correlations of variables in the sets with the canonical variates. In the tables labeled SQUARED MULTIPLE CORRELATIONS OF EACH VARIABLE . . . are R-SQUARED, ADJUSTED R-SQUARED, and an F STATISTIC where each variable in one set, in turn, serves as DV in multiple regression with the variables in the other set as IVs.

The final table (AVERAGE SQUARED LOADING . . .) shows, for each canonical variate, the proportion of variance extracted from each set of variables (AVERAGED SQUARED LOADING FOR EACH CANONICAL VARIABLE; as per Equations 6.11 and 6.12) and redundancy (AV. SQ. LOADINGS . . . ; as per Equation 6.13), as described earlier in this section and in the output. The squared canonical correlation is repeated for each canonical variate in the last column.

In SPSS MANOVA (Table 6.5), one set of variables is called DEPENDENT variables while the other is called COVARIATES. In the MANOVA instruction, the set listed before WITH is the DV set. The DISCRIM and PRINT statements provide information useful for canonical analysis.

After various tests of the within cells correlation matrix are several significance tests in the section labeled EFFECT . . . WITHIN CELLS Regression. The first three (Pillais, Hotellings, & Wilks) test all canonical variates together; Roys tests the first canonical variate by itself (see Section 9.5.1). Information from this table is repeated in subsequent sections. Eigenvalues and Canonical Correlations are printed in the next section. In SPSS MANOVA, the terms eigenvalue and canonical correlation are not used interchangeably—here the eigenvalue is $R^2/(1 - R^2)$. The Canon Cor. and Sq. (canonical) Cor are given for each canonical variate pair in rows labeled Root No. The percent of variance in the solution contributed by the canonical variate pair is listed in the column labeled Pct;

TABLE 6.4   SETUP AND SELECTED BMDP6M OUTPUT FOR CANONICAL CORRELATION ANALYSIS ON SAMPLE DATA IN TABLE 6.1

```
/INPUT      VARIABLES ARE 5.   FILE = SSCANON.
            FORMAT IS FREE.
            TITLE IS 'SMALL SAMPLE CANONICAL CORRELATION'.
/VARIABLES  NAMES ARE ID, TS, TC, BS, BC.
            LABEL IS ID.
/CANON      FIRST = TS, TC.   SECOND = BS, BC.
/END
```

UNIVARIATE SUMMARY STATISTICS

VARIABLE	MEAN	STANDARD DEVIATION	COEFFICIENT OF VARIATION	SMALLEST VALUE	LARGEST VALUE	SMALLEST STANDARD SCORE	LARGEST STANDARD SCORE	SKEWNESS	KURTOSIS
2 TS	5.20000	2.72134	0.523335	1.00000	7.00000	-1.54	0.66	-0.74	-1.46
3 TC	3.51250	2.75755	0.785069	1.00000	7.00000	-0.91	1.26	0.12	-2.09
4 BS	4.85000	2.84203	0.585986	1.00000	7.00000	-1.35	0.76	-0.46	-1.91
5 BC	2.95000	2.42133	0.820791	1.00000	7.00000	-0.81	1.67	0.56	-1.53

VALUES FOR KURTOSIS GREATER THAN ZERO INDICATE DISTRIBUTIONS
WITH HEAVIER TAILS THAN THE NORMAL DISTRIBUTION.

CORRELATIONS

		TS	TC	BS	BC
		2	3	4	5
TS	2	1.000			
TC	3	-0.161	1.000		
BS	4	0.758	0.110	1.000	
BC	5	-0.341	0.857	0.051	1.000

SQUARED MULTIPLE CORRELATIONS OF EACH VARIABLE IN
SECOND SET WITH ALL OTHER VARIABLES IN SECOND SET

VARIABLE NUMBER	NAME	R-SQUARED
4	BS	0.00261
5	BC	0.00261

208

TABLE 6.4 (CONTINUED)

SQUARED MULTIPLE CORRELATIONS OF EACH VARIABLE IN
FIRST SET WITH ALL OTHER VARIABLES IN FIRST SET

VARIABLE		
NUMBER	NAME	R-SQUARED
2	TS	0.02594
3	TC	0.02594

EIGENVALUE	CANONICAL CORRELATION	NUMBER OF EIGENVALUES	BARTLETT'S TEST FOR REMAINING EIGENVALUES		
			CHI-SQUARE	D.F.	TAIL PROB.
0.83566	0.91414		12.04	4	0.0170
0.58137	0.76247	1	3.92	1	0.0478

BARTLETT'S TEST ABOVE INDICATES THE NUMBER OF CANONICAL
VARIABLES NECESSARY TO EXPRESS THE DEPENDENCY BETWEEN THE
TWO SETS OF VARIABLES. THE NECESSARY NUMBER OF CANONICAL
VARIABLES IS THE SMALLEST NUMBER OF EIGENVALUES SUCH THAT
THE TEST OF THE REMAINING EIGENVALUES IS NOT SIGNIFICANT.
FOR EXAMPLE, IF A TEST AT THE .01 LEVEL WERE DESIRED,
THEN 0 VARIABLES WOULD BE CONSIDERED NECESSARY.
HOWEVER, THE NUMBER OF CANONICAL VARIABLES OF PRACTICAL
VALUE IS LIKELY TO BE SMALLER.

CANONICAL VARIABLE LOADINGS
(CORRELATIONS OF CANONICAL VARIABLES WITH ORIGINAL VARIABLES)
FOR FIRST SET OF VARIABLES

		CNVRF1	CNVRF2
		1	2
TS	2	-0.736	0.677
TC	3	0.787	0.617

SQUARED MULTIPLE CORRELATIONS OF EACH VARIABLE IN THE
FIRST SET WITH ALL VARIABLES IN THE SECOND SET.

TABLE 6.4 (CONTINUED)

VARIABLE	R-SQUARED	ADJUSTED R-SQUARED	F STATISTIC	DEGREES OF FREEDOM		P-VALUE
2 TS	0.719054	0.606675	6.40	2	5	0.0418
3 TC	0.738797	0.634316	7.07	2	5	0.0349

CANONICAL VARIABLE LOADINGS

(CORRELATIONS OF CANONICAL VARIABLES WITH ORIGINAL VARIABLES)
FOR SECOND SET OF VARIABLES

		CNVRS1 1	CNVRS2 2
BS	4	0.436	0.900
BC	5	0.876	0.482

SQUARED MULTIPLE CORRELATIONS OF EACH VARIABLE IN THE
SECOND SET WITH ALL VARIABLES IN THE FIRST SET.

VARIABLE	R-SQUARED	ADJUSTED R-SQUARED	F STATISTIC	DEGREES OF FREEDOM		P-VALUE
4 BS	0.629766	0.481672	4.25	2	5	0.0834
5 BC	0.776666	0.687332	8.69	2	5	0.0236

CANON. VAR.	AVERAGE SQUARED LOADING FOR EACH CANONICAL VARIABLE (1ST SET)	AV. SQ. LOADING TIMES SQUARED CANON. CORREL. (1ST SET)	AVERAGE SQUARED LOADING FOR EACH CANONICAL VARIABLE (2ND SET)	AV. SQ. LOADING TIMES SQUARED CANON. CORREL. (2ND SET)	SQUARED CANON. CORREL.
1	0.58028	0.48491	0.47917	0.40042	0.83566
2	0.41972	0.24401	0.52083	0.30279	0.58137

THE AVERAGE SQUARED LOADING MULTIPLIED BY THE SQUARED CANONICAL
CORRELATION IS THE AVERAGE SQUARED CORRELATION OF A
VARIABLE IN ONE SET WITH THE CANONICAL VARIABLE FROM
THE OTHER SET.   IT IS SOMETIMES CALLED A REDUNDANCY INDEX.

TABLE 6.5 SETUP AND SELECTED SPSS MANOVA OUTPUT FOR CANONICAL CORRELATOIN ANALYSIS OF SMALL SAMPLE DATA OF TABLE 6.1

```
DATA LIST   FILE = 'SSCANON.DAT'  FREE
            /ID, TS, TC, BS, BC.
MANOVA      BS BC WITH TS TC/
            DISCRIM = (RAW, STAN, ESTIM, COR)/
            PRINT = SIGNIF(EIGEN, DIMENR)/
            DESIGN.
```

* * Analysis of Variance * *

EFFECT .. WITHIN CELLS Regression
Multivariate Tests of Significance (S = 2, M = -1/2, N = 1)

Test Name	Value	Approx. F	Hypoth. DF	Error DF	Sig. of F
Pillais	1.41702	6.07669	4.00	10.00	.010
Hotellings	6.47355	4.85516	4.00	6.00	.043
Wilks	.06880	5.62496	4.00	8.00	.019
Roys	.83566				

Note.. F statistic for WILKS' Lambda is exact.

EFFECT .. WITHIN CELLS Regression (Cont.)
Eigenvalues and Canonical Correlations

Root No.	Eigenvalue	Pct.	Cum. Pct.	Canon Cor.	Sq. Cor
1	5.085	78.548	78.548	.914	.836
2	1.389	21.452	100.000	.762	.581

- - - - - - - - -

EFFECT .. WITHIN CELLS Regression (Cont.)
Dimension Reduction Analysis

Roots	Wilks L.	F	Hypoth. DF	Error DF	Sig. of F
1 TO 2	.06880	5.62496	4.00	8.00	.019
2 TO 2	.41863	6.94366	1.00	5.00	.046

- - - - - - - - -

EFFECT .. WITHIN CELLS Regression (Cont.)
Univariate F-tests with (2,5) D. F.

Variable	Sq. Mul. R	Adj. R-sq.	Hypoth. MS	Error MS	F
BS	.62977	.48167	17.80347	4.18661	4.25248
BC	.77667	.68733	15.93718	1.83313	8.69399

Variable	Sig. of F
BS	.083
BC	.024

- - - - - - - - -

Raw canonical coefficients for DEPENDENT variables
        Function No.

Variable	1	2
BS	-.170	.309
BC	.372	.180

- - - - - - - - -

TABLE 6.5   *(CONTINUED)*

Standardized canonical coefficients for DEPENDENT variables
        Function No.

Variable	1	2
BS	-.482	.878
BC	.901	.437

- - - - - - - - -

Correlations between DEPENDENT and canonical variables
        Function No.

Variable	1	2
BS	-.436	.900
BC	.876	.482

- - - - - - - - -

Variance explained by canonical variables of DEPENDENT variables

CAN. VAR.	Pct Var DE	Cum Pct DE	Pct Var CO	Cum Pct CO
1	47.917	47.917	40.042	40.042
2	52.083	100.000	30.279	70.322

- - - - - - - - -

Raw canonical coefficients for COVARIATES
        Function No.

COVARIATE	1	2
TS	-.230	.293
TC	.249	.270

Standardized canonical coefficients for COVARIATES
        CAN. VAR.

COVARIATE	1	2
TS	-.625	.797
TC	.686	.746

- - - - - - - - -

Correlations between COVARIATES and canonical variables
        CAN. VAR.

Covariate	1	2
TS	-.736	.677
TC	.787	.617

- - - - - - - - -

Variance explained by canonical variables of the COVARIATES

CAN. VAR.	Pct Var DE	Cum Pct DE	Pct Var CO	Cum Pct CO
1	48.491	48.491	58.028	58.028
2	24.401	72.893	41.972	100.000

```
TABLE 6.5      (CONTINUED)

- - - - - - - - - -
Regression analysis for WITHIN CELLS error term
--- Individual Univariate .9500 confidence intervals
Dependent variable .. BS

COVARIATE           B        Beta    Std. Err.    t-Value    Sig. of t

TS               .83168     .79636      .288       2.888        .034
TC               .24519     .23790      .284        .863        .428


COVARIATE     Lower -95%  CL- Upper

TS                .092      1.572
TC               -.485       .976
-------------------------------------------------------------------
Regression analysis for WITHIN CELLS error term   (Cont.)
Dependent variable .. BC

COVARIATE           B        Beta    Std. Err.    t-Value    Sig. of t

TS              -.18524    -.20819      .191       -.972        .376
TC               .72307     .82347      .188       3.845        .012

COVARIATE     Lower -95%  CL- Upper

TS               -.675       .305
TC                .240      1.206


-------------------------------------------------------------------
```

Cum. Pct. cumulates the percent of variance over the successive roots. In this example, the first canonical variate pair accounts for 79% of the variance in the solution. Both canonical variate pairs considered together account for 100% of the variance in the solution (because here there are only two pairs).

The Dimension Reduction Analysis follows the same "peel-off" procedure as in BMDP6M but tested as F and labeled a bit better. The first row tests both variate pairs (ROOTS 1 TO 2) and the second row (2 TO 2) tests variate pairs remaining after the first is peeled off. (The section on Univariate F Tests is of limited utility because it evaluates only the set of variables labeled DEPENDENT; remember this is a MANOVA program at heart).

Four tables are then given for each set of variables, labeled DEPENDENT variables and COVARIATES. The first two tables show Raw and Standardized canonical coefficients (see Equations 6.5 and 6.6). The loading matrix follows (Correlations between . . . and canonical variables). The last table is Variance explained by . . . and contains percent of variance (Pct Var same set) and redundancy measures (Pct Var other set). Considering the DEPENDENT variable set for the example, the first canonical variate accounts for 48% of the variance in that set (Pct Var DE) while redundancy is 40% (Pct Var CO). Cumulative percents of variance and redundancy over successive canonical variates are also given.

The final MANOVA table shows bivariate regression analyses (cf. Chapter 3) for each of the dependent variables with each of the COVARIATE variables.

Setup for SYSTAT GLM is fairly complex for canonical analysis, but the procedures are clearly spelled out in the manual. Basically you set up two multiple-multiple regression equations where each set of variables takes a turn as DVs (left side of the equation). As you might guess, this produces voluminous output with confusing labels. Comparison of results with the hand-worked example and other output helps clarify them, however.

```
OUTPUT CANON.OUT
USE SSCANON
MODEL TS TC = CONSTANT + BS + BC
  ESTIMATE
  PRINT = LONG
  HYPOTHESIS
  STANDARDIZE = TOTAL
  EFFECT = BS & BC
  TEST
MODEL BS BC = CONSTANT + TS + TC
  ESTIMATE
  PRINT = LONG
  HYPOTHESIS
  STANDARDIZE = TOTAL
  EFFECT = TS & TC
  TEST
```

Number of cases processed: 8
Dependent variable means

	TS	TC
	5.200	3.512

Regression coefficients $B = (X'X)^{-1} X'Y$

	TS	TC
CONSTANT	2.851	0.334
BS	0.744	0.064
BC	-0.428	0.972

Multiple correlations

	TS	TC
	0.848	0.860

Squared multiple correlations

	TS	TC
	0.719	0.739

Adjusted $R^2 = 1-(1-R^2)*(N-1)/DF$, where N = 8, and DF = 5

	TS	TC
	0.607	0.634

TABLE 6.6 *(CONTINUED)*

Test for effect called: BS

                          and
                             BC

Univariate F Tests

Effect	SS	DF	MS	F	P
TS	37.276	2	18.638	6.398	0.042
Error	14.564	5	2.913		
TC	39.325	2	19.663	7.071	0.035
Error	13.903	5	2.781		

Multivariate Test Statistics

Wilks' Lambda =	0.069						
F-Statistic =	5.625	DF =	4,	8	Prob =	0.019
Pillai Trace =	1.417						
F-Statistic =	6.077	DF =	4,	10	Prob =	0.010
Hotelling-Lawley Trace =	6.474						
F-Statistic =	4.855	DF =	4,	6	Prob =	0.043

THETA = 0.836 S = 2, M =-0.5, N = 1.0 Prob =   0.066

Test of Residual Roots

  Roots 1 through 2
    Chi-Square Statistic =   12.045   DF = 4

  Roots 2 through 2
    Chi-Square Statistic =   3.918   DF = 1

Canonical Correlations

1	2
0.914	0.762

Dependent variable canonical coefficients standardized
by sample standard deviations

	1	2
TS	-0.625	0.797
TC	0.686	0.746

Canonical loadings (correlations between dependent
variables and dependent canonical factors)

	1	2
TS	-0.736	0.677
TC	0.787	0.617

Number of cases processed: 8
Dependent variable means

BS	BC
4.850	2.950

TABLE 6.6   *(CONTINUED)*

Regression coefficients $B = (X'X)^{-1} X'Y$

	BS	BC
CONSTANT	-0.336	1.373
TS	0.832	-0.185
TC	0.245	0.723

Standardized regression coefficients

	BS	BC
CONSTANT	0.0	0.0
TS	0.796	-0.208
TC	0.238	0.823

Multiple correlations

BS	BC
0.794	0.881

Squared multiple correlations

BS	BC
0.630	0.777

Adjusted $R^2 = 1-(1-R^2)*(N-1)/DF$, where $N = 8$, and $DF = 5$

BS	BC
0.482	0.687

Test for effect called: TS

                       and
                           TC

Univariate F Tests

Effect	SS	DF	MS	F	P
BS	35.607	2	17.803	4.252	0.083
Error	20.933	5	4.187		
BC	31.874	2	15.937	8.694	0.024
Error	9.166	5	1.833		

Canonical Correlations

1	2
0.914	0.762

TABLE 6.6   *(CONTINUED)*

Dependent variable canonical coefficients standardized
by sample standard deviations

	1	2
BS	-0.482	0.878
BC	0.901	0.437

Canonical loadings (correlations between dependent
variables and dependent canonical factors)

	1	2
BS	-0.436	0.900
BC	0.876	0.482

The first segment of output shows means for the variables labeled Dependent. These are followed by multiple regression information for predicting each DV from the set IVS—unstandardized coefficients (B weights), multiple correlations, SMCs, adjusted SMCs, and univariate F tests for the significance of the prediction.

This is followed by Multivariate Test Statistics for the canonical correlation considering all variates together (compare with SPSS MANOVA section labeled EFFECT ... WITHIN CELLS Regression). The Test of Residual Roots provides the "peel off" chi square tests of the significance of pairs of canonical variates. Canonical correlations for each pair of canonical variates are then shown.

The next two segments of output give standardized canonical coefficients and the loading matrix for the DVs. The remaining output is based on the second MODEL statement in the input where the second set of variables serves as DVs. The output parallels that of the first MODEL although the Multivariate Test Statistics and Test of Residual Roots are deleted because they are exactly the same for both MODELs.

In SAS CANCORR (Table 6.7), the one set of variables (DVs) is listed in the input statement that begins VAR, the other set (IVs) in the statement that begins WITH. The RED instruction requests a redundancy analysis. The first segment of output contains the canonical correlations for each of the canonical variates (labeled 1 and 2), including adjusted and squared correlations as well as standard errors for the correlations. The next table shows the eigenvalues (calculated as for SPSS MANOVA above), the difference between eigenvalues, the proportion and the cumulative proportion of variance in the solution accounted for by each canonical variate pair. The Test of HO: ... table shows "peel off" significance tests for canonical variate pairs evaluated through approximate F followed in the next table by several multivariate significance tests. Matrices of raw and standardized canonical coefficients for each canonical variate labeled 'VAR' and 'WITH' in the setup follow; loading matrices are labeled Correlations Between the ... Variables and Their Canonical Variables. The portion labeled Canonical Structure is part of the redundancy analysis, and shows another type of loading matrices: the correlations between each set of variables and the canonical variates of the other set. Then the portion labeled Canonical Redundancy Analysis shows several tables in which proportion of variance for a set of variables as explained by their own and opposite canonical variates. It is in the Standardized Variance of the ... Variables table that the proportions of variance (*pvs*) are shown (Explained by Their Own Canonical Variables) as well as the redundancies (*rds*, Explained by The Opposite Canonical Variables). The final matrices show the SMCs for each of the individual IVs and DVs with their own and the opposite canonical variables.

TABLE 6.7   SETUP AND SAS CANCORR OUTPUT FOR CANONICAL CORRELATION ANALYSIS OF SAMPLE DATA OF TABLE 6.1

```
DATA SSAMPLE;
INFILE 'SSCANON.DAT';
INPUT ID TS TC BS BC;
PROC CANCORR RED;
  VAR BS BC;
  WITH TS TC;
```

Canonical Correlation Analysis

	Canonical Correlation	Adjusted Canonical Correlation	Approx Standard Error	Squared Canonical Correlation
1	0.914142	0.889541	0.062116	0.835656
2	0.762475	.	0.158228	0.581368

Eigenvalues of INV(E)*H
= CanRsq/(1-anRsq)

	Eigenvalue	Difference	Proportion	Cumulative
1	5.0848	3.6961	0.7855	0.7855
2	1.3887	.	0.2145	1.0000

Test of H0: The canonical correlations in the current row
and all that follow are zero

	Likelihood Ratio	Approx F	Num DF	Den DF	Pr > F
1	0.06879947	5.6250	4	8	0.0187
2	0.41863210	6.9437	1	5	0.0462

Multivariate Statistics and F Approximations

S=2   M=-0.5   N=1

Statistic	Value	F	Num DF	Den DF	Pr > F
Wilks' Lambda	0.06879947	5.6250	4	8	0.0187
Pillai's Trace	1.41702438	6.0767	4	10	0.0096
Hotelling-Lawley Trace	6.47354785	4.8552	4	6	0.0433
Roy's Greatest Root	5.08481559	12.7120	2	5	0.0109

NOTE: F Statistic for Roy's Greatest Root is an upper bound.
NOTE: F Statistic for Wilks' Lambda is exact.

Raw Canonical Coefficients for the 'VAR' Variables

	V1	V2
BS	-0.169694016	0.3087617628
BC	0.3721067975	0.1804101724

Raw Canonical Coefficients for the 'WITH' Variables

	W1	W2
TS	-0.229789498	0.2929561178
TC	0.2488132088	0.2703732438

TABLE 6.7   *(CONTINUED)*

Standardized Canonical Coefficients for the 'VAR' Variables

	V1	V2
BS	-0.4823	0.8775
BC	0.9010	0.4368

Standardized Canonical Coefficients for the 'WITH' Variables

	W1	W2
TS	-0.6253	0.7972
TC	0.6861	0.7456

Canonical Structure

Correlations Between the 'VAR' Variables and Their Canonical Variables

	V1	V2
BS	-0.4363	0.8998
BC	0.8764	0.4816

Correlations Between the 'WITH' Variables and Their Canonical Variables

	W1	W2
TS	-0.7358	0.6772
TC	0.7868	0.6172

Correlations Between the 'VAR' Variables and the
Canonical Variables of the 'WITH' Variables

	W1	W2
BS	-0.3988	0.6861
BC	0.8011	0.3672

Correlations Between the 'WITH' Variables and
the Canonical Variables of the 'VAR' Variables

	V1	V2
TS	-0.6727	0.5163
TC	0.7193	0.4706

Canonical Redundancy Analysis

Raw Variance of the 'VAR' Variables

	Explained by				
	Their Own Canonical Variables			The Opposite Canonical Variables	
	Proportion	Cumulative Proportion	Canonical R-Squared	Proportion	Cumulative Proportion
1	0.4333	0.4333	0.8357	0.3621	0.3621
2	0.5667	1.0000	0.5814	0.3295	0.6915

TABLE 6.7   (CONTINUED)

Raw Variance of the 'WITH' Variables

	Explained by				
	Their Own Canonical Variables			The Opposite Canonical Variables	
	Proportion	Cumulative Proportion	Canonical R-Squared	Proportion	Cumulative Proportion
1	0.5808	0.5808	0.8357	0.4853	0.4853
2	0.4192	1.0000	0.5814	0.2437	0.7291

Standardized Variance of the 'VAR' Variables

	Explained by				
	Their Own Canonical Variables			The Opposite Canonical Variables	
	Proportion	Cumulative Proportion	Canonical R-Squared	Proportion	Cumulative Proportion
1	0.4792	0.4792	0.8357	0.4004	0.4004
2	0.5208	1.0000	0.5814	0.3028	0.7032

Standardized Variance of the 'WITH' Variables

	Explained by				
	Their Own Canonical Variables			The Opposite Canonical Variables	
	Proportion	Cumulative Proportion	Canonical R-Squared	Proportion	Cumulative Proportion
1	0.5803	0.5803	0.8357	0.4849	0.4849
2	0.4197	1.0000	0.5814	0.2440	0.7289

Squared Multiple Correlations Between the 'VAR' Variables and
the First 'M' Canonical Variables of the 'WITH' Variables

M	1	2
BS	0.1590	0.6298
BC	0.6418	0.7767

Squared Multiple Correlations Between the 'WITH' Variables and
the First 'M' Canonical Variables of the 'VAR' Variables

M	1	2
TS	0.4525	0.7191
TC	0.5174	0.7388

6.5 ■ SOME IMPORTANT ISSUES

6.5.1 ■ Importance of Canonical Variates

As in most statistical procedures, establishing significance is usually the first step in evaluating a solution. Conventional statistical procedures apply to significance tests for number of reliable canonical variate pairs. The results of Equations 6.3 and 6.4, or a corresponding F test, are available in all programs reviewed in Section 6.6. But the number of statistically significant pairs of canonical variates is often larger than the number of interpretable pairs if N is at all sizable.

The only potential source of confusion is the meaning of the chain of significance tests. The first test is for all pairs taken together and is a test of independence between the two sets of variables. The second test is for all pairs of variates with the first and most important pair of canonical variates removed; the third is done with the first two pairs removed, and so forth. If the first test, but not the second, reaches significance, then only the first pair of canonical variates is interpreted.[5] If the first and second tests are significant but the third is not, then the first two pairs of variates are interpreted, and so on. Because canonical correlations are reported out in descending order of importance, usually only the first few pairs of variates are interpreted.

Once significance is established, amount of variance accounted for is of critical importance. Because there are two sets of variables, several assessments of variance are relevant. First, there is variance overlap between variates in a pair. Second is variance overlap between a variate and its own set of variables. Third is variance overlap between a variate and the other set of variables.

The first, and easiest, is the variance overlap between each significant set of canonical variate pairs. As indicated in Equation 6.2, the squared canonical correlation (or eigenvalue in BMDP) is the overlapping variance between a pair of canonical variates. Most researchers do not interpret pairs with a canonical correlation lower than .30, even if interpreted,[6] because r_c values of .30 or lower represent, squared, less than a 10% overlap in variance.

The next consideration is the variance a canonical variate extracts from its own set of variables. A pair of canonical variates may extract very different amounts of variance from their respective sets of variables. Equations 6.11 and 6.12 indicate that the variance extracted, pv, is the sum of squared loadings on a variate divided by the number of variables in the sets.[7] Because canonical variates are independent of one another (orthogonal), pvs are summed across all reliable variates to arrive at the total variance extracted from the variables by all the variates of the set.

The last consideration is the variance a variate from one set extracts from the variables in the other set, called redundancy (Stewart & Love, 1968; Miller & Farr, 1971). Equation 6.13 shows that redundancy is the percent of variance extracted by a canonical variate times the canonical correlation for the pair. A canonical variate from the IVs may be strongly correlated with the IVs, but weakly correlated with the DVs (and vice versa). Therefore the redundancies for a pair of canonical variates are usually not equal. Because canonical variates are orthogonal, redundancies for a set of variables are also added across canonical variates to get a total for the DVs relative to the IVs, and vice versa.

6.5.2 ■ Interpretation of Canonical Variates

Canonical correlation creates linear combinations of variables, canonical variates, that represent mathematically viable combinations of variables. However, although mathematically viable, they

---

[5] It is possible that the first canonical variate pair is not, by itself, significant, but rather achieves significance only in combination with the remaining canonical variate pairs. To date, there is no significance test for each pair by itself.

[6] Significance depends, to a large extent, on N.

[7] This calculation is identical to the one used in factor analysis for the same purpose, as shown in Table 13.4.

are not necessarily interpretable. A major task for the researcher is to discern, if possible, the meaning of pairs of canonical variates.

Interpretation of reliable pairs of canonical variates is based on the loading matrices, $\mathbf{A}_x$ and $\mathbf{A}_y$ (Equations 6.9 and 6.10, respectively). Each pair of canonical variates is interpreted as a pair, with a variate from one set of variables interpreted vis-à-vis the variate from the other set. A variate is interpreted by considering the pattern of variables highly correlated (loaded) with it. Because the loading matrices contain correlations, and because squared correlations measure overlapping variance, variables with correlations of .30 (9% of variance) and above are usually interpreted as part of the variate, while variables with loadings below .30 are not. Deciding on a cutoff for interpreting loadings is, however, somewhat a matter of taste, although guidelines are presented in Section 12.6.5.

6.6 ■ COMPARISON OF PROGRAMS

One program each in the SAS, SYSTAT, BMDP and SPSS packages are available for canonical analyses. Table 6.8 provides a comparison of important features of the programs. If available, the program of choice is BMDP6M. Second choices are SAS CANCORR and, with limitations, SPSS MANOVA.

6.6.1 ■ SPSS Package

The only program in the current SPSS series for canonical analysis is MANOVA (SPSS, Inc., 1990). A complete canonical analysis is available through SPSS MANOVA, which provides loadings, percents of variance, redundancy and much more. But problems arise with reading the results, because MANOVA is not designed specifically for canonical analysis and some of the labels are confusing.

Canonical analysis is requested through MANOVA by calling one set of variables DVs and the other set covariates; no IVs are listed. An example of the MANOVA setup for canonical analysis appears in Table 6.5. Interpretation of the output is facilitated if DISCRIM statistics are requested, as in Table 6.5, simply because there is less output and much of what there is pertains directly to canonical analysis. You request the dimension reduction analysis (peel off tests) and eigenvalues through the PRINT instruction. Remembering that one set of variables is identified as DVs and the other set as covariates assists in reading the output. Immediately following the first section, labeled Multivariate Tests of Significance, is output labeled Eigenvalues and Canonical Correlations. Both r_c (Canon Cor.) and r_c^2 (Sq. Cor) are given for each pair of canonical variates (Root No.). The Dimension Reduction Analysis that follows contains significance tests for the canonical correlations, evaluated as F. Also available are raw canonical coefficients and standardized canonical coefficients for dependent variables and covariates, as well as loadings: Correlations between DEPENDENT (COVARIATES) and canonical variables. Under Variance explained by canonical variates of the DEPENDENT variables one finds percent of variance (Pct Var DEP) and redundancy (Pct Var COV). These statistics are all repeated for the second set of variables, labeled COVARIATES.

Although SPSS MANOVA provides a rather complete canonical analysis, it does not calculate canonical variate scores, nor does it offer multivariate plots.

6.6.2 ■ BMDP Series

BMDP6M (Dixon, 1992) provides a complete canonical analysis and is the program of choice, if available. In addition to checks on accuracy of input are checks for univariate outliers (LARGEST

TABLE 6.8 COMPARISON OF SPSS, BMDP, SAS, AND SYSTAT PROGRAMS FOR CANONICAL CORRELATION

Feature	SPSS MANOVA[a]	BMDP6M	SAS CANCORR	SYSTAT GLM[a]
Input				
Correlation matrix	Yes	Yes	Yes	Yes
Covariance matrix	No	Yes	Yes	Yes
SSCP matrix	No	No	Yes	Yes
Number of canonical variates	No	Yes	Yes	No
Tolerance	No	Yes	Yes	Yes
Minimum canonical correlation	Specify alpha	CONSTANT	No	No
Labels for canonical variates	No	No	Yes	No
Error df if residuals input	No	No	Yes	No
Output				
Univariate:				
Means	Yes	Yes	Yes	Yes
Standard deviations	Yes	Yes	Yes	No
Coefficients of variation	No	Yes	No	No
Smallest and largest values	No	Yes	No	No
Smallest and largest standard scores	No	Yes	No	No
Skewness and kurtosis	No	Yes	No	No
Confidence intervals	Yes	No	No	No
Normal plots	Yes	No	No	No
Multivariate:				
Canonical correlations	Yes	Yes	Yes	Yes
Eigenvalues (r_c^2)	Yes	Yes	Yes	No
Significance test	F	χ^2	F	χ^2
Lambda	Yes	No	Yes	Yes
Additional test criteria	Yes	No	Yes	Yes
Correlation matrix	Yes	Yes	Yes	No
Covariance matrix	Yes	Yes	No	No
Loading matrix	Yes	Yes	Yes	Yes
Loading matrix for opposite set	No	No	Yes	No
Raw canonical coefficients	Yes	Yes	Yes	Yes
Standardized canonical coefficients	Yes	Yes	Yes	Yes
Canonical variate scores	No	Yes	Data file	Data file
Percent of variance	Yes	Yes	Yes	No
Redundancy	Yes	Yes	Yes	No
Plots: variable-variable, variable-variate, variate-variate	No	Yes	No	No
Numerical consistency	No	Yes	No	No
Multicollinearity (within-sets SMCs)	No	Yes	No	No
Between-sets SMCs	No	Yes	Yes	No
Multiple regression analyses	DVs only	No	Yes	No
Separate analyses by groups	No	No	Yes	No

[a]Additional features are listed in Section 8.6.

and SMALLEST VALUES), multicollinearity (WITHIN SET SMCs), linearity (variable-variable, variable-variate, and variate-variate PLOTS), and normality (SKEWNESS and KURTOSIS). Flexibility of input, control over the progression of the analysis, and flexibility of output are other desirable features of the program. Percent of variance and redundancy are given. And, as for many of the BMDP programs, the output is replete with helpful interpretive comments.

6.6.3 ■ SAS System

SAS CANCORR (SAS Institute, 1990) is a flexible program second only to BMDP6M in features and ease of interpretation. Along with the basics, you can specify easily interpretable labels for canonical variates and the program accepts several types of input matrices.

Multivariate output is more detailed than in BMDP6M, with several test criteria and voluminous redundancy analyses. Univariate output is minimal, however, and if plots are desired, case statistics such as canonical scores are written to a file to be analyzed by the SAS PLOT procedure. If requested, the program does separate multiple regressions with each variable predicted from the other set. You can also do separate canonical correlation analyses for different groups.

6.6.4 ■ SYSTAT System

Canonical analysis is done through the multivariate general linear model GLM program in SYS-TAT (Wilkinson, 1994). But to get all the output, the analysis must be done twice, once with the first set of variables defined as the DVs, and a second time with the other set of variables defined as DVs. Although tests of significance and canonical correlations are the same for both analyses, coefficients and loading matrices differ. And, of course, you must keep track of which set was called "DVs" in which analysis.

The program provides several test criteria for overall significance of the canonical analysis. However, few other amenities are available in this program. As for most SYSTAT procedures, additional output is produced by writing case statistics to a file and analyzing them through modules such as STAT for univariate statistics or GRAPH for plots.

6.7 ■ COMPLETE EXAMPLE OF CANONICAL CORRELATION

For an example of canonical correlation, variables are selected from among those made available by research described in Appendix B, Section B.1. The goal of analysis is to discover the dimensions, if any, along which certain attitudinal variables are related to certain health characteristics.

Selected attitudinal variables (Set 1) include attitudes toward the role of women (ATTROLE), toward locus of control (CONTROL), toward current marital status (ATTMAR), and toward self (ESTEEM). Larger numbers indicate increasingly conservative attitudes about the proper role of women, increasing feelings of powerlessness to control one's fate (external as opposed to internal locus of control), increasing dissatisfaction with current marital status, and increasingly poor self-esteem.

Selected health variables (Set 2) include mental health (MENHEAL), physical health (PHY-HEAL), number of visits to health professionals (TIMEDRS), attitude toward use of medication (ATTDRUG), and a frequency-duration measure of use of psychotropic drugs (DRUGUSE). Larger numbers reflect poorer mental and physical health, more visits, greater willingness to use drugs, and more use of them.

TABLE 6.9 SETUP AND SELECTED BMDP2D OUTPUT FOR INITIAL SCREENING OF CANONICAL CORRELATION DATA SET

```
/INPUT       VARIABLES = 10.  FILE = 'CANON.DAT'.  FORMAT IS FREE.
/VARIABLE    NAMES ARE SUBNO, TIMEDRS, ATTDRUG, PHYHEAL, MENHEAL, ESTEEM,
             CONTROL, ATTMAR, DRUGUSE, ATTROLE.

             MISSING ARE (7)0, 0.
             LABEL = SUBNO.
/PRINT       NO COUNT.
/END
```

```
************
*  ATTDRUG *
************
```

VARIABLE NUMBER	3	MAXIMUM	10.0000000
NUMBER OF DISTINCT VALUES . .	6	MINIMUM	5.0000000
NUMBER OF VALUES COUNTED. .	465	RANGE	5.0000000
NUMBER OF VALUES NOT COUNTED	0	VARIANCE	1.3365499

```
                                          ST.DEV.      1.1560925
                                          (Q3-Q1)/2    1.0000000
                                          MX.ST.SC.    2.00
                                          MN.ST.SC.   -2.32
```

	ESTIMATE	ST.ERROR		95% CONFIDENCE	
				LOWER	UPPER
MEAN	7.6860213	0.0536125		7.5806680	7.7913747
MEDIAN	8.0000000	0.0000000			
MODE	8.0000000				

```
                                                    H
                                                    H            H
                                        H   H       H   H
                                        H   H   H   H   H
                                        H   H   H   H   H
                                        H   H   H   H   H
                                        H   H   H   H   H
                                    H   H   H   H   H   H
                                    H   H   H   H   H   H
                                    H   H   H   H   H   H
                                L-----------------------U

                                    EACH '-' ABOVE  =      0.2000
                                                    L=      5.0000
                                                    U=     10.2000
                        CASE NO. OF MIN. VAL. =   15
                        CASE NO. OF MAX. VAL. =   20

                                          EACH 'H'
                                          REPRESENTS
                                                15
                                          COUNT(S)
```

	VALUE	VALUE/S.E.		
SKEWNESS	-0.12	-1.07	Q1=	7.0000000
KURTOSIS	-0.47	-2.05	Q3=	9.0000000
			S-=	6.5299287
			S+=	8.8421135

TABLE 6.9 (CONTINUED)

```
***********
*  ATTMAR  *
***********

VARIABLE NUMBER . . . . . .          8
NUMBER OF DISTINCT VALUES .         37
NUMBER OF VALUES COUNTED. .        460
NUMBER OF VALUES NOT COUNTED         5

                  ESTIMATE     ST.ERROR
MEAN            22.9804344    0.3998048
MEDIAN          21.0000000    0.5773506
MODE            16.0000000
```

```
MAXIMUM        58.0000000
MINIMUM        11.0000000
RANGE          47.0000000
VARIANCE       73.1608353
ST.DEV.         8.5534105
(Q3-Q1)/2       6.0000000
MX.ST.SC.       4.09
MN.ST.SC.      -1.40

          95% CONFIDENCE
      LOWER          UPPER
   22.1967239     23.7641449
```

```
                                                    H
                                                    H  H
                                                   HHHH
                                                   HHHH  H            EACH 'H'
                                                   HHHHHHH            REPRESENTS
                                                 HHHHHHHH  H  H            6
                                                 HHHHHHHHHH  H
                                                 HHHHHHHHHHH  H       COUNT(S)
                                               HHHHHHHHHHHH  HH  H  H
                                             HHHHHHHHHHHHHHHHHHH  H  H  H
                                        L---------------------------------U

                         EACH '-' ABOVE =    2.0000
                                         L=  10.0000
                                         U=  62.0000

               CASE NO. OF MIN. VAL. =    11
               CASE NO. OF MAX. VAL. =   279

                         VALUE   VALUE/S.E.        Q1=  16.0000000
             SKEWNESS     1.00      8.73           Q3=  28.0000000
             KURTOSIS     0.77      3.39           S-=  14.4270239
                                                   S+=  31.5338440
```

```
************
*  DRUGUSE  *
************

VARIABLE NUMBER . . . . . .          9
NUMBER OF DISTINCT VALUES .         42
NUMBER OF VALUES COUNTED. .        465
NUMBER OF VALUES NOT COUNTED         0

                  ESTIMATE     ST.ERROR
MEAN             9.0021505    0.4689813
MEDIAN           5.0000000    0.5773506
MODE             0.0000000
```

```
MAXIMUM        66.0000000
MINIMUM         0.0000000
RANGE          66.0000000
VARIANCE      102.2736970
ST.DEV.        10.1130457
(Q3-Q1)/2       6.5000000
MX.ST.SC.       5.64
MN.ST.SC.      -0.89

          95% CONFIDENCE
      LOWER          UPPER
    8.0805597      9.9237413
```

```
                                                    H
                                                    H
                                                    H
                                                    H
                                                    H
                                                   HH                 EACH 'H'
                                                   HHH                REPRESENTS
                                                 HHHHH  H                 19
                                             HHHHHHHHHHHHH  HH  H     COUNT(S)
                                        L----------------------------U

                         EACH '-' ABOVE =    3.0000
                                         L=   0.0000
                                         U=  78.0000

               CASE NO. OF MIN. VAL. =     2
               CASE NO. OF MAX. VAL. =    35

                         VALUE   VALUE/S.E.        Q1=   1.0000000
             SKEWNESS     1.75     15.40           Q3=  14.0000000
             KURTOSIS     4.17     18.36           S-=  -1.1108952
                                                   S+=  19.1151962
```

6.7.1 ■ Evaluation of Assumptions

6.7.1.1 ■ Missing Data

A screening run through BMDP2D, partially illustrated in Table 6.9, finds missing data for 6 of the 465 cases. One woman lacks a score on CONTROL, and five lack scores on ATTMAR. With deletion of these cases, remaining $N = 459$.

6.7.1.2 ■ Normality, Linearity, and Homoscedasticity

BMDP6M provides a particularly flexible scheme for assessing normality, linearity, and homoscedasticity between pairs of canonical variates, between pairs of variables, and between pairs of variables and variates. The PLOT paragraph allows one to request any number of scatterplots of pairs of variables and variates as well as to control the size of the plot.

Figure 6.1 shows two scatterplots produced by BMDP6M for the example using default size values for the plots. The scatterplots are between the first and second pairs of canonical variates, respectively. CNVRF1 is canonical variate scores, first set, first variate; CNVRS1 is canonical variate scores, second set, first variate. CNVRF2 is canonical variate scores, first set, second variate; CNVRS2 is canonical variate scores, second set, second variate.

The shapes of the scatterplots reflect the low canonical correlations for the solution (see section 6.7.2), particularly for the second pair of variates where the overall shape is nearly circular except for a few extreme values in the lower third of the plot. There are no obvious departures from linearity or homoscedasticity because the overall shapes do not curve and they are of about the same width throughout.

Deviation from normality is evident, however, for both pairs of canonical variates: on both plots the 0–0 point departs from the center of the vertical and horizontal axes. If the points are projected as a frequency distribution to the vertical or horizontal axes of the plots, there is further evidence of skewness. For the first plot there is a pileup of cases at low scores and a smattering of cases at high scores on both axes, indicating positive skewness. In plot 2, there are widely scattered cases with extremely low scores on CNVRS2, with no corresponding high scores, indicating negative skewness.

Departure from normality is confirmed by the output of BMDP2D, where Table 6.9 shows positive skewness for ATTMAR and, especially, DRUGUSE. Logarithmic transformation of these variables, plus TIMEDRS and PHYHEAL (output containing their distributions is not shown) results in variables that are far less skewed. The transformed variables are named LATTMAR, LDRUGUSE, LTIMEDRS, and LPHYHEAL. A second BMDP6M run provides univariate statistics for both transformed and untransformed variables. Compare the skewness and kurtosis of ATTMAR and DRUGUSE in Table 6.9 with that of LATTMAR and LDRUGUSE in Table 6.10. BMDP6M plots based on transformed variables (not shown) confirm improvement in normality with transformed variables, particularly for the second pair of canonical variates.

Note in the BMDP6M descriptive output of Table 6.10 that coefficients of variation, and kurtosis are also reasonable for a data set of this size.

---

[8] BMDP4M could also be used.

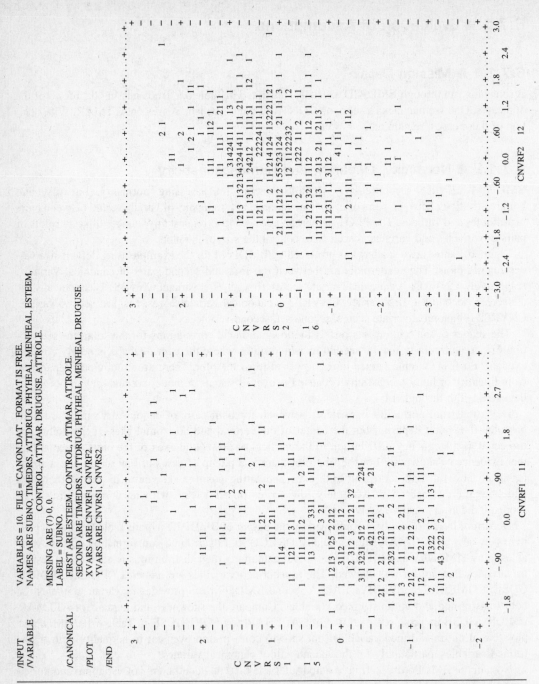

FIGURE 6.1 SETUP AND SELECTED BMDP6M OUTPUT SHOWING SCATTERPLOTS BETWEEN FIRST AND SECOND PAIRS OF CANONICAL VARIATES.

TABLE 6.10   SETUP AND SELECTED BMDP6M OUTPUT SHOWING DESCRIPTIVE STATISTICS

```
/INPUT       VARIABLES = 10.  FILE = 'CANON.DAT'.  FORMAT IS FREE.
/VARIABLE    NAMES ARE SUBNO, TIMEDRS, ATTDRUG, PHYHEAL, MENHEAL, ESTEEM,
                CONTROL, ATTMAR, DRUGUSE, ATTROLE.
             MISSING ARE (7)0, 0.
             LABEL = SUBNO.
/TRANSFORM   LTIMEDRS = LOG(TIMEDRS + 1).
             LPHYHEAL = LOG(PHYHEAL).
             LATTMAR = LOG(ATTMAR).
             LDRUGUSE = LOG(DRUGUSE + 1).
/CANONICAL   FIRST ARE ESTEEM, CONTROL, LATTMAR, ATTROLE.
             SECOND ARE LTIMEDRS, ATTDRUG, LPHYHEAL, MENHEAL, LDRUGUSE.
/PRINT       MATRICES ARE CORR, COEF, LOAD.
/PLOT        XVARS ARE CNVRF1, CNVRF2.
             YVARS ARE CNVRS1, CNVRS2.
/END
```

UNIVARIATE SUMMARY STATISTICS

VARIABLE	MEAN	STANDARD DEVIATION	COEFFICIENT OF VARIATION	SMALLEST VALUE	LARGEST VALUE	SMALLEST STANDARD SCORE	LARGEST STANDARD SCORE	SKEWNESS	KURTOSIS
6 ESTEEM	15.81699	3.94798	0.249604	8.00000	29.00000	-1.98	3.34	0.48	0.27
7 CONTROL	6.74946	1.27127	0.188352	5.00000	10.00000	-1.38	2.56	0.48	-0.44
13 LATTMAR	1.33374	0.15408	0.115522	1.04139	1.76343	-1.90	2.79	0.23	-0.61
10 ATTROLE	35.19608	6.75141	0.191823	18.00000	55.00000	-2.55	2.93	0.05	-0.42
11 LTIMEDRS	0.74139	0.41677	0.562148	0.00000	1.91381	-1.78	2.81	0.23	-0.21
3 ATTDRUG	7.67756	1.15433	0.150352	5.00000	10.00000	-2.32	2.01	-0.11	-0.45
12 LPHYHEAL	0.64766	0.20675	0.319231	0.30103	1.17609	-1.68	2.56	-0.01	-0.71
5 MENHEAL	6.11329	4.14532	0.678083	0.00000	18.00000	-1.47	2.87	0.57	-0.35
14 LDRUGUSE	0.76372	0.48867	0.639851	0.00000	1.82607	-1.56	2.17	-0.15	-1.10

VALUES FOR KURTOSIS GREATER THAN ZERO INDICATE DISTRIBUTIONS
WITH HEAVIER TAILS THAN THE NORMAL DISTRIBUTION.

6.7.1.3 ■ Outliers

UNIVARIATE SUMMARY STATISTICS in Table 6.10 provide information on univariate outliers. Note that the smallest and largest standard scores on the transformed variables are within a range anticipated in a sample of over 400 cases. The extremely large standard scores (MX.ST.SC. and MN.ST.SC) on ATTMAR and DRUGUSE (before transformation, Table 6.9) are not seen on LATTMAR and LDRUGUSE.

Analysis of multivariate outliers is not available through BMDP6M. Therefore BMDPAM is used to screen for multivariate outliers separately among the two sets of variables. Outliers appear in the BMDPAM program as cases with large Mahalanobis distances, shown as χ^2, from the centroid of the cases for original data. BMDPAM highlights with an asterisk any SIGNIFICANCE value for a case where $p < .001$. A segment of the output for the first set of variables appears in Table 6.11. There are no outliers in this segment or any other in either set of variables.

Cases with missing values on one variable show three rather than four degrees of freedom for χ^2. The full run confirms that there are indeed six cases with missing data, rather than fewer as there would be if a case were missing data on two variables.

6.7.1.4 ■ Multicollinearity and Singularity

BMDP6M provides a direct test of multicollinearity/singularity, as shown in Table 6.12. SMCs are reported where each variable in a set serves, in turn, as DV with the others as IVs. If R-SQUARED values become very large (say, .90 or above), then one variable in the set is a near-linear combination of others. In this case, low R-SQUARED values indicate absence of singularity or multicollinearity and, indeed, considerable heterogeneity in the sets of variables, particularly the attitudinal set.

6.7.2 ■ Canonical Correlation

The number and importance of canonical variates are determined using procedures from Section 6.5.1. Significance of the relationships between the sets of variables is reported directly by BMDP6M, as shown in Table 6.13. With all four canonical correlations included, $\chi^2(20) = 108.19$, $p < .001$. With the first and second canonical correlations removed, χ^2 values are not significant; $\chi^2(6) = 4.11$, $p = .66$. Therefore only significant relationships are in the first two pairs of canonical variates and these are interpreted.

Canonical correlations (r_c) and eigenvalues (r_c^2) are also in Table 6.13. The first canonical correlation is .38, representing 14% overlapping variance for the first pair of canonical variates (see Equation 6.2). The second canonical correlation is .27, representing 7% overlapping variance for the second pair of canonical variates. Although highly significant, neither of these two canonical correlations represents a substantial relationship between pairs of canonical variates. Interpretation of the second canonical correlation and its corresponding pair of canonical variates is marginal.

Loading matrices between canonical variates and original variables are in Table 6.14. Interpretation of the two significant pairs of canonical variates from loadings follows procedures mentioned in Section 6.5.2. Correlations between variables and variates (loadings) in excess of .3 are interpreted. Both the direction of correlations in the loading matrices and the direction of scales of measurement are considered when interpreting the canonical variates.

The first pair of canonical variates has high loadings on ESTEEM, CONTROL, and LATTMAR (.596, .784, and .730, respectively) on the attitudinal set and on LPHYHEAL and MENHEAL (.408

TABLE 6.11 SETUP AND SELECTED BMDPAM OUTPUT FOR IDENTIFICATION OF OUTLIERS FOR THE FIRST SET OF VARIABLES

```
/INPUT        VARIABLES = 10.   FILE = 'CANON.DAT'.   FORMAT IS FREE.
/VARIABLE     NAMES ARE SUBNO, TIMEDRS, ATTDRUG, PHYHEAL, MENHEAL, ESTEEM,
                    CONTROL, ATTMAR, DRUGUSE, ATTROLE.
              MISSING ARE (7)0, 0.
              LABEL = SUBNO.
              USE = 6, 7, 10, 13.
/TRANSFORM    LTIMEDRS = LOG(TIMEDRS + 1).
              LPHYHEAL = LOG(PHYHEAL).
              LATTMAR = LOG(ATTMAR).
              LDRUGUSE = LOG(DRUGUSE + 1).
/EST          METHOD = REGR.
/PRINT        MATR = DIS.
/END
```

MAHALANOBIS DISTANCES ARE COMPUTED FROM EACH CASE TO THE
CENTROID OF ITS GROUP. ONLY THOSE VARIABLES WHICH WERE
ORIGINALLY AVAILABLE ARE USED--ESTIMATED VALUES ARE NOT USED.
FOR LARGE MULTIVARIATE NORMAL SAMPLES, THE MAHALANOBIS
DISTANCES HAVE AN APPROXIMATELY CHI-SQUARE DISTRIBUTION WITH
THE NUMBER OF DEGREES OF FREEDOM EQUAL TO THE NUMBER OF
NONMISSING VARIABLES. SIGNIFICANCE LEVELS REPORTED BELOW
THAT ARE LESS THAN .001 ARE FLAGGED WITH AN ASTERISK (*).

ESTIMATES OF MISSING DATA, MAHALANOBIS D-SQUARED (CHI-SQUARED)
AND SQUARED MULTIPLE CORRELATIONS WITH AVAILABLE VARIABLES

CASE LABEL	CASE NUMBER	MISSING VARIABLE	ESTIMATE	R- SQUARED	GROUP	CHI-SQ	CHISQ/DF	D.F.	SIGNIFICANCE
121	101					0.572	0.143	4	0.9661
122	102					1.178	0.294	4	0.8818
123	103					1.105	0.276	4	0.8934
124	104					3.109	0.777	4	0.5397
125	105					2.130	0.533	4	0.7118
126	106					9.019	2.255	4	0.0606
127	107					2.459	0.615	4	0.6520
128	108					2.282	0.570	4	0.6841
129	109					4.520	1.130	4	0.3402
130	110					3.101	0.775	4	0.5410
131	111					1.026	0.256	4	0.9059
132	112					3.404	0.851	4	0.4926
133	113					4.624	1.541	3	0.2015
134	114					3.312	0.828	4	0.5070
135	115					3.073	0.768	4	0.5456

TABLE 6.12   SELECTED BMDP6M OUTPUT FOR ASSESSMENT OF MULTICOLLINEARITY.   SETUP SHOWN IN
TABLE 6.10.

SQUARED MULTIPLE CORRELATIONS OF EACH VARIABLE IN
SECOND SET WITH ALL OTHER VARIABLES IN SECOND SET

VARIABLE NUMBER	NAME	R-SQUARED
11	LTIMEDRS	0.36864
3	ATTDRUG	0.08002
12	LPHYHEAL	0.47204
5	MENHEAL	0.29564
14	LDRUGUSE	0.30872

SQUARED MULTIPLE CORRELATIONS OF EACH VARIABLE IN
FIRST SET WITH ALL OTHER VARIABLES IN FIRST SET

VARIABLE NUMBER	NAME	R-SQUARED
6	ESTEEM	0.22705
7	CONTROL	0.13316
13	LATTMAR	0.13184
10	ATTROLE	0.06751

TABLE 6.13   SELECTION PORTION OF BMDP6M OUTPUT SHOWING CANONICAL CORRELATIONS AND SIGNIFI-
CANCE LEVELS FOR SETS OF CANONICAL CORRELATIONS

EIGENVALUE	CANONICAL CORRELATION	NUMBER OF EIGENVALUES	CH-SQUARE	D.F.	TAIL PROB.
			108.19	20	0.0000
0.14358	0.37892	1	37.98	12	0.0002
0.07203	0.26839	2	4.11	6	0.6613
0.00787	0.08873	3	0.53	2	0.7661
0.00118	0.03429				

BARTLETT'S TEST ABOVE INDICATES THE NUMBER OF CANONICAL
VARIABLES NECESSARY TO EXPRESS THE DEPENDENCY BETWEEN THE
TWO SETS OF VARIABLES. THE NECESSARY NUMBER OF CANONICAL
VARIABLES IS THE SMALLEST NUMBER OF EIGENVALUES SUCH THAT
THE TEST OF THE REMAINING EIGENVALUES IS NOT SIGNIFICANT.
FOR EXAMPLE, IF A TEST AT THE .01 LEVEL WERE DESIRED,
THEN   2 VARIABLES WOULD BE CONSIDERED NECESSARY.
HOWEVER, THE NUMBER OF CANONICAL VARIABLES OF PRACTICAL
VALUE IS LIKELY TO BE SMALLER.

and .968) on the health side. Thus, low self esteem, external locus of control, and dissatisfaction with marital status are related to poor physical and mental health.

The second pair of canonical variates has high loadings on ESTEEM, LATTMAR and ATTROLE (.601, −.317, and .783) on the attitudinal side and LTIMEDRS, ATTDRUG, and LDRUGUSE (−.359, .559, and −.548) on the health side. Big numbers on ESTEEM, little numbers on LATTMAR, and big numbers on ATTROLE go with little numbers on LTIMEDRS, big numbers on

TABLE 6.14 SELECTED BMDP6M OUTPUT OF LOADING MATRICES FOR THE TWO SETS OF VARIABLES IN THE EXAMPLE

CANONICAL VARIABLE LOADINGS

(CORRELATIONS OF CANONICAL VARIABLES WITH ORIGINAL VARIABLES)
FOR FIRST SET OF VARIABLES

		CNVRF1	CNVRF2	CNVRF3	CNVRF4
		1	2	3	4
ESTEEM	6	0.596	0.601	-0.286	-0.450
CONTROL	7	0.784	0.148	-0.177	0.577
LATTMAR	13	0.730	-0.317	0.434	-0.422
ATTROLE	10	-0.094	0.783	0.605	0.113

CANONICAL VARIABLE LOADINGS

(CORRELATIONS OF CANONICAL VARIABLES WITH ORIGINAL VARIABLES)
FOR SECOND SET OF VARIABLES

		CNVRS1	CNVRS2	CNVRS3	CNVRS4
		1	2	3	4
LTIMEDRS	11	0.123	-0.359	-0.860	0.249
ATTDRUG	3	0.077	0.559	-0.033	0.405
LPHYHEAL	12	0.408	-0.048	-0.640	-0.505
MENHEAL	5	0.968	-0.143	-0.189	0.066
LDRUGUSE	14	0.276	-0.548	0.016	-0.005

TABLE 6.15 SELECTED BMDP6M OUTPUT SHOWING PERCENTS OF VARIANCE AND REDUNDANCY FOR FIRST AND SECOND SET OF CANONICAL VARIATES

CANON. VAR.	AVERAGE SQUARED LOADING FOR EACH CANONICAL VARIABLE (1ST SET)	AV. SQ. LOADING TIMES SQUARED CANON. CORREL. (1ST SET)	AVERAGE SQUARED LOADING FOR EACH CANONICAL VARIABLE (2ND SET)	AV. SQ. LOADING TIMES SQUARED CANON. CORREL. (2ND SET)	SQUARED CANON. CORREL.
1	0.37773	0.05424	0.24008	0.03447	0.14358
2	0.27392	0.01973	0.15294	0.01102	0.07203
3	0.16678	0.00131	0.23720	0.00187	0.00787
4	0.18157	0.00021	0.09702	0.00011	0.00118

THE AVERAGE SQUARED LOADING MULTIPLIED BY THE SQUARED CANONICAL
CORRELATION IS THE AVERAGE SQUARED CORRELATION OF A
VARIABLE IN ONE SET WITH THE CANONICAL VARIABLE FROM
THE OTHER SET. IT IS SOMETIMES CALLED A REDUNDANCY INDEX.

ATTDRUG, and little numbers on LDRUGUSE. That is, low self-esteem, satisfaction with marital status, and conservative attitudes toward the proper role of women in society go with few visits to physicians, favorable attitudes toward use of drugs, and little actual use of them. (Figure that one out!)

Loadings are converted to *pv* values by application of Equations 6.11 and 6.12. These values are shown in the BMDP6M output in columns labeled AVERAGE SQUARED LOADING FOR EACH CANONICAL VARIATE (Table 6.15). The values for the first pair of canonical variates are .38 for the first set of variables and .24 for the second set of variables. That is, the first canonical variate pair extracts 38% of variance from the attitudinal variables and 24% of variance from the health variables. The values for the second pair of canonical variates are .27 for the first set of variables

---

TABLE 6.16   SELECTED BMDP6M OUTPUT OF UNSTANDARDIZED AND STANDARDIZED CANONICAL VARIATE COEFFICIENTS

COEFFICIENTS FOR CANONICAL VARIABLES FOR FIRST SET OF VARIABLES

		CNVRF1 1	CNVRF2 2	CNVRF3 3	CNVRF4 4
ESTEEM	6	0.061949	0.155148	-0.158971	-0.172697
CONTROL	7	0.465185	0.021395	-0.086712	0.699597
LATTMAR	13	3.401715	-2.916037	4.756204	-2.413351
ATTROLE	10	-0.012933	0.091924	0.118283	0.030312

STANDARDIZED COEFFICIENTS FOR CANONICAL VARIABLES FOR FIRST SET OF VARIABLES

(THESE ARE THE COEFFICIENTS FOR THE STANDARDIZED VARIABLES)

		CNVRF1 1	CNVRF2 2	CNVRF3 3	CNVRF4 4
ESTEEM	6	0.245	0.613	-0.628	-0.682
CONTROL	7	0.591	0.027	-0.110	0.889
LATTMAR	13	0.524	-0.449	0.733	-0.372
ATTROLE	10	-0.087	0.621	0.799	0.205

COEFFICIENTS FOR CANONICAL VARIABLES FOR SECOND SET OF VARIABLES

		CNVRS1 1	CNVRS2 2	CNVRS3 3	CNVRS4 4
LTIMEDRS	11	-0.643182	-0.925366	-2.051052	1.873687
ATTDRUG	3	0.039664	0.673311	-0.039252	0.388091
LPHYHEAL	12	0.208193	2.159183	-2.144725	-5.739981
MENHEAL	5	0.256358	0.008586	0.036875	0.091726
LDRUGUSE	14	-0.122005	-1.693256	1.048614	-0.102427

STANDARDIZED COEFFICIENTS FOR CANONICAL VARIABLES FOR SECOND SET OF VARIABLES

(THESE ARE THE COEFFICIENTS FOR THE STANDARDIZED VARIABLES)

		CNVRS1 1	CNVRS2 2	CNVRS3 3	CNVRS4 4
LTIMEDRS	11	-0.268	-0.386	-0.855	0.781
ATTDRUG	3	0.046	0.777	-0.045	0.448
LPHYHEAL	12	0.043	0.446	-0.443	-1.187
MENHEAL	5	1.063	0.036	0.153	0.380
LDRUGUSE	14	-0.060	-0.827	-0.512	-0.050

---

and .15 for the second set; the second canonical variate pair extracts 27% of variance from the attitudinal variables and 15% of variance from the health variables. Together, the two canonical variates account for 65% of variance (38 plus 27%) in the attitudinal set, and 39% of variance (24% and 15%) in the health set.

Redundancies for the canonical variates are found in BMDP6M under the columns labeled AV. SQ. LOADING TIMES SQUARED CANON. CORREL. (Table 6.15). That is, the first health variate accounts for 5% of the variance in the attitudinal variables, while the second health variate accounts for 2% of the variance. Together, two health variates "explain" 7% of the variance in attitudinal

TABLE 6.17 CORRELATIONS, STANDARDIZED CANONICAL COEFFICIENTS, CANONICAL CORRELATIONS, PERCENTS OF VARIANCE, AND REDUNDANCIES BETWEEN ATTITUDINAL AND HEALTH VARIABLES AND THEIR CORRESPONDING CANONICAL VARIATES

	First canonical variate		Second canonical variate	
	Correlation	Coefficient	Correlation	Coefficient
Attitudinal set				
CONTROL	.78	.59	.15	.03
LATTMAR	.73	.52	−.32	−.45
ESTEEM	.60	.25	.60	.61
ATTROLE	−.09	−.09	.78	.62
Percent of Variance	.38		.27	Total = .65
Redundancy	.05		.02	Total = .07
Health set				
MENHEAL	.97	1.06	−.14	.04
LPHYHEAL	.41	.04	−.53	.45
LTIMEDRS	.12	−.27	−.36	−.39
ATTDRUG	.08	.05	.56	.78
LDRUGUSE	.28	−.06	−.55	−.83
Percent of variance	.24		.15	Total = .39
Redundancy	.03		.01	Total = .04
Canonical correlation	.38		.27	

TABLE 6.18 CHECKLIST FOR CANONICAL CORRELATION

1. Issues
    a. Missing data
    b. Normality, linearity, homoscedasticity
    c. Outliers
    d. Multicollinearity and singularity
2. Major analyses
    a. Significance of canonical correlations
    b. Correlations of variables and variates
    c. Variance accounted for
        (1) By canonical correlations
        (2) By same-set canonical variates
        (3) By other-set canonical variates (redundancy)
3. Additional analyses
    a. Canonical coefficients
    b. Canonical variates scores

variables. The first attitudinal variate accounts for 3% and the second 1% of the variance in the health set. Together the two attitudinal variates overlap the variance in the health set 4%.

If a goal of analysis is production of scores on canonical variates, coefficients for them are readily available. Table 6.16 shows both standardized and unstandardized coefficients for production of canonical variates. Scores on the variates themselves for each case are also produced by BMDP6M if CANV is requested in the /PRINT paragraph. A summary table of information appropriate for inclusion in a journal article appears in Table 6.17.

A checklist for canonical correlation appears in Table 6.18. An example of a Results section in journal format follows for the complete analysis described in Section 6.7.

Results

Canonical correlation was performed between a set of attitudinal variables and a set of health variables using BMDP6M (Dixon, 1992). The attitudinal set included attitudes toward the role of women, toward locus of control, toward current marital status, and toward self worth. The health set measured mental health, physical health, visits to health professionals, attitude toward use of medication, and use of psychotropic drugs. Increasingly large numbers reflected more conservative attitudes toward women's role, external locus of control, dissatisfaction with marital status, low self-esteem, poor mental health, poor physical health, more numerous health visits, favorable attitudes toward drug use, and more drug use.

To improve linearity of relationship between variables and normality of their distributions, logarithmic transformations were applied to attitude toward marital status, visits to health professionals, physical health, and drug use. No within-set multivariate outliers were identified at $p < .001$, although six cases were found to be missing data on locus of control or attitude toward marital status and were deleted, leaving $N = 459$. Assumptions regarding within-set multicollinearity were met.

The first canonical correlation was .38 (14% overlapping variance); the second was .27 (7% overlapping variance). The remaining two canonical correlations were effectively zero. With all four canonical correlations included, $\chi^2(20) = 108.19$, $p < .001$, and with the first canonical correlation removed, $\chi^2(12) = 37.98$, $p < .001$. Subsequent χ^2 tests were not statistically significant. The first two pairs of canonical variates, therefore, accounted for the significant relationships between the two sets of variables.

Data on the first two pairs of canonical variates appear in Table 6.17. Shown in the table are correlations between the variables and the canonical variates, standardized canonical variate coefficients, within-set variance accounted for by the canonical variates (percent of variance), redundancies, and canonical correlations. Total percent of variance and total redundancy indicate that the first pair of canonical variates was moderately related, but the second pair was only minimally related; interpretation of the second pair is questionable.

With a cutoff correlation of .3, the variables in the attitudinal set that were correlated with the first canonical variate were locus of control, (log of) attitude toward marital status, and self-esteem. Among the health variables, mental health and (log of) physical health correlated with the first canonical variate. The first pair of canonical variates indicate that those with external locus of control (.78), feelings of dissatisfaction toward marital status (.73), and lower self-esteem (.60) are associated with more numerous mental health symptoms (.97) and more numerous physical health symptoms (.41).

The second canonical variate in the attitudinal set was composed of attitude toward role of women, self-esteem, and negative of (log of) attitude toward marital status, while the corresponding canonical variate from the health set was composed of negative of (log of) number of physical health symptoms, negative of (log of) drug use, attitude toward drugs, and negative of (log of) visits to health professionals. Taken as a pair, these variates suggest that a combination of more conservative attitudes toward the role of women (.78), lower self-esteem (.60), but relative satisfaction with marital status (−.32) is associated with a combination of more favorable attitudes toward use of drugs (.56), but lower psychotropic drug use (−.55), fewer physical health symptoms (−.53), and fewer visits to health professionals (−.36).

6.8 ■ SOME EXAMPLES FROM THE LITERATURE

Wingard, Huba, and Bentler (1979) report a canonical analysis of relationships between personality variables and use of various licit and illicit drugs among junior high school students in a metropolitan area. The sample was sufficiently large ($N = 1634$) to be randomly divided into groups for assessment of the stability of the canonical solution. (When a large enough sample is available, cross validation between randomly selected halves of the sample is highly desirable.) At least two reliable pairs of canonical variates were discovered for both samples; the first pair, but not the second, was similar for the samples. The first canonical variate in the drug use set seemed to reflect early patterns of experimentation with the relatively less dangerous, more readily available legal and illegal drugs. The corresponding variate from the personality set reflected "non-abidance with the law, liberalism, leadership, extraversion, lack of diligence, and lack of deliberateness" (p. 139). An attempt to rotate the two dimensions to facilitate interpretation of the second pair of variates failed and also degraded the correspondence on the first dimension for the subsamples. Problems with variate skewness were noted as potentially responsible for difficulties in comparing the second pair of canonical variates for the two subsamples. Although canonical correlations and significance levels were high, redundancies were not.

Cohen, Gaughran, and Cohen (1979) examined the relationships between patterns of fertility across six different age groups (as DVs) and five different sets of demographic characteristics as IVs (education and occupation; income-labor force; ethnicity; marriage life cycle; and housing and occupancy) in five separate canonical analyses. The units of analysis were 338 New York City health areas. At least three of a possible six pairs of canonical variates were interpreted for each set of IVs, with the first pair showing substantial canonical correlations in all cases.

On the DV side, the first canonical variates represented substantial teenage and early twenties childbearing. On the IV side, the first canonical variates represented poorly educated, poverty-level, minority, unmarried persons living in overcrowded housing. Redundancy levels for predicting birthrates from each of the five sets of IVs were about 30%.

On the DV side, the second canonical variates were associated with low rates of childbearing in middle years (20 to 39). Canonical correlations were around .65 and redundancies around 14% for second pairs of variates. On the IV side, low middle-age childbearing was associated with constellations of demographic characteristics associated with an unmarried, affluent-singles life style and working, educated women. The third pairs of canonical variates had modest correlations and redundancies. Childbearing among the oldest group of women (over 40) was associated with certain ethnic and religious characteristics. Cohen and colleagues direct those who seek to understand and control population expansion to the relationships found in the first pairs of canonical variates, where the absolute level of childbearing was highest.

In a paper comparing canonical analysis with a method known as external single-set components analysis (ESSCA), Fornell (1979) describes relationships between a set of variables measuring characteristics of 128 consumer affairs departments and a set measuring the ability of the departments to influence management decision making. Two significant pairs of canonical variates were discovered, but the second pair had distinctly marginal redundancy (7%) in the direction of interest (predicting influence on decision making from characteristics of departments). Neither pair was deemed interpretable without rotation.

The rotated solution provided by ESSCA did, however, permit interpretation. The first component predicted impact on consumer service and information, and the second predicted impact on marketing decisions. Fornell recommends ESSCA, which maximizes the sum of squared loadings between IV variates and DV variables, over canonical analysis, which maximizes correlation between pairs of canonical variates, when the distinction between IV and DV is clear, so that only a set of variates from the IVs is required. ESSCA may also prove more interpretable when values in $\mathbf{R}_{xy}$ differ greatly in magnitude.

CHAPTER 7
Multiway Frequency Analysis

7.1 ■ GENERAL PURPOSE AND DESCRIPTION

Relationships among three or more discrete (categorical, qualitative) variables are studied through multiway frequency analysis or an extension of it called loglinear analysis. Relationships among two discrete variables, say area of psychology (clinical, general experimental, developmental) and average number of publications a year (0, 1, 2, 3, and 4 or more), are studied through the two-way χ^2 test of association. If a third variable is added, such as number of statistics courses taken (two or fewer vs. more than two), two- and three-way associations are sought through multiway frequency analysis. Is number of publications related to area of psychology and/or to number of statistics courses taken? Is number of statistics courses taken related to area of psychology? Is there a three-way relationship among number of publications, area of psychology, and number of statistics courses taken?

To do a multiway frequency analysis, tables are formed that contain the one-way, two-way, three-way, and higher-order associations. A linear model of (the logarithm of) expected cell frequencies is developed. The loglinear model starts with all of the one-, two-, three-, and higher-way associations and then eliminates as many of them as possible while still maintaining an adequate fit between expected and observed cell frequencies. In the example above, the three-way association between number of publications, area of psychology, and number of statistics courses is tested first and then eliminated if not statistically significant. Then the two-way associations (number of publications and area of psychology, number of publications and number of statistics courses, area of psychology and number of statistics courses) are tested and, if not significant, eliminated. Finally, there is a one-way test for each of the variables against the hypothesis that frequencies are equal in each cell (e.g., that there are equal numbers of psychologists in each area—a test analogous to equal frequency goodness-of-fit tests in χ^2 analysis).

The researcher may consider one of the variables a DV while the others are considered IVs. For example, a psychologist's success as a professional (successful vs. unsuccessful) is studied as a function of number of publications, area of psychology, number of statistics courses taken, and their interactions. Used this way, multiway frequency analysis is like a nonparametric analysis of variance with a discrete DV as well as discrete IVs. If some of the IVs are continuous, but the DV is discrete, the method of choice usually is logistic regression (Chapter 12).

7.2 ■ KINDS OF RESEARCH QUESTIONS

The purpose of multiway frequency analysis is to discover associations among discrete variables. Once a preliminary search for associations is complete, a model is fit that includes only the associations necessary to reproduce the observed frequencies. Each cell has its own combination of parameter estimates for the associations retained in the model. The parameter estimates are used to predict cell frequency, and they also reflect the importance of each effect to the frequency in that cell. If one of the variables is a DV, the odds that a case falls into one of its categories can be predicted from the cell's combination of parameter estimates. The following questions, then, are addressed by multiway frequency analysis.

7.2.1 ■ Associations among Variables

Which variables are associated with one another? By knowing which category a case falls into on one variable, can you predict the category it falls into on another? The procedures of Section 7.4 show, for a simple data set, how to determine statistically which variables are associated and how to decide on the level of complexity of associations necessary to describe the relationships.

As the number of variables increases so do the number of potential associations and their complexity. With three variables there are seven potential associations: one three-way association, three two-way associations, and three one-way associations. With four variables there is a potential four-way association, four three-way associations, and so on. With more variables, then, the highest-level associations are tested and eliminated if unreliable until a preliminary model is found with the fewest required associations.

In the example above, the three-way association between number of publications, number of statistics courses, and area in psychology might be ruled out in preliminary analysis. The set of two-way associations is then tested to see which of these might be ruled out. Number of statistics courses and number of publications might be associated, as well as area of psychology and number of statistics courses, but not area of psychology and number of publications. Finally, one-way "associations" are tested. For example, there is a one-way association for area of psychology if numbers of psychologists differ significantly among areas.

7.2.2 ■ Effect on a Dependent Variable

In the usual multiway frequency table, cell frequency is the DV that is influenced by one or more discrete variables and their associations. Sometimes, however, one of the variables is considered a DV. In this case, questions about association are translated into tests of main effects (associations between the DV and each IV) and interactions (association between the DV and the joint effects of two or more IVs).

For example, suppose we investigate a dichotomous measure of a psychologist's success (say, a rating by a supervisor as successful or not) as a function of number of statistics courses, number of publications, and area of psychology. The only associations of research interest are those that include the success variable. The analysis reveals the odds of being in various DV categories as a function of the levels of the IVs: what, for example, are the odds of success if a psychologist takes two or fewer vs. more than two statistics courses? This multiple regression-like application of multiway frequency analysis is called logit analysis and is discussed in Section 7.5.3.

7.2.3 ■ Parameter Estimates

What is the expected frequency for a particular combination of categories of variables? Or, given the level(s) of one or more variables, what are the odds of being at a particular level on another variable? First, reliable effects are identified, and then coefficients, called parameter estimates, are found for each level of all the reliable effects. Section 7.4.3.2 shows how to calculate parameter estimates and use them to find expected frequencies. Interpretation of parameter estimates as odds in logit analysis (where one variable is considered a DV) is described in Section 7.5.3. For example, given information about number of statistics courses and number of publications, what are the odds that a psychologist is successful?

7.2.4 ■ Importance of Effects

Because parameter estimates are developed for each level (or combinations of levels) of each reliable effect, the relative importance of each effect to the frequency in each cell can be evaluated. Effects with larger standardized parameter estimates are more important in predicting that cell's frequency than effects with smaller standardized parameter estimates. If, for instance, number of statistics courses has a higher standardized parameter estimate than number of publications in the cell for successful psychologists, it is the more important effect.

7.2.5 ■ Strength of Association

How well does a model fit the observed frequencies? Recall that in ANOVA, a statistical test gives little insight into the magnitude of the effect of the IV on the DV because the test depends on degrees of freedom as well as the strength of association between the IV and DV. In loglinear analysis, similarly, the goodness-of-fit statistic depends on sample size in addition to how well a model fits the observed frequencies.

Strength of association measures typically are not available in statistical packages used for loglinear analysis. However, two of them, entropy and concentration, are discussed in Section 7.5.2.1, in the context of tests of models. Additionally, Bonett and Bentler (1983) describe the use of a normed fit index. Like any strength of association measure, the normed fit index varies between 0 and 1, but because the index does not represent shared variance it is better termed effect size than strength of association. Although influenced by sample size, NFI still can give a better notion of how well a model fits observed frequencies than is available from formal goodness-of-fit tests. See Section 14.5.3.1 for further discussion of NFI and other indices of model fit. Finally, strength of association measures applicable to logistic regression analysis are described in Chapter 12, and can be applied to some loglinear analyses (i.e., logit models) as well.

7.2.6 ■ Specific Comparisons and Trend Analysis

If a significant association is found, it may be of interest to decompose the association to find its significant components. For example, if area of psychology and number of publications, both with several levels, are associated, which areas differ in number of publications? These questions are analogous to those of analysis of variance where a many-celled interaction is investigated in terms of simpler interactions or in terms of simple effects (cf. Section 10.5.1). Similarly, if the categories of one of the variables differ in quantity (e.g., number of publications), a trend analysis often helps one understand the nature of its relationship with other variables. Planned and post hoc comparisons are discussed in Section 7.5.5.

7.3 ■ LIMITATIONS TO MULTIWAY FREQUENCY ANALYSIS

7.3.1 ■ Theoretical Issues

As a nonparametric statistical technique with no assumptions about population distributions, multiway frequency analysis is remarkably free of limitations. The technique can be applied almost universally, even to continuous variables which fail to meet distributional assumptions of parametric statistics, if the variables are cut into discrete categories.

With the enormous flexibility of current programs for loglinear analysis, many of the questions posed by highly complex data sets can be answered. However, the greatest danger in the use of this analysis is inclusion of so many variables that interpretation boggles the mind—a danger frequently noted in multifactorial analysis of variance, as well.

In logit analyses where one of the variables is considered a DV, the usual cautions about causal inference apply. Validity of causal inference is determined by manipulation of IVs, random assignment of subjects to conditions, and experimental control, not by statistical technique.

7.3.2 ■ Practical Issues

The only limitations to using multiway frequency analysis are the requirement for independence, adequate sample size, and the size of the expected frequency in each cell. During interpretation, however, certain cells may turn out to be poorly predicted by the solution.

7.3.2.1 ■ Independence

Only between-subjects designs may be analyzed, so that the frequency in each cell is independent of the frequencies in all other cells. If the same case contributes values to more than one cell, those cells are not independent. *Verify that the total N is equal to the number of cases.*

Sometimes this restriction is circumvented by inclusion of a time variable, as in McNemar's test for two-way χ^2. A case is in a particular combination of cells over the time periods. Similarly, "yes-no" variables may be developed. For example, in a 2×2 design, a person attends karate classes but does not take piano lessons (yes on karate, no on piano), or does neither (no on both), or does both (yes on both) or takes piano lessons but not karate (no on karate, yes on piano). Each case is in only one of four cells, despite having "scores" on both karate and piano.

7.3.2.2 ■ Ratio of Cases to Variables

A number of problems may occur when there are too few cases relative to the number of variables. Loglinear analysis may fail to converge when combinations of variables result in too many cells with no cases. *You should have at least 5 times the number of cases as cells in your design.* In the example, area of psychology has 3 levels and number of publications 5 levels, so $3 \times 5 \times 5$ or 75 cases are needed.

A logit solution (one variable as DV) is impossible when there is perfect prediction of the DV, for example when all cases in one level of the DV have a particular value of one predictor. This is more likely to be the result of too small a sample than a discovery of the perfect predictor. An overly-good fit also occurs when there are too many variables relative to the number of cases in one level of the DV—leading to overfitting (cf. Section 5.3.2.1). See Section 12.3.2.1 for more discussion of this issue. If there is a problem with too many variables relative to the number of cases, *increase the number of cases or eliminate one or more predictors.*

7.3.2.3 ■ Adequacy of Expected Frequencies

The fit between observed and expected frequencies is an empirical question in tests of association among discrete variables. Sample cell sizes are observed frequencies; statistical tests compare them with expected frequencies derived from some hypothesis, such as independence between variables. The requirement in multiway frequency analysis is that expected frequencies are large enough. Two conditions produce expected frequencies that are too small: a small sample in conjunction with too many variables with too many levels (as discussed in Section 7.3.2.2), and rare events.

When events are rare, the marginal frequencies are not evenly distributed among the various levels of the variables. For example, there are likely to be few psychologists who average four or more publications a year. A cell from a low-probability row and/or a low-probability column will have a very low expected frequency. The best way to avoid low expected frequencies is to attempt to determine in advance of data collection which cells will be rare, and then sample until those cells are adequately filled.

In any event, *examine expected cell frequencies for all two-way associations to assure that all are greater than one, and that no more than 20% are less than five.* Inadequate expected frequencies generally do not lead to increased Type I error (except in some cases with use of the Pearson χ^2 statistic, cf. Section 7.5.2). But power can be so drastically reduced with inadequate expected frequencies that the analysis is worthless. Reduction of power becomes notable as expected frequencies for two-way associations drop below five in some cells (Milligan, 1980).

If low expected frequencies are encountered despite care in obtaining your sample, several choices are available. First, you can simply choose to accept reduced power for testing effects associated with low expected frequencies. Second, you can collapse categories for variables with more than two levels. For example, you could collapse the "three" and "four or more" categories for number of publications into one category of "three or more." The categories you collapse depend on theoretical considerations as well as practical ones because it is quite possible that associations will disappear as a result. Because this is equivalent to a complete reduction in power for testing those associations, nothing has been gained.

Finally, you can delete variables to reduce the number of cells. Care is taken to delete only variables that are not associated with the remaining variables. For example, in a three-way table you might consider deleting a variable if there is no three-way association and if at least one of the two-way associations with the variable is nonsignificant (Milligan, 1980). The common practice of adding a constant to each cell is not recommended because it has the effect of further reducing power. Its purpose is to stabilize Type I error rate, but as noted above, that is generally not the problem and when it is, other remedies are available (Section 7.5.2). Some of the programs, such as SPSS LOGLINEAR and HILOGLINEAR, add the constant by default anyway under circumstances that do not affect the outcome of the analysis.

Section 7.7.1. demonstrates procedures for screening a multidimensional frequency table for expected cell frequencies.

7.3.2.4 ■ Outliers in the Solution

Sometimes there are substantial differences between observed and expected frequencies derived from the best fitting model for some cells. If the differences are large enough, there may be no model that adequately fits the data. Levels of variables may have to be deleted or collapsed or new variables added before a model is fit. But whether or not a model is fit, examination of residuals in search of discrepant cells leads to better interpretation of the data set. Analysis of residuals is discussed in Sections 7.4.3.1 and 7.7.2.3.

7.4 ■ FUNDAMENTAL EQUATIONS FOR MULTIWAY FREQUENCY ANALYSIS

Analysis of multiway frequency tables typically requires three steps: (1) screening, (2) choosing and testing appropriate models, and (3) evaluating and interpreting the selected model. A small sample example of hypothetical data with three discrete variables is illustrated in Table 7.1. The first variable is type of preferred reading material, READTYP, with two levels: science fiction (SCIFI) and spy novels (SPY). The second variable is SEX; the third variable is three levels of profession, PROFESS: politicians (POLITIC), administrators (ADMIN), and belly dancers (BELLY).

In this section, the simpler calculations are illustrated in detail, while the more complex arithmetic is covered only enough to provide some idea of the methods used to model multidimension-

TABLE 7.1 SMALL SAMPLE OF HYPOTHETICAL DATA FOR ILLUSTRATION OF MULTIWAY FREQUENCY ANALYSIS

PROFESSION	SEX		Reading type		Total
			SCIFI	SPY	
Politicians	Male		15	15	30
	Female		10	15	25
		Total	25	30	55
Administrators	Male		10	30	40
	Female		5	10	15
		Total	15	40	55
Belly dancers	Male		5	5	10
	Female		10	25	35
		Total	15	30	45

al data sets. The computer packages used in this section are also the most straightforward. With real data sets, the various computer packages allow choice of strategy on the basis of utility rather than simplicity. Computer analyses of this data set through BMDP4F, SAS CATMOD, SPSS LOGLIN-EAR, GENLOG, and HILOGLINEAR, as well as SYSTAT LOGIN are in Section 7.4.4.

If only a single association is of interest, as is usually the case in the analysis of a two-way table, the familiar χ^2 statistic is used:

$$\chi^2 = \sum_{ij} \frac{(fo - Fe)^2}{Fe} \tag{7.1}$$

where fo represents observed frequencies in each cell of the table and Fe represents the expected frequencies in each cell under the null hypothesis of independence (no association) between the two variables. Summation is over all cells in the two-way table.

If the goodness-of-fit tests for the two marginal effects are also computed, the usual χ^2 tests for the two one-way and one two-way effects do not sum to total χ^2. This situation is similar to that of unequal-n ANOVA where F tests of main effects and interactions are not independent (cf. Chapter 8). Because overlapping variance cannot be unambiguously assigned to effects, and because overlapping variance is repeatedly analyzed, interpretation of results is not clear-cut. In multiway frequency tables, as in ANOVA, nonadditivity of χ^2 becomes more serious as additional variables produce higher-order (e.g., three-way and four-way) associations.

An alternative strategy is to use the likelihood ratio statistic, G^2. The likelihood ratio statistic is distributed as χ^2, so the χ^2 tables can be used to evaluate significance. However, under conditions to be described in section 7.4.2, G^2 has the property of additivity of effects. For example, in a two-way analysis,

$$G_T^2 = G_A^2 + G_B^2 + G_{AB}^2 \tag{7.2}$$

The test of overall association within a two-way table, G_T^2, is the sum of the first-order goodness-of-fit tests, G_A^2 and G_B^2, and the test of association, G_{AB}^2

G^2, like χ^2, has a single equation for its various manifestations which differ among themselves only in how the expected frequencies are found.

$$G^2 = 2\sum (fo) \ln\left(\frac{fo}{Fe}\right) \tag{7.3}$$

For each cell, the natural logarithm of the ratio of obtained to expected frequency is multiplied by the obtained frequency. These values are summed over cells, and the sum is doubled to produce the likelihood ratio statistics.

7.4.1 ■ Screening for Effects

Screening is done if the researcher is data snooping and wishes simply to identify reliable effects. Screening is also done if the researcher hypothesizes a full model, a model with all possible effects included. Screening is not done if the researcher has hypothesized an incomplete model, a model with some effects included and others eliminated; in this case the hypothesized model is tested and evaluated (see section 7.4.2) followed by, perhaps, post hoc analysis.

The first step in screening is to determine if there are any effects to investigate. If there are, then screening progresses to a computation of Fe for each effect, a test of the reliability (significance) of each effect (finding G^2 for the first-order effects, the second-order or two-way associations, the third-order or three-way associations, and so on), and an estimation of the size of the reliable effects. Because Equation 7.3 is used for all tests of the observed frequencies (fo), the trick is to find the Fe necessary to test the various hypotheses, as illustrated below using the data of Table 7.1.

7.4.1.1 ■ Total Effect

If done by hand, the process starts by calculation of overall G_T^2, which is used to test the hypothesis of no effects in the table (the hypothesis that all cells have equal frequencies). If this hypothesis cannot be rejected, there is no point to proceeding further. (Note that when all effects are tested simultaneously, as in computer programs, one can test either G_T^2 or G^2 for each of the effects, but not both, because degrees of freedom limit the number of hypotheses to be tested.)

For the test of total effect,

$$Fe = \frac{N}{rsp} \qquad\qquad (7.4)$$

Expected frequencies, Fe, for testing the hypothesis of no effects, are the same for each cell in the table and are found by dividing the total frequency (N) by the number of cells in the table, i.e., the number of levels of READTYP (represented by r) times the number of levels of SEX (s) times the number of levels of PROFESS (p).

For these data, then,

$$Fe = \frac{155}{(2)(2)(3)} = 12.9167$$

Applying Equation 7.3 for the test of overall effect,

$$G_T^2 = 2 \sum_{ijk} (fo) \ln(\frac{fo}{Fe}) \qquad\qquad df = rsp - 1$$

where $i = 1, 2, \ldots, r; j = 1, 2, \ldots, s;$ and $k = 1, 2, \ldots, p$. Filling in frequencies for each of the cells in Table 7.1, then,

$$G^2 = 2[15 \ln(\frac{15}{12.9167}) + 15 \ln(\frac{15}{12.9167}) + 10 \ln(\frac{10}{12.9167})$$

$$+ 15 \ln(\frac{15}{12.9167}) + 10 \ln(\frac{10}{12.9167}) + 30 \ln(\frac{30}{12.9167})$$

$$+ 5 \ln(\frac{5}{12.9167}) + \ldots + 25 \ln(\frac{25}{12.9167})]$$

$$= 2[2.243 + 2.243 + (-2.559) + 2.243 + (-2.559) + 25.280 + (-4.745)$$

$$+ (-2.559) + (-4.745) + (-4.745) + (-2.559) + 16.509]$$

$$= 48.09$$

With df $= 12 - 1 = 11$ and critical χ^2 at $\alpha = .05$ equal to 19.68 (cf. Table C.4 in Appendix C), there is a statistically reliable departure from equal frequencies among the 12 cells.[1] Further analysis is now required to screen the table for sources of this departure. In the normal course of data analysis the highest-order association is tested first, and so on. Because, however, of the greater complexity for finding expected frequencies with higher-order associations, the presentation here is in the reverse direction, from the first-order to highest-order associations.

7.4.1.2 ■ First-Order Effects

There are three first-order effects to test, one for each of the discrete variables. Starting with READ-TYP, a goodness-of-fit test evaluates the equality of preference for science fiction and spy novels. Only the marginal sums for the two types of reading material are relevant, producing the following observed frequencies:

fo

SCIFI	SPY
55	100

viding the total frequency by the number of relevant "cells," 5. The expected frequencies, then, are

Fe

SCIFI	SPY
77.5	77.5

$$(fo) \ln(\frac{fo}{Fe}) \qquad df = r - 1$$

$$5 \ln(\frac{55}{77.5}) + 100 \ln(\frac{100}{77.5})]$$

$$25 \qquad df = 1$$

Because critical χ^2 with df $= 1$ at $\alpha = .05$ is 3.84, a significant preference for spy novels is suggested. As in ANOVA, however, significant lower-order (main) effects cannot be interpreted unambiguously if there are higher-order (interaction) effects involving the same variable.

---

[1] Throughout this section calculations may differ slightly from those produced by computer programs due to rounding error.

Similar tests for main effects of SEX and PROFESS produce $G_S^2 = 0.16$ with 1 df and $G_P^2 = 1.32$ with 2 df, suggesting no statistically significant difference in the number of men (80) and women (75), nor a significant difference in the numbers of politicians (55), administrators (55), and belly dancers (45), and an interesting sampling strategy.

7.4.1.3 ■ Second-order Effects

Complications arise in how to compute intermediate level associations. In a three-way design, such as the present example, there are two methods of computing the two-way associations. The simpler method is the marginal (or unconditional) test, in which each two-way association is analyzed ignoring the other two-way and higher-order associations. A two-way table is formed by summing over the third factor and ignoring the presence or absence of other reliable effects.

In the more complicated partial (or conditional) test, each two-way association is adjusted for all the other associations. This second method of analysis requires computation of higher-order associations before lower-order ones, to provide for that adjustment.[2]

The two procedures test different hypotheses about the associations, akin to the various strategies for dealing with unequal sample sizes in ANOVA (see Section 8.5.2.2). There is no consensus as to which procedure is more valid, or even consensus regarding the circumstances under which one procedure is more appropriate than the other. One currently popular strategy (Brown, 1976) is to compute intermediate-order associations both ways and make decisions about them with a conservative statistical criterion to avoid inflated alpha levels due to multiple tests. Because these computations typically are only part of the initial screening process, the results of both sets of calculations are often useful in the second step of choosing and evaluating a model.

For the simpler *marginal* tests of association in the small sample data of Table 7.1, the three-way table is collapsed into three two-way tables, one for each two-way interaction. For the $R \times S$ (READTYP × SEX) association, for instance, the cells for each combination of reading type and sex are summed over the three levels of profession (P), forming as the observed frequencies:

	fo		
	SCIFI	SPY	
MEN	30	50	80
WOMEN	25	50	75
	55	100	155

The expected frequencies are found as in the usual way for a two-way χ^2 test of association:

$$\text{Cell } Fe = \frac{(\text{row sum})(\text{column sum})}{N} \tag{7.5}$$

for the appropriate row and column for each cell; that is, for the first cell, men preferring science fiction,

$$Fe = \frac{(80)(55)}{155} = 28.3871$$

---

[2] Recall sequential multiple regression (Chapter 5). A marginal effect is the test of a term entering the equation before other terms with lower priority; therefore the other terms are not partialed out of the effect. A partial effect is one that enters the equation after other terms with higher priority; therefore, effects of those other terms are partialed out.

After the computations are completed for the remaining cells, the following table of expected frequencies is found:

	Fe		
	SCIFI	SPY	
MEN	28.3871	51.6129	80
WOMEN	26.6129	48.3871	75
	55	100	155

The only difference between the usual two-way test of association and this test is use of the likelihood ratio statistic instead of χ^2.

$$G^2_{RS\,(marg)} = 2\sum_{ij} (fo)\ln(fo/Fe) \qquad df = (r-1)(s-1)$$

$$= 2\,[30\ln(\frac{30}{28.3871}) + 50\ln(\frac{50}{51.6129})$$

$$+ 25\ln(\frac{25}{26.6129}) + 50\ln(\frac{50}{48.3871})]$$

$$= 0.29 \qquad df = 1$$

The result is obviously not statistically significant.[3]

The same procedure for the remaining two-way associations leads to a nonsignificant result for the READTYP × PROFESS interaction with $G^2_{RP(marg)} = 4.09$ and df $= (r-1)(p-1) = 2$. For the SEX × PROFESS interaction, $G^2_{SP(marg)} = 26.77$ with df $= (s-1)(p-1) = 2$, which exceeds critical $\chi^2 = 5.99$ at $\alpha = .05$, suggesting a significant association. A look back at the original data in Table 7.1 reveals a dearth of male belly dancers and female administrators.

Tests of the more complicated *partial* associations use an iterative procedure to develop a full set of expected frequencies in which all marginal sums (except the one to be tested) match the observed marginal frequencies.[4] The first iteration uses Equation 7.5 to compute expected frequencies as in the marginal test. Once found, the expected frequencies are duplicated at each level of the other variable. The results of this iteration for the partial test of the $R \times S$ association appear in Table 7.2. Notice that the same expected frequencies as computed for the marginal test are repeated for politicians, administrators, and belly dancers.

All the entries are too large because the two-way table has simply been duplicated three times. That is, $N = 465$ instead of 155, there are 80 male politicians instead of 30, and so on. A second iteration is performed to adjust the values in Table 7.2 for another two-way association, in this case the $R \times P$ association. This iteration begins with the $R \times P$ table of observed frequencies and relevant marginal sums:

---

[3] The test just illustrated is the simplest marginal test of association using the loglinear model (Marascuilo & Levin, 1983) and, unlike the algorithms in some computer packages, does not use an iterative process.

[4] Other methods for finding partial association are based on differences in G^2 between hierarchical models.

TABLE 7.2 FIRST ITERATION ESTIMATES OF EXPECTED FREQUENCIES FOR THE PARTIAL TEST OF THE READTYP X SEX ASSOCIATION

			Reading type		
PROFESSION	SEX		SCIFI	SPY	Total
Politicians	Male		28.3871	51.6129	80
	Female		26.6129	48.3871	75
		Total	55	100	155
Administrators	Male		28.3871	51.6129	80
	Female		26.6129	48.3871	75
		Total	55	100	155
Belly dancers	Male		28.3871	51.6129	80
	Female		26.6129	48.3871	75
		Total	55	100	155

	fo	
	SCIFI	SPY
POLITIC	25	30
ADMIN	15	40
BELLY	15	30
	55	100

Note that the actual number of politicians preferring science fiction is 25, while after the first iteration (Table 7.2) the number is $(28.3871 + 26.6129) = 55$. The goal is to compute a proportion that, when applied to the relevant numbers in Table 7.2 (in this case, both male and female politicians who prefer science fiction), eliminates the effects of any $R \times P$ interaction:

$$\frac{fo}{Fe^{\#1}} = \frac{25}{55} = 0.45455$$

producing

$$Fe^{\#2} = Fe^{\#1}(0.45455) = (28.3871)(0.45455) = 12.9032$$

and

$$Fe^{\#2} = Fe^{\#1}(0.45455) = (26.6129)(0.45455) = 12.0968$$

for male and female politicians preferring science fiction, respectively.

To find second iteration expected frequency for female belly dancers preferring spy stories, the last cell in the table,

$$\frac{fo}{Fe^{\#1}} = \frac{30}{100} = 0.3$$

$$Fe^{\#2} = (48.3871)(0.3) = 14.5161$$

Table 7.3 shows the results of applying this procedure to all cells of the data matrix.

TABLE 7.3 SECOND ITERATION ESTIMATES OF EXPECTED FREQUENCIES FOR THE PARTIAL TEST OF THE READTYP BY SEX ASSOCIATION

PROFESSION	SEX		Reading type		
			SCIFI	SPY	Total
Politicians	Male		12.9032	15.4839	28.3871
	Female		12.0968	14.5161	26.6129
		Total	25	30	55
Administrators	Male		7.7419	20.6452	28.3871
	Female		7.2581	19.3548	26.6129
		Total	15	40	55
Belly dancers	Male		7.7419	15.4839	23.2258
	Female		7.2581	14.5161	21.7742
		Total	15	30	45

Notice that correct totals have been produced for overall N, for R, P, and S, and for $R \times P$, but that the $S \times P$ values are incorrect. The third and final iteration, then, adjusts the $S \times P$ expected values from the second iteration for the $S \times P$ matrix of observed values. These $S \times P$ matrices are:

	fo				Fe	
	Men	Women			Men	Women
POLITIC	30	25		POLITIC	28.3871	26.6129
ADMIN	40	15		ADMIN	28.3871	26.6129
BELLY	10	35		BELLY	23.22258	21.7742

For the first cell, male politicians preferring to read science fiction, the proportional adjustment (rounded off) is

$$\frac{fo}{Fe^{\#2}} = \frac{30}{28.3871} = 1.0568$$

to produce

$$Fe^{\#3} = Fe^{\#2}(1.0568) = (12.9032)(1.0568) = 13.6363$$

And for the last cell, female belly dancers who prefer spy stories,

$$\frac{fo}{Fe^{\#2}} = \frac{35}{21.7742} = 1.6074$$

$$Fe^{\#3} = (14.5161)(1.6074) = 23.3333$$

Following this procedure for the remaining 10 cells of the matrix produces the third iteration estimates as shown in Table 7.4. These values fulfill the requirement that all expected marginal frequencies are equal to observed marginal frequencies except for $R \times S$, the association to be tested.

At this point we have the Fe necessary to calculate G^2_{RS}

TABLE 7.4 THIRD ITERATION ESTIMATES OF EXPECTED FREQUENCIES FOR THE PARTIAL TEST OF THE READTYP BY SEX ASSOCIATION

PROFESSION	SEX		Reading type		Total
			SCIFI	SPY	
Politicians	Male		13.6363	16.3637	30
	Female		11.3637	13.6363	25
		Total	25	30	55
Administrators	Male		10.9090	29.0910	40
	Female		4.0909	10.9091	15
		Total	15	40	55
Belly dancers	Male		3.3333	6.6666	10
	Female		11.6667	23.3333	35
		Total	15	30	45

$$G^2_{RS} = 2 \sum_{ij} (fo) \ln(\frac{fo}{Fe})$$

$$= 2[(fo) \ln(\frac{fo}{Fe}) = 2[15 \ln(\frac{15}{13.6363}) + \ldots + 25 \ln(\frac{25}{23.3333})]$$

$$= 2.47$$

However, a final adjustment is made for the three-way association, G^2_{RSP} (as computed below). The partial likelihood ratio statistic for the association between READTYP and SEX, then, is

$$G^2_{RS(part)} = G^2_{RS} - G^2_{RSP} \qquad\qquad df = (r-1)(s-1)$$

$$= 2.47 - 1.85 = 0.62 \qquad\qquad df = 1$$

This partial test shows a lack of association, as does the corresponding marginal test of the $R \times S$ association.

The same process is (tediously) followed for the partial tests of the $R \times P$ and the $S \times P$ associations. The resultant partial likelihood ratio statistic for the $R \times P$ association is

$$G^2_{RP(part)} = 4.42 \qquad\qquad df = 2$$

showing lack of association between reading preferences and profession, a result consistent with that of the marginal test. For the $S \times P$ association, the partial likelihood ratio result is

$$G_{SP(part)} = 27.12 \qquad\qquad df = 2$$

a statistically significant association that is also consistent with the corresponding marginal test.

Corresponding marginal and partial tests of intermediate associations in this example produce the same conclusions and interpretation is clear-cut: there is a reliable association between sex and profession and no evidence of association between sex and reading preferences or between reading preferences and profession. In some situations, however, interpretation is more problematic because the results of marginal and partial tests differ. Procedures for dealing with such situations are discussed in Section 7.5.4.

7.4.1.4 ■ Third-order Effect

The test for the three-way $R \times S \times P$ association requires a much longer iterative process because all marginal expected frequencies must match observed frequencies (R, S, P, $R \times S$, $R \times P$, and $S \times P$). Ten iterations are required to compute the appropriate Fe for the 12 cells (not shown in the interests of brevity and avoidance of terminal boredom), producing

$$G_{RSP}^2 = 2 \sum_{ijk} (fo) \ln(fo/Fe) \qquad \qquad \text{df} = (r-1)(s-1)(p-1)$$

$$= 1.85 \qquad \qquad \qquad \text{df} = 2$$

The three-way association, then, shows no statistical significance.

A summary of the results of the calculations for all effects appears in Table 7.5. At the bottom of the table is the sum of all one-, two-, and three-way effects using both the marginal and partial methods for calculating G^2. As can be seen, neither of these matches G_T^2. If the simple marginal method is applied to the two-way associations, the sum is too large. If the more complicated method is applied, the sum is too small. Further, depending on the data, either over- or underadjustment of each effect may occur with either method. Therefore, Brown (1976) recommends attention to both marginal and partial results in subsequent modeling decisions (see Section 7.5.4).

7.4.2 ■ Modeling

In some applications of multiway frequency analysis, results of screening provide sufficient information for the researcher. In the current example, for instance, the results are clear-cut. One first-order

TABLE 7.5 SUMMARY OF SCREENING TESTS FOR SMALL SAMPLE EXAMPLE OF MULTIWAY FREQUENCY ANALYSIS

EFFECT	df	G^2	PROB	G^2	PROB
All (total)	11	48.09	<.05		
READTYP	1	13.25	<.05		
SEX	1	0.16	>.05		
PROFESS	2	1.32	>.05		
		(partial)		(marginal)	
$R \times S$	1	0.62	>.05	0.29	>.05
$R \times P$	2	4.42	>.05	4.09	>.05
$S \times P$	2	27.12	<.05	26.79	<.05
$R \times S \times P$	2	1.85	>0.5		
Sums	11	48.74		45.90	

effect, preference for reading type, is statistically significant, as is the sex by profession association. Often, however, the results are not so evident and consistent, and/or the goal is to find the best model for predicting frequencies in each cell of the design.

A loglinear model is developed where an additive regression-type equation is written for (the log of) expected frequency as a function of the effects in the design. The procedure is similar to multiple regression where a predicted DV is obtained by combining the effects of several IVs.

A full[5] model includes all possible effects in a multiway frequency analysis. The full model for the 3-way design of the example is

$$\ln Fe_{ijk} = \theta + \lambda_{A_i} + \lambda_{B_j} + \lambda_{C_k} + \lambda_{AB_{ij}} + \lambda_{AC_{ik}} + \lambda_{BC_{jk}} + \lambda_{ABC_{ijk}} \tag{7.6}$$

For each cell (the natural logarithm of) the expected frequency, $\ln Fe$, is an additive sum of the effect parameters, λs, and a constant, θ.

For each effect in the design, there are as many values of λ—called effect parameters—as there are levels in the effect, and these values sum to zero. In the example there are two levels of READ-TYP, so there is a value of λ_R for SCIFI and for SPY, and the sum of these two values is zero. For most cells, then, the expected frequency is derived from a different combination of effect parameters.

The full (saturated) model always provides a perfect fit to data so that expected frequencies exactly equal observed frequencies. The purpose of modeling is to find the *incomplete* model with the fewest effects that still closely mimics the observed frequencies. Screening is done to avoid the necessity of exploring all possible incomplete models, an inhumane effort with large designs, even with computers. Effects that are found to be unreliable during the screening process are often omitted during modeling.

Model fitting is accomplished by finding G^2 for a particular incomplete model and evaluating its significance. Because G^2 is a test of fit between observed and expected frequencies, a good model is one with a *nonsignificant* G^2. Because there are often many "good" models, however, there is a problem in choosing among them. The task is to compare nonsignificant models with one another.

Models come in two flavors, hierarchical and nonhierarchical. Hierarchical (nested) models include the highest-order reliable association and all its component parts; nonhierarchical models do not necessarily include all the components (see Section 7.5.1). For hierarchical models, the optimal model is one that is not significantly worse than the next most complex one. Therefore the choice among hierarchical models is made with reference to statistical criteria. There are no statistical criteria for choosing among nonhierarchical models and they are not recommended

Several methods for comparing models are available, as discussed in Section 7.5.4. In the simplest method, illustrated here, a few hierarchical models are selected on the basis of screening results and compared using the significance of the difference in G^2 between them. When the models are hierarchical, *the difference between the two G^2's is itself a G^2*. That is,

$$G^2_1 - G^2_2 = G^2 \tag{7.7}$$

---

[5] Full models are also called saturated models.

if Model 1 is a subset of Model 2 in which all the effects in Model 1 are included in Model 2. For the example, a Model 1 with $R \times P$, R, and P effects is nested within a Model 2 with $R \times S$, $R \times P$, R, S, and P effects.

To simplify description of models, Model 1 above is designated (RP) and Model 2 (RS,RP). This is a fairly standard notation for hierarchical models. Each association term (e.g., RS), implies that all lower-order effects (R and S) are included in the model. In the example, the most obvious model to choose is (SP,R), which includes the $S \times P$ association and all three first-order effects.

In practice, the first step is to evaluate the highest-order effect before sequentially testing lower-order effects. During screening on the example, the three-way association is ruled out but at least one of the two-way associations is reliable. Because there are only three effects in the design, it would not be difficult by computer to try out a model with all three two-way associations (RS,RP,SP), and compare that with models with all pairwise combinations of two-way associations. If there are discrepancies between marginal and partial tests of effects, models with and without the ambiguous effects are compared.

In the example, lack of significance for either the marginal or partial tests of the RP and RS effects would ordinarily preclude their consideration in the set of models to be tested. The RP effect is included in a model to be tested here for illustrative purposes only.

For each model to be tested, expected frequencies and G^2 are found. To obtain G^2 for a model, the G^2 for each of the effects is subtracted from total G^2 to yield a test of residual frequency that is not accounted for by effects in the model. If the residual frequencies are not significant, there is a good fit between obtained and expected frequencies from the reduced model.

For the example, G^2 values for the (SP,R) model are available from the screening tests shown in Table 7.5. For the two-way effects, the G^2 values from the marginal tests are used because in this case they are smaller and more conservative than the partial values. G^2 for the (SP,R) model is, then

$$G^2_{(SP,R)} = G^2_T - G^2_{SP} - G^2_S - G^2_P - G^2_R$$

$$= 48.09 - 26.79 - 0.16 - 1.32 - 13.25$$

$$= 6.57$$

Degrees of freedom are those associated with each of the effects as in Section 7.4.1, so that df $= 11 - 2 - 1 - 2 - 1 = 5$. Because residuals from this model are not statistically significant, the model is adequate.

For the example, a more complex model includes the $R \times P$ association. Following the above procedures, the (SP,RP) model produces $G^2 = 2.48$ with 3 df. The test of the difference between (SP,R) and (SP,RP) is simply the difference between G^2's (Equation 7.7) for the two models, using the difference between degrees of freedom to test for significance:

$$G^2_{(diff)} = G^2_{(SP,R)} - G^2_{(SP,RP)}$$

$$= 6.57 - 2.48 = 4.09 \text{ with df} = 5 - 3 = 2$$

a nonsignificant result. (In this simple case, G^2 for the difference is the same as G^2 for the partial test of the $S \times P$ association.) Because the difference between models is not statistically significant, the more parsimonious (SP,R) model is preferred over the more complex (SP,RP) model. The model of choice, then, is

$$\ln Fe = \theta = \lambda_R + \lambda_S + \lambda_P + \lambda_{SP}$$

7.4.3 ■ Evaluation and Interpretation

The optimal model, once chosen, is evaluated in terms of both the degree of fit to the overall data matrix (as discussed in the previous section) and the amount of deviation from fit in each cell.

7.4.3.1 ■ Residuals

Once a model is chosen, expected frequencies are computed for each cell and the deviation between the expected and observed frequencies in each cell (the residual) is used to assess the adequacy of the model for fitting the observed frequency in that cell. In some cases, a model predicts the frequencies in some cells well, and in others very poorly, to give an indication of the combination of levels of variables for which the model is and is not adequate.

For the example, the observed frequencies are in Table 7.1. Expected frequencies under the (SP,R) model, derived through an iterative procedure as demonstrated in Section 7.4.1.3, are shown in Table 7.6. Residuals are computed as the cell-by-cell differences between the values in the two tables.

Rather than trying to interpret raw differences, residuals usually are standardized by dividing the difference between observed and expected frequencies by the square root of the expected frequency to produce a z value. Both raw differences and standardized residuals for the example are in Table 7.7. The most deviant cell is for male politicians preferring science fiction, with 4.4 fewer cases expected than observed and a standardized residual of $z = 1.3$. Although the discrepancies for men are larger than those for women, none of the cells is terribly discrepant; so this seems to be an acceptable model.

7.4.3.2 ■ Parameter Estimates

There is a different linear combination of parameters for most cells and the sizes of the parameters in a cell reflect the contribution of each of the effects in the model to the frequency found in that

TABLE 7.6 EXPECTED FREQUENCIES UNDER THE MODEL (SP, R)

| | | Reading type | | |
		SCIFI	SPY	Total
PROFESSION	SEX			
Politicians	Male	10.6	19.4	30.0
	Female	8.9	16.1	25.0
	Total	19.5	35.5	55.0
Administrators	Male	14.2	25.8	40.0
	Female	5.3	9.7	15.0
	Total	19.5	35.5	55.0
Belly dancers	Male	3.5	6.5	10.0
	Female	12.4	22.6	35.0
	Total	16.0	29.0	45.0

cell. One can evaluate, for example, how important READTYP is to the number of cases found in the cell for female politicians who read science fiction.

Parameters are estimated for the model from the *Fe* in Table 7.6 in a manner that closely follows ANOVA. In ANOVA, the size of an effect for a cell is expressed as a deviation from the grand mean. Each cell has a different combination of deviations that correspond to the particular combination of levels of the reliable effects for that cell.

In MFA deviations are derived from natural logarithms of proportions: $\ln(P_{ijk})$. Expected frequencies for the model (Table 7.6) are converted to proportions by dividing *Fe* for each cell by $N = 155$, and then the proportions are changed to natural logarithms. For example, for the first cell, male politicians who prefer science fiction:

$$\ln(P_{ijk}) = \ln(\frac{Fe_{ijk}}{155})$$

$$= \ln(\frac{10.6}{155})$$

$$= -2.6825711$$

Table 7.8 gives all the resulting values.

The values in Table 7.8 are then used in a three-step process that culminates in parameter estimates, expressed in standard deviation units, for each effect for each cell. The first step is to find both the overall mean and the mean (in natural logarithm units) for each level of each of the effects in the model. The second step is to express each level of each effect as a deviation from the overall mean. The third step is to convert the deviations to standard scores to compare the relative contributions of various parameters to the frequency in a cell.

TABLE 7.7 RAW AND STANDARDIZED RESIDUALS FOR HYPOTHETICAL DATA SET UNDER MODEL (*SP, R*)

PROFESSION	SEX	Reading type	
		SCIFI	SPY
Raw residuals (*fo* − *Fe*):			
Politicians	Male	4.4	−4.4
	Female	1.1	− 1.1
Administrators	Male	− 4.2	4.2
	Female	− 0.3	0.3
Belly dancers	Male	1.5	− 1.5
	Female	− 2.4	2.4
Standardized residuals (*fo* − *Fe*)/$Fe^{1/2}$:			
Politicians	Male	1.3	− 1.0
	Female	0.4	− 0.3
Administrators	Male	− 1.1	0.8
	Female	− 0.1	0.1
Belly dancers	Male	0.8	− 0.6
	Female	− 0.7	0.5

TABLE 7.8 EXPECTED ln (P_{ijk}) FOR MODEL (SP,R)

PROFESSION	SEX	Reading type SCIFI	SPY
Politicians	Male	− 2.6825711	− 2.0781521
	Female	− 2.8573738	− 2.2646058
Administrators	Male	− 2.3901832	− 1.7930506
	Female	− 3.3757183	− 2.7712992
Belly dancers	Male	− 3.7906621	− 3.1716229
	Female	− 2.5257286	− 1.9254752

Note: $\ln(P_{ijk}) = \ln(F_{ijk}/155) = \ln(F_{ijk}) - \ln(155)$.

In the first step, various means are found by summing $\ln(P_{ijk})$ across appropriate cells and dividing each sum by the number of cells involved. For example, to find the overall mean,

$$\bar{x}_{...} = (1/rsp) \sum_{ijk} \ln(P_{ijk})$$

$$= (\frac{1}{12})[-2.6825711 + (-2.0781521) + (-2.8573738)$$

$$+ \ldots + (-1.9254752)]$$

$$= -2.6355346$$

To find the mean for SCIFI, the first level of READTYP:

$$\bar{x}_{1..} = (1/sp) \sum_{jk} \ln(P_{ijk})$$

$$= (\frac{1}{6}) [-2.6825711 + (-2.8573738) + (-2.3901832)$$

$$+ (-3.3757183) + (-3.7906621) + (-2.5257286)]$$

$$= -2.9370395$$

The mean for belly dancers is

$$\bar{x}_{..3} = (1/rs) \sum_{ij} \ln(P_{ijk})$$

$$= (\frac{1}{4}) [-3.7906621 + (-3.1716229) + (-2.5257286)$$

$$+ (-1.9254752)]$$

$$= -2.8533722$$

and so on for the first-order effects.

The means for second-order effects are found in a similar manner. For instance, for the $S \times P$ association, the mean for male politicians is

$$\bar{x}_{\cdot 11} = (1/r) \sum_i \ln(P_{ijk})$$

$$= (\frac{1}{2})\, [-2.6825711 + (-2.0781521)]$$

$$= -2.3803616$$

In the second step, parameter estimates are found by subtraction. For first-order effects, the overall mean is subtracted from the mean for each level. For example, λ_{R_1}, the parameter for SCIFI, the first level of READTYP is

$$\lambda_{R_1} = \bar{x}_{1\cdot\cdot} - \bar{x}_{\cdot\cdot\cdot}$$

$$= -2.9370395 - (-2.6355346)$$

$$= -.302$$

For belly dancers, the third level of PROFESS

$$\lambda_{P_3} = \bar{x}_{3\cdot\cdot} - \bar{x}_{\cdot\cdot\cdot}$$

$$= -2.8533722 - (-2.63555346)$$

$$= -.218$$

and so on.

To find λ for a cell in two-way effect, the two appropriate main effect means are subtracted from the two-way mean, and the overall mean is added (in a pattern that is also familiar from ANOVA). For example, $\lambda_{SP_{23}}$, the parameter for female belly dancers (second level of sex, third level of profession), is found by subtracting from the female belly dancer mean (averaged over the two types of reading material) the mean for women and the mean for belly dancers, and then adding the overall mean.

$$\lambda_{SP_{23}} = \bar{x}_{23\cdot} - \bar{x}_{\cdot\cdot 2} - \bar{x}_{\cdot\cdot 3} + \bar{x}_{\cdot\cdot\cdot}$$

$$= -2.2256019 - (-2.6200335) - (-2.8533722) + (-2.6355346)$$

$$= .612$$

All the λ values, as shown in Table 7.9, are found in a similar, if tedious, fashion. In the table, θ is the conversion of the overall mean from proportion to frequency units by addition of $\ln(N)$:

TABLE 7.9 PARAMETER ESTIMATES FOR MODEL (SP,R).
θ (MEAN) = 2.4079

EFFECT	LEVEL	λ	λ/SE
READTYP	SCIFI	−.302	−3.598
	SPY	.302	3.598
SEX	MALE	−.015	−0.186
	FEMALE	.015	0.186
PROFESSION	POLITICIAN	.165	2.045
	ADMINISTRATOR	.053	0.657
	BELLY DANCER	−.218	−2.702
SEX BY PROFESS	MALE POLITICIAN	.106	1.154
	FEMALE POLITICIAN	−.106	−1.154
	MALE ADMINISTRATOR	.506	5.510
	FEMALE ADMINISTRATOR	−.506	−5.510
	MALE BELLY DANCER	.612	7.200
	FEMALE BELLY DANCER	−.612	−7.200

$$\theta = \overline{x}\ldots + \ln(155)$$

$$= 2.4079$$

The expected frequency generated by the model for each cell is then expressed as a function of the appropriate parameters. For example, the expected frequency (19.40) for male politicians who read spy novels is

$$\ln Fe = \theta + \lambda_{R_2} + \lambda_{S_1} + \lambda_{P_1} + \lambda_{SP_{11}}$$

$$= 2.4079 + .302 + (-.015) + .165 + .106$$

$$= 2.9659 \approx \ln(19.40)$$

within rounding error.

These parameters are used to find expected frequencies for each cell but are not interpreted in terms of magnitude until step 3 is taken. During step 3, parameters are divided by their respective standard errors to form standard normal deviates that are interpreted according to their relative magnitudes. Therefore, the parameter values in Table 7.9 are given both in their λ form and after division by their standard errors.

Standard errors of parameters, SE, are found by squaring the reciprocal of the number of levels for the set of parameters, dividing by the observed frequencies, and summing over the levels. For example, for READTYP:

$$SE^2 = \sum (1/r_i)^2/fo$$

$$= (\tfrac{1}{2})^2/55 + (\tfrac{1}{2})^2/100$$

$$= (.25)/55 + (.25)/100$$

$$= .0070455$$

and

$$SE = .0839372$$

Note that this is the simplest method for finding SE (Goodman, 1978) and does not weight the number of levels by unequal marginal frequencies, as do methods such as the one used in BMDP4F.

To find the standard normal deviate for SCIFI (the first level of READTYP), λ for SCIFI is divided by its standard error

$$\lambda_{R_1}/SE = -.302/.0839372$$

$$= -3.598$$

This ratio is interpreted as a standard normal deviate (z) and compared with critical z to assess the contribution of an effect to a cell. The relative importance of the various effects to a cell is also derived from these values. For female belly dancers preferring spy novels, for example, the standard normal deviates for the parameters are 3.598 (SPY), 0.186 (FEMALE), -2.702 (BELLY), -7.200 (FEMALE BELLY). The most important influences on cell frequency are, in order, the sex by profession association, preferred type of reading material, and profession—all statistically significant at $p < .01$ because they exceed 2.58. Sex contributes little to the expected frequency in this cell and is not statistically significant.

Because of the large number of effects produced in typical loglinear models, a conservative criterion should be used if statistical significance is evaluated. A criterion z of 4.00 often is considered reasonable.

Further insights into interpretation are provided in Section 7.7.2.4. Conversion of parameters to odds when one variable is a DV is discussed in Section 7.5.3.2.

7.4.4 ■ Computer Analyses of Small Sample Example

Setup and selected output for computer analyses of the data in Table 7.1 appear in Tables 7.10 through 7.15. SPSS HILOGLINEAR, LOGLINEAR, and GENLOG are in Tables 7.10 through 7.12, respectively, BMDP4F in Table 7.13, SAS CATMOD in Table 7.14, and SYSTAT LOGLIN in Table 7.15.

The setup of SPSS HILOGLINEAR in Table 7.10 produces output appropriate for screening a hierarchical multiway frequency analysis. Additional instructions are necessary to test models. The instruction PRINT=FREQ produces the table of Observed, Expected Frequencies and Residuals. Because no model is specified in the setup, a full model (all effects included in the model) is produced in which expected and observed frequencies are identical. SPSS adds 0.5 to each observed

TABLE 7.10 MULTIWAY FREQUENCY ANALYSIS OF SMALL SAMPLE EXAMPLE THROUGH SPSS HILOGLINEAR
(SETUP AND SELECTED OUTPUT)

```
TITLE          SMALL SAMPLE MULTIWAY ANALYSIS.
DATA LIST      FILE='SSMFA.DAT'    FREE
               /PROFESS SEX READTYP FREQ.
WEIGHT BY FREQ.
VALUE LABELS   PROFESS 1 'POLITIC' 2 'ADMIN' 3 'BELLY'/
               SEX 1 'MALE' 2 'FEMALE'/
               READTYP 1 'SCIFI' 2 'SPY'.
HILOGLINEAR    PROFESS(1,3) SEX READTYP(1,2)/
               PRINT = FREQ, ASSOCIATION/
               DESIGN.
```

DATA   Information

```
    12 unweighted cases accepted.
     0 cases rejected because of out-of-range factor values.
     0 cases rejected because of missing data.
   155 weighted cases will be used in the analysis.
```

FACTOR Information

```
   Factor  Level  Label
   PROFESS    3
   SEX        2
   READTYP    2
```

* * * * * * * * * * * H I E R A R C H I C A L   L O G   L I N E A R * * * * * * * * * * *

DESIGN 1 has generating class

   PROFESS*SEX*READTYP

Note: For saturated models 0.500 has been added to all observed cells.
This value may be changed by using the CRITERIA = DELTA subcommand.

The Iterative Proportional Fit algorithm converged at iteration 1.
The maximum difference between observed and fitted marginal totals is   .000
and the convergence criterion is   .250

- -

Observed, Expected Frequencies and Residuals.

Factor	Code	OBS. count & PCT.	EXP. count & PCT.
PROFESS	POLITIC		
SEX	MALE		
READTYP	SCIFI	15.50 (10.00)	15.50 (10.00)
READTYP	SPY	15.50 (10.00)	15.50 (10.00)
SEX	FEMALE		
READTYP	SCIFI	10.50 (6.77)	10.50 (6.77)
READTYP	SPY	15.50 (10.00)	15.50 (10.00)
PROFESS	ADMIN		
SEX	MALE		
READTYP	SCIFI	10.50 (6.77)	10.50 (6.77)
READTYP	SPY	30.50 (19.68)	30.50 (19.68)
SEX	FEMALE		
READTYP	SCIFI	5.50 (3.55)	5.50 (3.55)
READTYP	SPY	10.50 (6.77)	10.50 (6.77)

TABLE 7.10   *(CONTINUED)*

PROFESS	BELLY		
SEX	MALE		
READTYP	SCIFI	5.50 (3.55)	5.50 (3.55)
READTYP	SPY	5.50 (3.55)	5.50 (3.55)
SEX	FEMALE		
READTYP	SCIFI	10.50 (6.77)	10.50 (6.77)
READTYP	SPY	25.50 (16.45)	25.50 (16.45)

- -

Goodness-of-fit test statistics

    Likelihood ratio chi square =   .00000   DF = 0 P = 1.000
                Pearson chi square =   .00000   DF = 0 P = 1.000

- -
* * * * * * * * * * H I E R A R C H I C A L   L O G   L I N E A R * * * * * * * * * * * * *

Tests that K-way and higher order effects are zero.

K	DF	L.R. Chisq	Prob	Pearson Chisq	Prob	Iteration
3	2	1.848	.3969	1.920	.3828	3
2	7	33.353	.0000	32.994	.0000	2
1	11	48.089	.0000	52.097	.0000	0

- -

Tests that K-way effects are zero.

K	DF	L.R. Chisq	Prob	Pearson Chisq	Prob	Iteration
1	4	14.737	.0053	19.103	.0008	0
2	5	31.505	.0000	31.073	.0000	0
3	2	1.848	.3969	1.920	.3828	0

* * * * * * * H I E R A R C H I C A L   L O G   L I N E A R * * * * * * * *

Tests of PARTIAL associations.

Effect Name	DF	Partial Chisq	Prob	Iter
PROFESS*SEX	2	27.122	.0000	2
PROFESS*READTYP	2	4.416	.1099	2
SEX*READTYP	1	.621	.4308	2
PROFESS	2	1.321	.5166	2
SEX	1	.161	.6879	2
READTYP	1	13.255	.0003	2

TABLE 7.11 MULTIWAY FREQUENCY ANALYSIS OF SMALL SAMPLE EXAMPLE THROUGH SPSS LOGLINEAR (SETUP AND SELECTED OUTPUT

```
TITLE          SMALL SAMPLE MULTIWAY ANALYSIS.
DATA LIST      FILE='SSMFA.DAT'    FREE
               /PROFESS SEX READTYP FREQ.
WEIGHT BY FREQ.
VALUE LABLES   PROFESS  1 'POLITIC', 2 'ADMIN'  3  'BELLY'/
               SEX 1 'MALE', 2 'FEMALE'/
               READTYP 1 'SCIFI'  2 'SPY'.
LOGLINEAR      PROFESS(1,3) SEX READTYP(1,2)/
               PRINT = EST1M.

>Note # 12717
>The last command is not a DESIGN/MODEL specification.  A saturated model is
>generated for this problem.

* * * */ * * * * * * * * * *   L O G   L I N E A R   A N A L Y S I S   * * * * * * * * * *

DATA   Information

    12  unweighted cases accepted.
     0  cases rejected because of  outofrange factor values.
     0  cases rejected because of  missing data.
   155  weighted cases will be used in the analysis.


FACTOR  Information

   Factor    Level Label
   PROFESS      3
   SEX          2
   READTYP      2


DESIGN Information

  1 Design/Model will be processed.

* * * * * * * * * * * * *   L O G   L I N E A R   A N A L Y S I S   * * * * * * * * * *

Correspondence Between Effects and  Columns  of  Design/Model 1

Starting  Ending
Column    Column    Effect Name

   1        2       PROFESS
   3        3       SEX
   4        4       READTYP
```

TABLE 7.11 (CONTINUED)

```
5    6    PROFESS BY SEX
7    8    PROFESS BY READTYP
9    9    SEX BY READTYP
10   11   PROFESS BY SEX BY READTYP
```

Note: for saturated models 0.500 has been added to all observed cells.
This value may be changed by using the CRITERIA = DELTA subcommand.
*** ML converged at iteration 3. The converge criterion = .00000

Observed, Expected Frequencies and Residuals

Factor	Code	OBS. count & PCT.	EXP. count & PCT.	Residual	Std. Resid.	Adj. Resid.
PROFESS	POLITIC					
SEX	MALE					
READTYP	SCIFI	15.50 (9.63)	15.50 (9.63)	.0000	.0000	.0000
READTYP	SPY	15.50 (9.63)	15.50 (9.63)	.0000	.0000	.0000
SEX	FEMALE					
READTYP	SCIFI	10.50 (6.52)	10.50 (6.52)	.0000	.0000	.0000
READTYP	SPY	15.50 (9.63)	15.50 (9.63)	.0000	.0000	.0000
PROFESS	ADMIN					
SEX	MALE					
READTYP	SCIFI	10.50 (6.52)	10.50 (6.52)	.0000	.0000	.0000
READTYP	SPY	30.50 (18.94)	30.50 (18.94)	.0000	.0000	.0000
SEX	FEMALE					
READTYP	SCIFI	5.50 (3.42)	5.50 (3.42)	.0000	.0000	.0000
READTYP	SPY	10.50 (6.52)	10.50 (6.52)	.0000	.0000	.0000
PROFESS	BELLY					
SEX	MALE					
READTYP	SCIFI	5.50 (3.42)	5.50 (3.42)	.0000	.0000	.0000
READTYP	SPY	5.50 (3.42)	5.50 (3.42)	.0000	.0000	.0000
SEX	FEMALE					
READTYP	SCIFI	10.50 (6.52)	10.50 (6.52)	.0000	.0000	.0000
READTYP	SPY	25.50 (15.84)	25.50 (15.84)	.0000	.0000	.0000

Goodness-of-Fit test statistics

```
Likelihood Ratio Chi Square =    .00000        DF = 0   P = 1.000
Pearson Chi Square          =    .00000        DF = 0   P = 1.000
```

TABLE 7.11 (CONTINUED)

* * * * * * * * * * * L O G   L I N E A R   A N A L Y S I S *

Estimates for Parameters

PROFESS

Parameter	Coeff.	Std. Err.	Z-Value	Lower 95 CI	Upper 95 CI
1	.1935846228	.11956	1.61918	-.04075	.42792
2	.0064171131	.12929	.04963	-.24700	.25983

SEX

Parameter	Coeff.	Std. Err.	Z-Value	Lower 95 CI	Upper 95 CI
3	-.0065095139	.09098	-.07155	-.18483	.17181

READTYP

Parameter	Coeff.	Std. Err.	Z-Value	Lower 95 CI	Upper 95 CI
4	-.2491455461	.09098	-2.73844	-.42747	-.07082

PROFESS BY SEX

Parameter	Coeff.	Std. Err.	Z-Value	Lower 95 CI	Upper 95 CI
5	.1038757056	.11956	.86884	-.13046	.33821
6	.4347541617	.12929	3.36255	.18134	.68817

PROFESS BY READTYP

Parameter	Coeff.	Std. Err.	Z-Value	Lower 95 CI	Upper 95 CI
7	.1517793544	.11956	1.26951	-.08255	.38611
8	-.1790991017	.12929	-1.38522	-.43251	.07432

SEX BY READTYP

Parameter	Coeff.	Std. Err.	Z-Value	Lower 95 CI	Upper 95 CI
9	.0714203084	.09098	.78501	-.10690	.24974

PROFESS BY SEX BY READTYP

Parameter	Coeff.	Std. Err.	Z-Value	Lower 95 CI	Upper 95 CI
10	.0259458833	.11956	.21702	-.20839	.26028
11	-.1763513737	.12929	-1.36397	-.42977	.07706

TABLE 7.12   MULTIWAY FREQUENCY ANALYSIS OF SMALL SAMPLE EXAMPLE THROUGH SPSS GENLOG (SETUP AND SELECTED OUTPUT)

```
DATA FILE   FILE='SSMFA.DAT' FREE
            /PROFESS SEX READTYP FREQ.
GENLOG   PROFESS READTYP SEX/
         CSTRUCTURE=FREQ/
         PRINT ESTIM FREQ/
         DESIGN.
```

>Warning # 19345 in column 80. Text: (End of Command)
>The DESIGN subcommand is empty. A saturated design will be generated.

- -
                        GENERALIZED LOGLINEAR ANALYSIS
- -

Data Information

        12 cases are accepted.
         0 cases are rejected because of missing data.
        12 weighted cases will be used in the analysis.
        12 cells are defined.
         0 structural zeros are imposed by design.
         0 sampling zeros are encountered.

- -

Variable Information

Factor   Levels   Value

PROFESS   3
                        1.00
                        2.00
                        3.00

READTYP   2
                        1.00
                        2.00

SEX   2
                        1.00
                        2.00

- -

Model and Design Information

 Model: Poisson
Design: Constant + PROFESS + READTYP + SEX + PROFESS*READTYP + PROFESS*SEX +
      READTYP*SEX + PROFESS*READTYP*SEX

- -

Correspondence Between Parameters and Terms of the Design

Parameter   Aliased   Term

        1   Constant
        2   [PROFESS = 1.00]
        3   [PROFESS = 2.00]
        4   x   [PROFESS = 3.00]
        5   [READTYP = 1.00]
        6   x   [READTYP = 2.00]
        7   [SEX = 1.00]
        8   x   [SEX = 2.00]
        9   [PROFESS = 1.00]*[READTYP = 1.00]
       10   x   [PROFESS = 1.00]*[READTYP = 2.00]
       11   [PROFESS = 2.00]*[READTYP = 1.00]
       12   x   [PROFESS = 2.00]*[READTYP = 2.00]
```
```

TABLE 7.12   *(CONTINUED)*

Parameter	Aliased	Term
13	x	[PROFESS = 3.00]*[READTYP = 1.00]
14	x	[PROFESS = 3.00]*[READTYP = 2.00]
15		[PROFESS = 1.00]*[SEX = 1.00]
16	x	[PROFESS = 1.00]*[SEX = 2.00]
17		[PROFESS = 2.00]*[SEX = 1.00]
18	x	[PROFESS = 2.00]*[SEX = 2.00]
19	x	[PROFESS = 3.00]*[SEX = 1.00]
20	x	[PROFESS = 3.00]*[SEX = 2.00]
21		[READTYP = 1.00]*[SEX = 1.00]
22	x	[READTYP = 1.00]*[SEX = 2.00]
23	x	[READTYP = 2.00]*[SEX = 1.00]
24	x	[READTYP = 2.00]*[SEX = 2.00]
25		[PROFESS = 1.00]*[READTYP = 1.00]*[SEX = 1.00]
26	x	[PROFESS = 1.00]*[READTYP = 1.00]*[SEX = 2.00]
27	x	[PROFESS = 1.00]*[READTYP = 2.00]*[SEX = 1.00]
28	x	[PROFESS = 1.00]*[READTYP = 2.00]*[SEX = 2.00]
29		[PROFESS = 2.00]*[READTYP = 1.00]*[SEX = 1.00]
30	x	[PROFESS = 2.00]*[READTYP = 1.00]*[SEX = 2.00]
31	x	[PROFESS = 2.00]*[READTYP = 2.00]*[SEX = 1.00]
32	x	[PROFESS = 2.00]*[READTYP = 2.00]*[SEX = 2.00]
33	x	[PROFESS = 3.00]*[READTYP = 1.00]*[SEX = 1.00]
34	x	[PROFESS = 3.00]*[READTYP = 1.00]*[SEX = 2.00]
35	x	[PROFESS = 3.00]*[READTYP = 2.00]*[SEX = 1.00]
36	x	[PROFESS = 3.00]*[READTYP = 2.00]*[SEX = 2.00]

Note: 'x' indicates an aliased (or a redundant) parameter.
      These parameters are set to zero.

- -

Convergence Information

Maximum number of iterations: 20
Relative difference tolerance: .001
Final relative difference: 9.81737E-15

Maximum likelihood estimation converged at iteration 1.

- -
                    GENERALIZED LOGLINEAR ANALYSIS
- -

Table Information

Factor	Value	Observed Count	%	Expected Count	%
PROFESS	1.00				
READTYP	1.00				
SEX	1.00	1.50 (	8.33)	1.50 (	8.33)
SEX	2.00	1.50 (	8.33)	1.50 (	8.33)
READTYP	2.00				
SEX	1.00	1.50 (	8.33)	1.50 (	8.33)
SEX	2.00	1.50 (	8.33)	1.50 (	8.33)
PROFESS	2.00				
READTYP	1.00				
SEX	1.00	1.50 (	8.33)	1.50 (	8.33)
SEX	2.00	1.50 (	8.33)	1.50 (	8.33)
READTYP	2.00				
SEX	1.00	1.50 (	8.33)	1.50 (	8.33)
SEX	2.00	1.50 (	8.33)	1.50 (	8.33)
PROFESS	3.00				
READTYP	1.00				
SEX	1.00	1.50 (	8.33)	1.50 (	8.33)
SEX	2.00	1.50 (	8.33)	1.50 (	8.33)
READTYP	2.00				
SEX	1.00	1.50 (	8.33)	1.50 (	8.33)
SEX	2.00	1.50 (	8.33)	1.50 (	8.33)

TABLE 7.12   *(CONTINUED)*

- -
Goodness-of-fit Statistics

	Chi-Square	DF	Sig.
Likelihood Ratio	.0000	0	.
Pearson	.0000	0	.

- -
GENERALIZED LOGLINEAR ANALYSIS
- -

Parameter Estimates

				Asymptotic 95% CI	
Parameter	Estimate	SE	Z-value	Lower	Upper
1	-2.8134	.8165	-3.45	-4.41	-1.21
2	.5108	1.1547	.44	-1.75	2.77
3	.9163	1.1547	.79	-1.35	3.18
4	.0000	.	.	.	.
5	.9163	1.1547	.79	-1.35	3.18
6	.0000	.	.	.	.
7	1.6094	1.1547	1.39	-.65	3.87
8	.0000	.	.	.	.
9	-.5108	1.6330	-.31	-3.71	2.69
10	.0000	.	.	.	.
11	-.2231	1.6330	-.14	-3.42	2.98
12	.0000	.	.	.	.
13	.0000	.	.	.	.
14	.0000	.	.	.	.
15	-1.6094	1.6330	-.99	-4.81	1.59
16	.0000	.	.	.	.
17	-2.7081	1.6330	-1.66	-5.91	.49
18	.0000	.	.	.	.
19	.0000	.	.	.	.
20	.0000	.	.	.	.
21	-.9163	1.6330	-.56	-4.12	2.28
22	.0000	.	.	.	.
23	.0000	.	.	.	.
24	.0000	.	.	.	.
25	.5108	2.3094	.22	-4.02	5.04
26	.0000	.	.	.	.
27	.0000	.	.	.	.
28	.0000	.	.	.	.
29	1.3218	2.3094	.57	-3.20	5.85
30	.0000	.	.	.	.
31	.0000	.	.	.	.
32	.0000	.	.	.	.
33	.0000	.	.	.	.
34	.0000	.	.	.	.
35	.0000	.	.	.	.
36	.0000	.	.	.	.

- -

>Note # 19373
>For a saturated model all residuals are zero. Therefore all plots are
>skipped.

TABLE 7.13   MULTIWAY FREQUENCY ANALYSIS OF SMALL SAMPLE EXAMPLE THROUGH BMDP4F (SETUP AND SELECTED OUTPUT)

```
/INPUT        VARIABLES = 4.  FORMAT IS FREE.    CASES = 12.
              FILE = 'SSMFA.DAT'.
/VARIABLES    NAMES ARE PROFESS, SEX, READTYP, FREQ.
/CATEGORY     CODES(SEX, READTYP) ARE 1, 2.
              NAMES(READTYP) ARE SCIFI, SPY.
              NAMES(SEX) ARE MALE, FEMALE.
              CODES(PROFESS) ARE 1 TO 3.
              NAMES(PROFESS) ARE POLITIC, ADMIN, BELLY.
/TABLE        INDICES ARE READTYP, SEX, PROFESS.
              COUNT = FREQ.
/FIT          ASSOCIATION IS 3.
/END
```

DESCRIPTIVE STATISTICS OF DATA

VARIABLE NO. NAME	TOTAL FREQ.	MEAN	STANDARD DEV.	ST.ERR OF MEAN	COEFF OF VAR	SMALLEST VALUE	LARGEST VALUE	RANGE
1 PROFESS	12	2.0000	.85280	.24618	.42640	1.0000	3.0000	2.0000
2 SEX	12	1.5000	.52223	.15076	.34816	1.0000	2.0000	1.0000
3 READTYP	12	1.5000	.52223	.15076	.34816	1.0000	2.0000	1.0000
4 FREQ	12	12.917	7.8214	2.2578	.60553	5.0000	30.000	25.000

```
****************************
*  TABLE PARAGRAPH 1  *
****************************
```

***** OBSERVED FREQUENCY TABLE 1

VARIABLE   4   FREQ   USED AS COUNT VARIABLE.
         ****

PROFESS	SEX	READTYP		
		SCIFI	SPY	TOTAL
POLITIC	MALE	15	15	30
	FEMALE	10	15	25
	TOTAL	25	30	55
ADMIN	MALE	10	30	40
	FEMALE	5	10	15
	TOTAL	15	40	55

TABLE 7.13   (CONTINUED)

BELLY

MALE	5	5	10
FEMALE	10	25	35
TOTAL	15	30	45

TOTAL OF THE OBSERVED FREQUENCY TABLE IS   155

ALL CASES HAD COMPLETE DATA FOR THIS TABLE.

***** THE RESULTS OF FITTING ALL KFACTOR MARGINALS.
SIMULTANEOUS TEST THAT ALL K+1 AND HIGHER FACTOR INTERACTIONS ARE ZERO.

K-FACTOR	D.F.	LR CHISQ	PROB.	PEARSON CHISQ	PROB.	ITERATION
0-MEAN	11	48.09	0.00000	52.10	0.00000	
1	7	33.35	0.00002	32.99	0.00003	2
2	2	1.85	0.39695	1.92	0.38249	6
3	0	0.00	1.00000	0.00	1.00000	1

*****SIMULTANEOUS TEST THAT ALL KFACTOR INTERACTIONS ARE SIMULTANEOUSLY ZERO.
THE CHISQUARES ARE DIFFERENCES IN THE ABOVE TABLE.

K-FACTOR	D.F.	LR CHISQ	PROB.	PEARSON CHISQ	PROB.
1	4	14.74	0.00528	19.10	0.00075
2	5	31.50	0.00001	31.07	0.00001
3	2	1.85	0.39695	1.92	0.38249

***** ASSOCIATION OPTION SELECTED FOR ALL TERMS OF ORDER LESS THAN OR EQUAL TO   3

EFFECT	D.F.	PARTIAL ASSOCIATION CHISQUARE	PROB	ITER	D.F.	MARGINAL ASSOCIATION CHISQUARE	PROB	ITER
R.	1	13.25	0.0003					
S.	1	0.16	0.6880					
P.	2	1.32	0.5166					
RS.	1	0.62	0.4308	2	1	0.29	0.5878	2
RP.	2	4.42	0.1099	2	2	4.09	0.1294	2
SP.	2	27.12	0.0000	2	2	26.79	0.0000	2
RSP.	2	1.85	0.3969	2				

TABLE 7.14 MULTIWAY FREQUENCY ANALYSIS OF SMALL SAMPLE EXAMPLE THROUGH SAS CATMOD (SETUP AND SELECTED OUTPUT)

```
DATA    SAMPLE;
INFILE  'SSMFA.DAT';
INPUT   PROFESS  SEX  READTYP  FREQ;
PROC    CATMOD;
WEIGHT  FREQ;
MODEL   PROFESS*SEX*READTYP=_RESPONSE_/
        PRED=FREQ ML NOGLS;
        LOGLIN PROFESS|SEX|READTYP;
```

CATMOD PROCEDURE

Response: PROFESS*SEX*READTYP	Response Levels (R) = 12
Weight Variable: FREQ	Populations (S) = 1
Data Set: SAMPLE	Total Frequency (N) = 155
Frequency Missing: 0	Observations (Obs) = 12

Sample	Sample Size
1	155

RESPONSE PROFILES

Response	PROFESS	SEX	READTYP
1	1	1	1
2	1	1	2
3	1	2	1
4	1	2	2
5	2	1	1
6	2	1	2
7	2	2	1
8	2	2	2
9	3	1	1
10	3	1	2
11	3	2	1
12	3	2	2

TABLE 7.14 *(CONTINUED)*

MAXIMUM-LIKELIHOOD ANALYSIS

Iteration	Sub Iteration	-2 Log Likelihood	Convergence Criterion
0	0	770.32106	1.0000
1	0	727.84402	0.0551
2	0	722.24874	0.007687
3	0	722.23169	0.0000236
4	0	722.23169	9.649E-10

Parameter Estimates

Iteration	1	2	3	4	5	6	7	8	9	10	11
0	0	0	0	0	0	0	0	0	0	0	0
1	0.0645	0.0645	0.0323	0.0645	0.4516	-0.2903	0.1935	-0.1935	0.0323	0.0645	-0.3226
2	0.2107	0.008986	-0.006229	0.1085	0.4609	-0.2610	0.1587	-0.1937	0.0744	0.0279	-0.1795
3	0.2081	0.005397	-0.008773	0.1101	0.4567	-0.2595	0.1581	-0.1885	0.0763	0.0250	-0.1778
4	0.2081	0.005378	-0.008780	0.1101	0.4567	-0.2595	0.1581	-0.1885	0.0764	0.0250	-0.1777

MAXIMUM-LIKELIHOOD ANALYSIS-OF-VARIANCE TABLE

Source	DF	Chi-Square	Prob
PROFESS	2	3.46	0.1777
SEX	1	0.01	0.9256
PROFESS*SEX	2	17.58	0.0002
READTYP	1	7.61	0.0058
PROFESS*READTYP	2	2.62	0.2691
SEX*READTYP	1	0.66	0.4168
PROFESS*SEX*READTYP	2	1.89	0.3894
LIKELIHOOD RATIO	0	.	.

TABLE 7.14 *(CONTINUED)*

ANALYSIS OF MAXIMUM-LIKELIHOOD ESTIMATES

Effect	Parameter	Estimate	Standard Error	Chi-Square	Prob
PROFESS	1	0.2081	0.1229	2.87	0.0903
	2	-0.00538	0.1337	0.00	0.9679
SEX	3	-0.00878	0.0940	0.01	0.9256
PROFESS*SEX	4	0.1101	0.1229	0.80	0.3700
	5	0.4567	0.1337	11.67	0.0006
READTYP	6	-0.2595	0.0940	7.61	0.0058
PROFESS*READTYP	7	0.1581	0.1229	1.66	0.1982
	8	-0.1885	0.1337	1.99	0.1586
SEX*READTYP	9	0.0764	0.0940	0.66	0.4168
PROFESS*SEX*READTYP	10	0.0250	0.1229	0.04	0.8387
	11	-0.1777	0.1337	1.77	0.1837

MAXIMUM-LIKELIHOOD PREDICTED VALUES FOR RESPONSE FUNCTIONS AND FREQUENCIES

Sample	PROFESS	SEX	READTYP	Function Number	Observed Function	Observed Standard Error	Predicted Function	Predicted Standard Error	Residual
1				1	-0.5108256	0.32659863	-0.5108256	0.32660074	0
				2	-0.5108256	0.32659863	-0.5108256	0.32660074	0
				3	-0.9162907	0.37416574	-0.9162907	0.37416963	2.4299E-10
				4	-0.5108256	0.32659863	-0.5108256	0.32660074	0
				5	-0.9162907	0.37416574	-0.9162907	0.37416963	2.4299E-10
				6	0.18232156	0.27080128	0.18232156	0.27079026	-4.6857E-9
				7	-1.6094379	0.48989795	-1.6094379	0.48989912	1.42673E-9
				8	-0.9162907	0.37416574	-0.9162907	0.37416963	2.4299E-10
				9	-1.6094379	0.48989795	-1.6094379	0.48989912	1.42673E-9
				10	-1.6094379	0.48989795	-1.6094379	0.48989912	1.42673E-9
				11	-0.9162907	0.37416574	-0.9162907	0.37416963	2.4299E-10
	1	1	1	F1	15	3.68081337	15	3.68090255	1.16293E-8
	1	1	2	F2	15	3.68081337	15	3.68090255	1.16293E-8
	1	2	1	F3	10	3.05856808	9.99999999	3.05864144	9.20842E-9
	1	2	2	F4	15	3.68081337	15	3.68090255	1.16293E-8
	2	1	1	F5	10	3.05856808	9.99999999	3.05864144	9.20842E-9
	2	1	2	F6	30	4.91869377	30.0000001	4.91848472	-1.2024E-7
	2	2	1	F7	5	2.19970673	4.99999999	2.19972473	1.05229E-8
	2	2	2	F8	10	3.05856808	9.99999999	3.05864144	9.20842E-9
	3	1	1	F9	5	2.19970673	4.99999999	2.19972473	1.05229E-8
	3	1	2	F10	5	2.19970673	4.99999999	2.19972473	1.05229E-8
	3	2	1	F11	10	3.05856808	9.99999999	3.05864144	9.20842E-9
	3	2	2	F12	25	4.5790547	25	4.57895107	1.69463E-8

```
USE SSMFA
OUTPUT SSMFA.OUT
PRINT SHORT / TERM
FREQ=FREQ
MODEL SEX*PROFESS*READTYP = SEX + PROFESS + READTYP + SEX*PPROFESS,
                            SEX*READTYP + PROFESS*READTYP,
                            SEX*PROFESS*READTYP
ESTIMATE
 Number of cells (product of levels):   12
                        Total count:   155
```

Observed Frequencies
====================

SEX	PROFESS		READTYP	
			1	2
1	1		15	15
	2		10	30
	3		5	5
2	1		10	15
	2		5	10
	3		10	25

```
Pearson ChiSquare     0.0000  df    0  Probability
    Lr ChiSquare      0.0000  df    0  Probability
   Raftery's BIC      0.0000
   Dissimilarity      0.0000
```

Lambda / SE(Lambda)
===================

```
      THETA
 -------------
      2.399
 -------------
```

SEX	
1	2
-0.093	0.093

PROFESS		
1	2	3
1.694	0.040	-1.506

READTYP	
1	2
-2.759	2.759

SEX		PROFESS	
	1	2	3
1	0.897	3.417	-3.999
2	-0.897	-3.417	3.999

SEX		READTYP	
	1	2	
1	0.812	-0.812	
2	-0.812	0.812	

TABLE 7.15   (CONTINUED)

```
PROFESS   |    READTYP
          | 1          2
----------+-------------------------------
1         |    1.287      -1.287
2         |    1.410       1.410
3         |    0.214      -0.214
----------+-------------------------------


SEX     PROFESS |    READTYP
                | 1          2
--------+-------+-------------------------------
1         1     |    0.204      -0.204
          2     |   -1.329       1.329
          3     |    1.077      -1.077
                +
2         1     |   -0.204       0.204
          2     |    1.329      -1.329
          3     |   -1.077       1.077
----------------+-------------------------------
```

```
 Model ln(MLE):    -25.525
Term tested             The model without the term    Removal of term from model
                  ln(MLE)  Chi-Sq  df  p-value          Chi-Sq  df  p-value
----------------  -------- -------- ---- ---------      -------- ---- ---------
SEX. . . . . .    -25.529    0.01   1   0.9256           0.01    1   0.9256
PROFESS. . . .    -27.304    3.56   2   0.1688           3.56    2   0.1688
READTYP. . . .    -29.374    7.70   1   0.0055           7.70    1   0.0055
SEX
* PROFESS. . .    -35.339   19.63   2   0.0001          19.63    2   0.0001
SEX
* READTYP. . .    -25.856    0.66   1   0.4157           0.66    1   0.4157
PROFESS
* READTYP. . .    -26.868    2.69   2   0.2611           2.69    2   0.2611
SEX
* PROFESS
* READTYP. . .    -26.449    1.85   2   0.3969           1.85    2   0.3969
```
The 3 most outlandish cells (based on FTD, stepwise):
===

```
                                    SEX
                                    | PROFESS
                                    | |  READTYP
  ln(MLE) LR_ChiSq p-value Frequency | |
  -------- -------- -------- --------- - -  -
```

frequency for a full model; however this has no effect on subsequent values. The table of Goodness-of-fit test statistics also reflects a perfectly fitting model.

The next three tables are produced by the ASSOCIATION instruction and consist of tests of all effects individually, effects combined at each order, and effects combined at each order and higher orders. The table labeled Tests of PARTIAL associations shows tests of each two-way and one-way effect. These values are the same as those of Table 7.5, produced by hand calculation. Tests of the combined associations at each order are presented in the table labeled Tests that K-way effects are zero. In the row labeled 2 is the test of the three two-way associations combined which, in this case, shows statistical significance using both the likelihood ratio (L.R.) and Pearson chi-square criteria. This output suggests that at least one of the two-way associations is significant by both criteria. The test of the single three-way association is also provided in this table when K = 3; it is not significant. In the table labeled Tests that K-way and higher order effects are zero, the row labeled 1 contains the test of the combination of all one-way, two-way, and three-way associations, significant in this case by both likelihood ratio and Pearson chi-square criteria. The row labeled 2 contains the test of the combination of all two- and three-way associations, and so on.

Table 7.11 shows the results of an unspecified, and therefore full (saturated), model run through SPSS LOGLINEAR, a general loglinear program. After some descriptive information about how the program internally processes the model, a table of Observed, Expected Frequencies and Residuals is printed out, similar to that of the HILOGLINEAR program but with the Residuals columns included along with Goodness-of-Fit test statistics. The remaining output reports tests of effects in a different way. Instead of a partial (or marginal) test for each effect, parameter estimates for the effect are tested by dividing each λ (Coeff.) by its standard error (Std. Err.) to produce a Z-Value and a 95% confidence interval (Lower 95 CI and Upper 95 CI).[6] Note that if an effect has more than 1 df, a single test for the effect is not provided because the parameter estimate for each df is tested separately.

Table 7.12 shows the results of an unspecified (full, saturated) model run through SPSS GENLOG, the newer general loglinear program in SPSS Version 6.1. Output begins with descriptive information, followed by specification of the (full) model. A table entitled Correspondence Between Parameters and Terms of the Design follows, which shows which parameter numbers are associated with each of the parameters estimated for the model, to interpret estimates which are shown at the end of the output. For example, parameters 2 and 3 are associated with the first and second df (two parameters because there are three levels) of the one-way effect of PROFESS, respectively; parameter 5 is for the effect of READTYP; parameter 21 is for READTYP by SEX, and so on. Terms that are Aliased (x) have no parameter estimates—they are redundant with terms that do have estimates.

This is followed by Convergence Information and then Table Information, which displays observed and expected frequencies (Count) and percentages (%) for each cell. Tests of effects are presented as single df Parameter Estimates, as per SPSS LOGLINEAR. Each estimated parameter is shown with its standard error (SE), as well as Z-value: the estimate divided by SE. The Asymptotic 95% CI (confidence interval) also is shown for each parameter estimate. However, the labeling of parameters is different from that of SPSS LOGLINEAR, and follows that of the table of Correspondence Between Parameters and Terms of the Design. Note that the difference between

---

[6] These parameter estimates differ somewhat from those produced by hand calculation—Table 7.9—because of the different algorithm used by this program.

the parameters for SPSS LOGLINEAR (which correspond to hand calculations of lambdas) and GENLOG are due to the different ways that the models are parameterized in the two programs.

BMDP4F output in Table 7.13 is similar to that of SPSS HILOGLINEAR. The program begins by printing out univariate descriptive statistics for each variable, useful for ensuring against miscoded values. The observed frequency table shows cell frequencies and marginal sums. The FIT paragraph with ASSOCIATION IS 3 provides the remaining output: tests of all terms, individually, combined at each order, and combined at each order and higher orders, as in SPSS HILOGLINEAR. The table labeled ***** ASSOCIATION OPTION SELECTED FOR ALL TERMS OF ORDER LESS THAN OR EQUAL TO 3 shows both marginal and partial significance tests of the individual effects. In the *****SIMULTANEOUS TEST .. ... . table, tests of the combined associations at each order are reported (i.e., the row labeled 2 shows statistical significance when all two-way associations are tested together). In the ***** THE RESULTS OF FITTING ALL K-FACTOR MARGINALS table, 0-MEAN contains output for testing whether there are any effects considering all associations together. The row labeled 1 contains output for testing whether any of the two- or higher-way associations are statistically significant, and so on. Tests of specific models require additional instructions.

SAS CATMOD setup and output for MFA appear in Table 7.14. The full model is specified by listing the three-way association, PROFESS*SEX*READTYP equal to RESPONSE, a keyword that induces a loglinear model. ML requests maximum likelihood estimates, and NOGLS suppresses unneeded output. The LOGLIN instruction is used unless a logit model is to be tested. The instructions in this setup specify that all variables—PROFESS, SEX, READTYP—are to be treated the same, that none is the DV.

The Response label in the RESPONSE PROFILES table refers to cells; levels 1 to 12 refer to the 12 ($2 \times 2 \times 3$) cells in the frequency table. The Weight Variable refers to the variable in the data set in which cell frequencies are encoded (FREQ). After information on description of the design, CATMOD provides details about the iterative process in fitting the model and estimating the parameters. The MAXIMUM LIKELIHOOD ANALYSIS-OF-VARIANCE TABLE contains likelihood ratio Chi-Square tests of each effect individually. Note that due to differences in the algorithms used, these estimates do not match those produced by hand calculation, SPSS HILOGLINEAR, or BMDP4F, all of which are based on a different testing strategy.

There are also tests of individual parameter estimates in the following section (ANALYSIS OF MAXIMUM-LIKELIHOOD ESTIMATES), although some of these differ from both the ones shown for hand calculation (Table 7.9) and those produced by SPSS LOGLINEAR and GENLOG (Tables 7.11 and 7.12). Chi-Square tests (rather than z) are given for each of the parameter estimates. The bottom part of the last table shows observed and predicted frequencies and residuals for each of the 12 cells. The relevant information appears at the bottom portion of the table. For the first cell (F1, first level of each of the variables), both the observed and predicted frequencies are 15 and the residual is effectively zero—again because this is a full, perfectly fitting model.

SYSTAT LOGLIN, set up for a full model, is illustrated in Table 7.15. The PRINT SHORT/ TERM instruction produces tests for individual effects. The observed frequencies are printed out for the 3-way table, followed by tests of the full (saturated) model. After printing out all relevant lambdas divided by their standard errors, individual effects are tested in two ways. First, the model without the term is tested, in which a small p-value designates a poorly fit model because its expected frequency significantly differs from the observed frequency. Then, a chi-square test is given for Removal of term from the model, in which a small p-value indicates a significant degradation of the tested model. The tests indicate that models without READTYP and SEX X PROFESS are inadequate. When dealing with a saturated model, results of both types of tests are identical. Additional default output providing information about outlying ("outlandish") cells is irrelevant for a full model.

7.5 ■ SOME IMPORTANT ISSUES

7.5.1 ■ Hierarchical and Nonhierarchical Models

A model is hierarchical, or nested, if it includes all the lower effects contained in the highest-order association that is retained in the model. A hierarchical model for a four-way design, *ABCD*, with a significant three-way association, *ABC*, is $A \times B \times C, A \times B, A \times C, B \times C$, and *A, B*, and *C*. The hierarchical model might or might not also include some of the other two-way associations and the *D* first-order effect. A nonhierarchical model derived from the same four-way design includes only the significant two-way associations and first-order effects along with the significant three-way association; that is, a nonsignificant $B \times C$ association is included in a hierarchical model that retains the *ABC* effect but is not included automatically in a nonhierarchical model.

In loglinear analysis of multiway frequency tables, hierarchical models are the norm (e.g., Goodman, 1978; Knoke & Burke, 1980). Nonhierarchical models are suspect because higher order effects are confounded with lower order components. Therefore, it is best to explicitly include component lower order associations when specifying models in general loglinear programs.

One major advantage of hierarchical models is the availability of a significance test for the difference between models, so that the most parsimonious adequately fitting model can be identified using inferential procedures. With nonhierarchical models, a statistical test for the difference between models is not available unless one of the candidate models happen to be nested in the other.

SPSS LOGLINEAR and GENLOG, SYSTAT LOGLIN and SAS CATMOD have the Newton-Raphson algorithm for assessing models and are considered general loglinear programs because they do not automatically impose hierarchical modeling. BMDP4F and SPSS HILOGLINEAR are restricted to hierarchical models.

7.5.2 ■ Statistical Criteria

Confusion is rampant in MFA because tests of models look for statistical *non*significance while tests of effects look for statistical significance. Worse, both kinds of tests commonly use the same statistics—forms of χ^2.

7.5.2.1 ■ Tests of Models

Both Pearson χ^2 and the likelihood ratio statistic G^2 are routinely available for screening for the complexity of model necessary to fit data and for testing overall fit of models. Of the two, consistency favors use of G^2 because it is available for testing overall fit, screening, and testing for differences among hierarchical models. Also, under some conditions, inadequate expected frequencies can inflate Type I error rate when Pearson χ^2 is used (Milligan, 1980).

In assessing goodness-of-fit for a model, you look for a nonsignificant G^2 where the frequencies estimated from the model are similar to the observed frequencies. Thus, retention of the null hypothesis is the desired outcome—an unhappy state of affairs for choosing an appropriate alpha level. In order to avoid finding too many "good" models, you need a less strict criterion for α, say .10 or .25.

Further, with very large samples, small discrepancies between expected and observed frequencies often result in statistical significance. A significant model, even at $\alpha = .05$, may actually have adequate fit. With very small samples, on the other hand, large discrepancies often fail to reach statistical significance so that a nonsignificant model, even at $\alpha = .25$, actually has a poor fit. Choice of a significance level, then, is a matter of considering both sample size and the nature of the test. With larger samples, smaller tail probability values are chosen.

For logit models, in which one of the variables is viewed as a DV, SPSS LOGLINEAR offers two additional tests based on Haberman (1982). Both are "analyses of dispersion," an analog to analysis of variance for linear models. As in ANOVA, dispersion and degrees of freedom are partitioned into sources due to fit (systematic "variance"), conditional dispersion (error "variance"), and total dispersion. An F ratio is formed in the usual fashion for ANOVA. Both tests have corresponding measures of association (effect size) equal to the proportion of total dispersion that is due to fit—a direct analogy to eta square.

The two measures of dispersion are entropy and concentration (the latter called concordance by Haberman). Shannon (1948) defined entropy as the negative of the sum of a set of probabilities times their logarithm. Gini (1912) defined concentration as one minus the sum of a set of squared probabilities. Both measures are used to construct an analysis of dispersion for cases in which one or more IVs are used to predict a discrete DV.

For these measures, a statistically significant result indicates that the model fits the observed frequencies better than expected by chance; i.e., at $\alpha = .05$, you expect this good a fit by chance only 5% of the time when the DV is unrelated to the IV(s). Significant dispersion tests, then, correspond to nonsignificant chi-square tests in evaluating a model. If dispersion due to fit is significant, measures of association for entropy and concentration indicate how good the fit is.

7.5.2.2 ■ Tests of Individual Effects

Two types of tests are available for testing individual effects in multiway frequency tables: chi-square tests of marginal and partial effects, and z tests for single df parameter estimates.

During screening with a full model, BMDP4F provides partial G^2 tests of all effects and both partial and marginal tests of all intermediate effects. Therefore statistically significant effects can be identified without modeling (though not always unambiguously, cf. Section 7.4.1). SPSS HILOGLINEAR provides only partial G^2 tests of all effects in a full model. SAS CATMOD provides a maximum likelihood chi-square test of association for all components whether or not the model is full. SYSTAT LOGLIN provides chi-square tests of the model without each term and for removal of the term itself, as well as tests for terms tested hierarchically. All of the tests are available for full as well as partial models. In addition, all four programs print parameter estimates and their standard errors, which are easily converted to z tests of parameters.

SPSS LOGLINEAR and GENLOG provide parameter estimates and their associated z tests, but no omnibus test for any effect that has more than one degree of freedom. If an effect has more than two levels, there is no single inferential test of that effect. Although one can attribute statistical significance to an effect if any of its single df tests is significant, no overall tail probability level is available. Also, an effect may be statistically significant even though none of its single df parameters reaches significance. With only the single df z tests of parameters, such an effect is not identified.

7.5.3 ■ One Variable as DV (Logit Analysis)

In many applications of multiway frequency analysis one of the discrete variables is considered a DV and the others IVs. When loglinear analysis is used with one variable considered a DV, it is called logit analysis. The questions are the same as in multiple regression, ANOVA, discriminant function analysis, or logistic regression; the significance of the relationship with the DV is assessed overall and for all IVs and their interactions.

If the DV has only two categories and the responses are split relatively evenly between the two categories (say no more extreme than 25/75%), either multiple regression/ANOVA or logit analysis is an appropriate analysis and the results of logit analysis are usually quite close to those of multiple regression/ANOVA with a dichotomous DV (Goodman, 1978). If, on the other hand, the splits are more extreme or there are more than two categories for the DV, logit analysis or logistic regression (Chapter 12) are the better choices because they have no parametric assumptions that are likely to be violated. Logistic regression allows continuous as well as discrete IVs; however, logistic regression programs lack some of the goodies of programs that do logit analysis.

When devising a model for a logit problem, the rationale behind picking terms is fairly straightforward if one thinks in terms of regression/ANOVA. The main effect of any IV in ANOVA is the IV-DV association in logit analysis. Take, for example, prediction of vegetarianism (yes or no) by gender, age category (older, younger), and geographic region (several levels representing west coast to east coast). The main effect of gender in ANOVA is tested by the vegetarianism by gender association in logit analysis. The gender by age interaction in ANOVA is the three-way association between gender, age, and vegetarianism, and so on. The test of equality of frequency for vegetarianism in logit analysis is the test of the constant or "correction for the mean" in ANOVA. In logit analysis, only effects with vegetarianism as a component are of research interest.

If modeling is a goal of logit analysis, reliable first- and higher-order effects of IVs are included, so that adjustments are made for them. Modeling follows the usual hierarchical strategy. Can the highest-order interaction (vegetarianism by gender by age by geographic region) be eliminated without significantly degrading the fit of the model to the data? Can any or all of the three-way associations be eliminated? And so on.

7.5.3.1 ■ Programs for Logit Analysis

SPSS LOGLINEAR and GENLOG, as well as SAS CATMOD are well suited to logit analysis with their simple instructions for requesting such models and provision for running contrasts on an effect with more than two levels (cf. Section 7.5.5). For example, if geographic region is associated with vegetarianism, one might want to look at trends (linear, quadratic, etc.) to identify the source of geographic differences. If vegetarianism is more prevalent along the coasts than in the interior, for instance, a quadratic trend is present.

Additionally, SPSS LOGLINEAR and GENLOG allow inclusion of continuous covariates, should they be of interest. If age is viewed as a nuisance variable, for instance, it could serve as a continuous covariate rather than as one of the IVs. Logit analysis can also be performed through logistic regression (Chapter 12), the more general procedure because it allows continuous as well as discrete IVs. BMDPLR, BMDPPR, SPSS LOGISTIC REGRESSION, SAS LOGISTIC, and SYSTAT LOGIT perform logistic regression.

The hierarchical loglinear programs—BMDP4F and SPSS HILOGLINEAR—are less convenient for logit analysis.

7.5.3.2 ■ Odds Ratios

Once a logit model is identified and the parameters for each cell computed, the parameters can be converted to odds. The term "logit," in fact, refers to interpretation of the parameters as the log of odds ratios. The general equation for converting a parameter, λ, to an odds ratio is

$$\text{Odds ratio} = e^{2\lambda} \tag{7.8}$$

In the research described above, for example, suppose that two effects are required for an optimal model, vegetarianism (the DV, required in all models) and the association between gender and vegetarianism (the main effect of gender on vegetarianism). Suppose the λ_V value for nonvegetarianism in that optimal model is 1.15. Applying Equation 7.8:

$$\text{Odds ratio} = e^{2(1.15)} = 9.97$$

The a priori odds for being a nonvegetarian, then, are approximately 10 to 1 (an odds ratio of 10).

Parameters for significant effects change the odds. If, for example, the parameter for gender by vegetarianism (the effect of gender on vegetarianism), λ_{SV}, is -0.2 for women, the odds ratio for that parameter is:

$$\text{Odds ratio (women)} = e^{2(-0.2)} = 0.67$$

To determine the odds associated with being a female nonvegetarian, one multiplies the odds for the level of the DV (10 for nonvegetarians) by the odds for the significant effect (.67 for the association between female and nonvegetarianism). (Odds for parameters are multiplicative since log odds are additive.) Thus, the odds against vegetarianism are about 6.7 to 1 for women. For men, the associated λ_{SV} of 0.2 leads to odds of about 15 to 1 against vegetarianism, since

$$\text{Odds ratio (men)} = e^{2(0.2)} = 1.50$$

$$\text{Odds ratio (nonvegetarians)} = e^{2(1.15)} = 9.97$$

and

$$\text{Odds ratio (male nonvegetarians)} = (1.50)(9.97) \approx 15$$

7.5.4 ■ Strategies for Choosing a Model

If you have one or more models hypothesized a priori, then there is no need for the strategies discussed in this section. The techniques in this section are used if you are building a model, or trying to find the most parsimonious incomplete model.

Strategies for choosing a model differ depending on whether you are using SPSS, SYSTAT, SAS, or BMDP. Options and features differ among programs. For example, while BMDP4F provides the information necessary to closely follow model screening strategies discussed by Benedetti and Brown (1978), other strategies are available for the remaining programs. You may find it handy to use one program to screen and another to evaluate models. Recall that hierarchical programs automatically include lower order components of higher order associations; general loglinear programs may require that you explicitly include lower order components when specifying candidate hierarchical models.

7.5.4.1 ■ BMDP4F (Hierarchical)

For the full model, this program provides a test of each individual effect (with both marginal and partial χ^2 reported where appropriate), simultaneous tests of all k-way effects (all one-way effects combined, all two-way effects combined, and so on), and simultaneous tests of all k- and higher-way effects (with a four-way model, all three- and four-way effects combined, all two-, three-, and four-way effects combined, and so on). Both Pearson and likelihood ratio χ^2 (G^2) are reported. A strategy that follows the recommendations of Benedetti and Brown (1978) proceeds as follows.

Consider the *ABC* effect in a four-way design with *ABCD*. First, look at the tests of all three-way effects combined and three-way and four-way effects combined because combined results take precedence over tests of individual effects. If both combined tests are nonsignificant, the *ABC* association is deleted regardless of its partial and marginal tests, unless this specific three-way interaction has been hypothesized beforehand. If both combined tests are significant, and the *ABC* effect is significant, the *ABC* effect is retained in the final model. If some of the tests are significant while others are not, further screening is recommended. This process is demonstrated in Section 7.7.2.1. (Recall that the cutoff p values for assessing significance depend on sample size. Larger samples are tested with smaller p values to avoid including statistically significant but trivial effects.)

Further screening of effects with mixed results proceeds stepwise. BMDP4F allows two ways of stepping, simple and multiple. In simple stepping, models are changed (whether by addition or deletion) one term at a time. For example, model (*AB,C*) with effects $A \times B, A, B,$ and C is compared with model (*AB*) with effects $A \times B, A,$ and B. In multiple stepping, models that differ by more than one effect are compared; model (*AB,C*) with effects $A \times B, A, B,$ and C is compared with model (*C*) with effect C only. Fine points of multiple and simple effects assessment are covered in Dixon (1992).

BMDP4F also provides two stepping options, forward and backward. Forward stepping starts with the simplest model that includes all the unambiguously significant effects and adds terms. The "best" term, whether simple or multiple, is added first where "best" is the term that produces the largest difference in G^2 between the simpler and the more complex model. Then other terms of the same order are added, with G^2 for difference reported as each one is added. Terms that add significantly to the model are retained in the final model.

In backward stepping, one starts with all the unambiguously significant effects plus all ambiguous effects from the initial screening of the full model. The term (either simple or multiple) that is least helpful to the model is deleted first, followed by assessment of the remaining terms of the same order. Again χ^2 for the difference between simpler and more complex models is reported. Terms that do not significantly degrade the model when deleted are excluded.

According to Dixon (1992) the safest procedure is a combination of backward stepping and simple deletion. This is also the most meaningful procedure if one hypothesizes a full model. However, if forward stepping is used, multiple rather than simple addition is recommended. Additional choices for stepping include specification of the criterion probability for significance of fit (default is .05) and of the maximum number of steps for adding or deleting effects.

Note that this stepwise procedure, like others, violates rules of hypothesis testing. Therefore, don't take the χ^2 and probability values produced by the stepping procedure too seriously. View this as a search for the most reasonable model, with χ^2 providing guidelines for choosing among models, as opposed to a stricter view that some models are truly significantly better or worse than others.

7.5.4.2 ■ SPSS HILOGLINEAR (Hierarchical)

There is less information and fewer choices for screening with SPSS HILOGLINEAR, but the strategy is basically the same as for BMDP4F. The simultaneous tests for k-way and k- and higher-way associations for the full model are identical to those of BMDP4F, but only partial tests of individual effects are reported. If all tests of an effect are unambiguously significant, the effect is retained. If all tests are unambiguously nonsignificant, the effect is deleted. Ambiguity for any effect leads to further stepwise screening.

Only one stepping method is available—backward deletion of simple effects. Therefore the starting model includes all ambiguous effects as well as those to be definitely retained. Rules for interpreting results of stepping are the same as when BMDP4F is used, including the caveat with respect to overinterpreting significance.

7.5.4.3 ■ SPSS LOGLINEAR and GENLOG (General Loglinear)

These programs provide neither simultaneous tests for associations nor a stepping algorithm. Therefore, the procedure for choosing an appropriate model is simpler but less flexible.

A preliminary run with a full model is used to identify effects whose parameters differ significantly from zero. Recall that each cell of a design has a parameter for each effect and that, if the effect has more than two levels, the size of the parameter for the same effect may be different in the different cells. If an effect has a parameter that is highly significant for any cell, the effect is retained. If all the parameters for an effect are clearly nonsignificant, the effect is deleted.

Ambiguous cases occur when some parameters are marginally significant. Subsequent runs are made with and without ambiguous effects. In these runs, the significance of parameters is assessed along with the fit of the overall model. The strategy of backward elimination of simple effects, as described above, is followed for the safest route to the most reasonable model.

7.5.4.4 ■ SAS CATMOD and SYSTAT LOGLIN (General Loglinear)

Although these programs have no provision for stepwise model building and no simultaneous tests of association for each order, they do provide separate tests for all effects in a model. A preliminary run with a full model, then, is used to identify candidates for model testing through the maximum likelihood chi-square test of association. Evaluation of models follows the spirit of backward elimination of simple effects as described in Section 7.5.4.1. This is facilitated in SYSTAT LOGLIN by consulting the tables describing the effects of removing a term from the model.

7.5.5 ■ Contrasts

Just as in ANOVA, you may be interested in trends across levels of an ordered discrete variable, or in associations among discrete variables when only some levels are considered, or when some levels are combined. These are often particularly interesting when one of your variables is considered a DV.

The distinction between planned and post hoc analysis applies to multiway frequency analysis as it does to ANOVA. If contrasts are not planned, adjustment is made for the possibility of capitalizing on chance associations. Unlike ANOVA, multiway frequency analysis has no cornucopia of adjustment strategies. None of the contrast procedures in the programs covered here provides for post hoc testing. About the best you can do is apply a Bonferroni-type adjustment (cf. Section 3.2.6.4). At the very least, set a strict criterion level for each contrast, say $\alpha = .01$.

SPSS LOGLINEAR offers extensive provision for contrasts and trend analysis. The logic and setup closely follow that of SPSS MANOVA. You can specify orthogonal polynomial contrasts (trend analysis) or a matrix of special contrasts, applying contrast coefficients as illustrated in Sections 3.2.6.1, 8.5.2.3, and 10.5.1.

For example, consider the gender by profession association found in the small sample data set of Section 7.4. We can decompose this association into two interesting single df comparisons: one where we combine politicians and administrators and contrast them with belly dancers, and a second where we contrast politicians with administrators. An SPSS LOGLINEAR run in which this is done is in Table 7.16.

The first line of the contrast matrix adjusts for the constant. The second line (contrast 1) pits the first two levels of profession, politician and administrator, against belly dancers. The third line (contrast 2) contrasts politicians and administrators. The design statement requests separate parameter estimates for the two contrasts.

The output shows a highly significant association between profession and gender when politicians and administrators are combined and contrasted with belly dancers, $z = 4.45$, $p < .01$. The second contrast barely exceeds the criterion level for $\alpha = .05$. This contrast is statistically significant if it is a planned comparison, but not if it is unplanned.

SAS CATMOD also requires that you specify matrices of coefficients. The manual (SAS, 1990), however, warns against using the same contrast coefficients you would use in an analogous general linear model, because of differences in the way parameters are estimated. The same caveat applies to SPSS GENLOG in setting up contrasts; consult a manual.

BMDP4F has extensive statistics outside the loglinear model for two-way tables, including, for example, linear trend analysis. These statistics might be applied to portions of a multiway frequency table if contrasts are planned and alpha levels are stringent. Otherwise, BMDP4F as well as SYSTAT LOGLIN and SPSS HILOGLINEAR have no procedures to test contrasts within the context of a loglinear model.

7.6 ■ COMPARISON OF PROGRAMS

Six programs are available in BMDP, SAS, SPSS, and SYSTAT for analysis of multiway frequency tables. There are two types of programs for loglinear analysis, those that deal exclusively with hierarchical models and general loglinear programs that can handle nonhierarchical models as well (cf. Section 7.5.1). SPSS LOGLINEAR and GENLOG, SYSTAT LOGLIN and SAS CATMOD are general programs for nonhierarchical as well as hierarchical models (cf. Section 7.5.1). SPSS HILOGLINEAR and BMDP4F deal only with hierarchical models, but include features for stepwise model building (cf. Section 7.5.4). All six programs provide observed and expected cell frequencies, tests of fit of incomplete models, and parameter estimates accompanied by their standard errors. Beyond that, the programs differ widely. Features of the six programs appear in Table 7.17.

TABLE 7.16 CONTRASTS ON SMALL SAMPLE DATA THROUGH SPSS LOGLINEAR (SETUP AND PARTIAL OUTPUT)

```
TITLE       CONTRASTS FOR SMALL SAMPLE MULTIWAY ANALYSIS.
DATA LIST   FILE='SSMFA.DAT'  FREE
            /PROFESS SEX READTYP FREQ.
WEIGHT BY FREQ.
VALUE LABLES  PROFESS 1 'POLITIC' 2 'ADMIN'  3 'BELLY'/
              SEX 1 'MALE' 2 'FEMALE'/
              READTYP 1 'SCIFI' 2 'SPY'./
LOGLINEAR   PROFESS(1,3) SEX READTYP (1,2)/
            PRINT=ESTIM/
            CONTRAST(PROFESS)  =  SPECIAL(1  1  1,
                                          1  1 -2,
                                          1 -1  0)/
            DESIGN=READTYP, SEX, PROFESS,
                   SEX BY PROFESS(1), SEX BY PROFESS(2).
```

Correspondence Between Effects and Columns of Design/Model 1

Starting Column	Ending Column	Effect Name
1	1	READTYP
2	2	SEX
3	4	PROFESS
5	5	SEX BY PROFESS(1)
6	6	SEX BY PROFESS(2)

- -

Estimates for Parameters

READTYP

Parameter	Coeff.	Std. Err.	Z-Value	Lower 95 CI	Upper 95 CI
1	-.2989185004	.08394	-3.56122	-.46344	-.13440

SEX

Parameter	Coeff.	Std. Err.	Z-Value	Lower 95 CI	Upper 95 CI
2	-.0149353598	.09030	-.16539	-.19193	.16206

TABLE 7.16 (CONTINUED)

PROFESS

Parameter	Coeff.	Std. Err.	Z-Value	Lower 95 CI	Upper 95 CI
3	.1084280461	.06868	1.57869	-.02619	.24305
4	.0557858878	.10155	.54934	-.14325	.25482

SEX BY PROFESS(1)

Parameter	Coeff.	Std. Err.	Z-Value	Lower 95 CI	Upper 95 CI
5	.3057230622	.06868	4.45126	.17111	.440 SEX BY PROFESS(2)

Parameter	Coeff.	Std. Err.	Z-Value	Lower 95 CI	Upper 95 CI
6	-.1996269241	.10155	-1.96579	-.39867	-.00059

TABLE 7.17 COMPARISONS OF PROGRAMS FOR MULTIWAY FREQUENCY TABLES

Feature	SPSS Genlog	SPSS LOGLINEAR	SPSS HILOGLINEAR	bmdp4f	SAS CATMOD	SYSTAT LOGLIN
Input						
Individual case data	Yes	Yes	Yes	Yes	Yes	Yes
Multiway frequency table	No	No	No	Yes	No	No
Cell frequencies and indices	CSTRUCTURE	WEIGHT	WEIGHT	COUNT	WEIGHT	FREQ
Cell weights (strucural zeroes)	CSTRUCTURE	CWEIGHT	CWEIGHT	EMPTY	Yes	ZERO CELL
Convergence criteria	CONVERGE	CONVERGE	CONVERGE	CONV	EPSILON	CONV,LCONV
Tolerance	No	No	No	No	No	TOL
Level of confidence interval	CIN	No	No	No	No	No
Epsilon value for redundancy checking	EPS	No	No	No	No	No
Specify maximum number of iterations	ITERATE	ITERATION	ITERATE	ITER	MAXITER	ITER
Maximum number of halvings	No	No	No	No	No	HALF
Stepping options	N.A.	N.A.	Yes	Yes	N.A.	N.A.
Specify maximum no. of steps	N.A.	N.A.	Yes	Yes	N.A.	N.A.
Specify significance level for adequate fit	N.A.	N.A.	P	PROB	N.A.	N.A.
Specify maximum order of terms	N.A.	N.A.	MAXORDER	No	N.A.	N.A.
Force terms into stepping model	N.A.	N.A.	No	INC	N.A.	N.A
Covariates (continous)	Yes	Yes	No	No	No	No
Logit model specification	Yes	Yes	No	No	Yes	No
Single df partitions & contrasts	Yes	Yes	No	No	Yes	No
Specify delta for each cell	DELTA	DELTA	DELTA	DELTA	ADDCELL	DELTA
Specify maximum no. of missing values to delete case	No	No	No	MMV	No	No
Include cases with user-missing values	INCLUDE	INCLUDE	INCLUDE	No	No	No
Specify a repeated measures factor	No	No	No	No	Yes	No
Specify ordered factor(s)	Nob	Yes	No	No	Yes	No
Poisson model	Default	No	No	No	No	No
Multinomial model	Yes	Yes	Yes	Yes	Yes	Yes
Specify weighted least-squares method	No	No	No	No	Yes	No
Output						
Nonhierarchical models	Yes	Yes	No	No	Yes	Yes
Tests of partial association	No	No	Yes[a]	Yes	No	No
Tests of marginal association	No	No	No	Yes	No	No
Tests of models with and without each item	No	No	No	No	No	Yes
Maximum likelihood ($x2$) tests of association (ANOVA)	No	No	No	No	Yes	No
Tests of k-way effects	No	No	Yes	Yes	No	No
Tests of k-way & higher effects	No	No	Yes	Yes	No	No
Pearson model tests	Yes	Yes	Yes	Yes	No	Yes
Likelihood ratio model tests	Yes	Yes	Yes	Yes	Yes	Yes
Observed & expected (predicted) frequencies	Yes	Yes	Yes	Yes	Yes	Yes

TABLE 7.17 *(CONTINUED)*

Feature	SPSS Genlog	SPSS LOGLINEAR	SPSS HILOGLINEAR	bmdp4f	SAS CATMOD	SYSTAT LOGLIN
Observed & expected probabilities or percentages	Yes	Yes	Yes	Yes	Yes	Yes
Cell percentages of rows & columns	No	No	No	Yes	No	No
Raw residuals	Yes	Yes	Yes	Yes	Yes	Yes
Standardized residuals	Yes	Yes	Yes	Yes	No	Yes
Deviation residuals	Yes	No	No	No	No	No
Generalized residuals	Yes	Yes	No	No	No	No
Adjusted residuals	Yes	Yes	No	No	No	No
Freeman-Turkey residuals	No	No	No	Yes	No	Yes
Pearson x^2 residuals	No	No	No	Yes	No	Yes
Likelihood ratio components	No	No	No	Yes	No	Yes
Contribution to log likelihood for each cell	No	No	No	Yes	No	Yes
Loglinear parameter estimates	Parameter	Coeff.	Coeff.[a]	LAMBDA	Estimate	Param
Standard error of parameter estimate	SE	Std.Err.	Std.Err.[a]	No error	Standard	SE(Param)
Ratio of parameter estimate to standard error (*z* or *t*)	Z-value	Z-value	Z-value[a]	Yes	No	Param/SE
Confidence limits for parameter estimates	Yes	Yes	Yes	No	No	No
Chi-square tests for parameter estimates	No	No	No	No	Yes	No
Multiplicative parameter estimates	No	No	No	Beta	No	Yes
Index of dissimilarity	No	No	No	No	No	Yes
Correlation matrix for parameter estimates	Yes	Yes	No	Yes	Yes	Yes
Covariance matrix for parameter estimates	Yes	No	No	Yes	Yes	Yes
Design matrix	Yes	Yes	No	No	Yes*	No
Marginal subtables	No	No	No	Yes	No	No
Table of reasons for excluded cases	No	No	No	Yes	No	No
Plots of standardized or adjusted residuals vs. observed and expected frequencies	Yes	Yes	Yes	No	No	No
Normal plots of adjusted residuals	Yes	Yes	Yes	No	No	No
Detrended normal plots of adjusted and deviance residuals	Yes	Yes	Yes	No	No	No[c]
Descriptive statistics	No	No	No	Yes	No	No
Raftery's BIC	No	No	No	No	No	Yes
Dissimilarity	No	No	No	No	No	Yes

[a]indirect procedures are available and described in manuals as continuation ratio logit models

[b]saturated model only

[c]available through PPLOT

7.6.1 ■ SPSS Package

Currently there are three programs for handling multiway frequency tables in the SPSS package: LOGLINEAR, a general purpose program for both hierarchical and nonhierarchical models, and HILOGLINEAR, which deals with only hierarchical models. GENLOG is a newer loglinear program available in version 6.1.

SPSS HILOGLINEAR is well suited to choosing among hierarchical models, with several options for controlling stepwise selection of effects. Simultaneous tests of all k-way effects and of all k-way and higher effects are available for a quick screening of the complexity of the model from which to start stepwise selection. Parameter estimates and partial tests of association are available, but only for full models. If your purpose is to discover which discrete variables are associated rather than to model, SPSS HILOGLINEAR provides direct and easily interpreted output (although there is some doubt as to the appropriateness of providing only partial tests of association, cf. Section 7.4.1).

SPSS LOGLINEAR and GENLOG do not provide stepwise selection of hierarchical models, although they can be used to compare user-specified models of any sort. Of the six programs, only these two allow adjustment for continuous covariates. Also available are a simple specification of a logit model (in which one factor is a DV). SPSS LOGLINEAR and GEN-LOG also provide single-degree-of-freedom partitions for contrasts among specified cells (cf. Section 7.5.5).

No inferential tests of model components are provided in either program. Parameter estimates and their z tests are available for any specified model, along with their 95% confidence intervals. However, the parameter estimates are reported by single degrees of freedom, so that a factor with more than two categories has no omnibus significance test reported for either its main effect or its association with other effects (cf. Section 7.5.2.2). Moreover, the parameterizations differ between LOGLINEAR and GENLOG, so this output doesn't match. No quick screening for k-way tests is available. Screening information can be gleaned from a full model run, but identifying an appropriate model may be tedious with a large number of factors. All three SPSS programs offer residuals plots. SPSS GENLOG is the only program offering specification of Poisson models, which do not require that the analysis be conditional on total sample size.

7.6.2 ■ BMDP Series

BMDP4F (Dixon, 1992) is a highly flexible program with useful features for evaluating model components, and for selecting and comparing hierarchical models (cf. Section 7.5.1).

Several stepping options are available. Logit models are dealt with by forcing specified terms into the stepping model. Screening is facilitated by use of tests of k-way and k-way and higher effects. Extensive output of parameter estimation is available for any selected model.

BMDP4F offers the most comprehensive evaluation of effects of any k-way model of association. Likelihood ratio statistics are given for each effect. For intermediate level effects both partial and marginal tests of association are given (cf. Section 7.4.1.3). If you are simply interested in associations among effects rather than modeling, BMDP4F provides the most direct output.

BMDP4F offers the most extensive descriptive output, with cell percentages of rows and columns, marginal subtables, and reasons for excluded cases available on request. Several forms of residuals analysis but no residuals plots are available.

7.6.3 ■ SAS System

SAS CATMOD is a general program for modeling discrete data, of which loglinear modeling is only one type. The program is primarily set up for logit analyses where one variable is the DV but provision is made for loglinear models where no such distinction is made. The program offers simple designation of logit models, contrasts, and single df tests of parameters as well as maximum likelihood tests of more complex components. The program lacks provision for continuous covariates and stepwise model building procedures.

SAS CATMOD uses different algorithms from the other three programs both for parameter estimation and model testing. Although parameter estimation uses the Newton-Raphson algorithm, as does SPSS LOGLINEAR, and iterative proportional fitting, as does BMDP4F, resulting parameters differ somewhat from those in SPSS and BMDP4F, although parameters generated by SPSS and BMDP4F agree. The test for model components as well as overall model fit similarly do not always agree with SPSS and BMDP4F, although the latter programs agree with each other. The output in Table 7.14 compared with that of Tables 7.10 to 7.13 demonstrates some of the inconsistencies.

This is the only program that allows specification of factors that are ordered or represent repeated measurement of the same variable.

7.6.4 ■ SYSTAT System

SYSTAT LOGLIN (Wilkinson & Hill, 1994) is a general program for loglinear analysis of categorical data unavailable in versions earlier than 6.0. The program uses its typical MODEL statement to set up the full, saturated, model (i.e., observed frequencies) on the left hand side of the equation, and the desired model to be tested on the right. Structural zeroes can be specified, and several options are available for controlling the iterative processing of model estimation. All of the usual descriptive and parameter estimate statistics are available, as well as multiple tests of effects in the model, both hierarchical and nonhierarchical. The program also prints outlying cells, designated "outlandish". Estimated frequencies and parameter estimates can be saved to a file.

7.7 ■ COMPLETE EXAMPLE OF MULTIWAY FREQUENCY ANALYSIS

Data to illustrate multiway frequency analysis were taken from the survey of clinical psychologists described in Appendix B, Section B.3. The example is a hierarchical analysis of five dichotomous variables: whether the therapists thought (1) that their clients were aware of the therapist's attraction to them (AWARE), (2) the attraction was beneficial to the therapy (BENEFIT), and (3) the attraction was harmful to the therapy (HARM), as well as whether the therapists had (4) sought consultation when attracted to a client (CONSULT), or (5) felt uncomfortable as a result of the attraction (DISCOMF).

7.7.1 ■ Evaluation of Assumptions: Adequacy of Expected Frequencies

There are 585 psychologists in the sample. Of these, 151 are excluded from the analysis because of missing data and because only therapists who had felt attraction to at least one client answered the questions used for the analysis. The usable sample, then, consists of 434 psychologists for the hierarchical analysis.

Sample sizes are adequate for the analysis. The $2 \times 2 \times 2 \times 2 \times 2$ data table contains 32 cells, for which a sample of 434 should be sufficient; more than five cases are expected per cell if the dichotomous splits are not too bad.

To determine the adequacy of expected frequencies, all two-way contingency tables are examined. Table 7.18 shows the setup and the two-way tables from the preliminary run through BMDP4F. Expected frequencies (Fe) are calculated in the usual way for χ^2 analysis:

$$\text{Cell } Fe = \frac{(\text{row sum})(\text{column sum})}{N}$$

The smallest $Fe = (128)(140)/434 = 41.29$ for a cell of the AWARE by BENEFIT association; all cells have $Fe > 5$.

BMDP4F provides printed output of expected frequencies for two-way tables. However, obtaining the expected frequencies by computer requires an additional run because special control language is needed. In light of the simplicity of the calculations, it might be as easy to perform them by hand.

Discussion of outliers in the solution appears in the section on adequacy of fit of the selected model that follows the section on selection of a model.

7.7.2 ■ Hierarchical Loglinear Analysis

7.7.2.1 ■ Preliminary Model Screening

The full model is proposed since there are no a priori reasons to eliminate any associations. Therefore, screening and model building are used to eliminate associations that do not contribute to observed cell frequencies. Table 7.19 contains the information needed to start the model-building procedure; the simultaneous tests for effects of each order, each order and higher, and the tests of individual association.

Both likelihood ratio and Pearson criteria are used to evaluate the k level and $k + 1$ and higher-level associations. Note that the probability levels for more than two-way associations are greater than 0.05 for the simultaneous tests of both k level and $k + 1$ and higher level. The two sets of simultaneous tests agree that variables are independent in three-way and higher-order effects. Thus the model need contain no associations greater than two-way.[7]

The final portion of the table provides the basis of a search for the best model of one- and two-way effects. Among the two-way effects, several associations are clearly significant ($p < .01$) by both the partial and marginal tests: BENEFIT by AWARE, BENEFIT by CONSULT, HARM by AWARE, HARM by DISCOMF, AWARE by CONSULT, DISCOMF by CONSULT. None of the

---

[7]Although one three-way effect, BENEFIT by HARM by AWARE, meets the $p < .01$ criterion by the marginal test and $p < .05$ by the partial test, the three-way associations are not considered for inclusion because the simultaneous tests take precedence over the component associations.

```
/INPUT          FILE='MFA.DAT'.   VARIABLES = 10.  FORMAT IS FREE.
/VARIABLE       NAMES ARE SEX, AGE, AWARE, BENEFIT, HARM, CONSULT,
                    DISCOMF, INTIMACY, GRADTRNG, IDNO.
                    LABEL IS IDNO.
                MISSING = 9*9.
/CATEGORY           CODES(SEX TO DISCOMF) ARE 1, 2.
                    NAMES(SEX) ARE MALE, FEMALE.
                    NAMES(AGE) ARE YOUNGER, OLDER.
                    NAMES(AWARE) ARE 'PROB NOT', YES.
                    NAMES(BENEFIT TO DISCOMF) ARE NEVER, YES.
                    CODES(INTIMACY) ARE 1 TO 4.
                    NAMES(INTIMACY) ARE NEVER,YES,YES,YES.
                    NAMES(GRADTRNG) ARE NONE, VLITTLE, SOME, SOME.
/TABLE          INDICES ARE BENEFIT, HARM, AWARE, DISCOMF, CONSULT.
/FIT            ASSOCIATION IS 5.
                ITERATION = 30.
/PRINT          MARGINALS ARE 4.
/END
```

```
*****   MARGINAL SUBTABLE -- TABLE 1
```

HARM	BENEFIT		
	NEVER	YES	TOTAL
NEVER	90	119	209
YES	50	175	225
TOTAL	140	294	434

```
*****   MARGINAL SUBTABLE -- TABLE 1
```

AWARE	BENEFIT		
	NEVER	YES	TOTAL
PROB NOT	129	177	306
YES	11	117	128
TOTAL	140	294	434

```
*****   MARGINAL SUBTABLE -- TABLE 1
```

DISCOMF	BENEFIT		
	NEVER	YES	TOTAL
NEVER	63	90	153
YES	77	204	281
TOTAL	140	294	434

```
*****   MARGINAL SUBTABLE -- TABLE 1
```

CONSULT	BENEFIT		
	NEVER	YES	TOTAL
NEVER	85	95	180
YES	55	199	254
TOTAL	140	294	434

TABLE 7.18   (CONTINUED)

```
*****    MARGINAL  SUBTABLE  --  TABLE   1

AWARE          HARM
-------        ------
               NEVER      YES      TOTAL
------------------------------------------
PROB  NOT       173       133  |    306
YES              36        92  |    128
------------------------------|-----------
TOTAL           209       225  |    434

*****    MARGINAL  SUBTABLE  --  TABLE   1

DISCOMF        HARM
-------        ------
               NEVER      YES      TOTAL
------------------------------------------
NEVER           107        46  |    153
YES             102       179  |    281
------------------------------|-----------
TOTAL           209       225  |    434

*****    MARGINAL  SUBTABLE  --  TABLE   1

CONSULT        HARM
-------        ------
               NEVER      YES      TOTAL
------------------------------------------
NEVER           114        66  |    180
YES              95       159  |    254
------------------------------|-----------
TOTAL           209       225  |    434

*****    MARGINAL  SUBTABLE  --  TABLE   1

DISCOMF        AWARE
-------        ------
             PROB  NOT    YES      TOTAL
------------------------------------------
NEVER           119        34  |    153
YES             187        94  |    281
------------------------------|-----------
TOTAL                     128  |    434

*****    MARGINAL  SUBTABLE  --  TABLE   1

CONSULT        AWARE
-------        ------
             PROB  NOT    YES      TOTAL
------------------------------------------
NEVER           155        25  |    180
 YES            151       103  |    254
------------------------------|-----------
TOTAL           306       128  |    434

*****    MARGINAL  SUBTABLE  --  TABLE   1

CONSULT        DISCOMF
-------        -------
               NEVER      YES      TOTAL
------------------------------------------
NEVER            94        86  |    180
YES              59       195  |    254
------------------------------|-----------
TOTAL           153       281  |    434
```

two-way effects is clearly nonsignificant ($p > .05$) by both partial and marginal tests. The four two-way effects that are ambiguous, BENEFIT by HARM, BENEFIT by DISCOMF, HARM by CONSULT, and AWARE by DISCOMF, are tested through a stepwise analysis.

All first-order effects need to be included in the final hierarchical model, most because they are highly significant, and HARM because it is part of a significant two-way association. Recall that in a hierarchical model a term automatically is included if it is a part of an included higher-order association.

7.7.2.2 ■ Stepwise Model Selection

Stepwise selection by simple deletion from the model with all two-way terms is illustrated in the BMDP4F run of Table 7.20. Although 10 steps are permitted by the instruction STEP = 10, the selection process stops after the fourth step because the criterion probability is reached.

Recall that each potential model generates a set of expected frequencies. The goal of model selection is to find the model with the smallest number of effects that still provides a fit between expected frequencies and observed frequencies. First, the optimal model must have a nonsignificant LIKELIHOOD-RATIO CHI-SQUARE value (cf. Section 7.5.2.1 for choice between Pearson and likelihood ratio values). Second, the selected model should not be significantly worse than the next more complicated model. That is, if an effect is deleted from a model, that model should not be significantly worse than the model with the term still in it.

Notice first in Table 7.20 that Model 1 includes all the effects, certain and ambiguous, that might be included. Model 1 is not significant, meaning that it provides an acceptable fit between expected and observed frequencies. At step 1, effects are deleted one at a time. The best model at step 1 is the model with the smallest CHI-SQUARE value because that is the model with the best fit between expected and observed frequencies. For these data, the best model at step 1 is the one with *AD* deleted.

The best model at step 1 becomes the basis of testing at step 2 in which, once again, effects are deleted one at a time. The best model at step 2 becomes the basis for tests at step 3 when effects are again deleted one at a time, and so forth.

In Table 7.20, all the models at step 4 are significant. That is, all the models with only six effects provide expected frequencies that differ significantly from observed frequencies. None of them is a good candidate for model of choice. The best model at step 3 has seven effects (*BH, BA, BC, HA, HD, AC, DC*) and, with a nonsignificant CHI-SQUARE value of 28.92, produces a good fit between observed and expected frequencies.

However, the second criterion is that the model should not be significantly different from the next more complicated model. The next more complicated model is the step 2 model that contains *HC*. Deletion of *HC* at step 3 results in a significant difference between the models, $\chi^2(1) = 4.37$, $p < .05$. Therefore, the model at step 3 is unsatisfactory because it is significantly worse than the next more complicated model. (Use of a more conservative alpha, for example $p < .01$, would lead to a decision in favor of the best model at step 3 with seven effects.)

The best model at step 2 (*BH, BA, BC, HA, HD, HC, AC, DC*) is satisfactory by all criteria. Observed and expected frequencies based on this model do not differ significantly, nor is there a significant difference between this model and the next more complex one because deletion of *BD* produces a nonsignificant CHI-SQUARE value of .20. Remember that this model includes all one-way effects because all variables are represented in one or more associations.

The model of choice for explaining the observed frequencies, then, includes all first-order effects and the two-way associations between benefit and harm, benefit and awareness, benefit and

TABLE 7.19 PARTIAL OUTPUT FOR BMDP4F PRELIMINARY RUN OF SIMULTANEOUS AND COMPONENT ASSOCIATIONS. (SEE TABLE 7.18 FOR SETUP)

***** THE RESULTS OF FITTING ALL K-FACTOR AND HIGHER FACTOR MARGINALS.
SIMULTANEOUS TEST THAT ALL K+1 AND HIGHER FACTOR INTERACTIONS ARE ZERO.

K-FACTOR	D.F.	LR CHISQ	PROB.	PEARSON CHISQ	PROB.	ITERATION
0-MEAN	31	436.15	0.00000	491.35	0.00000	
1	26	253.51	0.00000	364.21	0.00000	2
2	16	24.09	0.08764	21.82	0.14925	9
3	6	10.16	0.11801	11.15	0.08381	8
4	0	0.02	1.00000	0.01	1.00000	44
5	0	0.00	1.00000	0.00	1.00000	1

*****SIMULTANEOUS TEST THAT ALL KFACTOR INTERACTIONS ARE SIMULTANOUSLY ZERO.
THE CHISQUARES ARE DIFFERENCES IN THE ABOVE TABLE.

K-FACTOR	D.F.	LR CHISQ	PROB.	PEARSON CHISQ	PROB.
1	5	182.64	0.00000	127.13	0.00000
2	10	229.42	0.00000	342.40	0.00000
3	10	13.92	0.17646	10.66	0.38427
4	6	10.13	0.11921	11.14	0.08425
5	0	0.03	1.00000	0.02	1.00000

***** ASSOCIATION OPTION SELECTED FOR ALL TERMS OF ORDER LESS THAN OR EQUAL TO 5

EFFECT	D.F.	PARTIAL ASSOCIATION CHISQUARE	PROB	ITER	D.F.	MARGINAL ASSOCIATION CHISQUARE	PROB	ITER
B.	1	55.85	0.0000	8				
H.	1	0.59	0.4424	8				
A.	1	75.20	0.0000	8				
D.	1	38.32	0.0000	9				
C.	1	12.68	0.0004	8				
BH.	1	4.69	0.0304	8	1	21.73	0.0000	2
BA.	1	31.95	0.0000	8	1	54.14	0.0000	2
BD.	1	0.31	0.5760	8	1	8.47	0.0036	2
BC.	1	9.77	0.0018	9	1	31.40	0.0000	2
HA.	1	11.71	0.0006	8	1	30.00	0.0000	2
HD.	1	28.99	0.0000	7	1	45.79	0.0000	2
HC.	1	4.28	0.0385	8	1	28.67	0.0000	2
AD.	1	0.26	0.6081	8	1	6.18	0.0129	2
AC.	1	15.95	0.0001	7	1	38.40	0.0000	2
DC.	1	21.47	0.0000	7	1	38.81	0.0000	2

TABLE 7.19 (CONTINUED)

***** ASSOCIATION OPTION SELECTED FOR ALL TERMS OF ORDER LESS THAN OR EQUAL TO 5

EFFECT	PARTIAL ASSOCIATION				MARGINAL ASSOCIATION			
	D.F.	CHISQUARE	PROB	ITER	D.F.	CHISQUARE	PROB	ITER
BHA.	1	6.09	0.0136	7	1	7.32	0.0068	6
BHD.	1	1.06	0.3022	7	1	2.85	0.0916	5
BHC.	1	0.41	0.5210	7	1	0.41	0.5217	5
BAD.	1	0.16	0.6927	7	1	2.12	0.1453	5
BAC.	1	0.66	0.4166	7	1	0.02	0.8817	5
BDC.	1	0.42	0.5152	7	1	0.01	0.9218	4
HAD.	1	0.75	0.3879	7	1	2.16	0.1413	6
HAC.	1	0.61	0.4338	6	1	0.15	0.6949	5
HDC.	1	0.06	0.8135	7	1	0.30	0.5852	6
ADC.	1	2.20	0.1378	7	1	2.64	0.1040	5
BHAD.	1	3.86	0.0495	7	2	2.12	0.3461	5
BHAC.	1	3.33	0.0681	7	2	2.31	0.3148	6
BHDC.	1	0.49	0.4839	30	1	0.38	0.5384	5
BADC.	1	1.47	0.2260	5	1	0.04	0.8422	5
HADC.	1	2.08	0.1488	30	1	3.31	0.0690	5
BHADC.	0	0.03	1.0000					

TABLE 7.20 SETUP AND PARTIAL OUTPUT FROM BMDP4F MODEL SELECTION RUN FOR HIERARCHICAL LOGLINEAR ANALYSIS

```
/INPUT       FILE='MFA.DAT'.  VARIABLES = 10.  FORMAT IS FREE.
/VARIABLE    NAMES ARE SEX, AGE, AWARE, BENEFIT, HARM, CONSULT,
                DISCOMF, INTIMACY, GRADTRNG, IDNO.
             LABEL IS IDNO.
             MISSING = 9*9.
/CATEGORY    CODES(SEX  TO DISCOMF)  ARE 1, 2.
             NAMES(SEX)  ARE MALE, FEMALE.
             NAMES(AGE)  ARE YOUNGER, OLDER.
             NAMES(AWARE)  ARE 'PROB NOT', YES.
             NAMES(BENEFIT TO DISCOMF)  ARE NEVER, YES.
             CODES(INTIMACY)  ARE 1 TO 4.
             NAMES(INTIMACY)  ARE NEVER, YES, YES, YES.
             NAMES(GRADTRNG)  ARE NONE, VLITTLE, SOME, SOME.
/TABLE       INDICES ARE BENEFIT, HARM, AWARE, DISCOMF, CONSULT.
/FIT         MODEL = BH, BA, BD, BC, HA, HD, HC, AD, AC, DC.
             DELETE = SIMPLE.
             STEP = 10.
/END
```

```
****************
*  MODEL 1  *
****************
```

| | LIKELIHOOD-RATIO | | | PEARSON | | |
MODEL	D.F.	CHI-SQUARE	PROB	CHI-SQUARE	PROB	ITER.
BH,BA,BD,BC,HA,HD,HC,AD,AC,DC.	16	24.09	0.0876	21.82	0.1492	9

MODELS FORMED BY DELETING TERMS FROM MODEL --
BH,BA,BD,BC,HA,HD,HC,AD,AC,DC.

| | LIKELIHOOD-ATIO | | | PEARSON | | |
MODEL	D.F.	CHI-SQUARE	PROB	CHI-SQUARE	PROB	ITER.
BA,BD,BC,HA,HD,HC,AD,AC,DC.	17	28.77	0.0367	25.76	0.0790	8
DIFF. DUE TO DELETING BH.	1	4.69	0.0304			
BH,BD,BC,HA,HD,HC,AD,AC,DC.	17	56.04	0.0000	50.44	0.0000	8
DIFF. DUE TO DELETING BA.	1	31.95	0.0000			
BH,BA,BC,HA,HD,HC,AD,AC,DC.	17	24.40	0.1090	22.22	0.1766	8
DIFF. DUE TO DELETING BD.	1	0.31	0.5760			
BH,BA,BD,HA,HD,HC,AD,AC,DC.	17	33.86	0.0088	34.91	0.0064	9
DIFF. DUE TO DELETING BC.	1	9.77	0.0018			

TABLE 7.20 *(CONTINUED)*

MODEL	D.F.	LIKELIHOOD-RATIO CHI-SQUARE	PROB	PEARSON CHI-SQUARE	PROB	ITER.
BH,BA,BD,BC,HD,HC,AD,AC,DC.	17	35.79	0.0049	34.35	0.0076	8
DIFF. DUE TO DELETING HA.	1	11.71	0.0006			
BH,BA,BD,BC,HA,HC,AD,AC,DC.	17	53.07	0.0000	49.79	0.0000	7
DIFF. DUE TO DELETING HD.	1	28.99	0.0000			
BH,BA,BD,BC,HA,HD,AD,AC,DC.	17	28.37	0.0408	28.39	0.0406	8
DIFF. DUE TO DELETING HC.	1	4.28	0.0385			
BH,BA,BD,BC,HA,HD,HC,AC,DC.	17	24.35	0.1103	21.64	0.1991	8
DIFF. DUE TO DELETING AD.	1	0.26	0.6081			
BH,BA,BD,BC,HA,HD,HC,AD,DC.	17	40.03	0.0013	35.07	0.0061	7
DIFF. DUE TO DELETING AC.	1	15.95	0.0001			
BH,BA,BD,BC,HA,HD,HC,AD,AC.	17	45.56	0.0002	42.89	0.0005	7
DIFF. DUE TO DELETING DC.	1	21.47	0.0000			

STEP 1. BEST MODEL FOUND IS --
BH,BA,BD,BC,HA,HD,HC,AC,DC.

MODEL	D.F.	LIKELIHOOD-RATIO CHI-SQUARE	PROB	PEARSON CHI-SQUARE	PROB	ITER.
BA,BD,BC,HA,HD,HC,AC,DC.	18	29.22	0.0457	25.84	0.1035	7
DIFF. DUE TO DELETING BH.	1	4.87	0.0273			
BH,BD,BC,HA,HD,HC,AC,DC.	18	56.19	0.0000	49.88	0.0001	8
DIFF. DUE TO DELETING BA.	1	31.84	0.0000			
BH,BA,BC,HA,HD,HC,AC,DC.	18	24.55	0.1379	21.99	0.2326	8
DIFF. DUE TO DELETING BD.	1	0.20	0.6550			
BH,BA,BD,HA,HD,HC,AC,DC.	18	34.39	0.0113	34.83	0.0099	7
DIFF. DUE TO DELETING BC.	1	10.04	0.0015			
BH,BA,BD,BC,HD,HC,AC,DC.	18	35.95	0.0072	34.73	0.0102	8
DIFF. DUE TO DELETING HA.	1	11.60	0.0007			
BH,BA,BD,BC,HA,HC,AC,DC.	18	53.23	0.0000	50.19	0.0001	7
DIFF. DUE TO DELETING HD.	1	28.88	0.0000			
BH,BA,BD,BC,HA,HD,AC,DC.	18	28.86	0.0501	27.82	0.0648	7
DIFF. DUE TO DELETING HC.	1	4.51	0.0336			
BH,BA,BD,BC,HA,HD,HC,DC.	18	40.18	0.0020	35.15	0.0091	7
DIFF. DUE TO DELETING AC.	1	15.83	0.0001			

TABLE 7.20 (CONTINUED)

MODEL	D.F.	LIKELIHOOD-RATIO CHI-SQUARE	PROB	PEARSON CHI-SQUARE	PROB	ITER.
BH,BA,BD,BC,HA,HD,HC,AC.	18	45.71	0.0003	43.24	0.0007	7
DIFF. DUE TO DELETING DC.	1	21.36	0.0000			

STEP 2. BEST MODEL FOUND IS
BH,BA,BC,HA,HD,HC,AC,DC.

MODEL	D.F.	LIKELIHOOD-RATIO CHI-SQUARE	PROB	PEARSON CHI-SQUARE	PROB	ITER.
BA,BC,HA,HD,HC,AC,DC.	19	30.31	0.0480	27.32	0.0974	6
DIFF. DUE TO DELETING BH.	1	5.76	0.0164			
BH,BC,HA,HD,HC,AC,DC.	19	56.38	0.0000	50.65	0.0001	7
DIFF. DUE TO DELETING BA.	1	31.83	0.0000			
BH,BA,HA,HD,HC,AC,DC.	19	35.82	0.0111	36.72	0.0086	7
DIFF. DUE TO DELETING BC.	1	11.27	0.0008			
BH,BA,BC,HD,HC,AC,DC.	19	36.14	0.0102	34.80	0.0148	6
DIFF. DUE TO DELETING HA.	1	11.59	0.0007			
BH,BA,BC,HA,HC,AC,DC.	19	55.01	0.0000	52.13	0.0001	6
DIFF. DUE TO DELETING HD.	1	30.46	0.0000			
BH,BA,BC,HA,HD,AC,DC.	19	28.92	0.0673	28.05	0.0825	6
DIFF. DUE TO DELETING HC.	1	4.37	0.0366			
BH,BA,BC,HA,HD,HC,DC.	19	40.37	0.0029	35.28	0.0129	6
DIFF. DUE TO DELETING AC.	1	15.82	0.0001			
BH,BA,BC,HA,HD,HC,AC.	19	48.03	0.0003	45.55	0.0006	6
DIFF. DUE TO DELETING DC.	1	23.48	0.0000			

STEP 3. BEST MODEL FOUND IS --
BH,BA,BC,HA,HD,AC,DC.

MODEL	D.F.	LIKELIHOOD-RATIO CHI-SQUARE	PROB	PEARSON CHI-SQUARE	PROB	ITER.
BA,BC,HA,HD,AC,DC.	20	36.70	0.0127	35.14	0.0194	5
DIFF. DUE TO DELETING BH.	1	7.78	0.0053			
BH,BC,HA,HD,AC,DC.	20	59.39	0.0000	59.33	0.0000	5
DIFF. DUE TO DELETING BA.	1	30.47	0.0000			
BH,BA,HA,HD,AC,DC.	20	42.21	0.0026	43.91	0.0015	5
DIFF. DUE TO DELETING BC.	1	13.29	0.0003			

TABLE 7.20   (CONTINUED)

MODEL	D.F.	LIKELIHOOD-RATIO CHI-SQUARE	PROB	PEARSON CHI-SQUARE	PROB	ITER.
BH,BA,BC,HD,AC,DC.	20	43.91	0.0015	42.68	0.0023	5
DIFF. DUE TO DELETING HA.	1	14.99	0.0001			
BH,BA,BC,HA,AC,DC.	20	67.44	0.0000	64.06	0.0000	6
DIFF. DUE TO DELETING HD.	1	38.53	0.0000			
BH,BA,BC,HA,HD,DC.	20	48.14	0.0004	42.92	0.0021	4
DIFF. DUE TO DELETING AC.	1	19.23	0.0000			
BH,BA,BC,HA,HD,AC.	20	60.46	0.0000	58.09	0.0000	6
DIFF. DUE TO DELETING DC.	1	31.54	0.0000			

STEP 4.   BEST MODEL FOUND IS
  BA,BC,HA,HD,AC,DC.

consultation, harm and awareness, harm and discomfort, harm and consultation, awareness and consultation, and discomfort and consultation. Not required in the model are the two-way associations between benefit and discomfort or discomfort and harm.

7.7.2.3 ■ Adequacy of Fit

Overall evaluation of the model is made on the basis of the likelihood ratio or Pearson χ^2, both of which, as seen in Table 7.20, indicate a good fit between observed and expected frequencies. For the model of choice (*BH, BA, BC, HA, HD, HC, AC, DC*) the likelihood ratio value is 24.55 with $p = .1379$; χ^2 is 21.99 with $p = .2325$.

Assessment of fit of the model in individual cells proceeds through inspection of the standardized residuals for each cell (cf. Section 7.4.3.1). These residuals, as produced by BMDP4F, are shown in Table 7.21. The table displays the observed frequencies for each cell, the expected frequencies for each cell, the standardized deviates (the standardized residual values from which discrepancies are evaluated), and the differences between observed and expected frequencies.

Most of the standardized residual values are quite small; only one cell has a value that exceeds 1.96. Since the classification table has 32 cells, a residual value of 2.2 (the largest of the standardized residuals) for one of them is not unexpected; this cell is not deviant enough to be considered an outlier. However, the fit of the model is least effective for this cell, which contains therapists who felt their attraction to clients was beneficial to the therapy, who thought their clients aware of the attraction, and who felt uncomfortable about it, but who never felt it harmful to the therapy or sought consultation about it. As seen from the observed frequency table, seven of the 434 therapists responded in this way. The expected frequency table shows that, according to the model, only about three were predicted to provide this pattern of responses.

7.7.2.4 ■ Interpretation of the Selected Model

Two types of information are useful in interpreting the selected model: parameter estimates for the model and marginal observed frequency tables for all included effects.

The loglinear parameter estimate, lambda, and the ratio λ/SE (cf. Section 7.4.3.2) are shown in Table 7.22 for each effect included in the model. Because there are only two levels of each variable, each effect is summarized by a single parameter value where one level of the effect has the positive value of the parameter and the other the negative value of the parameter.

Especially useful for interpretation are the standardized parameter estimates. Effects with the largest standardized parameter estimates are the most important in influencing the frequency in a cell. If the effects are rank ordered by the sizes of their standardized parameter estimates, the relative importance of the various effects becomes apparent. With a standardized parameter estimate of 8.815, the strongest predictor of cell size is whether or not the therapist thought the client was aware of the therapist's attraction. The least predictive of all the effects in the model, with a standardized parameter estimate of .508, is whether the therapist's attraction to the client was believed to be harmful to the therapy. (Recall from Table 7.19 that this one-way effect is included in the hierarchical model only because it is a component of at least one two-way association; it was not statistically significant by itself.)

Parameter estimates are useful in determining the relative strength of effects and in creating a prediction equation, but they do not provide a simple view of the direction of effects. For interpretation

TABLE 7.21 SETUP AND PARTIAL OUTPUT OF BMDP4F RUN TO EVALUATE RESIDUALS

```
/INPUT       FILE='MFA.DAT'.       VARIABLES = 10,  FORMAT  IS  FREE.
/VARIABLE    NAMES  ARE  SEX,  AGE,  AWARE,  BENEFIT,  HARM,  CONSULT,
                  DISCOMF,  INTIMACY,  GRADTRNG,  IDNO.
             LABEL  IS  IDNO.
             MISSING = 9*9.
/CATEGORY    CODES(SEX    TO  DISCOMF)  ARE  1,  2.
             NAMES(SEX)   ARE  MALE,  FEMALE.
             NAMES(AGE)   ARE  YOUNGER,  OLDER.
             NAMES(AWARE) ARE  'PROB NOT',  YES.
             NAMES(BENEFIT   TO  DISCOMF)  ARE  NEVER,  YES.
             CODES(INTIMACY)  ARE  1  TO  4.
             NAMES(INTIMACY)  ARE  NEVER,YES,YES,YES.
             NAMES(GRADTRNG)  ARE  NONE,  VLITTLE,  SOME,  SOME.
/TABLE       INDICES  ARE  BENEFIT,  HARM,  AWARE,  DISCOMF,  CONSULT.
/FIT         MODEL = BH,  BA,  BC,  HA,  HD,  HC,  AC,  DC.
/PRINT       EXPECTED.  DIFF.  STANDARDIZED.  LAMBDA.
/END

     *********************
     *  TABLE  PARAGRAPH  1  *
     *********************
```

***** OBSERVED FREQUENCY TABLE 1

CONSULT	DISCOMF	AWARE	HARM	BENEFIT		
				NEVER	YES	TOTAL
NEVER	NEVER	PROB NOT	NEVER	43	27	70
			YES	4	11	15
			TOTAL	47	38	85
		YES	NEVER	0	3	3
			YES	1	5	6
			TOTAL	1	8	9
	YES	PROB NOT	NEVER	20	14	34
			YES	14	22	36
			TOTAL	34	36	70
		YES	NEVER	0	7	7
			YES	3	6	9
			TOTAL	3	13	16

TABLE 7.21   (CONTINUED)

				NEVER	YES	TOTAL
YES	NEVER	PROB NOT	NEVER	10	13	23
			YES	4	7	11
			TOTAL	14	20	34
		YES	NEVER	1	10	11
			YES	0	14	14
			TOTAL	1	24	25
YES	YES	PROB NOT	NEVER	16	30	46
			YES	18	53	71
			TOTAL	34	83	117
		YES	NEVER	0	15	15
			YES	6	57	63
			TOTAL	6	72	78

TOTAL OF THE OBSERVED FREQUENCY TABLE IS 434

151 CASES HAD INCOMPLETE DATA.

MODEL	D.F.	LIKELIHOOD-RATIO CHI-SQUARE	PROB	PEARSON CHI-SQUARE	PROB	ITER.
BH,BA,BC,HA,HD,HC,AC,DC.	18	24.55	0.1379	21.99	0.2326	8

***** EXPECTED VALUES USING ABOVE MODEL

CONSULT   DISCOMF   AWARE   HARM   BENEFIT

				NEVER	YES	TOTAL
NEVER	NEVER	PROB NOT	NEVER	37.1	28.0	65.1
			YES	7.5	9.8	17.4
			TOTAL	44.6	37.9	82.5
		YES	NEVER	1.3	5.4	6.6
			YES	0.6	4.3	4.9
			TOTAL	1.9	9.6	11.5

TABLE 7.21 (CONTINUED)

YES	PROB	NOT NEVER	YES	21.8	16.5	38.3
		YES		14.8	19.4	34.2
		TOTAL		36.6	35.9	72.5
	YES	NEVER		0.8	3.1	3.9
		YES		1.2	8.4	9.6
		TOTAL		1.9	11.6	13.5

NEVER	PROB	NOT NEVER	YES	9.8	16.0	25.8
		YES		3.2	9.0	12.2
		TOTAL		13.0	25.0	38.0
	YES	NEVER		0.9	8.5	9.4
		YES		0.7	10.9	11.6
		TOTAL		1.6	19.4	21.0

YES	PROB	NOT NEVER	YES	16.7	27.1	43.8
		YES		18.1	51.2	69.3
		TOTAL		34.8	78.3	113.0
	YES	NEVER		1.6	14.4	16.0
		YES		4.0	62.0	65.9
		TOTAL		5.6	76.4	82.0

TABLE 7.21 *(CONTINUED)*

***** STANDARDIZED DEVIATES = (OBS - EXP)/SQRT(EXP) FOR ABOVE MODEL

CONSULT	DISCOMF	AWARE	HARM	BENEFIT NEVER	BENEFIT YES
NEVER	NEVER	PROB NOT	NEVER	1.0	-0.2
			YES	-1.3	0.4
		YES	NEVER	-1.1	-1.0
			YES	0.5	0.3
	YES	PROB NOT	NEVER	-0.4	-0.6
			YES	-0.2	0.6
		YES	NEVER	-0.9	2.2
			YES	1.7	-0.8
YES	NEVER	PROB NOT	NEVER	0.1	-0.7
			YES	0.5	-0.7
		YES	NEVER	0.1	0.5
			YES	-0.8	0.9
	YES	PROB NOT	NEVER	-0.2	0.6
			YES	-0.0	0.3
		YES	NEVER	-1.3	0.2
			YES	1.0	-0.6

***** DIFFERENCES BETWEEN OBSERVED AND EXPECTED USING ABOVE MODEL

CONSULT	DISCOMF	AWARE	HARM	BENEFIT NEVER	BENEFIT YES
NEVER	NEVER	PROB NOT	NEVER	5.9	-1.0
			YES	-3.5	1.2
		YES	NEVER	-1.3	-2.4
			YES	0.4	0.7

TABLE 7.21 *(CONTINUED)*

YES	PROB NOT	NEVER		-1.8	-2.5
		YES		-0.8	2.6
	YES	NEVER		-0.8	3.9
		YES		1.8	-2.4
YES	NEVER	PROB NOT	NEVER	0.2	-3.0
			YES	0.8	-2.0
		YES	NEVER	0.1	1.5
			YES	-0.7	3.1
	YES	PROB NOT	NEVER	-0.7	2.9
			YES	-0.1	1.8
		YES	NEVER	-1.6	0.6
			YES	2.0	-5.0

ASYMPTOTIC STANDARD ERRORS OF THE PARAMETER ESTIMATES ARE COMPUTED BY INVERTING THE INFORMATION MATRIX.

TABLE 7.22   PARTIAL OUTPUT FOR BMDP4F RUN ON PARAMETER ESTIMATES (SETUP APPEARS IN TABLE 7.21)

```
*****   ESTIMATES OF THE LOG-LINEAR PARAMETERS (LAMBDA) IN THE MODEL ABOVE

              BENEFIT
              ------
        NEVER       YES
        ---------------------
        -0.617      0.617

*****        RATIO OF THE LOG-LINEAR PARAMETER ESTIMATE TO ITS STANDARD ERROR

                  BENEFIT
                  ------
            NEVER       YES
            -------------------
            -7.228      7.228

*****   ESTIMATES OF THE LOG-LINEAR PARAMETERS (LAMBDA) IN THE MODEL ABOVE

              HARM

        NEVER       YES
        ---------------------
        0.035      -0.035

*****        RATIO OF THE LOG-LINEAR PARAMETER ESTIMATE TO ITS STANDARD ERROR

                  HARM
                  -----
            NEVER       YES
            -------------------
            0.507      -0.507

*****   ESTIMATES OF THE LOG-LINEAR PARAMETERS (LAMBDA) IN THE MODEL ABOVE

HARM          BENEFIT
------          ------
              NEVER       YES
        ---------------------------
NEVER          0.138      -0.138
YES           -0.138       0.138

*****        RATIO OF THE LOG-LINEAR PARAMETER ESTIMATE TO ITS STANDARD ERROR

        HARM          BENEFIT
        ------          ------
                      NEVER       YES
        ----------------------------------------
            NEVER      2.395      -2.395
            YES       -2.395       2.395

*****   ESTIMATES OF THE LOG-LINEAR PARAMETERS (LAMBDA) IN THE MODEL ABOVE

              AWARE
              ------
        PROB NOT     YES
        ---------------------
        0.794      -0.794

*****        RATIO OF THE LOG-LINEAR PARAMETER ESTIMATE TO ITS STANDARD ERROR
                        AWARE
                        -------
            PROB NOT     YES
            -------------------
            8.815      -8.815
```

TABLE 7.22 *(CONTINUED)*

***** ESTIMATES OF THE LOG-LINEAR PARAMETERS (LAMBDA) IN THE MODEL ABOVE

```
AWARE       BENEFIT
------      ------
            NEVER      YES
--------------------------
PROB NOT    0.428    -0.428
 YES       -0.428     0.428
```

***** RATIO OF THE LOG-LINEAR PARAMETER ESTIMATE TO ITS STANDARD ERROR

```
    AWARE       BENEFIT
    ------      ------
                NEVER      YES
    --------------------------
    PROB NOT    4.947    4.947
    YES        -4.947    4.947
```

***** ESTIMATES OF THE LOG-LINEAR PARAMETERS (LAMBDA) IN THE MODEL ABOVE

```
AWARE       HARM
-----       -----
            NEVER      YES
--------------------------
PROB NOT    0.206    -0.206
YES        -0.206     0.206
```

***** RATIO OF THE LOG-LINEAR PARAMETER ESTIMATE TO ITS STANDARD ERROR

```
    AWARE       HARM
    ------      ------
                NEVER      YES
    --------------------------
    PROB NOT    3.355    -3.355
    YES        -3.355     3.355
```

***** ESTIMATES OF THE LOG-LINEAR PARAMETERS (LAMBDA) IN THE MODEL ABOVE

```
        DISCOMF

        NEVER      YES
    -------------------
        -0.302     0.302
```

***** RATIO OF THE LOG-LINEAR PARAMETER ESTIMATE TO ITS STANDARD ERROR

```
            DISCOMF
            ---------
            NEVER      YES
        -------------------
            -5.445     5.445
```

***** ESTIMATES OF THE LOG-LINEAR PARAMETERS (LAMBDA) IN THE MODEL ABOVE

```
DISCOMF     HARM
------      ------
            NEVER      YES
------------------------------
NEVER       0.302    -0.302
YES        -0.302     0.302
```

TABLE   7.22   *(CONTINUED)*

*****   RATIO OF THE LOG-LINEAR PARAMETER ESTIMATE TO ITS STANDARD ERROR

DISCOMF	HARM	
	NEVER	YES
NEVER	5.414	-5.414
YES	-5.414	5.414

*****   ESTIMATES OF THE LOG-LINEAR PARAMETERS (LAMBDA) IN THE MODEL ABOVE

CONSULT	
NEVER	YES
-0.166	0.166

*****   RATIO OF THE LOG-LINEAR PARAMETER ESTIMATE TO ITS STANDARD ERROR

CONSULT	
NEVER	YES
-2.319	2.319

*****   ESTIMATES OF THE LOG-LINEAR PARAMETERS (LAMBDA) IN THE MODEL ABOVE

CONSULT	BENEFIT	
	NEVER	YES
NEVER	0.192	-0.192
YES	-0.192	0.192

*****   RATIO OF THE LOG-LINEAR PARAMETER ESTIMATE TO ITS STANDARD ERROR

CONSULT	BENEFIT	
	NEVER	YES
NEVER	3.346	-3.346
YES	-3.346	3.346

*****   ESTIMATES OF THE LOG-LINEAR PARAMETERS (LAMBDA) IN THE MODEL ABOVE

CONSULT	HARM	
	NEVER	YES
NEVER	0.118	-0.118
YES	-0.118	0.118

*****   RATIO OF THE LOG-LINEAR PARAMETER ESTIMATE TO ITS STANDARD ERROR

CONSULT	HARM	
	NEVER	YES
NEVER	2.095	-2.095
YES	-2.095	2.095

TABLE 7.22   *(CONTINUED)*

*****   ESTIMATES OF THE LOG-LINEAR PARAMETERS (LAMBDA) IN THE MODEL ABOVE

CONSULT   AWARE
------   -------
              PROB NOT   YES

NEVER   0.256   -0.256
YES   -0.256   0.256

*****   RATIO OF THE LOG-LINEAR PARAMETER ESTIMATE TO ITS STANDARD ERROR

          CONSULT   AWARE
          ------   -------
                        PROB NOT   YES

          NEVER   3.844   -3.844
          YES   -3.844   3.844

*****   ESTIMATES OF THE LOG-LINEAR PARAMETERS (LAMBDA) IN THE MODEL ABOVE

CONSULT   DISCOMF
------   ---------
              NEVER   YES

NEVER   0.265   -0.265
YES   -0.265   0.265

*****   RATIO OF THE LOG-LINEAR PARAMETER ESTIMATE TO ITS STANDARD ERROR

          CONSULT   DISCOMF
                        NEVER   YES

          NEVER   4.812   -4.812
          YES   -4.812   4.812

of direction, the marginal tables of observed frequencies for each effect in the model are useful, as illustrated in Table 7.23.

The results as displayed in Table 7.23 are best interpreted as proportions of therapists responding in a particular way. For example, the BENEFIT marginal subtable shows that 32% (140/434) of the therapists believe that there was never any benefit to be gained from the therapist being attracted to a client. The BENEFIT by HARM marginal subtable shows that, among those who believe that there was no benefit, 64% (90/140) also believe there was no harm. Of those who believe there was at least some benefit, 59% (175/294) also believe there was at least some harm.

Table 7.24 summarizes the model in terms of significance tests and parameter estimates.

A checklist for hierarchical multiway frequency analysis appears in Table 7.25. A Results section, in journal format, follows for the analysis described.

TABLE 7.23   PARTIAL OUTPUT FROM BMDP4F RUN SHOWING OBSERVED FREQUENCIES FOR SIGNIFICANT
EFFECTS. (SETUP APPEARS IN TABLE 7.18)

```
*** MARGINAL SUBTABLE -- TABLE  1

            BENEFIT
            ------
        NEVER     YES    TOTAL
    ------------------------------
         140      294 |    434

*** MARGINAL SUBTABLE -- TABLE  1

            HARM
            ------
        NEVER     YES    TOTAL
    ------------------------------
         209      225 |    434

*** MARGINAL SUBTABLE -- TABLE  1

            AWARE
            -------
        PROB NOT   YES    TOTAL
    ------------------------------
         306      128     434

*** MARGINAL SUBTABLE -- TABLE  1

            DISCOMP
            ------
        NEVER     YES    TOTAL
    ------------------------------
         153      281 |    434

*** MARGINAL SUBTABLE -- TABLE  1

            CONSULT
            ------
        NEVER     YES    TOTAL
    ------------------------------
         180      254 |    434

*** MARGINAL SUBTABLE -- TABLE  1

    HARM        BENEFIT
    ------      ------
                NEVER     YES    TOTAL
    ---------------------------------------
    NEVER          90     119 |    209
    YES            50     175 |    225
    -----------------------------|---------
    TOTAL         140     294 |    434

*** MARGINAL SUBTABLE -- TABLE  1

    AWARE       BENEFIT
    ------      ------
                NEVER     YES    TOTAL
    ---------------------------------------
    PROB NOT      129     177 |    306
    YES            11     117 |    128
    -----------------------------|---------
    TOTAL         140     294 |    434
```

TABLE 7.23 *(CONTINUED)*

```
*** MARGINAL SUBTABLE -- TABLE  1
```

CONSULT   BENEFIT
------   ------

	NEVER	YES	TOTAL
NEVER	85	95	180
YES	55	199	254
TOTAL	140	294	434

```
*** MARGINAL SUBTABLE -- TABLE 1
```

AWARE   HARM
------   ------

	NEVER	YES	TOTAL
PROB NOT	173	133	306
YES	36	92	128
TOTAL	209	225	434

```
*** MARGINAL SUBTABLE -- TABLE 1
```

DISCOMF   HARM
------   ------

	NEVER	YES	TOTAL
NEVER	107	46	153
YES	102	179	281
TOTAL	209	225	434

```
*** MARGINAL SUBTABLE -- TABLE 1
```

CONSULT   HARM
------   ------

	NEVER	YES	TOTAL
NEVER	114	66	180
YES	95	159	254
TOTAL	209	225	434

```
*** MARGINAL SUBTABLE -- TABLE 1
```

CONSULT   AWARE
------   ------

	PROB NOT	YES	TOTAL
NEVER	155	25	180
YES	151	103	254
TOTAL	306	128	434

```
*** MARGINAL SUBTABLE -- TABLE 1
```

CONSULT   DISCOMF
------   ------

	NEVER	YES	TOTAL
NEVER	94	86	180
YES	59	195	254
TOTAL	153	281	434

TABLE 7.24 SUMMARY OF HIERARCHICAL MODEL OF THERAPISTS' ATTRACTION TO CLIENTS,

$N = 434$
Constant = 1.966

Effect	Partial association chi-square	Loglinear parameter estimate (lambda)		Lambda/*SE*	
First-order effects:					
		Prob. not	Yes	Prob. not	Yes
Aware	75.20**	0.794	− 0.794	8.815	− 8.815
		Never	Yes	Never	Yes
Benefit	55.85**	− 0.617	0.617	− 7.228	7.228
		Never	Yes	Never	Yes
Discomfort	38.32**	−0.302	0.302	−5.445	5.445
		Never	Yes	Never	Yes
Consult	12.68**	− 0.166	0.166	− 2.319	2.319
		Never	Yes	Never	Yes
Harm	0.59	0.035	− 0.035	0.508	− 0.508

Second-order effects:

Effect	Partial association chi-square		Prob.not	Yes		Prob.not	Yes
Benefit by aware	31.95**	Never	0.428	− 0.428	Never	4.947	− 4.947
		Yes	− 0.428	0.428	Yes	− 4.947	4.947
			Never	Yes		Never	Yes
Harm by discomfort	28.99**	Never	0.302	− 0.302	Never	5.414	− 5.414
		Yes	− 0.302	0.302	Yes	− 5.414	5.414
			Never	Yes		Never	Yes
Discomfort by consult	21.47**	Never	0.265	− 0.265	Never	4.812	− 4.812
		Yes	− 0.265	0.265	Yes	− 4.812	4.812

TABLE 7.24 (CONTINUED)

			Never	Yes		Never	Yes
Aware by consult	15.95**	Prob. not	0.256	−0.256	Prob.not	3.844	−3.844
		Yes	−0.256	0.256	Yes	−3.844	3.844
			Prob.not	Yes		Prob.not	Yes
Harm by aware	11.71**	Never	0.206	−0.206	Never	3.355	−3.355
		Yes	−0.206	0.206	Yes	−3.355	3.355
			Never	Yes		Never	Yes
Benefit by consult	9.77**	Never	0.191	− 0.191	Never	3.345	− 3.345
		Yes	− 0.191	0.191	Yes	− 3.345	3.345
			Never	Yes		Never	Yes
Benefit by harm	4.69*	Never	0.138	− 0.138	Never	2.397	− 2.397
		Yes	− 0.138	0.138	Yes	− 2.397	2.397
			Never	Yes		Never	Yes
Harm by consult	4.28*	Never	0.118	−0.118	Never	2.095	− 2.095
		Yes	− 0.118	0.118	Yes	− 2.095	2.095

Note:*$p<.05$.

**$p<.01$.

TABLE 7.25 CHECKLIST FOR HIERARCHICAL MULTI-WAY FREQUENCY ANALYSIS

1. Issues
   a. Adequacy of expected frequencies
   b. Outliers in the solution
2. Major analysis
   a. Model screening
   b. Model selection
   c. Evaluation of overall fit. If adequate:
      (1) Significance tests for each model effect
      (2) Parameter estimates
3. Additional analyses
   a. Interpretation via proportions
   b. Identifying extreme cells (if fit inadequate)

Results

A five-way frequency analysis was performed to develop a hierarchical loglinear model of attraction of therapists to clients. Dichotomous variables analyzed were whether the therapist (1) believed the attraction to be beneficial to the client, (2) believed the attraction to be harmful to the client, (3) thought the client was aware of the attraction, (4) felt discomfort, and (5) sought consultation as a result of the attraction.

Four hundred thirty-four therapists provided usable data for this analysis. All two-way contingency tables provided expected frequencies in excess of five. After the model was selected, none of the 32 cells was an outlier.

Stepwise selection by simple deletion of effects using BMDP4F produced a model that included all first-order effects and eight of the ten possible two-way associations. The model had a likelihood ratio $\chi^2(18) = 24.55$, $p = .14$, indicating a good fit between observed frequencies and expected frequencies generated by the model. A summary of the model with results of tests of significance (partial likelihood ratio χ^2) and loglinear parameter estimates in raw and standardized form appears in Table 7.24.

Most of the therapists (68%) reported that the attraction they felt for clients was at least occasionally beneficial to therapy, while a slight majority (52%) also reported that it was at least occasionally harmful. Seventy-one percent of the therapists thought that clients were probably aware of the attraction. Most therapists (65%) felt at least some discomfort about the attraction, and more than half (58%) sought consultation as a result of the attraction.

Of those therapists who thought the attraction beneficial to the therapy, 60% also thought it harmful. Of those who thought the attraction never beneficial, 36% thought it harmful. Perception of benefit was also related to client's awareness. Of those who thought their clients were aware of the attraction, 91% thought it beneficial. Among those who thought clients unaware, only 58% thought it beneficial.

Those who sought consultation were also more likely to see the attraction as beneficial. Of those seeking consultation, 78% judged the attraction beneficial. Of those not seeking consultation, 53% judged it beneficial.

Lack of harm was associated with lack of awareness. Fifty-seven percent of therapists who thought their clients unaware felt the attraction was never harmful. Only 28% of those who thought their clients

aware considered it never harmful. Discomfort was more likely to be
felt by those therapists who considered the attraction harmful to therapy (80%) than by those therapists who thought it was not harmful to
therapy (49%). Similarly, consultation was more likely to be sought by
those who felt the attraction harmful (71%) than by those who did not
feel it harmful (45%).

Seeking consultation was also related to client awareness and therapist discomfort. Therapists who thought clients were aware of the
attraction were more likely to seek consultation (80%) than those who
thought the client unaware (43%). Those who felt discomfort were more
likely to seek consultation (69%) than those who felt no such discomfort (39%).

No statistically significant two-way associations were found between
benefit and discomfort or between awareness and discomfort. None of the
higher-order associations reached statistical significance.

7.8 ■ SOME EXAMPLES FROM THE LITERATURE

Stability in types of errors made on math tests by boys and girls was studied by Marshall (1983).
Data were derived from the Survey of Basic Skills, Grade 6, a test given annually to all sixth grade
children in California. Chosen for study were the 160 math items on the survey that had four alternatives. Only incorrect responses were analyzed. A three-way contingency table was formed for
each item: a $2 \times 4 \times 4$ table of sex by year (1976, 1977, 1978, 1979) by distractor (three incorrect
alternatives plus no answer).

It was hypothesized that the best-fit model would include the distractor by sex association (indicating that boys and girls make different errors) but not the three-way association (which would indicate
that sex differences were not stable but varied from year to year). First-order effects were not of interest. The analysis was also to determine which of the other two-way interactions (sex by year, and distractor by year) were needed if the three-way effect and the distractor by sex effects were as predicted.

The clear winner among the hierarchical models was indeed limited to second-order effects; the
distractor by sex interaction was significant as hypothesized as was the distractor by year interaction. This was the best-fit model for 128 of the 160 items. The next best model included the three-
way interaction, which only emerged as best for 10 items. The author concluded that sex differences
in choice of distractor were indeed stable—boys and girls systematically select different incorrect
alternatives.

Porter and Albert (1977) studied attitudes toward women's role among South African and
American women. Attitude toward women's role was categorized as traditional or liberal on two
scales, familial roles and working mothers. Two $3 \times 2 \times 2 \times 2 \times 2$ frequency table were formed,
one for attitudes on the familial roles dimension and the other for attitudes on the working mother
measure. Predictors of attitudes were religion (liberal Christian, orthodox Christian, Jew), country

(South Africa, United States), whether the mother worked outside the home, and educational level (low, high).

Separate logit analysis was done for each of the two attitude scales. In addition, hierarchical modeling was used to find the best model to predict attitudes. This, then, was a combination of hierarchical modeling with logit analysis.

Attitude toward women's role on both scales was predicted by country and work status; the model for attitude toward familial roles also included the effect of education. The questions of interest were whether religion added predictability and whether religion interacted with the other predictors in enhancing prediction of attitudes.

For both attitude scales, religion significantly enhanced predictability over that of country, work status, and education while religion in association with the other three predictors did not significantly enhance prediction of attitude. For both scales, then, the best fit model was one that included the interaction of each predictor, including religion, with attitude but none of the higher-order interactions.

Relative magnitude of effects was evaluated through logits (log odds) of each effect as computed from effect parameters and then translated into probabilities. For example, South Africans were found to be 52% more likely to be traditional on attitudes toward familial roles than were Americans; this was the single strongest effect in both models. For attitudes toward working mothers, the strongest effects were associated with whether or not the respondent was a working woman. Also for this scale, a particularly good fit was obtained by including the work by religion interaction, although this was not significantly better than the model that excluded this interaction.

Probability of keeping appointments in a community mental health center was analyzed by Rock (1982) in a six-way frequency table. With outcome (show, no show) as the dependent variable, the five predictors were method of contact (in person, telephone), contact person (client, significant other), sex of client, time of day of intake appointment (A.M., P.M.), and number of days between first contact and intake appointment (0 to 7 days). As a result of fitting the full model, it was found that four of the six main effects were significant but that none of the interactions was. Although one of the main effects was the outcome, none of the interactions with the outcome reached statistical significance. There was, then, no logit model to test.

This analysis was followed by several bivariate chi-square analyses searching for significant relationships among those variables which produced significant main effects in the loglinear analysis. Since this is akin to performing post hoc interaction contrasts after finding nonsignificant omnibus interaction effects in analysis of variance, the statistical justification for these analyses is not clear. Nevertheless, three "significant" associations were found among the four tested; method of contact by outcome, latency of outcome, and method of contact by contact person. The author interpreted the results as evidence that individuals were more likely to show up for their appointments if contact was made by telephone and the appointment was scheduled relatively soon after initial contact. Also, contact was most likely to be made by telephone if made by a significant other.

Hierarchical modeling was used to investigate change in household composition of the elderly by Fillenbaum and Wallman (1984). Two separate analyses were done, one on a full sample of elderly respondents and another on a subsample of those who had no change in marital status in the 30-month period before first and second interviews. For both analyses, the dependent variable was whether there was change in household composition during the 30-month period. As predictors, both analyses used indices of whether there was change in economic status and self-care capacity. A third predictor was whether extensive or limited help was available. For the full sample, an additional predictor was whether there was change in marital status.

Second-order models, including associations of change in household composition with the remaining variables, were tested individually against a model that included only the main effect of household composition change. Among these, the only model with a single two-way association that provided improvement in fit was the one that included change in marital status. However, the fit was improved by considering a second two-way association—change in household composition by extent of help available. That is, the best fitting model included the associations of change in marital status with change in household composition, and extent of help with change in household composition. No mention was made of testing third-order models.

Change in household composition, then, is most likely by those elderly who experience change in marital status. A small gain in predictability is made by considering extent of help available. Those who perceive extensive availability of help are less likely to change their household composition. This interpretation is supported by separate analysis of the subsample which did not experience change in marital status. The only significant two-way association for this sample was between availability of help and change in household composition.

CHAPTER 8
Analysis of Covariance

8.1 ■ GENERAL PURPOSE AND DESCRIPTION

Analysis of covariance is an extension of analysis of variance in which main effects and interactions of IVs are assessed after DV scores are adjusted for differences associated with one or more covariates.[1] The major question for ANCOVA is essentially the same as for ANOVA: Are mean differences among groups on the adjusted DV likely to have occurred by chance? For example, is there a mean difference between a treated group and a control group on a posttest (the DV) after posttest scores are adjusted for differences in pretest scores (the covariate)?

Analysis of covariance is used for three major purposes. The first purpose is to increase the sensitivity of the test of main effects and interactions by reducing the error term; the error term is adjusted for, and hopefully reduced by, the relationship between the DV and the covariate(s). The second purpose is to adjust the means on the DV themselves to what they would be if all subjects scored equally on the covariate(s). The third use of ANCOVA occurs in MANOVA (Chapter 9) where the researcher assesses one DV after adjustment for other DVs that are treated as covariates.

The first use of ANCOVA is the most common. In an experimental setting, ANCOVA increases the power of an F test for a main effect or interaction by removing predictable variance associated with covariate(s) from the error term. That is, covariates are used to assess the "noise" where "noise" is undesirable variance in the DV (e.g., individual differences) that is estimated by scores on covariates (e.g., pretests). Although in most experiments subjects are randomly assigned to groups, random assignment in no way assures equality among groups—it only guarantees, within probability limits, that there are no systematic differences between groups to begin with. Random individual differences, however, can both spread out scores among subjects within groups (and increase the error term) and create differences among groups that are not associated with treatment.

---

[1] Strictly speaking, ANCOVA, like multiple regression, is not a multivariate technique because it involves a single DV. For the purposes of this book, however, it is convenient to consider it along with multivariate analyses.

Either of these effects makes it hard to show that differences among groups are due to different experimental treatments. One way to diminish the effect of individual differences is to adjust for them statistically. Individual differences are measured as covariates and ANCOVA is used to provide the adjustment. The effect of the adjustment is to provide a (usually) more powerful test of differences among groups uncontaminated with differences on covariates.

In this sense, ANCOVA is similar to a within-subjects (repeated measures) ANOVA (cf. Chapter 3), in which individual differences among subjects are estimated from consistencies within subjects over treatment and then the variance due to individual differences is removed from the error term. In ANCOVA, variance due to individual differences is assessed through one or more covariates, usually continuous variables, that are linearly related to the DV. Then the relationship between the DV and the covariate(s) is removed from the error term.

In the classical experimental use of ANCOVA, subjects are randomly assigned to levels of one or more IVs and are measured on one or more covariates before administration of treatment. A common covariate is a pretest score, measured the same as the DV but before manipulation of treatment. However, a covariate might also be some demographic characteristic (educational level, socioeconomic level, or whatever) or a personal characteristic (anxiety level, IQ, etc.) that is completely different from the DV. Manipulation of the IV(s) occurs and then the DV is measured.

For example, suppose an experiment is designed to investigate methods for reducing test anxiety in statistics courses. Volunteers enrolled in statistics courses are randomly assigned to one of three treatment groups: desensitization, relaxation training, or a control group (for whom anxiety reduction is offered after the experiment). The three treatment groups are the three levels of the IV. Before treatment, however, students in all three groups are given a standardized measure of test anxiety that serves as the covariate. Then treatment or "waiting-list" control is carried out for some specified time. After the treatment period, students in all three groups are tested again on the measure of test anxiety (preferably an alternate form to avoid memory of the pretest); this measure serves as the DV. The goal of statistical analysis is to test the null hypothesis that all three groups have the same mean test anxiety after treatment after adjusting for preexisting differences in test anxiety.

The second use of ANCOVA commonly occurs in nonexperimental situations where subjects cannot be randomly assigned to treatments. ANCOVA is used as a statistical matching procedure, although interpretation is fraught with difficulty, as discussed in Section 8.3.1. ANCOVA is used primarily to adjust group means to what they would be if all subjects scored identically on the covariate(s). Differences between subjects on covariates are removed so that, presumably, the only differences that remain are related to the effects of the grouping IV(s). (Differences could also, of course, be due to attributes that have not been used as covariates.)

This second application of ANCOVA is primarily for descriptive model building: the covariate enhances prediction of the DV, but there is no implication of causality. If the research question to be answered involves causality, ANCOVA is no substitute for running an experiment.

As an example, suppose we are looking at regional differences in political attitudes where the DV is some measure of liberalism-conservatism. Regions of the United States form the IV, say, Northeast, South, Midwest, and West. Two variables that are expected to vary with political attitude and with geographical region are socioeconomic status and age. These two variables serve as covariates. The statistical analysis tests the null hypothesis that political attitudes do not differ with geographical region after adjusting for socioeconomic status and age. However, if age and socioeconomic differences are inextricably tied to geography, adjustment for them is not realistic. And, of course, there is no implication that political attitudes are caused in any way by geographic region. Further, unreliability in measurement of the covariate and the DV-covariate relationship may lead to over- or underadjustment of scores and means and, therefore, to misleading results. These issues are discussed in greater detail throughout the chapter.

In the third major application of ANCOVA, discussed more fully in Chapter 9, ANCOVA is used to interpret IV differences when several DVs are used in MANOVA. After a multivariate analysis of variance, it is frequently desirable to assess the contribution of the various DVs to significant differences among IVs. One way to do this is to test DVs, in turn, with the effects of other DVs removed. Removal of the effects of other DVs is accomplished by treating them as covariates in a procedure called a stepdown analysis.

The statistical operations are identical in all three major applications of ANCOVA. As in ANOVA, variance in scores is partitioned into variance due to differences between groups and variance due to differences within groups. Squared differences between scores and various means are summed (see Chapter 3) and these sums of squares, when divided by appropriate degrees of freedom, provide estimates of variance attributable to different sources (main effects of IVs, interactions between IVs, and error). Ratios of variances then provide tests of hypotheses about the effects of IVs on the DV.

However, in ANCOVA, the regression of one or more covariates on the DV is estimated first. Then DV scores and means are adjusted to remove the linear effects of the covariate(s) before analysis of variance is performed on these adjusted values.

Lee (1975) presents an intuitively appealing illustration of the manner in which ANCOVA reduces error variance in a one-way between-subjects design with three levels of the IV (Figure 8.1). Note that the vertical axis on the right-hand side of the figure illustrates scores and group

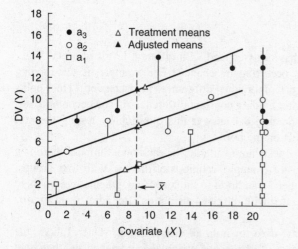

FIGURE 8.1 PLOT OF HYPOTHETICAL DATA. THE STRAIGHT LINES WITH COMMON SLOPE ARE THOSE THAT BEST FIT THE DATA FOR THE THREE TREATMENTS. THE DATA POINTS ARE ALSO PLOTTED ALONG THE SINGLE VERTICAL LINE ON THE RIGHT. (FROM *EXPERIMENTAL DESIGN AND ANALYSIS* BY WAYNE LEE. COPYRIGHT © 1975 BY W.H. FREEMAN AND COMPANY. USED WITH PERMISSION.)

means in ANOVA. The error term is computed from the sum of squared deviations of DV scores around their associated group means. In this case, the error term is substantial because there is considerable spread in scores within each group.

When the same scores are analyzed in ANCOVA, a regression line is found first that relates the DV to the covariate. The error term is based on the (sum of squared) deviations of the DV scores from the regression line running through each group mean instead of from the means themselves. Consider the score in the lower left-hand corner of Figure 8.1. The score is near the regression line (a small deviation for error in ANCOVA) but far from the mean for its own group (a large deviation for error in ANOVA). As long as the slope of the regression lines is not zero, ANCOVA produces a smaller sum of squares for error than ANOVA. If the slope is zero, error sum of squares is the same as in ANOVA but error mean square is larger because covariates use up degrees of freedom.

Covariates can be used in all ANOVA designs—factorial between-subjects, within-subjects,[2] mixed within-between, nonorthogonal, and so on. Analyses of these more complex designs are readily available in only a few programs, however. Similarly, specific comparisons and trend analysis of adjusted means are possible in ANCOVA but not always readily available through the programs.

8.2 ■ KINDS OF RESEARCH QUESTIONS

As with ANOVA, the question in ANCOVA is whether mean differences in the DV between groups are larger than expected by chance. In ANCOVA, however, one gets a more precise look at the IV-DV relationship after removal of the effect of covariate(s).

8.2.1 ■ Main Effects of IVs

Holding all else constant, are changes in behavior associated with different levels of an IV larger than expected through random fluctuations occurring by chance? For example, is test anxiety affected by treatment, after holding constant prior individual differences in test anxiety? Does political attitude vary with geographical region, after holding constant differences in socioeconomic status and age? The procedures described in Section 8.4 answer this question by testing the null hypothesis that the IV has no systematic effect on the DV.

With more than one IV, separate statistical tests are available for each one. Suppose there is a second IV in the political attitude example, for example, religious affiliation, with four groups: Protestant, Catholic, Jewish, and None-or-other. In addition to the test of geographic region, there is also a test of differences in attitudes associated with religious affiliation after adjustment for differences in socioeconomic status and age.

In experimental design, several devices are used to "hold all else constant." One of them, the topic for this chapter, is to measure the influence of extraneous variables (covariates) and then hold their influence constant by adjusting for differences on them statistically. A second procedure is to institute strict experimental controls over extraneous variables to hold them constant while levels of IV(s) are manipulated. A third procedure is to turn a potentially effective extraneous variable into another IV and hold it constant by crossing the levels of the more interesting IV with it in factorial

---

[2] If a covariate is measured only once, it does not provide adjustment to the levels of a within-subjects effect because each subject has the same value on the covariate for all levels of the effect. However, if a covariate is measured repeatedly along with the DV, the covariate provides adjustment to the within-subjects effect.

design. The last procedure is to randomize the assignment of levels of the extraneous variable (such as time of day) throughout the levels of the primary IV in the hopes of distributing the effects of the extraneous variable equally among treatments. A major consideration in experimental design is which of these procedures is more effective or more feasible for a given research program.

8.2.2 ■ Interactions among IVs

Holding all else constant, does change in behavior over levels of one IV depend on levels of another IV? That is, do IVs interact in their effect on behavior? (See Chapter 3 for a discussion of interaction.) For the political attitude example where religious affiliation is added as a second IV, are differences in attitudes over geographic region the same for all religions, after adjusting for socioeconomic status and age?

Tests of interactions, while interpreted differently from main effects, are statistically similar, as demonstrated in Section 8.7. With more than two IVs, numerous interactions are generated. Except for common error terms, each interaction is tested separately from other interactions and from main effects. All tests are independent when sample sizes in all groups are equal and the design is balanced.

8.2.3 ■ Specific Comparisons and Trend Analysis

When statistically significant effects are found in a design with more than two levels of a single IV, it is often desirable to pinpoint the nature of the differences. Which groups differ significantly from each other? Or, is there a simple trend over sequential levels of an IV? For the test anxiety example, we ask whether (1) the two treatment groups are more effective in reducing test anxiety than the waiting-list control, after adjusting for individual differences in test anxiety; and if (2) among the two treatment groups, desensitization is more effective than relaxation training in reducing test anxiety, again after adjusting for preexisting differences in test anxiety?

These two questions could be asked in planned comparisons instead of answering, through routine ANCOVA, the omnibus question of whether means are the same for all three levels of the IV. Or, with some loss in sensitivity, these two questions could be asked post hoc after finding a main effect of the IV in ANCOVA. Planned and post hoc comparisons are discussed in Section 8.5.2.3.

8.2.4 ■ Effects of Covariates

Analysis of covariance is based on a linear regression (Chapter 5) between covariate(s) and the DV, but there is no guarantee that the regression is reliable. The regression can be evaluated statistically by testing the covariate(s) as a source of variance in DV scores, as discussed in Section 8.5.3. For instance, consider the test anxiety example where the covariate is a pretest and the DV a posttest. To what extent is it possible to predict posttest anxiety from pretest anxiety, ignoring effects of differential treatment?

8.2.5 ■ Strength of Association

If a main effect or interaction of IVs is reliably associated with changes in the DV, the next logical question is, How much? How much of the variance in the adjusted DV scores—adjusted for the covariate(s)—is associated with the IV(s)? In the test anxiety example, if a main effect is found between the means for desensitization, relaxation training, and control group, one then asks, What proportion of variance in the adjusted test anxiety scores is attributed to the IV? Simple descriptions of strength of association are demonstrated in Sections 8.4 and 8.5.2.4.

8.2.6 ■ Adjusted Marginal and Cell Means

If any main effects or interactions are statistically significant, what is the estimated population parameter (adjusted mean) for each level of the IV? How do group scores differ, on the average, on the DV, after adjustment for covariates? For the test anxiety example, if there is a main effect of treatment, what is the average adjusted posttest anxiety score in each of the three groups? The reporting of parameter estimates is demonstrated in Section 8.7.

8.3 ■ LIMITATIONS TO ANALYSIS OF COVARIANCE

8.3.1 ■ Theoretical Issues

As with ANOVA, the statistical test in no way assures that changes in the DV were caused by the IV. The inference of causality is a logical rather than a statistical problem that depends on the manner in which subjects are assigned to levels of the IV(s), manipulation of levels of the IV(s) by the researcher, and the controls used in the research. The statistical test is available to test hypotheses from both nonexperimental and experimental research, but only in the latter case is attribution of causality justified.

Choice of covariates is a logical exercise as well. As a general rule, one wants a very small number of covariates, all correlated with the DV and none correlated with each other. The goal is to obtain maximum adjustment of the DV with minimum loss of degrees of freedom for error. Calculation of the regression of the DV on the covariates(s) results in the loss of one degree of freedom for error for each covariate. Thus *the gain in power from decreased sum of squares for error may be offset by the loss in degrees of freedom.* When there is a substantial correlation between the DV and a covariate, increased sensitivity due to reduced error variance offsets the loss of a degree of freedom for error. With multiple covariates, however, a point of diminishing returns is quickly reached, especially if the covariates correlate with one another (see Section 8.5.4).

In experimental work, a frequent caution is that the covariates must be independent of treatment. It is suggested that data on covariates be gathered before treatment is administered. Violation of this precept results in removal of some portion of the effect of the IV on the DV—that portion of the effect that is associated with the covariate. In this situation, adjusted group means may be closer together than unadjusted means. Further, the adjusted means may be difficult to interpret.

In nonexperimental work, adjustment for prior differences in means associated with covariates is appropriate. If the adjustment reduces mean differences on the DV, so be it—unadjusted differences reflect unwanted influences (other than the IV) on the DV. In other words, mean differences on a covariate associated with an IV are quite legitimately corrected for as long as the covariate differences are not caused by the IV (Overall & Woodward, 1977).

When ANCOVA is used to evaluate a series of DVs after MANOVA, independence of the "covariates" and the IV is not required. Because covariates are actually DVs, it is expected that they be dependent on the IV.

In all uses of ANCOVA, however, adjusted means must be interpreted with great caution because the adjusted mean DV score may not correspond to any situation in the real world. Adjusted means are the means that would have occurred if all subjects had the same scores on the covariates. Especially in nonexperimental work, such a situation may be so unrealistic as to make the adjusted values meaningless.

Sources of bias in ANCOVA are many and subtle, and can produce either under- or overadjustment of the DV. At best, the nonexperimental use of ANCOVA allows you to look at IV-DV relationships (noncausal) adjusted for the effects of covariates, as measured. If causal inference regarding effects is desired, there is no substitute for random assignment of subjects. Don't expect ANCOVA to permit causal inference of treatment effects with non-randomly assigned groups. If random assignment is absolutely impossible, or if it breaks down because of nonrandom loss of subjects, be sure to thoroughly ground yourself in the literature regarding use of ANCOVA in such cases, starting with Cook and Campbell (1979).

Limitations to generalizability apply to ANCOVA as they do to ANOVA or any other statistical test. One can generalize only to those populations from which a random sample is taken. ANCOVA may, in some limited sense, sometimes adjust for a failure to randomly assign the sample to groups, but it does not affect the relationship between the sample and the population to which one can generalize.

8.3.2 ■ Practical Issues

The ANCOVA model assumes reliability of covariates, linearity between pairs of covariates and between covariates and the DV, and homogeneity of regression, in addition to the usual ANOVA assumptions of normality and homogeneity of variance.

8.3.2.1 ■ Unequal Sample Sizes and Missing Data

If scores on the DV are missing in a between-subjects ANCOVA, this is reflected as the problem of unequal n because all IV levels or combinations of IV levels do not contain equal numbers of cases. Consult Section 8.5.2.2 for strategies for dealing with unequal sample sizes. If some subjects are missing scores on covariate(s), or if, in repeated measures ANOVA, some DV scores are missing for some subjects, this is more clearly a missing-data problem. Consult Chapter 4 for methods of dealing with missing data.

8.3.2.2 ■ Outliers

Within each group, univariate outliers can occur in the DV or any one of the covariates. Multivariate outliers can occur in the space of the DV and covariate(s). Multivariate outliers among DV and covariate(s) can produce heterogeneity of regression (Section 8.3.2.7), leading to rejection of ANCOVA or at least unreasonable adjustment of the DV. If the covariates are serving as a convenience in most analyses, rejection of ANCOVA because there are multivariate outliers is hardly convenient.

Consult Chapter 4 for methods dealing with univariate outliers in the DV or covariate(s) and multivariate outliers among the DV and covariate(s). Tests for univariate outliers within each group through BMDP7D and multivariate outliers within each group through BMDPAM are demonstrated in Section 8.7.1.

8.3.2.3 ■ Multicollinearity and Singularity

If there are multiple covariates, they should not be highly correlated with each other. *Highly correlated covariates should be eliminated,* both because they add no adjustment to the DV over that of other covariates and because of potential computational difficulties if they are singular or multicollinear. Programs for ANCOVA in the BMDP series automatically guard against multicollinearity or singularity of covariates, as do the general linear hypothesis programs of SAS and SYSTAT; but the SPSS programs do not. Consult Chapter 4 for methods of testing for multicollinearity and singularity among multiple covariates. A test for multicollinearity and singularity is demonstrated in Section 8.7.1.5.

8.3.2.4 ■ Normality

As in all ANOVA, it is assumed that the sampling distributions of means, as described in Chapter 3, are normal within each group. Note that it is the sampling distributions of means and not the raw scores within each cell that need to be normally distributed. Without knowledge of population values, or production of actual sampling distributions of means, there is no way to test this assumption. However, the central limit theorem suggests that, with large samples, sampling distributions are normal even if raw scores are not. *With relatively equal sample sizes in groups, no outliers, and two-tailed tests, robustness is expected with 20 degrees of freedom for error*. (See Chapter 3 for calculation of error degrees of freedom.)

Larger samples are necessary for one-tailed tests. With small, unequal samples or with outliers present, it may be necessary to consider data transformation (cf. Chapter 4).

8.3.2.5 ■ Homogeneity of Variance

Just as in ANOVA, it is assumed in ANCOVA that the variance of DV scores within each cell of the design is a separate estimate of the same population variance. Because of the robustness of the analysis to violation of this assumption, as long as there are no outliers, it is typically unnecessary to test for homogeneity of variance if samples are large and the following checks are met.

The following checks commonly are suggested to ensure robustness. *For two-tailed tests, sample sizes should preferably be equal, but in no event should the ratio of largest to smallest sample size for groups be greater than 4:1*. Then examine the variances (standard deviations squared) within each cell to *assure that the ratio between largest and smallest variance is no greater than approximately 10:1*. If one-tailed tests are used, use more conservative criteria because the test is less robust.

If any of these conditions is violated, a formal test for homogeneity of variance is called for as described in any standard ANOVA text (e.g., Keppel, 1991; Winer, 1971) and as available in some canned programs (see Table 8.14). Most of these tests are sensitive to nonnormality as well as heterogeneity of variance and can lead to overly conservative rejection of the use of ANCOVA. BMDP7D offers a test of homogeneity of variance that is typically not sensitive to departures from normality. Major heterogeneity should, of course, always be reported by providing within-cell standard deviations.

In any event, gross violations of homogeneity can usually be corrected by transformation of the DV scores (cf. Chapter 4). Interpretation, however, is then limited to the transformed scores. Add to this the difficulty of interpreting adjusted means, and interpretation becomes increasingly speculative. Another option is to use untransformed variables with a more stringent α level (for nominal $\alpha = .05$, use .025 with moderate violation and .01 with severe violation).

8.3.2.6 ■ Linearity

The ANCOVA model is based on the assumption that the relationship between each covariate and the DV and the relationships among pairs of covariates are linear. As with multiple regression (Chapter 5), violation of this assumption reduces the power of the statistical test; errors in statistical decision making are in a conservative direction. Error terms are not reduced as fully as they might be, optimum matching of groups is not achieved, and group means are incompletely adjusted.

To begin screening for linearity, produce residuals plots through BMDP1V, SAS GLM, or SPSS MANOVA, and interpret them as described in Section 5.3.2.4. *If there is indication of serious curvilinearity, examine within-cell scatterplots of the DV with each covariate and all covariates with one another.* Any of the programs that produce scatterplots (e.g., BMDP6D and SPSS PLOT) can be used to evaluate linearity. Relationships between the DV and each covariate can be produced as one of the sets of plots in BMDP1V.

Where curvilinearity is indicated,[3] it may be corrected by transforming some of the variables. Or, because of the difficulties in interpreting transformed variables, you may consider eliminating a covariate that produces nonlinearity. Or, a higher-order power of the covariate can be used to produce an alternative covariate that incorporates nonlinear influences.

8.3.2.7 ■ Homogeneity of Regression

Adjustment of scores in ANCOVA is made on the basis of an average within-cell regression coefficient. The assumption is that the slope of the regression between the DV and the covariate(s) within each cell is an estimate of the same population regression coefficient, that is, that the slopes are equal for all cells.

Heterogeneity of regression implies that there is a different DV-covariate(s) slope in some cells of the design, or, that there is an interaction between IV(s) and covariate(s). If IV and covariate(s) interact, the relationship between the covariate(s) and the DV is different at different levels of the IV(s), and the covariate adjustment that is needed for various cells is different. Figure 8.2 illustrates,

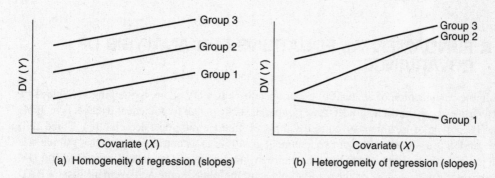

(a) Homogeneity of regression (slopes) (b) Heterogeneity of regression (slopes)

FIGURE 8.2 DV-COVARIATE REGRESSION LINES FOR THREE GROUPS PLOTTED ON THE SAME COORDINATES FOR CONDITIONS OF (a) HOMOGENEITY AND (b) HETEROGENEITY OF REGRESSION.

---

[3] Tests for deviation from linearity are available (cf. Winer, 1971), but there seems to be little agreement as to the appropriate test or the seriousness of significant deviation in the case of ANCOVA. Therefore formal tests are not recommended.

for three groups, perfect homogeneity of regression (equality of slopes) and extreme heterogeneity of regression (inequality of slopes).

If a between-subjects design is used, test the assumption of homogeneity of regression according to procedures described in Section 8.5.1. If any other experimental design is used, and interaction between IVs and covariates is suspected, ANCOVA is inappropriate. If there is no reason to suspect an IV-covariate interaction with complex designs, it is probably safe to proceed with ANCOVA on the basis of the robustness of the model. Alternatives to ANCOVA are discussed in Section 8.5.5.

8.3.2.8 ■ Reliability of Covariates

It is assumed in ANCOVA that covariates are measured without error; that they are perfectly reliable. In the case of such variables as sex and age, the assumption can usually be justified. With self-report of demographic variables, and with variables measured psychometrically, such assumptions are not so easily made. And variables such as attitude may be reliable at the point of measurement, but fluctuate over short periods.

In experimental research, unreliable covariates lead to loss of power and a conservative statistical test through underadjustment of the error term. In nonexperimental applications, however, unreliable covariates can lead to either under- or overadjustment of the means. Group means may be either spread too far apart (Type I error) or compressed too closely together (Type II error). The degree of error depends on how unreliable the covariates are. *In nonexperimental research, limit covariates to those that can be measured reliably ($r_{xx} > .8$).*

If fallible covariates are absolutely unavoidable, they can be adjusted for unreliability. However, there is no one procedure that produces appropriate adjustment under all conditions nor is there even agreement about which procedure is most appropriate for which application. Because of this disagreement, and because procedures for correction require use of sophisticated programs, they are not covered in this book. The interested reader is referred to Cohen and Cohen (1983, pp 406 – 413) or to procedures discussed by St. Pierre (1978).

8.4 ■ FUNDAMENTAL EQUATIONS FOR ANALYSIS OF COVARIANCE

In the simplest application of analysis of covariance there is a DV score, a grouping variable (IV), and a covariate score for each subject. An example of such a small hypothetical data set is in Table 8.1. The IV is type of treatment given to a sample of nine learning-disabled children. Three children are randomly assigned to one of two treatment groups or to a control group, so that sample size of each group is three. For each of the nine children, two scores are measured, covariate and DV. The covariate is a pretest score on the reading subtest of the Wide Range Achievement Test (WRAT-A), measured before the experiment begins. The DV is a posttest score on the same test measured at the end of the experiment.

The research question is, Does differential treatment of learning-disabled children affect reading scores, after adjusting for differences in the children's prior reading ability? The sample size is, of course, inadequate for a realistic test of the research question but is convenient for illustration of

TABLE 8.1 SMALL SAMPLE DATA FOR ILLUSTRATION OF ANALYSIS OF COVARIANCE

	Groups					
	Treatment 1		Treatment 2		Control	
	Pre	Post	Pre	Post	Pre	Post
	85	100	86	92	90	95
	80	98	82	99	87	80
	92	105	95	108	78	82
Sums	257	303	263	299	255	257

the techniques in ANCOVA. The reader is encouraged to follow this example with hand calculations. Computer analyses using four popular programs follow this section.

8.4.1 ■ Sums of Squares and Cross Products

Equations for ANCOVA are an extension of those for ANOVA, as discussed in Chapter 3. Averaged squared deviations from means—variances—are partitioned into variance associated with different levels of the IV (between-groups variance) and variance associated with differences in scores within groups (unaccounted for or error variance). Variance is partitioned by summing and squaring differences between scores and various means.

$$\sum_i \sum_j (Y_{ij} - GM)^2 = n \sum_j (\bar{Y}_j - GM)^2 + \sum_i \sum_j (Y_{ij} - \bar{Y}_j)^2 \tag{8.1}$$

or

$$SS_{total} = SS_{bg} + SS_{wg}$$

The total sum of squared differences between scores on Y (the DV) and the grand mean (GM) is partitioned into two components: sum of squared differences between group means (Y_j) and the grand mean (i.e., systematic or between-groups variability); and sum of squared differences between individual scores (Y_{ij}) and their respective group means (i.e., error).

In ANCOVA, there are two additional partitions. First, the differences in covariate scores are partitioned into between- and within-groups sums of squares:

$$SS_{total(x)} = SS_{bg(x)} + SS_{wg(x)} \tag{8.2}$$

The total sum of squared differences on the covariate (X) is partitioned into differences between groups and differences within groups.

Similarly, the covariance (the linear relationship between the DV and the covariate) is partitioned into sums of products associated with covariance between groups and sums of products associated with covariance within groups.

$$SP_{total} = SP_{bg} + SP_{wg} \tag{8.3}$$

A *sum of squares* involves taking deviations of scores from means (e.g., $X_{ij} - \overline{X}_j$ or $Y_{ij} - \overline{Y}_j$), squaring them, and then summing the squares over all subjects; a *sum of products* involves taking two deviations from the same subject (e.g., both $X_{ij} - \overline{X}_j$ and $Y_{ij} - \overline{Y}_j$), multiplying them together (instead of squaring), and then summing the products over all subjects. As discussed in Chapter 3, the means that are used to produce the deviations are different for the different sources of variance in the research design.

The partitions for the covariate (Equation 8.2) and the partitions for the association between the covariate and DV (Equation 8.3) are used to adjust the sums of squares for the DV according to the following equations:

$$SS'_{bg} = SS_{bg} - \left[\frac{(SP_{bg} + SP_{wg})^2}{SS_{bg(x)} + SS_{wg(x)}} - \frac{(SP_{wg})^2}{SS_{wg(x)}} \right] \tag{8.4}$$

The adjusted between-groups sum of squares (SS'_{bg}) is found by subtracting from the unadjusted between-groups sum of squares a term based on sums of squares associated with the covariate, X, and sums of products for the linear relationship between the DV and the covariate.

$$SS'_{wg} = SS_{wg} - \frac{(SP_{wg})^2}{SS_{wg(x)}} \tag{8.5}$$

The adjusted within-groups sum of squares (SS'_{wg}) is found by subtracting from the unadjusted within-groups sum of squares a term based on within-groups sums of squares and products associated with the covariate and with the linear relationship between the DV and the covariate.

This can be expressed in an alternate form. The adjustment for each score consists of subtracting from the deviation of that score from the grand mean a value that is based on the deviation of the corresponding covariate from the grand mean on the covariate, weighted by the regression coefficient for predicting the DV from the covariate. Symbolically, for an individual score:

$$(Y - Y') = (Y - GM_y) - B_{y.x}(X - GM_x) \tag{8.6}$$

The adjustment for any subject's score ($Y - Y'$) is obtained by subtracting from the unadjusted deviation score ($Y - GM_y$) the individual's deviation on the covariate ($X - GM_x$) weighted by the regression coefficient, $B_{y.x}$.

Once adjusted sums of squares are found, mean squares are found as usual by dividing by appropriate degrees of freedom. The only difference in degrees of freedom between ANOVA and ANCOVA is that in ANCOVA the error degrees of freedom are reduced by one for each covariate because a degree of freedom is used up in estimating each regression coefficient.

For computational convenience, raw score equations rather than deviation equations are provided in Table 8.2 for Equations 8.4 and 8.5. Note that these equations apply only to equal-n designs.

TABLE 8.2 COMPUTATION EQUATIONS FOR SUMS OF SQUARES AND CROSS PRODUCTS IN ONE-WAY BETWEEN-SUBJECTS ANALYSIS OF COVARIANCE

Source	Sum of squares for Y (DV)	Sum of squares for X (covariate)	Sum of products
Between groups	$SS_{bg} = \dfrac{\sum\limits^{k}\left(\sum\limits^{n} Y\right)^2}{n} - \dfrac{\left(\sum\limits^{k}\sum\limits^{n} Y\right)^2}{kn}$	$SS_{bg(x)} = \dfrac{\sum\limits^{k}\left(\sum\limits^{n} X\right)^2}{n} - \dfrac{\left(\sum\limits^{k}\sum\limits^{n} X\right)^2}{kn}$	$SP_{bg} = \dfrac{\sum\limits^{k}\left(\sum\limits^{n} Y\right)\left(\sum\limits^{n} X\right)}{n} - \dfrac{\left(\sum\limits^{k}\sum\limits^{n} Y\right)\left(\sum\limits^{k}\sum\limits^{n} X\right)}{kn}$
Within groups	$SS_{wg} = \sum\limits^{k}\sum\limits^{n} Y^2 - \dfrac{\sum\limits^{k}\left(\sum\limits^{n} Y\right)^2}{n}$	$SS_{wg(x)} = \sum\limits^{k}\sum\limits^{n} X^2 - \dfrac{\sum\limits^{k}\left(\sum\limits^{n} X\right)^2}{n}$	$SP_{wg} = \sum\limits^{k}\sum\limits^{n} (XY) - \dfrac{\sum\limits^{k}\left(\sum\limits^{n} Y\right)\left(\sum\limits^{n} X\right)}{n}$

Note: k = number of groups; n = number of subjects per group.

When applied to the data in Table 8.1, the six sums of squares and products are as follows:

$$SS_{bg} = \frac{(303)^2 + (299)^2 + (257)^2}{3} - \frac{(859)^2}{(3)(3)} = 432.889$$

$$SS_{wg} = (100)^2 + (98)^2 + (105)^2 + (92)^2 + (99)^2 + (108)^2$$

$$+ (95)^2 + (80)^2 + (82)^2 - \frac{(303)^2 + (299)^2 + (257)^2}{3} = 287.333$$

$$SS_{bg(x)} = \frac{(257)^2 + (263)^2 + (255)^2}{3} - \frac{(775)^2}{(3)(3)} = 11.556$$

$$SS_{wg(x)} = (85)^2 + (80)^2 + (92)^2 + (86)^2 + (82)^2 + (95)^2$$

$$+ (90)^2 + (87)^2 + (78)^2 - \frac{(257)^2 + (263)^2 + (255)^2}{3} = 239.333$$

$$SP_{bg} = \frac{(257)(303) + (263)(299) + (255)(257)}{3} - \frac{(775)(859)}{(3)(3)} = 44.889$$

$$SP_{wg} = (85)(100) + (80)(98) + (92)(105) + (86)(92) + (82)(99) + (95)(108) + (90)(105)$$

$$+ (87)(80) + (78)(82) - \frac{(257)(303) + (263)(299) + (255)(257)}{3} = 181.667$$

These values are conveniently summarized in a sum-of-squares and cross-products matrix (cf. Chapter 1). For the between-groups sums of squares and cross products,

$$\mathbf{S}_{bg} = \begin{bmatrix} 11.556 & 44.889 \\ 44.889 & 432.889 \end{bmatrix}$$

The first entry (first row and first column) is the sum of squares for the covariate and the last (second row, second column) is the sum of squares for the DV; the sum of products is shown in the off-diagonal portion of the matrix. For the within-groups sums of squares and cross products, arranged similarly,

$$\mathbf{S}_{wg} = \begin{bmatrix} 239.333 & 181.667 \\ 181.667 & 287.333 \end{bmatrix}$$

From these values, the adjusted sums of squares are found as per Equations 8.4 and 8.5.

$$SS'_{bg} = 432.889 - \left[\frac{(44.889 + 181.667)^2}{11.556 + 239.333} - \frac{(181.667)^2}{239.333} \right] = 366.202$$

$$SS'_{wg} = 287.333 - \frac{(181.667)^2}{239.333} = 149.438$$

8.4.2 ■ Significance Test and Strength of Association

These values are entered into a source table such as Table 8.3. Degrees of freedom for between-groups variance are $k - 1$, and for within-groups variance $N - k - c$. (N = total sample size, k = number of levels of the IV, and c = number of covariates.)

As usual, mean squares are found by dividing sums of squares by appropriate degrees of freedom. The hypothesis that there are no differences among groups is tested by the F ratio formed by dividing the adjusted mean square between groups by the adjusted mean square within groups.

$$F = \frac{183.101}{29.888} = 6.13$$

From a standard F table, we find that the obtained F of 6.13 exceeds the critical F of 5.79 at $\alpha = .05$ with 2 and 5 df. We therefore reject the null hypothesis of no change in WRAT reading scores associated with the three treatment levels, after adjustment for pretest reading scores.

If a statistically significant effect of the IV on the adjusted DV scores is found, the strength of association for the effect is assessed using η^2.

$$\eta^2 = \frac{SS'_{bg}}{SS'_{bg} + SS'_{wg}} \tag{8.7}$$

For the sample data,

$$\eta^2 = \frac{366.202}{366.202 + 149.438} = .71$$

We conclude that 71% of the variance in the adjusted DV scores (WRAT-R) is associated with treatment.

ANOVA for the same data appears in Table 8.4. Compare the results with those of ANCOVA in Table 8.3. ANOVA produces larger sums of squares, especially for the error term. There is also one more degree of freedom for error because there is no covariate. However, in ANOVA the null hypothesis is retained while in ANCOVA it is rejected. Thus, use of the covariate has reduced the "noise" in the error term for this example.

TABLE 8.3 ANALYSIS OF COVARIANCE FOR DATA OF TABLE 8.1

Source of variance	Adjusted SS	df	MS	F
Between groups	366.202	2	183.101	6.13*
Within groups	149.439	5	29.888	

*$p < .05$.

TABLE 8.4 ANALYSIS OF VARIANCE FOR DATA OF TABLE 8.1

Source of Variance	SS	df	MS	F
Between groups	432.889	2	216.444	4.52
Within groups	287.333	6	47.889	

ANCOVA extends to factorial and repeated-measures designs (Section 8.5.2.1), unequal n (Section 8.5.2.2), and multiple covariates (Section 8.5.3). In all cases, the analysis is done on adjusted, rather than raw, DV scores.

8.4.3 ■ Computer Analyses of Small Sample Example

Tables 8.5 through 8.8 demonstrate analyses of covariance of this small data set using BMDP2V, SPSS ANOVA, SYSTAT ANOVA, and SAS GLM. Minimal output is requested for each of the programs, although much more is available upon request.

In BMDP2V, a DESIGN paragraph is used to specify the DV (DEPENDENT) and the COVARIATE(s), as seen in Table 8.5. The GROUP paragraph specifies the IV(s), (VAR =) and defines their levels. Cell and marginal means, standard deviations, and sample sizes are given for all covariates and DVs. The variable is identified at the left of each of these tables—PRE for the covariate and POST for the DV in this example.

These descriptive statistics are followed by a standard ANOVA source table where the first effect is the IV, TREATMNT, adjusted for the covariate(s). The second source listed (PRE) is the covariate. The test is whether the covariate significantly adjusts the DV after effects of the IV(s) are removed. Along with the significance test, the unstandardized regression coefficient (B weight, see Equation 8.6) is provided. The ERROR term ends the table. The "1" to the left of that line indicates that this is the first (and in this case only) error term in the analysis.

Finally, adjusted cell means are provided. If required, adjusted marginal means (with appropriate adjustment for unequal sample sizes) can be calculated from these adjusted cell means for factorial designs, as illustrated in Section 8.7.2.

In SPSS ANOVA (Table 8.6) the DV (POST) is specified in the ANOVA paragraph followed BY the IV (TREATMNT, with the range of levels in parentheses) WITH the covariate (PRE). The output is an ANOVA source table with some redundancies. First, Covariates are pooled and tested as a group if there is more than one covariate. (In this example the overall test is also the test of the single covariate, PRE.) In BMDP2V, the test of the covariate(s) is adjusted for TREATMNT; in SPSS ANOVA, the test of the covariate(s) is not adjusted for TREATMNT if default values are used. In this case, SPSS produces a significant result for PRE while BMDP2V does not. Because there is only one IV in this example, the test of the single main effect for TREATMNT (adjusted for the covariate) is identical to the test for all Main Effects considered together. The source of variance labeled Explained provides a test of all known sources of variability: covariates, main effects and interactions. The source labeled Residual is the error term. The final source of variance is Total but beware: total is not adjusted for the covariate and is not appropriate for the denominator of η^2 (Equation 8.7). Subtract the sum of squares for the covariates to calculate the adjusted total sum of squares.

In SYSTAT ANOVA or GLM (Table 8.7) the DV is specified in the DEPEND instruction, the IV is specified in the CATEGORY instruction, and the covariate in the COVARIATE instruction. Output begins with multiple R and R^2 for predicting the DV from the combination of the IV and covariate. The standard ANOVA table shows tests of the IV and the covariate, each adjusted for the other.

SAS GLM requires the IV to be specified in the CLASS instruction. Then a MODEL instruction is set up, with the DV on the left side of the equation, and the IV and covariate on the right side as seen in Table 8.8. Two source tables are provided, one for the problem as a regression and the other for the problem as a more standard ANOVA, both with the same Error term.

The first test in the regression table asks if there is significant prediction of the DV by the combination of the IV and the covariate. The output resembles that of standard multiple regression

TABLE 8.5   SETUP AND SELECTED BMDP2V OUTPUT FOR ANALYSIS OF COVARIANCE ON SAMPLE DATA
            IN TABLE 8.1

```
/INPUT      VARIABLES ARE 4. FORMAT IS FREE.
            FILE = 'TAPE70.DAT'.
/VARIABLE   NAMES ARE SUBJNO, TREATMNT, PRE, POST.
/GROUP      VAR = TREATMNT.
            CODES(TREATMNT) ARE 1 TO 3.
            NAMES(TREATMNT) ARE TREAT1, TREAT2, CONTROL.
/DESIGN     DEPENDENT IS POST.
            COVARIATE IS PRE.
/END
```

GROUP STRUCTURE

TREATMNT	COUNT
TREAT1	3
TREAT2	3
CONTROL	3

CELL MEANS   FOR 1-ST COVARIATE

				MARGINAL
TREATMNT=	TREAT1	TREAT2	CONTROL	
PRE	85.66667	87.66667	85.00000	86.11111
COUNT	3	3	3	9

STANDARD DEVIATIONS FOR 1-ST COVARIATE

TREATMNT=	TREAT1	TREAT2	CONTROL
PRE	6.02771	6.65833	6.24500

CELL MEANS   FOR 1-ST DEPENDENT VARIABLE

				MARGINAL
TREATMNT=	TREAT1	TREAT2	CONTROL	
POST	101.00000	99.66667	85.66667	95.44444
COUNT	3	3	3	9

STANDARD DEVIATIONS FOR 1-ST DEPENDENT VARIABLE

TREATMNT=	TREAT1	TREAT2	CONTROL
POST	3.60555	8.02081	8.14453

TABLE 8.5   *(CONTINUED)*

A N A L Y S I S   O F   V A R I A N C E   FOR   THE   1-ST DEPENDENT VARIABLE

THE TRIALS ARE REPRESENTED BY THE VARIABLES:
POST

SOURCE	SUM OF SQUARES	D.F.	MEAN SQUARE	F	TAIL PROB.
TREATMNT	366.20123	2	183.10061	6.13	0.0452
PRE	137.89461	1	137.89461	4.61	0.0845
1 ERROR	149.43872		29.88774		

REG. COEFF.	ESTIMATE	STD. ERROR	T-VALUE	P-VALUE
PRE	0.75905	0.35338	2.15	0.0845

ADJUSTED CELL MEANS   FOR   1-ST DEPENDENT VARIABLE

			MARGINAL	
TREATMNT=	TREAT1	TREAT2	CONTROL	
POST	101.33736	98.48592	86.51006	95.44444
COUNT	3	3	3	9

STANDARD ERRORS OF ADJUSTED CELL MEANS FOR   1-ST DEPENDENT VARIABLE

TREATMNT=	TREAT1	TREAT2	CONTROL
POST	3.16026	3.20387	3.18068

(Chapter 5) and includes R-Square, the unadjusted Mean on POST (the DV), the Root MSE (square root of the error mean square), and C.V., the coefficient of variation (100 times the Root MSE divided by the mean of the DV).

In the ANOVA-like table, two forms of tests are given by default, labeled Type I SS and Type III SS. The sums of squares for TREATMNT are the same in both forms because both adjust treatment for the covariate. In Type I SS the sum of squares for the covariate (PRE) is not adjusted for the effect of treatment (like SPSS ANOVA) while in Type III SS the sum of squares for the covariate is adjusted for treatment (like BMDP2V).

8.5 ■ SOME IMPORTANT ISSUES

8.5.1 ■ Test for Homogeneity of Regression

The assumption of homogeneity of regression is that the slopes of the regression of the DV on the covariate(s) (the regression coefficients or *B* weights as described in Chapter 5) are the same for all cells of a design. Both homogeneity and heterogeneity of regression are illustrated in Figure 8.2. Because the average of the slopes for all cells is used to adjust the DV, it is assumed that the slopes do not differ significantly either from one another or from a single estimate of the population value.

TABLE 8.6 SETUP AND SPSS ANOVA OUTPUT FOR ANALYSIS OF COVARIANCE ON SAMPLE DATA IN TABLE 8.1

```
DATA LIST          FILE = 'TAPE70.DAT' FREE
                   /SUBJNO, TREATMNT, PRE, POST.
VALUE LABELS       TREATMNT 1 'TREAT1' 2 'TREAT2' 3 'CONTROL'.
ANOVA              POST BY TREATMNT (1,3) WITH PRE.
```

* * * A N A L Y S I S   O F   V A R I A N C E * * *
POST
by TREATMNT
with PRE

Source of Variation	Sum of Squares	DF	Mean Square	F	Sig of F
Covariates	204.582	1	204.582	6.845	.047
PRE	204.582	1	204.582	6.845	.047
Main Effects	366.201	2	183.101	6.126	.045
TREATMNT	366.201	2	183.101	6.126	.045
Explained	570.784	3	190.261	6.366	.037
Residual	149.439	5	29.888		
Total	720.222	8	90.028		

TABLE 8.7 SETUP AND SYSTAT GLM OUTPUT FOR ANALYSIS OF COVARIANCE ON SAMPLE DATA IN TABLE 8.1

```
USE TAPE70
DEPEND POST
CATEGORY TREATMNT
COVARIATE PRE
ESTIMATE
```

Categorical values encountered during processing are:
TREATMNT (3 levels)
  1,   2,   3

Dep Var: POST N: 9 Multiple R: 0.890 Squared multiple R: 0.793

Analysis of Variance

Source	Sum-of-Squares	DF	Mean-Square	F-Ratio	P
TREATMNT	366.201	2	183.101	6.126	0.045
PRE	137.895	1	137.895	4.614	0.084
Error	149.439	5	29.888		

If the null hypothesis of equality among slopes is rejected, the analysis of covariance is inappropriate and an alternative strategy as described in Sections 8.3.2.7 and 8.5.5 should be used.

Hand calculation of the test of homogeneity of regression (see, for instance, Keppel, 1991) is extremely tedious . However, for designs consisting solely of between-subjects IVs, computer programs (cf. Table 8.14) are available for testing the assumption, and their use is recommended. The most straightforward programs for testing homogeneity of regression are BMDP1V, SYSTAT ANOVA or GLM, and SPSS MANOVA.

TABLE 8.8   SETUP AND SAS GLM OUTPUT FOR ANALYSIS OF COVARIANCE ON SAMPLE DATA IN TABLE 8.1

```
DATA SSAMPLE;
INFILE TAPE70;
INPUT CASENO TREATMNT PRE POST;
PROC GLM;
        CLASS TREATMNT;
        MODEL POST = PRE TREATMNT;
```

General Linear Models Procedure
Class Level Information

Class	Levels	Values
TREATMNT	3	1 2 3

Number of observations in data set = 9

Dependent Variable: POST

Source	DF	Sum of Squares	Mean Square	F Value	Pr > F
Model	3	570.7835036	190.2611679	6.37	0.0369
Error	5	149.4387187	29.8877437		
Corrected Total	8	720.2222222			

R-Square	C.V.	Root MSE	POST Mean
0.792510	5.727906	5.466968	95.4444444

Dependent Variable: POST

Source	DF	Type I SS	Mean Square	F Value	Pr > F
PRE	1	204.5822754	204.5822754	6.85	0.0473
TREATMNT	2	366.2012282	183.1006141	6.13	0.0452

Source	DF	Type III SS	Mean Square	F Value	Pr > F
PRE	1	137.8946147	137.8946147	4.61	0.0845
TREATMNT	2	366.2012282	183.1006141	6.13	0.0452

BMDP1V automatically provides a test for homogeneity of regression, as shown in Table 8.9 for the data of Table 8.1.[4] The obtained F value of .0337 (with 2 and 3 df) for EQUALITY OF SLOPE falls far short of critical $F = 9.55$ at the .95 level of confidence. Therefore the slopes are not reliably different and the assumption of homogeneity of regression is tenable. The slopes themselves are shown for the three groups in the segment of output labeled ESTIMATES OF SLOPES WITHIN EACH GROUP in Table 8.9.[5] The test of EQUALITY OF ADJ. CELL MEANS is a test of the major ANCOVA hypothesis and the results are identical to those in Table 8.3. The test of the efficacy of the covariate, adjusted for treatment, in providing adjustment of the DV is given under ZERO SLOPES. Because the test does not reach significance, PRE is not providing a reliable adjustment to POST in this sample, owing, no doubt, to the very small sample size.

---

[4] Dave DeYoung (personal communication, January 6, 1994) reports that on occasion when there are multiple covariates, the BMDP1V program fails to provide the test of homogeneity of regression and erroneously reports, instead, singularity among the covariates. In this case, try a run with SPSS or SYSTAT which may provide the test.
[5] If contrasts of a certain form are desired, tests for homogeneity of regression for each group contrasted with all others are provided in the segment of output labeled EQUALITY OF SLOPES BETWEEN EACH GROUP AND ALL OTHERS.

| TABLE 8.9 | SETUP AND SELECTED OUTPUT FROM BMDP1V FOR SAMPLE DATA OF TABLE 8.1 |

```
/INPUT      VARIABLES ARE 4.  FORMAT IS FREE.
            FILE = 'TAPE70.DAT'.
/VARIABLE   NAMES ARE SUBJNO, TREATMNT, PRE, POST.
/GROUP      VAR = TREATMNT.
            CODES(TREATMNT) ARE 1 TO 3.
            NAMES(TREATMNT) ARE TREAT1, TREAT2, CONTROL.
/DESIGN     DEPENDENT IS POST.
            INDEPENDENT IS PRE.
            CONTRAST = 1, 0, - 1.
            CONTRAST = 1, - 2, 1.
            CONTRAST = 1, 1, - 2.
/END
```

ANALYSIS OF COVARIANCE TABLE
============================

SOURCE OF VARIATION	D.F.	SUM OF SQ.	MEAN SQ.	F-VALUE	PROB(TAIL)
EQUALITY OF ADJ. MEANS	2	366.2014	183.1007	6.1263	0.045

ZERO SLOPES

ALL COVARIATES	1	137.8947	137.8947	4.6138	0.084
ERROR	5	149.4387	29.8877		

EQUALITY OF SLOPES

ALL COVARIATES, ALL GROUPS	2	3.2868	1.6434	0.0337	0.967
ERROR	3	146.1519	48.7173		

BETWEEN EACH GROUP AND ALL OTHERS

GROUP 1 TREAT1 , ALL COVARIATES	1	2.9210	2.9210	0.0797	0.792
ERROR	4	146.5177	36.6294		
GROUP 2 TREAT2 , ALL COVARIATES	1	1.9243	1.9243	0.0522	0.831
ERROR	4	147.5144	36.8786		
GROUP 3 CONTROL , ALL COVARIATES	1	0.0612	0.0612	0.0016	0.970
ERROR	4	149.3775	37.3444		

*** N O T E *** THE TESTS FOR EQUALITY OF ADJUSTED MEANS AND ZERO SLOPES ARE BASED ON THE ASSUMPTION THAT THE SLOPES OF THE DEPENDENT VARIABLE ON THE COVARIATES ARE THE SAME IN EVERY GROUP. THUS, THE RESULTS OF THESE TESTS MAY BE MISLEADING IF ANY OF THE ABOVE TESTS FOR EQUALITY OF SLOPES IS SIGNIFICANT AT THE 5% LEVEL.

*** N O T E *** THE ABOVE TESTS FOR PARALLELISM DO NOT INDICATE A PROBLEM WITH EQUALITY OF SLOPES.

*** WARNING *** SINCE NONE OF THE TESTS FOR ZERO SLOPE IS SIGNIFICANT, THE USE OF THIS (THESE) COVARIATE(S) MAY NOT BE WARRANTED IN FUTURE STUDIES.

ESTIMATES OF SLOPES WITHIN EACH GROUP

		TREAT1 1	TREAT2 2	CONTROL 3
PRE	3	0.5917	0.8759	0.7821

But BMDP1V only provides analysis for one-way between subjects designs. For factorial designs it is necessary to create a variable that corresponds to the cells of the design and to code each case according to the cell in which it falls. For example, in a 3×3 factorial design, cells are coded 1 through 9, and analyzed as a one-way design with 9 levels of the IV. This type of recoding is demonstrated in Section 8.7.1.7, Table 8.19. The output contains the assessment of equality of slope in each cell but tests of main effects and interactions require additional runs on programs for factorial ANCOVA such as BMDP2V, BMDP4V, SPSS ANOVA and MANOVA, SYSTAT ANOVA and GLM, and SAS GLM.

Programs based on the general linear model test for homogeneity of regression by evaluating the covariate(s) by IV(s) interaction as the last effect entering a regression model (Chapter 5). This procedure is demonstrated in a straightforward manner in the ANCOVA sections of the SYSTAT manuals (Wilkinson & Hill, 1994a and b). The same procedure works for SAS GLM but is not shown in the manual.

BMDP2R and SPSS REGRESSION can be used to test homogeneity of regression through a slightly more complicated procedure. IVs and their interactions and the IV by covariate interaction are dummy variable coded with the IV by covariate interaction entered as the last step of analysis. Homogeneity of regression is tested as the difference between the last step and the preceding step as discussed in Section 5.6.2.3 and illustrated in Tabachnick and Fidell (1983).

A simpler procedure is available in the SPSS MANOVA program. Special language is provided for the test, as shown in Table 8.10. The ANALYSIS statement specifies the DV. Effects are listed in order in the DESIGN statement. First the covariate is listed, then the IV, and finally the IV by covariate interaction. The significance test for homogeneity of regression is the last one listed (PRE by TREATMNT) and is the same as the test of equality of slopes in Table 8.9. Section 8.7 illustrates this procedure with multiple IVs and covariates.

8.5.2 ■ Design Complexity

Extension of ANCOVA to factorial between-subjects designs is straightforward as long as sample sizes within cells are equal. Partitioning of sources of variance follows ANOVA (cf. Chapter 3) with "subjects nested within cells" as the simple error term. Sums of squares are adjusted for the aver-

---

TABLE 8.10 SETUP AND SELECTED OUTPUT FOR TEST OF HOMOGENEITY OF REGRESSION THROUGH SPSS MANOVA

```
DATA LIST         FILE = 'TAPE70.DAT'   FREE
                  /SUBJNO, TREATMNT, PRE, POST.
VALUE LABELS      TREATMNT 1 'TREAT1' 2 'TREAT2' 3 'CONTROL'.
MANOVA            PRE POST BY TREATMNT (1,3)/
                  PRINT = SIGNIF(BRIEF)/
                  ANALYSIS = POST/
                  METHOD = SEQUENTIAL/
                  DESIGN = PRE, TREATMNT, PRE BY TREATMNT.
```

Tests of Significance for POST using SEQUENTIAL Sums of Squares

Source of Variation	SS	DF	MS	F	Sig of F
WITHIN+1RESIDUAL	146.15	3	48.72		
PRE	204.58	1	204.58	4.20	.133
TREATMNT	366.20	2	183.10	3.76	.152
PRE BY TREATMNT	3.29	2	1.64	.03	.967

age association between the covariate and the DV in each cell, just as they are for the one-way design demonstrated in Section 8.4.

There are, however, two major design complexities that arise: within-subjects IVs, and unequal sample sizes in the cells of a factorial design. And, just as in ANOVA, contrasts are appropriate for a significant IV with more than two levels and assessment of strength of association is appropriate for any significant effect.

8.5.2.1 ■ Within-Subjects and Mixed Within-Between Designs

Just as a DV can be measured once or repeatedly, so also a covariate can be measured once or repeatedly. In fact, the same design may contain one or more covariates measured once and other covariates measured repeatedly.

A covariate that is measured only once does not provide adjustment to a within-subjects effect because it provides the same adjustment (equivalent to no adjustment) for each level of the effect. The covariate does, however, adjust any between-subjects effects in the same design. Thus ANCOVA with one or more covariates measured once is useful in a design with both between- and within-subjects effects for increasing the power of the test of between-subjects IVs. Both between- and within-subjects effects are adjusted for covariates measured repeatedly.

ANOVA and ANCOVA with repeated measures (within-subjects effects) are more complicated than designs with only between-subjects effects. One complication is that some programs cannot handle repeatedly measured covariates. A second complication is the assumption of homogeneity of covariance, as discussed in Chapter 10 (which also discusses alternatives in the event of violation of the assumption). A third complication (more computational than conceptual) is development of separate error terms for various segments of within-subjects effects.

BMDP2V handles designs with between-subjects effects, within-subjects effects, and both types of effects. Both types of covariates can be entered—those that are measured once and those that are measured repeatedly. BMDP2V provides adjusted cell means. BMDP4V also is useful for all varieties of complex ANCOVA but does not provide adjusted cell or marginal means. Both BMDP2V and BMDP4V provide corrections when the assumption of homogeneity of covariance is violated: the Greenhouse-Geisser correction and the Huynh-Feldt correction, both of which vary with the degree to which the homogeneity of covariance assumption is violated.[6]

Of the SPSS programs, only MANOVA can be used for complex designs. MANOVA provides adjusted means and a test for homogeneity of covariance and provides the Huynh-Feldt and Greenhouse-Geisser corrections for failure of the assumption. However, there is no provision for separate covariates at each level of the within-subjects IV.

SYSTAT ANOVA and GLM and SAS GLM handle complex designs and provide adjustments for violation of homogeneity of covariance as well as adjusted cell and marginal means. The SYSTAT Advanced Applications manual (Wilkinson & Hill, 1994a) shows how to set up a repeated measures analysis of covariance with covariates measured repeatedly. No such aid is available in the SAS manual.

---

[6] If there is reason to doubt homogeneity of covariance—if strong or differential carry-over effects are suspected from one level of the within-subjects IV to the other—a conservative adjustment of critical F for the within-subjects effect can be made where critical F (at a given α level) is assessed with 1 instead of $k - 1$ degrees of freedom. The correction is applied to the main effect of the within-subjects IV, all interactions that contain the within-subjects IV, and all error terms associated with it. Keppel (1991) refers to this as the Geisser-Greenhouse correction. However, this extremely conservative adjustment is not as desirable as the Huynh-Feldt or Greenhouse-Geisser corrections available in programs because of loss of power. The Huynh-Feldt correction loses less power than the Greenhouse-Geisser correction. See Section 10.5.2 for additional discussion of the corrections.

8.5.2.2 ■ Unequal Sample Sizes

In a factorial design, two problems arise if cells have unequal numbers of scores. First, there is ambiguity regarding a marginal mean from cells with unequal n. Is the marginal mean the mean of the means, or is the marginal mean the mean of the scores? Second, the total sums of squares for all effects is greater than SS_{total} and there is ambiguity regarding assignment of overlapping sums of squares to sources. The design has become nonorthogonal and tests for main effects and interactions are no longer independent (cf. Section 3.2.5.3). The problem generalizes directly to ANCOVA.

If equalizing cell sizes by random deletion of cases is undesirable, there are a number of strategies for dealing with unequal n. The choice among strategies depends on the type of research. Of the three major methods described by Overall and Spiegel (1969), Method 1 is usually appropriate for experimental research, Method 2 for survey or nonexperimental research, and Method 3 for research in which the researcher has clear priorities for effects.

Table 8.11 summarizes research situations calling for different methods and notes some of the jargon used by various sources. As Table 8.11 reveals, there are a number of ways of viewing these methods; the terminology associated with these viewpoints by different authors is quite different and, sometimes, seemingly contradictory.

Differences in these methods are easiest to understand from the perspective of multiple regression (Chapter 5). Method 1 is like standard multiple regression with each main effect and interaction assessed after adjustment is made for all other main effects and interactions, as well as for covariates. In SPSS this is called the regression approach.[7] The same hypotheses are tested as in the unweighted-means approach where each cell mean is given equal weight regardless of its sample size. This is the recommended approach for experimental research unless there are reasons for doing otherwise.

Reasons include a desire to give heavier weighting to some effects than others because of importance, or because unequal population sizes have resulted from treatments that occur naturally. (If a design intended to be equal-n ends up grossly unequal, the problem is not type of adjustment but, more seriously, differential dropout.)

Method 2 imposes a hierarchy of testing effects where main effects are adjusted for each other and for covariates, while interactions are adjusted for main effects, for covariates, and for same- and lower-level interactions. (The SAS implementation also makes adjustments for some higher order effects.) The order of priority for adjustment emphasizes main effects over interactions and lower-order interactions over higher-order interactions. It is called the classic experimental approach by SPSS although it is normally used in nonexperimental work when there is a desire to weight marginal means by the sizes of samples in cells from which they are computed. The adjustment assigns heavier weighting to cells with larger sample sizes when computing marginal means and lower-order interactions.

Method 3 allows the researcher to set up the sequence of adjustment of covariates, main effects, and interactions. In addition, BMDP4V gives the researcher the opportunity to weight cells by importance or by population size (or anything else) rather than by sample size.

All programs in the reviewed packages perform ANCOVA with unequal sample sizes. BMDP2V and SYSTAT ANOVA provide for complex designs and unequal n, but only with Method 1 adjustment (although SYSTAT provides guidelines for setting up other types of adjustment). SPSS ANOVA allows Methods 1 and 2, and many varieties of Method 3, but requires a between-subjects design. SPSS MANOVA, SYSTAT GLM, SAS GLM, and BMDP4V (for two-way designs) provide

---

[7] Note that the terms regression and classic experimental as used by SPSS do not imply that these approaches are most appropriately used for nonexperimental and experimental designs, respectively. If anything, the opposite is true.

TABLE 8.11 TERMINOLOGY FOR STRATEGIES FOR ADJUSTMENT FOR UNEQUAL CELL SIZE

Research type	Overall and Spiegel	SPSS	BMDP	SYSTAT	SAS
1. Experiments designed to to be equal-n, with random dropout. All cells equally important.	Method 1	Regression approach. option = 9 in ANOVA program. In MANOVA, METHOD = UNIQUE (default)	Equal cell weights for BMDP2V. Default for BMDP4V	Default	TYPE III and TYPE IV[a]
2. Nonexperimental research in which sample sizes reflect importance of cells. Main effects have equal priority.[b]	Method 2	Classic experimental approach. Default in ANOVA	N.A.	N.A.[e]	TYPE II
3. Like number 2 above, except all effects have unequal priority.	Method 3	Sequential approach. In MANOVA, METHOD = SEQUENTIAL	N.A.	N.A.[e]	TYPE I[f]
4. Like number 2 above, except main effects have unequal priority, interactions have equal priority.[c]	Method 3	Hierarchical approach. METHOD = HIERARCHICAL in ANOVA.	Cell size weights[d] for BMDP4V	N.A.[e]	N.A.
5. Research in which cells are given unequal weight on basis of prior knowledge.	N.A.	N.A.	User-defined cell weights for BMDP4V	N.A.	N.A.

[a] Type III and IV differ only if there are missing cells.
[b] The programs take different approaches to adjustment for interaction effects.
[c] The programs take different approaches to adjustment for main effects. For a two-way design, hierarchical = sequential.
[d] For a two-way design, output table in BMDP4V gives information for interpretation as either research type 2 or research type 3.
[e] Instructions are given in the SYSTAT manual to produce adjustments other than Method 1.
[f] Type III output provides values for covariates adjusted for main effects and interactions.

for design complexity and flexibility in adjustment for unequal n. If BMDP4V is used for higher-order designs, one main effect can be unadjusted (given highest priority) but all other effects must be fully adjusted. The SYSTAT advanced applications manual shows how to achieve SAS methods through SYSTAT GLM.

Unless you choose Method 1, output from various programs may disagree even though ostensibly the same method is chosen. This is because programs differ in algorithms used to generate sums of squares, and in the subtleties of adjustment for various effects. Because of these discrepancies, and of disagreements as to the best adjustment method in a research situation, some researchers advocate use of Method 1 always. Because Method 1 is the most conservative, you are unlikely to draw criticism by using it. On the other hand, with a nonexperimental design you risk loss of power and perhaps interpretability and generalizability by treating all cells as if they had equal sample sizes.

8.5.2.3 ■ Specific Comparisons and Trend Analysis

If there are more than two levels of an IV, the finding of a significant main effect or interaction in ANOVA or ANCOVA is often insufficient for full interpretation of the effects of IV(s) on the DV. The omnibus F test of a main effect or interaction gives no information as to which means are significantly different from which other means. With a qualitative IV (whose levels differ in kind) the researcher generates contrast coefficients to compare some adjusted mean(s) against other adjusted means. With a quantitative IV (whose levels differ in amount rather than in kind), trend coefficients are used to see if adjusted means of the DV follow a linear or quadratic pattern, say, over sequential levels of the IV.

As with ANOVA (Chapter 3), specific comparisons or trends can be either planned as part of the research design, or tested post hoc as part of a data-snooping procedure after omnibus analyses are completed. For planned comparisons, protection against inflated Type I error is achieved by running a small number of comparisons instead of omnibus F (where the number does not exceed the available degrees of freedom) and by working with an orthogonal set of coefficients. For post hoc comparisons, the probability of Type I error increases with the number of possible comparisons, so adjustment is made for inflated α error.

Comparisons are achieved by specifying coefficients and running analyses based on these coefficients. The comparisons can be simple (between two marginal or cell means with the other means left out) or complicated (where means for some cells or margins are pooled and contrasted with means for other cells or margins, or where coefficients for trend—linear, quadratic, etc.—are used). The difficulty of conducting comparisons depends on the complexity of the design and the effect to be analyzed. Comparisons are more difficult if the design has within-subjects IVs, either alone or in combination with between-subjects IVs, where problems arise from the need to develop a separate error term for each comparison. Comparisons are more difficult for interactions than for main effects because there are several approaches to comparisons for interactions. Some of these issues are reviewed in Section 10.5.1.

If all you want to do is pairwise comparisons of adjusted means (and if all IVs are between-subjects and cell sample sizes are equal[8]), they are easily obtained through BMDP1V. The design

---

[8] If cell sizes are unequal, BMDP4V can be used for a test of appropriately adjusted (weighted) means (see Section 8.6) but the program does not print out the adjusted means. Specific comparisons with unequal cell sizes and covariates are also available through SPSS MANOVA, as discussed in Chapter 9.

must be one-way or converted to one-way. The program automatically produces a T-TEST MATRIX FOR ADJUSTED GROUP MEANS where differences among all pairs of adjusted means are evaluated. For the data of Table 8.1, the matrix of t tests is shown in Table 8.12. Evaluation of the significance of any of the t values depends on whether the pairwise comparisons were planned or post hoc.

If only a few pairwise comparisons are planned (as many as degrees of freedom for the IV) and the rest are ignored or evaluated post hoc, then the significance of each planned comparison is tested as t with df $= N - k - c$, where N = total sample size (9 in the example), k = number of cells (3 in the example), and c = number of covariates (1 in the example). BMDP1V prints out both the degrees of freedom and the matrix of probability levels. A planned comparison between the adjusted means for TREAT1 and CONTROL, for instance, with an obtained t value of -3.3171, is statistically significant for a two-tailed test at $\alpha = .05$ because the p value of .0211 is less than .05.

If this same comparison is evaluated post hoc, a Scheffé adjustment (or some other adjustment) is made to counteract inflation of the Type I error rate caused by too many tests. First, obtained t is squared to produce obtained F. For the example, F obtained for TREAT1 versus CONTROL is $(-3.3171)^2 = 11.00$. F obtained is then compared with an adjusted critical F produced by multiplying the tabled F value (in this case 5.79 for 2 and 5 df, $\alpha = .05$) by the degrees of freedom associated with the number of cells, or $k - 1$. For this example, the adjusted critical F value is 2(5.79) = 11.58, and the difference between TREAT1 and CONTROL now fails to reach statistical significance.

Pairwise tests of adjusted means are also available through options in SPSS MANOVA, SYSTAT ANOVA and GLM and SAS GLM. In addition, SAS GLM as well as SYSTAT ANOVA and GLM provide several tests that incorporate post hoc adjustments, such as Scheffé or Tukey.

For more complicated comparisons, one set of coefficients is entered in the BMDP1V setup for each comparison. The program prints out the t-test value, the probability level P(T), and the coefficients for each comparison. As previously, the probability level is appropriate for a planned comparison but must be adjusted for a post hoc comparison.

TABLE 8.12   COMPARISONS OF ALL PAIRS OF CELL MEANS AS PRODUCED BY BMDP1V FOR SAMPLE DATA IN TABLE (SETUP APPEARS IN TABLE 8.9)

T-TEST MATRIX FOR ADJUSTED GROUP MEANS ON   5 DEGREES OF FREEDOM

		TREAT1 1	TREAT2 2	CONTROL 3
TREAT1	1	0.0000		
TREAT2	2	−0.6309	0.0000	
CONTROL	3	−3.3171	−2.6250	0.0000

PROBABILITIES FOR THE T-VALUES ABOVE

		TREAT1 1	TREAT2 2	CONTROL 3
TREAT	1	1.0000		
TREAT2	2	0.5558	1.0000	
CONTROL	3	0.0211	0.0468	1.0000

TABLE 8.13   CONTRASTS THROUGH BMDP1V FOR SAMPLE DATA IN TABLE 8.1 (SETUP IN TABLE 8.9)

T-VALUES FOR CONTRASTS IN ADJUSTED GROUP MEANS

CONTRAST NUMBER	T	P(T)	GROUP TREAT1	GROUP TREAT2	GROUP CONTROL
1	3.3171	0.0211	1.0000	0.0000	−1.0000
2	−1.1542	0.3006	1.0000	−2.0000	1.0000
3	3.4272	0.0187	1.0000	1.0000	−2.0000

Table 8.13 illustrates more complicated comparisons with the sample data. The first two sets of coefficients are for linear and quadratic trend, respectively.[9] (Because the IV in this sample is not quantitative, trend analysis is inappropriate; trend coefficients are used here for illustration only.) The third comparison contrasts the pooled adjusted means of the two treatment groups against the adjusted mean of the control group to test the null hypothesis of no difference between treatment and control. Consider the first two comparisons planned and the third post hoc. The first two are tested against a critical t with 5 df of 2.57; the linear trend is reliable, but the quadratic trend is not. The third comparison, tested post hoc, is reliable because obtained F of 11.75 exceeds adjusted critical F of 11.58. Thus, even by the conservative post hoc criterion, the adjusted mean or the control group differs from the pooled adjusted means of the two treated groups.

In designs with more than one IV, the size of the Scheffé adjustment to critical F for post hoc comparisons depends on the degrees of freedom for the effect being analyzed. For a two-way design, for example, with IVs A and B, adjusted critical F for A is tabled critical F for A multiplied by $(a - 1)$ (where a is the number of levels of A); adjusted critical F for B is tabled critical F for B multiplied by $(b - 1)$; adjusted critical F for the $A \times B$ interaction is tabled critical F for the interaction multiplied by $(a - 1)(b - 1)$.[10]

Procedures similar to these for BMDP1V are applied when using ANCOVA programs such as BMDP4V, SPSS MANOVA, SYSTAT GLM or ANOVA and SAS GLM. If appropriate programs are unavailable, hand calculations for specific comparisons are not particularly difficult, as long as sample sizes are equal for each cell. Equation 3.23 is for hand calculation of comparisons; to apply it, obtain the adjusted cell or marginal means and the error mean square from an omnibus ANCOVA program.

8.5.2.4 ■ Strength of Association

Once an effect is found reliable, the next logical question is, How important is the effect? Importance is usually assessed as the percentage of variance in the DV that is associated with the IV. For one-way designs, the strength of association between effect and DV for adjusted sums of squares is found using Equation 8.7. For factorial designs, one uses an extension of Equation 8.7.

The numerator for η^2 is the adjusted sum of squares for the main effect or interaction being evaluated; the denominator is the total adjusted sum of squares. The total adjusted sum of squares includes adjusted sums of squares for all main effects, interactions, and error terms but does not

---

[9] Coefficients for orthogonal polynomials are available in most standard ANOVA texts such as Keppel (1991) or Winer (1971).

[10] The adjusted critical F for an interaction is insufficient if a great many post hoc comparisons are undertaken; it should suffice, however, if a moderate number are performed. If a great many are undertaken, multiply by $ab - 1$ instead of $(a - 1)(b - 1)$.

include components for covariates or the mean, which are typically printed out by computer programs.[11] To find the strength of association between an effect and the adjusted DV scores, then,

$$\eta^2 = \frac{SS'_{effect}}{SS'_{total}} \tag{8.8}$$

In multifactorial designs, the size of η^2 for a particular effect is, in part, dependent on the strength of other effects in the design. In a design where several main effects and interactions are significant, η^2 for a particular effect is diminished because other significant effects increase the size of the denominator. An alternative method of computing η^2 (partial η^2) uses in the denominator only the adjusted sum of squares for the effect being tested and the adjusted sum of squares for the appropriate error term for that effect (see Chapter 3 for appropriate error terms).

$$\text{partial } \eta^2 = \frac{SS'_{effect}}{SS'_{effect} + SS'_{error}} \tag{8.9}$$

Because this alternative form is not standard, an explanatory footnote is appropriate whenever it is used in published results.

8.5.3 ■ Evaluation of Covariates

Covariates in ANCOVA can themselves be interpreted as predictors of the DV. From a sequential regression perspective (Chapter 5), each covariate is a high-priority, continuous IV with remaining IVs (main effects and interactions) evaluated after the relationship between the covariate and the DV is removed.

Significance tests for covariates test their utility in adjusting the DV. If a covariate is significant, it provides adjustment of the DV scores. For the example in Table 8.5, the covariate, PRE, does not provide significant adjustment to the DV, POST, with $F(1, 5) = 4.61, p > .05$. PRE is interpreted in the same way as any IV in multiple regression (Chapter 5).

With multiple covariates, all covariates enter the multiple regression equation at once and, as a set, are treated as a standard multiple regression (Section 5.5.1). Within the set of covariates, the significance of each covariate is assessed as if it entered the equation last; only the unique relationship between the covariate and the DV is tested for significance after overlapping variability with other covariates, in their relationship with the DV, is removed. Therefore, although a covariate may be significantly correlated with the DV when considered individually, it may add no significant adjustment to the DV when considered last. When interpreting the utility of a covariate, it is necessary to consider correlations among covariates, correlations between each covariate and the DV, and significance levels for each covariate as reported in ANCOVA source tables. Evaluation of covariates is demonstrated in Section 8.7.

Unstandardized regression coefficients, provided by most canned computer programs, have the same meaning as regression coefficients described in Chapter 5. However, with unequal n, interpretation of the coefficients depends on the method used for adjustment. When Method 1 (standard multiple regression—see Table 8.11 and Section 8.5.2.2) is used, the significance of the regression coefficients for covariates is assessed as if the covariate entered the regression equation after all

---

[11] If you are using SPSS ANOVA, such summary terms as main effects, two-way interactions, explained variance, and the like, are also omitted. Similarly, the total sum of squares is inappropriate because it is the sum of squares for original DV scores rather than adjusted DV scores.

main effects and interactions. With other methods, however, covariates enter the equation first, or after main effects but before interactions. The coefficients are evaluated at whatever point the covariates enter the equation.

8.5.4 ■ Choosing Covariates

One wants to use an optimal set of covariates if several are available. When too many covariates are used and they are correlated with each other, a point of diminishing returns in adjustment of the DV is quickly reached. Power is reduced because numerous correlated covariates subtract degrees of freedom from the error term while not removing commensurate sums of squares for error. Preliminary analysis of the covariates improves chances of picking a good set.

Statistically, the goal is to identify a small set of covariates that are uncorrelated with each other but correlated with the DV. Conceptually, one wants to select covariates that adjust the DV for predictable but unwanted sources of variability. It may be possible to pick the covariates on theoretical grounds or on the basis of knowledge of the literature regarding important sources of variability that should be controlled.

If theory is unavailable or the literature is insufficiently developed to provide a guide to important sources of variability in the DV, statistical considerations assist the selection of covariates. One strategy is to look at correlations among covariates and select one from among each of those groups of potential covariates that are substantially correlated with each other, perhaps by choosing the one with the highest correlation with the DV. A more sophisticated strategy is to use all subsets regression (cf. Chapter 5) to see which (small) set of covariates best predicts the DV, ignoring the IV. BMDP9R and SAS REG do all subsets regression.

If N is large and power is not a problem, it may still be worthwhile to find a small set of covariates for the sake of parsimony. Useless covariates are identified in the first ANCOVA run. Then further runs are made, each time eliminating covariates, until a small set of useful covariates is found. The analysis with the smallest set of covariates is reported, but mention is made in the Results section of the discarded covariate(s) and the fact that the pattern of results did not change when they were eliminated.

8.5.5 ■ Alternatives to ANCOVA

Because of the stringent limitations to ANCOVA and potential ambiguity in interpreting results of ANCOVA, alternative analytical strategies are often sought. The availability of alternatives depends on such issues as the scale of measurement of the covariate(s) and the DV, the time that elapses between measurement of the covariate and assignment to treatment, and the difficulty of interpreting results.

When the covariate(s) and the DV are measured on the same scale, two alternatives are available: use of difference scores and conversion of the pretest and posttest scores into a within-subjects IV. In the first alternative, the difference between a pretest score (the previous covariate) and a posttest score (the previous DV) is computed for each subject and used as the DV in ANOVA. If the research question is phrased in terms of "change," then difference scores provide the answer. For example, suppose self-esteem is measured both before and after a year of either belly dance or aerobic dance classes. If difference scores are used, the research question is, Does one year of belly dance training change self-esteem scores more than participation in aerobic dance classes? If

ANCOVA is used, the research question is, Do belly dance classes produce greater self-esteem than aerobic dance classes after adjustment for pretreatment differences in self-esteem?

A problem with difference scores is ceiling and floor effects (or, in general, a problem of regression toward the mean). A difference may be small because the pretest score is very near the end of the scale and no treatment effect can change it very much, or it may be small because the effect of treatment is small—the DV is the same in either case and the researcher is hard pressed to decide between them. A second problem is that use of difference scores assumes that covariate and DV are measured with perfect reliability. If either ANCOVA or ANOVA with difference scores could be used, ANCOVA is usually the better approach.

The second alternative, if measurement scales are the same, is to convert pre- and posttest scores into two levels of a within-subjects IV. The problem with this strategy is that the effect of the IV of interest is no longer assessed as a main effect, but rather as an interaction between the IV and the pre- vs. posttest IV.

When covariates are measured on any continuous scale, other alternatives are available: randomized blocks and blocking. In the randomized block design, subjects are matched into blocks—equated—on the basis of scores on what would have been the covariate(s). Each block has as many subjects as the number of levels of the IV in a one-way design or number of cells in a larger design (cf. Section 3.2.3). Subjects within each block are randomly assigned to levels or cells of the IV(s) for treatment. In the analytic phase, subjects in the same block are treated as if they were the same person, in a repeated-measures analysis.

Disadvantages to this approach are the strong assumption of homogeneity of covariance of a within-subjects analysis and the loss of degrees of freedom for error without commensurate loss of sums of squares for error if the variables used to block is not highly related to the DV. In addition, implementation of the randomized block design requires the added step of equating subjects before randomly assigning them to treatment, a step that may be inconvenient, if not impossible, in some applications.

The last alternative is use of blocking. Subjects are measured on potential covariate(s) and then grouped according to their scores (e.g., into groups of high, medium, and low self-esteem on the basis of pretest scores). The groups of subjects become the levels of another full-scale IV that are crossed with the levels of the IV(s) of interest in factorial design. Interpretation of the main effect of the IV of interest is straightforward and variation due to the potential covariate(s) is removed from the estimate of error variance and assessed as a separate main effect. Furthermore, if the assumption of homogeneity of regression would have been violated in ANCOVA, it shows up as an interaction between the blocking IV and the IV of interest.

Blocking has several advantages over ANCOVA and the other alternatives listed here. First, it has none of the assumptions of ANCOVA or within-subjects ANOVA. Second, the relationship between the potential covariate(s) and the DV need not be linear (blocking is less powerful when the covariate-DV relationship *is* linear); curvilinear relationships can be captured in ANOVA when three (or more) levels of an IV are analyzed. Blocking, then, is preferable to ANCOVA in many situations, and particularly for experimental, rather than correlational, research.

Blocking can also be expanded to multiple covariates. That is, several new IVs, one per covariate, can be developed through blocking and, with some difficulty, crossed in factorial design. However, as the number of IVs increases, the design rapidly becomes very large and cumbersome to implement.

For some applications, however, ANCOVA is preferable to blocking. When the relationship between the DV and the covariate is linear, ANCOVA is more powerful than blocking. And, if the assumptions of ANCOVA are met, conversion of a continuous covariate to a discrete IV can result in loss of information. Finally, practical limitations may prevent measurement of potential covariate(s) sufficiently in advance of treatment to accomplish random assignment of equal numbers of subjects to the cells of the design. When blocking is attempted after treatment, sample sizes within cells are likely to be highly discrepant, leading to the problems of unequal n.

In some applications, a combination of blocking and ANCOVA may turn out to be best. Some potential covariates are used to create new IVs, while others are analyzed as covariates.

8.6 ■ COMPARISON OF PROGRAMS

For the novice, there is a bewildering array of canned computer programs in the BMDP (BMDP1V, 2R, 2V, and 4V), SPSS (REGRESSION, ANOVA, and MANOVA), and SYSTAT (ANOVA, GLM, and REGRESS) packages for ANCOVA. For our purposes, the programs based on regression (BMD2R, SYSTAT REGRESS and SPSS REGRESSION) are not discussed because they offer little advantage over the other, more easily used programs.

SAS has a single general linear model program designed for use with both discrete and continuous variables. This program deals well with ANCOVA. Features of eight programs are described in Table 8.14.

8.6.1 ■ BMDP Series

BMDP1V is designed for one-way, between-subjects ANOVA and ANCOVA but offers features unavailable in programs for factorial designs. Most notably, it tests both homogeneity of regression (equality of slopes) and user-designated comparisons of adjusted means. As discussed in Sections 8.5.1 and 8.5.2.3, data from factorial between-subjects designs can be recoded into one-way designs to take advantage of these features and then analyzed through factorial programs.

For factorial ANCOVA, the best all-around program in the BMDP series is BMDP2V. It allows for complex designs (both between- and within-subjects IVs and unbalanced layouts) and unequal n (but only with Method 1 adjustment). Both singly and repeatedly measured covariates can be analyzed. The test for homogeneity of covariance in within-subjects designs, labeled compound symmetry, is available within the program; only the portion of the compound-symmetry test relevant to homogeneity of covariance is evaluated. If the assumption of homogeneity of covariance is violated, two forms of adjustment (Huynh-Feldt and Greenhouse-Geisser) are available. There is full orthogonal decomposition (trend analysis) of within-subjects variables for equal or unequal spacing between levels of the within-subjects variable(s). Specific comparisons (contrasts) can be specified for within- and between-subjects effects. Box-Cox diagnostic plots are available to suggest appropriate transformations for violation of homogeneity of variance.

BMDP4V, described more fully in Section 8.7, is more comprehensive but is also more difficult to use. It provides a variety of methods for dealing with unequal n. In addition, specific comparisons of both between- and within-subjects IVs are available, as well as adjustment for violation of homogeneity of covariance in within-subjects designs. A disadvantage of the program is the failure to provide adjusted cell or marginal means, or regression coefficients for covariates.

The combination of BMDP1V (to test for equality of slope and user-specified comparisons in between-subjects designs), BMDP2V (to analyze complex designs and to test for homogeneity of

TABLE 8.14 COMPARISON OF SELECTED PROGRAMS FOR ANALYSIS OF COVARIANCE

Feature	BMDP4V[a]	BMDP1V	BMDP2V	SPSS ANOVA	SPSS MANOVA[a]	SAS GLM[a]	SYSTAT ANOVA[a]	SYSTAT GLM[a]
Input								
Maximum number of IVs	No limit	One	No limit	10	10	No limit	No limit	No limit
Choice of unequal-n adjustment	Yes	N.A.	No	Yes	Yes	Yes	No	Yes
Within-subjects IVs	Yes	No	Yes	No	Yes	Yes	Yes	Yes
Specify tolerance	No	No	Yes	No	No	SINGULAR	Yes	Yes
Specify separate variance error term for contrasts	No	No	No	N.A.	No	No	Yes	Yes
Output								
Source table	Yes	Yes	Yes	Yes	Yes	Yes	Yes	Yes
Unadjusted cell means	Yes	Yes	Yes	Yes	Yes	Yes	No	No
Confidence interval for unadjusted cell means	No	No	Yes	No	Yes	No	No	No
Unadjusted marginal means	Yes	N.A.	No	Yes	Yes	Yes	No	No
Cell standard deviations	Yes	No	Yes	No	Yes	Yes	No	No
Adjusted cell means	No	Yes	Yes	No	PMEANS	LS MEANS	PRINT MEDIUM	PRINT MEDIUM
Confidence intervals for adjusted cell means	No	No	Yes	No	No	No	No	No
Standard errors for adjusted cell means	No	No	Yes	No	No	Data file	PRINT MEDIUM	PRINT MEDIUM
Adjusted marginal means	No	N.A.	No	No	Yes[b]	LS MEANS	PRINT MEDIUM	PRINT MEDIUM
Standard error for adjusted marginal means	No	N.A.	No	No	No	STDERR	PRINT MEDIUM	PRINT MEDIUM
Grand means and adjusted marginal deviation	No	No	No	Yes	No	No	No	No
Skewness	No	No	Yes	No	No	No	No	No
Power analysis	No	No	No	No	POWER	No	No	No
Effect sizes	No	No	No	No	POWER	No	No	No
Test for equality of slope (homogeneity of regression)	No	Yes	No	No	Yes	No	Yes	Yes
Within-groups regression coefficients	No	Yes	No	No	No	No	No	No
t test for all pairs of cell means	No	Yes	No	No	Yes	Yes	No	No
Post hoc tests with adjustment	No	No	Yes[c]	No	Yes[d]	Yes	Yes	Yes

TABLE 8.14 (CONTINUED)

Feature	BMDP4V[a]	BMDP1V	BMDP2V	SPSS ANOVA	SPSS MANOVA[a]	SAS GLM[a]	SYSTAT ANOVA[a]	SYSTAT GLM[a]
User-specified contrasts	Yes	Yes	Yes	No	Yes	Yes	Yes	Yes
Hypothesis SSCP matrices	Yes	No	No	No	Yes	Yes	No	Yes
Pooled within-cell error SSCP matrices	Yes	No	No	No	Yes	Yes	No	Yes
Hypothesis covariance matrices	No	Yes	No	No	Yes	No	Yes	Yes
Pooled within-cell covariance matrix	No	Yes	No	No	Yes	No	Yes	Yes
Group covariance matrices	No	Yes	No	No	Yes	No	Yes	No
Pooled within-cell correlation matrix	No	No	No	No	Yes	Yes	No	Yes
Group correlation matrices	No	Yes	No	No	Yes	No	No	No
Correlation matrix for adjusted group means	No	Yes	No	No	No	No	No	No
Covariance matrix for adjusted group means	No	No	No	No	No	Data file	No	No
Correlation matrix for regression coefficients	No	Yes	No	No	No	No	No	No
Regression coefficient for each covariate	No	Yes	Yes	Yes	Yes	No	PRINT LONG	PRINT LONG
Regression coefficient for each cell	No	No	No	No	No	Yes	No	No
Regression coefficient for each main effect	No	No	No	Yes	No	No	PRINT LONG	PRINT LONG
Multiple R and/or R^2	No	No	No	Yes	Yes	Yes	Yes	Yes
Observed maximums and minimums	Yes	Yes	Yes	No	No	No	No	No
Test for homogeneity of variance	No	No	No	No	Yes	No	No	No
Box-Cox diagnostic plot for homogeneity of variance	No	No	Yes	No	No	No	No	No
Test for homogeneity of covariance	No	N.A.	Sphericity	N.A.	Yes	Yes	No	No
Adjustment for heterogeneity of covariance	Yes	N.A.	Yes	N.A.	Yes	Yes	Yes	Yes

TABLE 8.14 *(CONTINUED)*

Feature	BMDP4V[a]	BMDP1V	BMDP2V	SPSS ANOVA	SPSS MANOVA[a]	SAS GLM[a]	SYSTAT ANOVA[a]	SYSTAT GLM[a]
Predicted values and residuals	No	No	Yes	No	Yes	Yes	Data file	Data file
Plots of means	No	No	Yes	No	No	No	No	No
Mahalanobis distance for outliers	No	No	No	No	No	No	Data file	Data file
Other influence and/or leverage statistics	No	No	No	No	No	Data file	Data file	Data file

[a] Additional features described in Chapter 9 (MANOVA)

[b] Available through the CONSPLUS procedure

[c] Bonferroni contrasts for pairwise contrasts among within-subjects factor levels only

[d] Bonferroni and Scheffé confidence intervals

covariance and trend in within-subjects IVs), BMDP4V (to analyze complex designs, provide a variety of adjustments for unequal n, and give complete flexibility in comparisons), and BMDPAM or BMDP7M (to search for outliers in each group) gives the user ability to handle almost any ANCOVA problem.

8.6.2 ■ SPSS Package

Two SPSS programs perform ANCOVA: MANOVA, and ANOVA. Of the two, MANOVA is far richer. Because the MANOVA program is so flexible, it takes more time to master than SPSS ANOVA but is well worth the effort. SPSS MANOVA is useful for nonorthogonal and complex designs, and shines in its ability to test assumptions such as homogeneity of regression (see Section 8.5.1) and of variance and covariance. It provides adjusted cell and marginal means; specific comparisons and trend analysis are readily available. However, SPSS does not facilitate the search for multivariate outliers among the DV and covariate(s) in each group.

SPSS ANOVA has nowhere near the flexibility of BMDP2V or of SPSS MANOVA but with between-subjects designs and an assist from SPSS MANOVA, a fairly large subset of ANCOVA problems can be analyzed. The program only works for between-subjects IVs but does offer a wide variety of adjustments for unequal n (see Section 8.5.2.2). Tests of assumptions and specific comparisons, including trend analysis, cannot be accomplished through SPSS ANOVA but are available through SPSS MANOVA. Another major limitation is lack of adjusted cell means, although adjusted marginal means can be computed from marginal deviations.

8.6.3 ■ SYSTAT System

SYSTAT ANOVA (Wilkinson & Hill, 1994b) is an easily used program which handles all types of ANOVA and ANCOVA. In addition, SYSTAT GLM (Wilkinson & Hill, 1994a) is a multivariate general linear program that does everything that the ANOVA program does, and more, but is not always quite as easy to set up. The ANOVA program handles repeated measures and post hoc comparisons, and provides a great deal of control over error terms and comparisons. Cell means adjusted for covariates are saved to a file. The program also provides Mahalanobis distance (actually the square root of the value usually reported) as a test for multivariate outliers. The program provides adjustment for violation of homogeneity of covariance in within-subject designs.

The major advantage in using SYSTAT GLM over ANOVA is the greater flexibility with unequal-n designs. Only method 1 is available in the ANOVA program. GLM allows specification of a MEANS model, which, when WEIGHTS are applied to cell means, provides a weighted means analysis. The advanced applications manual (Wilkinson & Hill, 1994a) also describes the multiple models that can be estimated to provide sums of squares that correspond to the four SAS types of adjustment for unequal n. GLM is also the only SYSTAT program set up for simple effects and such designs as Latin square, nesting, and incomplete blocks.

8.6.4 ■ SAS System

SAS GLM is a program for univariate and multivariate analysis of variance and covariance. The setup is similar to that of SYSTAT GLM.

SAS GLM offers analysis of complex designs, several adjustments for unequal n, a test for homogeneity of covariance for within-subjects IVs, a full array of descriptive statistics (upon request), and a wide variety of post hoc tests in addition to user-specified comparisons and trend analysis. Although there is no example of a test for homogeneity of regression in the SAS manual, the procedures described in the SYSTAT manual can be followed.

8.7 ■ COMPLETE EXAMPLE OF ANALYSIS OF COVARIANCE

The research described in Appendix B, Section B.1, provides the data for this illustration of ANCO-VA. The research question is whether attitudes toward drugs are associated with current employment status and/or religious affiliation.

Attitude toward drugs (ATTDRUG) serves as the DV, with increasingly high scores reflecting more favorable attitudes. The two IVs, factorially combined, are current employment status (EMPLMNT) with two levels, (1) employed and (2) unemployed; and religious affiliation (RELIGION) with four levels, (1) None-or-other, (2) Catholic, (3) Protestant, and (4) Jewish.

In examining other data for this sample of women, three variables stand out that could be expected to relate to attitudes toward drugs and might obscure effects of employment status and religion. These variables are general state of physical health, mental health, and the use of psychotropic drugs. In order to control for the effects of these three variables on attitudes toward drugs, they are treated as covariates. Covariates, then, are physical health (PHYHEAL), mental health (MENHEAL), and sum of all psychotropic drug uses, prescription and over-the-counter (PSYDRUG). For all three covariates, larger scores reflect increasingly poor health or more use of drugs.

The 2×4 analysis of covariance, then, provides a test of the effects of employment status, religion, and their interaction on attitudes toward drugs after adjustment for differences in physical health, mental health, and use of psychotropic drugs. Note that this is a form of ANCOVA in which no causal inference can be made.

8.7.1 ■ Evaluation of Assumptions

These variables are examined with respect to practical limitations of ANCOVA as described in Section 8.3.2.

8.7.1.1 ■ Unequal *n* and Missing Data

BMDP7D provides an initial screening run to look at histograms and descriptive statistics for DV and covariates for the eight groups. Output for PSYDRUG is shown in Table 8.15. Three women out of 465 failed to provide information on religious affiliation. Because RELIGION is one of the IVs for which cell sizes are unequal in any event, the three cases are dropped from analysis.

The cell-size approach (Method 3 of Section 8.5.2.2) for dealing with unequal *n* is chosen for this study. This method weights cells by their sample sizes, which, in this study, are meaningful because they represent population sizes for the groups.

8.7.1.2 ■ Normality

The histograms in Table 8.15 reveal obvious positive skewness for some variables. Because the assumption of normality applies to the sampling distribution of means, and not to the raw scores,

TABLE 8.15   SETUP AND PARTIAL OUTPUT OF SCREENING RUN FOR UNIVARIATE OUTLIERS USING BMDP7D

```
/INPUT      VARIABLES ARE 7. FORMAT IS FREE.
            FILE = 'ANCOVA.DAT'.
/VARIABLE   NAMES ARE SUBJNO, ATTDRUG, PHYHEAL, MENHEAL,
                     PSYDRUG, EMPLMNT, RELIGION.
            MISSING = (7)9.
            LABEL = SUBJNO.
/GROUP      CODES(EMPLMNT) ARE 1, 2
            NAMES(EMPLMNT) ARE EMPLYD, UNEMPLYD.
            CODES(RELIGION) ARE 1 TO 4.
            NAMES(RELIGION) ARE NONOTHER,CATHOLIC,PROTEST,JEWISH.
/HISTOGRAM  GROUPING = EMPLMNT, RELIGION.
/END

    ************                        ************
HISTOGRAM OF * PSYDRUG *  5)  GROUPED BY  * EMPLMNT *   ( 6)
    ************               AND        * RELIGION *  ( 7)
                                          ************
```

TABLE 8.15   (CONTINUED)

CASES WITH
UNUSED
VALUES FOR

MIDPOINTS	EMPLYD NONOTHER	EMPLYD CATHOLIC	EMPLYD PROTEST	EMPLYD JEWISH	UNEMPLYD NONOTHER	UNEMPLYD CATHOLIC	UNEMPLYD PROTEST	UNEMPLYD JEWISH	EMPLMNT RELIGION
45.0000)									
43.2000)									
41.4000)									
39.6000)									
37.8000)									
36.0000)									
34.2000)									
32.4000)*							*		
30.6000)									
28.8000)									
27.0000)			**				*	*	
25.2000)						*		*	
23.4000)**		*	*				**	**	
21.6000)								****	
19.8000)		*	*			*	*		
18.0000)									
16.2000)**		***	***	***	*	*	**	**	
14.4000)***		******	******	*	*		*****	****	
12.6000)*		*	*	*	*	**	**	***	
10.8000)*		***	*******	**	*	***	*****	*	
9.0000)*		*	***	***	*	***	**		
7.2000)***		******	*********		**	***	*********	********	
5.4000)M*****		****	***	M****		*	M****	M*	
3.6000)*		**********	M******	***	*****	M**	*****	*****	
1.8000)***	M***	M***	**		M	*	*		
0.0000)*********36	*********23	*********50	*********24	*********20	*********37	*********47	*********21 M***		

LEGEND FOR GROUP MEANS: M - MEAN COINCIDES WITH AN ASTERISK
                          N - MEAN DOES NOT COINCIDE WITH ANY ASTERISK

	EMPLYD NONOTHER	EMPLYD CATHOLIC	EMPLYD PROTEST	EMPLYD JEWISH	UNEMPLYD NONOTHER	UNEMPLYD CATHOLIC	UNEMPLYD PROTEST	UNEMPLYD JEWISH	EMPLMNT RELIGION
MEAN	5.348	2.349	4.185	4.932	2.067	3.232	5.096	6.083	0.000
STD.DEV.	7.657	3.413	5.736	7.212	3.629	5.579	8.641	7.582	0.000
S.E.M.	1.129	0.430	0.598	1.087	0.663	0.746	0.949	1.094	0.000
MAXIMUM	32.000	14.000	23.000	27.000	13.000	25.000	43.000	25.000	0.000
MINIMUM	0.000	0.000	0.000	0.000	0.000	0.000	0.000	0.000	0.000
CASES EXCL.	(0)	(0)	(0)	(0)	(0)	(0)	(0)	(0)	(0)
CASES INCL.	46	63	92	44	30	56	83	48	3

skewness by itself poses no problem. With the large sample size and use of two-tailed tests, normality of sampling distributions of means is anticipated.

8.7.1.3 ■ Linearity

There is no reason to expect curvilinearity considering the variables used and the fact that the variables, when skewed, are all skewed in the same direction. Had there been reason to suspect curvilinearity, plots of residuals would have been examined through the BMDP1V run for homogeneity of regression (see below).

8.7.1.4 ■ Outliers

The histograms as well as maximum values in the BMDP7D run of Table 8.15 show that, although no outliers are evident for the DV, several cases are univariate outliers for two of the covariates, PHYHEAL and PSYDRUG (note the histogram for the UNEMPLYD PROTEST group in Table 8.15). Positive skewness is also visible for these variables.

To facilitate the decision between transformation of variables and deletion of outliers, the groups are recoded into a one-way, eight level design and analyzed by BMDPAM. A portion of the output is shown in Table 8.16. Significance levels for skewness are given for each variable in each group. The two covariates in question have significant skewness in the group of unemployed, Protestant women. Skewness is evident for these variables in the other groups as well.

In addition, five cases are multivariate outliers in their respective groups as identified by extreme Mahalanobis distance from group means. The Mahalanobis distance is evaluated as χ^2 with degrees of freedom equal to the number of variables (in this case the three covariates plus the DV). BMDPAM prints significance values for each case, and places an asterisk for each case that exceeds $\alpha = .001$ for Mahalanobis distance. Notice the two outliers in Table 8.16. The first, with $\chi^2 = 25.968$ is from the seventh group (unemployed Protestant women) and the second, with $\chi^2 = 20.657$ departs significantly from the eighth group (unemployed Jewish women). As it turns out, all five outliers are from these two groups—three in group 7 and two in group 8—so that about 4% of the cases in each group are outliers.

This is a borderline case in terms of whether to transform variables or delete outliers—somewhere between "few" and "many." A log transform of the two skewed variables is undertaken to see if outliers remain after transformation. LPSYDRUG is created as the logarithm of PSYDRUG (incremented by 1 since many of the values are at zero) and LPHYHEAL as the logarithm of PHYHEAL.

A second run of BMDPAM with the four variables (the DV plus three covariates, two of them transformed) reveals no outliers at $\alpha = .001$ (Table 8.17). All five former outliers are within acceptable distance from their groups once the two covariates are transformed. The decision is to proceed with the analysis using the two transformed covariates rather than to delete cases, although the alternative decision is also acceptable in this situation.

8.7.1.5 ■ Multicollinearity and Singularity

BMDPAM provides squared multiple correlations for each variable as a DV with the remaining variables acting as IVs. This is helpful for detecting the presence of multicollinearity and singularity among the DV-covariate set, as seen in Table 8.18 for the transformed variables. Because the largest $R^2 = .30$, there is no danger of multicollinearity or singularity.

TABLE 8.16 TEST FOR MULTIVARIATE OUTLIERS. BMDPAM SETUP AND SELECTED OUTPUT

```
/INPUT      VARIABLES ARE 7. FORMAT IS FREE.
            FILE = 'ANCOVA.DAT'.
/VARIABLE   NAMES ARE SUBJNO, ATTDRUG, PHYHEAL, MENHEAL,
                  PSYDRUG, EMPLMNT, RELIGION.
            MISSING = (7)9.
            LABEL = SUBJNO.
            USE = 2 TO 5, CELL.
/GROUP      VAR = CELL.
            CODES(CELL) ARE 1 TO 8.
/TRANSFORM  IF (EMPLMNT EQ 1 AND RELIGION EQ 1) THEN CELL = 1.
            IF (EMPLMNT EQ 1 AND RELIGION EQ 2) THEN CELL = 2.
            IF (EMPLMNT EQ 1 AND RELIGION EQ 3) THEN CELL = 3.
            IF (EMPLMNT EQ 1 AND RELIGION EQ 4) THEN CELL = 4.
            IF (EMPLMNT EQ 2 AND RELIGION EQ 1) THEN CELL = 5.
            IF (EMPLMNT EQ 2 AND RELIGION EQ 2) THEN CELL = 6.
            IF (EMPLMNT EQ 2 AND RELIGION EQ 3) THEN CELL = 7.
            IF (EMPLMNT EQ 2 AND RELIGION EQ 4) THEN CELL = 8.
/EST        METHOD = REGR.
/PRINT      MATR = DIS.
/END
```

UNIVARIATE SUMMARY STATISTICS

GROUP IS *2   SIZE IS   63

VARIABLE	SAMPLE SIZE	MEAN	STANDARD DEVIATION	COEFFICIENT OF VARIATION	SMALLEST VALUE	LARGEST VALUE	SMALLEST STANDARD SCORE	LARGEST STANDARD SCORE	SKEWNESS	KURTOSIS
2 ATTDRUG	63	7.66667	0.98374	0.128314	6.00000	10.00000	-1.69	2.37	0.29	-0.54
3 PHYHEAL	63	4.61905	2.47848	0.536578	2.00000	14.00000	-1.06	3.78	1.36	2.18
4 MENHEAL	63	5.84127	4.75261	0.813626	0.00000	18.00000	-1.23	2.56	0.79	20.40
5 PSYDRUG	63	2.34921	3.41325	1.452938	0.00000	14.00000	-0.69	3.41	1.39	1.11

ESTIMATES OF MISSING DATA, MAHALANOBIS D-SQUARED (CHI-SQUARED)
AND SQUARED MULTIPLE CORRELATIONS WITH AVAILABLE VARIABLES

CASE LABEL	CASE NUMBER	MISSING VARIABLE	ESTIMATE	R-SQUARED	GROUP	CHI-SQ	CHISQ/D.F.	D.F.	SIGNIFICANCE
127	107				*2	8.943	2.236	4	0.0625
128	108				*7	25.968	6.492	4	0.0000*
129	109				*3	1.889	0.472	4	0.7562
130	110				*6	5.043	1.261	4	0.2829

TABLE 8.16 (CONTINUED)

CASE LABEL	CASE NUMBER	MISSING VARIABLE	ESTIMATE	R-SQUARED	GROUP	CHI-SQ	CHISQ/D.F.	D.F.	SIGNIFICANCE
131	111				*6	3.051	0.763	4	0.5494
132	112				*1	1.553	0.388	4	0.8172
133	113				*2	12.571	3.143	4	0.0136
134	114				*8	3.871	0.968	4	0.4238
135	115				*5	2.090	0.522	4	0.7193
136	116				*3	0.212	0.053	4	0.9948
137	117				*3	12.940	3.235	4	0.0116
138	118				*6	1.841	0.460	4	0.7650
139	119				*8	4.073	1.018	4	0.3962
140	120				*7	6.660	1.665	4	0.1550
141	121				*3	1.782	0.445	4	0.7759
142	122				*2	2.517	0.629	4	0.6417
143	123				*2	1.639	0.410	4	0.8017
144	124				*4	1.119	0.280	4	0.8913
145	125				*4	11.153	2.788	4	0.0249
146	126				*8	20.657	5.164	4	0.0004*
148	127				*6	5.246	1.311	4	0.2630

362

TABLE 8.17 CHECK FOR MULTIVARIATE OUTLIERS WITH TRANSFORMED VARIABLES. SETUP AND PARTIAL OUTPUT FROM BMDPAM.

```
/INPUT      VARIABLES ARE 7. FORMAT IS FREE.
            FILE = 'ANCOVA.DAT'.
/VARIABLE   NAMES ARE SUBJNO, ATTDRUG, PHYHEAL, MENHEAL,
              PSYDRUG, EMPLMNT, RELIGION.
            MISSING = (7)9.
            LABEL = SUBJNO.
/GROUP      VAR = CELL.
            USE = 2, 4, CELL, LPHYHEAL, LPSYDRUG.
            CODES(CELL) ARE 1 TO 8.
/TRANSFORM  IF (EMPLMNT EQ 1 AND RELIGION EQ 1) THEN CELL = 1.
            IF (EMPLMNT EQ 1 AND RELIGION EQ 2) THEN CELL = 2.
            IF (EMPLMNT EQ 1 AND RELIGION EQ 3) THEN CELL = 3.
            IF (EMPLMNT EQ 1 AND RELIGION EQ 4) THEN CELL = 4.
            IF (EMPLMNT EQ 2 AND RELIGION EQ 1) THEN CELL = 5.
            IF (EMPLMNT EQ 2 AND RELIGION EQ 2) THEN CELL = 6.
            IF (EMPLMNT EQ 2 AND RELIGION EQ 3) THEN CELL = 7.
            IF (EMPLMNT EQ 2 AND RELIGION EQ 4) THEN CELL = 8.
            LPHYHEAL=LOG(PHYHEAL).
            LPSYDRUG=LOG(PSYDRUG11).
/EST        METHOD = REGR.
/PRINT      MATR = CORR, DIS.
/END
```

CASE LABEL	CASE NUMBER	MISSING VARIABLE	ESTIMATE	R-SQUARED	GROUP	CHI-SQ	CHISQ/D.F.	D.F.	SIGNIFICANCE
127	107				*2	8.169	2.042	4	0.0856
128	108				*7	11.375	2.844	4	0.0227
129	109				*3	2.408	0.602	4	0.6612
130	110				*6	5.401	1.350	4	0.2486
131	111				*6	3.199	0.800	4	0.5251
132	112				*1	2.196	0.549	4	0.6997
133	113				*2	13.731	3.433	4	0.0082
134	114				*8	4.181	1.045	4	0.3821
135	115				*5	2.510	0.628	4	0.6428
136	116				*3	0.743	0.186	4	0.9460
137	117				*3	9.712	2.428	4	0.0456
138	118				*6	2.711	0.678	4	0.6072
139	119				*8	2.851	0.713	4	0.5831
140	120				*7	4.771	1.193	4	0.3116
141	121				*3	2.696	0.674	4	0.6100
142	122				*2	3.647	0.912	4	0.4558
143	123				*2	2.429	0.607	4	0.6575
144	124				*4	1.649	0.412	4	0.7999
145	125				*4	11.776	2.944	4	0.0191
146	126				*8	13.582	3.395	4	0.0088
148	127				*6	4.340	1.085	4	0.3619

TABLE 8.18   CHECK FOR MULTICOLLINEARITY THROUGH BMDPAM. SETUP APPEARS IN TABLE 8.17.

SQUARED MULTIPLE CORRELATIONS OF EACH VARIABLE WITH ALL OTHER VARIABLES

(MEASURES OF MULTICOLLINEARITY OF VARIABLES)
AND TESTS OF SIGNIFICANCE OF MULTIPLE REGRESSION
DEGREES OF FREEDOM FOR F-STATISTICS ARE   3 AND   451

VARIABLE		R-SQUARED	F-STATISTIC	SIGNIFICANCE
NO.	NAME			(P LESS THAN)
2	ATTDRUG	0.093267	15.46	0.00000
4	MENHEAL	0.286702	60.42	0.00000
9	LPHYHEAL	0.303041	65.37	0.00000
10	LPSYDRUG	0.229175	44.70	0.00000

8.7.1.6 ■ Homogeneity of Variance

Sample variances are available from the full BMDPAM run in the section on UNIVARIATE SUM-MARY STATISTICS (cf. Table 8.17). For the DV, find the largest and smallest standard deviations over the groups and square them to get variances. For example, the variance for ATTDRUG in the EMPLYD CATHOLIC group is $(.98374)^2 = .967$ (the smallest variance). The largest variance is for group 1 (EMPLYD NONOTHER), with $s^2 = 1.825$ (not shown). The variance ratio $(F_{max}) = 1.89$, well below the criterion of 10:1. There is no need for a formal test of homogeneity of variance with this variance ratio since the ratio of sample sizes is less than 4:1 (92/30 = 3.07, as seen in Table 8.15), and there are no outliers.

8.7.1.7 ■ Homogeneity of Regression

Because this is a between-subjects design, the test of homogeneity of regression is readily available through BMDP1V when the two-way factorial is recoded into a one-way design with 8 levels (Table 8.19). The obtained $F(21, 430) = 1.0367$ for equality of slopes indicates that regression coefficients for the eight groups are homogeneous at the .05 level of significance. In addition, none of the specific tests for equality of slopes shows significant deviation from equality.

8.7.1.8 ■ Reliability of Covariates

The three covariates, MENHEAL, PHYHEAL, and PSYDRUG, were measured as counts of symptoms or drug use—"have you ever . . . ?" It is assumed that people are reasonably consistent in reporting the presence or absence of symptoms and that high reliability is likely. Therefore no adjustment in ANCOVA is made for unreliability of covariates.

8.7.2 ■ Analysis of Covariance

The program chosen for the major two-way analysis of covariance is BMDP4V. The cell size weights (Table 8.11, number 4, Method 3) approach to adjustment of unequal n is chosen for this set of survey data. Ease of use, then, makes BMDP4V a convenient program for this unequal-n data set.

Selected output from application of BMDP4V to these data appears in Table 8.20 that shows unadjusted marginal and cell means and sample sizes, and the source table. Sources of variance are the combined covariates; the main effects for RELIGION and EMPLMNT, unadjusted (R and E, respectively) and adjusted for each other (R|E and E|R, respectively); the interaction adjusted for both main effects, RE; and the error term, ERROR.

TABLE 8.19   SETUP AND SELECTED OUTPUT FOR BMDP1V TEST OF HOMOGENEITY OF REGRESSION

```
/INPUT      VARIABLES ARE 7.  FORMAT IS FREE.
            FILE = 'ANCOVA.DAT'.
/VARIABLE   NAMES ARE SUBJNO, ATTDRUG, PHYHEAL, MENHEAL,
                PSYDRUG, EMPLMNT, RELIGION.
            MISSING = (7)9.
            LABEL = SUBJNO.
/GROUP      VAR = CELL.
            CODES(CELL) ARE 1 TO 8.
/TRANSFORM      IF (EMPLMNT EQ 1 AND RELIGION EQ 1) THEN CELL = 1.
                IF (EMPLMNT EQ 1 AND RELIGION EQ 2) THEN CELL = 2.
                IF (EMPLMNT EQ 1 AND RELIGION EQ 3) THEN CELL = 3.
                IF (EMPLMNT EQ 1 AND RELIGION EQ 4) THEN CELL = 4.
                IF (EMPLMNT EQ 2 AND RELIGION EQ 1) THEN CELL = 5.
                IF (EMPLMNT EQ 2 AND RELIGION EQ 2) THEN CELL = 6.
                IF (EMPLMNT EQ 2 AND RELIGION EQ 3) THEN CELL = 7.
                IF (EMPLMNT EQ 2 AND RELIGION EQ 4) THEN CELL = 8.
            LPHYHEAL = LOG(PHYHEAL).
            LPSYDRUG = LOG(PSYDRUG + 1).
/DESIGN     DEPENDENT IS ATTDRUG.
            INDEPENDENT ARE LPHYHEAL, MENHEAL, LPSYDRUG.
/END
```

ANALYSIS OF COVARIANCE TABLE
=============================

SOURCE OF VARIATION	D.F.	SUM OF SQ.	MEAN SQ.	F-VALUE	PROB(TAIL)
EQUALITY OF ADJ. MEANS	7	22.9146	3.2735	2.7382	0.009

ZERO SLOPES

ALL COVARIATES	3	55.4604	18.4868	15.4634	0.000
VAR. 4 MENHEAL	1	1.4289	1.4289	1.1952	0.275
VAR. 9 LPHYHEAL	1	0.6300	0.6300	0.5270	0.468
VAR. 10 LPSYDRUG	1	46.7374	46.7374	39.0939	0.000
ERROR	451	539.1788	1.1955		

EQUALITY OF SLOPES

ALL COVARIATES, ALL GROUPS	21	25.9840	1.2373	1.0367	0.417
ERROR	430	513.19	1.1935		

FOR EACH VARIABLE SEPARATELY

VAR. 4 MENHEAL , ALL GROUPS	7	7.4122	1.0589	0.8841	0.519
ERROR	444	531.7665	1.1977		
VAR. 9 LPHYHEAL, ALL GROUPS	7	6.5384	0.9341	0.7786	0.606
ERROR	444	532.6404	1.1996		
VAR. 10 LPSYDRUG, ALL GROUPS	7	10.6053	1.5150	1.2726	0.262
ERROR	444	528.5734	1.1905		

TABLE 8.19   *(CONTINUED)*

BETWEEN EACH GROUP AND ALL OTHERS

GROUP 1 *1 ,					
ALL COVARIATES	3	0.1874	0.0625	0.0519	0.984
ERROR	448	538.9914	1.2031		
GROUP 2 *2 ,					
ALL COVARIATES	3	8.2559	2.7520	2.3222	0.074
ERROR	448	530.9229	1.1851		
GROUP 3 *3 ,					
ALL COVARIATES	3	5.2828	1.7609	1.4776	0.220
ERROR	448	533.8959	1.1917		
GROUP 4 *4 ,					
ALL COVARIATES	3	0.6694	0.2231	0.1856	0.906
ERROR	448	538.5094	1.2020		
GROUP 5 *5 ,					
ALL COVARIATES	3	3.6113	1.2038	1.0069	0.389
ERROR	448	535.5675	1.1955		
GROUP 6 *6 ,					
ALL COVARIATES	3	1.9523	0.6508	0.5427	0.653
ERROR	448	537.2264	1.1992		
GROUP 7 *7 ,					
ALL COVARIATES	3	3.1080	1.0360	0.8658	0.459
ERROR	448	536.0708	1.1966		
GROUP 8 *8 ,					
ALL COVARIATES	3	6.2889	2.0963	1.7624	0.154
ERROR	448	532.8899	1.1895		

BMDP4V provides the researcher with results of all options for adjustment and the researcher chooses among them. In this example, the researcher can choose the main effect of RELIGION adjusted for the effect of EMPLMNT (R|E) or the main effect of EMPLMNT adjusted for RELIGION (E|R). In this case the decision is made to give RELIGION priority over EMPLMNT because for most people religious affiliation precedes employment status. The sources used, then, are RELIGION (R) adjusted for covariates, EMPLMNT adjusted for RELIGION (E|R) and covariates, and the interaction RE adjusted for all lower-order effects and covariates.

The significance level is provided for each computed F, so one need not consult a table of critical F; any p less than .05 is statistically significant with $\alpha = .05$. In this example, only the main effect of RELIGION reaches statistical significance.

Entries in the sum of squares column of the source table are used to calculate η^2 as a measure of strength of association for any significant main effect and interaction (Sections 8.4 and 8.5.2.4). Because there is only one significant effect, the more standard equation (Equation 8.7) is preferable.

$$\eta^2 = \frac{10.178}{(10.178 + 3.787 + 8.871) + 539.179} = .02$$

Significance tests of individual covariates assess the utility of covariates, as described in Section 8.5.3. Tests of individual covariates are unavailable in BMDP4V but are in the BMDP1V run of Table 8.19 and reproduced in Table 8.21. In this run, covariates are adjusted for each other and for main effects and interactions. The table of pooled within-cell correlations between the DV

```
/INPUT      VARIABLES ARE 7.  FORMAT IS FREE.
            FILE = 'ANCOVA.DAT'.
/VARIABLE   NAMES ARE SUBJNO, ATTDRUG, PHYHEAL, MENHEAL,
                 PSYDRUG, EMPLMNT, RELIGION.
            MISSING = (7)9.
            LABEL = SUBJNO.
            USE = ATTDRUG, MENHEAL, EMPLMNT, RELIGION, LPHYHEAL,
                LPSYDRUG.
/TRANSFORM      LPHYHEAL = LOG(PHYHEAL).
                LPSYDRUG = LOG(PSYDRUG + 1).
/BETWEEN    FACTORS ARE RELIGION, EMPLMNT.
            CODES(EMPLMNT) ARE 1, 2
            NAMES(EMPLMNT) ARE EMPLYD, UNEMPLYD.
            CODES(RELIGION) ARE 1 TO 4.
            NAMES(RELIGION) ARE NONOTHER,CATHOLIC,PROTEST,JEWISH.
/WEIGHTS    BETWEEN ARE SIZES.
/END
ANALYSIS PROCEDURE IS FACTORIAL.
            COVARIATES ARE LPHYHEAL, MENHEAL, LPSYDRUG./
END/
```

SUMMARY STATISTICS FOR VARIATE(S):

VARIATE	COUNT	MEAN	STDERROR	STD_DEV	WTD_MEAN	MAXIMUM	MINIMUM
ATTDRUG	462	7.684	0.5386E-01	1.158	7.684	10.00	5.000
MENHEAL	462	6.136	0.1954	4.199	6.136	18.00	0.0000
LPHYHEAL	462	0.6496	0.9587E-02	0.2061	0.6496	1.176	0.3010
LPSYDRUG	462	0.4159	0.2331E-01	0.5009	0.4159	1.643	0.0000

LEVEL   1 MARGINALS

FACTOR	LEVEL	VARIATE	COUNT	MEAN	STDERROR	STD_DEV	WTD_MEAN	MAXIMUM	MINIMUM
RELIGION	NONOTHER	ATTDRUG	76	7.4474	0.1503	1.3104	7.4474	10.0000	5.0000
		MENHEAL	76	5.9342	0.4582	3.9944	5.9342	17.0000	0.0000
		LPHYHEAL	76	0.6564	0.0192	0.1675	0.6564	1.0000	0.3010
		LPSYDRUG	76	0.4011	0.0571	0.4975	0.4011	1.5185	0.0000
	CATHOLIC	ATTDRUG	119	7.8403	0.0962	1.0495	7.8403	10.0000	6.0000
		MENHEAL	119	6.1345	0.4270	4.6576	6.1345	18.0000	0.0000
		LPHYHEAL	119	0.6303	0.0190	0.2067	0.6303	1.1461	0.3010
		LPSYDRUG	119	0.3271	0.0403	0.4395	0.3271	1.4150	0.0000
	PROTEST	ATTDRUG	175	7.6686	0.0867	1.1467	7.6686	10.0000	5.0000
		MENHEAL	175	6.0857	0.3100	4.1005	6.0857	16.0000	0.0000
		LPHYHEAL	175	0.6618	0.0167	0.2209	0.6618	1.1139	0.3010
		LPSYDRUG	175	0.4348	0.0390	0.5164	0.4348	1.6435	0.0000
	JEWISH	ATTDRUG	92	7.7065	0.1212	1.1630	7.7065	10.0000	5.0000
		MENHEAL	92	6.4022	0.4142	3.9726	6.4022	18.0000	0.0000
		LPHYHEAL	92	0.6458	0.0215	0.2060	0.6458	1.1761	0.3010
		LPSYDRUG	92	0.5070	0.0559	0.5359	0.5070	1.4472	0.0000
EMPLMNT	EMPLYD	ATTDRUG	245	7.5959	0.0712	1.1144	7.5959	10.0000	5.0000
		MENHEAL	245	6.1020	0.2622	4.1044	6.1020	18.0000	0.0000
		LPHYHEAL	245	0.6396	0.0130	0.2039	0.6396	1.1461	0.3010
		LPSYDRUG	245	0.4206	0.0314	0.4913	0.4206	1.5185	0.0000
	UNEMPLYD	ATTDRUG	217	7.7834	0.0814	1.1996	7.7834	10.0000	5.0000
		MENHEAL	217	6.1751	0.2928	4.3126	6.1751	18.0000	0.0000
		LPHYHEAL	217	0.6608	0.0141	0.2083	0.6608	1.1761	0.3010
		LPSYDRUG	217	0.4105	0.0348	0.5127	0.4105	1.6435	0.0000

TABLE 8.20   *(CONTINUED)*

CELL STATISTICS

```
==========================================================
FACTOR      LEVEL
RELIGION    NONOTHER
==>
```

FACTOR	LEVEL	VARIATE	COUNT	MEAN	STDERROR	STD_DEV	WTD_MEAN	MAXIMUM	MINIMUM
EMPLMNT	EMPLYD	ATTDRUG	46	7.6739	0.1992	1.3508	7.6739	10.0000	5.0000
		MENHEAL	46	6.5435	0.5977	4.0536	6.5435	17.0000	0.0000
		LPHYHEAL	46	0.6733	0.0251	0.1702	0.6733	0.9542	0.3010
		LPSYDRUG	46	0.4882	0.0789	0.5355	0.4882	1.5185	0.0000
	UNEMPLYD	ATTDRUG	30	7.1000	0.2163	1.1847	7.1000	10.0000	5.0000
		MENHEAL	30	5.0000	0.6898	3.7783	5.0000	12.0000	0.0000
		LPHYHEAL	30	0.6304	0.0297	0.1626	0.6304	1.0000	0.3010
		LPSYDRUG	30	0.2676	0.0741	0.4060	0.2676	1.1461	0.0000

```
==========================================================
FACTOR      LEVEL
RELIGION    CATHOLIC
==>
```

FACTOR	LEVEL	VARIATE	COUNT	MEAN	STDERROR	STD_DEV	WTD_MEAN	MAXIMUM	MINIMUM
EMPLMNT	EMPLYD	ATTDRUG	63	7.6667	0.1239	0.9837	7.6667	10.0000	6.0000
		MENHEAL	63	5.8413	0.5988	4.7526	5.8413	18.0000	0.0000
		LPHYHEAL	63	0.6095	0.0275	0.2186	0.6095	1.1461	0.3010
		LPSYDRUG	63	0.3275	0.0509	0.4039	0.3275	1.1761	0.0000
	UNEMPLYD	ATTDRUG	56	8.0357	0.1463	1.0949	8.0357	10.0000	6.0000
		MENHEAL	56	6.4643	0.6105	4.5685	6.4643	18.0000	0.0000
		LPHYHEAL	56	0.6537	0.0256	0.1918	0.6537	1.0414	0.3010
		LPSYDRUG	56	0.3267	0.0642	0.4802	0.3267	1.4150	0.0000

```
==========================================================
FACTOR      LEVEL
RELIGION    PROTEST
==>
```

FACTOR	LEVEL	VARIATE	COUNT	MEAN	STDERROR	STD_DEV	WTD_MEAN	MAXIMUM	MINIMUM
EMPLMNT	EMPLYD	ATTDRUG	92	7.5109	0.1130	1.0843	7.5109	10.0000	5.0000
		MENHEAL	92	5.8804	0.3879	3.7206	5.8804	16.0000	0.0000
		LPHYHEAL	92	0.6521	0.0223	0.2139	0.6521	1.0792	0.3010
		LPSYDRUG	92	0.4333	0.0522	0.5010	0.4333	1.3802	0.0000
	UNEMPLYD	ATTDRUG	83	7.8434	0.1311	1.1943	7.8434	10.0000	5.0000
		MENHEAL	83	6.3133	0.4935	4.4964	6.3133	16.0000	0.0000
		LPHYHEAL	83	0.6724	0.0252	0.2292	0.6724	1.1139	0.3010
		LPSYDRUG	83	0.4364	0.0588	0.5361	0.4364	1.6435	0.0000

```
==========================================================
FACTOR      LEVEL
RELIGION    JEWISH
==>
```

FACTOR	LEVEL	VARIATE	COUNT	MEAN	STDERROR	STD_DEV	WTD_MEAN	MAXIMUM	MINIMUM
EMPLMNT	EMPLYD	ATTDRUG	44	7.5909	0.1668	1.1064	7.5909	9.0000	5.0000
		MENHEAL	44	6.4773	0.6026	3.9970	6.4773	18.0000	0.0000
		LPHYHEAL	44	0.6214	0.0289	0.1917	0.6214	0.9542	0.3010
		LPSYDRUG	44	0.4568	0.0803	0.5329	0.4568	1.4472	0.0000
	UNEMPLYD	ATTDRUG	48	7.8125	0.1753	1.2144	7.8125	10.0000	6.0000
		MENHEAL	48	6.3333	0.5761	3.9911	6.3333	15.0000	0.0000
		LPHYHEAL	48	0.6680	0.0315	0.2179	0.6680	1.1761	0.3010
		LPSYDRUG	48	0.5530	0.0780	0.5402	0.5530	1.4150	0.0000

TABLE 8.20 *(CONTINUED)*

--- ANALYSIS SUMMARY ---

THE FOLLOWING EFFECTS ARE COMPONENTS OF THE SPECIFIED
LINEAR MODEL FOR THE BETWEEN DESIGN. ESTIMATES AND TESTS
OF HYPOTHESES FOR THESE EFFECTS CONCERN PARAMETERS OF THAT MODEL.

```
              COVARIATES
              OVALL: GRAND MEAN
              RE
              R|E
              E|R
```

THE FOLLOWING EFFECTS INVOLVE WEIGHTED COMBINATIONS OF
REGRESSION INTERCEPTS (ADJUSTED MEANS FOR CONTRAST EFFECTS).
THEY ARE NOT COMPONENTS OF THE LINEAR MODEL FOR THE BETWEEN DESIGN,
SO ESTIMATES AND TESTS MAY NOT BE MEANINGFUL.

```
              R: RELIGION
              E: EMPLMNT
```

EFFECT	VARIATE	STATISTIC		F	DF		P
COVARIATES							
	ATTDRUG						
		SS=	55.460423				
		MS=	18.486808	15.46	3,	451	0.0000
OVALL: GRAND MEAN							
	ATTDRUG						
		SS=	2225.713856				
		MS=	2225.713856	1861.71	1,	451	0.0000
R: RELIGION							
	ATTDRUG						
		SS=	10.178336				
		MS=	3.392779	2.84	3,	451	0.0377
E: EMPLMNT							
	ATTDRUG						
		SS=	4.189173				
		MS=	4.189173	3.50	1,	451	0.0619
RE							
	ATTDRUG						
		SS=	8.870919				
		MS=	2.956973	2.47	3,	451	0.0611
R|E							
	ATTDRUG						
		SS=	9.774202				
		MS=	3.258067	2.73	3,	451	0.0438
E|R							
	ATTDRUG						
		SS=	3.787008				
		MS=	3.787008	3.17	1,	451	0.0758
ERROR							
	ATTDRUG						
		SS=	539.17869323				
		MS=	1.19551817				

ATTDRUG and the covariates, as provided in BMDPAM in Table 8.22 (and reproduced in Table 8.23), is also needed for a full assessment of the covariates. Pooled within-cell correlations adjust the correlations for main effects and interactions, but not for the DV and other covariates.

The correlations provide a measure of the total relationship between DV and covariates (independent of factors in the design) and the significance tests provide a measure of the unique contribution of each covariate, adjusted for the others, to adjustment of the DV. As seen in Table 8.23, both LPHYHEAL and LPSYDRUG are reliably related to the DV, but, as seen in Table 8.21, only LPSYDRUG provides reliable unique adjustment of the DV after all other effects are considered. According to the criteria of Section 8.5.4, then, use of MENHEAL as a covariate in future research is not warranted (it has, in fact, lowered the power of this analysis) and use of LPHYHEAL is questionable.

Adjusted marginal and cell means are likewise not available through BMDP4V, but again are provided by BMDP1V. Output showing adjusted cell means from the setup in Table 8.19 appears in Table 8.24. The adjusted cell means are appropriate to report for a significant interaction.

Because only the main effect of RELIGION is significant here, adjusted marginal means for the four religious groups are needed. Adjusted marginal means are found by weighting adjusted cell means by sample sizes, summing across appropriate cells, and then dividing by total sample size on the margin. Groups 1 and 5 comprise the none-or-other RELIGION group, as seen from the coding in the setup portion of Table 8.19. Multiply each ADJ.GRP.MEAN by its cell size to find the adjusted cell sum:

TABLE 8.21 ANALYSIS OF COVARIANCE OF ATTITUDE TOWARD DRUGS

Source of variance	Adjusted SS	df	MS	F
Religion	10.178	3	3.393	2.84*
Employment status (adjusted for Religion)	3.787	1	3.787	3.17
Interaction	8.871	3	2.957	2.47
Covariates	55.460	3	18.487	15.46
Physical health (LOG)	0.630	1	0.630	0.53
Mental health	1.429	1	1.429	1.20
Drug uses (LOG)	46.737	1	46.737	39.09**
Error	539.179	451	1.196	

$*p < .05$

$**p < .01$

TABLE 8.22 OUTPUT FROM BMDPAM SHOWING POOLED WITHIN-CELL CORRELATIONS AMONG COVARIATES AND THE DEPENDENT VARIABLE. SETUP APPEARS IN TABLE 8.17

```
CORRELATIONS
------------
(COMPUTED FROM POOLED WITHIN-GROUPS COVARIANCE MATRIX)


          ATTDRUG  MENHEAL  LPHYHEAL LPSYDRUG

              2       4        9        10
ATTDRUG    2  1.000
MENHEAL    4  0.064    1.000
LPHYHEAL   9  0.121    0.510    1.000
LPSYDRUG  10  0.301    0.333    0.365    1.000
```

$$\text{Group 1: } (46)(7.62434) = 350.72$$

$$\text{Group 5: } (30)(7.19055) = 215.72$$

Add the adjusted cell sums and divide by the sum of sample sizes to get the adjusted marginal mean:

$$\text{Weighted mean} = \frac{350.72 + 215.72}{46 + 30} = 7.453$$

Application of this procedure to the other three religious groups results in the adjusted marginal means shown in Table 8.25. The table includes the unadjusted marginal means from Table 8.20 for comparison.

Because no a priori hypotheses about differences among religious groups were generated, planned comparisons are not appropriate. A glance at the four adjusted means in Table 8.25 suggests a straightforward interpretation; the None-or-other group has the least favorable attitude toward use of psychotropic drugs, the Catholic group the most favorable attitude, and the Protestant and Jewish groups intermediate attitude. In the absence of specific questions about differences

TABLE 8.23 POOLED WITHIN-CELL INTERCORRELATIONS AMONG THREE COVARIATES AND THE DEPENDENT VARIABLE, ATTITUDE TOWARDS DRUGS

	Physical health (LOG)	Mental health	Drug uses (LOG)
Attitude toward drugs	.121*	.064	.301*
Physical health (LOG)		.510*	.365*
Mental health			.333*

*$p < .01$

TABLE 8.24 OUTPUT FROM BMDP1V SHOWING CELL SIZES AND ADJUSTED CELL MEANS (SETUP APPEARS IN TABLE 8.19)

GROUP	N	GP.MEAN	ADJ.GP.MEAN	STD.ERR.
*1	46.	7.67391	7.62434	0.16143
*2	63.	7.66667	7.73281	0.13827
*3	92.	7.51087	7.49404	0.11409
*4	44.	7.59091	7.57361	0.16533
*5	30.	7.10000	7.19055	0.20050
*6	56.	8.03571	8.10272	0.14660
*7	83.	7.84337	7.82679	0.12015
*8	48.	7.81250	7.71522	0.15847

TABLE 8.25 ADJUSTED AND UNADJUSTED MEAN ATTITUDE TOWARD DRUGS FOR FOUR CATEGORIES OF RELIGION

Religion	Adjusted mean	Unadjusted mean
None-or-other	7.45	7.45
Catholic	7.91	7.84
Protestant	7.65	7.67
Jewish	7.65	7.71

between means, there is no compelling reason to evaluate post hoc the significance of these differences, although they certainly provide a rich source of speculation for future research.

A checklist for analysis of covariance appears as Table 8.26. An example of a Results section, in
journal format, follows for the analysis described above.

TABLE 8.26 CHECKLIST FOR ANALYSIS OF COVARIANCE

1. Issues
   a. Unequal sample size and missing data
   b. Within-cell outliers
   c. Normality
   d. Homogeneity of variance
   e. Within-cell linearity
   f. Homogeneity of regression
   g. Reliability of covariates
2. Major analyses
   a. Main effect(s) or planned comparison. If significant:
      (1) Adjusted marginal means
      (2) Strength of association
   b. Interactions or planned comparisons. If significant:
      (1) Adjusted cell means (in table or interaction graph)
      (2) Strength of association
3. Additional analyses
   a. Evaluation of covariate effects
   b. Evaluation of intercorrelations
   c. Post hoc comparisons (if appropriate)
   d. Unadjusted marginal and/or cell means (if significant main effect and/or interaction) if nonexperimental
      application

```
                        Results

    A 2 × 4 between-subjects analysis of covariance was performed on
attitude toward drugs. Independent variables consisted of current
employment status (employed and unemployed) and religious identifica-
tion (None-or-other, Catholic, Protestant, and Jewish), factorially
combined. Covariates were physical health, mental health, and the sum
of psychotropic drug uses. Analyses were performed by BMDP4V, weighting
cells by their sample sizes to adjust for unequal n.

    Results of evaluation of the assumptions of normality of sampling
distributions, linearity, homogeneity of variance, homogeneity of
regression, and reliability of covariates were satisfactory. Presence
of outliers led to transformation of two of the covariates. Logarithmic
transforms were made of physical health and the sum of psychotropic
drug uses. No outliers remained after transformation. The original sam-
```

ple of 465 was reduced to 462 by three women who did not provide information as to religious affiliation.

After adjustment by covariates, attitude toward drugs varied significantly with religious affiliation, as summarized in Table 8.21, with $F(3, 451) = 2.84$, $p < .05$. The strength of the relationship between adjusted attitudes toward drugs and religion was weak, however, with $\eta^2 = .02$. The adjusted marginal means, as displayed in Table 8.25, show that the most favorable attitudes toward drugs were held by Catholic women, and least favorable attitudes by women who either were unaffiliated with a religion or identified with some religion other than the major three. Attitudes among Protestant and Jewish women were almost identical, on average, for this sample, and fell between those of the two other groups.

No statistically significant main effect of current employment status was found. Nor was there a significant interaction between employment status and religion after adjustment for covariates.

Pooled within-group correlations among covariates and attitude toward drugs are shown in Table 8.23. Two of the covariates, logarithm of physical health and logarithm of drug use, were significantly associated with the dependent variable. However, only logarithm of drug use uniquely adjusted the attitude scores, $F(1, 451) = 39.09$, $p < .01$, after covariates were adjusted for other covariates, main effects, and interaction. The remaining two covariates, mental health and logarithm of physical health, provided no reliable unique adjustment.

8.8 ■ SOME EXAMPLES FROM THE LITERATURE

A straightforward, classic experimental application of ANCOVA is demonstrated in a study of Hall and colleagues (1977). Obese members of a weight reduction club were randomly assigned to five experimental treatments: self-management, external management, self-management plus external management, psychotherapy, and no-treatment control. Using pretreatment body weight as a covariate, a one-way ANCOVA was done on posttreatment weight. After a 10-week treatment period, the groups differed significantly in weight after adjustment for the covariate. A Tukey HSD test showed that the three behavioral conditions did not differ significantly from one another but did differ from the two remaining conditions. The remaining conditions, psychotherapy and no-treatment control, did not differ from each other. ANCOVAs performed on weight at 3- and 6-month follow-ups showed that differences among groups disappeared.

Merrill and Towle (1976) investigated the effects of the availability of objectives in a graduate course in programmed instruction. Analyses of covariance were used on both student performance and anxiety level (as DVs) to evaluate the effect of presence versus absence of course objectives. The course consisted of 12 cognitive units. For the first half of the course, consisting of six units, half of the 32 students received objectives and half did not.

Pretest scores of knowledge of the material served as the covariate for posttest performance (scores on tests after the first six units and the final); no significant effect of course objectives was found. For anxiety as DV, the two covariates were A-Trait and A-State scales of the State-Trait Anxiety Inventory administered to graduate students during the first class session. The DV was the average of the short-form A-State scale administered after each unit test for the first six units. Availability of objectives significantly decreased A-State scores after adjustment for covariates. Six separate analyses of covariance were then done on the A-State scores, one for each of the first six units.[12] These analyses showed that anxiety was significantly reduced only for the first three units when objectives were given.

O'Kane and coworkers (1977) studied the relationship between anticipated social mobility and social political ideology. A total of 307 male Catholic adolescents were divided into four groups. They were first separated into either middle or working class on the basis of the Duncan Index score of the heads of households. These two groups were then further subdivided on the basis of predicted future class, either same as class of origin or different—upward or downward depending on class of origin. The four groups, then, were middle-middle, working-middle, working-working, and middle-working. IQ served as the covariate for analyses of covariance on three measures of ideology: economic liberalism, noneconomic liberalism, and ethnocentrism.[13]

For economic liberalism, no significant effects were found for class of origin, class of destination, or the interaction between origin and destination classes. For both noneconomic liberalism and ethnocentrism, significant main effects were found for class of destination. Those who expected to join (or remain in) the working class showed more conservative scores on the scale of noneconomic liberalism, and scored higher in ethnocentrism, than those who expected to join (or remain in) the middle class. For these two measures, no significant effects of class of origin were found, nor was there a significant interaction between class of origin and class of destination. Additional *t* tests, presumably on the basis of planned comparisons, showed that the two stable groups differed from each other, in the expected direction, on measures of noneconomic liberalism and ethnocentrism, but not on economic liberalism.

---

[12] This is a questionable practice because of inflated Type I error rates with multiple testing, especially when the DVs are correlated. Stepdown analysis as described in Chapter 9, or profile analysis as described in Chapter 10 is more appropriate. Or, if assumptions are met, "unit" could act as a six-level within-subjects IV in ANCOVA with both between- and within-subjects IVs.

[13] Because these three measures are probably correlated, multivariate analysis of covariance followed by stepdown analysis as described in Chapter 9 is a better analytic strategy.

CHAPTER 9
Multivariate Analysis of Variance and Covariance

9.1 ■ GENERAL PURPOSE AND DESCRIPTION

Multivariate analysis of variance (MANOVA) is a generalization of ANOVA to a situation in which there are several DVs. For example, suppose a researcher is interested in the effect of different types of treatment on several types of anxiety: test anxiety, anxiety in reaction to minor life stresses, and so-called free-floating anxiety. The IV is different treatment with three levels (desensitization, relaxation training, and a waiting-list control). After random assignment of subjects to treatments and a subsequent period of treatment, subjects are measured for test anxiety, stress anxiety, and free-floating anxiety; scores on all three measures for each subject serve as DVs. MANOVA is used to ask whether a combination of the three anxiety measures varies as a function of treatment.

ANOVA tests whether mean differences among groups on a single DV are likely to have occurred by chance. MANOVA tests whether mean differences among groups on a combination of DVs are likely to have occurred by chance. In MANOVA, a new DV that maximizes group differences is created from the set of DVs. The new DV is a linear combination of measured DVs, combined so as to separate the groups as much as possible. ANOVA is then performed on the newly created DV. As in ANOVA, hypotheses about means in MANOVA are tested by comparing variances—hence multivariate analysis of variance.

In factorial or more complicated MANOVA, a different linear combination of DVs is formed for each main effect and interaction. If gender of subject is added to the example as a second IV, one combination of the three DVs maximizes the separation of the three treatment groups, a second combination maximizes separation of women and men, and a third combination maximizes separation of the cells of the interaction. Further, if the treatment IV has more than two levels, the DVs can be recombined in yet other ways to maximize the separation of groups formed by comparisons.[1]

MANOVA has a number of advantages over ANOVA. First, by measuring several DVs instead of only one, the researcher improves the chance of discovering what it is that changes as a result of

_____
[1] The linear combinations themselves are of interest in discriminant function analysis (Chapter 11).

different treatments and their interactions. For instance, desensitization may have an advantage over relaxation training or waiting-list control, but only on test anxiety; the effect is missing if test anxiety isn't one of your DVs. A second advantage of MANOVA over a series of ANOVAs when there are several DVs is protection against inflated Type I error due to multiple tests of (likely) correlated DVs.

Another advantage of MANOVA is that, under certain, probably rare conditions, it may reveal differences not shown in separate ANOVAs. Such a situation is shown in Figure 9.1 for a one-way design with two levels. In this figure, the axes represent frequency distributions for each of two DVs, Y_1 and Y_2. Notice that from the point of view of either axis, the distributions are sufficiently overlapping that a mean difference might not be found in ANOVA. The ellipses in the quadrant, however, represent the distributions of Y_1 and Y_2 for each group separately. When responses to two DVs are considered in combination, group differences become apparent. Thus, MANOVA, which considers DVs in combination, may occasionally be more powerful than separate ANOVAs.

But there are no free lunches in statistics, just as there are none in life. MANOVA is a substantially more complicated analysis than ANOVA. There are several important assumptions to consider, and there is often some ambiguity in interpretation of the effects of IVs on any single DV. Further, the situations in which MANOVA is more powerful than ANOVA are quite limited; often MANOVA is considerably less powerful than ANOVA. Thus, our recommendation is to think very carefully about the need for more than one DV in light of the added complexity and ambiguity of analysis (see also Section 9.5.2.).[2]

Multivariate analysis of covariance (MANCOVA) is the multivariate extension of ANCOVA (Chapter 8). MANCOVA asks if there are statistically reliable mean differences among groups after adjusting the newly created DV for differences on one or more covariates. For the example, suppose that before treatment subjects are pretested on test anxiety, minor stress anxiety, and free-floating anxiety. When pretest scores are used as covariates, MANCOVA asks if mean anxiety on the composite score differs in the three treatment groups, after adjusting for preexisting differences in the three types of anxiety.

MANCOVA is useful in the same ways as ANCOVA. First, in experimental work, it serves as a noise-reducing device where variance associated with the covariate(s) is removed from error variance; smaller error variance provides a more powerful test of mean differences among groups. Second, in nonexperimental work, MANCOVA provides statistical matching groups when random assignment to groups is not possible. Prior differences among groups are accounted for by adjusting DVs as if all subjects scored the same on the covariate(s). (But review Chapter 8 for a discussion of the logical difficulties of using covariates this way.)

ANCOVA is used after MANOVA (or MANCOVA) in Roy-Bargmann stepdown analysis where the goal is to assess the contributions of the various DVs to a significant effect. One asks whether, after adjusting for differences on higher-priority DVs serving as covariates, there is any significant mean difference among groups on a lower-priority DV. That is, does a lower-priority DV provide additional separation of groups beyond that of the DVs already used? In this sense, ANCOVA is used as a tool in interpreting MANOVA results.

Although computing procedures and programs for MANOVA and MANCOVA are not as well developed as for ANOVA and ANCOVA, there is in theory no limit to the generalization of the model, despite complications that arise. There is no reason why all types of designs—one-way, factorial, repeated measures, nonorthogonal, and so on—cannot be extended to research with several

---

[2] In several years of working with students, we have found that, once the possibility of measuring several DVs is raised, they each and every one become necessary. That is, we have been royally unsuccessful at talking our students out of MANOVA, despite its disadvantage.

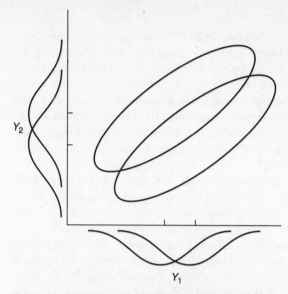

FIGURE 9.1 ADVANTAGE OF MANOVA, WHICH COM-
BINES DVS, OVER ANOVA. EACH AXIS
REPRESENTS A DV; FREQUENCY DIS-
TRIBUTIONS PROJECTED TO AXES
SHOW CONSIDERABLE OVERLAP,
WHILE ELLIPSES, SHOWING DVS IN
COMBINATION, DO NOT.

DVs. Questions of strength of association, and specific comparisons and trend analysis are equally interesting with MANOVA. In addition, there is the question of importance of DVs—that is, which DVs are affected by the IVs and which are not.

MANOVA developed in the tradition of ANOVA. Typically, MANOVA is applied to experimental situations where all, or at least some, IVs are manipulated and subjects are randomly assigned to groups, usually with equal cell sizes. Discriminant function analysis (Chapter 11) developed in the context of nonexperimental research where groups are formed naturally and are not usually the same size. MANOVA asks if mean differences among groups on the combined DV are larger than expected by chance; discriminant function analysis asks if there is some combination of variables that reliably separates groups. But it should be noted that there is no mathematical distinction between MANOVA and discriminant function analysis. At a practical level, computer programs for discriminant analysis are more informative but are also, for the most part, limited to one-way designs. Therefore, analysis of one-way MANOVA is deferred to Chapter 11 and the present chapter covers factorial MANOVA and MANCOVA.

9.2 ■ KINDS OF RESEARCH QUESTIONS

The goal of research using MANOVA is to discover whether behavior, as reflected by the DVs, is changed by manipulation (or other action) of the IVs. Statistical techniques are currently available for answering the types of questions posed in Sections 9.2.1 through 9.2.8.

9.2.1 ■ Main Effects of IVs

Holding all else constant, are mean differences in the composite DV among groups at different levels of an IV larger than expected by chance? The statistical procedures described in Sections 9.4.1 and 9.4.3 are designed to answer this question, by testing the null hypothesis that the IV has no systematic effect on the optimal linear combination of DVs.

As in ANOVA "holding all else constant" refers to a variety of procedures: (1) controlling the effects of other IVs by "crossing over" them in a factorial arrangement, (2) controlling extraneous variables by holding them constant (e.g., running only women as subjects), counterbalancing their effects, or randomizing their effects, or (3) using covariates to produce an "as if constant" state by statistically adjusting for differences on covariates.

In the anxiety-reduction example, the test of main effect asks, Are there mean differences in anxiety—measured by test anxiety, stress anxiety, and free-floating anxiety—associated with differences in treatment? With addition of covariates, the question is: Are there differences in anxiety associated with treatment, after adjustment for individual differences in anxiety prior to treatment?

When there are two or more IVs, separate tests are made for each IV. Further, when sample sizes are equal in all cells, the separate tests are independent of one another (except for use of a common error term) so that the test of one IV in no way predicts the outcome of the test of another IV. If the example is extended to include gender of subject as an IV, and if there are equal numbers of subjects in all cells, the design produces tests of the main effect of treatment and of gender of subject, the two tests independent of each other.

9.2.2 ■ Interactions among IVs

Holding all else constant, does change in the DV over levels of one IV depend on the level of another IV? The test of interaction is similar to the test of main effect, but interpreted differently, as discussed more fully in Chapter 3 and in Sections 9.4.1 and 9.4.3. In the example, the test of interaction asks, Is the pattern of response to the three types of treatment the same for men as it is for women? If the interaction is significant, it indicates that one type of treatment "works best" for women while another type "works best" for men.

With more than two IVs, there are multiple interactions. Each interaction is tested separately from tests of other main effects and interactions, and these tests (but for a common error term) are independent when sample sizes in all cells are equal.

9.2.3 ■ Importance of DVs

If there are significant differences for one or more main effects or interactions, the researcher usually asks which of the DVs are changed and which are unaffected by the IVs. If the main effect of treatment is significant, it may be that only test anxiety is changed while stress anxiety and free-floating anxiety do not differ with treatment. As mentioned in Section 9.1, Roy-Bargmann stepdown analysis is often used where each DV is assessed in ANCOVA with higher-priority DVs serving as covariates. Stepdown analysis and other procedures for assessing importance of DVs appear in Section 9.5.2.

9.2.4 ■ Adjusted Marginal and Cell Means

Ordinarily, marginal means are the best estimates of population parameters for main effects and cell means are the best estimates of population parameters for interactions. But when Roy-Bargmann stepdown analysis is used to test the importance of the DVs, the means that are tested are adjusted means rather than sample means. In the example, suppose free-floating anxiety is given first, stress anxiety second, and test anxiety third priority. Now suppose that a stepdown analysis shows that only test anxiety is affected by differential treatment. The means that are tested for test anxiety are not sample means, but sample means adjusted for stress anxiety and free-floating anxiety. In MAN-COVA, additional adjustment is made for covariates. Interpretation and reporting of results is based on both adjusted and sample means, as illustrated in Section 9.7.

9.2.5 ■ Specific Comparisons and Trend Analysis

If an interaction or a main effect for an IV with more than two levels is significant, you probably want to ask which levels of main effect or cells of interaction are different from which others. If, in the example, treatment, with three levels, is significant, the researcher would be likely to want to ask if the pooled average for the two treated groups is different from the average for the waiting-list control, and if the average for relaxation training is different from the average for desensitization. Indeed, the researcher may have planned to ask these questions instead of the omnibus F questions about treatment. Similarly, if the interaction of gender of subject and treatment is significant, you may want to ask if there is a significant difference in the average response of women and men to, for instance, desensitization.

Specific comparisons and trend analysis are discussed more fully in Sections 9.5.3, 3.2.6, 8.5.2.3, and 10.5.1.

9.2.6 ■ Strength of Association

If a main effect or interaction reliably affects behavior, the next logical question is, How much? What proportion of variance of the linear combination of DV scores is attributable to the effect? You can determine, for instance, the proportion of the variance in the linear combination of anxiety scores that is associated with differences in treatment. These procedures are described in Section 9.4.1. Procedures are also available for finding the strength of association between the IV and an individually significant DV, as demonstrated in Section 9.7.

9.2.7 ■ Effects of Covariates

When covariates are used, the researcher normally wants to assess their utility. Do the covariates provide reliable adjustment and what is the nature of the DV-covariate relationship? For example, when pretests of test, stress, and free-floating anxiety are used as covariates, to what degree does each covariate adjust the composite DV? Assessment of covariates is demonstrated in Section 9.7.3.1.

9.2.8 ■ Repeated-Measures Analysis of Variance

MANOVA is an alternative to repeated-measures ANOVA in which responses to the levels of the within-subjects IV are simply viewed as separate DVs. Suppose, in the example, that measures of test anxiety are taken four times (instead of measuring three different kinds of anxiety once), before, immediately after, 3 months after, and 6 months after treatment. Results could be analyzed as a two-way ANOVA, with treatment as a between-subjects IV and tests as a within-subject IV, or as a one-way MANOVA with treatment as a between-subjects IV and the four testing occasions as four DVs.

As discussed in Sections 8.5.2.1 and 3.2.3, repeated measures ANOVA has the often-violated assumption of homogeneity of covariance. When the assumption is violated, significance tests are too liberal and some alternative to ANOVA is necessary. Other alternatives are adjusted tests of the significance of the within-subjects IV (e.g., Huynh-Feldt), decomposition of the repeated-measures IV into an orthogonal series of single degree of freedom tests (e.g., trend analysis), and profile analysis of repeated measures (Chapter 10).

9.3 ■ LIMITATIONS TO MULTIVARIATE ANALYSIS OF VARIANCE AND COVARIANCE

9.3.1 ■ Theoretical Issues

As with all other procedures, attribution of causality to IVs is in no way assured by the statistical test. This caution is especially relevant because MANOVA, as an extension of ANOVA, stems from experimental research where IVs are typically manipulated by the experimenter and desire for causal inference provides the reason behind elaborate controls. But the statistical test is available whether or not IVs are manipulated, subjects randomly assigned, and controls implemented. Therefore the inference that significant changes in the DVs are caused by concomitant changes in the IVs is a logical exercise, not a statistical one.

Choice of variables is also a question of logic and research design rather than of statistics. Skill is required in choosing IVs and levels of IVs, as well as DVs that have some chance of showing effects of the IVs. A further consideration in choice of DVs is the extent of likely correlation among them. The best choice is a set of DVs that are uncorrelated with each other because they each measure a separate aspect of the influence of the IVs. When DVs are correlated, they measure the same or similar facets of behavior in slightly different ways. What is gained by inclusion of several measures of the same thing? Might there be some way of combining DVs or deleting some of them so that the analysis is simpler?

In addition to choice of number and type of DVs is choice of the order in which DVs enter a stepdown analysis if Roy-Bargmann stepdown F is the method chosen to assess the importance of DVs (see Section 9.5.2.2). Priority is usually given to more important DVs or to DVs that are considered causally prior to others in theory. The choice is not trivial because the significance of a DV may well depend on how high a priority it is given, just as in sequential multiple regression the significance of an IV is likely to depend on its position in the sequence.

When MANCOVA is used, the same limitations apply as in ANCOVA. Consult Sections 8.3.1, 8.5.3, and 8.5.4 for a review of some of the hazards associated with interpretation of designs that include covariates.

Finally, the usual limits to generalizability apply. The results of MANOVA and MANCOVA generalize only to those populations from which the researcher has randomly sampled. And, although MANCOVA may, in some very limited situations, adjust for failure to randomly assign subjects to groups, MANCOVA does not adjust for failure to sample from segments of the population to which one wishes to generalize.

9.3.2 ■ Practical Issues

In addition to the theoretical and logical issues discussed above, the statistical procedure demands consideration of some practical matters.

9.3.2.1 ■ Unequal Sample Sizes and Missing Data

Problems associated with unequal cell sizes are discussed in Section 9.5.4.2. Problems caused by incomplete data (and solutions to them) are discussed in Chapters 4 and 8 (particularly Section 8.3.2.1). The discussion applies to MANOVA and, in fact, may be even more relevant because, as experiments are complicated by numerous DVs and, perhaps, covariates, the probability of missing data increases.

In addition, when using MANOVA, it is necessary to have more cases than DVs in every cell. With numerous DVs this requirement can become burdensome, especially when the design is complicated and there are numerous cells. There are two reasons for the requirement. First, the power of the analysis is lowered unless there are more cases than DVs in every cell because of reduced degrees of freedom for error. One likely outcome of reduced power is a nonsignificant multivariate F, but one or more significant univariate Fs (and a very unhappy researcher).

The second reason for having more cases than DVs in every cell is associated with the assumption of homogeneity of variance-covariance matrices (see Section 9.3.2.4). If a cell has more DVs than cases, the cell becomes singular and the assumption is untestable. If the cell has only one or two more cases than DVs, the assumption is likely to be rejected. Thus MANOVA as an analytic strategy may be discarded because of a failed assumption when the assumption failed because the cases-to-DVs ratio is too low.

9.3.2.2 ■ Multivariate Normality

Significance tests for MANOVA, MANCOVA, and other multivariate techniques are based on the multivariate normal distribution. Multivariate normality implies that the sampling distributions of means of the various DVs in each cell and all linear combinations of them are normally distributed. With univariate F and large samples, the central limit theorem suggests that the sampling distribution of means approaches normality even when raw scores do not. Univariate F is robust to modest violations of normality as long as the violations are not due to outliers.

Mardia (1971) shows that MANOVA is also robust to modest violation of normality if the violation is created by skewness rather than by outliers. *A sample size that produces 20 degrees of freedom*[3] *for error in the univariate case should ensure robustness of the test, as long as sample sizes are equal and two-tailed tests are used.* Even with unequal n and only a few DVs, a sample size of about 20 in the smallest cell should ensure robustness. With small, unequal samples, normality is assessed by reliance on judgment. Are the individual DVs expected to be fairly normally distributed in the population? If not, would some transformation be expected to produce normality (cf. Chapter 4)?

9.3.2.3 ■ Outliers

One of the more serious limitations of MANOVA (and ANOVA) is its sensitivity to outliers. Especially worrisome is that an outlier can produce either a Type I or a Type II error, with no clue in the analysis as to which is occurring. Therefore, it is highly recommended that a test for outliers accompany any use of MANOVA.

Several programs are available for screening for univariate and multivariate outliers (cf. Chapter 4). *Run tests for univariate and multivariate outliers separately for each cell of the design and*

---

[3] See Chapter 3 for a review of calculation of degrees of freedom for univariate *F*.

transform or eliminate them. Report the transformation or deletion of outlying cases. Screening runs for within-cell univariate and multivariate outliers are shown in Sections 8.7.1.4 and 9.7.1.4.

9.3.2.4 ■ Homogeneity of Variance-Covariance Matrices

The multivariate generalization of homogeneity of variance for individual DVs is homogeneity of variance-covariance matrices.[4] The assumption is that variance-covariance matrices within each cell of the design are sampled from the same population variance-covariance matrix and can reasonably be pooled to create a single estimate of error.[5] If the within-cell error matrices are heterogeneous, the pooled matrix is misleading as an estimate of error variance.

The following guidelines for testing this assumption in MANOVA are based on a generalization of a Monte Carlo test of robustness in T^2 (Hakstian, Roed, & Lind, 1979). If sample sizes are equal, robustness of significance tests is expected; disregard the outcome of Box's M test, a notoriously sensitive test of homogeneity of variance-covariance matrices available through SPSS MANOVA.

However, if sample sizes are unequal and Box's M test is significant at $p < .001$, then robustness is not guaranteed. The more numerous the DVs and the greater the discrepancy in cell sample sizes, the greater the potential distortion of alpha levels. Look at both sample sizes and the sizes of the variances and covariances for the cells. If cells with larger samples produce larger variances and covariances, the alpha level is conservative so that null hypotheses can be rejected with confidence. If, however, cells with smaller samples produce larger variances and covariances, the significance test is too liberal. Null hypotheses may be retained with confidence but indications of mean differences are suspect. Use Pillai's criterion instead of Wilks' Lambda (see Section 9.5.1) to evaluate multivariate significance (Olson, 1979). Or equalize sample sizes by random deletion of cases, if power can be maintained at reasonable levels.

9.3.2.5 ■ Linearity

MANOVA and MANCOVA assume linear relationships among all pairs of DVs, all pairs of covariates, and all DV-covariate pairs in each cell. Deviations from linearity reduce the power of the statistical tests because (1) the linear combinations of DVs do not maximize the separation of groups for the IVs, and (2) covariates do not maximize adjustment for error. *With a small number of DVs and covariates, examine all within-cell scatterplots between pairs of DVs, pairs of covariates, and pairs of DV-covariate combinations* through SPSS PLOT, SAS PLOT, SYSTAT PLOT or BMDP6D.

With a large number of DVs and/or covariates, however, a complete check is unwieldy; spot check the bivariate relationships for which nonlinearity is likely. Or, use the procedure for checking linearity with large numbers of variables as available through BMDP1V and described in Section 8.3.2.6. One DV (preferably the one with the lowest priority for entry) is designated DEPENDENT, and the remainder are designated INDEPENDENT (covariates) in a BMDP1V run and plots of residuals are examined as described in Section 5.3.2.4. If serious curvilinearity is found with a covariate, consider deletion; if curvilinearity is found with a DV, consider transformation— provided, of course, that increased difficulty in interpretation of a transformed DV is worth the increase in power.

---

[4] In MANOVA, homogeneity of variance for each of the DVs is also assumed. See Section 8.3.2.5 for discussion and recommendations.

[5] Don't confuse this assumption with the assumption of homogeneity of covariance that is relevant to repeated measures ANOVA or MANOVA, as discussed in Section 8.5.2.1.

9.3.2.6 ■ Homogeneity of Regression

In Roy-Bargmann stepdown analysis (Section 9.5.2.2) and in MANCOVA (Section 9.4.3) it is assumed that the regression between covariates and DVs in one group is the same as the regression in other groups so that using the average regression to adjust for covariates in all groups is reasonable.

In both MANOVA and MANCOVA, if Roy-Bargmann stepdown analysis is used, the importance of a DV in a hierarchy of DVs is assessed in ANCOVA with higher-priority DVs serving as covariates. Homogeneity of regression is required for each step of the analysis, as each DV, in turn, joins the list of covariates. If heterogeneity of regression is found at a step, the rest of the stepdown analysis is uninterpretable. Once violation occurs, the IV-"covariate" interaction is itself interpreted and the DV causing violation is eliminated from further steps.

In MANCOVA (like ANCOVA) heterogeneity of regression implies that there is interaction between the IV(s) and the covariates and that a different adjustment of DVs for covariates is needed in different groups. If interaction between IVs and covariates is suspected, MANCOVA is an inappropriate analytic strategy, both statistically and logically. Consult Sections 8.3.2.7 and 8.5.5 for alternatives to MANCOVA where heterogeneity of regression is found.

For MANOVA, test for stepdown homogeneity of regression and for MANCOVA, test for overall and stepdown homogeneity of regression. These procedures are demonstrated in Section 9.7.1.6.

9.3.2.7 ■ Reliability of Covariates

In MANCOVA as in ANCOVA, the F test for mean differences is more powerful if covariates are reliable. If covariates are not reliable, either increased Type I or Type II errors can occur. Reliability of covariates is discussed more fully in Section 8.3.2.8.

In Roy-Bargmann stepdown analysis where all but the lowest-priority DV act as covariates in assessing other DVs, unreliability of any of the DVs (say $r_{yy} < .8$) raises questions about stepdown analysis as well as about the rest of the research effort. When DVs are unreliable, use another method for assessing the importance of DVs (Section 9.5.2) and report known or suspected unreliability of covariates and high-priority DVs in your Results section.

9.3.2.8 ■ Multicollinearity and Singularity

When correlations among DVs are high, one DV is a near-linear combination of other DVs; the DV provides information that is redundant to the information available in one or more of the other DVs. It is both statistically and logically suspect to include all the DVs in analysis and *the usual solution is deletion of the redundant DV*. But if there is some compelling theoretical reason to retain all DVs, a principal components analysis (cf. Chapter 13) is done on the pooled within-cell correlation matrix, and component scores are entered as an alternative set of DVs.

BMDP4V, SAS GLM and SYSTAT GLM protect against multicollinearity and singularity through computation of pooled within-cell tolerance $(1 - SMC)$ for each DV; DVs with insufficient tolerance are deleted from analysis. In SPSS MANOVA, singularity or multicollinearity may be present when the determinant of the within-cell correlation matrix insure zero (say, less than .0001). The redundant DV is identified by using the within-cell correlation matrix (from MANOVA) as input to the REGRESSION program and then performing several regressions, with each DV in turn serving as DV, with all other DVs serving as IVs in analysis. If R^2 approaches .90 for some DV, that DV is redundant. Alternatively, the within-cell correlation matrix is input to BMDP4M or to the SPSS FACTOR program where SMCs (called Communality in SPSS) are reported.

9.4 ■ FUNDAMENTAL EQUATIONS FOR MULTIVARIATE ANALYSIS OF VARIANCE AND COVARIANCE

9.4.1 ■ Multivariate Analysis of Variance

A minimum data set for MANOVA has one or more IVs, each with two or more levels, and two or more DVs for each subject within each combination of IVs. A fictitious small sample with two DVs and two IVs is illustrated in Table 9.1. The first IV is degree of disability with three levels—mild, moderate, and severe—and the second is treatment with two levels—treatment and no treatment. These two IVs in factorial arrangement produce six cells; three children are assigned to each cell so there are 3×6 or 18 children in the study. Each child produces two DVs, score on the reading subtest of the Wide Range Achievement Test (WRAT-R) and score on the arithmetic subtest (WRAT-A). In addition an IQ score is given in parentheses for each child to be used as a covariate in Section 9.4.3.

The test of the main effect of treatment asks, Disregarding degree of disability, does treatment affect the composite score created from the two subtests of the WRAT? The test of interaction asks, Does the effect of treatment on a difference composite score from the two subtests differ as a function of degree of disability?

The test of the main effect of disability is automatically provided in the analysis but is trivial in this example. The question is, Are scores on the WRAT affected by degree of disability? Because degree of disability is at least partially defined by difficulty in reading and/or arithmetic, a significant effect provides no useful information. On the other hand, the absence of this effect would lead us to question the adequacy of classification.

The sample size of three children per cell is probably inadequate for a realistic test but serves to illustrate the techniques of MANOVA. Additionally, if causal inference is intended, the researcher should randomly assign children to the levels of treatment. The reader is encouraged to analyze these data by hand and by computer. Setup and selected output for this example appear in Section 9.4.2 for several appropriate programs.

MANOVA follows the model of ANOVA where variance in scores is partitioned into variance attributable to difference among scores within groups and to differences among groups. Squared differences between scores and various means are summed (see Chapter 3); these sums of squares, when divided by appropriate degrees of freedom, provide estimates of variance attributable to different sources (main effects of IVs, interactions among IVs, and error). Ratios of variances provide tests of hypotheses about the effects of IVs on the DV.

In MANOVA, however, each subject has a score on each of several DVs. When several DVs for each subject are measured, there is a matrix of scores (subjects by DVs) rather than a simple set of

TABLE 9.1 SMALL SAMPLE DATA FOR ILLUSTRATION OF MULTIVATRIATE ANALYSIS OF VARIANCE

	Mild			Moderate			Severe		
	WRAT-R	WRAT-A	(IQ)	WRAT-R	WRAT-A	(IQ)	WRAT-R	WRAT-A	(IQ)
Treatment	115	108	(110)	100	105	(115)	89	78	(99)
	98	105	(102)	105	95	(98)	100	85	(102)
	107	98	(100)	95	98	(100)	90	95	(100)
Control	90	92	(108)	70	80	(100)	65	62	(101)
	85	95	(115)	85	68	(99)	80	70	(95)
	80	81	(95)	78	82	(105)	72	73	(102)

DVs within each group. Matrices of difference scores are formed by subtracting from each score an appropriate mean; then the matrix of differences is squared. When the squared differences are summed, a sum-of-squares-and-cross-products matrix, an S matrix, is formed, analogous to a sum of squares in ANOVA. Determinants[6] of the various S matrices are found, and ratios between them provide tests of hypotheses about the effects of the IVs on the linear combination of DVs. In MANCOVA, the sums of squares and cross products in the S matrix are adjusted for covariates, just as sums of squares are adjusted in ANCOVA (Chapter 8).

The MANOVA equation for equal n is developed below through extension of ANOVA. The simplest partition apportions variance to systematic sources (variance attributable to differences between groups) and to unknown sources of error (variance attributable to differences in scores within groups). To do this, differences between scores and various means are squared and summed.

$$\sum_i \sum_j (Y_{ij} - GM)^2 = n \sum_j (\overline{Y}_j - GM)^2 + \sum_i \sum_j (Y_{ij} - \overline{Y}_{ij})^2 \tag{9.1}$$

The total sum of squared differences between scores on Y (the DV) and the grand mean (GM) is partitioned into sum of squared differences between group means ($\overline{Y}_{ij}$) and the grand mean (i.e., systematic or between-groups variability), and sum of squared difference between individual scores (Y_{ij}) and their respective group means.

or

$$SS_{total} = SS_{bg} + SS_{wg}$$

For designs with more than one IV, SS_{bg} is further partitioned into variance associated with the first IV (e.g, degree of disability, abbreviated D), variance associated with the second IV (treatment, or T), and variance associated with the interaction between degree of disability and treatment (or DT).

$$n_{km} \sum_k \sum_m (DT_{km} - GM)^2 = n_k \sum_k (D_k - GM)^2 + n_m \sum_m (T_m - GM)^2 \tag{9.2}$$

$$+ \left[n_{km} \sum_k \sum_m (DT_{km} - GM)^2 - n_k \sum_k (D_k - GM)^2 - n_m \sum_m (T_m - GM)^2 \right]$$

The sum of squared differences between cell (DT_{km}) means and the grand mean is partitioned into (1) sum of squared differences between means associated with different levels of disability (D_k) and the grand mean; (2) sum of squared differences between means associated with different levels of treatment (T_m) and the grand mean; and (3) sum of squared differences associated with combinations of treatment and disability (DT_{km}) and the grand mean, from which differences associated with D_k and T_m are subtracted. Each n is the number of scores composing the relevant marginal or cell mean.

or

$$SS_{bg} = SS_D + SS_T + SS_{DT}$$

---

[6] A determinant, as described in Appendix A, can be viewed as a measure of generalized variance for a matrix.

The full partition for this factorial between-subjects design is

$$\sum_i \sum_k \sum_m (Y_{ikm} - GM)^2 = n_k \sum_k (D_k - GM)^2 + n_m \sum_m (T_m - GM)^2$$

$$+ \left[n_{km} \sum_k \sum_m (DT_{km} - GM)^2 - n_k \sum_k (D_k - GM)^2 - n_m \sum_m (T_m - GM)^2 \right]$$

$$+ \sum_i \sum_k \sum_m (Y_{ikm} - DT_{km})^2 \qquad (9.3)$$

For MANOVA, there is no single DV but rather a column matrix (or vector) of Y_{ikm} values, of scores on each DV. For the example in Table 9.1, column matrices of Y scores for the three children in the first cell of the design (mild disability with treatment) are

$$\mathbf{Y}_{i11} = \begin{bmatrix} 115 \\ 108 \end{bmatrix} \begin{bmatrix} 98 \\ 105 \end{bmatrix} \begin{bmatrix} 107 \\ 98 \end{bmatrix}$$

Similarly, there is a column matrix of disability—D_k—means for mild, moderate, and severe levels of D, with one mean in each matrix for each DV.

$$\mathbf{D}_1 = \begin{bmatrix} 95.83 \\ 96.50 \end{bmatrix} \qquad \mathbf{D}_2 = \begin{bmatrix} 88.83 \\ 88.00 \end{bmatrix} \qquad \mathbf{D}_3 = \begin{bmatrix} 82.67 \\ 77.17 \end{bmatrix}$$

where 95.83 is the mean on WRAT-R and 96.50 is the mean on WRAT-A for children with mild disability, averaged over treatment and control groups.

Matrices for treatment—T_m—means, averaged over children with all levels of disability are

$$\mathbf{T}_1 = \begin{bmatrix} 99.89 \\ 97.11 \end{bmatrix} \qquad \mathbf{T}_2 = \begin{bmatrix} 78.33 \\ 78.11 \end{bmatrix}$$

Similarly, there are six matrices of cell means (DT_{km}) averaged over the three children in each group.

Finally, there is a single matrix of grand means **(GM)**, one for each DV, averaged over all children in the experiment.

$$\mathbf{GM} = \begin{bmatrix} 89.11 \\ 87.22 \end{bmatrix}$$

As illustrated in Appendix A, differences are found by simply subtracting one matrix from another, to produce difference matrices. The matrix counterpart of a difference score, then, is a difference matrix. To produce the error term for this example, the matrix of grand means **(GM)** is subtracted from each of the matrixes of individual scores (Y_{ikm}). Thus for the first child in the example:

$$(\mathbf{Y}_{111} - \mathbf{GM}) = \begin{bmatrix} 115 \\ 108 \end{bmatrix} - \begin{bmatrix} 89.11 \\ 87.22 \end{bmatrix} = \begin{bmatrix} 25.89 \\ 20.78 \end{bmatrix}$$

In ANOVA, difference scores are squared. The matrix counterpart of squaring is multiplication by a transpose. That is, each column matrix is multiplied by its corresponding row matrix (see

Appendix A for matrix transposition and multiplication) to produce a sum-of-squares and cross-products matrix. For example, for the first child in the first group of the design:

$$(\mathbf{Y}_{111} - \mathbf{GM})(\mathbf{Y}_{111} - \mathbf{GM})' = \begin{bmatrix} 25.89 \\ 20.78 \end{bmatrix} [25.89 \quad 20.78] = \begin{bmatrix} 670.29 & 537.99 \\ 537.99 & 431.81 \end{bmatrix}$$

These matrices are then summed over subjects and over groups, just as squared differences are summed in univariate ANOVA. The order of summing and squaring is the same in MANOVA as in ANOVA for a comparable design. The resulting matrix (**S**) is called by various names: sum-of-squares and cross-products, cross-products, or sum-of-products. The MANOVA partition of sums-of-squares and cross-products for our factorial example is represented below in a matrix form of Equation 9.3:

$$\sum_i \sum_k \sum_m (\mathbf{Y}_{ikm} - \mathbf{GM})(\mathbf{Y}_{ikm} - \mathbf{GM})' = n_k \sum_k (\mathbf{D}_k - \mathbf{GM})(\mathbf{D}_k - \mathbf{GM})'$$

$$+ n_m \sum_m (\mathbf{T}_m - \mathbf{GM})(\mathbf{T}_m - \mathbf{GM})'$$

$$+ \left[n_{km} \sum_k \sum_m (\mathbf{DT}_{km} - \mathbf{GM})(\mathbf{DT}_{km} - \mathbf{GM})' \right.$$

$$- n_k \sum_k (\mathbf{D}_k - \mathbf{GM})(\mathbf{D}_k - \mathbf{GM})'$$

$$\left. - n_m \sum_m (\mathbf{T}_m - \mathbf{GM})(\mathbf{T}_m - \mathbf{GM})' \right]$$

$$+ \sum_i \sum_k \sum_m (\mathbf{Y}_{ikm} - \mathbf{DT}_{km})(\mathbf{Y}_{ikm} - \mathbf{DT}_{km})'$$

or

$$\mathbf{S}_{\text{total}} = \mathbf{S}_D + \mathbf{S}_T + \mathbf{S}_{DT} + \mathbf{S}_{S(DT)}$$

The total cross-products matrix ($\mathbf{S}_{\text{total}}$) is partitioned into cross-products matrices for differences associated with degree of disability, with treatment, with the interaction between disability and treatment, and for error—subjects within groups ($\mathbf{S}_{S(DT)}$).

For the example in Table 9.1, the four resulting cross-products matrices[7] are

$$\mathbf{S}_D = \begin{bmatrix} 570.29 & 761.72 \\ 761.72 & 1126.78 \end{bmatrix} \quad \mathbf{S}_T = \begin{bmatrix} 2090.89 & 1767.56 \\ 1767.56 & 1494.22 \end{bmatrix}$$

$$\mathbf{S}_{DT} = \begin{bmatrix} 2.11 & 5.28 \\ 5.28 & 52.78 \end{bmatrix} \quad \mathbf{SS}_{S(DT)} = \begin{bmatrix} 544.00 & 31.00 \\ 31.00 & 539.33 \end{bmatrix}$$

Notice that all these matrices are symmetric, with the elements top left to bottom right diagonal representing sums of squares (that, when divided by degrees of freedom, produce variances), and

---

[7] Numbers producing these matrices were carried to 8 digits before rounding.

with the off-diagonal elements representing sums of cross products (that, when divided by degrees of freedom, produce covariances). In this example, the first element in the major diagonal (top left to bottom right) is the sum of squares for the first DV, WRAT-R, and the second element is the sum of squares for the second DV, WRAT-A. The off-diagonal elements are the sums of cross products between WRAT-R and WRAT-A.

In ANOVA, sums of squares are divided by degrees of freedom to produce variances, or mean squares. In MANOVA, the matrix analog of variance is a determinant (see Appendix A); the determinant is found for each cross-products matrix. In ANOVA, ratios of variances are formed to test main effects and interactions. In MANOVA, ratios of determinants are formed to test main effects and interactions when using Wilks' Lambda (see Section 9.5.1 for additional criteria). These ratios follow the general form

$$\Lambda = \frac{|\mathbf{S}_{\text{error}}|}{|\mathbf{S}_{\text{effect}} + \mathbf{S}_{\text{error}}|} \tag{9.4}$$

Wilks' Lambda (Λ) is the ratio of the determinant of the error cross-products matrix to the determinant of the sum of the error and effect cross-products matrices.

To find Wilks' Λ, the within-groups matrix is added to matrices corresponding to main effects and interactions before determinants are found. For the example, the matrix produced by adding the $\mathbf{S}_{DT}$ matrix for interaction to the $\mathbf{S}_{S(DT)}$ matrix for subjects within groups (error) is

$$\mathbf{S}_{DT} + \mathbf{S}_{S(DT)} = \begin{bmatrix} 2.11 & 5.28 \\ 5.28 & 52.78 \end{bmatrix} + \begin{bmatrix} 544.00 & 31.00 \\ 31.00 & 539.33 \end{bmatrix} = \begin{bmatrix} 546.11 & 36.28 \\ 36.28 & 592.11 \end{bmatrix}$$

For the four matrices needed to test main effect of disability, main effect of treatment, and the treatment-disability interaction, the determinants are

$$|\mathbf{S}_{S(DT)}| = 292436.52$$

$$|\mathbf{S}_D + \mathbf{S}_{S(DT)}| = 1228124.71$$

$$|\mathbf{S}_T + \mathbf{S}_{S(DT)}| = 2123362.49$$

$$|\mathbf{S}_{DT} + \mathbf{S}_{S(DT)}| = 322040.95$$

At this point a source table, similar to the source table for ANOVA, is useful, as presented in Table 9.2. The first column lists sources of variance, in this case the two main effects and the interaction. The error term does not appear. The second column contains the value of Wilks' Lambda.

Wilks' Lambda is a ratio of determinants, as described in Equation 9.4. For example, for the interaction between disability and treatment, Wilks' Lambda is

$$\Lambda = \frac{|\mathbf{S}_{S(DT)}|}{|\mathbf{S}_{DT} + \mathbf{S}_{S(DT)}|} = \frac{292436.52}{322040.95} = .908068$$

Tables for evaluating Wilks' Lambda directly are not common, though they do appear in Harris (1975). However, an approximation to F has been derived that closely fits Lambda. The

TABLE 9.2 MULTIVARIATE ANALYSIS OF VARIANCE OF WRAT-R AND WRAT-A SCORES

Source of variance	Wilks' Lambda	df_1	df_2	Multivariate F
Treatment	.13772	2.00	11.00	34.43570**
Disability	.25526	4.00	22.00	5.38602*
Treatment by Disability	.90807	4.00	22.00	0.27170

*$p < .01$
**$p < .001$

last three columns of Table 9.2, then, represent the approximate F values and their associated degrees of freedom.

The following procedure for calculating approximate F (Rao, 1952) is based on Wilks' Lambda and the various degrees of freedom associated with it.

$$\text{Approximate } F(df_1, df_2) = \left(\frac{1-y}{y}\right)\left(\frac{df_2}{df_1}\right) \qquad (9.5)$$

where df_1 and df_2 are defined below as the degrees of freedom for testing the F ratio, and y is

$$y = \Lambda^{1/s} \qquad (9.6)$$

Λ is defined in Equation 9.4, and s is[8]

$$s = \sqrt{\frac{p^2(df_{effect})^2 - 4}{p^2 + (df_{effect})^2 - 5}} \qquad (9.7)$$

where p is the number of DVs, and df_{effect} is the degrees of freedom for the effect being tested. And

$$df_1 = p(df_{effect})$$

and

$$df_2 = s\left[(df_{error}) - \frac{p - df_{effect} + 1}{2}\right] - \left[\frac{p(df_{effect}) - 2}{2}\right]$$

where df_{error} are the degrees of freedom associated with the error term.

For the test of interaction in the sample problem, we have

$p = 2$ the number of DVs

$df_{effect} = 2$ the number of treatment levels minus 1 times the number of disability levels minus 1 or $(t - 1)(d - 1)$

$df_{error} = 12$ the number of treatment levels times the number of disability levels times the quantity $n - 1$ (where n is the number of scores per cell for each DV)—that is, $df_{error} = dt(n - 1)$

---

[8] When $df_{effect} = 1$, $s = 1$. When $p = 1$, we have univariate ANOVA.

Thus

$$s = \sqrt{\frac{(2)^2(2)^2 - 4}{(2)^2 + (2)^2 - 5}} = 2$$

$$y = .908068^{1/2} = .952926$$

$$df_1 = 2(2) = 4$$

$$df_2 = 2\left[12 - \frac{2 - 2 + 1}{2}\right] - \left[\frac{2(2) - 2}{2}\right] = 22$$

$$\text{Approximate } F(4,22) = \left(\frac{.047074}{.952926}\right)\left(\frac{22}{4}\right) = 0.2717$$

This approximate F value is tested for significance by using the usual tables of F at selected α. In this example, the interaction between disability and treatment is not statistically significant with 4 and 22 df, because the observed value of 0.2717 does not exceed the critical value of 2.82 at $\alpha = .05$.

Following the same procedures, the effect of treatment is statistically significant, with the observed value of 34.44 exceeding the critical value of 3.98 with 2 and 11 df, $\alpha = .05$. The effect of degree of disability is also statistically significant, with the observed value of 5.39 exceeding the critical value of 2.82 with 4 and 22 df, $\alpha = .05$. (As noted previously, this main effect is not of research, interest, but does serve to validate the classification procedure.) In Table 9.2, significance is indicated at the highest level of α reached, following standard practice.

A measure of strength of association is readily available from Wilks' Lambda.[9] For MANOVA:

$$\eta^2 = 1 - \Lambda \qquad\qquad (9.8)$$

This equation represents the variance accounted for by the best linear combination of DVs as explained below.

In a one-way analysis, according to Equation 9.4, Wilks' Lambda is the ratio of (the determinant of) the error matrix and (the determinant of) the total sum-of-squares and cross-products matrix. The determinant of the error matrix—Λ—is the variance not accounted for by the combined DVs so $1 - \Lambda$ is the variance that is accounted for.[10]

Thus, for each statistically significant effect, the proportion of variance accounted for is easily calculated using Equation 9.6. For example, the main effect of treatment:

$$\eta^2_T = 1 - \Lambda_T = 1 - .137721 = .862279$$

In the example, 86% of the variance in the best linear combination of WRAT-R and WRAT-A scores is accounted for by assignment to levels of treatment. The square root of η^2 ($\eta = .93$) is a form of correlation between WRAT scores and assignment to treatment.

---

[9] An alternative measure of strength of association is canonical correlation, printed out by some computer programs. Canonical correlation is the correlation between the optimal linear combination of IV levels and the optimal linear combination of DVs where optimal is chosen to maximize the correlation between combined IVs and DVs. Canonical correlation as a general procedure is discussed in Chapter 6, and the relation between canonical correlation and MANOVA is discussed briefly in Chapter 14.

[10] BMDP4V does not show Wilks' Lambda (LRATIO) for an effect when $s = 1$. However, by requesting EVALues (eigenvalues) in the ANALYSIS paragraph, η^2 is printed out, labeled HT EVALS.

However, unlike η^2 in the analogous ANOVA design, the sum of η^2 for all effects in MANOVA may be greater than 1.0 because DVs are recombined for each effect. This lessens the appeal of an interpretation in terms of proportion of variance accounted for, although the size of η^2 is still a measure of the relative importance of an effect.

Another difficulty in using this form of η^2 is that effects tend to be much larger in the multivariate than in the univariate case. Therefore, a recommended alternative, when $s > 1$ is

$$\text{partial } \eta^2 = 1 - \Lambda^{1/s} \tag{9.9}$$

Estimated strength of association is reduced to .63 with the use of partial η^2 for the current data, a more reasonable assessment.

9.4.2 ■ Computer Analyses of Small Sample Example

Tables 9.3 through 9.6 show setup and selected minimal output for SPSS MANOVA, BMDP4V, SYSTAT GLM, and SAS GLM, respectively.

In SPSS MANOVA (Table 9.3) simple MANOVA source tables, resembling those of ANOVA, are printed out when PRINT=SIGNIF(BRIEF) is requested. After interpretive material (including the design matrix) is printed, the source table is shown, labeled Tests of Significance for WITHIN CELLS using UNIQUE sums of squares.[11] WITHIN CELLS refers to the pooled within-cell error SSCP matrix (Section 9.4.1), the error term chosen by default for MANOVA. For the example, the two-way MANOVA source table consists of the two main effects and the interaction. For each source, you are given Wilks' Lambda, multivariate F with numerator and denominator degrees of freedom (Hypoth. DF and Error DF, respectively), and the probability level achieved for the significance test.

BMDP4V (Table 9.4) requires a USE= instruction if there are more variables in your data set than are used in the MANOVA. After the IVs alerted in the BETWEEN instruction (and in the WITHIN instruction, if applicable) the remaining variables serve as DVs. There is a great deal of output showing the program's interpretation of the data and the analysis, and the cell and marginal means, most of which has been deleted from Table 9.4. Shown is a detailed source table starting with the tests for the grand mean. Results for each effect are reported in a common format: the multivariate test for -ALL—(the DVs combined) is followed by univariate tests for each DV. For the example, the multivariate test of T: TREATMNT is a form of Hotelling's T^2 (because TREATMNT has only two levels), for which you are given the value of TSQ, the associated F ratio (34.44), numerator and denominator degrees of freedom (2, 11), and the significance level attained (rounded to four digits). For each DV, BMDP4V prints out sum of squares and mean square (SS and MS) for the effect, the univariate F ratio, numerator and denominator df, and significance level.

For an effect with more than one df (e.g., D: DISABLTY and TD, the interaction), several multivariate tests of significance are provided: Wilks' Lambda (LRATIO), Hotelling-Lawley TRACE, a form of Hotelling's T^2 labeled TZSQ, and Roy's gcr (MXROOT), described in Section 9.5.1. Because of the small n in this example, significance values for TZSQ and MXROOT are not printed out. Note that, unlike other programs reviewed here, the same F ratio is estimated for Wilks' Lambda and Hotelling-Lawley TRACE.

---

[11] Note that recent versions of SPSS MANOVA use UNIQUE rather than SEQUENTIAL SUMS OF SQUARES as default to correct for unequal-*n*, cf. Section 8.5.2.2.

TABLE 9.3   MANOVA ON SMALL SAMPLE EXAMPLE THROUGH SPSS MANOVA (SETUP AND OUTPUT)

DATA LIST   FILE='TAPE81.DAT'   FREE
                 /SUBJNO, TREATMNT, DISABLTY, WRATR, WRATA, IQ,
                 CELLNO.

VALUE LABELS   TREATMNT 1 'TREATMENT' 2 'CONTROL'/
                 DISABLTY 1 'MILD' 2 'MODERATE' 3 'SEVERE'.

MANOVA   WRATR, WRATA BY TREATMNT(1,2) DISABLTY(1,3)/
                 PRINT=SIGNIF(BRIEF)/
                 DESIGN.

* * * * * * * * * * * * * * * A N A L Y S I S   O F   V A R I A N C E *

    18 cases accepted.
     0 cases rejected because of out-of-range factor values.
     0 cases rejected because of missing data.
     6 non-empty cells.

     1 design will be processed.

- -
* * * * * * * * * * * * * * * A N A L Y S I S   O F   V A R I A N C E - DESIGN   1 * * * * * * * * * * * * * * * * * * *
Tests of significance for WITHIN CELLS using UNIQUE sums of squares
Source of Variation   Wilks Lambda   Mult. F   Hypoth. DF   Error DF   Sig. of F

TREATMNT   .13772   34.43570   2.00   11.00   .000
DISABLTY   .25526   5.38602   4.00   22.00   .004
TREATMNT BY DISABLTY   .90807   .27170   4.00   22.00   .893

```
/INPUT          VARIABLES ARE 7.  FILE IS 'TAPE81.DAT'.  FORMAT IS FREE.
/VARIABLE       NAMES ARE SUBJNO, TREATMNT, DISABLTY, WRATR,
                          WRATA, IQ, CELLNO.
                USE = TREATMNT, DISABLTY, WRATA, WRATR.
/BETWEEN        FACTORS ARE TREATMNT, DISABLTY.
                CODES(1) ARE 1, 2.
                NAMES(1) ARE TREATMNT, CONTROL.
                CODES(2) ARE 1 TO 3.
                NAMES(2) ARE MILD, MODERATE, SEVERE.
/WEIGHTS        BETWEEN ARE EQUAL.
/END
```

```
 --- ANALYSIS SUMMARY ---

THE FOLLOWING EFFECTS ARE COMPONENTS OF THE SPECIFIED
LINEAR MODEL FOR THE BETWEEN DESIGN.  ESTIMATES AND TESTS
OF HYPOTHESES FOR THESE EFFECTS CONCERN PARAMETERS OF THAT MODEL.

        OVALL: GRAND MEAN
        T: TREATMNT
        D: DISABLTY
        TD
```

```
=========================================================================

=========================================================================
EFFECT   VARIATE      STATISTIC          F        DF        P
-------------------------------------------------------------------------
 OVALL: GRAND MEAN
       -ALL----
                TSQ=      5864.25       2687.78    2,   11  0.0000
        WRATA
                SS=    136938.888889
                MS=    136938.888889    3046.85    1,   12  0.0000
        WRATR
                SS=    142934.222222
                MS=    142934.222222    3152.96    1,   12  0.0000
 T: TREATMNT
       -ALL----
                TSQ=        75.1324       34.44    2,   11  0.0000
        WRATA
                SS=      1494.222222
                MS=      1494.222222      33.25    1,   12  0.0001
        WRATR
                SS=      2090.888889
                MS=      2090.888889      46.12    1,   12  0.0000
 D: DISABLTY
       -ALL----
                LRATIO=  0.255263          5.39    4,   22.00 0.0035
                TRACE=   2.89503
                TZSQ=   34.7404
                 P APPROXIMATION FOR ABOVE STATISTIC UNAVAILABLE - DF IS ZERO
                MXROOT=  0.742748
                 P APPROXIMATION FOR ABOVE STATISTIC UNAVAILABLE
        WRATA
                SS=      1126.777778
                MS=       563.388889      12.54    2,   12  0.0012
        WRATR
                SS=       520.777778
                MS=       260.388889       5.74    2,   12  0.0178
```

TABLE 9.4 *(CONTINUED)*

```
TD
        -ALL----
                LRATIO=   0.908068              0.27    4,  22.00 0.8930
                TRACE=    0.100954
                TZSQ=     1.21144
                 P APPROXIMATION FOR ABOVE STATISTIC UNAVAILABLE - DF IS ZERO
                MXROOT=   0.892854E01
                 P APPROXIMATION FOR ABOVE STATISTIC UNAVAILABLE
        WRATA
                SS=       52.777778
                MS=       26.388889             0.59    2,   12  0.5711
        WRATR
                SS=       2.111111
                MS=       1.055556              0.02    2,   12  0.9770


ERROR
        WRATA
                SS=       539.33333333
                MS=       44.94444444

        WRATR
                SS=       544.00000000
                MS=       45.33333333
```

The last source of variance in the source table is ERROR. This section contains error SS and MS for each DV, in turn. These mean square values are the denominators for the univariate F tests above.

In SYSTAT GLM (Table 9.5) the CATEGORY instruction defines the IVs. An equation is formed in the MODEL statement with DVs on the left of the equation and effects on the right.

Separate source tables are provided for each effect in the output, including CONSTANT. Each source table begins with Univariate F tests for each of the DVs (giving the usual SS, DF, MS, F, and P level) followed by four fully labeled multivariate tests. The last, THETA, is Roy's gcr, with its three df parameters.

In SAS GLM (Table 9.6) IVs are defined in a CLASS instruction while the MODEL instruction defines the DVs and the effects to be considered. The NOUNI instruction suppresses printing of descriptive statistics and univariate F tests. MANOVA H = _ALL_ requests tests of all main effects and interactions listed in the MODEL instruction, and SHORT condenses the printout.

The output begins with some interpretative information followed by separate sections for TREATMNT, DISABLTY, and TREATMNT*DISABLTY. Each source table is preceded by information about characteristic roots and vectors of the error SSCP matrix (as discussed in Chapters 6, 12, and 13; note that in each case the first Characteristic Root is equal to Roy's Greatest Root), and the three df parameters (Section 9.4.1). Each source table shows results of four multivariate tests, fully labeled.

9.4.3 ■ Multivariate Analysis of Covariance

In MANCOVA, the linear combination of DVs is adjusted for differences in the covariates. The adjusted linear combination of DVs is the combination that would be obtained if all participants had the same scores on the covariates. For this example, preexperimental IQ scores (listed in parentheses in Table 9.1) are used as covariates.

TABLE 9.5 MANOVA ON SMALL SAMPLE EXAMPLE THROUGH SYSTAT GLM (SETUP AND SELECTED OUTPUT)

```
USE TAPE81
OUTPUT SSMANOVA.OUT
CATEGORY TREATMNT DISABLTY
MODEL WRATR WRATA = CONSTANT + TREATMNT + DISABLTY + TREATMNT*DISABLTY
ESTIMATE
```

```
 Categorical values encountered during processing are:
 TREATMNT (2 levels)
          1,        2
 DISABLTY (3 levels)
          1,        2,        3
```

```
 Number of cases processed: 18
 Dependent variable means
```

	WRATR	WRATA
	89.111	87.222

Test for effect called: CONSTANT

Univariate F Tests

Effect	SS	DF	MS	F	P
WRATR	142934.222	1	142934.222	3152.961	0.000
Error	544.000	12	45.333		
WRATA	136938.889	1	136938.889	3046.848	0.000
Error	539.333	12	44.944		

Multivariate Test Statistics

Wilks' Lambda =	0.002					
F-Statistic =	2687.779	DF =	2, 11	Prob =	0.000
Pillai Trace =	0.998					
F-Statistic =	2687.779	DF =	2, 11	Prob =	0.000
Hotelling-Lawley Trace =	488.687					
F-Statistic =	2687.779	DF =	2, 11	Prob =	0.000

Test for effect called: TREATMNT

Univariate F Tests

Effect	SS	DF	MS	F	P
WRATR	2090.889	1	2090.889	46.123	0.000
Error	544.000	12	45.333		
WRATA	1494.222	1	1494.222	33.246	0.000
Error	539.333	12	44.944		

TABLE 9.5 *(CONTINUED)*

Multivariate Test Statistics

```
        Wilks' Lambda =      0.138
          F-Statistic =     34.436   DF =   2,  11    Prob =     0.000

        Pillai Trace =       0.862
          F-Statistic =     34.436   DF =   2,  11    Prob =     0.000

  Hotelling-Lawley Trace =   6.261
          F-Statistic =     34.436   DF =   2,  11    Prob =     0.000
```

Test for effect called: DISABLTY

Univariate F Tests

Effect	SS	DF	MS	F	P
WRATR	520.778	2	260.389	5.744	0.018
Error	544.000	12	45.333		
WRATA	1126.778	2	563.389	12.535	0.001
Error	539.333	12	44.944		

Multivariate Test Statistics

```
        Wilks' Lambda =      0.255
          F-Statistic =      5.386   DF =   4,  22    Prob =     0.004

        Pillai Trace =       0.750
          F-Statistic =      3.604   DF =   4,  24    Prob =     0.019

  Hotelling-Lawley Trace =   2.895
          F-Statistic =      7.238   DF =   4,  20    Prob =     0.001

            THETA = 0.743 S =  2, M =-0.5, N =  4.5 Prob =       0.002
```

Test for effect called: TREATMNT*DISABLTY

Univariate F Tests

Effect	SS	DF	MS	F	P
WRATR	2.111	2	1.056	0.023	0.977
Error	544.000	12	45.333		
WRATA	52.778	2	26.389	0.587	0.571
Error	539.333	12	44.944		

Multivariate Test Statistics

```
        Wilks' Lambda =      0.908
          F-Statistic =      0.272   DF =   4,  22    Prob =     0.893

        Pillai Trace =       0.092
          F-Statistic =      0.290   DF =   4,  24    Prob =     0.882

  Hotelling-Lawley Trace =   0.101
          F-Statistic =      0.252   DF =   4,  20    Prob =     0.905

            THETA = 0.089 S =  2, M =-0.5, N =  4.5 Prob =       0.760
```

```
DATA        SSAMPLE;
INFILE      'TAPE81.DAT';
INPUT       SUBJNO TREATMNT DISABLTY WRATR WRATA IQ CELLNO;
PROC GLM;
     CLASS TREATMNT DISABLTY;
     MODEL WRATR WRATA = TREATMNT DISABLTY TREATMNT*DISABLTY/NOUNI;
     MANOVA H=_ALL_/SHORT;
```

General Linear Models Procedure
Class Level Information

Class	Levels	Values
TREATMNT	2	1 2
DISABLTY	3	1 2 3

Number of observations in data set = 18

General Linear Models Procedure
Multivariate Analysis of Variance

Characteristic Roots and Vectors of: E Inverse * H, where
H = Type III SS&CP Matrix for TREATMNT   E = Error SS&CP Matrix

Characteristic Root	Percent	Characteristic Vector V'EV=1	
		WRATR	WRATA
6.2610363687	100.00	0.03206535	0.02680051
0.0000000000	0.00	-0.02856728	0.03379300

Manova Test Criteria and Exact F Statistics for
the Hypothesis of no Overall TREATMNT Effect
H = Type III SS&CP Matrix for TREATMNT   E = Error SS&CP Matrix

S=1   M=0   N=4.5

Statistic	Value	F	Num DF	Den DF	Pr > F
Wilks' Lambda	0.13772139	34.4357	2	11	0.0001
Pillai's Trace	0.86227861	34.4357	2	11	0.0001
Hotelling-Lawley Trace	6.26103637	34.4357	2	11	0.0001
Roy's Greatest Root	6.26103637	34.4357	2	11	0.0001

Characteristic Roots and Vectors of: E Inverse * H, where
H = Type III SS&CP Matrix for DISABLTY   E = Error SS&CP Matrix

Characteristic Root	Percent	Characteristic Vector V'EV=1	
		WRATR	WRATA
2.8872408474	99.73	0.02260839	0.03531017
0.0077932220	0.27	-0.03651215	0.02476743

TABLE 9.6 *(CONTINUED)*

Manova Test Criteria and F Approximations for
the Hypothesis of no Overall DISABLTY Effect
H = Type III SS&CP Matrix for DISABLTY   E = Error SS&CP Matrix

S=2   M=-0.5   N=4.5

Statistic	Value	F	Num DF	Den DF	Pr > F
Wilks' Lambda	0.25526256	5.3860	4	22	0.0035
Pillai's Trace	0.75048108	3.6037	4	24	0.0195
Hotelling-Lawley Trace	2.89503407	7.2376	4	20	0.0009
Roy's Greatest Root	2.88724085	17.3234	2	12	0.0003

NOTE: F Statistic for Roy's Greatest Root is an upper bound.
NOTE: F Statistic for Wilks' Lambda is exact.

Characteristic Roots and Vectors of: E Inverse * H, where
H = Type III SS&CP Matrix for TREATMNT*DISABLTY   E = Error SS&CP Matrix

Characteristic Root	Percent	Characteristic Vector V'EV=1	
		WRATR	WRATA
0.0980388337	97.11	0.00187535	0.04291087
0.0029147009	2.89	0.04290407	-0.00434641

General Linear Models Procedure
Multivariate Analysis of Variance

Manova Test Criteria and F Approximations for
the Hypothesis of no Overall TREATMNT*DISABLTY Effect
H = Type III SS&CP Matrix for TREATMNT*DISABLTY   E = Error SS&CP Matrix

S=2   M=-0.5   N=4.5

Statistic	Value	F	Num DF	Den DF	Pr > F
Wilks' Lambda	0.90806786	0.2717	4	22	0.8930
Pillai's Trace	0.09219163	0.2899	4	24	0.8816
Hotelling-Lawley Trace	0.10095353	0.2524	4	20	0.9048
Roy's Greatest Root	0.09803883	0.5882	2	12	0.5706

NOTE: F Statistic for Roy's Greatest Root is an upper bound.
NOTE: F Statistic for Wilks' Lambda is exact.

In MANCOVA the basic partition of variance is the same as in MANOVA. However, all the matrices—$\mathbf{Y}_{ikm}$, $\mathbf{D}_k$, $\mathbf{T}_m$, $\mathbf{DT}_{km}$, and $\mathbf{GM}$—have three entries in our example; the first entry is the covariate (IQ score) and the second two entries are the two DV scores (WRAT-R and WRAT-A). For example, for the first child with mild disability and treatment, the column matrix of covariate and DV scores is

$$\mathbf{Y}_{111} = \begin{bmatrix} 110 \\ 115 \\ 108 \end{bmatrix} \quad \begin{matrix} \text{(IQ)} \\ \text{(WRAT-R)} \\ \text{(WRAT-A)} \end{matrix}$$

As in MANOVA, difference matrices are found by subtraction, and then the squares and cross-products matrices are found by multiplying each difference matrix by its transpose to form the **S** matrices.

At this point another departure from MANOVA occurs. The **S** matrices are partitioned into sections corresponding to the covariates, the DVs, and the cross products of covariates and DVs. For the example, the cross-products matrix for the main effect of treatment is

$$\mathbf{S}_T = \begin{bmatrix} [2.00] & [64.67 & 54.67] \\ 64.67 & 2090.89 & 1767.56 \\ 54.67 & 1767.56 & 1494.22 \end{bmatrix}$$

The lower right-hand partition is the $\mathbf{S}_T$ matrix for the DVs (or $\mathbf{S}_T^{(Y)}$), and is the same as the $\mathbf{S}_T$ matrix developed in Section 9.4.1. The upper left matrix is the sum of squares for the covariate (or $\mathbf{S}_T^{(X)}$). (With additional covariates, this segment becomes a full sum-of-squares and cross-products matrix.) Finally, the two off-diagonal segments contain cross products of covariates and DVs (or $\mathbf{S}_T^{(XY)}$).

Adjusted or **S*** matrices are formed from these segments. The **S*** matrix is the sums-of-squares and the cross-products of DVs adjusted for effects of covariates. Each sum of squares and each cross product is adjusted by a value that reflects variance due to differences in the covariate.

In matrix terms, the adjustment is

$$\mathbf{S}^* = \mathbf{S}^{(Y)} - \mathbf{S}^{(YX)}(\mathbf{S}^{(X)})^{-1}\mathbf{S}^{(XY)}$$

The adjusted cross-products matrix **S*** is found by subtracting from the unadjusted cross-products matrix of DVs ($\mathbf{S}^{(Y)}$) a product based on the cross-products matrix for covariate(s) ($\mathbf{S}^{(X)}$) and cross-products matrices for the relation between the covariates and the DVs ($\mathbf{S}^{(YX)}$ and $\mathbf{S}^{(XY)}$).

The adjustment is made for the regression of the DVs (Y) on the covariates (X). Because $\mathbf{S}^{(XY)}$ is the transpose of $\mathbf{S}^{(YX)}$, their multiplication is analogous to a squaring operation. Multiplying by the inverse of $\mathbf{S}^{(X)}$ is analogous to division. As shown in Chapter 3 for simple scalar numbers, the regression coefficient is the sum of cross products between X and Y, divided by the sum of squares for X.

An adjustment is made to each **S** matrix to produce **S*** matrices. The **S*** matrices are 2×2 matrices, but their entries are usually smaller than those in the original MANOVA **S** matrices. For the example, the reduced **S*** matrices are

$$\mathbf{S}_D^* = \begin{bmatrix} 388.18 & 500.49 \\ 500.49 & 654.57 \end{bmatrix} \qquad \mathbf{S}_T^* = \begin{bmatrix} 2059.50 & 1708.24 \\ 1708.24 & 1416.88 \end{bmatrix}$$

$$\mathbf{S}_{DT}^* = \begin{bmatrix} 2.06 & 0.87 \\ 0.87 & 19.61 \end{bmatrix} \qquad \mathbf{S}_{S(DT)}^* = \begin{bmatrix} 528.41 & -26.62 \\ -26.62 & 324.95 \end{bmatrix}$$

Note that, as in the lower right-hand partition, cross-products matrices may have negative values for entries other than the major diagonal which contains sums of squares.

Tests appropriate for MANOVA are applied to the adjusted S* matrices. Ratios of determinants are formed to test hypotheses about main effects and interactions by using Wilks' Lambda criterion (Equation 9.4). For the example, the determinants of the four matrices needed to test the three hypotheses (two main effects and the interaction) are

$$|\mathbf{S}_{S(DT)}{}^*| = 171032.69$$

$$|\mathbf{S}_D{}^* + \mathbf{S}_{S(DT)}{}^*| = 673383.31$$

$$|\mathbf{S}_T{}^* + \mathbf{S}_{S(DT)}{}^*| = 1680076.69$$

$$|\mathbf{S}_{DT}{}^* + \mathbf{S}_{S(DT)}{}^*| = 182152.59$$

The source table for MANCOVA, analogous to that produced for MANOVA, for the sample data is in Table 9.7.

One new item in this source table that is not in the MANOVA table of Section 9.4.1 is the variance in the DVs due to the covariate. (With more than one covariate, there is a line for combined covariates and a line for each of the individual covariates.) As in ANCOVA, one degree of freedom for error is used for each covariate so that df_2 and s of Equation 9.5 are modified. For MANCOVA, then,

$$s = \sqrt{\frac{(p + q)^2(df_{effect})^2 - 4}{(p + q)^2 + (df_{effect})^2 - 5}} \qquad (9.10)$$

where q is the number of covariates and all other terms are defined as in Equation 9.7. And

$$df_2 = s\left[(df_{error}) - \frac{(p + q) - df_{effect} + 1}{2}\right] - \left[\frac{(p + q)(df_{effect} - 2)}{2}\right]$$

Approximate F is used to test the significance of the covariate-DV relationship as well as main effects and interactions. If a significant relationship is found, Wilks' Lambda is used to find the strength of association as shown in Equations 9.6 or 9.9.

9.5 ■ SOME IMPORTANT ISSUES

9.5.1 ■ Criteria for Statistical Inference

Several multivariate statistics are available in MANOVA programs to test significance of main effects and interactions: Wilks' Lambda, Hotelling's trace criterion, Pillai's criterion, as well as Roy's gcr criterion. When an effect has only two levels (1 df, $s = 1$), the F tests for Wilks' Lambda, Hotelling's trace, and Pillai's criterion are identical. And usually when an effect has more than two levels (df > 1), the F values are slightly different but all three statistics are either significant or all are nonsignificant. Occasionally, however, some of the statistics are significant while others are not, and the researcher is left wondering which result to believe.

When there is only one degree of freedom for effect there is only one way to combine the DVs to separate the two groups from each other. However, when there is more than one degree of free-

TABLE 9.7 MULTIVARIATE ANALYSIS OF COVARIANCE OF WRAT-R AND WRAT-A SCORES

Source of variance	Wilks' Lambda	df_1	df_2	Multivariate F
Covariate	.58485	2.00	10.00	3.54913
Treatment	.13772	2.00	10.00	44.11554**
Disability	.25526	4.00	22.00	4.92112*
Treatment by Disability	.90807	4.00	22.00	0.15997

*$p < .01$.
**$p < .001$.

dom for effect, there is more than one way to combine DVs to separate groups. For example, with three groups, one way of combining DVs may separate the first group from the other two while the second way of combining DVs separates the second group from the third. Each way of combining DVs is a dimension along which groups differ (as described in gory detail in Chapter 11) and each generates a statistic.

When there is more than one degree of freedom for effect, Wilks' Lambda, Hotelling's trace criterion, and Pillai's criterion pool the statistics from each dimension to test the effect; Roy's gcr criterion uses only the first dimension (in our example, the way of combining DVs that separates the first group from the other two) and is the preferred test statistic for a few researchers (Harris, 1975). Most researchers, however, use one of the pooled statistics to test the effect (Olson, 1976).

Wilks' Lambda, defined in Equation 9.4 and Section 9.4.1, is a likelihood ratio statistic that tests the likelihood of the data under the assumption of equal population mean vectors for all groups against the likelihood under the assumption that population mean vectors are identical to those of the sample mean vectors for the different groups. Wilks' Lambda is the pooled ratio of error variance to effect variance plus error variance. Hotelling's trace is the pooled ratio of effect variance to error variance. Pillai's criterion is simply the pooled effect variances.

Wilks' Lambda, Hotelling's trace, and Roy's gcr criterion are often more powerful than Pillai's criterion when there is more than one dimension but the first dimension provides most of the separation of groups; they are less powerful when separation of groups is distributed over dimensions. But Pillai's criterion is said to be more robust than the other three (Olson, 1979). As sample size decreases, unequal n's appear, and the assumption of homogeneity of variance-covariance matrices is violated (Section 9.3.2.4), the advantage of Pillai's criterion in terms of robustness is more important. When the research design is less than ideal, then, Pillai's criterion is the criterion of choice.

In terms of availability, all the MANOVA programs reviewed here provide Wilks' Lambda, as do most research reports, so that Wilks' Lambda is the criterion of choice unless there is reason to use Pillai's criterion. Programs differ in the other statistics provided (see Section 9.6).

In addition to potentially conflicting significance tests for multivariate F is the irritation of a nonsignificant multivariate F but a significant univariate F for one of the DVs. If the researcher measures only one DV—the right one—the effect is significant, but because more DVs are measured, it is not. Why doesn't MANOVA combine DVs with a weight of 1 for the significant DV and a weight of zero for the rest? In fact, MANOVA comes close to doing just that, but multivariate F is often not as powerful as univariate or stepdown F, and significance can be lost. If this happens, about the best one can do is report the nonsignificant multivariate F and offer the univariate and/or stepdown result as a guide to future research.

9.5.2 ■ Assessing DVs

When a main effect or interaction is significant in MANOVA, the researcher has usually planned to pursue the finding to discover which DVs are affected. But the problems of assessing DVs in significant multivariate effects are similar to the problems of assigning importance to IVs in multiple regression (Chapter 5). First, there are multiple significance tests so some adjustment is necessary for inflated Type I error. Second, if DVs are uncorrelated, there is no ambiguity in assignment of variance to them, but if DVs are correlated, assignment of overlapping variance to DVs is problematical.

9.5.2.1 ■ Univariate F

If pooled within-group correlations among DVs are zero, univariate ANOVAs, one per DV, give the relevant information about their importance. Using ANOVA for uncorrelated DVs is analogous to assessing importance of IVs in multiple regression by the magnitude of their individual correlations with the DV. The DVs that have significant univariate Fs are the important ones, and they can be ranked in importance by strength of association. However, because of inflated Type I error rate due to multiple testing, more stringent alpha levels are required.

Because there are multiple ANOVAs, a Bonferroni type adjustment is made for inflated Type I error. The researcher assigns alpha for each DV so that alpha for the set of DVs does not exceed some critical value.

$$\alpha = 1 - (1 - \alpha_1)(1 - \alpha_2)\ldots(1 - \alpha_p) \tag{9.11}$$

The Type I error rate (α) is based on the error rate for testing the first DV (α_1), the second DV (α_2), and all other DVs to the pth, or last, DV (α_p).

All the alphas can be set at the same level, or more important DVs can be given more liberal alphas. For example, if there are four DVs and α for each DV is set at .01, the overall alpha level according to Equation 9.11 is .039, acceptably below .05 overall. Or, if α is set at .02 for 2 DVs, and at .001 for the other 2 DVs, overall α is .042, also below .05.

When DVs are correlated, there are two problems with univariate Fs. First, correlated DVs measure overlapping aspects of the same behavior. To say that two of them are both "significant" mistakenly suggests that the IV affects two different behaviors. For example, if the two DVs are Stanford-Binet IQ and WISC IQ, they are so highly correlated that an IV that affects one surely affects the other. The second problem with reporting univariate Fs for correlated DVs is inflation of Type I error rate; with correlated DVs, the univariate Fs are not independent and no straightforward adjustment of the error rate is possible.

Although reporting univariate F for each DV is a simple tactic, the report should also contain the pooled within-group correlations among DVs so the reader can make necessary interpretive adjustments. The pooled within-group correlation matrix is provided by SPSS MANOVA, SAS GLM, and SYSTAT GLM.

In the example of Tables 9.1 and 9.2, there is a significant multivariate effect of treatment (and of disability, although, as previously noted, it is not interesting in this example). It is appropriate to ask which of the two DVs is affected by treatment. Univariate ANOVAs for WRAT-R and WRAT-A are in Tables 9.8 and 9.9, respectively. The pooled within-group correlation between WRAT-R and WRAT-A is .057 with 12 df. Because the DVs are uncorrelated, univariate F with adjustment of α for multiple tests is appropriate. There are two DVs, so each is set at alpha .025.[12] With 2 and

TABLE 9.8 UNIVARIATE ANALYSIS OF VARIANCE OF WRAT-R SCORES

Source	SS	df	MS	F
D	520.7778	2	260.3889	5.7439
T	2090.8889	1	2090.8889	46.1225
DT	2.1111	2	1.0556	0.0233
S(DT)	544.0000	12	45.3333	

TABLE 9.9 ANALYSIS OF VARIANCE OF WRAT-A SCORES

Source	SS	df	MS	F
D	1126.7778	2	563.3889	12.5352
T	1494.2222	1	1494.2222	33.2460
DT	52.7778	2	26.3889	0.5871
S(DT)	539.5668	12	44.9444	

12 df, critical F is 5.10; with 1 and 12 df, critical F is 6.55. There is a main effect of treatment (and disability) for both WRAT-R and WRAT-A.

9.5.2.2 ■ Roy-Bargmann Stepdown Analysis[13]

The problem of correlated univariate F tests with correlated DVs is resolved by stepdown analysis (Bock, 1966; Bock & Haggard, 1968). Stepdown analysis of DVs is analogous to testing the importance of IVs in multiple regression by sequential analysis. Priorities are assigned to DVs according to theoretical or practical considerations.[14] The highest-priority DV is tested in univariate ANOVA, with appropriate adjustment of alpha. The rest of the DVs are tested in a series of ANCOVAs; each successive DV is tested with higher-priority DVs as covariates to see what, if anything, it adds to the combination of DVs already tested. Because successive ANCOVAs are independent, adjustment for inflated Type I error due to multiple testing is the same as in Section 9.5.2.1.

For the example, we assign WRAT-R scores higher priority since reading problems represent the most common presenting symptoms for learning disabled children. To keep overall alpha below .05, individual alpha levels are set at .025 for each of the two DVs. WRAT-R scores are analyzed through univariate ANOVA, as displayed in Table 9.8. Because the main effect of disability is not interesting and the interaction is not statistically significant in MANOVA (Table 9.2), the only effect of interest is treatment. The critical value for testing the treatment effect (6.55 with 1 and 12 df at $\alpha = .025$) is clearly exceeded by the obtained F of 46.1225.

WRAT-A scores are analyzed in ANCOVA with WRAT-R scores as covariate. The results of this analysis appear in Table 9.10.[15] For the treatment effect, critical F with 1 and 11 df at $\alpha = .025$ is

---

[12] When the design is very complicated and generates many main effects and interactions, further adjustment of α is necessary in order to keep overall α under .15 or so, across the ANOVAs for the DVs.

[13] Stepdown analysis can be run in lieu of MANOVA where a significant stepdown F is interpreted as a significant multivariate effect for the main effect or interaction.

[14] It is also possible to assign priority on the basis of statistical criteria such as univariate F, but the analysis suffers all the problems inherent in stepwise regression, discussed in Chapter 5.

[15] A full stepdown analysis is produced as an option through SPSS MANOVA. For illustration, however, it is helpful to show how the analysis develops.

TABLE 9.10 ANALYSIS OF COVARIANCE OF WRAT-A SCORES, WITH WRAT-R SCORES AS THE COVARIATE

Source	SS	df	MS	F
Covariate	1.7665	1	1.7665	0.0361
D	538.3662	2	269.1831	5.5082
T	268.3081	1	268.3081	5.4903
DT	52.1344	2	26.0672	0.5334
S(DT)	537.5668	11	48.8679	

6.72. This exceeds the obtained F of 5.49. Thus, according to stepdown analysis, the significant effect of treatment is represented in WRAT-R scores, with nothing added by WRAT-A scores.

Note that WRAT-A scores show significant univariate but not stepdown F. Because WRAT-A scores are not significant in stepdown analysis does not mean they are unaffected by treatment but rather that no unique variability is shared with treatment after adjustment for differences in WRAT-R.

This procedure can be extended to sets of DVs through MANCOVA. If the DVs fall into categories, such as scholastic variables and attitudinal variables, one can ask whether there is any change in attitudinal variables as a result of an IV, after adjustment for differences in scholastic variables. The attitudinal variables serve as DVs in MANCOVA while the scholastic variables serve as covariates. This type of analysis is demonstrated in the sequential discriminant function analysis of Chapter 11.

9.5.2.3 ■ Choosing among Strategies for Assessing DVs

You may find the procedures of Sections 11.6.3 and 11.6.4 more useful than univariate or stepdown F for assessing DVs when you have a significant multivariate main effect with more than two levels. Similarly, you may find the procedures described in Section 10.5.1 helpful for assessment of DVs if you have a significant multivariate interaction.

The choice between univariate and stepdown F is not always easy, and often you want to use both. When there is very little correlation among the DVs, univariate F with adjustment for Type I error is acceptable. When DVs are correlated, stepdown F is preferable on grounds of statistical purity, but you have to prioritize the DVs and the results can be difficult to interpret.

If DVs are correlated and there is some compelling priority ordering of them, stepdown analysis is clearly called for, with univariate Fs and pooled within-cell correlations reported simply as supplemental information. For significant lower-priority DVs, marginal and/or cell means adjusted for higher-priority DVs are reported and interpreted.

If the DVs are correlated but the ordering is somewhat arbitrary, an initial decision in favor of stepdown analysis is made. If the pattern of results from stepdown analysis makes sense in the light of the pattern of univariate results, interpretation takes both patterns into account with emphasis on DVs that are significant in stepdown analysis. If, for example, a DV has a significant univariate F but a nonsignificant stepdown F, interpretation is straightforward: the variance the DV shares with the IV is already accounted for through overlapping variance with one or more higher-priority DVs. This is the interpretation of WRAT-A in the preceding section and the strategy followed in Section 9.7.

But if a DV has a nonsignificant univariate F and a significant stepdown F, interpretation is much more difficult. In the presence of higher-order DVs as covariates, the DV suddenly takes on "importance." In this case, interpretation is tied to the context in which the DVs entered the stepdown analysis. It may be worthwhile at this point, especially if there is only a weak basis for ordering DVs, to forgo evaluation of statistical significance of DVs and resort to simple description.

After finding a significant multivariate effect, unadjusted marginal and/or cell means are reported for DVs with high univariate Fs but significance levels are not given.

An alternative to attempting interpretation of either univariate or stepdown F is interpretation of discriminant functions, as discussed in Section 11.6.3. This process is facilitated when SPSS MANOVA, SAS GLM, or SYSTAT GLM is used because information about the discriminant functions is provided as a routine part of the output.

Another perspective is whether DVs differ significantly in the effects of IVs on them. For example, Does treatment affect reading significantly more than it affects arithmetic? Tests for contrasts among DVs have been developed in the context of meta-analysis with its emphasis on comparing effect sizes. Rosenthal and Rubin (1986) demonstrate these techniques.

9.5.3 ■ Specific Comparisons and Trend Analysis

When there are more than two levels in a significant multivariate main effect and when a DV is important to the main effect, the researcher often wants to perform specific comparisons or trend analysis of the DV to pinpoint the source of the significant difference. Similarly, when there is a significant multivariate interaction and a DV is important to the interaction, the researcher follows up the finding with comparisons on the DV.[16] Review Sections 3.2.6, 8.5.2.3, and 10.5.1 for examples and discussions of comparisons. The issues and procedures are the same for individual DVs in MANOVA as in ANOVA.

Comparisons are either planned (performed in lieu of omnibus F) or post hoc (performed after omnibus F to snoop the data). When comparisons are post hoc, an extension of the Scheffé procedure is used to protect against inflated Type I error due to multiple tests. The procedure is very conservative but allows for an unlimited number of comparisons. Following Scheffé for ANOVA (see Section 3.2.6), the tabled critical value of F is multiplied by the degrees of freedom for the effect being tested to produce an adjusted, and much more stringent, F. If marginal means for a main effect are being contrasted, the degrees of freedom are those associated with the main effect. If cell means are being contrasted, our recommendation is to use the degrees of freedom associated with the interaction.

9.5.4 ■ Design Complexity

In between-subjects designs with more than two IVs, extension of MANOVA is straightforward as long as sample sizes are equal within each cell of the design. The partition of variance continues to follow ANOVA, with a variance component computed for each main effect and interaction. The pooled variance-covariance matrix due to differences among subjects within cells serves as the single error term. Assessment of DVs and comparisons proceed as described in Sections 9.5.2 and 9.5.3.

Two major design complexities that arise, however, are inclusion of within-subjects IVs and unequal sample sizes in cells.

9.5.4.1 ■ Within-Subjects and Between-Within Designs

The simplest design with repeated measures is a one-way within-subjects design where the same subjects are measured on a single DV on several different occasions. The design can be complicated by

---

[16] Comparisons and trend analysis can also be performed on a composite DV as illustrated in Tabachnick and Fidell (1983) Section 8.5.2 but the results for a composite DV are less interpretable.

addition of between-subjects IVs or more within-subjects IVs. Consult Chapters 3 and 8 for discussion of some of the problems that arise in ANOVA with repeated measures.

Repeated measures extends to MANOVA when the researcher measures several DVs on several different occasions. The occasions can be viewed in two ways. In the traditional sense, occasions produce a within-subjects IV with as many levels as occasions (Chapter 3). Alternatively, occasions can be treated as separate DVs—one DV per occasion (Section 9.2.8). In this latter view, if there is more than one DV measured on each occasion, the design is said to be doubly multivariate—multiple DVs are measured on multiple occasions. (There is no distinction between the two views when there are only two levels of the within-subjects IV.) Section 10.5.3 has an example of a doubly multivariate analysis of a small data set with a between-subjects IV (PROGRAM), a within-subjects IV (MONTH), and two DVs (WTLOSS and ESTEEM), both measured three times. Generalizations from the example are reasonably straightforward.

9.5.4.2 ■ Unequal Sample Sizes

When cells in a factorial ANOVA have an unequal number of scores, the sum of squares for effect plus error no longer equals the total sum of squares, and tests of main effects and interactions are correlated. There are a number of ways to adjust for overlap in sums of squares (cf. Woodward & Overall, 1975), as discussed in some detail in Section 8.5.2.2, particularly Table 8.11. Both the problem and the solutions generalize to MANOVA.

All the MANOVA programs described in Section 9.6 adjust for unequal n. SPSS MANOVA offers both Method 1 adjustment (METHOD = UNIQUE) and Method 3 adjustment (METHOD = SEQUENTIAL). Method 3 adjustment with survey data through SPSS MANOVA is shown in Section 9.7.2. BMDP4V allows very flexible specification of method through choice of cell-weighting procedure. In SAS GLM Method 1 (called TYPE III or TYPE IV) is the default among four options available. Default is also Method 1 for SYSTAT GLM; alternative methods require specification of custom sums of squares for error terms or MEANS and WEIGHTS instructions.

9.5.5. ■ MANOVA vs. ANOVAs

MANOVA works best with moderately correlated DVs. For example, two DVs, such as time to complete a task and number of errors, might be expected to have a moderate correlation. MANOVA is less attractive if correlations among DVs are very high or very low.

Using very highly correlated IVs in MANOVA is wasteful. For example, the effects of the Head Start program might be tested in a MANOVA with the WISC and Stanford-Binet as DVs. The overall multivariate test works well, but after the highest priority DV is entered in stepdown analysis, tests of remaining DVs are ambiguous. Once that DV becomes a covariate, there is no variance remaining in the lower priority DVs to be related to IV main effects or interactions. Univariate tests also are highly misleading, because they suggest effects on different behaviors when actually there is one behavior being measured repeatedly. Better strategies are to pick a single DV (preferably the most reliable) or to create a composite score (an average if the DVs are commensurate or a principal component score if they are not) for use in ANOVA.

MANOVA also is wasteful if DVs are uncorrelated. The multivariate test may lack power, and there is little difference between univariate and stepdown results. The only advantage to MANOVA over separate ANOVAs on each DV is control of familywise Type I error. However, this error rate can be controlled by applying a Bonferroni correction (cf. Equation 9.11) to each test in a set of separate ANOVAs on each DV.

Sometimes there is a mix of correlated and uncorrelated DVs. For example, there may be a set of moderately correlated DVs related to performance on a task and another set of moderately correlated DVs related to attitudes. Separate MANOVAs on each of the two sets of moderately correlated DVs are likely to produce the most interesting interpretations. Or, one set might serve as covariates in a single MANCOVA.

9.6 ■ COMPARISON OF PROGRAMS

SPSS, BMDP, SAS, and SYSTAT all have highly flexible and full-featured MANOVA programs, as seen in Table 9.11. One-way between-subjects MANOVA is also available through discriminant function programs, as discussed in Chapter 11.

9.6.1 ■ SPSS Package

SPSS MANOVA offers several methods of adjustment for unequal n and is the only program that performs Roy-Bargmann stepdown analysis as an option (Section 9.5.2.2). Should Pillai's criterion be desired as the multivariate significance test (Section 9.5.1), it and several other alternatives are available through SPSS MANOVA. The program is also reasonably simple to use for comparisons on both between- and within-subjects IVs (see Section 10.5.1).

Except for outliers, tests of assumptions are readily available in MANOVA. Bivariate collinearity and homogeneity of variance-covariance matrices are readily tested through within-cell correlations and homogeneity of dispersion matrices, respectively. Multicollinearity is assessed through the determinant of the within-cells correlation matrix (cf. Section 4.1.7).

For between-subjects designs, Bartlett's test of sphericity tests the null hypothesis that correlations among DVs are zero; if they are, univariate F (with Bonferroni adjustment) is used instead of stepdown F to test the importance of DVs (Section 9.5.2.1). In repeated measures designs, the sphericity test also evaluates the homogeneity of covariance assumption; if the assumption is rejected (that is, if the test is significant), one of the alternatives to repeated measures ANOVA—MANOVA, for instance—is appropriate (Section 8.5.5). There are also the Greenhouse-Geisser, Huynh-Feldt and lower-bound epsilons for adjustment of df for homogeneity of covariance. Optionally, SPSS MANOVA will do the adjustment and provide significance levels for the effects with adjusted df.

For designs with covariates or when stepdown F is used, homogeneity of regression is readily tested through setup language illustrated in Table 9.14. If the assumption is violated, the manuals describe procedures for ANCOVA with separate regression estimates, if that is your choice. Adjusted cell and marginal means can be obtained, although in different ways. The PMEANS procedure provides adjusted cell means. For adjusted marginal means, however, the CONSPLUS procedure is used, illustrated in Sections 9.7.2 and 9.7.3.

Finally, a principal components analysis can be performed on the DVs, as described in the manuals. In the case of multicollinearity or singularity among DVs (see Chapter 4), principal components analysis can be used to produce composite variables that are orthogonal to one another. However, the program still performs MANOVA on the raw DV scores, not the component scores.

TABLE 9.11 COMPARISON OF PROGRAMS FOR MULTIVARIATE ANALYSIS OF VARIANCE AND COVARIANCE[a]

Feature	BMDP4V	SPSS MANOVA	SYSTAT GLM	SAS GLM
Input				
Variety of strategies for unequal n	Yes	Yes	Yes	Yes
Alternative languages or procedures to specify design	Yes	Yes	Yes	Yes
Specify tolerance	No	No	Yes	SINGULAR
Output				
Standard source table for Wilks' Lambda	No	PRINT = SIGNIF(BRIEF)	No	No
Cell covariance matrices	No	Yes	No	No
Cell covariance matrix determinants	No	Yes	No	No
Cell correlation matrices	No	Yes	No	No
Cell SSCP matrices	No	Yes	No	No
Cell SSCP determinants	No	Yes	No	No
Marginal standard deviation for factorial design	Yes	No	No	No
Unadjusted marginal means for factorial design	Yes	Yes	No	Yes
Weighted marginal means	Yes	Yes	No	No
Unadjusted cell means	Yes	Yes	No	Yes
Unadjusted cell standard deviations	Yes	Yes	No	Yes
Cell and marginal variance	Yes	No	No	No
Confidence interval around unadjusted cell means	No	Yes	No	No
Unadjusted cell and marginal minimum and maximum values	Yes	No	No	No
Adjusted cell means	No	PMEANS	PRINT MEDIUM	LSMEANS
Standard errors for adjusted cell means	No	No	PRINT MEDIUM	LSMEANS
Adjusted marginal means	No	Yes[b]	PRINT MEDIUM	LSMEANS
Standard errors for adjusted marginal means	No	No	PRINT MEDIUM	LSMEANS
Wilks' Lambda (U statistic) with approximate F statistic	LRATIO	Yes	Yes	Yes
Criteria other than Wilks'	Yes	Yes	Yes	Yes
Mahalanobis distance for multivariate outliers	No	No	Data file	No
Other influence/leverage statistics	No	No	Data file	Data file
Canonical (discriminant function) statistics[c]	EVALues & VECTors	Yes	Yes	Yes
Univariate F tests	Yes	Yes	Yes	Yes
Averaged univariate F tests	No	Yes	No	No
Stepdown F tests (DVs)	No	Yes	No	No
Sphericity test for homogeneity of covariance	No	Yes	No	No
Greenhouse-Geisser and Huynh-Feldt adjustment for heterogeneity of covariance	Yes	Yes	Yes	Yes
Tests for univariate homogeneity of variance	No	Yes	No	No
Test for homogeneity of covariance matrices	No	Box's M	No	No
Principal component analysis of residuals	No	Yes	Yes	No
Hypothesis SSCP matrices	Yes	No	Yes	Yes
Hypothesis covariance matrices	No	Yes	No	No

TABLE 9.11 *(CONTINUED)*

Feature	BMDP4V	SPSS MANOVA	SYSTAT GLM	SAS GLM
Inverse of hypothesis SSCP matrices	No	No	No	Yes
Pooled within-cell error SSCP matrix	Yes	Yes	Yes	Yes
Pooled within-cell covariance matrix	No	Yes	Yes	No
Pooled within-cell correlation matrix	No	Yes	Yes	Yes
Total SSCP matrix	No	No	Yes	No
Determinants of pooled within-cell correlation matrix	No	Yes	No	No
Covariance matrix for adjusted cell means	No	No	No	Data file
Effect size for univariate tests	No	POWER	No	No
Power analysis	No	POWER	No	Yes
SMCs with effects for each DV	No	No	Yes	No
Confidence intervals for multivariate tests	No	Yes	No	No
Post hoc tests with adjustment	No	Yes[d]	Yes	Yes
Specific comparisons	Yes	Yes	Yes	Yes
Tests of simple effects (complete)	Yes	Yes	No	No
Homogeneity of regression	No	Yes	Yes	No
ANCOVA with separate regression estimates	No	Yes	No	No
Regression coefficient for each covariate	No	Yes	PRINT LONG	No
Regression coefficient for each cell	No	No	No	Yes
R^2 for model	No	No	No	Yes
R^2 for each DV	No	No	Yes	No
Coefficient of variation	No	No	No	Yes
Normalized plots for each DV and covariate	No	Yes	No	No
Predicted values and residuals for each case	No	Yes	Data file	Yes
Confidence limits for predicted values	No	No	No	Yes
Residuals plots	No	Yes	No	No

[a]Additional features are discussed in Chapter 8 (ANCOVA).

[b]Available through CONSPLUS procedure, see Section 9.7.

[c]Discussed more fully in Chapter 11.

[d]Bonferroni and Scheffé confidence intervals

If MANOVA for component scores is desired, use the results of PCA and the COMPUTE facility to generate component scores for use as DVs.

9.6.2 ■ BMDP Series

The MANOVA program in the BMDP series is 4V. The program was originally known as URWAS (University of Rochester Weighted Analysis of Variance System). It is unique among the programs reviewed here in that it uses a cell weighting system to adjust for unequal *n* and to specify hypotheses to be tested. Cell weighting can be used for situations where population sizes are unequal even though sample sizes are equal.

For repeated measures designs, both the Huynh-Feldt and the Greenhouse-Geisser-Imhof corrections for degrees of freedom are available to adjust for violation of the assumption of homogeneity of covariance. Output includes significance levels for effects with adjusted df.

Statistics appropriate for doubly multivariate analysis (in which levels of a within-subjects IV are treated as separate DVs) are automatically provided when multiple DVs are specified (cf.

Chapter 10). Weighted marginal means are given that are useful if weights other than sample sizes are used (see Section 8.5.2.2). Adjusted (for covariates) cell and marginal means are unavailable, and some other program such as BMDP2V or BMDP1V must be used to find them. Three multi-variate test criteria, as well as univariate F tests, are automatically provided. Specific comparisons include a simple effects procedure as described in Section 10.5.1.

For a factorial design in which tests of all main effects and interactions are desired, a simple statement so indicates. For complex and/or nonorthogonal designs, Model Description Language is available, in which the user specifies the design matrix of contrast coefficients and/or other design goodies. Finally, a structural-equation procedure allows the user to write out an equation that specifies the design and desired tests.

Neither stepdown analysis nor tests for homogeneity of regression are available in BMDP4V, although MANOVA stepdown analysis can be done through BMDP2V or 1V, and tests for homogeneity of regression can be done through BMDP1V, as described in Chapter 8. No direct evidence is available to evaluate multicollinearity, but the program does deal with the problem. If a sum-of-squares and products matrix is multicollinear, it is conditioned and a pseudoinverse calculated. And if there is any problem with the accuracy of a statistic, warning messages regarding precision are printed out.

9.6.3 ■ SYSTAT System

In SYSTAT, the MGLH (multivariate general linear model) program has been renamed GLM, and specialized programs have been excerpted for ANOVA, REGRESS, and DISCRIM. SYSTAT ANOVA may be used for simple, fully factorial MANOVA, however GLM is recommended for its numerous features and flexibility, and because it is not much more difficult to set up.

Model 1 adjustment for unequal n is provided by default, along with a strong argument as to its benefits. Other options are available, however, by specification of error terms or a series of sequential regression analyses. Several criteria are provided for tests of multivariate hypotheses, along with a great deal of flexibility in specifying these hypotheses. Mahalanobis distance is provided in an output data set for evaluation of multivariate outliers, but the values are the square root of those typically given. To apply a χ^2 criterion to evaluate significance of outliers, you must first square the distance value produced by SYSTAT.

For MANCOVA or stepdown F, tests are provided for homogeneity of regression. The program provides cell and marginal least squares means and their standard errors, adjusted for covariates, if any. Other univariate statistics are not provided in the program, but they can be obtained through the STATS module.

The program provides both multivariate and univariate (with Greenhouse-Geisser and Huynh-Feldt adjustment) output for designs with repeated measures, along with an automatic trend analysis for the within-subjects factor(s).

Like SPSS MANOVA, principal components analysis can be done on the pooled within-cell correlation matrix. But also like the SPSS program, the MANOVA is performed on the original scores.

9.6.4 ■ SAS System

MANOVA in SAS is done through the PROC GLM. This general linear model program, like SYSTAT GLM, has great flexibility in testing models and specific comparisons. Four types of adjustment for unequal-n are available, called TYPE I through TYPE IV estimable functions (cf. 8.5.2.2);

this program is considered by some to have provided the archetypes of the choices available for unequal-*n* adjustment. Adjusted cell and marginal means are printed out with the LSMEANS instruction. SAS has no test for multivariate outliers, however.

SAS GLM provides Greenhouse-Geisser and Huynh-Feldt adjustments to degrees of freedom and significance tests for effects using adjusted df. There is no explicit test for homogeneity of regression, but because this program can be used for any form of multiple regression, the assumption can be tested as a regression problem where the interaction between the covariate(s) and IV(s) is an explicit term in the regression equation (Section 8.5.1).

Of the programs reviewed, GLM has the most extensive selection of options for testing contrasts. In addition to a CONTRAST procedure for specifying comparisons, numerous post hoc procedures are available.

There is abundant information about residuals, as expected from a program that can be used for multiple regression. Should you want to plot residuals, however, a run through the PLOT procedure is required. As with most SAS programs, the output requires a fair amount of effort to decode until you become accustomed to the style.

9.7 ■ COMPLETE EXAMPLES OF MULTIVARIATE ANALYSIS OF VARIANCE AND COVARIANCE

In the research described in Appendix B, Section B.1, there is interest in whether the means of several of the variables differ as a function of sex role identification. Are there differences in self-esteem, introversion-extraversion, neuroticism, and so on, associated with a woman's masculinity and femininity?

Sex role identification is defined by the masculinity and femininity scales of the Bem Sex Role Inventory (Bem, 1974). Each scale is divided at its median to produce two levels of masculinity (high and low), two levels of femininity (high and low), and four groups: Undifferentiated (low femininity, low masculinity), Feminine (high femininity, low masculinity), Masculine (low femininity, high masculinity), and Androgynous (high femininity, high masculinity). The design produces a main effect of masculinity, a main effect of femininity, and a masculinity-femininity interaction.[17]

DVs for this analysis are self-esteem (ESTEEM), internal versus external locus of control (CONTROL), attitudes toward women's role (ATTROLE), socioeconomic level (SEL2), introversion-extroversion (INTEXT), and neuroticism (NEUROTIC).

Omnibus MANOVA (Section 9.7.2) asks whether these DVs are associated with the two IVs (femininity and masculinity) or their interaction. Then Roy-Bargmann stepdown analysis, in conjunction with the univariate *F* values, allows us to examine the pattern of relationships between DVs and each IV.

In a second example (Section 9.7.3), MANCOVA is performed with SEL2, CONTROL, and ATTROLE used as covariates and ESTEEM, INTEXT, and NEUROTIC used as DVs. The research question is whether the three personality DVs vary as a function of sex role identification (the two IVs and their interaction) after adjusting for differences in socioeconomic status, attitudes toward women's role, and beliefs regarding locus of control of reinforcements.

---

[17] Some would argue with the wisdom of considering masculinity and femininity separate IVs, and of performing a median split on them to create groups. This example is used for didactic purposes.

9.7.1 ■ Evaluation of Assumptions

Before proceeding with MANOVA and MANCOVA, we assess the variables with respect to practical limitations of the techniques.

9.7.1.1 ■ Unequal Sample Sizes and Missing Data

SPSS FREQUENCIES is run with SORT and SPLIT FILE to divide cases into the four groups. Data and distributions for each DV within each group are inspected for missing values, shape, and variance (see Table 9.12 for output on the CONTROL variable). The run reveals the presence of a case for which the CONTROL score is missing. No datum is missing on any of the other DVs for the 369 women who were administered the Bem Sex Role Inventory. Deletion of the case with the missing value, then, reduces the available sample size to 368.

Sample sizes are quite different in the four groups: there are 71 Undifferentiated, 172 Feminine, 36 Masculine, and 89 Androgynous women in the sample. Because it is assumed that these differ-

---

TABLE 9.12 SETUP AND SELECTED SPSS FREQUENCIES OUTPUT FOR MANOVA VARIABLES SPLIT BY GROUPS

```
DATA LIST         FILE='MANOVA.DAT'
                  /CASENO 1-4 ANDRM 5-7 FEM 8-10 MASC 11-13 ESTEEM 14-16
                         CONTROL 17-19 ATTROLE 20-22 SEL2 23-30 (5)
                         INTEXT 31-34 (1) NEUROTIC 35-37.
VALUE LABELS      ANDRM 1 'UNDIFF' 2 'FEMNINE' 3 'MASCLNE' 4'ANDROGNS'/
                  FEM,MASC 1 'LOW' 2 'HIGH'.
MISSING VALUES    CONTROL(0).
SORT CASES BY ANDRM.
SPLIT FILE BY ANDRM.
FREQUENCIES       VARIABLES=ESTEEM TO NEUROTIC/
                  FORMAT = NOTABLE/
                  HISTOGRAM = NORMAL/
                  STATISTICS = ALL.


CONTROL

     Count      Value    One symbol equals approximately  1.20 occurrences

       26        5.00   |*****************:*****
       54        6.00   |*************************************:********
       51        7.00   |*******************************************:
       18        8.00   |***************
       20        9.00   |*********:*******
        3       10.00   |*:*
                        +----+----+----+----+----+----+----+----+----+----+
                        0    12        24        36        48        60
                                    Histogram frequency


ANDRM:    2   FEMNINE

CONTROL


Mean          6.773     Std err       .097     Median        7.000
Mode          6.000     Std dev      1.266     Variance      1.603
Kurtosis      -.381     S E Kurt      .368     Skewness       .541
S E Skew       .185     Range        5.000     Minimum       5.000
Maximum      10.000     Sum       1165.000


Valid cases     172     Missing cases        1
```

ences in sample size reflect real processes in the population, the sequential approach to adju
for unequal n is used with FEM (femininity) given priority over MASC (masculinity), and F
MASC (interaction between femininity and masculinity).

9.7.1.2 ■ Multivariate Normality

The sample size of 368 includes over 35 cases for each cell of the 2×2 between-subjects design, far
more than the 20 df for error suggested to assure multivariate normality of the sampling distribution
of means, even with unequal sample sizes; there are far more cases than DVs in the smallest cell.
Further, the distributions for the full run (of which CONTROL in Table 9.12 is a part) produce no
cause for alarm. Skewness is not extreme and, when present, is roughly the same for the DVs.

Two-tailed tests are automatically performed by the computer programs used. That is, the F test
looks for differences between means in either direction.

9.7.1.3 ■ Linearity

The full output for the run of Table 9.12 reveals no cause for worry about linearity. All DVs in each
group have reasonably balanced distributions so there is no need to examine scatterplots for each
pair of DVs within each group. Had scatterplots been necessary, SPSS PLOT would have been used
with the SORT and SPLIT FILE setup in Table 9.12.

9.7.1.4 ■ Outliers

BMDPAM is used to check for univariate and multivariate outliers (cf. Table 8.16) within each of
the four groups. Using a cutoff criterion of $p < .001$, no outliers are found.

9.7.1.5 ■ Homogeneity of Variance-Covariance Matrices

As a preliminary check for robustness, sample variances (in the full run of Table 9.12) for each DV
are compared across the four groups. For no DV does the ratio of largest to smallest variance
approach 10:1. As a matter of fact, the largest ratio is about 1.5:1 for the Undifferentiated versus
Androgynous groups on CONTROL.

Sample sizes are widely discrepant, with a ratio of almost 5:1 for the Feminine to Masculine
groups. However, with very small differences in variance and two-tailed tests, the discrepancy in sample sizes does not invalidate use of MANOVA. The very sensitive Box's M test for homogeneity of
dispersion matrices (performed through SPSS MANOVA as part of the major analysis in Table 9.15)
produces $F(63, 63020) = 1.07, p > .05$, confirming homogeneity of variance-covariance matrices.

9.7.1.6 ■ Homogeneity of Regression

Because Roy-Bargmann stepdown analysis is planned to assess the importance of DVs after
MANOVA, a test of homogeneity of regression is necessary for each step of the stepdown analysis.
Table 9.13 shows the SPSS MANOVA setup for tests of homogeneity of regression where each DV,
in turn, serves as DV on one step and then becomes a covariate on the next and all remaining steps.

Table 9.13 also contains output for the last two steps where CONTROL serves as DV with
ESTEEM, ATTROLE, NEUROTIC, and INTEXT as covariates, and then SEL2 is the DV with

TABLE 9.13   TEST FOR HOMOGENEITY OF REGRESSION FOR LARGE SAMPLE MANOVA STEPDOWN ANALYSIS
(SETUP AND SELECTED OUTPUT FOR LAST TWO TESTS FROM SPSS MANOVA)

```
DATA LIST         FILE = 'MANOVA.DAT'
                  /CASENO 1-4 ANDRM 5-7 FEM 8-10 MASC 11-13 ESTEEM 14-16
                       CONTROL 17-19 ATTROLE 20-22 SEL2 23-30 (5)
                       INTEXT 31-34 (1) NEUROTIC 35-37.
VALUE LABELS      ANDRM 1 'UNDIFF' 2 'FEMNINE' 3 'MASCLNE' 4'ANDROGNS'/
                  FEM,MASC 1 'LOW' 2 'HIGH'.
MISSING VALUES    CONTROL(0).
MANOVA            ESTEEM,ATTROLE,NEUROTIC,INTEXT,CONTROL,SEL2 BY FEM,MASC(1,2)/
                  PRINT=SIGNIF(BRIEF)/
        ANALYSIS=ATTROLE/
        DESIGN=ESTEEM,FEM,MASC,FEM BY MASC, ESTEEM BY FEM BY MASC/
        ANALYSIS=NEUROTIC/
        DESIGN=ATTROLE,ESTEEM,FEM,MASC,FEM BY MASC, POOL(ATTROLE,ESTEEM)
              BY FEM + POOL(ATTROLE,ESTEEM) BY MASC + POOL(ATTROLE,
              ESTEEM)BY FEM BY MASC/
        ANALYSIS=INTEXT/
        DESIGN=NEUROTIC,ATTROLE,ESTEEM,FEM,MASC,FEM BY MASC, POOL(NEUROTIC,
              ATTROLE,ESTEEM) BY FEM + POOL(NEUROTIC,ATTROLE,ESTEEM)
              BY MASC + POOL(NEUROTIC,ATTROLE,ESTEEM)BY FEM BY MASC/
        ANALYSIS=CONTROL/
        DESIGN=INTEXT,NEUROTIC,ATTROLE,ESTEEM,FEM,MASC,FEM BY MASC,
              POOL(INTEXT,NEUROTIC,ATTROLE,ESTEEM) BY FEM +
              POOL(INTEXT,NEUROTIC,ATTROLE,ESTEEM) BY MASC +
              POOL(INTEXT,NEUROTIC,ATTROLE,ESTEEM)BY FEM BY MASC/
        ANALYSIS=SEL2/
        DESIGN=CONTROL,INTEXT,NEUROTIC,ATTROLE,ESTEEM,FEM,MASC,FEM BY MASC,
              POOL(CONTROL,INTEXT,NEUROTIC,ATTROLE,ESTEEM) BY FEM +
              POOL(CONTROL,INTEXT,NEUROTIC,ATTROLE,ESTEEM) BY MASC +
              POOL(CONTROL,INTEXT,NEUROTIC,ATTROLE,ESTEEM)BY FEM BY MASC.
```

Tests of Significance for CONTROL using UNIQUE sums of squares

Source of Variation	SS	DF	MS	F	Sig of F
WITHIN+RESIDUAL	442.61	348	1.27		
INTEXT	2.19	1	2.19	1.72	.190
NEUROTIC	42.16	1	42.16	33.15	.000
ATTROLE	.67	1	.67	.52	.470
ESTEEM	14.52	1	14.52	11.42	.001
FEM	2.80	1	2.80	2.20	.139
MASC	3.02	1	3.02	2.38	.124
FEM BY MASC	.00	1	.00	.00	.995
POOL(INTEXT NEUROTIC ATTROLE ESTEEM) BY FEM + POOL(INTEXT NEUROTIC ATTROLE ESTEEM) BY MASC + POOL(INTEXT NEUROTIC ATTROLE ESTEEM) BY FEM BY MASC	19.78	12	1.65	1.30	.219

TABLE 9.13 *(CONTINUED)*

Tests of Significance of SEL2 using UNIQUE sums of squares

Source of Variation	SS	DF	MS	F	Sig of F
WITHIN+RESIDUAL	220340.10	344	640.52		
CONTROL	1525.23	1	1525.23	2.38	.124
INTEXT	.99	1	.99	.00	.969
NEUR	262.94	1	262.94	.41	.522
ATT	182.98	1	182.98	.29	.593
EST	157.77	1	157.77	.25	.620
FEM	1069.23	1	1069.23	1.67	.197
MASC	37.34	1	37.34	.06	.809
FEM BY MASC	1530.73	1	1530.73	2.39	.123
POOL (CONTROL INTEXT NEUROTIC ATTROLE EST EEM) BY FEM + POOL(C ONTROL INTEXT NEUROT IC ATTROLE ESTEEM) B Y MASC + POOL(CONTRO L INTEXT NEUROTIC AT TROLE ESTEEM) BY FEM BY MASC)	14017.22	15	934.48	1.46	.118

ESTEEM, ATTROLE, NEUROTIC, INTEXT, and CONTROL as covariates. At each step, the relevant effect is the one appearing last in the column labeled Source of Variation, so that for SEL2 the F value for homogeneity of regression is $F(15, 344) = 1.46, p > .01$. (The more stringent cutoff is used here because robustness is expected.) Homogeneity of regression is established for all steps.

For MANCOVA, an overall test of homogeneity of regression is required, in addition to stepdown tests. Setups for all tests are shown in Table 9.14. The ANALYSIS sentence with three DVs specifies the overall test, while the ANALYSIS sentences with one DV each are for stepdown analysis. Output for the overall test and the last stepdown test are also shown in Table 9.14. Multivariate output is printed for the overall test because there are three DVs; univariate results are give for the stepdown tests. All runs show sufficient homogeneity of regression for this analysis.

9.7.1.7 ■ Reliability of Covariates

For the stepdown analysis in MANOVA, all DVs except ESTEEM must be reliable because all act as covariates. Based on the nature of scale development and data collection procedures, there is no reason to expect unreliability of a magnitude harmful to covariance analysis for ATTROLE, NEUROTIC, INTEXT, CONTROL, and SEL2. These same variables act as true or stepdown covariates in the MANCOVA analysis.

9.7.1.8 ■ Multicollinearity and Singularity

The log-determinant of the pooled within-cells correlation matrix is found (through SPSS MANOVA setup in Table 9.15) to be $-.4336$, yielding a determinant of 2.71. This is sufficiently different from zero that multicollinearity is not judged to be a problem.

9.7.2 ■ Multivariate Analysis of Variance

Setup and partial output of omnibus MANOVA produced by SPSS MANOVA appear in Table 9.15. The order of IVs listed in the MANOVA statement together with METHOD = SEQUENTIAL sets

```
DATA LIST        FILE = 'MANOVA.DAT'
            /CASENO 1-4 ANDRM 5-7 FEM 8-10 MASC 11-13 ESTEEM 14-16
                    CONTROL 17-19 ATTROLE 20-22 SEL2 23-30 (5)
                    INTEXT 31-34 (1) NEUROTIC 35-37.
VALUE LABELS     ANDRM 1 'UNDIFF' 2 'FEMNINE' 3 'MASCLNE' 4'ANDROGNS'/
                 FEM,MASC 1 'LOW' 2 'HIGH'.
MISSING VALUES   CONTROL(0).
MANOVA       ESTEEM,ATTROLE,NEUROTIC,INTEXT,CONTROL,SEL2 BY FEM,MASC(1,2)/
        PRINT=SIGNIF(BRIEF)/
      ANALYSIS=ESTEEM,INTEXT,NEUROTIC/
      DESIGN=CONTROL,ATTROLE,SEL2,FEM,MASC,FEM BY MASC,
            POOL(CONTROL,ATTROLE,SEL2) BY FEM +
            POOL(CONTROL,ATTROLE,SEL2) BY MASC +
            POOL(CONTROL,ATTROLE,SEL2) BY FEM BY MASC/
      ANALYSIS=ESTEEM/
      DESIGN=CONTROL,ATTROLE,SEL2,FEM,MASC,FEM BY MASC,
            POOL(CONTROL,ATTROLE,SEL2) BY FEM +
            POOL(CONTROL,ATTROLE,SEL2) BY MASC +
            POOL(CONTROL,ATTROLE,SEL2) BY FEM BY MASC/
      ANALYSIS=INTEXT/
      DESIGN=ESTEEM,CONTROL,ATTROLE,SEL2,FEM,MASC,FEM BY MASC,
            POOL(ESTEEM,CONTROL,ATTROLE,SEL2) BY FEM +
            POOL(ESTEEM,CONTROL,ATTROLE,SEL2) BY MASC +
            POOL(ESTEEM,CONTROL,ATTROLE,SEL2) BY FEM BY MASC/
      ANALYSIS=NEUROTIC/
      DESIGN=INTEXT,ESTEEM,CONTROL,ATTROLE,SEL2,FEM,MASC,FEM BY MASC,
            POOL(INTEXT,ESTEEM,CONTROL,ATTROLE,SEL2) BY FEM +
            POOL(INTEXT,ESTEEM,CONTROL,ATTROLE,SEL2) BY MASC +
            POOL(INTEXT,ESTEEM,CONTROL,ATTROLE,SEL2) BY FEM BY MASC.


WARNING 12176
The Cholesky approach is used for estimation with the observation model.

* * ANALYSIS  OF  VARIANCE -- DESIGN  1 * *

Tests of significance for WITHIN+RESIDUAL using UNIQUE sums of squares
Source of Variation        Wilks  Approx F  Hyp. DF  Error DF  Sig of F

CONSTANT                    .815    26.422    3.00    350.000     .000
CONTROL                    .814    26.656    3.00    350.000     .000
ATTROLE                    .973     3.221    3.00    350.000     .023
SEL2                       .999      .105    3.00    350.000     .957
FEM                        .992      .949    3.00    350.000     .417
MASC                       .993      .824    3.00    350.000     .481
FEM BY MASC                .988     1.414    3.00    350.000     .238
POOL(CONTROL ATTROL        .933      .911   27.00   1022.823     .596
E SEL2) BY FEM + POO
L(CONTROL ATTROLE SE
L2) BY MASC + POOL(C
ONTROL ATTROLE SEL2)
 BY FEM BY MASC
```

TABLE 9.14 *(CONTINUED)*

```
* * ANALYSIS  OF  VARIANCE - DESIGN   4 * *
```

Tests of Significance for NEUROTIC using UNIQUE sums of squares

Source of Variation	SS	DF	MS	F	Sig of F
WITHIN+RESIDUAL	6662.67	344	19.37		
CONSTANT	230.29	1	230.29	11.89	.001
INTEXT	82.19	1	82.19	4.24	.040
ESTEEM	308.32	1	308.32	15.92	.000
CONTROL	699.04	1	699.04	36.09	.000
ATTROLE	1.18	1	1.18	.06	.805
SEL2	2.24	1	2.24	.12	.734
FEM	.07	1	.07	.00	.952
MASC	74.66	1	74.66	3.85	.050
FEM BY MASC	1.65	1	1.65	.09	.770
POOL(INTEXT ESTEEM C	420.19	15	28.01	1.45	.124
ONTROL ATTROLE SEL2)					
BY FEM + POOL(INTEX					
T ESTEEM CONTROL ATT					
ROLE SEL2) BY MASC +					
POOL(INTEXT ESTEEM					
CONTROL ATTROLE SEL2					
) BY FEM BY MASC					

up the priority for testing FEM before MASC in this unequal-*n* design. Results are reported for FEM by MASC, MASC, and FEM, in turn. Tests are reported out in order of adjustment where FEM by MASC is adjusted for both MASC and FEM, and MASC is adjusted for FEM.

Four multivariate statistics are reported for each effect. Because there is only one degree of freedom for each effect, three of the tests—Pillai's, Hotelling's, and Wilks'—produce the same F.[18] Both main effects are highly significant, but there is no statistically reliable interaction. If desired, strength of association for the composite DV for each main effect is found using Equation 9.8 or 9.9.

Because omnibus MANOVA shows significant main effects, it is appropriate to investigate further the nature of the relationships among the IVs and DVs. Correlations, univariate Fs, and stepdown Fs help clarify the relationships.

The degree to which DVs are correlated provides information as to the independence of behaviors. Pooled within-cell correlations, adjusted for IVs, as produced by SPSS MANOVA through PRINT = ERROR(COR), appear in Table 9.16. (Diagonal elements are pooled standard deviations.) Correlations among ESTEEM, NEUROTIC, and CONTROL are in excess of .30 so stepdown analysis is appropriate.

Even if stepdown analysis is the primary procedure, knowledge of univariate Fs is required to correctly interpret the pattern of stepdown Fs. And, although the statistical significance of these F values is misleading, investigators frequently are interested in the ANOVA that would have been produced if each DV had been investigated in isolation. These univariate analyses are produced automatically by SPSS MANOVA and shown in Table 9.17 for the three effects in turn: FEM by MASC, MASC, and FEM. F values are substantial for all DVs except SEL2 for MASC and ESTEEM, ATTROLE, and INTEXT for FEM.

---

[18] For more complex designs, a single source table containing all effects can be obtained through PRINT=SIGNIF(BRIEF) but the table displays only Wilks' Lambda.

TABLE 9.15 MULTIVARIATE ANALYSIS OF VARIANCE OF ESTEEM, CONTROL, ATTROLE, SEL2, INTEXT, AND NEUROTIC, AS A FUNCTION OF (TOP TO BOTTOM) FEMININITY BY MASCULINITY INTERACTION, MASCULINITY, AND FEMININITY. SELECTED OUTPUT FROM SPSS MANOVA. (SETUP IN TABLE 9.15)

```
DATA LIST       FILE = 'MANOVA.DAT'
                /CASENO 1-4 ANDRM 5-7 FEM 8-10 MASC 11-13 ESTEEM 14-16
                 CONTROL 17-19 ATTROLE 20-22 SEL2 23-30 (5)
                 INTEXT 31-34 (1) NEUROTIC 35-37.

VALUE LABELS    ANDRM 1 'UNDIFF' 2 'FEMININE' 3 'MASCLNE' 4 'ANDROGNS'/
                FEM,MASC 1 'LOW' 2 'HIGH'.

MISSING VALUES  CONTROL(0).

MANOVA          ESTEEM,ATTROLE,NEUROTIC,INTEXT,CONTROL,SEL2 BY FEM,MASC(1,2)/
                PRINT=SIGNIF(STEPDOWN), ERROR(COR)/
                HOMOGENEITY(BARTLETT,COCHRAN,BOXM)/
                METHOD=SEQUENTIAL/
                DESIGN.
```

EFFECT .. FEM BY MASC
Multivariate Tests of Significance (S = 1, M = 2 , N = 178 1/2)

Test Name	Value	Exact F	Hypoth. DF	Error DF	Sig. of F
Pillais	.00816	.49230	6.00	359.00	.814
Hotellings	.00823	.49230	6.00	359.00	.814
Wilks	.99184	.49230	6.00	359.00	.814
Roys	.00816				

Note.. F statistics are exact.

EFFECT .. MASC
Multivariate Tests of Significance (S = 1, M = 2 , N = 178 1/2)

Test Name	Value	Exact F	Hypoth. DF	Error DF	Sig. of F
Pillais	.24363	19.27301	6.00	359.00	.000
Hotellings	.32211	19.27301	6.00	359.00	.000
Wilks	.75637	19.27301	6.00	359.00	.000
Roys	.24363				

Note.. F statistics are exact.

EFFECT .. FEM
Multivariate Tests of Significance (S = 1, M = 2 , N = 178 1/2)

Test Name	Value	Exact F	Hypoth. DF	Error DF	Sig. of F
Pillais	.08101	5.27423	6.00	359.00	.000
Hotellings	.08815	5.27423	6.00	359.00	.000
Wilks	.91899	5.27423	6.00	359.00	.000
Roys	.08101				

Note.. F statistics are exact.

TABLE 9.16 POOLED WITHIN-CELL CORRELATIONS AMONG SIX DVs (SELECTED OUTPUT FROM SPSS MANOVA--SEE TABLE 9.15 FOR SETUP)

WITHIN CELLS Correlations with Std. Devs. on Diagonal

	ESTEEM	ATTROLE	NEUROTIC	INTEXT	CONTROL	SEL2
ESTEEM	3.53338					
ATTROLE	.14529	6.22718				
NEUROTIC	.35772	.05070	4.96517			
INTEXT	-.16444	.01111	-.00923	3.58729		
CONTROL	-.34789	-.03051	.38737	-.08315	1.26680	
SEL2	-.03510	.01635	-.01501	.05526	-.08417	25.50097

419

TABLE 9.17 UNIVARIATE ANALYSES OF VARIANCE OF SIX DVs FOR EFFECTS OF (TOP TO BOTTOM) FEM BY MASC INTERACTION, MASCULINITY, AND FEMININITY (SELECTED OUTPUT FROM SPSS MANOVA--SEE TABLE 9.15 FOR SETUP)

EFFECT .. FEM BY MASC (Cont.)
Univariate F-tests with (1,364) D. F.

Variable	Hypoth. SS	Error SS	Hypoth. MS	Error MS	F	Sig. of F
ESTEEM	17.48685	4544.44694	17.48685	12.48474	1.40066	.237
ATTROLE	36.79594	14115.12120	36.79594	38.77781	.94889	.331
NEUROTIC	.20239	8973.67662	.20239	24.65296	.00821	.928
INTEXT	.02264	4684.17900	.02264	12.86862	.00176	.967
CONTROL	.89539	584.14258	.89539	1.60479	.55795	.456
SEL2	353.58143	236708.96634	353.58143	650.29936	.54372	.461

EFFECT .. MASC (Cont.)
Univariate F-tests with (1,364) D. F.

Variable	Hypoth. SS	Error SS	Hypoth. MS	Error MS	F	Sig. of F
ESTEEM	979.60086	4544.44694	979.60086	12.48474	78.46383	.000
ATTROLE	1426.75675	14115.12120	1426.75675	38.77781	36.79313	.000
NEUROTIC	179.53396	8973.67662	179.53396	24.65296	7.28245	.007
INTEXT	327.40797	4684.17900	327.40797	12.86862	25.44235	.000
CONTROL	11.85923	584.14258	11.85923	1.60479	7.38991	.007
SEL2	1105.38196	236708.96634	1105.38196	650.29936	1.69980	.193

EFFECT .. FEM (Cont.)
Univariate F-tests with (1,364) D. F.

Variable	Hypoth. SS	Error SS	Hypoth. MS	Error MS	F	Sig. of F
ESTEEM	101.46536	4544.44694	101.46536	12.48474	8.12715	.005
ATTROLE	610.88860	14115.12120	610.88860	38.77781	15.75356	.000
NEUROTIC	44.05442	8973.67662	44.05442	24.65296	1.78698	.182
INTEXT	87.75996	4684.17900	87.75996	12.86862	6.81968	.009
CONTROL	2.83106	584.14258	2.83106	1.60479	1.76414	.185
SEL2	9.00691	236708.96634	9.00691	650.29936	.01385	.906

Finally, Roy-Bargmann stepdown analysis, produced by PRINT=SIGNIF(STEPDOWN), allows a statistically pure look at the significance of DVs, in context, with Type I error rate controlled. For this study, the following priority order of DVs is developed, from most to least important: ESTEEM, ATTROLE, NEUROTIC, INTEXT, CONTROL, SEL2. Following the procedures for stepdown analysis (Section 9.5.2.2), the highest-priority DV, ESTEEM, is tested in univariate ANOVA. The second-priority DV, ATTROLE, is assessed in ANCOVA with ESTEEM as the covariate. The third-priority DV, NEUROTIC, is tested with ESTEEM and ATTROLE as covariates, and so on, until all DVs are analyzed. Stepdown analyses for the interaction and both main effects are in Table 9.18.

For purposes of journal reporting, critical information from Tables 9.17 and 9.18 is consolidated into a single table with both univariate and stepdown analyses, as shown in Table 9.19. The alpha level established for each DV is reported along with the significance levels for stepdown F. For the main effect of FEM, ESTEEM and ATTROLE are significant. (INTEXT would be significant in ANOVA but its variance is already accounted for through overlap with ESTEEM, as noted in the pooled within-cell correlation matrix.) For the main effect of MASC, ESTEEM, ATTROLE, and INTEXT are significant. (NEUROTIC and CONTROL would be significant in ANOVA, but their variance is also already accounted for through overlap with ESTEEM, ATTROLE, and, in the case of CONTROL, NEUROTIC and INTEXT.)

For the DVs significant in stepdown analysis, the relevant adjusted marginal means are needed for interpretation. Marginal means are needed for ESTEEM for FEM, and for MASC adjusted for FEM. Also needed are marginal means for ATTROLE with ESTEEM as a covariate for both FEM, and MASC adjusted for FEM; lastly, marginal means are needed for INTEXT with ESTEEM, ATTROLE, and NEUROTIC as covariates for MASC adjusted for FEM. Table 9.20 contains setup and selected output for these marginal means as produced through SPSS MANOVA. In the table, level of effect is identified under PARAMETER and mean is under COEFF. Thus, the mean for ESTEEM at level 1 of FEM is 16.57. Marginal means for effects with univariate, but not stepdown, differences are shown in Table 9.21 where means for NEUROTIC and CONTROL are found for the main effect of MASC adjusted for FEM.

Strength of association between the main effect and each DV with which it has a significant stepdown relationship is evaluated as η^2 (Equations 3.25, 3.26, 8.7, 8.8, or 8.9). The information you need for calculation of η^2 is available in SPSS MANOVA stepdown tables (see Table 9.18) but not in a convenient form; mean squares are given in the tables but you need sums of squares for calculation of η^2. Sums of squares are found by multiplying a mean square by its associated degrees of freedom. For this example, appropriate information is summarized in Table 9.22. Once sums of squares are computed, total sum of squares for a DV is calculated by adding the sums of squares for all effects and error.

For example, ATTROLE is significantly related to FEM (F). For ATTROLE, the adjusted sum of squares is the mean square for hypothesis (Hypoth. MS) times the degrees of freedom (Hypoth. DF) from the FEM section of Table 9.16.

$$SS'_F = (728.76735)(1) = 728.77$$

Adjusted sums of squares for the remaining effects and error are necessary for computation of sum of squares total; these values are also available, but scattered around through Table 9.18. The adjusted sum of squares for MASC (middle part of Table 9.18) is

$$SS'_M = (728.51682)(1) = 728.52$$

TABLE 9.18 STEPDOWN ANALYSES OF SIX ORDERED DVs FOR (TOP TO BOTTOM) FEM BY MASC INTERACTION, MASCULINITY, AND FEMININITY (SELECTED OUTPUT FROM SPSS MANOVA—SEE TABLE 9.15 FOR SETUP)

Roy-Bargman Stepdown F - tests

Variable	Hypoth. MS	Error MS	StepDown F	Hypoth. DF	Error DF	Sig. of F
ESTEEM	17.48685	12.48474	1.40066	1	364	.237
ATTROLE	24.85653	38.06383	.65302	1	363	.420
NEUROTIC	2.69735	21.61699	.12478	1	362	.724
INTEXT	.26110	12.57182	.02077	1	361	.885
CONTROL	.41040	1.28441	.31952	1	360	.572
SEL2	297.09000	652.80588	.45510	1	359	.500

Roy-Bargman Stepdown F - tests

Variable	Hypoth. MS	Error MS	StepDown F	Hypoth. DF	Error DF	Sig. of F
ESTEEM	979.60086	12.48474	78.46383	1	364	.000
ATTROLE	728.51682	38.06383	19.13935	1	363	.000
NEUROTIC	4.14529	21.61699	.19176	1	362	.662
INTEXT	139.98354	12.57182	11.13471	1	361	.001
CONTROL	.00082	1.28441	.00064	1	360	.980
SEL2	406.59619	652.80588	.62284	1	359	.431

Roy-Bargman Stepdown F - tests

Variable	Hypoth. MS	Error MS	StepDown F	Hypoth. DF	Error DF	Sig. of F
ESTEEM	101.46536	12.48474	8.12715	1	364	.005
ATTROLE	728.76735	38.06383	19.14593	1	363	.000
NEUROTIC	2.21946	21.61699	.10267	1	362	.749
INTEXT	47.98941	12.57182	3.81722	1	361	.052
CONTROL	.05836	1.28441	.04543	1	360	.831
SEL2	15.94930	652.80588	.02443	1	359	.876

and for FEM BY MASC (top part of Table 9.18) is

$$SS'_{F\,by\,M} = (24.85653)(1) = 24.86$$

Error mean square and degrees of freedom are duplicated in all three segments of Table 9.18. Adjusted sum of squares for error, then, is

$$SS'_{error} = (38.06383)(363) = 13817.177$$

Finally, η^2 for ATTROLE is found by computing

$$\eta^2 = \frac{SS'_F}{SS'_{total}} = \frac{SS'_F}{SS'_F + SS'_M + SS'_{F\,by\,M} + SS'_{error}}$$

$$= \frac{728.77}{728.77 + 728.52 + 24.86 + 13817.17} = \frac{728.77}{15299.32} = .05$$

A checklist for MANOVA appears in Table 9.23. An example of a Results section, in journal format, follows for the study just described.

TABLE 9.19   TESTS OF FEMININITY, MASCULINITY, AND THEIR INTERACTION

IV	DV	Univariate F	df	Stepdown F	df	α
Femininity	ESTEEM	8.13[a]	1/364	8.13**	1/364	.01
	ATTROLE	15.75[a]	1/364	19.15**	1/363	.01
	NEUROTIC	1.79	1/364	0.10	1/362	.01
	INTEXT	6.82[a]	1/364	3.82	1/361	.01
	CONTROL	1.76	1/364	0.05	1/360	.01
	SEL2	0.01	1/364	0.02	1/359	.001
Masculinity	ESTEEM	78.46[a]	1/364	78.46**	1/364	.01
	ATTROLE	36.79[a]	1/364	19.14**	1/363	.01
	NEUROTIC	7.28[a]	1/364	0.19	1/362	.01
	INTEXT	25.44[a]	1/364	11.13**	1/361	.01
	CONTROL	7.39[a]	1/364	0.00	1/360	.01
	SEL2	1.70	1/364	0.62	1/359	.001
Femininity by	ESTEEM	1.40	1/364	1.40	1/364	.01
masculinity	ATTROLE	0.95	1/364	0.65	1/363	.01
interaction	NEUROTIC	0.01	1/364	0.12	1/362	.01
	INTEXT	0.00	1/364	0.02	1/361	.01
	CONTROL	0.56	1/364	0.32	1/360	.01
	SEL2	0.54	1/364	0.46	1/359	.001

[a]Significance level cannot be evaluated but would reach $p < .01$ in univariate context.

**p < .01

TABLE 9.20   ADJUSTED MARGINAL MEANS FOR ESTEEM; ATTROLE WITH ESTEEM AS A COVARIATE; AND INTEXT WITH ESTEEM, ATTROLE, AND NEUROTIC AS COVARIATES (SETUP AND SELECTED OUTPUT FROM SPSS MANOVA)

```
DATA LIST        FILE = 'MANOVA.DAT'
                 /CASENO 1-4 ANDRM 5-7 FEM 8-10 MASC 11-13 ESTEEM 14-16
                       CONTROL 17-19 ATTROLE 20-22 SEL2 23-30 (5)
                       INTEXT 3134 (1) NEUROTIC 35-37.
VALUE LABELS     ANDRM 1 'UNDIFF' 2 'FEMNINE' 3 'MASCLNE' 4'ANDROGNS'/
                 FEM,MASC 1 'LOW' 2 'HIGH'.
MISSING VALUES   CONTROL(0).
MANOVA           ESTEEM,ATTROLE,NEUROTIC,INTEXT,CONTROL,SEL2 BY FEM,MASC(1,2)/
                 PRINT=PARAMETERS(ESTIM)/
         ANALYSIS=ESTEEM/DESIGN=CONSPLUS FEM/
                 DESIGN=FEM,CONSPLUS MASC/
         ANALYSIS=ATTROLE WITH ESTEEM/DESIGN=CONSPLUS FEM/
                 DESIGN=FEM, CONSPLUS MASC/
         ANALYSIS=INTEXT WITH ESTEEM,ATTROLE,NEUROTIC/
                 DESIGN=FEM, CONSPLUS MASC.
```

Estimates for ESTEEM

CONSPLUS FEM

Parameter	Coeff.
1	16.5700934579
2	15.4137931034

Estimates for ESTEEM

CONSPLUS MASC

Parameter	Coeff.
2	17.1588559924
3	13.7138143624

Estimates for ATTROLE adjusted for 1 covariate

CONSPLUS FEM

Parameter	Coeff.
1	32.7152449841
2	35.8485394127

Estimates for ATTROLE adjusted for 1 covariate

CONSPLUS MASC

Parameter	Coeff.
2	35.3913780733
3	32.1251994692

Estimates for INTEXT adjusted for 3 covariates

CONSPLUS MASC

Parameter	Coeff.
2	11.0021422402
3	12.4772020098

Note: Coeff. = adjusted marginal mean; first parameter=low, second parameter=high.

TABLE 9.21 UNADJUSTED MARGINAL MEANS FOR NEUROTIC AND CONTROL (SETUP AND SELECTED OUTPUT FROM SPSS MANOVA)

```
DATA LIST          FILE = 'MANOVA.DAT'
                   /CASENO 1-4 ANDRM 5-7 FEM 8-10 MASC 11-13 ESTEEM 14-16
                       CONTROL 17-19 ATTROLE 20-22 SEL2 23-30 (5)
                       INTEXT 31-34 (1) NEUROTIC 35-37.
VALUE LABELS       ANDRM 1 'UNDIFF' 2 'FEMNINE' 3 'MASCLNE' 4'ANDROGNS'/
                   FEM,MASC 1 'LOW' 2 'HIGH'.
MISSING VALUES     CONTROL(0).
MANOVA             ESTEEM,ATTROLE,NEUROTIC,INTEXT,CONTROL,SEL2 BY FEM,MASC(1,2)/
                   PRINT=PARAMETERS(ESTIM)/
                   ANALYSIS=NEUROTIC/DESIGN=FEM, CONSPLUS MASC/
                   ANALYSIS=CONTROL/ DESIGN=FEM, CONSPLUS MASC.


Estimates for NEUROTIC

  CONSPLUS MASC

  Parameter          Coeff.

       2         9.3709383047
       3         7.8961041130


Estimates for CONTROL

  CONSPLUS MASC

  Parameter          Coeff.

       2         6.8916331030
       3         6.5125816027
```

Note: Coeff. = unadjusted marginal mean; first parameter=low, second parameter=high.

TABLE 9.22 SUMMARY OF ADJUSTED SUMS OF SQUARES AND η^2 FOR EFFECTS OF FEMININITY, MASCULINITY, AND THEIR INTERACTION ON ESTEEM; ATTROLE WITH ESTEEM AS A COVARIATE; AND INTEXT ESTEEM, ATTROLE, AND NEUROTIC AS COVARIATES

Source of variance	ESTEEM		ATTROLE		INTEXT	
	SS'	η^2	SS'	η^2	SS'	η^2
F	101.47	.02	728.77	.05	47.99	—
M	979.60	.17	728.52	.05	139.98	.03
$F \times M$	17.49	—	24.86	—	0.26	—
Error	4544.45		13817.17		4538.43	
Total	5643.01		15299.32		4726.67	

Note: See Table 9.18 for MS values. SS = (MS)(df).

TABLE 9.23   CHECKLIST FOR MULTIVARIATE ANALYSIS OF VARIANCE

1. Issues
   a. Unequal sample sizes and missing data
   b. Normality of sampling distributions
   c. Outliers
   d. Homogeneity of variance-covariance matrices
   e. Linearity
   f. In stepdown, when DVs act as covariates
      (1) Homogeneity of regression
      (2) Reliability of DVs
   g. Multicollinearity and singularity
2. Major analyses: Planned comparisons or omnibus F, when significant
   a. Multivariate strength of association
   b. Importance of DVs
      (1) Within-cell correlations, stepdown F, univariate F
      (2) Strength of association for significant F
      (3) Means or adjusted marginal and/or cell means for significant F
3. Additional analyses
   a. Post hoc comparisons

Results

A 2 × 2 between-subjects multivariate analysis of variance was performed on six dependent variables: self-esteem, attitude toward the
role of women, neuroticism, introversion-extraversion, locus of control, and socioeconomic level. Independent variables were masculinity
(low and high) and femininity (low and high).

SPSS MANOVA was used for the analyses with the sequential adjustment
for nonorthogonality. Order of entry of IVs was femininity, then masculinity. Total $\underline{N}$ of 369 was reduced to 368 with the deletion of a case
missing a score on locus of control. There were no univariate or multivariate within-cell outliers at $\underline{p}$ = .001. Results of evaluation of
assumptions of normality, homogeneity of variance-covariance matrices,
linearity, and multicollinearity were satisfactory.

With the use of Wilks' criterion, the combined DVs were significant-
ly affected by both masculinity, $\underline{F}$(6, 359) = 19.27, $\underline{p}$ < .001, and femininity, $\underline{F}$(6, 359) = 5.27, $\underline{p}$ < .001, but not by their interaction,
$\underline{F}$(6,359) = 0.49, $\underline{p}$ > .05. The results reflected a modest association
between masculinity scores (low vs. high) and the combined DVs, partial
η^2 = .13. The association was even less substantial between femininity
and the DVs, partial η^2 = .04. [*F and lambda are from Table 9.15; partial η^2 is calculated according to Equation 9.9.*]

To investigate the impact of each main effect on the individual DVs,
a Roy-Bargmann stepdown analysis was performed on the prioritized DVs.

All DVs were judged to be sufficiently reliable to warrant stepdown
analysis. In stepdown analysis each DV was analyzed, in turn, with
higher-priority DVs treated as covariates and with the highest-priority
DV tested in a univariate ANOVA. Homogeneity of regression was achieved
for all components of the stepdown analysis. Results of this analysis
are summarized in Table 9.19. An experimentwise error rate of 5% was
achieved by the apportionment of alpha as shown in the last column of
Table 9.19 for each of the DVs.

A unique contribution to predicting differences between those low and
high on femininity was made by self-esteem, stepdown $F(1, 364) = 8.13$,
$p < .01$, $\eta^2 = .02$. Self-esteem was scored inversely, so women with higher femininity scores showed greater self-esteem (mean self-esteem =
15.41) than those with lower femininity (mean self-esteem = 16.57).
After the pattern of differences measured by self-esteem was entered, a
difference was also found on attitude toward role of women, stepdown
$F(1, 363) = 19.15$, $p < .01$, $\eta^2 = .05$. Women with higher femininity
scores had more conservative attitudes toward women's role (adjusted
mean attitude = 35.85) than those lower in femininity (adjusted mean
attitude = 32.72). Although a univariate comparison revealed that those
higher in femininity also were more extroverted, univariate $F(1, 364) =$
6.82, this difference was already represented in the stepdown analysis
by higher-priority DVs.

Three DVs—self-esteem, attitude toward role of women, and introvert-
extrovert—made unique contributions to the composite DV that best distinguished between those high and low in masculinity. The greatest
contribution was made by self-esteem, the highest-priority DV, stepdown
$F(1, 364) = 78.46$, $p < .01$, $\eta^2 = 17$. Women scoring high in masculinity
had higher self-esteem (mean self-esteem = 13.71) than those scoring
low (mean self-esteem = 17.16). With differences due to self-esteem
already entered, attitudes toward the role of women made a unique contribution, stepdown $F(1, 363) = 19.14$, $p < .01$, $\eta^2 = .05$. Women scoring
lower in masculinity had more conservative attitudes toward the proper
role of women (adjusted mean attitude = 35.39) than those scoring
higher (adjusted mean attitude = 32.12). Introversion-extroversion,
adjusted by self-esteem, attitudes toward women's role, and neuroticism
also made a unique contribution to the composite DV, stepdown $F(1, 361)$
= 11.13, $p < .01$, $\eta^2 = .03$. Women with higher masculinity were more
extroverted (mean adjusted introversion-extraversion score = 12.48)
than lower masculinity women (mean adjusted introversion-extraversion
score = 11.00). Univariate analyses revealed that women with higher
masculinity scores were also less neurotic, univariate $F(1, 364) =$
7.28, and had a more internal locus of control, univariate $F(1, 364) =$

```
    7.39, differences that were already accounted for in the composite DV
    by higher-priority DVs. [Means adjusted for main effects and for other
    DVs for stepdown interpretation are from Table 9.20, η² values are from
    Table 9.22. Means adjusted for main effects but not other DVs for uni-
    variate interpretation are in Table 9.21.]

      High-masculinity women, then, have greater self-esteem, less conserv-
    ative attitudes toward the role of women, and more extraversion than
    women scoring low on masculinity. High femininity is associated with
    greater self-esteem and more conservative attitudes toward women's role
    than low femininity. Of the five effects, however, only the association
    between masculinity and self-esteem shows even a moderate proportion of
    shared variance.

      Pooled within-cell correlations among DVs are shown in Table 9.16.
```

9.7.3 ■ Multivariate Analysis of Covariance

For MANCOVA the same six variables are used as for MANOVA but ESTEEM, INTEXT, and NEUROTIC are used as DVs and CONTROL, ATTROLE, and SEL2 are used as covariates. The research question is whether there are personality differences associated with femininity, mas-culinity, and their interaction after adjustment for differences in attitudes and socioeconomic status.

Setup and partial output of omnibus MANCOVA as produced by SPSS MANOVA appear in Table 9.24. As in MANOVA, Method 3 adjustment for unequal n is used with MASC adjusted for FEM. And, as in MANOVA, both main effects are highly significant but there is no interaction.

9.7.3.1 ■ Assessing Covariates

Under EFFECT..WITHIN CELLS REGRESSION is the multivariate significance test for the rela-tionship between the set of DVs (ESTEEM, INTEXT, and NEUROTIC) and the set of covariates (CONTROL, ATTROLE, and SEL2). Because there is multivariate significance, it is useful to look at the three multiple regression analyses of each DV in turn, with covariates acting as IVs (see Chapter 5). The setup of Table 9.24 automatically produces these regressions. They are done on the pooled within-cell correlation matrix, so that effects of the IVs are eliminated.

The results of the DV-covariate multiple regressions are shown in Table 9.25. At the top of Table 9.25 are the results of the univariate and stepdown analysis, summarizing the results of multiple regressions for the three DVs independently and then in priority order (see Section 9.7.3.2). At the bottom of Table 9.25 under Regression analysis for WITHIN CELLS error term are the separate regressions for each DV with covariates as IVs. For ESTEEM, two covariates, CONTROL and ATTROLE, are significantly related but SEL2 is not. None of the three covariates is related to INTEXT. Finally, for NEUROTIC, only CONTROL is significantly related. Because SEL2 pro-vides no adjustment to any of the DVs, it could be omitted from future analyses.

TABLE 9.24   MULTIVARIATE ANALYSIS OF COVARIANCE OF ESTEEM, INTEXT, AND NEUROTIC AS A FUNCTION OF (TOP TO BOTTOM) FEM BY MASC INTERACTION, MASCULINITY, FEMININITY; COVARIATES ARE ATTROLE, CONTROL, AND SEL2 (SETUP AND SELECTED OUTPUT FROM SPSS MANOVA)

```
DATA LIST        FILE = 'MANOVA.DAT'
                 /CASENO 1-4 ANDRM 5-7 FEM 8-10 MASC 11-13 ESTEEM 14-16
                    CONTROL 17-19 ATTROLE 20-22 SEL2 23-30 (5)
                    INTEXT 31-34 (1) NEUROTIC 35-37.
VALUE LABELS    ANDRM 1 'UNDIFF' 2 'FEMNINE' 3 'MASCLNE' 4'ANDROGNS'/
                FEM,MASC 1 'LOW' 2 'HIGH'.
MISSING VALUES  CONTROL(0).
MANOVA          ESTEEM,ATTROLE,NEUROTIC,INTEXT,CONTROL,SEL2 BY FEM,MASC(1,2)/
                ANALYSIS=ESTEEM,INTEXT,NEUROTIC WITH CONTROL,ATTROLE,SEL2/
                PRINT=SIGNIF(STEPDOWN), ERROR(COR),
                    HOMOGENEITY(BARTLETT,COCHRAN,BOXM)/
                METHOD=SEQUENTIAL/
                DESIGN.
```

EFFECT .. WITHIN CELLS Regression
Multivariate Tests of Significance (S = 3, M = -1/2, N = 178 1/2)

Test Name	Value	Approx. F	Hypoth. DF	Error DF	Sig. of F
Pillais	.23026	10.00372	9.00	1083.00	.000
Hotellings	.29094	11.56236	9.00	1073.00	.000
Wilks	.77250	10.86414	9.00	873.86	.000
Roys	.21770				

EFFECT .. FEM BY MASC
Multivariate Tests of Significance (S = 1, M = 1/2, N = 178 1/2)

Test Name	Value	Exact F	Hypoth. DF	Error DF	Sig. of F
Pillais	.00263	.31551	3.00	359.00	.814
Hotellings	.00264	.31551	3.00	359.00	.814
Wilks	.99737	.31551	3.00	359.00	.814
Roys	.00263				

Note.. F statistics are exact.

EFFECT .. MASC
Multivariate Tests of Significance (S = 1, M = 1/2, N = 178 1/2)

Test Name	Value	Exact F	Hypoth. DF	Error DF	Sig. of F
Pillais	.14683	20.59478	3.00	359.00	.000
Hotellings	.17210	20.59478	3.00	359.00	.000
Wilks	.85317	20.59478	3.00	359.00	.000
Roys	.14683				

Note.. F statistics are exact.

EFFECT .. FEM
Multivariate Tests of Significance (S = 1, M = 1/2, N = 178 1/2)

Test Name	Value	Exact F	Hypoth. DF	Error DF	Sig. of F
Pillais	.03755	4.66837	3.00	359.00	.003
Hotellings	.03901	4.66837	3.00	359.00	.003
Wilks	.96245	4.66837	3.00	359.00	.003
Roys	.03755				

Note.. F statistics are exact.

TABLE 9.25 UNIVARIATE, STEPDOWN, AND MULTIPLE REGRESSION ANALYSES FOR THREE DVs WITH THREE COVARIATES (SELECTED OUTPUT FROM SPSS MANOVA--SEE TABLE 9.26 FOR SETUP)

EFFECT .. WITHIN CELLS Regression (Cont.)
Univariate F-tests with (3,361) D. F.

Variable	Sq. Mul. R	Mul. R	Adj. R-sq.	Hypoth. MS	Error MS	F	Sig. of F
ESTEEM	.14542	.38134	.13832	220.28068	10.75791	20.47616	.000
INTEXT	.00932	.09655	.00109	14.55535	12.85461	1.13231	.336
NEUROTIC	.15425	.39274	.14722	461.38686	21.02359	21.94615	.000

Roy-Bargman Stepdown F - tests

Variable	Hypoth. MS	Error MS	StepDown F	Hypoth. DF	Error DF	Sig. of F
ESTEEM	220.28068	10.75791	20.47616	3	361	.000
INTEXT	6.35936	12.60679	.50444	3	360	.679
NEUROTIC	239.94209	19.72942	12.16164	3	359	.000

Regression analysis for WITHIN CELLS error term
--- Individual Univariate .9500 confidence intervals
Dependent variable .. ESTEEM

COVARIATE	B	Beta	Std. Err.	t-Value	Sig. of t	Lower -95%	CL- Upper
CONTROL	.9817325741	.3519752398	.13625	7.20542	.000	.71379	1.24967
ATTROLE	.0886059312	.1561581235	.02762	3.20773	.001	.03428	.14293
SEL2	-.0011127953	-.0080312308	.00677	-.16446	.869	-.01442	.01219

Dependent variable .. INTEXT

COVARIATE	B	Beta	Std. Err.	t-Value	Sig. of t	Lower -95%	CL- Upper
CONTROL	-.2232208679	-.0788274507	.14894	-1.49877	.135	-.51611	.06967
ATTROLE	.0045581881	.0079125744	.03019	.15096	.880	-.05482	.06394
SEL2	.0068216006	.0484927597	.00740	.92231	.357	-.00772	.02137

Dependent variable .. NEUROTIC

COVARIATE	B	Beta	Std. Err.	t-Value	Sig. of t	Lower -95%	CL- Upper
CONTROL	1.5312818992	.3906873702	.19047	8.03955	.000	1.15671	1.90585
ATTROLE	.0497076488	.0623419410	.03861	1.28727	.199	-.02623	.12565
SEL2	.0032824106	.0168583493	.00946	.34703	.729	-.01532	.02188

9.7.3.2 ■ Assessing DVs

Procedures for evaluating DVs, now adjusted for covariates, follow those specified in Section 9.7.2 for MANOVA. Correlations among all DVs, among covariates, and between DVs and covariates are informative so all the correlations in Table 9.16 are still relevant.[19]

Univariate Fs are now adjusted for covariates. The univariate ANCOVAs produced by the SPSS MANOVA run specified in Table 9.24 are shown in Table 9.26. Although significance levels are misleading, there are substantial F values for ESTEEM and INTEXT for MASC (adjusted for FEM) and for FEM.

For interpretation of effects of IVs on DVs adjusted for covariates, comparison of stepdown Fs with univariate Fs again provides the best information. The priority order of DVs for this analysis is ESTEEM, INTEXT, and NEUROTIC. ESTEEM is evaluated after adjustment only for the three covariates. INTEXT is adjusted for effects of ESTEEM and the three covariates; NEUROTIC is adjusted for ESTEEM and INTEXT and the three covariates. In effect, then, INTEXT is adjusted for four covariates and NEUROTIC is adjusted for five.

Stepdown analysis for the interaction and two main effects is in Table 9.27. The results are the same as those in MANOVA except that there is no longer a main effect of FEM on INTEXT after adjustment for four covariates. The relationship between FEM and INTEXT is already represented by the relationship between FEM and ESTEEM. Consolidation of information from Tables 9.26 and 9.27, as well as some information from Table 9.25, appears in Table 9.28, along with apportionment of the .05 alpha error to the various tests.

For the DVs associated with significant main effects, interpretation requires associated marginal means. Table 9.29 contains setup and adjusted marginal means for ESTEEM and for INTEXT (which is adjusted for ESTEEM as well as covariates) for FEM and for MASC adjusted for FEM. Setup and marginal means for the main effect of FEM on INTEXT (univariate but not stepdown effect) appear in Table 9.30. A checklist for MANCOVA appears in Table 9.31. An example of a Results section, as might be appropriate for journal presentation, follows.

---

[19] For MANCOVA, SPSS MANOVA prints pooled within-cell correlations among DVs (called criteria) adjusted for covariates. To get a pooled within-cell correlation matrix for covariates as well as DVs, you need a run in which covariates are included in the set of DVs.

TABLE 9.26 UNIVARIATE ANALYSES OF COVARIANCE OF THREE DVs ADJUSTED FOR THREE COVARIATES FOR (TOP TO BOTTOM) FEM BY MASC INTERACTION, MASCULINITY, AND FEMININITY (SELECTED OUTPUT FROM SPSS MANOVA--SEE TABLE 9.26 FOR SETUP)

EFFECT .. FEM BY MASC (Cont.)
Univariate F-tests with (1,361) D. F.

Variable	Hypoth. SS	Error SS	Hypoth. MS	Error MS	F	Sig. of F
ESTEEM	7.21931	3883.60490	7.21931	10.75791	.67107	.413
INTEXT	.02590	4640.51295	.02590	12.85461	.00202	.964
NEUROTIC	1.52636	7589.51604	1.52636	21.02359	.07260	.788

EFFECT .. MASC (Cont.)
Univariate F-tests with (1,361) D. F.

Variable	Hypoth. SS	Error SS	Hypoth. MS	Error MS	F	Sig. of F
ESTEEM	533.61774	3883.60490	533.61774	10.75791	49.60237	.000
INTEXT	264.44545	4640.51295	264.44545	12.85461	20.57204	.000
NEUROTIC	35.82929	7589.51604	35.82929	21.02359	1.70424	.193

EFFECT .. FEM (Cont.)
Univariate F-tests with (1,361) D. F.

Variable	Hypoth. SS	Error SS	Hypoth. MS	Error MS	F	Sig. of F
ESTEEM	107.44454	3883.60490	107.44454	10.75791	9.98749	.002
INTEXT	74.94312	4640.51295	74.94312	12.85461	5.83006	.016
NEUROTIC	26.81431	7589.51604	26.81431	21.02359	1.27544	.259

432

TABLE 9.27 STEPDOWN ANALYSES OF THREE ORDERED DVs ADJUSTED FOR THREE COVARIATES FOR (TOP TO BOTTOM) FEM BY MASC INTERACTION, MASCULINITY, FEMININITY (SELECTED OUTPUT FROM SPSS MANOVA--SEE TABLE 9.26 FOR SETUP)

Roy-Bargman Stepdown F - tests

Variable	Hypoth. MS	Error MS	StepDown F	Hypoth. DF	Error DF	Sig. of F
ESTEEM	7.21931	10.75791	.67107	1	361	.413
INTEXT	.35520	12.60679	.02817	1	360	.867
NEUROTIC	4.94321	19.72942	.25055	1	359	.617

Roy-Bargman Stepdown F - tests

Variable	Hypoth. MS	Error MS	StepDown F	Hypoth. DF	Error DF	Sig. of F
ESTEEM	533.61774	10.75791	49.60237	1	361	.000
INTEXT	137.74436	12.60679	10.92621	1	360	.001
NEUROTIC	1.07421	19.72942	.05445	1	359	.816

Roy-Bargman Stepdown F - tests

Variable	Hypoth. MS	Error MS	StepDown F	Hypoth. DF	Error DF	Sig. of F
ESTEEM	107.44454	10.75791	9.98749	1	361	.002
INTEXT	47.36159	12.60679	3.75683	1	360	.053
NEUROTIC	4.23502	19.72942	.21466	1	359	.643

433

TABLE 9.28 TESTS OF COVARIATES, FEMININITY, MASCULINITY, AND INTERACTION

Effect	DV	Univariate F	df	Stepdown F	df	α
Covariates	ESTEEM	20.48[a]	3/361	20.48**	3/361	.02
	INTEXT	1.13	3/361	0.50	3/360	.02
	NEUROTIC	21.95[a]	3/361	12.16**	3/359	.01
Femininity	ESTEEM	9.99[a]	1/361	9.99**	1/361	.02
	INTEXT	5.83[a]	1/361	3.76	1/360	.02
	NEUROTIC	1.28	1/361	0.21	1/359	.01
Masculinity	ESTEEM	49.60[a]	1/361	49.60**	1/361	.02
	INTEXT	20.57[a]	1/361	10.93**	1/360	.02
	NEUROTIC	1.70	1/361	0.05	1/359	.01
Femininity by	ESTEEM	0.67	1/361	0.67	1/361	.02
masculinity	INTEXT	0.00	1/361	0.03	1/360	.02
interaction	NEUROTIC	0.07	1/361	0.25	1/359	.01

[a]Significance level cannot be evaluated but would reach $p < .01$ in univariate context.

**$p < .01$.

TABLE 9.29 ADJUSTED MARGINAL MEANS FOR ESTEEM ADJUSTED FOR THREE COVARIATES, AND INTEXT ADJUSTED FOR ESTEEM PLUS THREE COVARIATES (SETUP AND SELECTED OUTPUT FROM SPSS MANOVA)

```
DATA LIST         FILE = 'MANOVA.DAT'
                  /CASENO 1-4 ANDRM 5-7 FEM 8-10 MASC 11-13 ESTEEM 14-16
                       CONTROL 17-19 ATTROLE 20-22 SEL2 23-30 (5)
                       INTEXT 31-34 (1) NEUROTIC 35-37.
VALUE LABELS      ANDRM 1 'UNDIFF' 2 'FEMNINE' 3 'MASCLNE' 4'ANDROGNS'/
                  FEM,MASC 1 'LOW' 2 'HIGH'.
MISSING VALUES    CONTROL(0).
MANOVA            ESTEEM,ATTROLE,NEUROTIC,INTEXT,CONTROL,SEL2 BY FEM,MASC(1,2)/
                  PRINT=PARAMETERS(ESTIM)/
             ANALYSIS=ESTEEM WITH CONTROL,ATTROLE,SEL2/
             DESIGN=CONSPLUS FEM/DESIGN=FEM,CONSPLUS MASC/
             ANALYSIS=INTEXT WITH CONTROL,ATTROLE,SEL2,ESTEEM/
             DESIGN=FEM, CONSPLUS MASC.


Estimates for ESTEEM adjusted for 3 covariates

CONSPLUS FEM

  Parameter        Coeff.

       1       16.7151721171
       2       15.3543164117


Estimates for ESTEEM adjusted for 3 covariates

  CONSPLUS MASC

  Parameter        Coeff.

       2       16.9175545469
       3       14.2243700126


Estimates for INTEXT adjusted for 4 covariates

CONSPLUS MASC

  Parameter        Coeff.

       2       11.0058841297
       3       12.4718466538
```

Note: Coeff. = adjusted marginal mean; first parameter=low, second parameter=high.

TABLE 9.30	MARGINAL MEANS FOR INTEXT ADJUSTED FOR THREE COVARIATES ONLY (SETUP AND SELECTED OUTPUT FROM SPSS MANOVA)

```
DATA LIST          FILE = 'MANOVA.DAT'
                   /CASENO 1-4 ANDRM 5-7 FEM 8-10 MASC 11-13 ESTEEM 14-16
                        CONTROL 17-19 ATTROLE 20-22 SEL2 23-30 (5)
                        INTEXT 31-34 (1) NEUROTIC 35-37.
VALUE LABELS       ANDRM 1 'UNDIFF' 2 'FEMNINE' 3 'MASCLNE' 4'ANDROGNS'/
                   FEM,MASC 1 'LOW' 2 'HIGH'.
MISSING VALUES     CONTROL(0).
MANOVA             ESTEEM,ATTROLE,NEUROTIC,INTEXT,CONTROL,SEL2 BY FEM,MASC(1,2)/
                   PRINT=PARAMETERS(ESTIM)/
                   ANALYSIS=INTEXT WITH CONTROL, ATTROLE, SEL2, ESTEEM/
                   DESIGN=CONSPLUS FEM.


Estimates for INTEXT adjusted for 4 covariates

CONSPLUS FEM

   Parameter          Coeff.

       1          11.1014710555
       2          11.9085923259
```

Note: Coeff. = adjusted marginal mean; first parameter=low, second parameter=high.

TABLE 9.31   CHECKLIST FOR MULTIVARIATE ANALYSIS OF COVARIANCE

1. Issues
   a. Unequal sample sizes and missing data
   b. Normality of sampling distributions
   c. Outliers
   d. Homogeneity of variance-covariance matrices
   e. Linearity
   f. Homogeneity of regression
      (1) Covariates
      (2) DVs for stepdown analysis
   g. Reliability of covariates (and DVs for stepdown)
   h. Multicollinearity and singularity
2. Major analyses: Planned comparisons or omnibus F; when significant:
   a. Multivariate strength of association
   b. Importance of DVs
      (1) Within-cell correlations, stepdown F, univariate F
      (2) Strength of association for significant F
      (3) Adjusted marginal and/or cell means for significant F
3. Additional analyses
   a. Assessment of covariates
   b. Interpretation of IV-covariates interaction (if homogeneity of regression violated for stepdown analysis)
   c. Post hoc comparisons

Results

A 2 × 2 between-subjects multivariate analysis of covariance was performed on three dependent variables associated with personality of respondents: self-esteem, introversion-extraversion, and neuroticism. Adjustment was made for three covariates: attitude toward role of women, locus of control, and socioeconomic status. Independent variables were masculinity (high and Low) and femininity (high and low).

SPSS MANOVA was used for the analyses with the hierarchical adjustment for nonorthogonality. Order of entry of IVs was femininity, then masculinity. Total N = 369 was reduced to 368 with the deletion of a case missing a score on locus of control. There were no univariate or multivariate within-cell outliers at α = .001. Results of evaluation of assumptions of normality, homogeneity of variance-covariance matrices, linearity, and multicollinearity were satisfactory. Covariates were judged to be adequately reliable for covariance analysis.

With the use of Wilks' criterion, the combined DVs were significantly related to the combined covariates, approximate $F(9, 873)$ = 10.86, p < .01, to femininity, $F(3, 359)$ = 4.67, p < .01, and to masculinity, $F(3, 359)$ = 20.59, p < .001 but not to the interaction, $F(3, 359)$ = 0.31, p > .05. There was a modest association between DVs and covariates, with partial η^2 = .12. A somewhat smaller association was found between combined DVs and the main effect of masculinity, η^2 = .08, and the association between the main effect of femininity and the combined DVs was smaller yet, η^2 = .02. [F and lambda are from Table 9.24; partial η^2 is calculated from Equation 9.9.]

To investigate more specifically the power of the covariates to adjust dependent variables, multiple regressions were run for each DV in turn, with covariates acting as multiple predictors. Two of the three covariates, locus of control and attitudes toward women's role, provided significant adjustment to self-esteem. The β value of .35 for locus of control was significantly different from zero, $t(361)$ = 7.21, p < .001, as was the η^2 value of .16 for attitudes toward women's role, $t(361)$ = 3.21, p < .01. None of the covariates provided adjustment to the introversion-extraversion scale. For neuroticism, only locus of control reached statistical significance, with β = .39, $t(361)$ = 8.04, p < .001. For none of the DVs did socioeconomic status provide significant adjustment.

Effects of masculinity and femininity on the DVs after adjustment for covariates were investigated in univariate and Roy-Bargmann step-down analysis, in which self-esteem was given the highest priority, introversion-extraversion second priority (so that adjustment was made for self-esteem as well as for the three covariates), and neuroticism third priority (so that adjustment was made for self-esteem and intro-version-extraversion as well as for the three covariates). Homogeneity of regression was satisfactory for this analysis, and DVs were judged to be sufficiently reliable to act as covariates. Results of this analysis are summarized in Table 9.28. An experimentwise error rate of 5% for each effect was achieved by apportioning alpha according to the values shown in the last column of the table.

After adjusting for differences on the covariates, self-esteem made a significant contribution to the composite of the DVs that best distin-guishes between women who were high or low in femininity, stepdown $F(1, 361) = 9.99$, $p < .01$, $\eta^2 = .02$. With self-esteem scored inversely, women with higher femininity scores showed greater self-esteem after adjustment for covariates (adjusted mean self-esteem = 15.40) than those scoring lower on femininity (adjusted mean self-esteem = 16.61). Univariate analysis revealed that a reliable difference was also pre-sent on the introversion-extraversion measure, with higher-femininity women more extraverted, univariate $F(1, 361) = 5.83$, a difference already accounted for by covariates and the higher-priority DV. [*Adjusted means are from Tables 9.29 and 9.30; η^2 is calculated as in Section 9.7.2.*]

Lower- versus higher-masculinity women differed in self-esteem, the highest-priority DV, after adjustment for covariates, stepdown $F(1, 361) = 49.60$, $p <, .01$, $\eta^2 = .10$. Greater self-esteem was found among higher-masculinity women (adjusted mean = 14.22) than among lower-mas-culinity women (adjusted mean = 16.92). The measure of introversion and extraversion, adjusted for covariates and self-esteem, was also related to differences in masculinity, stepdown $F(1, 360) = 10.93$, $p < .01$, $\eta^2 = .03$. Women scoring higher on the masculinity scale were more extraverted (adjusted mean extraversion 12.47) than those showing lower masculinity (adjusted mean extraversion = 11.01).

High-masculinity women, then, are characterized by greater self-esteem and extraversion than low-masculinity women when adjustments are made for differences in socioeconomic status, attitudes toward women's role, and locus of control. High-femininity women show greater self-esteem than low-femininity women with adjustment for those covariates.

Pooled within-cell correlations among dependent variables and covari-ates are shown in Table 9.16.

9.8 ■ SOME EXAMPLES FROM THE LITERATURE

9.8.1 ■ Examples of MANOVA

Wade and Baker (1977) surveyed clinical psychologists on their reasons for decisions to use tests. Psychologists were divided into low and high test users, as the two levels of the IV. Six "reasons," each rated on a 5-point scale of importance, served as the DVs. Separate MANOVAs were performed on users of objective and projective tests. After finding significant multivariate effects—that is, that high and low test users rated importance of reasons differently—for both objective and projective users, the researchers performed univariate ANOVAs on each of the reasons.

They found that for objective test users, the distinguishing reasons were reliability and validity, and agency requirements, for which high users gave higher ratings of importance than did low users. For the projective test users, high users rated as more important than low users graduate training experience, previous experience with tests, and reliability and validity. A MANOVA testing reasons for use as a function of eight major therapeutic orientations was not statistically significant.

In a study of romantic attraction, Giarrusso (1977) investigated the effects of physical attractiveness (low vs. high) and similarity of attitudes (similar vs. different) of a target date in a 2 × 2 between-groups MANOVA. Male students rated the target date on degree of liking, desire to date, comparability with previous dates, goodness of personality, and whether the target date liked them. These five ratings served as DVs. A significant multivariate effect of physical attractiveness was found, but not of similarity of attitudes or of the interaction between attractiveness and attitudes. After an *a priori* ordering of DVs, it was found that only the highest-priority variable, degree of liking, was significantly affected by physical attractiveness. That is, the more attractive the target date, the more she was liked. The remaining variables showed no statistically significant differences once the effect on degree of liking was partialed out.

Junior high school students were evaluated in terms of their behavioral changes after being exposed in class for one week to materials on "responsibility," in a study by Singh, Greer, and Hammond (1977). IVs were (1) whether or not materials were presented; (2) grade level: 7, 8, and 9; and (3) ability levels, three levels based on the School and College Ability Tests. The three DVs consisted of an attitude test designed to accompany the "responsibility" materials, an experimenter-designed essay, and objective tests, which tapped comprehension of various aspects of responsibility. With the use of a 2 × 3 × 3 between-groups multivariate analysis of variance on gain scores, only the main effect of program was found to be statistically significant. Bonferroni-type confidence intervals constructed for the three DVs, used instead of stepdown analysis, indicated that the attitude test alone accounted for the significant multivariate effect, but the authors felt that the magnitude of change was insufficient to support the program as implemented in the study.

Concerned that policy and decision makers seldom use evaluation information, Brown, Braskamp, and Newman (1978) investigated effects on readers' reactions to use of jargon and to objective vs. subjective statements in evaluation reports. After reading one of four statements (with presence vs. absence of jargon, and subjective vs. objective statements factorially combined), readers judged the reports on two dependent measures, difficulty and technicality, and judged the writer of the evaluation on nine measures (including such variables as thoroughness and believability). Separate MANOVAs were performed on two types of DVs—judgment of report and judgment of writer. Significant multivariate effects were interpreted in terms of cell means for DVs with significant univariate *F*s, rather than with the combination of univariate and stepdown analysis recommended here.

The only significant multivariate effect was that of jargon on ratings of the report. Both report DVs, technicality and difficulty, produced significant univariate Fs, indicating that reports high in jargon were considered to be more technical and more difficult. Ratings of the report writer were not affected by jargon, objectivity, or the interaction. Further, a univariate ANOVA revealed that extent of agreement with the recommendations of the evaluation report was independent of use of jargon, objectivity, or the interaction.

Jakubczak (1977) was interested in age differences in regulation of caloric intake of rats. For the first experiment, he used as IVs three age levels, three levels of dilution of food with cellulose, two levels of previous food deprivation experience, and six daysets, a within-subjects factor. DVs were caloric intake and body weight. Significant multivariate main effects were found on all factors, in addition to a number of two- and three-way interactions. Univariate ANOVAs, rather than stepdown and univariate analyses, were then used for interpreting effects on individual DVs. An age-related decrement in caloric intake and body weight was found in response to dilution with cellulose. Previous deprivation experience did not affect this relationship.

In a second experiment, two age levels, two levels of dilution by water, and nine daysets were used as factors. As an additional IV, rats were or were not given quinine in their diets. Older rats behaved like younger rats in response to dilution of the diet by water, maintaining caloric intake and weight. With addition of quinine, however, older rats decreased caloric intake and body weight to a greater degree than younger rats.

For both experiments, MANCOVAs run with initial levels of caloric intake and body weight as covariates revealed no difference in decisions with respect to within-subjects effects, as could be expected (cf. Section 8.5.2.1). Jakubczak made no mention of changes in between-subjects effects with the use of covariates.

Willis and Wortman (1976) studied public acceptance of randomized control-group designs in medical experimentation. Factors were gender of subject, three levels of science emphasis in describing the medical experiment, and three levels of treatment scarcity situations. For the last IV, descriptions of the medical experiment explicitly mentioned nonscarcity (i.e., the proposed treatment was available to all), mentioned scarcity (the treatment was expensive and scarce), or failed to mention anything about scarcity. Dependent measures were nine opinion questions, answered on bipolar scales. MANOVA revealed significant multivariate effects for two of the three IVs, scarcity of treatment and gender of subject. None of the interactions was statistically significant. DVs were analyzed in terms of cell means for scarcity and gender associated with those DVs with significant univariate Fs. For post hoc comparisons among the three scarcity levels, multivariate Fs were appropriately adjusted for inflated alpha error. It was concluded that the most favorable reactions occurred in situations in which scarcity was not mentioned at all, and that women tended to be less accepting than men of placebos and withholding of treatment.

9.8.2 ■ Examples of MANCOVA

Cornbleth (1977) studied the effect of a protected hospital ward on geriatric patients. IVs were ward assignment (protected or unprotected) and identification of the patient as a wanderer or nonwanderer. These two IVs were factorially combined, forming a 2×2 between-subjects design. Separate analyses were run for two periods in time (rather than including time period as a within-subjects factor). DVs were two physical measures, five cognitive measures, and six psychosocial measures. Separate MANOVAs were done for each of these three categories of functioning. In additional analyses, preassessment measures on the appropriate DVs were used as covariates. Therefore, there were three MANOVAs and three MANCOVAs. Contribution of the DVs to multivariate effects were evaluated through examination of standardized discriminant function coefficients (cf. Chapter 11). In addition, means adjusted for covariates were reported for all DVs.

On physical variables, a significant multivariate interaction was found (marginal at time 1, $p <$ .05 at time 2). The protected ward produced greater range of motion for wanderers, while the regular nonprotected ward facilitated performance for nonwanderers. In addition, independent of ward assignment, wanderers showed a lower level of psychosocial functioning than nonwanderers at both time points.

Modification of perceived locus of control in high school students was studied by Bradley and Gaa (1977). Subjects were blocked on gender and on two levels of prior achievement, and then randomly assigned to one of three treatment groups, goal-setting, conference (serving as a placebo), and control. Five variables measuring aspects of perceived locus of control were used as DVs. The covariate was a teacher-developed achievement pretest. In the report of results, no mention was made of main effects of the blocking IVs or any interactions between them and the manipulated IV.

Three separate analyses were reported, reflecting specific comparisons among groups rather than a single omnibus analysis of the three groups. These three comparisons were not mutually orthogonal, and they required more degrees of freedom than the omnibus test. However, no mention was made of adjusting (by Bonferroni procedures or others) for inflated alpha produced by multiple testing. Further, although it was reported that three MANCOVAs were performed, no multivariate statistics were presented. Instead, univariate Fs for each DV were reported for each of the three analyses. It was concluded that goal-setting treatment was effective in promoting more internal orientation among students, but only on measures related to locus of control in an academic situation.

CHAPTER 10

Profile Analysis of Repeated Measures

10.1 ■ GENERAL PURPOSE AND DESCRIPTION

Profile analysis is a special application of multivariate analysis of variance (MANOVA) in which several DVs are measured and they are all measured on the same scale.[1] A classic example of profile analysis is the comparison of personality profiles on, say, the Profile of Mood States (POMS), for different groups of people—psychoanalysts and behavior therapists. The DVs are the various scales of the POMS, tension-anxiety, vigor, anger-hostility, and the like.

Profile analysis is also used in research where subjects are measured repeatedly on the same DV. In this case, profile analysis is an alternative to univariate repeated measures ANOVA (see Chapter 3). For example, math achievement tests are given at various points during a semester to test the effects of alternative educational programs such as traditional classroom vs. computer assisted instruction. The choice between profile analysis (called the multivariate approach) and univariate repeated measures ANOVA depends on sample size, power, and whether statistical assumptions of repeated measured ANOVA are met. These issues are discussed fully in Section 10.5.2.

Current computer programs allow application of profile analysis to complex designs where, for instance, groups are classified along more than one dimension to create multiple IVs as in ANOVA. For example, POMS profiles are examined for both male and female psychoanalysts and behavior therapists. Or changes in math achievement during a semester as a result of either traditional education or computer assisted instruction are evaluated for both elementary vs. junior high school students. In addition, "multivariate-multivariate," or doubly multivariate, tests are available, in which multiple DVs are measured repeatedly. For example, the POMS is administered to psychoanalysts and behavior therapists at the beginning of professional training, at the end of training, and after some period of time in the profession. A discussion of types of doubly multivariate analysis and an example of one type appear in Section 10.5.3.

---

[1]The term profile analysis is also applied to techniques for measuring resemblance among profile patterns through cluster analysis rather than the MANOVA strategy described in this chapter. The current usage was popularized by Harris (1975).

10.2 ■ KINDS OF RESEARCH QUESTIONS

The major question to be answered by profile analysis is whether profiles of groups differ on a set of measures. To apply profile analysis, all measures must have the same range of possible scores, with the same score value having the same meaning on all the measures. The restriction on scaling of the measures is present because in two of the major tests of profile analysis (parallelism and flatness) the numbers that are actually tested are difference scores between DVs measured on adjacent occasions. Difference scores are called segments in profile analysis.

10.2.1 ■ Parallelism of Profiles

Do different groups have parallel profiles? This is commonly known as the test of parallelism and is the primary question addressed by profile analysis. Using the therapist example, do psychoanalysts and behavior therapists have the same pattern of highs and lows on the various mood states measured by the POMS?

When using profile analysis as a substitute for univariate repeated measures ANOVA, the parallelism test is the test of interaction. For example, do traditional and computer assisted instruction lead to the same pattern of gains in achievement over the course of testing? Put another way, do changes in achievement depend on which method of instruction is used?

10.2.2 ■ Overall Difference among Groups

Whether or not groups produce parallel profiles, does one group, on average, score higher on the collected set of measures than another? For example, does one type of therapist have reliably higher scores on the set of states measured by the POMS than the other? Or, using the repeated measures example, does one method of instruction lead to greater overall math achievement than the other method?

In regular ANOVA, this question is answered by test of the "groups" hypothesis; in profile analysis jargon, this is the "levels" hypothesis. It addresses the same question as the between-subjects main effect in repeated-measures ANOVA.

10.2.3 ■ Flatness of Profiles

The third question addressed by profile analysis concerns the similarity of response to all DVs, independent of groups. Do the DVs all elicit the same average response? In profile jargon, this tests the "flatness" hypothesis. This question is typically relevant only if the profiles are parallel. If the profiles are not parallel, then at least one of them is necessarily not flat. Although it is conceivable that nonflat profiles from two or more groups could cancel each other out to produce, on average, a flat profile, this result is usually not of research interest.

Using the therapist example, if psychoanalysts and behavior therapists have the same pattern of mood states on the POMS (that is, they have parallel profiles), one might ask whether the groups combined are notably high or low on any of the states. In the instructional example, the flatness test evaluates whether achievement changes over the period of testing. In this context the flatness test evaluates the same hypothesis as the within-subjects main effect in repeated-measures ANOVA.

10.2.4 ■ Contrasts Following Profile Analysis

With more than two groups or more than two measures, differences in parallelism, flatness, and/or level can result from a variety of sources. For example, if a group of client-centered therapists is added to the therapist study, and the parallelism or levels hypothesis is rejected, it is not obvious whether it is the behavior therapists who differ from the other two groups, the psychoanalysts who differ from the client-centered therapists, or exactly which group differs from which other group or groups. Contrasts following profile analysis are demonstrated in Sections 10.5.1 and 10.7.2.

10.2.5 ■ Marginal/Cell Means and Plots

Whenever statistically significant differences are found between groups or measures, parameters are estimated. For profile analysis, the major description of results is typically a plot of profiles where means for each of the DVs are plotted for each of the groups. In addition, if the null hypothesis regarding levels is rejected, assessment of group means and group standard deviations is helpful. And if the null hypothesis regarding flatness is rejected and the finding is of interest, a plot of scores combined over groups is instructive. Profile plots are demonstrated in Sections 10.4 and 10.7.

10.2.6 ■ Strength of Association

As with all statistical techniques, a finding of statistically significant differences is accompanied by description of the strength of association between the relevant IVs and DVs. Such measures are demonstrated in Sections 10.4 and 10.7.

10.2.7 ■ Treatment Effects in Multiple Time-Series Designs

In multiple time-series designs, two groups—experimental and control—are tested on several pretreatment and several posttreatment occasions to provide multiple DVs. In these types of designs, profile analysis is available as a substitute for univariate repeated-measures ANOVA.

Random assignment to groups is rare in this design. Therefore, the levels test of overall group difference is usually not of interest, for the two groups may well differ prior to treatment. However, as Algina and Swaminathan (1979) suggest, the test of parallelism, or interaction, sometimes reflects treatment effects. Acceptance of parallelism suggests no treatment effects because the course of the time series is the same whether or not treatment occurs. Rejection of parallelism suggests that there may be treatment effects but further analysis, as described by Algina and Swaminathan, is required to reveal whether the deviation from parallelism supports interpretation as treatment effects.

10.3 ■ LIMITATIONS TO PROFILE ANALYSIS

10.3.1 ■ Theoretical Issues

Choice of DVs is more limited in profile analysis than in usual applications of multivariate statistics because DVs must be commensurate. That is, they must all have been subjected to the same scaling techniques.

In applications where profile analysis is used as an alternative to univariate repeated-measures ANOVA, this requirement is met because all DVs are literally the same measure. In other applications of profile analysis, however, careful consideration of measurement of the DVs is required to assure that the units of measurement are the same. One way to measure commensurability is to use standardized scores such as z scores, instead of raw scores for the DVs. In this case, each DV is standardized using the pooled within-groups standard deviation (the square root of the error mean square for the DV) provided by univariate one-way between-subjects ANOVA for the DV. Or, more commonly, commensurate DVs are subscales of standardized tests, in which subtests are scaled in the same manner.

If, and only if, groups are formed by random assignment, levels of IVs manipulated, and proper experimental controls maintained, are differences among profiles causally attributed to differences in treatments among groups. Causality, as usual, is not addressed by the statistical test. Generalizability, as well, is influenced by sampling strategy, not choice of statistical tests. That is, the results of profile analysis generalize only to the populations from which cases are randomly sampled.

As described in Section 10.2 and derived in Section 10.4, the DVs in profile analysis are differences (segments) between the scores for adjacent levels of the within-subjects IV. Creating difference scores is one of the ways to equate the number of DVs and the degrees of freedom for the within-subjects IV. Although different programs use different transformations, the resulting analysis is insensitive to the transform. Technically, then, limitations should be assessed with regard to segments; however, it is reasonable (and lots simpler) to assess the DVs in their original form. That is, for purposes of assessing limitations, scores for the levels of the within-subjects IV are treated as the set of DVs.

10.3.2 ■ Practical Issues

10.3.2.1 ■ Sample Size and Missing Data

The sample size in each group is an important issue in profile analysis, as in MANOVA, because *there must be more research units in the smallest group than there are DVs*. This is required both because of considerations of power and for evaluation of the assumption of homogeneity of variance-covariance matrices (cf. Section 9.3.2.1). In the choice between univariate repeated-measures ANOVA and profile analysis, sample size per group is often the deciding factor.

Unequal sample sizes typically provide no special difficulty in profile analysis because each hypothesis is tested as if in a one-way design and, as discussed in Section 8.5.2.2, unequal n creates difficulties in interpretation only in designs with more than one IV. However, unequal n sometimes has implications for evaluating homogeneity of variance-covariance matrices, as discussed in Section 10.3.2.4.

If some measures are missing from some cases, the usual problems and solutions discussed in Chapter 4 apply.

10.3.2.2 ■ Multivariate Normality

Profile analysis is as robust to violation of normality as other forms of MANOVA (cf. Section 9.3.2.2). So *unless there are fewer cases than DVs in the smallest group and highly unequal* n, *deviation from normality of sampling distributions is not expected*. In the unhappy event of small, unequal samples, however, a look at the distributions of DVs for each group is in order. If distributions of DVs show marked, highly significant skewness, some normalizing transformations might be investigated (cf. Chapter 4).

10.3.2.3 ■ Outliers

As in all MANOVA, profile analysis is extremely sensitive to outliers. *Tests for univariate and multivariate outliers*, detailed in Chapter 4, *are applied to DVs*. These tests are demonstrated in Section 10.7.1.

10.3.2.4 ■ Homogeneity of Variance-Covariance Matrices

If sample sizes are equal, evaluation of homogeneity of variance-covariance matrices is not necessary. However, if sample sizes are notably discrepant, Box's M test is available through SPSS MANOVA as a test of the homogeneity of the variance-covariance matrices. Box's M is too sensitive for use at routine α levels but if the test of homogeneity is rejected at highly significant levels, the guidelines in Section 9.3.2.4 are appropriate.

Univariate homogeneity of variance is also assumed but the robustness of ANOVA generalizes to profile analysis. Unless sample sizes are highly divergent or there is evidence of strong heterogeneity (maximum variance ratio of 10:1) of the DVs (cf. Section 8.3.2.5) this assumption is probably safely ignored.

10.3.2.5 ■ Linearity

For the parallelism and flatness tests, linearity of the relationships among DVs is assumed. This assumption is evaluated by examining scatterplots between all pairs of DVs through SPSS PLOT, BMDP6D, SYSTAT PLOT, or SAS CORR or PLOT. Because the major consequence of failure of linearity is loss of power in the parallelism test, violation is somewhat mitigated by large sample sizes. Therefore, *with many symmetrically distributed DVs and large sample sizes, the issue may be ignored.* On the other hand, if distributions are notably skewed in different directions, check a few scatterplots for the variables with the most discrepant distributions to assure that the assumption is not too badly violated.

10.3.2.6 ■ Multicollinearity and Singularity

BMDP4V protects against multicollinearity and singularity by calculating pooled within-cell tolerance and dealing with tolerances that are too low. SPSS MANOVA provides the determinant of the within-cell correlation matrix. If the determinant is less than .0001, then SMCs for all original DVs are examined, most easily through BMDPAM, SYSTAT GLM, or a factor analysis program (cf. Section 9.3.2.8).

10.4 ■ FUNDAMENTAL EQUATIONS FOR PROFILE ANALYSIS

In Table 10.1 is a hypothetical data set appropriate for profile analysis used as an alternative to repeated measures ANOVA. The three groups (the IV) whose profiles are compared are belly dancers, politicians, and administrators (or substitute your favorite scapegoat). The five respondents

TABLE 10.1 SMALL SAMPLE OF HYPOTHETICAL DATA FOR ILLUSTRATION OF PROFILE ANALYSIS

Group		Case No.	Activity				Combined activities
			Read	Dance	TV	Ski	
Belly dancers		1	7	10	6	5	7.00
		2	8	9	5	7	7.25
		3	5	10	5	8	7.00
		4	6	10	6	8	7.50
		5	7	8	7	9	7.75
	Mean		6.60	9.40	5.80	7.40	7.30
Politicians		6	4	4	4	4	4.00
		7	6	4	5	3	4.50
		8	5	5	5	6	5.25
		9	6	6	6	7	6.25
		10	4	5	6	5	5.00
	Mean		5.00	4.80	5.20	5.00	5.00
Administrators		11	3	1	1	2	1.75
		12	5	3	1	5	3.50
		13	4	2	2	5	3.25
		14	7	1	2	4	3.50
		15	6	3	3	3	3.75
	Mean		5.00	2.00	1.80	3.80	3.15
Grand mean			5.53	5.40	4.27	5.40	5.15

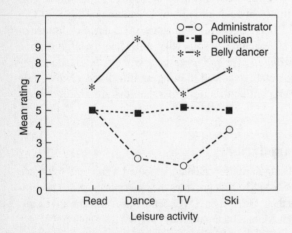

FIGURE 10.1 PROFILES OF LEISURE-TIME RATINGS FOR THREE OCCUPATIONS.

in each of these occupational groups participate in four leisure activities (the DVs) and, during each, are asked to rate their satisfaction on a 10-point scale. The leisure activities are reading, dancing, watching TV, and skiing. The profiles are illustrated in Figure 10.1, where mean ratings of each group for each activity are plotted.

Profile analysis tests of parallelism and flatness are multivariate and involve sum-of-squares and cross-products matrices. But the levels test is a univariate test, equivalent to the between-subjects main effect in repeated-measures ANOVA.[2]

10.4.1 ■ Differences in Levels

For the example, the levels test examines differences between the means of the three occupational groups combined over the four activities. Are the group means of 7.30, 5.00, and 3.15 significantly different from each other?

The relevant equation for partitioning variance is adapted from Equation 3.8 as follows:

$$\sum_i \sum_j (Y_{ij} - GM)^2 = np \sum_i \sum_j (\bar{Y}_j - GM)^2 + p \sum_i \sum_j (Y_{ij} - \bar{Y}_j)^2 \tag{10.1}$$

where n is the number of subjects in each group and p is the number of measures, in this case the number of ratings made by each respondent.

The partition of variance in Equation 10.1 produces the total, within-groups, and between-groups sums of squares, respectively, as in Equation 3.9. Because the score for each subject in the levels test is the subject's average score over the four activities, degrees of freedom follow Equations 3.10 through 3.13, with N equal to the total number of subjects and k equal to the number of groups.

For the hypothetical data of Table 10.1:

$$SS_{bg} = (5)(4)[(7.30 - 5.15)^2 + (5.00 - 5.15)^2 + (3.15 - 5.15)^2]$$

and

$$SS_{wg} = (4)[(7.00 - 7.30)^2 + (7.25 - 7.30)^2 + \ldots + (3.75 - 3.15)^2]$$

$$df_{bg} = k - 1 = 2$$

$$df_{wg} = N - k = 12$$

The levels test for the example produces a standard ANOVA source table for a one-way univariate test, as summarized in Table 10.2. There is a statistically significant difference between occupational groups in average rating of satisfaction during the four leisure activities.

TABLE 10.2 ANOVA SUMMARY TABLE FOR TEST OF LEVELS EFFECT FOR SMALL SAMPLE EXAMPLE OF TABLE 10.1

Source of variance	SS	df	MS	F
Between groups	172.90	2	86.45	44.14*
Within groups	23.50	12	1.96	

*$p < .001$

[2]Timm (1975) offers a multivariate form of this test, which he recommends when the parallelism hypothesis is rejected. It has been omitted here because it is not readily available through computer programs.

Standard η^2 is used to evaluate the strength of association between occupational groups and averaged satisfaction ratings:

$$\eta^2 = \frac{SS_{bg}}{SS_{bg} + SS_{wg}} = \frac{172.90}{172.90 + 23.50} = .88 \tag{10.2}$$

10.4.2 ■ Parallelism

Tests of parallelism and flatness are conducted through hypotheses about adjacent segments of the profiles. The test of parallelism, for example, asks if the difference (segment) between reading and dancing is the same for belly dancers, politicians, and administrators? How about the difference between dancing and watching TV?

The most straightforward demonstration of the parallelism test begins by converting the data matrix into difference scores. For the example, the four DVs are turned into three differences, as shown in Table 10.3. The difference scores are created from adjacent pairs of activities but in this example, as in many uses of profile analysis, the order of the DVs is arbitrary. In profile analysis it is often true that segments are formed from arbitrarily ordered DVs and have no intrinsic meaning. As discussed later, this sometimes creates difficulty in interpreting statistical findings.

In Table 10.3 the first entry for the first case is the difference between READ and DANCE scores, that is, $7 - 10 = -3$. Her second score is the difference in ratings between DANCE and TV: $10 - 6 = 4$, and so on.

TABLE 10.3   SCORES FOR ADJACENT SEGMENTS FOR SMALL SAMPLE HYPOTHETICAL DATA

Group		Case no.	Read vs. dance	Dance vs. TV	TV vs. ski
				Segment	
Belly dancers		1	−3	4	1
		2	−1	4	−2
		3	−5	5	−3
		4	−4	4	−2
		5	−1	1	−2
	Mean		−2.8	3.6	−1.6
Politicians		6	0	0	0
		7	2	−1	2
		8	0	0	−1
		9	0	0	−1
		10	−1	−1	1
	Mean		0.2	−0.4	0.2
Administrators		11	2	0	−1
		12	2	2	−4
		13	2	0	−3
		14	6	−1	−2
		15	3	0	0
	Mean		3.0	0.2	−2.0
Grand mean			0.13	1.13	−1.13

A one-way MANOVA on the segments tests the parallelism hypothesis. Because each segment represents a slope between two original DVs, if there is a multivariate difference between groups then one or more slopes are different and the profiles are not parallel.

Using procedures developed in Chapter 9, the total sum-of-squares and cross-products matrix (S_{tot}) is partitioned into the between-groups matrix S_{bg} and the within-groups or error matrix S_{wg}.[3] To produce the within-groups matrix, each person's score matrix, Y_{ikm}, has subtracted from it the mean matrix for that group, M_m. The resulting difference matrix is multiplied by its transpose to create the sum-of-squares and cross-products matrix. For the first belly dancer:

$$(\mathbf{Y}_{111} - \mathbf{M}_1) = \begin{bmatrix} -3 \\ 4 \\ 1 \end{bmatrix} - \begin{bmatrix} -2.8 \\ 3.6 \\ -1.6 \end{bmatrix} = \begin{bmatrix} -0.2 \\ 0.4 \\ 2.6 \end{bmatrix}$$

and

$$(\mathbf{Y}_{111} - \mathbf{M}_1)(\mathbf{Y}_{111} - \mathbf{M}_1)' = \begin{bmatrix} -0.2 \\ 0.4 \\ 2.6 \end{bmatrix} [-0.2 \quad 0.4 \quad 2.6]$$

$$= \begin{bmatrix} 0.04 & -0.08 & -0.52 \\ -0.08 & 0.16 & 1.04 \\ -0.52 & 1.04 & 6.76 \end{bmatrix}$$

This is the sum-of-squares and cross-products matrix for the first case. When these matrices are added over all cases and groups, the result is the error matrix S_{wg}:

$$\mathbf{S}_{wg} = \begin{bmatrix} 29.6 & -13.2 & 6.4 \\ -13.2 & 15.2 & -6.8 \\ 6.4 & -6.8 & 26.0 \end{bmatrix}$$

To produce the between-groups matrix, S_{bg}, the grand mean matrix, GM, is subtracted from each group mean matrix M_k, to form a difference matrix for each group. The mean matrix for each group in the example is

$$\mathbf{M}_1 = \begin{bmatrix} -2.8 \\ 3.6 \\ -1.6 \end{bmatrix} \qquad \mathbf{M}_2 = \begin{bmatrix} 0.2 \\ -0.4 \\ 0.2 \end{bmatrix} \qquad \mathbf{M}_3 = \begin{bmatrix} 3.0 \\ 0.2 \\ -2.0 \end{bmatrix}$$

and the grand mean matrix is

$$\mathbf{GM} = \begin{bmatrix} 0.13 \\ 1.13 \\ -1.13 \end{bmatrix}$$

The between-groups sum-of-squares and cross-products matrix S_{bg} is formed by multiplying each group difference matrix by its transpose, and then adding the three resulting matrices. After multiplying each entry by $n = 5$, to provide for summation over subjects,

---

[3] Other methods of forming **S** matrices can be used to produce the same result.

$$\mathbf{S}_{bg} = \begin{bmatrix} 84.133 & -50.067 & -5.133 \\ -50.067 & 46.533 & -11.933 \\ -5.133 & -11.933 & 13.733 \end{bmatrix}$$

Wilks' Lambda (Λ) tests the hypothesis of parallelism by evaluating the ratio of the determinant of the within-groups cross-products matrix to the determinant of the matrix formed by the sum of the within- and between-groups cross-products matrices:

$$\Lambda = \frac{|\mathbf{S}_{wg}|}{|\mathbf{S}_{wg} + \mathbf{S}_{bg}|} \tag{10.3}$$

For the example, Wilks' Lambda for testing parallelism is

$$\Lambda = \frac{6325.2826}{6325.2826 + 76598.7334} = .076279$$

By applying the procedures of Section 9.4.1, one finds an approximate $F(6, 20) = 8.74$, $p < .001$, leading to rejection of the hypothesis of parallelism. That is, the three profiles of Figure 10.1 are not parallel. Strength of association is measured as partial η^2:[4]

$$\text{partial } \eta^2 = 1 - \Lambda^{1/s} \tag{10.4}$$

For this example, then,

$$\text{partial } \eta^2 = 1 - .076279^{1/2} = .72$$

Seventy-two percent of the variance in the segments as combined for this test is accounted for by the difference in shape of the profiles for the three groups. Recall from Chapter 9, however, that segments are combined here to maximize group differences for parallelism. A different combination of segments is used for the test of flatness.

10.4.3 ■ Flatness

Because the hypothesis of parallelism is rejected for this example, the test of flatness is irrelevant; the question of flatness of combined profiles of Figure 10.1 makes no sense because at least one of them (and in this case probably two) is not flat. The flatness test is computed here to conclude the demonstration of this example.

Statistically, the test is whether, with groups combined, the three segments of Table 10.3 deviate from zero. That is, if segments are interpreted as slopes in Figure 10.1, are any of the slopes for the combined groups different from zero (nonhorizontal)? The test subtracts a set of hypothesized grand means, representing the null hypothesis, from the matrix of actual grand means:

$$(\mathbf{GM} - 0) = \begin{bmatrix} 0.13 \\ 1.13 \\ -1.13 \end{bmatrix} - \begin{bmatrix} 0 \\ 0 \\ 0 \end{bmatrix} = \begin{bmatrix} 0.13 \\ 1.13 \\ -1.13 \end{bmatrix}$$

---

[4] Partial η^2 is available through SPSS MANOVA.

The test of flatness is a multivariate generalization of the one-sample t test demonstrated in Chapter 3. Because it is a one-sample test, it is most conveniently evaluated through Hotelling's T^2, or trace:[5]

$$T^2 = N(\mathbf{GM} - 0)'\mathbf{S}_{wg}^{-1}(\mathbf{GM} - 0) \tag{10.5}$$

where N is the total number of cases and $\mathbf{S}_{wg}^{-1}$ is the inverse of the within-groups sum-of-squares and cross-products matrix developed in Section 10.4.2.
For the example:

$$T^2 = (15)[0.13 \quad 1.13 \quad -1.13] \begin{bmatrix} .05517 & .04738 & -.00119 \\ .04738 & .11520 & .01847 \\ -.00119 & .01847 & .04358 \end{bmatrix} \begin{bmatrix} 0.13 \\ 1.13 \\ -1.13 \end{bmatrix}$$

$$= 2.5825$$

From this is found F, with $p - 1$ and $N - k - p + 2$ degrees of freedom, where p is the number of original DVs (in this case 4), and k is the number of groups (3).

$$F = \frac{N - k - p + 2}{p - 1}(T^2) \tag{10.6}$$

so that

$$F = \frac{15 - 3 - 4 + 2}{4 - 1}(2.5825) = 8.608$$

with 3 and 10 degrees of freedom, $p < .01$ and the test shows significant deviation from flatness.

A measure of strength of association is found through Hotelling's T^2 that bears a simple relationship to lambda.

$$\Lambda = \frac{1}{1 + T^2} = \frac{1}{1 + 2.5825} = .27913$$

Lambda, in turn, is used to find η^2 (note that there is no difference between η^2 and partial η^2 because $s = 1$):

$$\eta^2 = 1 - \Lambda = 1 - .27913 = .72$$

showing that 72% of the variance in this combination of segments is accounted for by nonflatness of the profile collapsed over groups.

10.4.4 ■ Computer Analyses of Small Sample Example

Tables 10.4 through 10.7 show setup and selected output for computer analyses of the data in Table 10.1. Table 10.4 illustrates SPSS MANOVA with brief output. BMDP4V is illustrated in Table 10.5. Table 10.6 demonstrates profile analysis through SAS GLM, with SHORT printout requested. SYSTAT GLM is illustrated in Table 10.7. All programs are set up as repeated-measures ANOVA, which automatically produces both univariate and multivariate results.

---

[5] This is sometimes referred to as Hotelling's T.

TABLE 10.4 PROFILE ANALYSIS OF SMALL SAMPLE EXAMPLE THROUGH SPSS MANOVA (SETUP AND OUTPUT)

```
DATA LIST        FILE='tape42.dat'  FREE
                 /RESPOND, OCCUP, READ, DANCE, TV, SKI.
VAR LABELS       OCCUP 'OCCUPATION'.
VALUE LABELS     OCCUP 1 'BELLY DANCER' 2 'POLITICIAN' 3 'ADMINISTRATOR'.
MANOVA           READ TO SKI BY OCCUP(1,3)/
                 WSFACTOR=ACTIVITY(4)/
                 WSDESIGN=ACTIVITY/
                 RENAME=OVERALL, RDVSDAN, DANVSTV, TVVSSKI/
                 PRINT=SIGNIF(BRIEF)/
                 DESIGN.
```

The default error term in MANOVA has been changed from WITHIN CELLS to
WITHIN+RESIDUAL. Note that these are the same for all full factorial designs.

    15 cases accepted.
    0 cases rejected because of outofrange factor values.
    0 cases rejected because of missing data.
    3 nonempty cells.

    1 design will be processed.

* * * * * * * * * * * * * * * A n a l y s i s   o f   V a r i a n c e -- Design   1 * * * * * * * * * * * * * * * *

Tests of Between-Subjects Effects.

Tests of Significance for OVERALL using UNIQUE sums of squares
Source of Variation   SS   DF   MS   F   Sig of F

WITHIN+RESIDUAL   23.50   12   1.96
OCCUP   172.90   2   86.45   44.14   .000

* * * * * * * * * * * * * * * A n a l y s i s   o f   V a r i a n c e -- Design   1 * * * * * * * * * * * * * * * *

Tests involving 'ACTIVITY' Within-Subject Effect.

Mauchly sphericity test, W =   .66214
Chisquare approx. =   4.42047 with 5 D. F.
Significance =   .491

Greenhouse-Geisser Epsilon =   .79856
Huynh-Feldt Epsilon =   1.00000
Lower-bound Epsilon =   .33333
```
```

TABLE 10.4 *(CONTINUED)*

AVERAGED Tests of Significance that follow multivariate tests are equivalent to univariate or splitplot or mixedmodel approach to repeated measures. Epsilons may be used to adjust d.f. for the AVERAGED results.

* * * * * * * * * * * * * * * A n a l y s i s   o f   V a r i a n c e -- Design   1 *

Multivariate Tests of Significance
Tests using UNIQUE sums of squares and WITHIN+RESIDUAL error term

Source of Variation	Wilks Lambda	Mult. F	Hypoth. DF	Error DF	Sig. of F
ACTIVITY	.27914	8.60822	3.00	10.00	.004
OCCUP BY ACTIVITY	.07628	8.73584	6.00	20.00	.000

* * * * * * * * * * * * * * * * A n a l y s i s   o f   V a r i a n c e -- Design   1 *

Tests involving 'ACTIVITY' Within-Subject Effect.

AVERAGED Tests of Significance for MEAS.1 using UNIQUE sums of squares

Source of Variation	SS	DF	MS	F	Sig of F
WITHIN+RESIDUAL	40.10	36	1.11		
ACTIVITY	15.78	3	5.26	4.72	.007
OCCUP BY ACTIVITY	55.37	6	9.23	8.28	.000

```
/INPUT              VARIABLES = 6.  FORMAT IS FREE.  FILE = 'TAPE42.DAT'.
/VARIABLES          NAMES ARE RESPOND, OCCUP, READ, DANCE, TV, SKI.
                    LABEL IS RESPOND.
/BETWEEN            FACTOR IS OCCUP.  CODES ARE 1 TO 3.
                    NAMES ARE 'BELLY DANCER', 'POLITICIAN',
                        'ADMINISTRATOR'.
/WITHIN             FACTOR IS ACTIVITY.  CODES ARE 1 TO 4.
                    NAMES ARE READ, DANCE, TV, SKI.
/WEIGHTS            WITHIN ARE EQUAL.  BETWEEN ARE EQUAL.
/END
ANALYSIS            PROC = FACTORIAL.  EVAL./
END/
```

FACTORIAL PROCEDURE

THE FOLLOWING STATEMENTS HAVE BEEN GENERATED:

DESIGN TYPE IS BETWEEN, CONTRAST. MODEL.
  CODE IS CONST. NAME IS 'OVALL: GRAND MEAN'./

DESIGN FACTOR IS OCCUP.
  CODE IS EFFECT. NAME IS 'O: OCCUP'./

DESIGN TYPE IS WITHIN, CONTRAST. MODEL.
  CODE IS CONST. NAME IS 'OBS: WITHIN CASE MEAN'./

DESIGN FACTOR IS ACTIVITY.
  CODE IS EFFECT. NAME IS 'A: ACTIVITY'./

ANALYSIS /

END OF PROCEDURE-GENERATED STATEMENTS.
YOUR ANALYSIS INSTRUCTIONS RESUME AFTER ANALYSIS OUTPUT.

===

  --- ANALYSIS SUMMARY ---

THE FOLLOWING EFFECTS ARE COMPONENTS OF THE SPECIFIED
LINEAR MODEL FOR THE BETWEEN DESIGN. ESTIMATES AND TESTS
OF HYPOTHESES FOR THESE EFFECTS CONCERN PARAMETERS OF THAT MODEL.

        OVALL: GRAND MEAN
        O: OCCUP

THE FOLLOWING EFFECTS ARE COMPONENTS OF THE SPECIFIED
LINEAR MODEL FOR THE WITHIN DESIGN. ESTIMATES AND TESTS
OF HYPOTHESES FOR THESE EFFECTS CONCERN PARAMETERS OF THAT MODEL.

        OBS: WITHIN CASE MEAN
        A: ACTIVITY

EFFECTS CONCERNING PARAMETERS OF THE COMBINED BETWEEN AND
WITHIN MODELS ARE THE COMBINATIONS (INTERACTIONS) OF EFFECTS
IN BOTH MODELS.

454

TABLE 10.5   *(CONTINUED)*

```
========================================================================

========================================================================
WITHIN EFFECT: OBS: WITHIN CASE MEAN

EFFECT   VARIATE      STATISTIC              F         DF        P

  OVALL: GRAND MEAN
         DEP_VAR
                 SS=       1591.350000
                 MS=       1591.350000     812.60     1,    12  0.0000
  O: OCCUP
         DEP_VAR
                 SS=        172.900000
                 MS=         86.450000      44.14     2,    12  0.0000

  ERROR
         DEP_VAR
                 SS=         23.50000000
                 MS=          1.95833333


========================================================================
WITHIN EFFECT: A: ACTIVITY

EFFECT   VARIATE      STATISTIC              F         DF        P

  A
         DEP_VAR
                 S=    1,T=   3,DFH=   1,DFE=    12
                 HT EVALS= 0.72086265
                 HE EVALS=   2.5824657
                 TSQ=       30.9896         8.61      3,    10  0.0040
             WCP SS=        15.783333
             WCP MS=         5.261111       4.72      3,    36  0.0070
             GREENHOUSE-GEISSER ADJ. DF     4.72  2.40,  28.75  0.0126
             HUYNH-FELDT ADJUSTED DF        4.72  3.00,  36.00  0.0070
  (A) X (O: OCCUP)
         DEP_VAR
                 S=    2,T=   3,DFH=   2,DFE=    12
                 HT EVALS= 0.77976478, 0.65364964
                 HE EVALS=   3.5405999   ,    1.8872498
                 LRATIO=  0.762785E01     8.74      6,   20.00  0.0001
                 TRACE=    5.42785
                 TZSQ=    59.7063
                 P APPROXIMATION FOR ABOVE STATISTIC UNAVAILABLE - DF IS ZERO
                 MXROOT=  0.779765
                 P APPROXIMATION FOR ABOVE STATISTIC UNAVAILABLE
             WCP SS=        55.366667
             WCP MS=         9.227778      8.28      6,    36  0.0000
             GREENHOUSE-GEISSER ADJ. DF    8.28  4.79,  28.75  0.0001
             HUYNH-FELDT ADJUSTED DF       8.28  6.00,  36.00  0.0000

  ERROR
         DEP_VAR
             WCP SS=        40.10000000
             WCP MS=         1.11388889

             GGI EPSILON    0.79856
             H-F EPSILON    1.00000

------------------------------------------------------------------------

========================================================================
```

TABLE 10.6 PROFILE ANALYSIS OF SMALL SAMPLE EXAMPLE THROUGH SAS GLM (SETUP AND SELECTED OUTPUT)

```
DATA SSAMPLE;
INFILE 'TAPE42.DAT';
INPUT RESPOND OCCUP READ DANCE TV SKI;
PROC GLM;
      CLASS OCCUP;
      MODEL READ DANCE TV SKI = OCCUP/NOUNI;
      REPEATED ACTIVITY 4 PROFILE/SHORT;
```

General Linear Models Procedure
Class Level Information

Class	Levels	Values
OCCUP	3	1 2 3

Number of observations in data set = 15

Repeated Measures Analysis of Variance
Repeated Measures Level Information

Dependent Variable	READ	DANCE	TV	SKI
Level of ACTIVITY	1	2	3	4

Manova Test Criteria and Exact F Statistics for
the Hypothesis of no ACTIVITY Effect
H = Type III SS&CP Matrix for ACTIVITY   E = Error SS&CP Matrix

S=1   M=0.5   N=4

Statistic	Value	F	Num DF	Den DF	Pr > F
Wilks' Lambda	0.27913735	8.6082	3	10	0.0040
Pillai's Trace	0.72086265	8.6082	3	10	0.0040
Hotelling-Lawley Trace	2.58246571	8.6082	3	10	0.0040
Roy's Greatest Root	2.58246571	8.6082	3	10	0.0040

Manova Test Criteria and F Approximations for
the Hypothesis of no ACTIVITY*OCCUP Effect
H = Type III SS&CP Matrix for ACTIVITY*OCCUP   E = Error SS&CP Matrix

S=2   M=0   N=4

Statistic	Value	F	Num DF	Den DF	Pr > F
Wilks' Lambda	0.07627855	8.7358	6	20	0.0001
Pillai's Trace	1.43341443	9.2764	6	22	0.0001
Hotelling-Lawley Trace	5.42784967	8.1418	6	18	0.0002
Roy's Greatest Root	3.54059987	12.9822	3	11	0.0006

NOTE: F Statistic for Roy's Greatest Root is an upper bound.
NOTE: F Statistic for Wilks' Lambda is exact.

Tests of Hypotheses for Between Subjects Effects

Source	DF	Type III SS	Mean Square	F Value	Pr > F
OCCUP	2	172.9000000	86.4500000	44.14	0.0001

TABLE 10.6   *(CONTINUED)*

Error	12	23.5000000	1.9583333

Univariate Tests of Hypotheses for Within Subject Effects

Source: ACTIVITY

					Adj	Pr > F
DF	Type III SS	Mean Square	F Value	Pr > F	G - G	H - F
3	15.78333333	5.26111111	4.72	0.0070	0.0126	0.0070

Source: ACTIVITY*OCCUP

					Adj	Pr > F
DF	Type III SS	Mean Square	F Value	Pr > F	G - G	H - F
6	55.36666667	9.22777778	8.28	0.0001	0.0001	0.0001

Source: Error(ACTIVITY)

DF	Type III SS	Mean Square
36	40.10000000	1.11388889

Greenhouse-Geisser Epsilon = 0.7986
Huynh-Feldt Epsilon = 1.1778

The four programs differ substantially in setup and presentation of the three tests. To set up SPSS MANOVA for profile analysis, the DVs (levels of the within-subject effect) READ TO SKI are followed in the MANOVA statement by the keyword BY and the grouping variable with its levels— OCCUP(1,3). The DVs are combined for profile analysis in the WSFACTOR instruction and labeled ACTIVITY(4) to indicate four levels for the within-subjects factor. The RENAME instruction makes the output easier to read: OVERALL labels the test of the combined DVs, and the remaining three variables specify the segments. For the example, RDVSDAN labels the READ vs. DANCE segment (see Table 10.4).

In the SPSS MANOVA output, the levels test for differences among groups is the test of OCCUP in the section labeled Tests of Significance for OVERALL . . . This is followed by information about tests and adjustments for homogeneity of covariance (cf. Section 10.5.2). The flatness and parallelism tests appear in the section labeled Tests using UNIQUE sums of squares and WITHIN + RESIDUAL for ACTIVITY and OCCUP BY ACTIVITY, respectively. Output is limited in this example by the PRINT=SIGNIF(BRIEF) instruction. Without this statement, separate source tables are printed for flatness and parallelism, each containing several multivariate tests. Finally, the output concludes with univariate tests for the within-subjects main effect (ACTIVITY) and the interaction in the section labeled AVERAGED Tests of Significance for MEAS. 1. . . .

In BMDP4V, the BETWEEN instruction specifies the grouping variable and the WITHIN instruction indicates the multiple DVs (repeated measures). The EVAL instruction in the ANALYSIS paragraph requests eigenvalues, as shown in Table 10.5.

There are two separate source tables in the output, one for the levels test and another for the flatness and parallelism tests. The first source table is labeled OBS: WITHIN CASE MEAN and the levels effect is O: OCCUP. In the second table, labeled A: ACTIVITY, the flatness test is simply called A. The parallelism test is called (A) X (O: OCCUP). For both these tests, both univariate and multivariate tests are printed out. For flatness, TSQ, a form of Hotelling's T^2, is used. HT EVALS is η^2, helpful for finding strength of association in the absence of Wilks' Lambda. For parallelism,

```
USE TAPE42
OUTPUT SSPROFIL.OUT
CATEGORY OCCUP
MODEL READ DANCE TV SKI = CONSTANT + OCCUP / REPEAT,
     NAME = 'ACTIVITY'
ESTIMATE
```

Categorical values encountered during processing are:
OCCUP (3 levels)
          1,   2,   3

Number of cases processed: 15
Dependent variable means

	READ	DANCE	TV	SKI
	5.533	5.400	4.267	5.400

Univariate and Multivariate Repeated Measures Analysis

Between Subjects

Source	SS	DF	MS	F	P
OCCUP	172.900	2	86.450	44.145	0.000
Error	23.500	12	1.958		

Within Subjects

Source	SS	DF	MS	F	P	G-G	H-F
ACTIVITY	15.783	3	5.261	4.723	0.007	0.013	0.007
ACTIVITY *OCCUP	55.367	6	9.228	8.284	0.000	0.000	0.000
Error	40.100	36	1.114				

Greenhouse-Geisser Epsilon: 0.7986
Huynh-Feldt Epsilon   :   1.0000

Multivariate Repeated Measures Analysis

Test of: ACTIVITY		Hypoth. DF	Error DF	F	P
Wilks' Lambda=	0.279	3	10	8.608	0.004
Pillai Trace =	0.721	3	10	8.608	0.004
H-L Trace =	2.582	3	10	8.608	0.004

Test of: ACTIVITY *OCCUP		Hypoth. DF	Error DF	F	P
Wilks' Lambda=	0.076	6	20	8.736	0.000
Pillai Trace =	1.433	6	22	9.276	0.000
H-L Trace =	5.428	6	18	8.142	0.000
Theta = 0.780 S = 2, M = 0.0, N = 4.0 P = 0.004					

there are four multivariate tests, Wilks' Lambda (LRATIO), Hotelling's TRACE, TZSQ, and MXROOT (Roy's greatest characteristic root, cf. Section 9.5.1). The F ratios for WCP are the univariate within-subjects tests, along with Greenhouse-Geisser and Huynh-Feldt adjustments to univariate F for violation of homogeneity of covariance, if necessary. All cell and marginal means are also printed out (not shown).

In SAS GLM, the CLASS instruction identifies OCCUP as the grouping variable. The MODEL instruction shows the DVs on the left of an equation and the IV on the right. Profile analysis is distinguished from ordinary MANOVA by the instructions in the line beginning REPEATED, as seen in Table 10.6.

The results are presented in two multivariate tables and two univariate tables. The first table, labeled . . . Hypothesis of no ACTIVITY Effect, shows four fully-labeled multivariate tests of flatness. The second table shows the same four multivariate tests of parallelism, labeled . . . Hypothesis of no ACTIVITY*OCCUP Effect. The test of levels is the test for OCCUP in the third table, labeled Test of Hypotheses for Between Subjects Effects. The Univariate Tests of Hypotheses for Within Subjects Effects table provides the univariate ANOVA for the within-subjects main effect (ACTIVITY) and the within-between interaction (ACTIVITY*OCCUP). Greenhouse-Geisser (G-G) and Huynh-Feldt (H-F) adjustments for violation of homogeneity of covariance are also given for these univariate repeated-measures tests. More extensive output is available if the SHORT instruction is omitted.

SYSTAT GLM setup is similar to SAS except that CONSTANT is added to the IV side of the MODEL equation, as seen in Table 10.7. The program first prints out the means for each leisure time activity (but not means for each group). The levels test follows in the Between Subjects effects section. The following Within Subjects section shows the univariate repeated-measures tests for ACTIVITY and the ACTIVITY*OCCUP interaction, along with Greenhouse-Geisser and Huynh-Feldt adjustments, if necessary. A full trend analysis follows (not shown). Finally, the section labeled Multivariate Repeated Measures Analysis begins with the flatness test, labeled ACTIVITY, with three multivariate tests. The same tests are given for the parallelism test, labeled ACTIVITY*OCCUP. Roy's gcr, labeled Theta, is an additional test given for parallelism (because there is only one df for the flatness test, Roy's gcr provides no additional useful information).

10.5 ■ SOME IMPORTANT ISSUES

Issues discussed here are unique to profile analysis or affect profile analysis differently from traditional MANOVA. Issues such as choice among statistical criteria, for instance, are identical whether the DVs are analyzed directly (as in MANOVA) or converted into segments (as in profile analysis). Therefore, the reader is referred to Section 9.5 for consideration of these matters.

10.5.1 ■ Contrasts in Profile Analysis

When there are more than two levels of a significant effect in profile analysis, it is often desirable to perform contrasts to pinpoint sources of variability. For instance, because there is an overall difference between administrators, belly dancers, and politicians in their ratings of satisfaction with leisure time activities (see Section 10.4), contrasts are needed to discover which groups differ from which other groups. Are belly dancers the same as administrators? Politicians? Neither?

It is probably easiest to think of contrasts following profile analysis as coming from a regular ANOVA design with (at least) one grouping variable and one repeated measure, even when the application of the technique is to multiple, commensurate DVs. That is, the most interpretable contrasts following profile analysis are likely to be the ones that would also be appropriate after a mixed within- and between-subjects ANOVA.

There are, of course, numerous contrast procedures and the choice among them depends on what makes most sense in a given research setting. With a single control group, Dunnett's procedure often makes most sense. Or, if all pairwise comparisons are desired, the Tukey test is most appropriate. Or, if there are numerous repeated measures, and/or normative data are available, a confidence interval procedure, such as that used in Section 10.7.2, may make the most sense. With relatively few repeated measures, a Scheffé type procedure is probably the most general (if also the most conservative) and is the procedure illustrated in this section.

It is important to remember that the contrasts recommended here explore differences in original DV scores while the significance tests in profile analysis for parallelism and flatness typically evaluate segments. Although there is a logical problem with following up a significance test based on segments with a contrast based on the original scores, performing contrasts on segments or some other transformation of the variables seems even worse because of difficulty interpreting the results.

Contrasts in repeated-measures ANOVA with both grouping variables and repeated measures is not the easiest of topics, as you probably recall. First, when parallelism (interaction) is significant, there is the choice between a simple effects analysis and an interaction contrasts analysis. Second, there is a need in some cases to develop separate error terms for some of the contrasts. Third is the need to apply an adjustment such as Scheffé to the F test to avoid too liberal rejection of the null hypothesis. The researcher who is fascinated by these topics is referred to Keppel (1991, Chapters 11, 12, 17 and 18) for a detailed and reasonably clear discussion of them. The present effort is to illustrate several possible approaches through BMDP and SPSS and to recommend guidelines for when each is likely to be appropriate.

The most appropriate contrast to perform depends on which effect or combination of effects—levels, flatness, or parallelism—is significant. If either levels or flatness is significant, but parallelism (interaction) is not, contrasts are performed on marginal means. If the test for levels is significant, contrasts are formed on marginal values for the grouping variable. If the test for flatness is significant, contrasts are formed on the repeated-measures marginal values. Because contrasts formed on marginal values "fall out" of computer runs for interaction contrasts, they are illustrated in Section 10.5.1.3.

Sections 10.5.1.1 and 10.5.1.2 describe simple effects analyses, appropriate if parallelism is significant. In simple effects analysis one variable is held constant at some value while mean differences are examined on the levels of the other variable, as seen in Figure 10.2. For instance, the level of group is held constant at belly dancer while mean differences are examined among the leisure time activities [Figure 10.2(a)]. The researcher asks if belly dancers have mean differences in satisfaction with different leisure activities. Or, leisure activity is held constant at dance while mean differences are explored between administrators, politicians, and belly dancers [Figure 10.2(c)]. The researcher asks whether the three groups have different mean satisfaction while dancing.

Section 10.5.1.1 illustrates a simple effects analysis followed by simple contrasts [Figures 10.2(c) and (d)] for the case where parallelism and flatness effects are both significant, but the levels effect is not. Section 10.5.1.2 illustrates a simple effects analysis followed by simple contrasts [Figures 10.2(a) and (b)] for the case where parallelism and levels are both significant, but the flatness effect

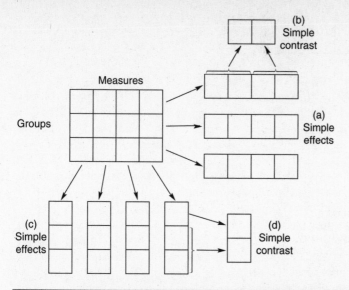

FIGURE 10.2 SIMPLE EFFECTS ANALYSIS EXPLORING: (a) DIFFER-
ENCES AMONG MEASURES FOR EACH GROUP, FOL-
LOWED BY (b) A SIMPLE CONTRAST BETWEEN
MEASURES FOR ONE GROUP: AND (c) DIFFERENCES
AMONG GROUPS FOR EACH MEASURE, FOLLOWED
BY (d) A SIMPLE CONTRAST BETWEEN GROUPS FOR
ONE MEASURE.

is not. This particular pattern of simple effects analysis is recommended because of the confounding inherent in analyzing simple effects.

The analysis is confounded because when the groups (levels) effect is held constant to analyze the repeated measure in a one-way within-subjects ANOVA, both the sum of squares for interaction and the sum of squares for the repeated measure are partitioned. When the repeated measure is held constant so the groups (levels) effect is analyzed in a one-way between-subjects ANOVA, both the sum of squares for interaction and the sum of squares for the group effect are partitioned. Because in simple effects analyses the interaction sum of squares is confounded with one or the other of the main effects, it seems best to confound it with a nonsignificant main effect where possible. This recommendation is followed in Sections 10.5.1.1 and 10.5.1.2.

Section 10.5.1.3 describes an interaction contrasts analysis. In such an analysis, an interaction between two IVs is examined through one or more smaller interactions (Figure 10.3). For instance, the significant interaction between the three groups on four leisure activities in the example might be reduced to examination of the difference between two groups on only two of the activities. One could, for instance, ask if there is a significant interaction in satisfaction between belly dancers and administrators while watching TV vs. dancing. Or, one could pool the results for administrator and politician and contrast them with belly dancer for one side of the interaction, while pooling the results for the two sedentary activities (watching TV and reading) against the results for the two active activities (dancing and skiing) as the other side of the interaction. The researcher asks

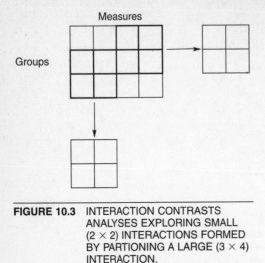

FIGURE 10.3 INTERACTION CONTRASTS
ANALYSES EXPLORING SMALL
(2 × 2) INTERACTIONS FORMED
BY PARTIONING A LARGE (3 × 4)
INTERACTION.

whether there is an interaction between dancers and the other professionals in their satisfaction while engaged in sedentary vs. active leisure time activities.

An interaction contrasts analysis is not a confounded analysis; only the sum of squares for interaction is partitioned. Thus, it is appropriate whenever the interaction is significant and regardless of the significance of the other two effects. However, because simple effects are generally easier to understand and explain, it seems better to perform them when possible. For this reason, we recommend an interaction contrasts analysis to explore the parallelism effect only when both the levels and flatness effects are also significant.

10.5.1.1 ■ Parallelism and Flatness Significant, Levels Not Significant (Simple Effects Analysis)

When parallelism and flatness are both significant, a simple effects analysis is recommended where differences among means for groups are examined separately at each level of the repeated measure [Figure 10.2(c)]. For the example, differences in means among politicians, administrators, and belly dancers are sought first in the reading variable, then in dance, then in TV, and finally in skiing. (Not all these effects need to be examined, of course, if they are not of interest.)

Both BMDP and SPSS have setup language to request simple effects analyses. In BMDP one specifies PROCEDURE IS SIMPLE and HOLD IS (variable name) in the ANALYSIS paragraph. In SPSS the keyword MWITHIN is used in the MANOVA paragraph. Table 10.8 shows control language and partial output from both BMDP and SPSS for performing simple effects analysis on groups with repeated measures held constant.

Table 10.8(a) shows the control language and output produced by BMDP. The simple effect of groups at READ is identified as (OBS.R) X (O: OCCUP) (sum of squares = 8.533333, mean square = 4.266667, and F = 2.67) in the output. The simple effect of groups at DANCE is (OBS.D) X (O: OCCUP) in the output, while simple effects of TV is (OBS.T) X (O: OCCUP) and SKI is (OBS.S) X (O: OCCUP).

(a) SETUP AND SELECTED OUTPUT FROM BMDP4V

```
/INPUT          VARIABLES = 6.  FORMAT IS FREE.  FILE = 'TAPE42.DAT'.
/VARIABLES      NAMES ARE RESPOND, OCCUP, READ, DANCE, TV, SKI.
                LABEL IS RESPOND.
/BETWEEN        FACTOR IS OCCUP.  CODES ARE 1 TO 3.
                NAMES ARE 'BELLY DANCER', 'POLITICIAN',
                    'ADMINISTRATOR'.
/WITHIN         FACTOR IS ACTIVITY.  CODES ARE 1 TO 4.
                NAMES ARE READ, DANCE, TV, SKI.
/WEIGHTS        WITHIN ARE EQUAL.  BETWEEN ARE EQUAL.
/END
ANALYSIS        PROC = SIMPLE.
                HOLD IS ACTIVITY./
END/
```

WITHIN EFFECT: OBS.R: MEAN AT READ

EFFECT	VARIATE	STATISTIC		F	DF		P
OBS.R							
	DEP_VAR						
		SS=	459.266667				
		MS=	459.266667	287.04	1,	12	0.0000
(OBS.R) X (O: OCCUP)							
	DEP_VAR						
		SS=	8.533333				
		MS=	4.266667	2.67	2,	12	0.1101
ERROR							
	DEP_VAR						
		SS=	19.20000000				
		MS=	1.60000000				

WITHIN EFFECT: OBS.D: MEAN AT DANCE

EFFECT	VARIATE	STATISTIC		F	DF		P
OBS.D							
	DEP_VAR						
		SS=	437.400000				
		MS=	437.400000	524.88	1,	12	0.0000
(OBS.D) X (O: OCCUP)							
	DEP_VAR						
		SS=	139.600000				
		MS=	69.800000	83.76	2,	12	0.0000
ERROR							
	DEP_VAR						
		SS=	10.00000000				
		MS=	0.83333333				

WITHIN EFFECT: OBS.T: MEAN AT TV

EFFECT	VARIATE	STATISTIC		F	DF		P
OBS.T							
	DEP_VAR						
		SS=	273.066667				
		MS=	273.066667	390.10	1,	12	0.0000
(OBS.T) X (O: OCCUP)							
	DEP_VAR						
		SS=	46.533333				
		MS=	23.266667	33.24	2,	12	0.0000

TABLE 10.8   *(CONTINUED)*

```
ERROR
      DEP_VAR
                    SS=            8.40000000
                    MS=            0.70000000

------------------------------------------------------------------------


WITHIN EFFECT: OBS.S: MEAN AT SKI

EFFECT    VARIATE       STATISTIC              F         DF        P
------------------------------------------------------------------------
  OBS.S
      DEP_VAR
                    SS=          437.400000
                    MS=          437.400000    201.88    1,    12   0.0000
  (OBS.S) X (O: OCCUP)
      DEP_VAR
                    SS=           33.600000
                    MS=           16.800000      7.75    2,    12   0.0069

  ERROR
      DEP_VAR
                    SS=           26.00000000
                    MS=            2.16666667


------------------------------------------------------------------------

========================================================================
```

(b) SETUP AND SELECTED OUTPUT FROM SPSS MANOVA

```
DATA LIST         FILE='TAPE42.DAT'  FREE
                  /RESPOND, OCCUP, READ, DANCE, TV, SKI.
VAR LABELS        OCCUP 'OCCUPATION'.
VALUE LABELS      OCCUP 1 'BELLY DANCER' 2 'POLITICIAN' 3 'ADMINISTRATOR'.
MANOVA            READ TO SKI BY OCCUP(1,3)/
                  WSFACTOR=ACTIVITY(4)/
                  WSDESIGN=MWITHIN ACTIVITY(1) MWITHIN ACTIVITY(2)
                       MWITHIN ACTIVITY(3) MWITHIN ACTIVITY(4)/
                  RENAME=READ, DANCE, TV, SKI/
                  DESIGN.
```

Tests involving 'MWITHIN ACTIVITY(1)' Within-Subject Effect.

Tests of Significance for READ using UNIQUE sums of squares

Source of Variation	SS	DF	MS	F	Sig of F
WITHIN + RESIDUAL	19.20	12	1.60		
MWITHIN ACTIVITY(1)	459.27	1	459.27	287.04	.000
OCCUP BY MWITHIN ACTIVITY(1)	8.53	2	4.27	2.67	.110

Tests involving 'MWITHIN ACTIVITY(2)' Within-Subject Effect.

Tests of Significance for DANCE using UNIQUE sums of squares

Source of Variation	SS	DF	MS	F	Sig of F
WITHIN + RESIDUAL	10.00	12	.83		
MWITHIN ACTIVITY(2)	437.40	1	437.40	524.88	.000
OCCUP BY MWITHIN ACTIVITY(2)	139.60	2	69.80	83.76	.000

TABLE 10.8 *(CONTINUED)*

```
Tests involving 'MWITHIN ACTIVITY(3)' Within-Subject Effect.

Tests of Significance for TV using UNIQUE sums of squares
Source of Variation        SS       DF       MS        F   Sig of F

WITHIN + RESIDUAL         8.40      12      .70
MWITHIN ACTIVITY(3)     273.07       1   273.07    390.10    .000
OCCUP BY MWITHIN ACT     46.53       2    23.27     33.24    .000
IVITY(3)

Tests involving 'MWITHIN ACTIVITY(4)' Within-Subject Effect.

Tests of Significance for SKI using UNIQUE sums of squares
Source of Variation        SS       DF       MS        F   Sig of F

WITHIN + RESIDUAL        26.00      12     2.17
MWITHIN ACTIVITY(4)     437.40       1   437.40    201.88    .000
OCCUP BY MWITHIN ACT     33.60       2    16.80      7.75    .007
IVITY(4)
```

Table 10.8(b) shows the control language and output produced by SPSS. The simple effect of groups at READ is identified as OCCUP BY MWITHIN ACTIVITY(1) in the output. Simple effects of DANCE, TV, and SKI are identified as OCCUP BY MWITHIN ACTIVITY(2), ACTIVITY(3), and ACTIVITY(4), respectively.

The same separate error terms are developed by both programs for each of the simple effects. The error term formed by subjects nested within groups at READ is used to test for differences between groups when the activity is READ. Similar error terms are developed for DANCE, TV, and SKI. In the BMDP output, the error terms are called ERROR, in SPSS, WITHIN+RESIDUAL (which is the same as WITHIN CELLS as long as OCCUP and ACTIVITY both are analyzed). For example, both BMDP and SPSS compute the sum of squares for subjects nested within groups at READ as 19.2000 (mean square = 1.6000).

To evaluate the significance of the simple effects, a Scheffé adjustment (see Section 3.2.6.1) is applied to unadjusted critical *F* under the assumptions that these tests are performed post hoc and that the researcher wants to control for familywise Type I error. For these contrasts the Scheffé adjustment is

$$F_s = (k - 1) F_{(k - 1), k(n - 1)}   \tag{10.7}$$

where k is the number of groups and n is the number of subjects in each group. For the example, using $\alpha = 0.05$

$$F_s = (3 - 1) F_{(2, 12)} = 7.76$$

By this criterion, there are not reliable mean differences between the groups when the DV is READ or SKI, but there are reliable differences when the DV is DANCE or TV.

Because there are three groups, these findings are still ambiguous. Which group or groups are different from which other group or groups? To pursue the analysis, simple contrasts are performed [Figure 10.2(d)]. Contrast coefficients are applied to the levels of the grouping variable to determine the source of the difference. For the example, contrast coefficients compare the mean for belly dancers with the mean for the other two groups combined for DANCE. Control language and partial output from both BMDP and SPSS are shown in Table 10.9.

In BMDP [Table 10.9(a)], the control language is appropriate to an ANOVA with a between-subjects IV. DANCE is identified as the DV in the ANALYSIS paragraph. (Note that the WITHIN paragraph has been omitted.) The sum of squares for the contrast is identified by its name as specified in the DESIGN paragraph (BD vs OTHR—belly dance vs. other).

In SPSS [Table 10.9(b)], the CONTRAST procedure is used.[6] The sum of squares for the contrast is labeled OCCUP(1). Both programs use the error term formed by subjects nested within groups for the particular DV. This error term is called ERROR in BMDP and WITHIN CELLS in SPSS. The WITHIN CELLS error term is explicitly requested in SPSS in DESIGN = OCCUP (1) VS WITHIN.

For this analysis, the sum of squares and mean square for the contrast is 120.000, error mean square is .83333, and F is 144.00. This F exceeds the F_s adjusted critical value of 7.76; it is no surprise to find there is a reliable difference between belly dancers and others in their satisfaction while engaging in DANCE.

10.5.1.2 ■ Parallelism and Levels Significant, Flatness Not Significant (Simple Effects Analysis)

This combination of findings occurs rarely because if parallelism and levels are significant, flatness is nonsignificant only if profiles for different groups are mirror images that cancel each other out.

The simple effects analysis recommended here examines mean differences among the various DVs in series of one-way within-subjects ANOVAs with each group in turn held constant [Figure 10.2(a)]. For the example, mean differences between READ, DANCE, TV, and SKI are sought first for belly dancers, then for politicians, and then for administrators. The researcher inquires whether each group, in turn, is more satisfied during some activities than during others.

The same setups that were used in Section 10.5.1.1 are appropriate here except that the other effect is specified. Table 10.10(a) contains control language and partial output for BMDP; Table 10.10(b) shows the same analyses through SPSS.

For BMDP the simple effect of repeated measures for belly dancers is labeled (A) X (OVALL.B: MEAN AT BELLY DA); the simple effect of repeated measures for politicians is (A) X (OVALL.P: MEAN AT POLITICI) and for administrators (A) X (OVALL.B: MEAN AT ADMINIST). For SPSS the simple effect for belly dancers is MWITHIN OCCUP(1) BY ACTIVITY, for politicians MWITHIN OCCUP(2) BY ACTIVITY, and for administrators MWITHIN OCCUP(3) BY ACTIVITY. The sum of squares for the simple effect at belly dancers is 35.80.

Both BMDP and SPSS use the error term from the full analysis (interaction between repeated measures and subjects nested within groups) as the error term for each of the contrasts. BMDP calls the term ERROR and SPSS calls it WITHIN+RESIDUAL (same as WITHIN CELLS here); the sum of squares is 40.10.

Some feel that it is best to develop a separate error term for each portion of an analysis, that is, that the error term for the simple effect for belly dancers should be the repeated measures by subjects interaction for that group alone. If you are of this opinion, judicious use of the USE = (OCCUP EQ 1)

---

[6] This could also be done in conjunction with the MWITHIN procedure, but the setup is more complicated and the labeling in the output misleading. Note that when contrasts are specified in SPSS MANOVA, a full rank set is required: a set of ones, and then as many contrasts as there are df for the effect.

TABLE 10.9 SIMPLE COMPARISONS ON OCCUPATIONS, HOLDING ACTIVITY CONSTANT: (a) BMDP AND (b) SPSS

(a) SETUP AND SELECTED OUTPUT FROM BMDP4V

```
/INPUT          VARIABLES = 6.  FORMAT IS FREE.  FILE = 'TAPE42.DAT'.
/VARIABLES      NAMES ARE RESPOND, OCCUP, READ, DANCE, TV, SKI.
                LABEL IS RESPOND.
/BETWEEN        FACTOR IS OCCUP.  CODES ARE 1 TO 3.
                NAMES ARE 'BELLY DANCER', 'POLITICIAN',
                          'ADMINISTRATOR'.
/WEIGHTS        BETWEEN ARE EQUAL.
/END
DESIGN          FACTOR IS OCCUP.
                TYPE = BETWEEN, REGRESSION.
                CODE = READ.
                VALUES = 2, -1, -1.
                NAME IS 'BD VS OTHER'./

ANALYSIS        DEPENDENT IS DANCE.
                PROC = FACT./
END/
```

==

EFFECT VARIATE	STATISTIC		F	DF		P
BD VS OTHER						
DANCE						
	SS=	120.000000				
	MS=	120.000000	144.00	1,	12	0.0000
OVALL: GRAND MEAN						
DANCE						
	SS=	437.400000				
	MS=	437.400000	524.88	1,	12	0.0000
O: OCCUP						
DANCE						
	SS=	139.600000				
	MS=	69.800000	83.76	2,	12	0.0000
ERROR						
DANCE						
	SS=	10.00000000				
	MS=	0.83333333				

==

(b) SETUP AND SELECTED OUTPUT FROM SPSS MANOVA

```
DATA LIST       FILE='TAPE42.DAT'  FREE
                /RESPOND, OCCUP, READ, DANCE, TV, SKI.
VAR LABELS      OCCUP 'OCCUPATION'.
VALUE LABELS    OCCUP 1 'BELLY DANCER' 2 'POLITICIAN' 3 'ADMINISTRATOR'.
MANOVA          DANCE BY OCCUP(1,3)/
                PARTITION(OCCUP)/
                CONTRAST(OCCUP)=SPECIAL ( 1  1  1,
                                          2 -1 -1,
                                          0  1 -1)/
                DESIGN=OCCUP(1) VS WITHIN.
```

Tests of Significance for DANCE using UNIQUE sums of squares

Source of Variation	SS	DF	MS	F	Sig of F
WITHIN CELLS	10.00	12	.83		
OCCUP(1)	120.00	1	120.00	144.00	.000

(a) SETUP AND SELECTED OUTPUT FROM BMDP4V

```
/INPUT           VARIABLES = 6.  FORMAT IS FREE.  FILE = 'TAPE42.DAT'.
/VARIABLES       NAMES ARE RESPOND, OCCUP, READ, DANCE, TV, SKI.
                 LABEL IS RESPOND.
/BETWEEN         FACTOR IS OCCUP.  CODES ARE 1 TO 3.
                 NAMES ARE 'BELLY DANCER', 'POLITICIAN',
                     'ADMINISTRATOR'.
/WITHIN          FACTOR IS ACTIVITY.  CODES ARE 1 TO 4.
                 NAMES ARE READ, DANCE, TV, SKI.
/WEIGHTS         WITHIN ARE EQUAL.  BETWEEN ARE EQUAL.
/END
ANALYSIS         PROCEDURE IS SIMPLE.
                 HOLD IS OCCUP./
END/
```

WITHIN EFFECT: A: ACTIVITY

EFFECT VARIATE STATISTIC	F	DF	P

(A) X (OVALL.B: MEAN AT BELLY DA)
DEP_VAR

	STATISTIC	F	DF	P
TSQ=	51.5102	14.31	3, 10	0.0006
WCP SS=	35.800000			
WCP MS=	11.933333	10.71	3, 36	0.0000
GREENHOUSE-GEISSER ADJ. DF		10.71	2.40, 28.75	0.0002
HUYNH-FELDT ADJUSTED DF		10.71	3.00, 36.00	0.0000

ERROR
DEP_VAR

WCP SS=	40.10000000
WCP MS=	1.11388889
GGI EPSILON	0.79856
H-F EPSILON	1.00000

EFFECT VARIATE STATISTIC	F	DF	P

(A) X (OVALL.P: MEAN AT POLITICI)
DEP_VAR

	STATISTIC	F	DF	P
TSQ=	0.705068	0.20	3, 10	0.8968
WCP SS=	0.400000			
WCP MS=	0.133333	0.12	3, 36	0.9479
GREENHOUSE-GEISSER ADJ. DF		0.12	2.40, 28.75	0.9177
HUYNH-FELDT ADJUSTED DF		0.12	3.00, 36.00	0.9479

ERROR
DEP_VAR

WCP SS=	40.10000000
WCP MS=	1.11388889
GGI EPSILON	0.79856
H-F EPSILON	1.00000

WITHIN EFFECT: A: ACTIVITY

EFFECT VARIATE STATISTIC	F	DF	P

(A) X (OVALL.B: MEAN AT ADMINIST)
DEP_VAR

	STATISTIC	F	DF	P
TSQ=	43.9086	12.20	3, 10	0.0011
WCP SS=	34.950000			
WCP MS=	11.650000	10.46	3, 36	0.0000
GREENHOUSE-GEISSER ADJ. DF		10.46	2.40, 28.75	0.0002
HUYNH-FELDT ADJUSTED DF		10.46	3.00, 36.00	0.0000

TABLE 10.10 *(CONTINUED)*

```
ERROR
       DEP_VAR
              WCP SS=              40.10000000
              WCP MS=              1.11388889

              GGI EPSILON       0.79856
              H-F EPSILON       1.00000
```

(b) SETUP AND SELECTED OUTPUT FROM SPSS MANOVA

```
DATA LIST           FILE='TAPE42.DAT'   FREE
                    /RESPOND, OCCUP, READ, DANCE, TV, SKI..
VAR LABELS          OCCUP 'OCCUPATION'.
VALUE LABELS        OCCUP 1 'BELLY DANCER' 2 'POLITICIAN' 3 'ADMINISTRATOR'.
MANOVA              READ TO SKI BY OCCUP(1,3)/
                    WSFACTOR=ACTIVITY(4)/
                    PRINT=SIGNIF(BRIEF)/
                    DESIGN=MWITHIN OCCUP(1), MWITHIN OCCUP(2),
                        MWITHIN OCCUP(3).
```

Tests involving 'ACTIVITY' WithinSubject Effect.

AVERAGED Tests of Significance for MEAS.1 using UNIQUE sums of squares

Source of Variation	SS	DF	MS	F	Sig of F
WITHIN+RESIDUAL	40.10	36	1.11		
MWITHIN OCCUP(1) BY ACTIVITY	35.80	3	11.93	10.71	.000
MWITHIN OCCUP(2) BY ACTIVITY	.40	3	.13	.12	.948
MWITHIN OCCUP(3) BY ACTIVITY	34.95	3	11.65	10.46	.000

feature in BMDP or the SELECT IF (OCCUP EQ 1) procedure in SPSS is used to separate out the group on whose data a one-way within-subject ANOVA is then performed.

For these simple effects, the Scheffé adjustment to critical F is

$$F_s = (p - 1) F_{(p - 1), k(p - 1)(n - 1)} \tag{10.8}$$

where p is the number of repeated measures, n is the number of subjects in each group, and k is the number of groups. For the example

$$F_s = (4 - 1) F_{(3, 36)} = 8.76$$

Two of the F values for simple effects (10.71 and 10.46) exceed adjusted critical F; the researcher concludes that there are reliable differences in mean satisfaction during different activities among belly dancers and among administrators.

These findings are also ambiguous because there are more than two activities. Contrast coefficients are therefore applied to the levels of the repeated measure to examine the pattern of differences in greater detail [Figure 10.2(b)].

Table 10.11 shows setups and partial outputs from BMDP and SPSS for a simple contrast. The contrast that is illustrated compares the pooled mean for the two sedentary activities (READ and

TV) against the pooled mean for the two active activities (DANCE and SKI) for (you guessed it) belly dancers.

For BMDP [Table 10.11(a)] the contrast is defined and named in the DESIGN paragraph. The contrast is calculated for only the first group because of the USE = (OCCUP EQ 1) sentence in the TRANSFORM paragraph and because the BETWEEN paragraph is omitted. The error term that is developed by BMDP for this contrast is the interaction of subjects with the contrast on the repeated measure for the group. The F value of 15.37 exceeds F_s of 8.76 and indicates that belly dancers have a reliable mean difference in their satisfaction while engaging in active vs. sedentary activities.

For SPSS [Table 10.11(b)] the contrast is specified in the WSFACTOR portion of the MANOVA paragraph while group and appropriate error term are selected by DESIGN = MWITHIN OCCUP(1) VS WITHIN. SPSS differs from BMDP in computation of the error term. SPSS uses the interaction of subjects with the contrast on the repeated measure for all groups combined as the error term. In this case the F value of 16.98246 also exceeds F_s of 8.76 and indicates that belly dancers have reliable mean differences in their satisfaction during active vs. sedentary activities.

10.5.1.3 ■ Parallelism, Levels, and Flatness Significant (Interaction Contrasts)

When all three effects are significant, an interaction contrasts analysis is often most appropriate. This analysis partitions the sum of squares for interaction into a series of smaller interactions (Figure 10.3). Smaller interactions are obtained by deleting or combining groups or measures with use of appropriate contrast coefficients.

For the example, illustrated in Table 10.12, means for administrators and politicians are combined and compared with the mean of belly dancers, while means for TV and READ are combined and compared with the combined mean of DANCE and SKI. The researcher asks whether belly dancers and others have the same pattern of satisfaction during sedentary vs. active leisure activities.

In BMDP [Table 10.12(a)] two DESIGN paragraphs are used, one to specify the contrast on the within-subjects portion of the design (the repeated measures) and the other to specify the contrast on the between-subjects portion (the grouping effect). In the heavily edited output, the interaction contrast is labeled (SEDENT.) X (BD VS OTHER), sum of squares = 21.675. The marginal test of the repeated measure is labeled SEDENT., sum of squares = 3.75. Both of these are tested against ERROR, sum of squares = 17.10. The error term is the interaction of subjects with the contrast applied to the repeated measure nested with groups.

The marginal test of the groupings variable is labeled BD VS OTHER, sum of squares = 138.675. It is tested against subjects nested within groups. The two marginal tests are themselves of interest in examination of significant levels or flatness effects. The marginal value of the coefficients applied to the repeated measures is a contrast appropriate as followup of a significant flatness effect; the marginal value of the coefficients applied to groups is a contrast appropriate as a follow-up of a significant levels effect. Both marginal contrasts are automatically provided during a run for interaction contrasts.

SPSS gives less voluminous output, as seen in Table 10.12(b). The interaction contrast is identified as OCCUP(1) BY ACTIVITY(1) (sum of squares = 21.68) with its corresponding WITHIN CELLS error term (sum of squares = 17.10). The second WITHIN CELLS error term is the appropriate one for the interaction contrast, and it is the one used by SPSS with the request for ERROR = WITHIN. It is the interaction of subjects with contrast applied to the repeated measure nested in groups.

TABLE 10.11	SIMPLE COMPARISONS ANALYSIS ON ACTIVITY, HOLDING OCCUPATION CONSTANT: (a) BMDP AND (b) SPSS

(a) SETUP AND SELECTED OUTPUT FROM BMDP4V

```
/INPUT          VARIABLES = 6.  FORMAT IS FREE.  FILE = 'TAPE42.DAT'.
/VARIABLES      NAMES ARE RESPOND, OCCUP, READ, DANCE, TV, SKI.
                LABEL IS RESPOND.
                USE = READ TO SKI.
/TRANSFORM      USE = OCCUP EQ 1.
/WITHIN         FACTOR IS ACTIVITY.  CODES ARE 1 TO 4.
                NAMES ARE READ, DANCE, TV, SKI.
/WEIGHTS        WITHIN ARE EQUAL.
/END
DESIGN          FACTOR IS ACTIVITY.
                TYPE = WITHIN, REGRESSION.
                CODE = READ.
                VALUES = -1, 1, -1, 1.
                NAME IS 'SEDENT. VS ACTIVE'./
ANALYSIS        PROCEDURE IS FACT./
END/
```

```
========================================================================
WITHIN EFFECT: SEDENT. VS ACTIVE
```

EFFECT	VARIATE	STATISTIC		F	DF		P
SEDENT.							
	DEP_VAR						
		SS=	24.200000				
		MS=	24.200000	15.37	1,	4	0.0173
ERROR							
	DEP_VAR						
		SS=	6.30000000				
		MS=	1.57500000				

```
------------------------------------------------------------------------


========================================================================
```

(b) SETUP AND SELECTED OUTPUT FROM SPSS MANOVA

```
DATA LIST       FILE='TAPE42.DAT'  FREE
                /RESPOND, OCCUP, READ, DANCE, TV, SKI.
VAR LABELS      OCCUP 'OCCUPATION'.
VALUE LABELS    OCCUP 1 'BELLY DANCER' 2 'POLITICIAN' 3 'ADMINISTRATOR'.
MANOVA          READ TO SKI BY OCCUP(1,3)/
                WSFACTOR=ACTIVITY(4)/
                PARTITION(ACTIVITY)/
                CONTRAST(ACTIVITY)=SPECIAL ( 1  1  1  1,
                                            -1  1 -1  1,
                                            -1  0  1  0,
                                             0 -1  0  1)/
                WSDESIGN=ACTIVITY(1)/
                RENAME=OVERALL, SEDVSACT, READVSTV, DANVSSK/
                PRINT=SIGNIF(BRIEF)/
                DESIGN=MWITHIN OCCUP(1) VS WITHIN.
```

Tests involving 'ACTIVITY(1)' Within-Subject Effect.

Tests of Significance for SEDVSACT using UNIQUE sums of squares

Source of Variation	SS	DF	MS	F	Sig of F
WITHIN CELLS	17.10	12	1.43		
MWITHIN OCCUP(1) BY ACTIVITY(1)	24.20	1	24.20	16.98	.001

TABLE 10.12	INTERACTION CONTRASTS, BELLY DANCERS VS. OTHERS AND ACTIVE VS. SEDENTARY ACTIVITIES: (a) BMDP AND (b) SPSS

(a) SETUP AND SELECTED OUTPUT FROM BMDP4V

```
/INPUT          VARIABLES = 6.  FORMAT IS FREE.  FILE = 'TAPE42.DAT'.
/VARIABLES      NAMES ARE RESPOND, OCCUP, READ, DANCE, TV, SKI.
                LABEL IS RESPOND.
/BETWEEN        FACTOR IS OCCUP.  CODES ARE 1 TO 3.
                NAMES ARE 'BELLY DANCER', 'POLITICIAN',
                    'ADMINISTRATOR'.
/WITHIN         FACTOR IS ACTIVITY.  CODES ARE 1 TO 4.
                NAMES ARE READ, DANCE, TV, SKI.
/WEIGHTS        WITHIN ARE EQUAL.  BETWEEN ARE EQUAL.
/END
DESIGN          FACTOR IS ACTIVITY.
                TYPE = WITHIN, REGRESSION.
                CODE = READ.
                VALUES = -1, 1, -1, 1.
                NAME IS 'SEDENT. VS ACTIVE'./

DESIGN          FACTOR IS OCCUP.
                TYPE = BETWEEN, REGRESSION.
                VALUES = 2, -1, -1.
                NAME = 'BD VS OTHER'./

ANALYSIS        PROCEDURE IS FACT./
```

==
WITHIN EFFECT: SEDENT. VS ACTIVE

EFFECT	VARIATE	STATISTIC	F	DF		P
(SEDENT.) X (BD VS OTHER)						
	DEP_VAR					
		SS= 21.675000				
		MS= 21.675000	15.21	1,	12	0.0021
SEDENT.						
	DEP_VAR					
		SS= 3.750000				
		MS= 3.750000	2.63	1,	12	0.1307
(SEDENT.) X (O: OCCUP)						
	DEP_VAR					
		SS= 21.900000				
		MS= 10.950000	7.68	2,	12	0.0071
ERROR						
	DEP_VAR					
		SS= 17.10000000				
		MS= 1.42500000				

==
WITHIN EFFECT: OBS: WITHIN CASE MEAN

EFFECT	VARIATE	STATISTIC	F	DF		P
BD VS OTHER						
	DEP_VAR					
		SS= 138.675000				
		MS= 138.675000	70.81	1,	12	0.0000
OVALL: GRAND MEAN						
	DEP_VAR					
		SS= 1591.350000				
		MS= 1591.350000	812.60	1,	12	0.0000
O: OCCUP						
	DEP_VAR					
		SS= 172.900000				
		MS= 86.450000	44.14	2,	12	0.0000

TABLE 10.12 *(CONTINUED)*

```
ERROR
      DEP_VAR
              SS=          23.50000000
              MS=           1.95833333
```

(b) SETUP AND SELECTED OUTPUT FROM SPSS MANOVA

```
DATA LIST          FILE='TAPE42.DAT'  FREE
                   /RESPOND, OCCUP, READ, DANCE, TV, SKI.
VAR LABELS         OCCUP 'OCCUPATION'.
VALUE LABELS       OCCUP 1 'BELLY DANCER' 2 'POLITICIAN' 3 'ADMINISTRATOR'.
MANOVA             READ TO SKI BY OCCUP(1,3)/
                   WSFACTOR=ACTIVITY(4)/
                   PARTITION(ACTIVITY)/
                   CONTRAST(ACTIVITY)=SPECIAL ( 1  1  1  1,
                                               -1  1 -1  1,
                                               -1  0  1  0,
                                                0 -1  0  1)/
                   WSDESIGN=ACTIVITY(1)/
                   RENAME=OVERALL, SEDVSACT, READVSTV, DANVSSK/
                   PARTITION(OCCUP)/
                   CONTRAST(OCCUP)=SPECIAL ( 1  1  1,
                                            2 -1 -1,
                                            0  1 -1)/
                   PRINT=SIGNIF(BRIEF)/ERROR=WITHIN/
                   DESIGN=OCCUP(1) VS WITHIN.
```

Tests of Between-Subjects Effects.

Tests of Significance for OVERALL using UNIQUE sums of squares

Source of Variation	SS	DF	MS	F	Sig of F
WITHIN CELLS	23.50	12	1.96		
OCCUP(1)	138.68	1	138.68	70.81	.000

Tests involving 'ACTIVITY(1)' Within-Subject Effect.

Tests of Significance for SEDVSACT using UNIQUE sums of squares

Source of Variation	SS	DF	MS	F	Sig of F
WITHIN CELLS	17.10	12	1.43		
ACTIVITY(1)	3.75	1	3.75	2.63	.131
OCCUP(1) BY ACTIVITY (1)	21.68	1	21.68	15.21	.002

The first WITHIN CELLS error term (sum of squares = 23.50, also produced by ERROR = WITHIN) is subjects nested in groups, the error term that accompanies OCCUP(1) (sum of squares = 138.68), the marginal test of contrast applied to groups and the test appropriate as follow-up of a significant levels effect. The effect labeled ACTIVITY(1) (sum of squares = 3.75) is the marginal test of contrast applied to repeated measures and is the test appropriate as follow-up of a significant flatness effect. This test is evaluated against the second error term.

Interaction contrasts also need Scheffé adjustment to critical F to hold down the rate of family-wise error. For an interaction, the Scheffé adjustment is

$$F_s = (p - 1)(k - 1) F_{(p-1)(k-1), k(p-1)(n-1)} \qquad (10.9)$$

where p is the number of repeated measures, k is the number of groups, and n is the number of subjects in each group. For the example

$$F_s = (4 - 1)(3 - 1)\, F_{(6,\ 36)} = 14.52$$

Because the F value for the interaction contrast is 15.21, a value that exceeds F_s, there is an interaction between belly dancers vs. others in their satisfaction during sedentary vs. active leisure time activities. A look at the means in Table 10.1 reveals that belly dancers favor active leisure time activities to a greater extent than others.

10.5.1.4 ■ Only Parallelism Significant

If the only significance is in the interaction of groups with repeated measures, any of the analyses in Section 10.5.1 is appropriate. The decision between simple effects analysis and interaction contrasts is based on which is more informative and easier to explain. Writers and readers seem likely to have an easier time explaining results of procedures in Section 10.5.1.2.[7]

10.5.2 ■ Multivariate Approach to Repeated Measures

Research where the same cases are repeatedly measured with the same instrument is common in many sciences. Longitudinal or developmental studies, research that requires follow-up, studies where changes in time are of interest—all involve repeated measurement. Further, many studies of short-term phenomena have repeated measurement of the same subjects under several experimental conditions, resulting in an economical research design.

When there are repeated measures, a variety of analytical strategies are available, all with advantages and disadvantages. Choice among the strategies depends upon details of research design and conformity between the data and the assumptions of analysis.

Univariate repeated-measures ANOVA with more than 1 df for the repeated-measure IV requires homogeneity of covariance. Although the test for homogeneity of covariance is fairly complicated, the notion is conceptually simple. All pairs of levels of the within-subjects variable need to have equivalent correlations. For example, consider a longitudinal study in which children are measured yearly from ages 5 to 10. If there is homogeneity of covariance the correlation between scores on the DV for ages 5 and 6 should be about the same as the correlation between scores between ages 5 and 7, or 5 and 8, or 6 and 10, etc. In applications like these, however, the assumption is almost surely violated. Things measured closer in time tend to be more highly correlated than things measured farther away in time; the correlation between scores measured at ages 5 and 6 is likely to be much higher than the correlation between scores measured at ages 5 and 10. Thus, whenever time is a within-subjects IV, the assumption of homogeneity of covariance is likely to be violated, leading to increased Type I error. All four packages now routinely provide information about homogeneity of covariance directly in their output: BMDP2V and SPSS MANOVA each show a sphericity test for the significance of departure from homogeneity of covariance.

If there is violation of homogeneity of covariance, several alternatives are available, as also discussed in Section 8.5.2.1. The first is to use one of the significance tests that is adjusted for viola-

---

[7] If you managed to read this far, go have a beer.

tion of the assumption, such as Greenhouse-Geisser or Huynh-Feldt. In BMDP2V both Greenhouse-Geisser (G-G) and Huynh-Feldt (H-F) adjustments are provided.[8] BMDP4V, SYSTAT GLM and SAS GLM also give adjusted significance levels. These four programs plus SPSS MANOVA and BMDP2V show epsilon values for the two tests (G-G and H-F), which when multiplied by the effect and error df, yield the corrected df for the univariate, repeated measures tests. In the absence of a formal test, take a look at the results with and without adjustment. If they are the same, report the unadjusted results and note the lack of difference. If the results differ, report the adjusted results.

A second strategy, available through all four of the programs, is a more stringent adjustment of the statistical criterion leading to a more honest Type I error rate, but lower power. This strategy has the advantage of simplicity of interpretation (because familiar main effects and interactions are evaluated) and simplicity of decision-making (you decide on one of the strategies before performing the analysis and then take your chances with respect to power).

For all of the multivariate programs, however, results of profile analysis are also printed out, and you have availed yourself of the third strategy, whether you meant to or not. Profile analysis, called the multivariate approach to repeated measures, is a statistically acceptable alternative to repeated-measures ANOVA because (transformed) multiple DVs replace the within-subjects IV and the assumption of homogeneity of covariance is no longer required. Other requirements such as homogeneity of variance-covariance matrices and absence of multicollinearity and singularity must be met, but they are less likely to be violated.

Profile analysis requires more cases than univariate repeated-measures ANOVA—certainly more cases than DVs in the smallest group. If the sample is too small, the choice between multivariate and univariate approaches is automatically resolved in favor of the univariate approach, with adjustment for failure of homogeneity of covariance, as necessary.

Sometimes, however, the choice is not so simple, and you find yourself with two sets of results. If the conclusions from both sets of results are the same, it often is easier to report the univariate solution, while noting that the multivariate solution is similar. But if conclusions differ between the two sets of results, you have a dilemma. Choice between conflicting results requires attention to the details of the research design. Clean, counterbalanced experimental designs "fit" better within the univariate model, while nonexperimental or contaminated designs often require the multivariate model that is more forgiving statistically but more ambiguous to interpret.

The best solution, the fourth alternative, is often to perform trend analysis (or some other set of single df contrasts) instead of either profile analysis or repeated-measures ANOVA if it makes conceptual sense within the context of the research design. Many longitudinal, follow-up, and other time-related studies lend themselves beautifully to interpretation in terms of trends. Because statistical tests of trends and other contrasts each use a single degree of freedom of the within-subjects IV, there is no possibility of violation of homogeneity of covariance. Furthermore, none of the assumptions of the multivariate approach need be met. BMDP2V offers simple, straightforward setup and output for polynomial decomposition of the within-subjects IV, in which results for each trend are printed out (1 = linear, 2 = quadratic, 3 = cubic, etc.). SYSTAT GLM automatically prints out a full trend analysis for a repeated-measures analysis.

A fifth alternative is straightforward MANOVA where DVs are treated directly (cf. Chapter 9), without transformation. The design becomes a one-way between-subjects analysis of the grouping

---

[8] See Keppel (1991, pp 352–353) for a discussion of the differences between the two types of adjustment (referring to the Huynh-Feldt procedure as the Box correction). Even greater insights are available through consultation with the original sources: Greenhouse & Geisser (1959) as well as Huynh and Feldt (1976).

variable with the repeated measures used simply as multiple DVs. There are two problems with conversion of repeated measures to MANOVA. First, because the design is now one-way between-subjects, MANOVA does not produce the interaction (parallelism) test most often of interest in a repeated-measures design. Second, MANOVA allows a Roy-Bargmann stepdown analysis, but not a trend analysis of DVs after finding a multivariate effect.

In summary then, if the levels of the IV differ along a single dimension such as time or dosage and trend analysis makes sense, use it. Or, if the design is a clean experiment where cases have been randomly assigned to treatment and there are expected to be no carry-over effects, the univariate repeated-measures approach is probably justified. (But just to be on the safe side, use a program that tests and adjusts for violation of homogeneity of covariance.) If, however, the levels of the IV do not vary along a single dimension but violation of homogeneity of covariance is likely, and if there are lots more cases than DVs, it is probably a good idea to choose either profile analysis or MANOVA.

10.5.3 ■ Doubly Multivariate Designs

10.5.3.1 Kinds of Doubly Multivariate Analysis

There are two ways that an analysis can be doubly multivariate. In the first, commensurate DVs are administered to groups more than once, creating repeated measures over time. For example, the POMS is administered to psychoanalysts and behavior therapists before training, immediately after training, and 1 year later. This is analogous to a univariate between-within-within design with time as one within-subjects factor and the set of DVs as the other. The analysis produces tests of three main effects (levels, flatness over time, and flatness over DVs), three two-way interactions (parallel groups over time, parallel groups over DVs, and doubly parallel time by DVs), as well as the mind-boggling three-way interaction.

In the second and more common doubly multivariate design, noncommensurate DVs are repeatedly measured. For example, children in classrooms with either traditional or computer assisted instruction are measured at several points over the semester on reading achievement, general information, and math achievement. There are two ways to conceptualize the analysis. If treated in a singly multivariate fashion, this is a between-within design (groups by time) with multiple DVs. The time effect, however, has the assumption of homogeneity of covariance. To circumvent the assumption, the analysis becomes doubly multivariate where both the within-subjects part of the design and the multiple DVs are analyzed multivariately. The between-subject effect is singly multivariate; the within-subject effects and interactions are doubly multivariate.

For both commensurate and noncommensurate applications, the limiting factor is the large number of cases needed for analysis. The number of measures for each case is the number of DVs times the number of repetitions. With 10 DVs measured three times, there are 30 measures per case. To avoid singularity of variance-covariance matrices (cf. Section 10.3.2) more than 30 cases per group are needed.

BMDP4V is particularly well set up to handle analysis of noncommensurate DVs. The multiple DVs are designated VARIATES, and an appropriate name is given to the within-subjects IV treated multivariately. This is illustrated in the example in Section 10.5.3.2.

For SPSS MANOVA, the procedure is not much more difficult. However, the fully documented example in the SPSS[x] manual (SPSS Inc, 1986) is no longer available in more recent manuals. The SAS manual now also has an example of a doubly multivariate design in the GLM chapter.

For repeatedly measured commensurate DVs, none of the programs provi̇tion for analysis. The procedures offered by the multivariate programs resu̇the tests of flatness and parallelism apply to the repeated-measures factorthe higher-order interaction or of parallelism or flatness of the commensurate design matrices have to be devised, a topic beyond the scope of this book.

10.5.3.2 ■ Example of a Doubly Multivariate Analysis of Variance

In this example, a small hypothetical data set with repeatedly measured noncommensurate DVs is used. The between-subjects IV is three weight-loss programs (PROGRAM): a control group (CONTROL), a group that diets (DIET), and a group that both diets and exercises (DIET-EX). Each group has 12 participants, so total $N = 36$. The major DV is weight loss (WTLOSS) and a secondary DV is self-esteem (ESTEEM). The DVs are measured at the end of the first, second, and third months of treatment. The within-subject IV treated multivariately, then, is MONTH that the measures are taken. That is, the commensurate DVs are MONTH1, MONTH2, and MONTH3. The data set is in Table 10.13. Both BMDP4V and SPSS MANOVA analyses are demonstrated there.

The setup and results from BMDP4V as applied to this data set appear in Table 10.14. Note that in the WITHIN paragraph, VARIATES[9] appears first, indicating that in the data set the three values for the first DV appear before the three values of the second DV. Months "change" fastest in this arrangement of data.

The first effect of interest is that of PROGRAM, the between-subject IV. Note that there is no significant difference between groups when the DVs are combined (-ALL–), with Wilks' Lambda (LRATIO) = .773006, $F(4, 64) = 2.20$, $p = .0791$. Therefore, the results of the main effect of program for the individual DVs are ignored.

The second effect that appears is M (MONTH). The section labeled -ALL– presents the doubly multivariate test of MONTH; the test is multivariate in the DVs and multivariate for MONTH. Wilks' Lambda is also used to test this effect and $F(4, 130) = 59.23$. There is, then, a significant difference over the 3 months in the combination of weight loss and self-esteem measures. The singly multivariate test of this same effect, multivariate in the DVs but univariate for MONTH, is WCP LRATIO $F(4, 130) = 40.40$.

Below this section appear the tests of differences over months for the individual DVs (the MONTH × flatness of DVs interaction). For each DV there are both the multivariate (TSQ) and univariate (WCP) results, the latter with the two forms of adjustment for violation of homogeneity of covariance. According to the multivariate results both the DVs show highly significant effects, with $F(2, 32) = 120.62$ for weight loss and $F(2, 32) = 12.19$ for self-esteem.

BMDP4V offers no automatic stepdown analysis. To test changes in self-esteem over the 3 months after adjustment for differences in weight loss, a second run with self-esteem as the DV and weight loss as a covariate is needed.

The final doubly multivariate test is of the PROGRAM by MONTH interaction, highly significant with Wilks' criterion (LRATIO) $F(8, 60) = 4.43$. The singly multivariate tests follow, labeled WCP LRATIO. According to multivariate tests of the individual DVs, weight loss shows significant change as a joint function of PROGRAM and MONTH, with multivariate $F(4, 64) = 8.31$ for LRATIO, but there is no significant multivariate effect for self-esteem, $F(4, 64) = 1.71$. Univariate

---

[9] VARIATES is a BMDP4V key word which must always be used to specify the multiple (noncommensurate) DVs.

TABLE 10.13 HYPOTHETICAL DATA SET TO DEMONSTRATE DOUBLY MULTIVARIATE ANALYSIS

DV	Weight loss			Self-esteem		
Month	1	2	3	1	2	3
Group Control	4	3	3	14	13	15
	4	4	3	13	14	17
	4	3	1	17	12	16
	3	2	1	11	11	12
	5	3	2	16	15	14
	6	5	4	17	18	18
	6	5	4	17	16	19
	5	4	1	13	15	15
	5	4	1	14	14	15
	3	3	2	14	15	13
	4	2	2	16	16	11
	5	2	1	15	13	16
Diet	6	3	2	12	11	14
	5	4	1	13	14	15
	7	6	3	17	11	18
	6	4	2	16	15	18
	3	2	1	16	17	15
	5	5	4	13	11	15
	4	3	1	12	11	14
	4	2	1	12	11	11
	6	5	3	17	16	19
	7	6	4	19	19	19
	4	3	2	15	15	15
	7	4	3	16	14	18
Diet + exercise	8	4	2	16	12	16
	3	6	3	19	19	16
	7	7	4	15	11	19
	4	7	1	16	12	18
	9	7	3	13	12	17
	2	4	1	16	13	17
	3	5	1	13	13	16
	6	5	2	15	12	18
	6	6	3	15	13	18
	9	5	2	16	14	17
	7	9	4	16	16	19
	8	6	1	17	17	17

results for individual DVs are consistent for weight loss but questionable for self-esteem. This finding needs to be verified in a run in which self-esteem scores are adjusted for differences in weight loss acting as a covariate.

Table 10.15 shows the setup and results of an ANCOVA run through BMDP2V with self-esteem as the DV and weight loss as a covariate. BMDP2V is more convenient to use than BMDP4V because adjusted cell means for self-esteem are available. The results for the two programs are the same because, in the case of a covariate changing with each level of the single DV, BMDP4V prints only univariate ANCOVA.

```
/INPUT          VARIABLES ARE 7.  FORMAT IS FREE.  FILE=DBLDAT.
/VARIABLES      NAMES ARE PROGRAM, WTLOSS1, WTLOSS2, WTLOSS3,
                             ESTEEM1, ESTEEM2, ESTEEM3.
/BETWEEN        FACTOR IS PROGRAM.
                CODES(PROGRAM) ARE 1, 2, 3.
                NAMES(PROGRAM) ARE CONTROL, DIET, 'DIET+EX'.
/WITHIN         FACTORS = VARIATES, MONTH.
                CODES(VARIATES) = 1, 2.
                NAMES(VARIATES) = WTLOSS, ESTEEM.
                CODES(MONTH) = 1, 2, 3.
                NAMES(MONTH) = MONTH1, MONTH2, MONTH3.
/WEIGHTS        BETWEEN ARE EQUAL.  WITHIN ARE EQUAL.
/END
ANALYSIS  PROCEDURE IS FACTORIAL./
END/
```

===
WITHIN EFFECT: OBS: WITHIN CASE MEAN

EFFECT	VARIATE	STATISTIC		F	DF		P
OVALL: GRAND MEAN							
	-ALL----						
		TSQ=	2637.25	1278.67	2,	32	0.0000
	WTLOSS						
		SS=	1688.231481				
		MS=	1688.231481	429.01	1,	33	0.0000
	ESTEEM						
		SS=	24390.083333				
		MS=	24390.083333	2608.75	1,	33	0.0000
P: PROGRAM							
	-ALL----						
		LRATIO=	0.773006	2.20	4,	64.00	0.0791
		TRACE=	0.292626				
		TZSQ=	9.65665				
		CHISQ =	9.68			17.481	0.0704
		MXROOT=	0.224252				0.0582
	WTLOSS						
		SS=	36.907407				
		MS=	18.453704	4.69	2,	33	0.0161
	ESTEEM						
		SS=	13.722222				
		MS=	6.861111	0.73	2,	33	0.4877
ERROR							
	WTLOSS						
		SS=	129.86111111				
		MS=	3.93518519				
	ESTEEM						
		SS=	308.52777778				
		MS=	9.34932660				

===

WITHIN EFFECT: M: MONTH

EFFECT	VARIATE	STATISTIC		F	DF		P
M							
	-ALL----						
		TSQ=	260.614	59.23	4,	30	0.0000
		WCP LRATIO=	0.198756	40.40	4,	130.00	0.0000
		WCP TRACE=	3.20173				
		WCP TZSQ=	211.314				
		CHISQ =	172.06			3.159	0.0000
		WCP MXROOT=	0.744727				0.0000

TABLE 10.14   *(CONTINUED)*

```
        WTLOSS
                TSQ=        248.783        120.62    2,    32  0.0000
            WCP SS=        181.351852
            WCP MS=         90.675926      88.37     2,    66  0.0000
            GREENHOUSE-GEISSER ADJ. DF     88.37  1.56,  51.34  0.0000
            HUYNH-FELDT ADJUSTED DF        88.37  1.72,  56.68  0.0000
        ESTEEM
                S=    1,T=    2,DFH=    1,DFE=    33
            HT EVALS= 0.43235806
            HE EVALS= 0.76167392
                TSQ=         25.1352        12.19    2,    32   0.0001
            WCP SS=         86.722222
            WCP MS=         43.361111       18.78    2,    66  0.0000
            GREENHOUSE-GEISSER ADJ. DF      18.78  1.58,  52.07  0.0000
            HUYNH-FELDT ADJUSTED DF         18.78  1.74,  57.56  0.0000
(M) X (P: PROGRAM)
        -ALL----
                LRATIO=  0.395418           4.43    8,  60.00  0.0003
                TRACE=   1.47509
                TZSQ=   45.7278
                    CHISQ =    6.48                23.712   0.0002
                MXROOT=  0.589762                            0.0001
            WCP LRATIO=  0.679124           3.47    8, 130.00  0.0012
            WCP TRACE=   0.464989
            WCP TZSQ=   30.6893
                    CHISQ =   22.97                 5.800   0.0007
            WCP MXROOT=  0.309520                            0.0737
        WTLOSS
                LRATIO=  0.433300           8.31    4,  64.00  0.0000
                TRACE=   1.28611
                TZSQ=   42.4415
                    CHISQ =    2.90                17.481   0.0000
                MXROOT=  0.559268                            0.0000
            WCP SS=         20.925926
            WCP MS=          5.231481        5.10    4,    66  0.0012
            GREENHOUSE-GEISSER ADJ. DF       5.10  3.11,  51.34  0.0033
            HUYNH-FELDT ADJUSTED DF          5.10  4.00,  66.00  0.0012
        ESTEEM
                LRATIO=  0.816616           1.71    4,  64.00  0.1597
                TRACE=   0.224443
                TZSQ=    7.40661
                    CHISQ =   11.53                17.481   0.1506
                MXROOT=  0.182936                            0.1206
            WCP SS=         25.555556
            WCP MS=          6.388889        2.77    4,    66  0.0344
            GREENHOUSE-GEISSER ADJ. DF       2.77  3.16,  52.07  0.0483
            HUYNH-FELDT ADJUSTED DF          2.77  4.00,  66.00  0.0344


ERROR
        WTLOSS
            WCP SS=         67.72222222
            WCP MS=          1.02609428

            GGI EPSILON     0.77784
            H-F EPSILON     1.00000
        ESTEEM
            WCP SS=        152.38888889
            WCP MS=          2.30892256

            GGI EPSILON     0.78896
            H-F EPSILON     1.00000
```

As seen in Table 10.15, the ANCOVA results are consistent with those of MANOVA. For adjusted self-esteem scores, there is a significant difference due to MONTH, $F(2, 65) = 11.96$, but no significant PROGRAM by MONTH interaction, $F(4, 65) = 2.03$.

To interpret the significant effects, cell and marginal means for the DVs are needed. These are most easily found in the BMDP2V run, although means for weight loss are available through the BMDP4V run. Table 10.16 shows the cell and marginal means for weight loss (called the 1-ST COVARIATE) and adjusted cell means for self-esteem.

For weight loss with all three groups combined, there is a decrease over the 3 months, with an average of 5.28 pounds lost the first month, 4.39 pounds the second month, and 2.19 pounds the third month. A plot of the PROGRAM by MONTH interaction based on the cell means for weight loss is helpful to show how the weight loss over the 3 months differs for the three program groups.

The adjusted cell means for self-esteem need to be averaged in order to see the progress over the 3 months for the combined groups. Because all cells have equal sample sizes, a simple unweighted averaging is appropriate. For example, for the third month,

$$\text{Mean}_{\text{Month3}} = \frac{(15.58007 + 16.36914 + 17.78581)}{3}$$

$$= 16.58$$

For the second month, the adjusted average self-esteem score is 13.80 and for the first month it is 14.70.

Strength of association measures for multivariate effects are, as usual, based on Wilks' Lambda (Equation 10.4) and for univariate effects, η^2 for adjusted scores (Equation 8.7).

Table 10.17 shows the same doubly multivariate analysis done through SPSS MANOVA. Univariate and Roy-Bargmann stepdown analyses are produced by request (PRINT=SIGNIF(UNIV, STEPDOWN)). The order of listing DVs in the MANOVA paragraph is critical: all repeated measures of one DV must be listed before the next DV, and so on. In this example, months *must* be the fastest changing index. Effects are a little harder to find in the output, and there is no univariate test of individual DVs where months is tested multivariately, but with a little patience, and perhaps a glass of wine, interpretation is possible.

The output begins with the levels test, labeled EFFECT .. PROGRAM. Four multivariate tests are provided, with Wilks' = .77301 corresponding to the value from BMDP4V. This is followed by univariate and Roy-Bargmann stepdown analyses for weight loss (labeled T1) and self esteem (labeled T4). The tests for EFFECT .. CONSTANT are printed out, followed by tests with adjustments for homogeneity of covariance, but both sections are omitted from output in Table 10.17.

The next portion of output of interest is the doubly multivariate parallelism test, labeled EFFECT .. PROGRAM BY MONTH, where multiple DVs are tested in repeated measures interacting with the grouping variable. The univariate and stepdown tests that follow, however, are of no interest. Next is the doubly multivariate flatness test, labeled EFFECT .. MONTH, again followed by uninteresting univariate and stepdown tests.

The remaining portions treat (MONTH) univariately. First, there is the multivariate test (both DVs) for the PROGRAM by MONTH interaction, followed by the univariate and stepdown tests for each of the DVs (appropriately labeled). Finally, there is the multivariate test for MONTH, again followed by univariate and stepdown tests of weight loss and self esteem. The CONSPLUS

TABLE 10.15   SETUP AND RESULTS OF BMDP2V ANALYSIS OF SELF-ESTEEM ADJUSTED FOR WEIGHT LOSS

```
/INPUT       VARIABLES ARE 7.  FORMAT IS FREE.  FILE=DBLDAT.
/VARIABLES   NAMES ARE PROGRAM, WTLOSS1, WTLOSS2, WTLOSS3,
                       ESTEEM1, ESTEEM2, ESTEEM3.
/GROUP       VAR IS PROGRAM.
             CODES(PROGRAM) ARE 1, 2, 3.
             NAMES(PROGRAM) ARE CONTROL, DIET, 'DIET+EX'.
/DESIGN      DEPENDENT ARE ESTEEM1 TO ESTEEM3.
             COVARIATES ARE WTLOSS1 TO WTLOSS3.
             LEVEL=3.
             NAME IS MONTH.
/END
```

A N A L Y S I S   O F   V A R I A N C E   FOR   THE   1-ST DEPENDENT VARIABLE

THE TRIALS ARE REPRESENTED BY THE VARIABLES:
ESTEEM1 ESTEEM2 ESTEEM3

SOURCE	SUM OF SQUARES	D.F.	MEAN SQUARE	F	TAIL PROB.	GREENHOUSE GEISSER PROB.	HUYNH FELDT PROB.	REGRESSION COEFFICIENTS
PROGRAM	1.71025	2	0.85512	0.12	0.8894			
1-ST COVAR	75.98203	1	75.98203	10.46	0.0028			0.7649
1 ERROR	232.54575	32	7.26705					
MONTH	56.03087	2	28.01544	11.96	0.0000	0.0002	0.0001	
MP	19.01707	4	4.75427	2.03	0.1006	0.1180	0.1093	
1-ST COVAR	0.12562	1	0.12562	0.05	0.8176			-0.0431
2 ERROR	152.26327	65	2.34251					

EPSILON FACTORS FOR DEGREES OF FREEDOM ADJUSTMENT

ERROR TERM	GREENHOUSE-GEISSER	HUYNH-FELDT
2	0.7933	0.8915

TABLE 10.16   CELL AND MARGINAL MEANS FROM BMDP2V (SETUP APPEARS IN TABLE 10.15)

CELL MEANS   FOR   1-ST COVARIATE

	MONTH	PROGRAM = CONTROL	DIET	DIET+EX	MARGINAL
WTLOSS1	1	4.50000	5.33333	6.00000	5.27778
WTLOSS2	2	3.33333	3.91667	5.91667	4.38889
WTLOSS3	3	2.08333	2.25000	2.25000	2.19444
MARGINAL		3.30556	3.83333	4.72222	3.95370
COUNT		12	12	12	36

ADJUSTED CELL MEANS   FOR   1-ST DEPENDENT VARIABLE

	MONTH	PROGRAM = CONTROL	DIET	DIET+EX	MARGINAL
ESTEEM1	1	14.60491	14.46693	15.03987	14.70391
ESTEEM2	2	14.49809	13.75984	13.14534	13.80109
ESTEEM3	3	15.58007	16.36914	17.78581	16.57834
MARGINAL		14.89436	14.86530	15.32367	15.02778
COUNT		12	12	12	36

procedure described in Section 9.7.2 and Table 9.20 is used to find means for self-esteem adjusted for weight loss.

10.5.4 ■ Classifying Profiles

A procedure typically available in programs designed for discriminant function analysis is classification of cases into groups on the basis of a best-fit statistical function. Although classification is done on the basis of scores rather than segments, the principle of classification is often of interest in research where profile analysis is appropriate. If it is found that groups differ on their profiles, it could be useful to classify new cases into groups according to their profiles.

For example, given profile of scores for different groups on a standardized test such as the Illinois Test of Psycholinguistic Abilities, one might use the profile of a new child to see if that child more closely resembles a group of children who have difficulty reading or a group who does not show such difficulty. If reliable profile differences were available before the age at which children are taught to read, classification according to profiles could provide a powerful diagnostic tool.

Note that this is no different from using classification procedures in discriminant function analysis. It is simply mentioned here because choice of profile analysis as the initial vehicle for testing group differences does not preclude use of classification. To use a discriminant function program such as SPSS DISCRIMINANT or BMDP7M for classification, one simply defines the levels of the IV as "groups" and the DVs as "predictors."

TABLE 10.17 SETUP AND RESULTS OF DOUBLY MULTIVARIATE ANALYSIS THROUGH SPSS MANOVA

```
DATA LIST      FILE='DBLDAT.DAT'  FREE
               /PROGRAM, WTLOSS1, WTLOSS2, WTLOSS3,
               ESTEEM1, ESTEEM2, ESTEEM3.

VALUE LABELS   PROGRAM 1 'CONTROL' 2 'DIET' 3 'DIET+EX'.  Gender 1 'Male' 2 'Female'
MANOVA         WTLOSS1 TO ESTEEM3 BY PROGRAM(1,3)/          Gender(1,2)
               WSFACTOR=MONTH(3)/
               MEASURES=WTLOSS, ESTEEM/
               WSDESIGN=MONTH/
               PRINT=SIGNIF(UNIV, STEPDOWN)/
               DESIGN=PROGRAM.
```

* * * * * * * * * * * * * * A N A L Y S I S   O F   V A R I A N C E -- DESIGN   1 * * * * * * * * * * * * * * *

Tests involving Between-Subjects Effects.

EFFECT .. PROGRAM
Multivariate Tests of Significance (S = 2, M = -1/2, N = 15)

Test Name	Value	Approx. F	Hypoth. DF	Error DF	Sig. of F
Pillais	.22779	2.12079	4.00	66.00	.088
Hotellings	.29263	2.26785	4.00	62.00	.072
Wilks	.77301	2.19821	4.00	64.00	.079
Roys	.22425				

Note.. F statistic for WILK'S Lambda is exact.

- -

EFFECT .. PROGRAM (Cont.)
Univariate F-tests with (2,33) D. F.

Variable	Hypoth. SS	Error SS	Hypoth. MS	Error MS	F	Sig. of F
T1	36.90741	129.86111	18.45370	3.93519	4.68941	.016
T4	13.72222	308.52778	6.86111	9.34933	.73386	.488

- -

Roy-Bargman Stepdown F - tests

Variable	Hypoth. MS	Error MS	StepDown F	Hypoth. DF	Error DF	Sig. of F
T1	18.45370	3.93519	4.68941	2	33	.016
T4	.85512	7.26705	.11767	2	32	.889
```
```

TABLE 10.17 (CONTINUED)

```
* * * * * * * * * * * * * * * * A N A L Y S I S   O F   V A R I A N C E -- DESIGN   1 * * * * * * * * * * * * * * * *

EFFECT .. PROGRAM BY MONTH
Multivariate Tests of Significance (S = 2, M = 1/2, N = 14 )
```

Test Name	Value	Approx. F	Hypoth. DF	Error DF	Sig. of F
Pillais	.62589	3.53001	8.00	62.00	.002
Hotellings	1.47509	5.34720	8.00	58.00	.000
Wilks	.39542	4.42706	8.00	60.00	.000
Roys	.58976				

Note.. F statistic for WILK'S Lambda is exact.

```
- - - - - - - - - - - - - - - - - - - - - - - - - - - - - - - - - - - - - - - - - - - - - - - - - - - -

EFFECT .. PROGRAM BY MONTH   (Cont.)
Univariate F-tests with (2,33) D. F.
```

Variable	Hypoth. SS	Error SS	Hypoth. MS	Error MS	F	Sig. of F
T2	5.33333	42.04167	2.66667	1.27399	2.09316	.139
T3	15.59259	25.68056	7.79630	.77820	10.01839	.000
T5	6.02778	60.91667	3.01389	1.84596	1.63269	.211
T6	19.52778	91.47222	9.76389	2.77189	3.52247	.041

```
- - - - - - - - - - - - - - - - - - - - - - - - - - - - - - - - - - - - - - - - - - - - - - - - - - - -

Roy-Bargman Stepdown F - tests
```

Variable	Hypoth. MS	Error MS	StepDown F	Hypoth. DF	Error DF	Sig. of F
T2	2.66667	1.27399	2.09316	2	33	.139
T3	10.20965	.60884	16.76892	2	32	.000
T5	.40057	1.82513	.21947	2	31	.804
T6	2.73201	2.26243	1.20755	2	30	.313

```
* * * * * * * * * * * * * * * * A N A L Y S I S   O F   V A R I A N C E -- DESIGN   1 * * * * * * * * * * * * * * * *

EFFECT .. MONTH
Multivariate Tests of Significance (S = 1, M = 1, N = 14 )
```

Test Name	Value	Exact F	Hypoth. DF	Error DF	Sig. of F
Pillais	.88761	59.23050	4.00	30.00	.000
Hotellings	7.89740	59.23050	4.00	30.00	.000
Wilks	.11239	59.23050	4.00	30.00	.000
Roys	.88761				

Note.. F statistics are exact.

```
- - - - - - - - - - - - - - - - - - - - - - - - - - - - - - - - - - - - - - - - - - - - - - - - - - - -
```

TABLE 10.17 *(CONTINUED)*

EFFECT .. MONTH (Cont.)
Univariate F-tests with (1,33) D. F.

Variable	Hypoth. SS	Error SS	Hypoth. MS	Error MS	F	Sig. of F
T2	171.12500	42.04167	171.12500	1.27399	134.32210	.000
T3	10.22685	25.68056	10.22685	.77820	13.14170	.001
T5	20.05556	60.91667	20.05556	1.84596	10.86457	.002
T6	66.66667	91.47222	66.66667	2.77189	24.05102	.000

Roy-Bargman Stepdown F - tests

Variable	Hypoth. MS	Error MS	StepDown F	Hypoth. DF	Error DF	Sig. of F
T2	171.12500	1.27399	134.32210	1	33	.000
T3	13.32786	.60884	21.89046	1	32	.000
T5	.07883	1.82513	.04319	1	31	.837
T6	2.75131	2.26243	1.21608	1	30	.279

* * * * * * * * * * * * * * * A N A L Y S I S   O F   V A R I A N C E -- DESIGN   1 * * * * * * * * * * * * * * *

Tests involving 'MONTH' WithinSubject Effect.

EFFECT .. PROGRAM BY MONTH
AVERAGED Multivariate Tests of Significance (S = 2, M = 1/2, N = 31 1/2)

Test Name	Value	Approx. F	Hypoth. DF	Error DF	Sig. of F
Pillais	.32597	3.21286	8.00	132.00	.002
Hotellings	.46499	3.71992	8.00	128.00	.001
Wilks	.67912	3.46872	8.00	130.00	.001
Roys	.30952				

Note.. F statistic for WILK'S Lambda is exact.

EFFECT .. PROGRAM BY MONTH (Cont.)
Univariate F-tests with (4,66) D. F.

Variable	Hypoth. SS	Error SS	Hypoth. MS	Error MS	F	Sig. of F
WTLOSS	20.92593	67.72222	5.23148	1.02609	5.09844	.001
ESTEEM	25.55556	152.38889	6.38889	2.30892	2.76704	.034

TABLE 10.17 *(CONTINUED)*

Roy-Bargman Stepdown F - tests

Variable	Hypoth. MS	Error MS	StepDown F	Hypoth. DF	Error DF	Sig. of F
WTLOSS	5.23148	1.02609	5.09844	4	66	.001
ESTEEM	4.75427	2.34251	2.02956	4	65	.101

* * * * * * * * * * * * * * * A N A L Y S I S   O F   V A R I A N C E -- DESIGN   1 * * * * * * * * * * * * * * *

EFFECT .. MONTH
AVERAGED Multivariate Tests of Significance (S = 2, M = -1/2, N = 31 1/2)

Test Name	Value	Approx. F	Hypoth. DF	Error DF	Sig. of F
Pillais	.96612	30.83743	4.00	132.00	.000
Hotellings	3.20173	51.22767	4.00	128.00	.000
Wilks	.19876	40.39924	4.00	130.00	.000
Roys	.74473				

Note.. F statistic for WILK'S Lambda is exact.

EFFECT .. MONTH (Cont.)
Univariate F-tests with (2,66) D. F.

Variable	Hypoth. SS	Error SS	Hypoth. MS	Error MS	F	Sig. of F
WTLOSS	181.35185	67.72222	90.67593	1.02609	88.36998	.000
ESTEEM	86.72222	152.38889	43.36111	2.30892	18.77980	.000

Roy-Bargman Stepdown F - tests

Variable	Hypoth. MS	Error MS	StepDown F	Hypoth. DF	Error DF	Sig. of F
WTLOSS	90.67593	1.02609	88.36998	2	66	.000
ESTEEM	28.01544	2.34251	11.95957	2	65	.000

10.6 ■ COMPARISON OF PROGRAMS

Programs for MANOVA are covered in detail in Chapter 9. Therefore, this section is limited to those features of particular relevance to profile analysis. SYSTAT, SPSS, BMDP each have a program useful for profile analysis. SAS has an additional program that can be used for profile analysis, but it is limited to equal-n designs. The BMDP and SAS manuals show by example how to set up doubly multivariate designs. No such help is available in the SYSTAT or SPSS manuals, unless you happen to have an old (1986) SPSS^x manual handy. All programs now provide output for within-in-subjects effects in both univariate and multivariate form. All programs also provide information useful for deciding between multivariate and univariate approaches, with Greenhouse-Geisser and Huynh-Feldt adjustments for violation of homogeneity of covariance in the univariate approach. Features of the programs appear in Table 10.18.

10.6.1 ■ SPSS Package

Profile analysis in SPSS MANOVA currently is run like any other repeated-measures design. Output includes both univariate and multivariate tests for all within-subjects effects (flatness) and mixed interactions (parallelism). The MWITHIN feature provides for testing simple effects when repeated measures are analyzed. And the MEASURES command allows the multiple DVs to be given a generic name in doubly multivariate designs, making the output more readable. Also, several repeated-measures analyses can be specified within a single run.

The full output for SPSS MANOVA consists of three separate source tables, one each for parallelism, levels, and flatness. Specification of PRINT= SIGNIF(BRIEF) is used to simplify the multivariate output.

The determinant of the within-cells correlation matrix is used as an aid to deciding whether or not further investigation of multicollinearity is needed; the residuals plots provide evidence of skewness and outliers. And Box's M test is available as an ultrasensitive test of homogeneity of variance-covariance matrices. But tests of linearity and homoscedasticity are not directly available within SPSS or SPSS MANOVA and a test for multivariate outliers requires use of SPSS REGRESSION separately for each group.

10.6.2 ■ BMDP Series

Although the BMDP manual (Dixon, 1992) is helpful in interpreting the output for BMDP4V, a few additional hints might be in order. The first main effect shown is the test for levels, easily interpreted in any case because it is a univariate test. The second main effect is the test for flatness based on Hotelling's T^2 and labeled TSQ. (Note that η^2, labeled HT EVALS, can be found in the absence of Wilks' Lambda by specifying EVAL in the ANALYSIS paragraph.) Finally, the parallelism test is shown as an interaction effect, and several multivariate statistics are given, including LRATIO for Wilks' Lambda.

There is no provision for evaluation of homogeneity of variance-covariance matrices directly within BMDP4V. Nor can multicollinearity, linearity, skewness, heteroscedasticity of variance, or outliers be evaluated within that program. Instead, it is necessary to apply other programs (cf. Section 10.4.2). Other programs are necessary for adjusted means, as well. Unadjusted cell and marginal means and standard deviations are automatically printed, along with maximum and minimum values, standard errors, and weighted means, should some weighting scheme be used other than cell sizes.

TABLE 10.18 COMPARISON OF PROGRAMS FOR PROFILE ANALYSIS[a]

	SPSS MANOVA	BMDP4V	SAS GLM and ANOVA	SYSTAT GLM
Input				
Variety of strategies for unequal n	Yes	Yes	Yes[b]	Yes
Special specification for doubly multivariate analysis	Yes	Yes	Yes	No
Special specification for simple effects	Yes	Yes	No	No
Output				
Single source table for Wilks' Lambda	PRINT= SIGNIF(BRIEF)	Yes	No	No
Specific comparisons	Yes	Yes	Yes	Yes
Within-cells correlation matrix	Yes	No	Yes	Yes
Determinant of within-cells correlation matrix	Yes	No	No	No
Cell means and standard deviations	PRINT= CELLINFO(MEANS)		LSMEANS	PRINT MEDIUM
Marginal means	OMEANS	Yes	LSMEANS	No
Marginal standard deviations	No	Yes	No	No
Cell and marginal variances	No	Yes	No	No
Cell and marginal minimum and maximum values	No	Yes	No	No
Confidence intervals around cell means	Yes	No	No	No
Wilks' Lambda and F for parallelism	Yes	LRATIO	Yes	Yes
Hotelling's T^2 for flatness	No	TSQ	No	No
Pillai's criterion	Yes	No	Yes	Yes
Additional statistical criteria	Yes	Yes	Yes	Yes
Test for homogeneity of covariance	Yes	No	No	No
Greenhouse-Geisser epsilon and adjusted p	Yes	Yes	Yes	Yes
Huynh-Feldt epsilon and adjusted p	Yes	Yes	Yes	Yes
Predicted values and residuals for each case	Yes	No	Yes	Data file
Residuals plot	Yes	No	No	No
Homogeneity of variance-covariance matrices	Box's M	No	No	No
Test for multivariate outliers	No	No	No	Data file
Effect sizes (strength of association)	POWER	No	No	No
Observed power values	POWER		No	No

[a] Additional features of these programs appear in Chapter 9 (MANOVA).

[b] SAS ANOVA requires equal n

10.6.3 ■ SAS System

Profile analysis is available through the GLM (general linear model) procedure of SAS or through ANOVA if the groups have equal sample sizes. The two procedures use very similar setup conventions and provide the same output. In GLM and ANOVA, profile analysis is treated as a special case of repeated-measures ANOVA. PROFILE is an explicit statement that generates the segments between adjacent levels of the within-subjects factor.

Both univariate and multivariate results are provided by default, with the multivariate output providing the tests of parallelism and flatness. However, the output is not particularly easy to read. Each multivariate effect appears separately but instead of a single table for each effect, there are multiple sections filled with cryptic symbols. These symbols are defined above the output for that effect. Only the test of levels appears in a familiar source table. Specification of SHORT within the REPEATED statement provides condensed output for the multivariate tests of flatness and parallelism, but the tests for each effect are still separated. The manual shows an example of a doubly multivariate setup, not an easy feat to accomplish.

SAS GLM and ANOVA have no provision for evaluating the assumptions of profile analysis. The superior data-management capabilities of SAS are useful for creating matrices and plotting exotic functions, but the statistical procedures such as GLM and ANOVA have minimal built-in capabilities for testing assumptions.

10.6.4 ■ SYSTAT System

The GLM program in SYSTAT handles profile analysis through the REPEAT format. (SYSTAT ANOVA also does profile analysis, but with few features and less flexibility.) There are three forms for printing output—long, medium, and short. The short form is the default option. The long form provides such extras as error correlation matrices and canonical analysis. Cell means and standard errors for the DVs are available through PRINT=MEDIUM or LONG. Additional statistics are available through the STATS program but the data set must be sorted by groups. SYSTAT GLM automatically prints out a full trend analysis on the within-subjects variable. There is no example to follow for a doubly multivariate analysis.

Multivariate outliers are found by applying the discriminant function procedure detailed in the SYSTAT manual. Mahalanobis distances (the square root of values given by other programs) for each group are saved to a file. Additional assumptions are not directly tested through the GLM procedure although assumptions such as linearity, and homogeneity of variance can be evaluated through STATS and GRAPH. No test for homogeneity of variance-covariance matrices is available.

10.7 ■ COMPLETE EXAMPLE OF PROFILE ANALYSIS

Variables are chosen from among those in the learning disabilities data bank described in Appendix B, Section B.2 to illustrate the application of profile analysis. Three groups are formed on the basis of the preference of learning-disabled children for age of playmates (AGEMATE): children whose parents report that they have (1) preference for playmates younger than themselves, (2) preference for playmates older than themselves, and (3) preference for playmates the same age as themselves or no preference.

DVs are the 11 subtests of the Wechsler Intelligence Scale for Children given either in its original or revised (WISC-R) form, depending on the date of administration of the test. The subtests are information (INFO), comprehension (COMP), arithmetic (ARITH), similarities (SIMIL), vocabulary (VOCAB), digit span (DIGIT), picture completion (PICTCOMP), picture arrangement (PARANG), block design (BLOCK), object assembly (OBJECT), and CODING.

The primary question is whether profiles of learning-disabled children on the WISC subscales differ if the children are grouped on the basis of their choice of age of playmates (the parallelism test). Secondary questions are whether preference for age of playmates is associated with overall IQ (the levels test), and whether the subtest pattern of the combined group of learning-disabled children is flat (the flatness test), as it is for the population on which the WISC was standardized.

10.7.1 ■ Evaluation of Assumptions

Assumptions and limitations of profile analysis are evaluated as described in Section 10.3.2.

10.7.1.1 ■ Unequal Sample Sizes and Missing Data

From the sample of 177 learning-disabled children given the WISC or WISC-R, 168 could be grouped according to preferred age of playmates. Among the 168 children who could be grouped, a preliminary run of BMDPAM (Table 10.19) is used to reveal the extent and pattern of missing data.

Missing data are sought among the DVs (subtests, levels of the within-subjects IV) for cases grouped by AGEMATE as indicated in the GROUP paragraph. The 11 subtests of the WISC yield 11 df, so cases with complete data have 11 df in the case by case listing of the MAHALANOBIS D-SQUARED test for outliers in the bottom portion of Table 10.19. Cases with fewer than 11 df are those with missing data. Four children are identified through the BMDPAM run, among them case 59 from the group preferring younger playmates. Because so few cases have missing data, and the missing variables are scattered over groups and DVs, as seen in the upper portion of Table 10.19, it is decided to delete them from analysis, leaving $N = 164$. Other strategies for dealing with missing data are discussed in Chapter 4.

Of the remaining 164 children, 45 are in the group preferring younger playmates, 54 older playmates, and 65 same age playmates or no preference. This leaves 4.5 times as many cases as DVs in the smallest group, posing no problems for multivariate analysis.

10.7.1.2 ■ Multivariate Normality

Groups are large and not notably discrepant in size. Therefore, the central limit theorem should assure acceptably normal sampling distributions of means for use in profile analysis. Other portions of the BMDPAM output show all of the DVs to be well-behaved; univariate summary statistics for the YOUNGER group, for example, are in Table 10.20. Although cases with missing data are not yet deleted in this run, resulting in slightly different sample sizes for different segments, skewness and kurtosis values are acceptable and unlikely to be much influenced by deletion of so few cases.

The levels test is based on the average of the DVs. However, this should pose no problem; since the individual DVs are so well-behaved, there is no reason to expect problems with the average of them. Had the DVs shown serious departures from normality, an "average" variable could have been created through a transformation and tested through the usual procedures of Section 4.2.2.1.

TABLE 10.19 IDENTIFICATION OF MISSING DATA (SETUP AND SELECTED OUTPUT FROM BMDPAM)

```
/INPUT       VARIABLES ARE 13.   FILE='PROFILE.DAT'.   FORMAT IS FREE.
/VARIABLES   NAMES ARE CLIENT, AGEMATE, INFO, COMP, ARITH, SIMIL,
                VOCAB, DIGIT, PICTCOMP, PARANG, BLOCK, OBJECT,
                CODING.
             MISSING = 13*0.
             LABEL = CLIENT.
/GROUP       VAR = AGEMATE.
             CODES(AGEMATE) ARE 1 TO 3.
             NAMES(AGEMATE) ARE 'YOUNGER', 'OLDER', 'SAME-NO PREF'.
/ESTIMATE    METHOD=REGR.
/PRINT       MATR=DIS.
/END
```

```
NUMBER OF CASES READ. . . . . . . . . . . . . . . . .    177
* CASES WITH GROUPING VALUES NOT USED. . . . . . . . .      1
    REMAINING NUMBER OF CASES . . . . . . . . . . . .     176
** CASES WITH GROUPING VARIABLE MISSING OR BEYOND LIMITS    8
    REMAINING NUMBER OF CASES . . . . . . . . . . . .     168

NUMBER OF CASES WITH NO DATA MISSING AND WITH
    POSITIVE CASE WEIGHT. . . . . . . . . . . . . . .     164

*  THESE CASES ARE EXCLUDED FROM ALL COMPUTATIONS.
** THESE CASES ARE NOT USED IN ESTIMATING MISSING DATA.
```

TABLE OF SAMPLE SIZES

(PERCENTAGES OF MISSING DATA INCLUDE CASES WITH ANY VARIABLE
MISSING OR BEYOND MAXIMUM OR MINIMUM LIMITS)

GROUP	SIZE	COMPLETE CASES	PERCENT MISSING
YOUNGER	46	45	2.2
OLDER	55	54	1.8
SAME-NO	67	65	3.0

TABLE 10.19 *(CONTINUED)*

TABLE OF SAMPLE SIZES, NUMBER MISSING AND PERCENT
MISSING FOR ALL GROUPS TAKEN TOGETHER. NUMBER AND
PERCENT FOR MISSING INCLUDES VALUES OUTSIDE LIMITS.

VARIABLE NO.	NAME	SAMPLE SIZE	NUMBER MISSING	PERCENT MISSING
3	INFO	168	0	0.0
4	COMP	166	2	1.2
5	ARITH	168	0	0.0
6	SIMIL	168	0	0.0
7	VOCAB	168	0	0.0
8	DIGIT	167	1	0.6
9	PICTCOMP	168	0	0.0
10	PARANG	168	0	0.0
11	BLOCK	168	0	0.0
12	OBJECT	168	0	0.0
13	CODING	167	1	0.6

ESTIMATES OF MISSING DATA, MAHALANOBIS D-SQUARED (CHI-SQUARED)
AND SQUARED MULTIPLE CORRELATIONS WITH AVAILABLE VARIABLES

CASE LABEL	CASE NUMBER	MISSING VARIABLE	ESTIMATE	R-SQUARED	GROUP	CHI-SQ	CHISQ/DF	D.F.	SIGNIFICANCE
68	54				YOUNGER	7.172	0.652	11	0.7850
69	55				SAME-NO	7.095	0.645	11	0.7914
73	57				YOUNGER	8.864	0.806	11	0.6345
75	59				OLDER	3.569	0.357	10	0.9647
76	60				OLDER	11.622	1.057	11	0.3928
78	61				SAME-NO	5.541	0.504	11	0.9022
79	62				SAME-NO	3.499	0.318	11	0.9824
80	63				OLDER	7.801	0.709	11	0.7310
82	64				YOUNGER	8.112	0.737	11	0.7032
83	65				OLDER	15.471	1.406	11	0.1619
84	66				YOUNGER	6.216	0.565	11	0.8585
85	67				SAME-NO	29.952	2.723	11	0.0016
86	68				OLDER	6.999	0.636	11	0.7992
87	69				OLDER	6.979	0.634	11	0.8008
90	70				SAME-NO	6.502	0.591	11	0.8379

TABLE 10.20   SAMPLE UNIVARIATE STATISTICS FOR FIRST GROUP PRODUCED BY BMDPAM.   SEE TABLE 10.18 FOR SETUP

UNIVARIATE SUMMARY STATISTICS

GROUP IS YOUNGER   SIZE IS   46

VARIABLE	SAMPLE SIZE	MEAN	STANDARD DEVIATION	COEFFICIENT OF VARIATION	SMALLEST VALUE	LARGEST VALUE	SMALLEST STANDARD SCORE	LARGEST STANDARD SCORE	SKEWNESS	KURTOSIS
3 INFO	46	8.95652	3.37281	0.376576	4.00000	19.00000	-1.47	2.98	0.54	0.08
4 COMP	46	9.54348	2.87291	0.301034	3.00000	18.00000	-2.28	2.94	0.76	1.19
5 ARITH	46	9.17391	2.70230	0.294563	5.00000	19.00000	-1.54	3.64	0.97	2.06
6 SIMIL	46	9.73913	3.33623	0.342559	4.00000	19.00000	-1.72	2.78	0.65	0.07
7 VOCAB	46	10.26087	3.44761	0.335996	2.00000	19.00000	-2.40	2.53	0.54	0.56
8 DIGIT	45	8.53333	2.68498	0.314646	3.00000	16.00000	-2.06	2.78	0.52	0.20
9 PICTCOMP	46	11.13043	2.68004	0.240785	5.00000	17.00000	-2.29	2.19	-0.19	-0.46
10 PARANG	46	10.00000	2.42212	0.242212	5.00000	15.00000	-2.06	2.06	0.20	-0.47
11 BLOCK	46	10.00000	2.95898	0.295898	3.00000	19.00000	-2.37	3.04	0.32	0.97
12 OBJECT	46	10.41304	2.60407	0.250077	3.00000	14.00000	-2.85	1.38	-0.71	-0.08
13 CODING	46	8.73913	2.57712	0.294894	4.00000	14.00000	-1.84	2.04	0.21	-0.82

VALUES FOR KURTOSIS GREATER THAN ZERO INDICATE DISTRIBUTIONS
WITH HEAVIER TAILS THAN THE NORMAL DISTRIBUTION.

10.7.1.3 ■ Linearity

Considering the well-behaved nature of these DVs and the known linear relationship among subtests of the WISC, no threats to linearity are anticipated.

10.7.1.4 ■ Outliers

A second run of BMDPAM, after deletion of cases with missing values, provides information for evaluation of several issues, including univariate and multivariate outliers among DVs. As seen in the univariate summary statistics of Table 10.21 for the various groups, one standard score (ARITH for the YOUNGER group) is 3.6, suggesting a univariate outlier.

A later portion of the run in Table 10.21 (not shown), identical in format to the final portion of Table 10.19, also reveals no multivariate outliers with a criterion of $p = .001$. The decision is made to retain the univariate outlier since the subtest score of 19 is acceptable and trial analyses with and without the outlier removed made no difference in the results (cf. Section 4.1.4.3).

10.7.1.5 ■ Homogeneity of Variance-Covariance Matrices

With relatively equal sample sizes and no gross discrepancy in within-cell variances, there is no need to consult the overly sensitive Box's M test available in SPSS MANOVA. Evidence for relatively equal variances is available from the BMDPAM run of Table 10.21, where standard deviations are given for each variable within each group. All the variances (squared standard deviations) are quite close in value; for no variable is there a between-group ratio of largest to smallest variance approaching 10:1.

10.7.1.6 ■ Multicollinearity and Singularity

BMDPAM provides evidence of multicollinearity and/or singularity by printing out squared multiple correlations of each variable with remaining variables. As seen in the last portion of Table 10.21, none of the R-SQUARED values approaches the .99 level which could pose danger to matrix inversion in performing profile analysis.

10.7.2 ■ Profile Analysis

Setup and major output for profile analysis of the 11 WISC subtests for the three groups as produced by SPSS MANOVA appear in Table 10.22. After cell statistics, significance tests are shown, in turn, for levels, parallelism, and flatness.

The levels test for AGEMATE in the portion of output called the Test of Significance for OVERALL... (the average of all subtests) shows no significant univariate effect of the three AGEMATE groups on the combined subtests, $F(2, 161) = 0.81$. That is, the groups do not differ on their combined WISC subtest scores, on their IQ.

The parallelism test, called the test of the AGEMATE BY SUBTEST effect, shows significantly different profiles for the three AGEMATE groups. The various multivariate tests of parallelism produce slightly different probability levels for α, all less than 0.05. Because there is no reason to doubt compliance with multivariate assumptions in this data set, Wilks' Lambda (.78398) is used for statistical evaluation and strength of association using Equation 10.4. The

TABLE 10.21   EVALUATION OF OUTLIERS AND HOMOGENEITY OF VARIANCE (SETUP AND SELECTED OUTPUT FROM BMDPAM)

```
/INPUT       VARIABLES ARE 13.  FILE='PROFILE.DAT'.   FORMAT IS FREE.
/VARIABLES   NAMES ARE CLIENT, AGEMATE, INFO, COMP, ARITH, SIMIL,
                VOCAB, DIGIT, PICTCOMP, PARANG, BLOCK, OBJECT,
                CODING.
             MISSING = 13*0.
             LABEL = CLIENT.
/TRANSFORM   DELETE = 59, 123, 129, 130.
/GROUP       VAR = AGEMATE.
             CODES(AGEMATE) ARE 1 TO 3.
             NAMES(AGEMATE) ARE 'YOUNGER', 'OLDER', 'SAMENO PREF'.
/ESTIMATE    METHOD=REGR.
/PRINT       MATR=DIS.
/END
```

UNIVARIATE SUMMARY STATISTICS

GROUP IS YOUNGER   SIZE IS   45

VARIABLE	SAMPLE SIZE	MEAN	STANDARD DEVIATION	COEFFICIENT OF VARIATION	SMALLEST VALUE	LARGEST VALUE	SMALLEST STANDARD SCORE	LARGEST STANDARD SCORE	SKEWNESS	KURTOSIS
3 INFO	45	9.06667	3.32620	0.366861	4.00000	19.00000	-1.52	2.99	0.55	0.14
4 COMP	45	9.51111	2.89688	0.304579	3.00000	18.00000	-2.25	2.93	0.79	1.17
5 ARITH	45	9.22222	2.71267	0.294145	5.00000	19.00000	-1.56	3.60	0.94	2.01
6 SIMIL	45	9.86667	3.25856	0.330259	5.00000	19.00000	-1.49	2.80	0.73	0.11
7 VOCAB	45	10.28889	3.48126	0.338352	5.00000	19.00000	-2.38	2.50	0.52	0.48
8 DIGIT	45	8.53333	2.68498	0.314646	3.00000	16.00000	-2.06	2.78	0.52	0.20
9 PICTCOMP	45	11.20000	2.66799	0.238214	5.00000	17.00000	-2.32	2.17	-0.24	-0.38
10 PARANG	45	10.08889	2.37240	0.235150	5.00000	15.00000	-2.15	2.07	0.22	-0.41
11 BLOCK	45	10.04444	2.97685	0.296367	3.00000	19.00000	-2.37	3.01	0.28	0.94
12 OBJECT	45	10.46667	2.60768	0.249141	3.00000	14.00000	-2.86	1.35	-0.77	0.02
13 CODING	45	8.64444	2.52403	0.291982	4.00000	14.00000	-1.84	2.12	0.23	-0.73

GROUP IS OLDER   SIZE IS   54

VARIABLE	SAMPLE SIZE	MEAN	STANDARD DEVIATION	COEFFICIENT OF VARIATION	SMALLEST VALUE	LARGEST VALUE	SMALLEST STANDARD SCORE	LARGEST STANDARD SCORE	SKEWNESS	KURTOSIS
3 INFO	54	10.18519	3.27411	0.321458	3.00000	19.00000	-2.19	2.69	0.16	-0.21
4 COMP	54	10.42593	2.87213	0.275480	3.00000	18.00000	-2.59	2.64	-0.09	-0.09
5 ARITH	54	8.79630	2.25187	0.256002	4.00000	14.00000	-2.13	2.31	0.08	-0.76
6 SIMIL	54	11.20370	2.98031	0.266011	5.00000	18.00000	-2.08	2.28	0.16	-0.62
7 VOCAB	54	11.46296	2.80641	0.244825	6.00000	19.00000	-1.95	2.69	0.24	-0.34
8 DIGIT	54	9.01852	2.53646	0.281250	4.00000	16.00000	-1.98	2.75	0.53	-0.12
9 PICTCOMP	54	9.79630	3.33297	0.340227	2.00000	19.00000	-2.34	2.76	0.05	0.42

TABLE 10.21   (CONTINUED)

	SAMPLE SIZE	MEAN	STANDARD DEVIATION	COEFFICIENT OF VARIATION	SMALLEST VALUE	LARGEST VALUE	SMALLEST STANDARD SCORE	LARGEST STANDARD SCORE	SKEWNESS	KURTOSIS
10 PARANG	54	10.70370	2.95009	0.275614	2.00000	17.00000	-2.95	2.13	-0.24	0.13
11 BLOCK	54	10.29630	2.89195	0.280873	2.00000	17.00000	-2.87	2.32	-0.36	0.52
12 OBJECT	54	10.90741	2.87651	0.263720	3.00000	16.00000	-2.75	1.77	-0.43	-0.06
13 CODING	54	8.81481	2.97203	0.337163	2.00000	15.00000	-2.29	2.08	-0.10	-0.70

GROUP IS SAME-NO   SIZE IS   65

VARIABLE	SAMPLE SIZE	MEAN	STANDARD DEVIATION	COEFFICIENT OF VARIATION	SMALLEST VALUE	LARGEST VALUE	SMALLEST STANDARD SCORE	LARGEST STANDARD SCORE	SKEWNESS	KURTOSIS
3 INFO	65	9.36923	2.54073	0.271178	3.00000	15.00000	-2.51	2.22	-0.32	-0.33
4 COMP	65	10.12308	2.88047	0.284545	6.00000	18.00000	-1.43	2.73	0.62	-0.30
5 ARITH	65	9.13846	2.49297	0.272800	5.00000	16.00000	-1.66	2.75	0.70	-0.13
6 SIMIL	65	10.75385	3.44615	0.320458	2.00000	18.00000	-2.54	2.10	-0.21	-0.07
7 VOCAB	65	10.36923	2.75323	0.265519	3.00000	17.00000	-2.68	2.41	0.18	-0.04
8 DIGIT	65	8.73846	2.62367	0.300244	4.00000	14.00000	-1.81	2.01	0.20	-0.89
9 PICTCOMP	65	11.15385	2.77956	0.249202	6.00000	18.00000	-1.85	2.46	0.15	-0.38
10 PARANG	65	10.38462	2.59623	0.250007	4.00000	15.00000	-2.46	1.78	-0.43	-0.26
11 BLOCK	65	10.64615	2.45860	0.230938	5.00000	18.00000	-2.30	2.99	0.04	0.56
12 OBJECT	65	11.07692	2.92782	0.264317	5.00000	19.00000	-2.08	2.71	0.33	0.16
13 CODING	65	8.20000	2.80736	0.342361	2.00000	15.00000	-2.21	2.42	0.22	-0.56

VALUES FOR KURTOSIS GREATER THAN ZERO INDICATE DISTRIBUTIONS
WITH HEAVIER TAILS THAN THE NORMAL DISTRIBUTION.

SQUARED MULTIPLE CORRELATIONS OF EACH VARIABLE WITH ALL OTHER VARIABLES

(MEASURES OF MULTICOLLINEARITY OF VARIABLES)
AND TESTS OF SIGNIFICANCE OF MULTIPLE REGRESSION
DEGREES OF FREEDOM FOR F-STATISTICS ARE   10 AND   151

VARIABLE NO.	NAME	R-SQUARED	F-STATISTIC	SIGNIFICANCE (P LESS THAN)
3	INFO	0.533228	17.25	0.00000
4	COMP	0.545236	18.10	0.00000
5	ARITH	0.404673	10.26	0.00000
6	SIMIL	0.467074	13.23	0.00000
7	VOCAB	0.535717	17.42	0.00000
8	DIGIT	0.184073	3.41	0.00048
9	PICTCOMP	0.351333	8.18	0.00000
10	PARANG	0.195119	3.66	0.00021
11	BLOCK	0.339068	7.75	0.00000
12	OBJECT	0.295201	6.32	0.00000
13	CODING	0.086320	1.43	0.17356

```
DATA LIST       FILE='PROFILE.dat'    FREE
                /CLIENT,AGEMATE,INFO,COMP,ARITH,SIMIL,VOCAB,DIGIT,
                PICTCOMP,PARANG,BLOCK,OBJECT,CODING.
MISSING VALUES   ALL(0).
VAR LABELS       AGEMATE 'PREFERRED AGE OF PLAYMATES'.
VALUE LABELS     AGEMATE 1 'YOUNGER' 2 'OLDER' 3 'SAME OR NO PREFER.'.
MANOVA               INFO TO CODING BY AGEMATE(1,3)/
               WSFACTOR=SUBTEST(11)/
               WSDESIGN=SUBTEST/
               PRINT=CELLINFO(MEANS), HOMOGENEITY(BOXM)/
               DESIGN=AGEMATE.
```

Cell Means and Standard Deviations
Variable .. INFO

FACTOR	CODE	Mean	Std. Dev.	N
AGEMATE	YOUNGER	9.067	3.326	45
AGEMATE	OLDER	10.185	3.274	54
AGEMATE	SAME OR	9.369	2.541	65
For entire sample		9.555	3.036	164

Variable .. COMP

FACTOR	CODE	Mean	Std. Dev.	N
AGEMATE	YOUNGER	9.511	2.897	45
AGEMATE	OLDER	10.426	2.872	54
AGEMATE	SAME OR	10.123	2.880	65
For entire sample		10.055	2.887	164

Variable .. ARITH

FACTOR	CODE	Mean	Std. Dev.	N
AGEMATE	YOUNGER	9.222	2.713	45
AGEMATE	OLDER	8.796	2.252	54
AGEMATE	SAME OR	9.138	2.493	65
For entire sample		9.049	2.471	164

Variable .. SIMIL

FACTOR	CODE	Mean	Std. Dev.	N
AGEMATE	YOUNGER	9.867	3.259	45
AGEMATE	OLDER	11.204	2.980	54
AGEMATE	SAME OR	10.754	3.446	65
For entire sample		10.659	3.270	164

Variable .. VOCAB

FACTOR	CODE	Mean	Std. Dev.	N
AGEMATE	YOUNGER	10.289	3.481	45
AGEMATE	OLDER	11.463	2.806	54
AGEMATE	SAME OR	10.369	2.753	65
For entire sample		10.707	3.015	164

Variable .. DIGIT

FACTOR	CODE	Mean	Std. Dev.	N
AGEMATE	YOUNGER	8.533	2.685	45
AGEMATE	OLDER	9.019	2.536	54
AGEMATE	SAME OR	8.738	2.624	65
For entire sample		8.774	2.603	164

Variable .. PICTCOMP

FACTOR	CODE	Mean	Std. Dev.	N
AGEMATE	YOUNGER	11.200	2.668	45
AGEMATE	OLDER	9.796	3.333	54
AGEMATE	SAME OR	11.154	2.780	65
For entire sample		10.720	2.998	164

TABLE 10.22   *(CONTINUED)*

Variable .. PARANG

FACTOR	CODE	Mean	Std. Dev.	N
AGEMATE	YOUNGER	10.089	2.372	45
AGEMATE	OLDER	10.704	2.950	54
AGEMATE	SAME OR	10.385	2.596	65
For entire sample		10.409	2.656	164

Variable .. BLOCK

FACTOR	CODE	Mean	Std. Dev.	N
AGEMATE	YOUNGER	10.044	2.977	45
AGEMATE	OLDER	10.296	2.892	54
AGEMATE	SAME OR	10.646	2.459	65
For entire sample		10.366	2.747	164

Variable .. OBJECT

FACTOR	CODE	Mean	Std. Dev.	N
AGEMATE	YOUNGER	10.467	2.608	45
AGEMATE	OLDER	10.907	2.877	54
AGEMATE	SAME OR	11.077	2.928	65
For entire sample		10.854	2.820	164

Variable .. CODING

FACTOR	CODE	Mean	Std. Dev.	N
AGEMATE	YOUNGER	8.644	2.524	45
AGEMATE	OLDER	8.815	2.972	54
AGEMATE	SAME OR	8.200	2.807	65
For entire sample		8.524	2.786	164

- - - - - - - - - -

Cell Number .. 1

Determinant of Covariance matrix of dependent variables =   68287979.65217
LOG(Determinant) =
                                        18.03924
- - - - - - - - - -

Cell Number .. 2

Determinant of Covariance matrix of dependent variables = 333847926.18883
LOG(Determinant) =
                                             19.62620

- - - - - - - - - -

Cell Number .. 3

Determinant of Covariance matrix of dependent variables = 105220916.10364
LOG(Determinant) =
                                             18.47157

- - - - - - - - - -

Determinant of pooled Covariance matrix of dependent vars. = 300201402.79794
LOG(Determinant) =
                                             19.51996

- - - - - - - - - -

Multivariate test for Homogeneity of Dispersion matrices

Boxs M =   126.61843
F WITH (132,59321) DF =   .86399, P =   .868 (Approx.)
Chi-Square with 132 DF =   114.32900, P =   .864 (Approx.)

TABLE 10.22 *(CONTINUED)*

```
Tests of Between-Subjects Effects.

Tests of Significance for T1 using UNIQUE sums of squares
Source of Variation          SS       DF        MS         F   Sig of F

WITHIN+RESIDUAL           4916.23     161      30.54
AGEMATE                     49.61       2      24.81       .81     .446

EFFECT .. AGEMATE BY SUBTEST
Multivariate Tests of Significance (S = 2, M = 3 1/2, N = 75 )

Test Name          Value  Approx. F Hypoth. DF   Error DF  Sig. of F

Pillais           .22243  1.91452      20.00      306.00     .011
Hotellings        .26735  2.01851      20.00      302.00     .007
Wilks             .78398  1.96682      20.00      304.00     .009
Roys              .18838
Note.. F statistic for WILKS' Lambda is exact.

EFFECT .. SUBTEST
Multivariate Tests of Significance (S = 1, M = 4 , N = 75 )

Test Name          Value   Exact F Hypoth. DF   Error DF  Sig. of F

Pillais           .46444  13.18150     10.00      152.00     .000
Hotellings        .86720  13.18150     10.00      152.00     .000
Wilks             .53556  13.18150     10.00      152.00     .000
Roys              .46444
Note.. F statistics are exact.
```

test shows that there are reliable differences among the three AGEMATE groups in their profiles on the WISC. The profiles are illustrated in Figure 10.4. Mean values for the plots are found in the cell means portion of the output in Table 10.22, produced by the statement PRINT=CELLINFO(MEANS).

For interpretation of the nonparallel profiles, a contrast procedure is needed to determine which WISC subtests separate the three groups of children. Because there are so many subtests, however, the procedure of Section 10.5.1 is unwieldy. The decision is made to evaluate profiles in terms of subtests on which group averages fall outside the confidence interval of the pooled profile. Cell statistics in Table 10.22 provide the 95% confidence intervals for the pooled profile (confidence interval for entire sample for each variable) and group means for each variable (cell means).

In order to compensate for multiple testing, a wider confidence interval is developed for each test to reflect an experimentwise 95% confidence interval. Alpha rate is set at .0015 for each test to account for the 33 comparisons available—3 groups·at each of 11 subtests—generating a 99.85% confidence interval. Because an N of 164 produces a t distribution similar to z, it is appropriate to base the confidence interval on $z = 3.19$.

For the first subtest, INFO,

$$P(\overline{Y} - zs_m < \mu < \overline{Y} + zs_m) = 99.85 \tag{10.10}$$

$$P(9.55488 - 3.19(3.03609)/\sqrt{164} < \mu < 9.55488 + 3.19(3.03609)/\sqrt{164}) = 99.85$$

$$P(8.79860 < \mu < 10.31116) = 99.85$$

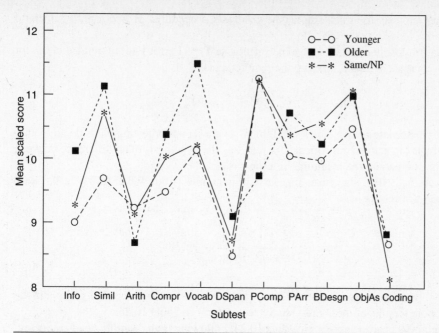

FIGURE 10.4 PROFILES OF WISC SCORES FOR THREE AGEMATE GROUPS.

Because none of the group means on INFO falls outside this interval for the INFO subtest, profiles are not differentiated on the basis of the information subtest of the WISC. It is not necessary to calculate intervals for any variable for which none of the groups deviate from the 95% confidence interval, because they cannot deviate from a wider interval. Therefore, intervals are calculated only for SIMIL, COMP, VOCAB, and PICTCOMP. Applying Equation 10.10 to these variables, significant profile deviation is found for vocabulary and picture completion. (The direction of differences is given in the Results section that follows.)

The final omnibus test produced by SPSS MANOVA is the SUBTEST effect, for which the flatness hypothesis is rejected. All multivariate criteria show essentially the same result, but Hotelling's criterion, with approximate $F(10, 152) = 13.18$, $p < .001$, is most appropriately reported because it is a test of a single group (all groups combined). Strength of association, however, is best computed through Wilks' Lambda, as shown in Section 10.4.3.

Although not usually of interest when the hypothesis of parallelism is rejected, the flatness test is interesting in this case because it reveals differences between learning-disabled children (our three groups combined) and the sample used for standardizing the WISC, for which the profile is necessarily flat. (The WISC was standardized so that all subtests produce the same mean value.) Any sample that differs from a flat profile, that is, has different mean values on various subtests, diverges from the standard profile of the WISC.

Appropriate contrasts for the flatness test in this example are simple one-sample z tests (cf. Section 3.2.1) against the standardized population values for each subtest with mean = 10.0 and standard deviation = 3.0. In this case we are less interested in how the subtests differ from one another than in how they differ from the normative population. (Had we been interested in differences among subtests for this sample, the contrasts procedures of Section 10.5.1 could have been applied.)

As a correction for post hoc inflation of experimentwise Type I error rate, individual alpha for each of the 11 z tests is set at .0045, meeting the requirements of

$$\alpha_{ew} = 1 - (1 - .0045)^{11} < .05$$

as per Equation 9.11. Because most z tables (cf. Table C.1) are set up for testing one-sided hypotheses, critical α is divided in half to find critical z for rejecting the hypothesis of no difference between our sample and the population; the resulting criterion z is ± 2.845.

For the first subtest, INFO, the mean for our entire sample (from Table 10.22) is 9.55488. Application of the z test results in

$$z = \frac{\overline{Y} - \mu}{\sigma/\sqrt{N}} = \frac{9.55488 - 10}{3.0/\sqrt{164}} = -1.900$$

For INFO, then, there is no significant difference between the learning disabled group and the normative population. Results of these individual z tests appear in Table 10.23.

A checklist for profile analysis appears in Table 10.24. Following is an example of a Results section in APA journal format.

TABLE 10.23   RESULTS OF Z-TESTS COMPARING EACH
SUBTEST WITH WISC POPULATION MEAN
(ALPHA = .0045, TWO-TAILED TEST)

Subtest	Mean for entire sample	z for comparison with population mean
Information	9.55488	−1.90
Similarities	10.65854	2.81
Arithmetic	9.04878	−4.06*
Comprehension	10.05488	0.23
Vocabulary	10.70732	3.02*
Digit span	8.77439	−5.23*
Picture completion	10.71951	3.07*
Picture arrangement	10.40854	1.74
Block design	10.36585	1.56
Object assembly	10.85366	3.64*
Coding	8.52439	−6.30*

*$p < .0045$

TABLE 10.24 CHECKLIST FOR PROFILE ANALYSIS

1. Issues
   a. Unequal sample sizes and missing data
   b. Normality of sampling distributions
   c. Outliers
   d. Homogeneity of variance-covariance matrices
   e. Linearity
   f. Multicollinearity and singularity

2. Major analysis
   a. Tests for parallelism. If significant:
      (1) Strength of association
      (2) Figure showing profile for deviation from parallelism
   b. Test for differences among levels, if appropriate. If significant:
      (1) Strength of association
      (2) Marginal means for groups
   c. Test for deviation from flatness, if appropriate. If significant:
      (1) Strength of association
      (2) Means for measures

3. Additional analyses
   a. Planned comparisons
   b. Post hoc comparisons appropriate for significant effect(s)
      (1) Comparisons among groups
      (2) Comparisons among measures
      (3) Comparisons among measures within groups
   c. Power analysis for nonsignificant effects

```
                                Results

    A profile analysis was performed on 11 subtests of the Wechsler
Intelligence Scale for Children (WISC): information, similarities,
arithmetic, comprehension, vocabulary, digit span, picture completion,
picture arrangement, block design, object assembly, and coding. The
grouping variable was preference for age of playmates, divided into
children who (1) prefer younger playmates, (2) prefer older playmates,
and (3) those who have no preference or prefer playmates the same age
as themselves.

    BMDPAM and BMDP7D were used for data screening. Four children in the
original sample, scattered through groups and DVs, had missing data on
one or more subtest, reducing the sample size to 164. No univariate or
multivariate outliers were detected among these children, with p =
.001. After deletion of cases with missing data, assumptions regarding
normality of sampling distributions, homogeneity of variance-covariance
matrices, linearity, and multicollinearity were met.
```

SPSS MANOVA was used for the major analysis. Using Wilks' criterion, the profiles, seen in Figure 10.4, deviated significantly from parallelism, $F(20, 304) = 1.97$, $p = .009$, partial $\eta^2 = .11$ [cf. Equation 10.4]. For the levels test, no reliable differences were found among groups when scores were averaged over all subtests, $F(2, 161) = 0.81$, $p = .45$. When averaged over groups, however, subtests were found by Hotelling's criterion to deviate significantly from flatness, $F(10, 152) = 13.18$, $p < .001$, $\eta^2 = .46$.

To evaluate deviation from parallelism of the profiles, confidence limits were calculated around the mean of the profile for the three groups combined. Alpha error for each confidence interval was set at .0015 to achieve an experimentwise error rate of 5%. Therefore, 99.85% limits were evaluated for the pooled profile. For two of the subtests, one or more groups had means that fell outside these limits. Children who preferred older playmates had a reliably higher mean on the vocabulary subtest (mean = 11.46) than that of the pooled groups (where the 99.85% confidence limits were 9.956 to 11.458); children who preferred older playmates had reliably lower scores on the picture completion subtest (mean = 9.80) than that of the pooled groups (99.85% confidence limits were 9.973 to 11.466).

Deviation from flatness was evaluated by identifying which subtests differed from those of the standardization population of the WISC, with mean = 10 and standard deviation = 3 for each subtest. Experimentwise α = .05 was achieved by setting α for each test at .0045. As seen in Table 10.23, learning-disabled children had significantly lower scores than the WISC normative population in arithmetic, digit span, and coding. On the other hand, these children had significantly higher than normal performance on vocabulary, picture completion, and object assembly.

10.8 ■ SOME EXAMPLES FROM THE LITERATURE

Gray-Toft (1980) used profile analysis to evaluate the effectiveness of a counseling program for hospice nurses. Two groups of nurses participated in the program at different times, with each group acting as both treatment (during participation) and control. One group consisted of day shift nurses, while the other group was made up of nurses on evening and night shifts. (Because groups participated as a whole, the use of individual subjects for calculation of error variance is questionable.) Variables were background characteristics, seven subscales of a nursing stress scale, and five subscales of a job satisfaction index.

Profile analysis in this study was limited to evaluation of pretest scores. For the nursing stress scale, profiles were parallel and the groups did not differ in level. There were, however, differences in subscales (flatness), which were evaluated through stepdown analysis. Because subscales were based on factor analysis of a larger set of questions, priority for stepdown analysis was presumably based on percentage of variance accounted for by successive factors. Nurses experienced significantly more stress associated with death and dying and with their workload than with other sources, and significantly less stress associated with conflict with other nurses than with other sources of stress. There was no difference in stress associated with conflict with physicians, inadequate preparation, need for support, and uncertainty concerning treatment.

On the job satisfaction index, profiles were also parallel. However, significant differences were found in both levels and subscales. Day shift nurses showed greater satisfaction than evening or night shift nurses, and stepdown analysis showed nurses to be more satisfied with supervision than with pay, promotions, and coworkers. (No rationale for ordering of subscales was provided for the satisfaction index.)

To study career-making decisions, Mitchell and Krumboltz (1984) performed separate profile analyses for several time periods, during and after university students participated in one of three groups: training in cognitive restructuring, decision skills intervention, or a test-only control. Students were administered, among other instruments, a Vocational Exploratory Behavior Inventory (VEBI), which measures four components of appropriate career decision-making behavior.

No differences in profiles (parallelism) were reported between groups, nor were differences between components noted (flatness). Group (levels) differences were, however, reported at a number of the time periods. These differences began to emerge after the third week of treatment with the treatment groups performing more appropriate behaviors than controls during the treatment period. After treatment had ended, the cognitive restructuring group generally performed more vocational exploratory behaviors than the other two groups. That is, the cognitive restructuring treatment was more effective in maintaining vocational exploratory behavior. Specific differences between groups were explored via contrast analysis.

CHAPTER 11
Discriminant Function Analysis

11.1 ■ GENERAL PURPOSE AND DESCRIPTION

The goal of discriminant function analysis is to predict group membership from a set of predictors. For example, can a differential diagnosis between a group of normal children, a group of children with learning disability, and a group with emotional disorder be made reliably from a set of psychological test scores? The three groups are normal children, children with learning disability, and children with emotional disorder. The predictors are a set of psychological test scores such as the Illinois Test of Psycholinguistic Ability, subtests of the Wide Range Achievement Test, Figure Drawing tests, and the Wechsler Intelligence Scale for Children.

Discriminant function analysis (DISCRIM) is MANOVA turned around. In MANOVA, we ask whether group membership is associated with reliable mean differences on a combination of DVs. If the answer to that question is yes, then the combination of variables can be used to predict group membership—the DISCRIM perspective. In univariate terms, a significant difference between groups implies that, given a score, you can predict (imperfectly, no doubt) which group it comes from.

Semantically, however, confusion arises between MANOVA and DISCRIM because in MANOVA the IVs are the groups and the DVs predictors while in DISCRIM the IVs are the predictors and the DVs groups. We have tried to avoid confusion here by always referring to IVs as "predictors" and to DVs as "groups" or "grouping variables."[1]

Mathematically, MANOVA and DISCRIM are the same, although the emphases often differ. The major question in MANOVA is whether group membership is associated with reliable mean differences in combined DV scores, analogous in DISCRIM to the question of whether predictors can be combined to predict group membership reliably. In many cases, DISCRIM is carried to the point of actually putting cases into groups in a process called classification.

---

[1] Many texts also refer to IVs or predictors as discriminating variables and to DVs or groups as classification variables. However, there are also discriminant functions and classification functions to contend with, so the terminology becomes quite confusing. We have tried to simplify it by using only the terms predictors and groups.

Classification is a major extension of DISCRIM over MANOVA. Most computer programs for DISCRIM evaluate the adequacy of classification. How well does the classification procedure do? How many learning-disabled kids in the original sample, or a cross-validation sample, are classified correctly? When errors occur, what is their nature? Are learning-disabled kids more often confused with normal kids or with kids suffering emotional disorder?

A second difference involves interpretation of differences among the predictors. In MANOVA, there is frequently an effort to decide which DVs are associated with group differences, but rarely an effort to interpret the pattern of differences among the DVs as a whole. In DISCRIM, there is often an effort to interpret the pattern of differences among the predictors as a whole in an attempt to understand the dimensions along which groups differ.

Complexity arises with this attempt, however, because with more than two groups there may be more than one way to combine the predictors to differentiate among groups. There may, in fact, be as many dimensions that discriminate among groups as there are degrees of freedom for groups. For instance, if there are three groups, there may be two dimensions that discriminate among groups: a dimension that separates the first group from the second and third groups, and a dimension that separates the second group from the third group, for example. This process is conceptually similar to contrasts in ANOVA except that in DISCRIM the divisions among groups are almost never as neat and clean.

In our example of three groups of children (normal, learning-disabled, and emotionally disordered) given a variety of psychological measures, one way of combining the psychological test scores may tend to separate the normal group from the two groups with disorders, while a second way of combining the test scores may tend to separate the group with learning disability from the group with emotional disorder. The researcher attempts to understand the "message" in the two ways of combining test scores to separate groups differently. What is the meaning of the combination of scores that separates normal from disordered kids, and what is the meaning of the different combination of scores that separates kids with one kind of disorder from kids with another? This attempt is facilitated by statistics available in many of the canned computer programs for DISCRIM that are not printed in some programs for MANOVA.

Thus there are two facets of DISCRIM, and one or both may be emphasized in any given research application. The researcher may simply be interested in a decision rule for classifying cases where the number of dimensions and their meaning is irrelevant. Or, the emphasis may be on interpreting the results of DISCRIM in terms of the combinations of predictors—called discriminant functions—that separate various groups from each other.

A DISCRIM version of covariates analysis (MANCOVA) is available, because DISCRIM can be set up in a sequential manner. When sequential DISCRIM is used, the covariate is simply a predictor that is given top priority. For example, a researcher might consider the score on the Wechsler Intelligence Scale for Children a covariate and ask how well the Wide Range Achievement Test, the Illinois Test of Psycholinguistic Ability, and Figure Drawings differentiate between normal, learning-disabled, and emotionally disordered children after differences in IQ are accounted for.

If groups are arranged in a factorial design, it is frequently best to rephrase research questions so that they are answered within the framework of MANOVA. (However, DISCRIM can in some circumstances be directly applied to factorial designs as discussed in Section 11.6.5.) Similarly, DISCRIM programs make no provision for within-subjects variables. If a within-subjects analysis is desired, the question is also rephrased in terms of MANOVA or Profile Analysis. For this reason the emphasis in this chapter is on one-way between-subjects DISCRIM.

11.2 ■ KINDS OF RESEARCH QUESTIONS

The primary goals of DISCRIM are to find the dimension or dimensions along which groups differ and to find classification functions to predict group membership. The degree to which these goals are met depends, of course, on choice of predictors. Typically, the choice is made either on the basis of theory about which variables should provide information about group membership, or on the basis of pragmatic considerations such as expense, convenience, or unobtrusiveness.

It should be emphasized that the same data are profitably analyzed through either MANOVA or DISCRIM programs, and frequently both, depending on the kinds of questions you want to ask. If group sizes are very unequal, and/or distributional assumptions are untenable, logistic regression also answers most of the same questions. In any event, statistical procedures are readily available within canned computer programs for answering the following types of questions generally associated with DISCRIM.

11.2.1 ■ Significance of Prediction

Can group membership be predicted reliably from the set of predictors? For example, can we do better than chance in predicting whether children are learning-disabled, emotionally disordered, or normal on the basis of the set of psychological test scores? This is the major question of DISCRIM that the statistical procedures described in Section 11.6.1 are designed to answer. The question is identical to the question about "main effects of IVs" for a one-way MANOVA.

11.2.2 ■ Number of Significant Discriminant Functions

Along how many dimensions do groups differ reliably? For the three groups of children in our example, two discriminant functions are possible, and neither, one, or both may be statistically reliable. For example, the first function may separate the normal group from the other two while the second, which would separate the group with learning disability from the group with emotional disorder, is not reliable. This pattern of results indicates the predictors can differentiate normal from abnormal kids, but cannot separate learning disabled kids from kids with emotional disorder.

In DISCRIM, like canonical correlation (Chapter 6), the first discriminant function provides the best separation among groups. Then a second discriminant function, orthogonal to the first, is found that best separates groups on the basis of associations not used in the first discriminant function. This procedure of finding successive orthogonal discriminant functions continues until all possible dimensions are evaluated. The number of possible dimensions is either one fewer than the number of groups or equal to the number of predictor variables, whichever is smaller. Typically, only the first one or two discriminant functions reliably discriminate among groups; remaining functions provide no additional information about group membership and are better ignored. Tests of significance for discriminant functions are discussed in Section 11.6.2.

11.2.3 ■ Dimensions of Discrimination

How can the dimensions along which groups are separated be interpreted? Where are groups located along the discriminant functions and how do predictors correlate with the discriminant functions? In our example, if two significant discriminant functions are found, which predictors correlate highly with each function? What pattern of test scores discriminates between normal children and the other two groups (first discriminant function)? And what pattern of scores discriminates between children with learning disability and children with emotional disorder (second discriminant function)? These questions are discussed in Section 11.6.3.

11.2.4 ■ Classification Functions

What linear equation(s) can be used to classify new cases into groups? For example, suppose we have the battery of psychological test scores for a group of new, undiagnosed children. How can we combine (weight) their scores to achieve the most reliable diagnosis? Procedures for deriving and using classification functions are discussed in Sections 11.4.2 and 11.6.6.[2]

11.2.5 ■ Adequacy of Classification

Given classification functions, what proportion of cases is correctly classified? When errors occur, how are cases misclassified? For instance, what proportion of learning-disabled children is correctly classified as learning-disabled, and, among those who are incorrectly classified, are they more often put into the group of normal children or into the group of emotionally disordered children?

Classification functions are used to predict group membership for new cases and to check the adequacy of classification for cases in the same sample through cross-validation. If the researcher knows that some groups are more likely to occur, or if some kinds of misclassification are especially undesirable, the classification procedure can be modified. Procedures for deriving classification functions and modifying them are discussed in Section 11.4.2; procedures for testing them are discussed in Section 11.6.6.

11.2.6 ■ Strength of Association

What is the degree of relationship between group membership and the set of predictors? If the first discriminant function separates the normal group from the other two groups, how much does the variance for groups overlap the variance in combined test scores? If the second discriminant function separates learning-disabled from emotionally disordered children, how much does the variance for these groups overlap the combined test scores for this discriminant function? This is basically a question of percent of variance accounted for and, as seen in Section 11.4.1, is answered through canonical correlation. A canonical correlation is found for each discriminant function that, when squared, indicates the proportion of variance shared between groups and predictors on that function.

11.2.7 ■ Importance of Predictor Variables

Which predictors are most important in predicting group membership? Which test scores are helpful for separating normal children from children with disorders, and which are helpful for separating learning-disabled from emotionally disordered children?

---

[2] Discriminant function analysis provides classification of cases into groups where group membership is known, at least for the sample from whom the classification equations are derived. Cluster analysis is a similar procedure except that group membership is not known. Instead, the analysis develops groups on the basis of similarities among cases.

Questions about importance of predictors are analogous to those of importance of DVs in MANOVA, to those of IVs in multiple regression, and to those of IVs and DVs in canonical correlation. One procedure in DISCRIM is to interpret the correlations between the predictors and the discriminant functions, as discussed in Section 11.6.3.2. A second procedure is to evaluate predictors by how well they separate each group from all the others, as discussed in Section 11.6.4. (Or importance can be evaluated as in MANOVA, Section 9.5.2.)

11.2.8 ■ Significance of Prediction with Covariates

After statistically removing the effects of one or more covariates, can one reliably predict group membership from a set of predictors? In DISCRIM, as in MANOVA, the ability of some predictors to promote group separation can be assessed after adjustment for prior variables. If scores on the Wechsler Intelligence Scale for Children (WISC) are considered the covariate and given first entry in DISCRIM, do scores on the Illinois Test of Psycholinguistic Ability (ITPA), the Wide Range Achievement Test, and Figure Drawings contribute to prediction of group membership when they are added to the equation?

Rephrased in terms of sequential discriminant function analysis, the question becomes, Do scores on the ITPA, the Wide Range Achievement Test, and Figure Drawings provide significantly better classification among the three groups than that afforded by scores on the WISC alone? Sequential DISCRIM is discussed in Section 11.5.2. A test for contribution of added predictors is given in Section 11.6.6.3.

11.2.9 ■ Estimation of Group Means

If predictors discriminate among groups, it is important to report just how the groups differ on those variables. The best estimate of central tendency in a population is the sample mean. If, for example, the ITPA discriminates between groups with learning disability and emotional disorder, it is worthwhile to compare and report the mean ITPA score for learning-disabled children and the mean ITPA score for emotionally disordered children.

11.3 ■ LIMITS TO DISCRIMINANT FUNCTION ANALYSIS

11.3.1 ■ Theoretical Issues

Because DISCRIM is typically used to predict membership in naturally occurring groups rather than groups formed by random assignment, questions such as why we can reliably predict group membership, or what causes differential membership are often not asked. If, however, group membership has occurred by random assignment, inferences of causality are justifiable as long as proper experimental controls have been instituted. The DISCRIM question then becomes, Does treatment following random assignment to groups produce enough difference in the predictors that we can now reliably separate groups on the basis of those variables?

As implied, limitations to DISCRIM are the same as limitations to MANOVA. The usual difficulties of generalizability apply to DISCRIM. But the cross-validation procedure described in Section 11.6.6.1 gives some indication of the generalizability of a solution.

11.3.2 ■ Practical Issues

Practical issues for DISCRIM are basically the same as for MANOVA. Therefore, they are discussed here only to the extent of identifying the similarities between MANOVA and DISCRIM and identifying the situations in which assumptions for MANOVA and DISCRIM differ.

Classification makes fewer statistical demands than does inference. If classification is the primary goal, then, most of the following requirements (except for outliers and homogeneity of variance-covariance matrices) are relaxed. If, for example, you achieve 95% accuracy in classification, you hardly worry about the shape of distributions. Nevertheless, DISCRIM is optimal under the same conditions where MANOVA is optimal; and, if the classification rate is unsatisfactory, it may be because of violation of assumptions or limitations.

11.3.2.1 ■ Unequal Sample Sizes and Missing Data

As DISCRIM is typically a one-way analysis, *no special problems are posed by unequal sample sizes in groups.*[3] In classification, however, a decision is required as to whether you want the a priori probabilities of assignment to groups to be influenced by sample size. That is, do you want the probability with which a case is assigned to a group to reflect the fact that the group itself is more (or less) probable in the sample? Section 11.4.2 discusses this issue, and use of unequal a priori probabilities is demonstrated in Section 11.8.

Regarding missing data (absence of scores on predictors for some cases), consult Section 8.3.2.1 and Chapter 4 for a review of problems and potential solutions.

As discussed in Section 9.3.2.1, *the sample size of the smallest group should exceed the number of predictor variables.* Although sequential and stepwise DISCRIM avoid the problems of multicollinearity and singularity by a tolerance test at each step, overfitting (producing results so close to the sample they don't generalize to other samples) occurs with all forms of DISCRIM if the number of cases does not notably exceed the number of predictors in the smallest group.

11.3.2.2 ■ Multivariate Normality

When using statistical inference in DISCRIM, the assumption of multivariate normality is that scores on predictors are independently and randomly sampled from a population, and that the sampling distribution of any linear combination of predictors is normally distributed. No tests are currently feasible for testing the normality of all linear combinations of sampling distributions of means of predictors.

However, DISCRIM, like MANOVA, is robust to failures of normality if violation is caused by skewness rather than outliers. *A sample size that would produce 20 df for error in the univariate ANOVA case should ensure robustness with respect to multivariate normality, as long as sample sizes are equal and two-tailed tests are used.* (Calculation of df for error in the univariate case is discussed in Section 3.2.1.)

Because tests for DISCRIM typically are two-tailed, this requirement poses no difficulty. Sample sizes, however, are often not equal for applications of DISCRIM because naturally occurring groups rarely occur or are sampled with equal numbers of cases in groups. As differences in sample size

---

[3] Actually a problem does occur because discriminant functions may be nonorthogonal with unequal *n* (cf. Chapter 13), but rotation of axes is uncommon in discriminant function analysis. Also, highly unequal sample sizes are better handled by logistic regression (Chapter 12) than by discriminant function analysis.

among groups increase, larger overall sample sizes are necessary to assure robustness. As a conservative recommendation, robustness is expected with 20 cases in the smallest group if there are only a few predictors (say, five or fewer).

If samples are both small and unequal in size, assessment of normality is a matter of judgment. Are predictors expected to have normal sampling distributions in the population being sampled? If not, transformation of one or more predictors (cf. Chapter 4) may be worthwhile.

11.3.2.3 ■ Outliers

DISCRIM, like MANOVA, is highly sensitive to inclusion of outliers. Therefore, *run a test for univariate and multivariate outliers for each group separately, and transform or eliminate significant outliers before DISCRIM* (see Chapter 4).

11.3.2.4 ■ Homogeneity of Variance-Covariance Matrices

In inference, when sample sizes are equal or large, DISCRIM, like MANOVA (Section 9.3.2.4) is robust to violation of the assumption of equality of within-group variance-covariance (dispersion) matrices. However, when sample sizes are unequal and small, results of significance testing may be misleading if there is heterogeneity of the variance-covariance matrices.

Although inference is usually robust with respect to heterogeneity of variance-covariance matrices with decently sized samples, classification is not. Cases tend to be overclassified into groups with greater dispersion. If classification is an important goal of analysis, test for homogeneity of variance-covariance matrices.

Homogeneity of variance-covariance matrices is assessed through procedures of Section 9.3.2.4, or by *inspection of scatterplots of scores on the first two canonical discriminant functions produced separately for each group*. These scatterplots are available through BMDP7M or SPSS DISCRIMINANT. *Rough equality in overall size of the scatterplots is evidence of homogeneity of variance-covariance matrices (cf. Section 11.8.1.5).*

Anderson's test, available in BMDP5M or SAS DISCRIM (POOL=TEST) assesses homogeneity of variance-covariance matrices, but is also sensitive to nonnormality. Another overly sensitive test, Box's *M*, is available in SPSS MANOVA and DISCRIMINANT.

If heterogeneity is found, one can transform predictors, use separate covariance matrices during classification, use quadratic discriminant function analysis, or use nonparametric classification. Transformation of predictors follows procedures of Chapter 4. Classification on the basis of separate covariance matrices, the second remedy, is available through SPSS DISCRIMINANT and SAS DISCRIM. Because this procedure often leads to overfitting, it should be used only when the sample is large enough to permit cross-validation (Section 11.6.6.1). Quadratic discrimination function analysis, the third remedy, is available in BMDP5M, SYSTAT DISCRIM, and SAS DISCRIM. This procedure avoids overclassification into groups with greater dispersion, but performs poorly with small samples (Norušis, 1990).

SAS DISCRIM uses separate matrices and computes quadratic discriminant functions with the instruction POOL=NO. With the instruction POOL=TEST, SAS DISCRIM uses the pooled variance-covariance matrix only if heterogeneity of variance-covariance matrices is not significant. With small samples, nonnormal predictors, and heterogeneity of variance-covariance matrices, SAS DISCRIM offers a fourth remedy, nonparametric classification methods, which avoid overclassification into groups with greater dispersion and are robust to nonnormality.

Therefore, *transform variables if there is significant departure from homogeneity, samples are small and unequal, and inference is the major goal. If the emphasis is on classification and dispersions are unequal, use (1) separate covariance matrices and/or quadratic discriminant analysis if samples are normal and large and (2) nonparametric classification methods if variables are non-normal and/or samples are small.*

11.3.2.5 ■ Linearity

The DISCRIM model assumes linear relationships among all pairs of predictors within each group. The assumption is less serious (from some points of view) than others, however, in that violation leads to reduced power rather than increased Type I error. The procedures in Section 8.3.2.6 may be applied to test for and improve linearity and to increase power.

11.3.2.6 ■ Multicollinearity and Singularity

Multicollinearity or singularity may occur with highly redundant predictors, making matrix inversion unreliable. Fortunately, most computer programs for DISCRIM protect against this possibility by testing tolerance. Predictors with insufficient tolerance are excluded.

Guidelines for assessing multicollinearity and singularity for programs that do not include tolerance tests, and for dealing with multicollinearity or singularity when it occurs, are in Section 9.3.2.8. Note that analysis is done on predictors, not "DVs" in DISCRIM.

11.4 ■ FUNDAMENTAL EQUATIONS FOR DISCRIMINANT FUNCTION ANALYSIS

Hypothetical scores on four predictors are given for three groups of learning-disabled children for demonstration of DISCRIM. Scores for three cases in each of the three groups are shown in Table 11.1.

The three groups are MEMORY (children whose major difficulty seems to be with tasks related to memory), PERCEPTION (children who show difficulty in visual perception), and COMMUNICATION (children with language difficulty). The four predictors are PERF (Performance Scale IQ of the WISC), INFO (Information subtest of the WISC), VERBEXP (Verbal Expression subtest of the ITPA), and AGE (chronological age in years). The grouping variable, then, is type of learning disability, and the predictors are selected scores from psychodiagnostic instruments and age.

Fundamental equations are presented for two major parts of DISCRIM: discriminant functions and classification equations. Setup and selected output for this example appear in Section 11.4.3 for BMDP7M, SPSS DISCRIMINANT, SAS DISCRIM, and SYSTAT DISCRIM.

11.4.1 ■ Derivation and Test of Discriminant Functions

The fundamental equations for testing the significance of a set of discriminant functions are the same as for MANOVA, discussed in Chapter 9. Variance in the set of predictors is partitioned into two sources: variance attributable to differences between groups and variance attributable to differences within groups. Through procedures shown in Equations 9.1 to 9.3, cross-products matrices are formed.

TABLE 11.1 HYPOTHETICAL SMALL DATA FOR ILLUSTRATION OF DISCRIMINANT FUNCTION ANALYSIS

Group	Predictors			
	PERF	INFO	VERBEXP	AGE
MEMORY	87	5	31	6.4
	97	7	36	8.3
	112	9	42	7.2
PERCEPTION	102	16	45	7.0
	85	10	38	7.6
	76	9	32	6.2
COMMUNICATION	120	12	30	8.4
	85	8	28	6.3
	99	9	27	8.2

$$\mathbf{S}_{\text{total}} = \mathbf{S}_{bg} + \mathbf{S}_{wg} \tag{11.1}$$

The total cross-products matrix ($\mathbf{S}_{\text{total}}$) is partitioned into a cross-products matrix associated with differences between groups ($\mathbf{S}_{bg}$) and a cross-products matrix of differences within groups ($\mathbf{S}_{wg}$).

For the example in Table 11.1, the resulting cross-products matrices are

$$\mathbf{S}_{bg} = \begin{bmatrix} 314.89 & -71.56 & -180.00 & 14.49 \\ -71.56 & 32.89 & 8.00 & -2.22 \\ -180.00 & 8.00 & 168.00 & -10.40 \\ 14.49 & -2.22 & -10.40 & 0.74 \end{bmatrix}$$

$$\mathbf{S}_{wg} = \begin{bmatrix} 1286.00 & 220.00 & 348.33 & 50.00 \\ 220.00 & 45.33 & 73.67 & 6.37 \\ 348.33 & 73.67 & 150.00 & 9.73 \\ 50.00 & 6.37 & 9.73 & 5.49 \end{bmatrix}$$

Determinants[4] for these matrices are

$$|\mathbf{S}_{wg}| = 4.70034789 \times 10^{13}$$

$$|\mathbf{S}_{bg} - \mathbf{S}_{wg}| = 448.63489 \times 10^{13}$$

Following procedures in Equation 9.4, Wilks' Lambda[5] for these matrices is

$$\Lambda = \frac{|\mathbf{S}_{wg}|}{|\mathbf{S}_{bg} + \mathbf{S}_{wg}|} = .010477$$

---

[4] A determinant, as described in Appendix A, can be viewed as a measure of generalized variance of a matrix.

[5] Alternative statistical criteria are discussed in Section 11.6.1.1.

To find the approximate F ratio, as per Equation 9.5, the following values are used:

$p = 4$ the number of predictor variables

$df_{effect} = 2$ the number of groups minus one, or $k - 1$

$df_{error} = 6$ the number of groups times the quantity $n - 1$, where n is the number of cases per group. Because n is often not equal for all groups in DISCRIM, an alternative equation for df_{error} is $N - k$, where N is the total number of cases in all groups—9 in this case.

Thus we obtain

$$s = \sqrt{\frac{(4)^2(2)^2 - 4}{2(4)^2 + (2)^2 - 5}} = 2$$

$$y = (.010477)^{\frac{1}{2}} = .102357$$

$$df_1 = 4(2) = 8$$

$$df_2 = (2)\left[6 - \frac{4 - 2 + 1}{2}\right] - \left[\frac{4(2) - 2}{2}\right] = 6$$

$$\text{Approximate } F(8, 6) = \left(\frac{1 - .102357}{.102357}\right)\left(\frac{6}{8}\right) = 6.58$$

Critical F with 8 and 6 df at $\alpha = 0.05$ is 4.15. Because obtained F exceeds critical F, we conclude that the three groups of children can be distinguished on the basis of the combination of the four predictors.

This is a test of overall relationship between groups and predictors. It is the same as the overall test of a main effect in MANOVA. In MANOVA, this result is followed by an assessment of the importance of the various DVs to the main effect. In DISCRIM, however, when an overall relationship is found between groups and predictors, the next step is to examine the discriminant functions that compose the overall relationship.

The maximum number of discriminant functions is either (1) the number of predictors or (2) the degrees of freedom for groups, whichever is smaller. Because there are three groups (and four predictors) in this example, there are potentially two discriminant functions contributing to the overall relationship. And, because the overall relationship is reliable, at least the first discriminant function is very likely to be reliable, and both may be reliable.

Discriminant functions are like regression equations; a discriminant function score for a case is predicted from the sum of the series of predictors, each weighted by a coefficient. There is one set of discriminant function coefficients for the first discriminant function, a second set of coefficients for the second discriminant function, and so forth. Subjects get separate discriminant function scores for each discriminant function when their own scores on predictors are inserted into the equations.

To solve for the (standardized) discriminant function score for the ith function, Equation 11.2 is used.

$$D_i = d_{i1}z_1 + d_{i2}z_2 + \ldots + d_{ip}z_p \tag{11.2}$$

A child's standardized score on the ith discriminant function (D_i) is found by multiplying the standardized score on each predictor (z) by its standardized discriminant function coefficient (d_i) and then adding the products for all predictors.

Discriminant function coefficients are found in the same manner as are coefficients for canonical variates (Section 6.4.2). In fact, DISCRIM is basically a problem in canonical correlation with group membership on one side of the equation and predictors on the other, where successive canonical variates (here called discriminant functions) are computed. In DISCRIM, d_i are chosen to maximize differences between groups relative to differences within groups.

Just as in multiple regression, Equation 11.2 can be written either for raw scores or for standardized scores. A discriminant function score for a case, then, can also be produced by multiplying the raw score on each predictor by its associated unstandardized discriminant function coefficient, adding the products over all predictors, and adding a constant to adjust for the means. The score produced in this way is the same D_i as produced in Equation 11.2.

The mean of each discriminant function over all cases is zero, because the mean of each predictor, when standardized, is zero. The standard deviation of each D_i is 1.

Just as D_i can be calculated for each case, a mean value of D_i can be calculated for each group. The members of each group considered together have a mean score on a discriminant function that is the distance of the group, in standard deviation units, from the zero mean of the discriminant function. Group means on D_i are typically called centroids in reduced space, the space having been reduced from that of the p predictors to a single dimension, or discriminant function.

A canonical correlation is found for each discriminant function following procedures in Chapter 6. Successive discriminant functions are evaluated for significance, as discussed in Section 11.6.2. Also discussed in subsequent sections are loading matrices and group centroids.

If there are only two groups, discriminant function scores can be used to classify cases into groups. A case is classified into one group if its D_i score is above zero, and into the other group if the D_i score is below zero. With numerous groups, classification is possible from the discriminant functions, but it is simpler to use the procedure in the following section.

11.4.2 ■ Classification

To assign cases into groups, a classification equation is developed for each group. Three classification equations are developed for the example in Table 11.1, where there are three groups. Data for each case are inserted into each classification equation to develop a classification score for each group for the case. The case is assigned to the group for which it has the highest classification score.

In its simplest form, the basic classification equation for the jth group ($j = 1, 2, \ldots, k$) is

$$C_j = c_{j0} + c_{j1}X_1 + c_{j2}X_2 + \ldots c_{jp}X_p \tag{11.3}$$

A score on the classification function for group j (C_j) is found by multiplying the raw score on each predictor (X) by its associated classification function coefficient (c_j), summing over all predictors, and adding a constant c_{j0}.

Classification coefficients, c_j, are found from the means of the p predictors and the pooled within-group variance-covariance matrix, $\mathbf{W}$. The within-group covariance matrix is produced by dividing each element in the cross-products matrix, $\mathbf{S}_{wg}$, by the within-group degrees of freedom, $N - k$. In matrix form,

$$\mathbf{C}_j = \mathbf{W}^{-1}\mathbf{M}_j \qquad (11.4)$$

The row matrix of classification coefficients for group j ($\mathbf{C}_j = c_{j1}, c_{j2}, \ldots, c_{jp}$) is found by multiplying the inverse of the within-group variance-covariance matrix ($\mathbf{W}^{-1}$) by a column matrix of means for group j on the p variables ($\mathbf{M}_j = \bar{X}_{j1}, \bar{X}_{j2}, \ldots \bar{X}_{jp}$).

The constant for group j, c_{j0}, is found as follows:

$$c_{j0} = (-\frac{1}{2})\mathbf{C}_j\mathbf{M}_j \qquad (11.5)$$

The constant for the classification function for group j (c_{j0}) is formed by multiplying the row matrix of classification coefficients for group j ($\mathbf{C}_j$) by the column matrix of means for group j ($\mathbf{M}_j$).

For the sample data, each element in the $\mathbf{S}_{wg}$ matrix from Section 11.4.1 is divided by $df_{wg} = df_{error} = 6$ to produce the within-group variance-covariance matrix:

$$\mathbf{W}_{bg} = \begin{bmatrix} 214.33 & 36.67 & 58.06 & 8.33 \\ 36.67 & 7.56 & 12.28 & 1.06 \\ 58.06 & 12.28 & 25.00 & 1.62 \\ 8.33 & 1.06 & 1.62 & 0.92 \end{bmatrix}$$

The inverse of the within-group variance-covariance matrix is

$$\mathbf{W}^{-1} = \begin{bmatrix} 0.04362 & -0.20195 & 0.00956 & -0.17990 \\ -0.21095 & 1.62970 & -0.37037 & 0.60623 \\ 0.00956 & -0.37037 & 0.20071 & -0.01299 \\ -0.17990 & 0.60623 & -0.01299 & 2.05006 \end{bmatrix}$$

Multiplying $\mathbf{W}^{-1}$ by the column matrix of means for the first group gives the matrix of classification coefficients for that group, as per Equation 11.4.

$$\mathbf{C}_1 = \mathbf{W}^{-1} \begin{bmatrix} 98.67 \\ 7.00 \\ 36.33 \\ 7.30 \end{bmatrix} = [1.92, -17.56, 5.55, 0.99]$$

The constant for group 1, then, according to Equation 11.5, is

$$c_{1,0} = (-\frac{1}{2}) [1.92, -17.56, 5.55, 0.99] \begin{bmatrix} 98.67 \\ 7.00 \\ 36.33 \\ 7.30 \end{bmatrix} = -137.83$$

(Values used in these calculations were carried to several decimal places before rounding.) When these procedures are repeated for groups 2 and 3, the full set of classification equations is produced, as shown in Table 11.2.

In its simplest form, classification proceeds as follows for the first case in group 1. Three classification scores, one for each group, are calculated for the case by applying Equation 11.3:

$$C_1 = -137.83 + (1.92)(87) + (-17.56)(5) + (5.55)(31) + (0.99)(6.4) = 119.80$$

$$C_2 = -71.29 + (0.59)(87) + (-8.70)(5) + (4.12)(31) + (5.02)(6.4) = 96.39$$

$$C_3 = -71.24 + (1.37)(87) + (-10.59)(5) + (2.97)(31) + (2.91)(6.4) = 105.69$$

Because this child has the highest classification score in group 1, the child is assigned to group 1, a correct classification in this case.

This simple classification scheme is most appropriate when equal group sizes are expected in the population. If unequal group sizes are expected, the classification procedure can be modified by setting a priori probabilities to group size. Although a number of highly sophisticated classification schemes have been suggested (e.g., Tatsuoka, 1975), the most straightforward involves adding to each classification equation a term that adjusts for group size.[6] The classification equation for group j (C_j) then becomes

$$C_j = c_{j0} + \sum_{i=1}^{p} c_{ji}X_i + \ln(n_j/N) \tag{11.6}$$

where n_j = size of group j and N = total sample size.

TABLE 11.2 CLASSIFICATION FUNCTION COEFFICIENTS FOR SAMPLE DATA OF TABLE 11.1

	Group 1: MEMORY	Group 2: PERCEP	Group 3: COMMUN
PERF	1.92420	0.58704	1.36552
INFO	-17.56221	-8.69921	-10.58700
VERBEXP	5.54585	4.11679	2.97278
AGE	0.98723	5.01749	2.91135
(CONSTANT)	-137.82892	-71.28563	-71.24188

---

[6] Output from computer programs reflects this adjustment.

It should be reemphasized that the classification procedures are highly sensitive to heterogeneity of variance-covariance matrices. Cases are more likely to be classified into the group with the greatest dispersion—that is, into the group for which the determinant of the within-group covariance matrix is greatest. Section 11.3.2.4 provides suggestions for dealing with this problem.

Uses of classification procedures are discussed more fully in Section 11.6.6.

11.4.3 ■ Computer Analyses of Small Sample Example

Setup and selected output for computer analyses of the data in Table 11.1, using the simplest methods, are in Tables 11.3 through 11.6, for BMDP7M, SPSS DISCRIMINANT, SAS DISCRIM, and SYSTAT DISCRIM, respectively. There is less consensus than usual in these programs regarding the statistics that are reported owing, in part, to the complexity of discriminant function analysis.

Table 11.3 shows a BMDP7M run with all but the most relevant output suppressed or omitted. Direct DISCRIM is specified in the LEVEL instruction. The first two zeros delete the first two predictors named in the VARIABLE paragraph from the problem. The remaining four predictors, with level 1, are to be used as predictors. FORCE=1 forces all predictors with level 1 into the equation. Because all predictors to be included have the same level, this is a direct rather than a stepwise or sequential analysis.

In the output, PRIOR PROBABILITIES of group membership are set equal by default. At STEP NUMBER 0 univariate ANOVAs for each predictor considered separately, with df = 2 and 6, are given; *F* ratios for each predictor appear in the column labeled F TO ENTER. TOLERNCE is 1-SMC for each predictor already in the equation. Because there are no predictors in the equation at this point, SMCs are zero. The NO STEP instruction suppresses printing of steps 2 and 3.

At STEP NUMBER 4 all predictors have entered the equation. F TO REMOVE (with 2 and 3 df) tests the significance of the reduction in prediction if a predictor is removed from the equation. Both Wilks' Lambda and the APPROXIMATE F-STATISTIC for the multivariate effect (cf. Sections 9.4.1 and 11.4.1) are provided as a test of the discriminant functions considered together. The F-MATRIX, with 4 and 3 df, shows multivariate pairwise comparisons among the three groups. For example, multivariate $F(4, 3) = 9.70$ for the difference between groups 1 and 2.

In the CLASSIFICATION FUNCTIONS section, the CLASSIFICATION MATRIX shows all cases to be perfectly classified. Rows represent actual group membership; columns show the group into which cases are classified, following procedures in Section 11.4.2. With JACKKNIFED CLASSIFICATION, in which classification equations are developed for each case with that case deleted (cf. Section 11.6.6.2), one case in the third group is misclassified into the second group, reducing the PERCENT CORRECT for group 3 to 66.7 and the overall percentage to 88.9.

The EIGENVALUES show the relative proportion of variance contributed by the two discriminant functions, followed by the CUMULATIVE PROPORTION OF TOTAL DISPERSION (variance) in the solution accounted for by the functions—the first discriminant function accounts for 70.699 percent of the variance in the solution while the second accounts for the remaining variance. This is followed by the results of four multivariate tests, described in Section 9.5.1.

The remaining output describes the two discriminant functions (CANONICAL VARIABLES) produced for the three groups. CANONICAL CORRELATIONS are the multiple correlations between the predictors and the discriminant functions. COEFFICIENTS FOR CANONICAL VARIABLES are weighting coefficients for canonical correlation as described in Chapter 6, and STANDARDIZED . . . . COEFFICIENTS FOR CANONICAL VARIABLES are raw score discriminant function coefficients computed using the pooled within-cell variance-covariance matrix.

TABLE 11.3   SETUP AND SELECTED BMDP7M OUTPUT FOR DIRECT DISCRIMINANT FUNCTION ANALYSIS ON
              SAMPLE DATA IN TABLE 11.1

```
/INPUT      VARIABLES ARE 6.  FORMAT IS FREE.  FILE = 'TAPE41.DAT'.
/VARIABLE   NAMES ARE SUBJNO,GROUP,PERF,INFO,VERBEXP,AGE.
            LABEL IS SUBJNO.
/GROUP      VAR IS GROUP.
            CODES(GROUP) ARE 1, 2, 3.
/DISCRIM    LEVEL=2*0, 4*1.  FORCE=1.
/PRINT      NO STEP.
/END

PRIOR PROBABILITIES. . . .    0.33333    0.33333    0.33333

*****************************************************************************

STEP NUMBER    0

   VARIABLE     F TO   FORCE TOLERNCE *   VARIABLE     F TO   FORCE TOLERNCE
                REMOVE LEVEL           *                ENTER  LEVEL
        DF =  2    7                   *       DF =  2    6
                                       *    3 PERF        0.73    1   1.00000
                                       *    4 INFO        2.18    1   1.00000
                                       *    5 VERBEXP     3.36    1   1.00000
                                       *    6 AGE         0.40    1   1.00000


*****************************************************************************

STEP NUMBER    4
VARIABLE ENTERED   6 AGE

   VARIABLE     F TO   FORCE TOLERNCE *   VARIABLE     F TO   FORCE TOLERNCE
                REMOVE LEVEL           *                ENTER  LEVEL
        DF =  2    3                   *       DF =  2    2
    3 PERF        6.89    1   0.10696  *
    4 INFO       18.84    1   0.08121  *
    5 VERBEXP     9.74    1   0.19930  *
    6 AGE         0.27    1   0.53278  *

U-STATISTIC(WILKS' LAMBDA) 0.0104766   DEGREES OF FREEDOM    4    2       6
APPROXIMATE F-STATISTIC        6.577   DEGREES OF FREEDOM    8.00     6.00

  F - MATRIX        DEGREES OF FREEDOM =     4    3

              *1        *2
 *2         9.70
 *3         7.19       4.57

CLASSIFICATION FUNCTIONS

        GROUP =    *1            *2            *3
VARIABLE
   3 PERF          1.92420       0.58704       1.36552
   4 INFO        -17.56221      -8.69921     -10.58700
   5 VERBEXP       5.54585       4.11679       2.97278
   6 AGE           0.98723       5.01749       2.91135

CONSTANT       -138.91107      -72.38436     -72.34032

CLASSIFICATION MATRIX

GROUP     PERCENT   NUMBER OF CASES CLASSIFIED INTO GROUP
          CORRECT
                     *1      *2      *3
 *1       100.0       3       0       0
 *2       100.0       0       3       0
 *3       100.0       0       0       3
TOTAL     100.0       3       3       3
```

TABLE 11.3 *(CONTINUED)*

JACKKNIFED CLASSIFICATION

GROUP	PERCENT CORRECT	NUMBER OF CASES CLASSIFIED INTO GROUP		
		*1	*2	*3
*1	100.0	3	0	0
*2	100.0	0	3	0
*3	66.7	0	1	2
TOTAL	88.9	3	4	2

EIGENVALUES

　　　　　13.48590　　　5.58923

CUMULATIVE PROPORTION OF TOTAL DISPERSION

　　　　　　0.70699　　　1.00000

MULTIVARIATE TESTS

STATISTICS	VALUE	F	F APPROXIMATION		
			D.F.	D.F.	PVALUE
WILKS' LAMBDA	0.01048	6.57742	8	6.00	0.0169
D.F.　4　2　6					
PILLAI'S TRACE	1.77920	8.05816	8	8	0.0039
HOTELLING-LAWLEY TRACE	19.07512	APPROXIMATION NOT CALCULABLE			
ROY'S MAXIMUM ROOT	13.48590				

　　TABLES ARE AVAILABLE FOR ROY'S TEST:　MORRISON (1976)

CANONICAL CORRELATIONS

　　　　　　0.96487　　　0.92100

AVERAGE SQUARED CANONICAL CORRELATION　　　　　0.88960

VARIABLE	COEFFICIENTS FOR CANONICAL VARIABLES	
3 PERF	0.17100	-0.10069
4 INFO	-1.26953	-0.10325
5 VERBEXP	0.26493	0.35776
6 AGE	-0.52541	0.24690

VARIABLE	STANDARDIZED (BY POOLED WITHINGROUP VARIANCES) COEFFICIENTS FOR CANONICAL VARIABLES	
3 PERF	2.50352	-1.47406
4 INFO	-3.48961	-0.28380
5 VERBEXP	1.32466	1.78881
6 AGE	-0.50273	0.23625
CONSTANT	-9.67374	-3.45293

GROUP	CANONICAL VARIABLES EVALUATED AT GROUP MEANS	
*1	4.10234	0.69097
*2	-2.98068	1.94169
*3	-1.12166	-2.63265

TABLE 11.4 SETUP AND SELECTED SPSS DISCRIMINANT OUTPUT FOR DISCRIMINANT FUNCTION ANALYSIS OF SAMPLE DATA IN TABLE 11.1

```
DATA LIST        FILE='TAPE41.DAT'    FREE
                 /SUBJNO GROUP PERF INFO VERBEXP AGE.
DISCRIMINANT     GROUPS = GROUP(1,3)/
                 VARIABLES=PERF TO AGE/
                 ANALYSIS=PERF TO AGE/
                 METHOD = DIRECT/
                 STATISTICS = TABLE.
```

Prior probability for each group is .33333

                    Canonical Discriminant Functions

		Pct of	Cum	Canonical	After	Wilks'			
Fcn	Eigenvalue	Variance	Pct	Corr	Fcn	Lambda	Chisquare	DF	Sig
					: 0	.0105	20.514	8	.0086
1*	13.4859	70.70	70.70	.9649	: 1	.1518	8.484	3	.0370
2*	5.5892	29.30	100.00	.9210	:				

  * marks the   2 canonical discriminant functions remaining in the analysis.

Standardized Canonical Discriminant Function Coefficients

	FUNC 1	FUNC 2
PERF	-2.50352	-1.47406
INFO	3.48961	-.28380
VERBEXP	-1.32466	1.78881
AGE	.50273	.23625

Structure Matrix:

Pooled-withingroups correlations between discriminating variables
                              and canonical discriminant functions
(Variables ordered by size of correlation within function)

	FUNC 1	FUNC 2
INFO	.22796*	.06642
VERBEXP	-.02233	.44630*
PERF	-.07546	-.17341*
AGE	-.02786	-.14861*

Canonical Discriminant Functions evaluated at Group Means (Group Centroids)

Group	FUNC 1	FUNC 2
1	-4.10234	.69097
2	2.98068	1.94169
3	1.12166	-2.63265

Classification Results -

	No. of	Predicted Group Membership		
Actual Group	Cases	1	2	3
Group 1	3	3	0	0
		100.0%	.0%	.0%
Group 2	3	0	3	0
		.0%	100.0%	.0%
Group 3	3	0	0	3
		.0%	.0%	100.0%

Percent of "grouped" cases correctly classified: 100.00%

Classification Processing Summary
        9 Cases were processed.
        0 Cases were excluded for missing or outofrange group codes.
        0 Cases had at least one missing discriminating variable.
        9 Cases were used for printed output.

```
DATA SSAMPLE;
INFILE 'TAPE41.DAT';
INPUT SUBJNO GROUP PERF INFO VERBEXP AGE;
PROC DISCRIM;
    CLASS GROUP;
    VAR PERF INFO VERBEXP AGE;
```

Discriminant Analysis

9 Observations	8 DF Total	
4 Variables	6 DF Within Classes	
3 Classes	2 DF Between Classes	

Class Level Information

GROUP	Frequency	Weight	Proportion	Prior Probability
1	3	3.0000	0.333333	0.333333
2	3	3.0000	0.333333	0.333333
3	3	3.0000	0.333333	0.333333

Discriminant Analysis   Pooled Covariance Matrix Information

Covariance Matrix Rank	Natural Log of the Determinant of the Covariance Matrix
4	6.50509421

Discriminant Analysis   Pairwise Generalized Squared Distances Between Groups

$$D^2(i|j) = (\bar{X}_i - \bar{X}_j)' \, COV^{-1} \, (\bar{X}_i - \bar{X}_j)$$

Generalized Squared Distance to GROUP

From GROUP	1	2	3
1	0	51.73351	38.33673
2	51.73351	0	24.38053
3	38.33673	24.38053	0

Discriminant Analysis   Linear Discriminant Function

$$\text{Constant} = -.5 \, \bar{X}_j' \, COV^{-1} \, \bar{X}_j \qquad \text{Coefficient Vector} = COV^{-1} \, \bar{X}_j$$

GROUP

	1	2	3
CONSTANT	-137.81247	-71.28575	-71.24170
PERF	1.92420	0.58704	1.36552
INFO	-17.56221	-8.69921	-10.58700
VERBEXP	5.54585	4.11679	2.97278
AGE	0.98723	5.01749	2.91135

TABLE 11.5   *(CONTINUED)*

Discriminant Analysis   Classification Summary for Calibration Data: WORK.SSAMPLE

Resubstitution Summary using Linear Discriminant Function

Generalized Squared Distance Function: Posterior Probability of Membership in each GROUP:

$$D_j^2(X) = (X-\bar{X}_j)' \, COV^{-1} \, (X-\bar{X}_j) \qquad\qquad Pr(j|X) = exp(-.5 \, D_j^2(X)) \, / \, SUM_k \, exp(-.5 \, D_k^2(X))$$

Number of Observations and Percent Classified into GROUP:

From GROUP	1	2	3	Total
1	3	0	0	3
	100.00	0.00	0.00	100.00
2	0	3	0	3
	0.00	100.00	0.00	100.00
3	0	0	3	3
	0.00	0.00	100.00	100.00
Total	3	3	3	9
Percent	33.33	33.33	33.33	100.00
Priors	0.3333	0.3333	0.3333	

Error Count Estimates for GROUP:

	1	2	3	Total
Rate	0.0000	0.0000	0.0000	0.0000
Priors	0.3333	0.3333	0.3333	

(Note that the CONSTANT values are for the standardized equations.) Finally, the section labeled CANONICAL VARIABLES EVALUATED AT GROUP MEANS shows the group centroids, average discriminant function score for each group for each function.

SPSS DISCRIMINANT[7] (Table 11.4) also assigns equal prior probability for each group by default. The output summarizing the Canonical Discriminant Functions appears in two parallel tables separated by colons(:). At the left are shown Eigenvalue, Pct of Variance and Cum Pct of variance accounted for by each function, and Canonical Corr for each discriminant function. At the right are the "peel off" significance tests of successive discriminant functions. When After Fcn is 0, (Chisquare = 20.514) all discriminant functions are tested together (0 removed). After Fcn 1 is removed, Chisquare is still statistically significant at $\alpha = .05$ because Sig = .0370. This means that the second discriminant function is significant as well as the first. If not, the second would not have been marked as one of the discriminated functions remaining in the analysis.

---

[7] Note that in the PC+ implementation of SPSS, the keyword is DSCRIMINANT (drop the first I).

TABLE 11.6   SETUP AND SELECTED SYSTAT DISCRIM OUTPUT FOR DISCRIMINANT FUNCTION ANALYSIS OF SMALL SAMPLE DATA IN TABLE 11.1

```
USE TAPE41
OUTPUT SSDISC.OUT
MODEL  GROUP = PERF INFO VERBEXP AGE
PRINT MEDIUM
ESTIMATE
```

Group frequencies

	1	2	3
Frequencies	3	3	3

Group means

	1	2	3
PERF	98.667	87.667	101.333
INFO	7.000	11.667	9.667
VERBEXP	36.333	38.333	28.333
AGE	7.300	6.933	7.633

Between groups Fmatrix - df =   4   3

	1	2	3
1	0.0		
2	9.700	0.0	
3	7.188	4.571	0.0

Wilk's lambda
    Lambda =   0.0105   df =   4   2   6
    Approx. F= 6.5774   df =   8   6   prob = 0.0169

Classification functions

	1	2	3
Constant	-138.911	-72.384	-72.340
PERF	1.924	0.587	1.366
INFO	-17.562	-8.699	-10.587
VERBEXP	5.546	4.117	2.973
AGE	0.987	5.017	2.911

Variable	F-to-remove	Tolerance	Variable	F-to-enter	Tolerance
3 PERF	6.89	0.106964			
4 INFO	18.84	0.081213			
5 VERBEXP	9.74	0.199297			
6 AGE	0.27	0.532781			

Classification matrix (cases in row categories classified into columns)

	1	2	3	%correct
1	3	0	0	100
2	0	3	0	100
3	0	0	3	100
Total	3	3	3	100

Jacknifed classification matrix

	1	2	3	%correct
1	3	0	0	100
2	0	3	0	100
3	0	1	2	67
Total	3	4	2	89

TABLE 11.6   *(CONTINUED)*

Eigen values	Canonical correlations	Cumulative proportion of total dispersion
13.486	0.965	0.707
5.589	0.921	1.000

```
        Wilk's lambda=      0.010
             Approx.F=      6.577  DF=    8,       6  p-tail=  0.0169

      Pillai's trace=       1.779
             Approx.F=      8.058  DF=    8,       8  p-tail=  0.0039

Lawley-Hotelling trace=    19.075
             Approx.F=      4.769  DF=    8,       4  p-tail=  0.0742
```

Canonical discriminant functions

	1	2
Constant	-9.674	3.453
PERF	0.171	0.101
INFO	-1.270	0.103
VERBEXP	0.265	-0.358
AGE	-0.525	-0.247

Canonical discriminant functions -- standardized by within variances
----

	1	2
PERF	2.504	1.474
INFO	-3.490	0.284
VERBEXP	1.325	-1.789
AGE	-0.503	-0.236

Canonical scores of group means

1	4.102	-.691
2	-2.981	-1.942
3	-1.122	2.633

Standardized Canonical Discriminant Function Coefficients (Equation 11.2) are given for deriving discriminant function scores from standardized predictors. Correlations (loadings) between predictors and discriminant functions are given in the Structure Matrix. These are ordered so that predictors loading on the first discriminant function are listed first, and those loading on the second discriminant function second. Then Group Centroids are shown, indicating the average discriminant score for each group on each function.

In the Classification Results table, produced by the TABLE instruction in the STATISTICS paragraph (STATISTICS 13 is used for the PC version), rows represent actual group membership and columns represent predicted group membership. Within each cell, the number and percent of cases correctly classified are shown. For this example, all of the diagonal cells show 100% correction classification (100.0%).

SAS DISCRIM output (Table 11.5) begins with a summary of input and degrees of freedom, followed by a summary of GROUPs, their Frequency (number of cases), Weight (sample sizes in this case), Proportion of cases in each group, and Prior Probability (set equal by default). Information is then provided about the Pooled Covariance Matrix, which can signal problems in multicollinearity/singularity if the rank of the matrix is not equal to the number of predictors. An equation is provided by which distances between groups are derived and then presented in the Mahalanobis D^2 matrix labeled Pairwise Generalized Squared Distances Between Groups. For example the greatest distance is between groups 1 and 2 (51.73351) while the smallest distance is between groups 2 and 3 (24.38053). Equations for classification functions and classification coefficients (Equation 11.6) are given in the following matrix, labeled Linear Discriminant Function. Finally, results of classification are presented in the table labeled Number of Observations and Percent Classified into GROUP where, as usual, rows represent actual group and columns represent predicted group. Cell values show number of cases classified and percentage correct. Number of erroneous classifications for each group are presented, and Prior probabilities are repeated at the bottom of this table.

SYSTAT DISCRIM (Table 11.6), after requesting PRINT MEDIUM, begins with the frequencies and the means for each of the predictors. The Between groups Fmatrix then shows the F-ratio for the multivariate test comparing each group with each other group. Wilks' Lambda is then provided, followed by Classification functions for classifying new cases into groups. This is followed by the F-to-remove, showing the consequences of dropping each predictor, by itself, from the prediction equation.

The next two matrices show classification results, with and without jackknifing, followed by Eigenvalues (relative proportion of variance contributed by each predictor), Canonical correlations (multiple correlations between predictors and groups), and Cumulative proportion of total dispersion in the solution accounted for by the functions, as per BMDP7M. Wilks' Lambda is then repeated, along with Pillai's trace and Lawley-Hotelling trace, as alternative multivariate criteria for evaluated significance of prediction of groups by predictor variables. Canonical discriminant functions (raw discriminant function coefficients) are then shown, followed by coefficients standardized by pooled within-groups variances. The Canonical scores of group means are the centroids for the three groups on the two discriminant functions. Plots (not shown) of discriminant functions, with centroids and cases for all groups are then produced by SYSTAT DISCRIM.

11.5 ■ TYPES OF DISCRIMINANT FUNCTION ANALYSIS

The three types of discriminant function analysis—standard (direct), sequential, and statistical (stepwise)—are analogous to the three types of multiple regression discussed in Section 5.5. Criteria for choosing among the three strategies are the same as those discussed in Section 5.5.4 for multiple regression.

11.5.1 ■ Direct Discriminant Function Analysis

In standard (direct) DISCRIM, like standard multiple regression, all predictors enter the equations at once and each predictor is assigned only the unique association it has with groups. Variance shared among predictors contributes to the total relationship, but not to any one predictor.

The overall test of relationship between predictors and groups in direct DISCRIM is the same as the test of main effect in MANOVA where all discriminant functions are combined and DVs are considered simultaneously. Direct DISCRIM is the model demonstrated in Section 11.4.1.

All the computer programs described in Table 11.11 perform direct DISCRIM; the use of some of them for that purpose is shown in Tables 11.3 through 11.6.

11.5.2 ■ Sequential Discriminant Function Analysis

Sequential (or, as some prefer to call it, hierarchical) DISCRIM is used to evaluate contributions to prediction of group membership by predictors as they enter the equations in an order determined by the researcher. The researcher assesses improvement in classification when a new predictor is added to a set of prior predictors. Does classification of cases into groups improve reliably when the new predictor or predictors are added (cf. Section 11.6.6.3)?

If predictors with early entry are viewed as covariates and an added predictor is viewed as a DV, DISCRIM is used for analysis of covariance. Indeed, sequential DISCRIM can be used to perform stepdown analysis following MANOVA (cf. Section 9.5.2.2) because stepdown analysis is a sequence of ANCOVAs.

Sequential DISCRIM also is useful when a reduced set of predictors is desired and there is some basis for establishing a *priority* order among them. If, for example, some predictors are easy or inexpensive to obtain and they are given early entry, a useful, cost-effective set of predictors may be found through the sequential procedure.

Sequential DISCRIM through BMDP7M for the data of Section 11.4 is shown in Table 11.7. Age is given first priority because it is one of the easiest pieces of information to obtain about a child, and, furthermore, may be useful as a covariate. The two WISC scores are given the same priority 2 entry level and they compete with each other in a stepwise fashion for entry (see Section 11.5.3). Third priority is assigned the VERBEXP score because it is the least likely to be available without special testing.

Significance of discrimination is assessed at each step, and classification functions are provided. Adequacy of classification is shown automatically at the last step but can be requested after any step.

Classification information in SPSS DISCRIMINANT is provided only at the last step, however other useful information, unavailable in other programs, is given at each step. Particularly handy for stepdown analysis is the "change in Rao's V" at every step, as discussed in Section 11.6.1.1.[8]

In SPSS, sequential DISCRIM differs from sequential multiple regression in the way predictors with the same priority enter. In DISCRIM, predictors with the same priority compete with one another stepwise, and only one predictor enters at each step. In multiple regression, several IVs can enter in a single step. Further, the test for the significance of improvement in prediction is tedious in the absence of very large samples (Section 11.6.6.3). This makes DISCRIM less flexible than multiple regression. If you have only two groups and sample sizes are approximately equal, you might consider performing DISCRIM through regression where the DV is a dichotomous variable representing group membership, with groups coded 0 and 1. If classification is desired, preliminary multiple regression analysis with fully flexible entry of predictors could be followed by DISCRIM to provide classification.

In BMDP, the less flexible method (where predictors with the same priority compete for entry stepwise and only one enters at a time) is used for sequential DISCRIM. SYSTAT DISCRIM provides

---

[8] In using DISCRIM for stepdown analysis following MANOVA, change in Rao's *V* can replace stepdown *F* as the criterion for evaluating significance of successive DVs. Rao's *V* statistics are available through an option in SPSS DISCRIMINANT.

TABLE 11.7	SETUP AND SELECTED BMDP7M OUTPUT FOR SEQUENTIAL DISCRIMINANT FUNCTION ANALYSIS OF SAMPLE DATA IN TABLE 11.1

```
/INPUT    VARIABLES ARE 6.  FORMAT IS FREE.  FILE = 'TAPE41.DAT'.
/VARIABLE  NAMES ARE SUBJNO,GROUP,PERF,INFO,VERBEXP,AGE.
           LABEL IS SUBJNO.
/GROUP    VAR IS GROUP.
          CODES(GROUP) ARE 1, 2, 3.
          NAMES(GROUP) ARE MEMORY, PERCEPT, COMMUN.
/DISCRIM   LEVEL=2*0, 2, 2, 3, 1.  FORCE=3.
/END
```

****

```
STEP NUMBER   1
VARIABLE ENTERED   6 AGE

  VARIABLE       F TO   FORCE TOLERNCE  *   VARIABLE      F TO    FORCE TOLERNCE
                REMOVE  LEVEL           *                ENTER   LEVEL
        DF =  2     6                   *          DF =  2    5
  6 AGE          0.40     1  1.00000    *  3 PERF          0.31    2   0.64611
                                        *  4 INFO          2.47    2   0.83723
                                        *  5 VERBEXP       3.49    3   0.88503

U-STATISTIC(WILKS' LAMBDA) 0.8819122   DEGREES OF FREEDOM   1   2      6
APPROXIMATE F-STATISTIC        0.402   DEGREES OF FREEDOM   2.00      6.00

 F - MATRIX        DEGREES OF FREEDOM =    1     6

           MEMORY    PERCEPT
PERCEPT     0.22
COMMUN      0.18      0.80

CLASSIFICATION FUNCTIONS

        GROUP =  MEMORY       PERCEPT       COMMUN
VARIABLE
  6 AGE          7.97330      7.57282       8.33738

CONSTANT       -30.20117    -27.35104     -32.91961
```

****

```
STEP NUMBER   2
VARIABLE ENTERED   4 INFO

  VARIABLE       F TO   FORCE TOLERNCE  *   VARIABLE      F TO    FORCE TOLERNCE
                REMOVE  LEVEL           *                ENTER   LEVEL
        DF =  2     5                   *          DF =  2    4
  4 INFO         2.47     2  0.83723    *  3 PERF          9.30    2   0.10809
  6 AGE          0.77     1  0.83723    *  5 VERBEXP      13.13    3   0.20140

U-STATISTIC(WILKS' LAMBDA) 0.4435523   DEGREES OF FREEDOM   2   2      6
APPROXIMATE F-STATISTIC        1.254   DEGREES OF FREEDOM   4.00     10.00

 F - MATRIX        DEGREES OF FREEDOM =    2     5

           MEMORY    PERCEPT
PERCEPT     2.65
COMMUN      0.59      1.12

CLASSIFICATION FUNCTIONS

        GROUP =  MEMORY       PERCEPT       COMMUN
VARIABLE
  4 INFO        -0.23089      0.57401       0.12959
  6 AGE          8.24090      6.90755       8.18718

CONSTANT       -30.36978    -28.39318     -32.97273
```

TABLE 11.7   *(CONTINUED)*

```
**************************************************************************

STEP NUMBER    3
VARIABLE ENTERED    3 PERF

   VARIABLE     F TO    FORCE TOLERNCE *   VARIABLE      F TO    FORCE TOLERNCE
                REMOVE  LEVEL          *                ENTER    LEVEL
        DF =  2    4                   *        DF =  2    3
    3 PERF        9.30     2   0.10809 *   5 VERBEXP     9.74     3    0.19930
    4 INFO       18.01     2   0.14007 *
    6 AGE         0.44     1   0.53300 *

U-STATISTIC(WILKS' LAMBDA) 0.0784812   DEGREES OF FREEDOM    3    2    6
APPROXIMATE F-STATISTIC        3.426   DEGREES OF FREEDOM    6.00    8.00

  F - MATRIX          DEGREES OF FREEDOM =    3    4

           MEMORY   PERCEPT
PERCEPT    13.85
COMMUN      1.78     5.95

CLASSIFICATION FUNCTIONS

        GROUP =   MEMORY       PERCEPT       COMMUN
VARIABLE
   3 PERF          1.65998      0.39091       1.22389
   4 INFO         -7.31838     -1.09502      -5.09593
   6 AGE           1.34609      5.28388       3.10372

CONSTANT         -62.29002    -30.16335     -50.32441

**************************************************************************

STEP NUMBER    4
VARIABLE ENTERED    5 VERBEXP

   VARIABLE     F TO    FORCE TOLERNCE *   VARIABLE      F TO    FORCE TOLERNCE
                REMOVE  LEVEL          *                ENTER    LEVEL
        DF =  2    3                   *        DF =  2    2
    3 PERF        6.89     2   0.10696 *
    4 INFO       18.84     2   0.08121 *
    5 VERBEXP     9.74     3   0.19930 *
    6 AGE         0.27     1   0.53278 *

U-STATISTIC(WILKS' LAMBDA) 0.0104766   DEGREES OF FREEDOM    4    2    6
APPROXIMATE F-STATISTIC        6.577   DEGREES OF FREEDOM    8.00    6.00

  F - MATRIX          DEGREES OF FREEDOM =    4    3

           MEMORY   PERCEPT
PERCEPT     9.70
COMMUN      7.19     4.57

CLASSIFICATION FUNCTIONS

        GROUP =   MEMORY       PERCEPT       COMMUN
VARIABLE
   3 PERF          1.92420      0.58704       1.36552
   4 INFO        -17.56221     -8.69921     -10.58700
   5 VERBEXP       5.54585      4.11679       2.97278
   6 AGE           0.98723      5.01749       2.91135

CONSTANT        -138.91107    -72.38436     -72.34032
```

TABLE 11.7   *(CONTINUED)*

```
CLASSIFICATION MATRIX
GROUP        PERCENT     NUMBER OF CASES CLASSIFIED INTO GROUP -
             CORRECT
                         MEMORY    PERCEPT   COMMUN
   MEMORY    100.0         3          0        0
   PERCEPT   100.0         0          3        0
   COMMUN    100.0         0          0        3

TOTAL        100.0         3          3        3

JACKKNIFED CLASSIFICATION

GROUP        PERCENT     NUMBER OF CASES CLASSIFIED INTO GROUP -
             CORRECT
                         MEMORY    PERCEPT   COMMUN
   MEMORY    100.0         3          0        0
   PERCEPT   100.0         0          3        0
   COMMUN     66.7         0          1        2

TOTAL         88.9         3          4        2

   SUMMARY TABLE
```

STEP NO.	VARIABLE ENTERED REMOVED	F VALUE TO ENTER REMOVE	NO. OF VARIAB. INCLUDED	U-STATISTIC	APPROXIMATE F-STATISTIC	DEGREES OF FREEDOM	
1	6 AGE	0.402	1	0.8819	0.402	2.0	6.0
2	4 INFO	2.471	2	0.4436	1.254	4.0	10.0
3	3 PERF	9.303	3	0.0785	3.426	6.0	8.0
4	5 VERBEXP	9.737	4	0.0105	6.577	8.0	6.0

interactive entry of variables for sequential analysis. One or more variables are entered at each interactive step. Sequential DISCRIM is accomplished through a series of runs in SAS DISCRIM.

If you have more than two groups or your group sizes are very unequal, sequential logistic regression is the procedure of choice, and, as seen in Chapter 12, most programs classify cases.

11.5.3 ■ Stepwise (Statistical) Discriminant Function Analysis

When the researcher has no reasons for assigning some predictors higher priority than others, statistical criteria can be used to determine order of entry. That is, if a researcher wants a reduced set of predictors but has no preferences among them, stepwise DISCRIM can be used to produce the reduced set. Entry of predictors is determined by user-specified statistical criteria, of which several are available as discussed in Section 11.6.1.2.

Stepwise DISCRIM has the same controversial aspects as stepwise procedures in general (see Section 5.5.3). Order of entry may be dependent on trivial differences in relationships among predictors in the sample that do not reflect population differences. However, this bias is reduced if cross-validation is used (cf. Section 11.8). Costanza and Afifi (1979) recommend a probability to enter criterion more liberal than .05. They suggest a choice in the range of .15 to .20 to ensure entry of important variables.

Application of stepwise analysis to our small sample example through SPSS DISCRIMINANT is illustrated in Table 11.8. One of five statistical criteria for entry of predictors was selected for this example. Notice that AGE is dropped as a predictor by the statistical criterion and the analysis has only three steps. Compared with output in Table 11.7 where AGE is included, F for discrimination has improved ($F = 6.58$ with AGE included, $F = 10.67$ without AGE). That is, AGE is worthless as a predictor in this example; its inclusion only decreases within-groups degrees of freedom while not increasing between-groups sums of squares. If this result replicates, AGE should be dropped from consideration when classifying future cases.

Progression of the stepwise analysis is summarized in the Summary Table section of Table 11.8. The summary table is followed by discriminant function coefficients, the loading matrix, and group centroids. Moreover, a great deal of classification information is available, not shown in this segment of the output but discussed in Section 11.6.6.

Stepwise as well as sequential DISCRIM are produced by BMDP7M. Format of output for stepwise analysis is identical to that of sequential analysis, as shown in Table 11.7. The user can select among four statistical criteria for entry of predictors. In SAS, stepwise discriminant analysis is provided through a separate program—STEPDISC. Three entry methods are available (cf. Section 11.6.1.2), as well as additional statistical criteria for two of them. SYSTAT provides forward and backward stepwise DISCRIM, with statistical criteria for entry and removal.

11.6 ■ SOME IMPORTANT ISSUES

11.6.1 ■ Statistical Inference

Section 11.6.1.1 contains a discussion of criteria for evaluating the overall statistical significance of a set of predictors for predicting group membership. Section 11.6.1.2 summarizes methods for directing the progression of stepwise DISCRIM and statistical criteria for entry of predictors.

11.6.1.1 ■ Criteria for Overall Statistical Significance

Criteria for evaluating overall statistical reliability in DISCRIM are the same as those in MANO-VA. The choice between Wilks' Lambda, Roy's gcr, Hotelling's trace, and Pillai's criterion is based on the same considerations as discussed in Section 9.5.1. Different statistics are available in different programs, as noted in Section 11.7.

Two additional statistical criteria, Mahalanobis' D^2 and Rao's V, are especially relevant to stepwise DISCRIM. Mahalanobis' D^2 is based on distance between pairs of group centroids which is then generalizable to distances over multiple pairs of groups. Rao's V is another generalized distance measure that attains its largest value when there is greatest overall separation among groups.

These two criteria are available both to direct the progression of stepwise DISCRIM and to evaluate the reliability of a set of predictors to predict group membership. Like Wilks' Lambda, Mahalanobis' D^2 and Rao's V are based on all discriminant functions rather than one. Note that Lambda, D^2, and V are descriptive statistics; they are not, themselves, inferential statistics, although inferential statistics are applied to them.

11.6.1.2 ■ Stepping Methods

Related to criteria for statistical inference is the choice among methods to direct the progression of entry of predictors in stepwise DISCRIM. Different methods of progression maximize group differences

```
DATA LIST            FILE='TAPE41.DAT'    FREE
                     /SUBJNO TYPE PERF INFO VERBEXP AGE.
DISCRIMINANT         GROUPS = TYPE (1,3)/
                     VARIABLES=PERF TO AGE/
                     ANALYSIS=PERF TO AGE/
                     METHOD = WILKS.
```

Prior probability for each group is .33333
---------------- Variables not in the analysis after step   0 ----------------

		Minimum		
Variable	Tolerance	Tolerance	F to enter	Wilks' Lambda
PERF	1.0000000	1.0000000	.73458	.80330
INFO	1.0000000	1.0000000	2.1765	.57955
VERBEXP	1.0000000	1.0000000	3.3600	.47170
AGE	1.0000000	1.0000000	.40170	.88191

At step   1, VERBEXP was included in the analysis.

		Degrees of Freedom			Signif.	Between Groups
Wilks' Lambda	.47170	1	2	6.0		
Equivalent F	3.36000		2	6.0	.1050	

---------------- Variables in the analysis after step   1 ----------------

Variable	Tolerance	F to remove	Wilks' Lambda
VERBEXP	1.0000000	3.3600	

---------------- Variables not in the analysis after step   1 ----------------

		Minimum		
Variable	Tolerance	Tolerance	F to enter	Wilks' Lambda
PERF	.3709896	.3709896	5.4218	.14886
INFO	.2019444	.2019444	13.134	.07543
AGE	.8850270	.8850270	.70229	.36825

At step   2, INFO   was included in the analysis.

		Degrees of Freedom			Signif.	Between Groups
Wilks' Lambda	.07543	2	2	6.0		
Equivalent F	6.60264		4	10.0	.0072	

---------------- Variables in the analysis after step   2 ----------------

Variable	Tolerance	F to remove	Wilks' Lambda
INFO	.2019444	13.134	.47170
VERBEXP	.2019444	16.708	.57955

---------------- Variables not in the analysis after step   2 ----------------

		Minimum		
Variable	Tolerance	Tolerance	F to enter	Wilks' Lambda
PERF	.1676353	.0912506	10.231	.01233
AGE	.8349769	.1905241	.57354	.05862

At step   3, PERF   was included in the analysis.

		Degrees of Freedom			Signif.	Between Groups
Wilks' Lambda	.01233	3	2	6.0		
Equivalent F	10.6724		6	8.0	.0019	

TABLE 11.8 *(CONTINUED)*

```
--------------- Variables in the analysis after step   3 ---------------

Variable  Tolerance  F to remove  Wilks' Lambda
PERF       .1676353   10.231           .07543 INFO      .0912506   22.138          .14886
VERBEXP    .1993789   13.555           .09593

------------ Variables not in the analysis after step   3 ---------------

                     Minimum
Variable  Tolerance  Tolerance  F to enter   Wilks' Lambda
AGE        .5327805   .0812130    .26592          .01048

F level or tolerance or VIN insufficient for further computation.
```

Summary Table

	Action		Vars	Wilks'		
Step	Entered	Removed	In	Lambda	Sig.	Label
1	VERBEXP		1	.47170	.1050	
2	INFO		2	.07543	.0072	
3	PERF		3	.01233	.0019	

Canonical Discriminant Functions

		Pct of	Cum	Canonical	After	Wilks'			
Fcn	Eigenvalue	Variance	Pct	Corr	Fcn	Lambda	Chisquare	DF	Sig
					: 0	.0123	21.977	6	.0012
1*	11.7179	68.55	68.55	.9599	: 1	.1569	9.262	2	.0097
2*	5.3751	31.45	100.00	.9182	:				

  * marks the   2 canonical discriminant functions remaining in the analysis.

Standardized Canonical Discriminant Function Coefficients

	FUNC 1	FUNC 2
PERF	1.88474	-1.42704
INFO	-3.30213	-.07042
VERBEXP	1.50580	1.64500

Structure Matrix:

Pooled-withingroups correlations between discriminating variables
                              and canonical discriminant functions
(Variables ordered by size of correlation within function)

	FUNC 1	FUNC 2
INFO	-.23964*	.09887
VERBEXP	.05067	.45030*
AGE	.29955	-.31955*
PERF	.07023	-.18655*

Canonical Discriminant Functions evaluated at Group Means (Group Centroids)

Group	FUNC 1	FUNC 2
1	3.89650	.44986
2	-2.52348	2.06052
3	-1.37301	-2.51039

along different statistical criteria, as indicated in the Purpose column of Table 11.9. The variety of stepping methods, and the computer programs in which they are available, are presented in Table 11.9.

Selection of stepping method depends on the availability of programs and choice of statistical criterion. If, for example, the statistical criterion is Wilks' Lambda, it is beneficial to choose the stepping method that minimizes Lambda. (In SPSS DISCRIMINANT Lambda is the least expensive method, and is recommended in the absence of contrary reasons.) Or, if the statistical criterion is "change in Rao's *V*," the obvious choice of stepping method is RAO.

Statistical criteria also can be used to modify stepping. For example, the user can modify minimum *F* for a predictor to enter, minimum *F* to avoid removal, and so on. SAS allows forward, backward and "stepwise" stepping (cf. Section 5.5.3). Either partial R^2 or significance level is chosen for variables or enter (forward stepping) or stay (backward stepping) in the model. Tolerance (the proportion of variance for a potential predictor that is not already accounted for by predictors in the equation) can be modified in BMDP, SAS, and SPSS stepwise programs. Comparison of programs with respect to these stepwise statistical criteria is provided in Table 11.12.

11.6.2 ■ Number of Discriminant Functions

In DISCRIM with more than two groups, a number of discriminant functions are extracted. The maximum number of functions is the lesser of either degrees of freedom for groups or, as in canonical correlation, principal components analysis and factor analysis, equal to the number of predictors. As in these other analyses, some functions often carry no worthwhile information. It is frequently the case that the first one or two discriminant functions account for the lion's share of discriminating power, with no additional information forthcoming from the remaining functions.

Many of the programs evaluate successive discriminant functions. For the SPSS DISCRIMINANT example of Table 11.4, note that eigenvalues, percents of variance, and canonical correlations are given for each discriminant function for the small sample data of Table 11.1. To the right, with both functions included, the $\chi^2(8)$ of 20.514 indicates a highly reliable relationship between groups and predictors. With the first discriminant function removed, there is still a reliable relationship between groups and predictors as indicated by the $\chi^2(3) = 8.484$, $p = .037$. This finding indicates that the second discriminant function is also reliable.

How much between-group variability is accounted for by each discriminant function? The eigenvalues associated with discriminant functions indicate the relative proportion of between-group variability accounted for by each function. In the small sample example of Table 11.4, 70.70% of the between-group variability is accounted for by the first discriminant function and 29.30% by the second.

SPSS DISCRIMINANT offers the most flexibility with regard to number of discriminant functions. The user can choose the number of functions, the critical value for proportion of variance accounted for (with succeeding discriminant functions dropped once that value is exceeded), or the significance level of additional functions. SYSTAT GLM (but not DISCRIM) provides tests of successive functions, as do SAS DISCRIM and CANDISC (but not STEPDISC). Neither of the BMDP programs provide tests of successive functions.

11.6.3 ■ Interpreting Discriminant Functions

If a primary goal of analysis is to discover and interpret the combinations of predictors (the discriminant functions) that separate groups in various ways, then the next two sections are relevant. Section 11.6.3.1 reveals how groups are spaced out along the various discriminant functions. Section 11.6.3.2 discusses correlations between predictors and the discriminant functions.

TABLE 11.9 METHODS FOR DIRECTING STEPWISE DISCRIMINANT FUNCTION ANALYSIS

Label	Purpose	Program and Option
WILKS	Produces smallest value of Wilks' Lambda (therefore largest multivariate F)	SPSS:WILKS BMDP:METHOD=1.
MAHAL	Produces largest distance (D^2) for two closest groups	SPSS:MAHAL
MAXMINF	Maximizes the smallest F between pairs of groups	SPSS:MAXMINF
MINRESID	Produces smallest average residual variance ($1 - R^2$) between variables and pairs of groups	SPSS:MINRESI
RAO	Produces at each step largest increase in distance between groups as measured by Rao's V	SPSS:RAO
F TO ENTER	At each step picks variables with largest F TO ENTER	BMDP:METHOD=2.
Partial R^2	Specifies the partial R^2 (p) to enter and/or stay (remove) at each step (forward/backward stepping)	SAS:PR2ENTRY=p, PR2STAY=p
Significance	Specifies the significance level (p) to enter and/or stay (remove) at each step (forward/backward stepping)	SAS:SLENTRY=p, SLSTAY=p SYSTAT:Enter=p, Remove=p SPSS:PIN=p, POUT=p
F-enter/remove limit	Specifies the F-to-enter and/or remove at each step (forward/backward stepping)	SYSTAT:FEnter=f, FRemove=f BMDP:ENTER=f, REMOVE=f SPSS:FIN=f, FOUT=f
Maximum steps	Specifies the maximum number of steps	SAS:MAXSTEP=n SPSS:MAXSTEPS=n BMDP:STEP=n
Number of variables	Specifies the number of variables in the final model	SAS:STOP=n
CONTRAST	Stepping procedure based on groups as defined by contrasts	BMDP7M:CONTRAST. SYSTAT:CONTRAST

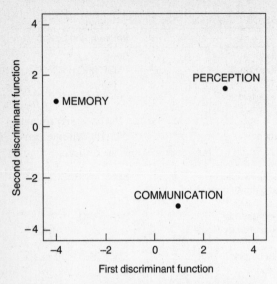

FIGURE 11.1   CENTROIDS OF THREE LEARNING
DISABILITY GROUPS ON THE TWO
DISCRIMINANT FUNCTIONS
DERIVED FROM SAMPLE DATA OF
TABLE 11.1.

11.6.3.1 ■ Discriminant Function Plots

Groups are spaced along the various discriminant functions according to their centroids. Recall from Section 11.4.1 that centroids are mean discriminant scores for each group on a function. Discriminant functions form axes and the centroids of the groups are plotted along the axes. If there is a big difference between the centroid of one group and the centroid of another along a discriminant function axis, the discriminant function separates the two groups. If there is not a big distance, the discriminant function does not separate the two groups. Many groups can be plotted along a single axis.

An example of a discriminant function plot is illustrated in Figure 11.1 for the data of Section 11.4. Centroids are obtained from the section called CANONICAL DISCRIMINANT FUNCTIONS EVALUATED AT GROUP MEANS in Table 11.3.

The plot emphasizes the utility of both discriminant functions in separating the three groups. On the first discriminant function (X axis), the MEMORY group is some distance from the other two groups, but the COMMUNICATION and PERCEPTION groups are close together. On the second function (Y axis) the COMMUNICATION group is far from the MEMORY and PERCEPTION groups. It takes both discriminant functions, then, to separate the three groups from each other.

If there are four or more groups and, therefore, more than two reliable discriminant functions, then pairwise plots of axes are used. One discriminant function is the X axis and another is the Y axis. Each group has a centroid for each discriminant function; paired centroids are plotted with respect to their values on the X and Y axes. Because centroids are only plotted pairwise, three significant discriminant functions require three plots (function 1 vs. function 2; function 1 vs. function 3; and function 2 vs. function 3) and so on.

BMDP7M, SYSTAT DISCRIM and SPSS DISCRIMINANT provide a plot of group centroids for the first pair of discriminant functions (called canonical variates in BMDP7M and canonical scores in SYSTAT DISCRIM). Cases as well as means are plotted in SPSS DISCRIMINANT and

SYSTAT DISCRIM, making separations among groups harder to see than with simpler plots, but facilitating evaluation of classification. Simplified plots of group means as well as plots including cases are available in BMDP7M.

Plots of centroids on additional pairs of reliable discriminant functions have to be prepared by hand, or discriminant scores can be passed to a "plotting" program such as BMDP6D. SAS and SYSTAT pass the discriminant scores to plotting programs.

With factorial designs (Section 11.6.5) separate sets of plots are required for each significant main effect and interaction. Main effect plots have the same format as Figure 11.1, with one centroid per group per margin. Interaction plots have as many centroids as cells in the design.

11.6.3.2 ■ Loading Matrices

Plots of centroids tell you how groups are separated by a discriminant function, but they do not reveal the meaning of the discriminant function. The meaning of the function is inferred by a researcher from the pattern of correlations between the function and the predictors.[9] Correlations between predictors and functions are called loadings in both discriminant function analysis and factor analysis (see Chapter 13). If predictors X_1, X_2, and X_3 load (correlate) highly with the function but predictors X_4 and X_5 do not, the researcher attempts to understand what X_1, X_2, and X_3 have in common with each other that is different from X_4 and X_5; the meaning of the function is determined by this understanding. (Read Section 13.6.5 for further insights into the art of interpreting loadings.)

Mathematically, the matrix of loadings is the pooled within-group correlation matrix multiplied by the matrix of standardized discriminant function coefficients.[10]

$$A = R_wD \tag{11.7}$$

The loading matrix of correlations between predictors and discriminant functions, A, is found by multiplying the matrix of within-group correlations among predictors, R_w, by a matrix of standardized discriminant function coefficients, D (standardized using pooled within-group standard deviations).

For the example of Table 11.1, the loading matrix, called POOLED WITHIN-GROUPS CORRELATIONS BETWEEN DISCRIMINATING VARIABLES AND CANONICAL DISCRIMINANT FUNCTIONS by SPSS DISCRIMINANT, appears as the middle matrix in Table 11.4.

Loading matrices are read in columns; the column is the discriminant function (FUNC 1 and FUNC 2), the rows are predictors (INFO to AGE), and the entries in the column are correlations. For this example, the first discriminant function correlates most highly with INFO (WISC Information scores, $r = .22796$), while the second function correlates most highly with VERBEXP (ITPA Verbal Expression scale, $r = .44630$).

These findings are related to discriminant function plots (e.g., Figure 11.1) for full interpretation. The first discriminant function is largely a measure of INFOrmation, and it separates the group with MEMORY problems from the groups with PERCEPTION and COMMUNICATION problems. The second discriminant function is largely a measure of VERBEXP (verbal expression) and it separates the group with COMMUNICATION problems from the groups with PERCEPTION and MEMORY

---

[9] Some researchers interpret standardized discriminant function coefficients, however they suffer from the same difficulties in interpretation as standardized regression coefficients, discussed in Section 5.6.1.

[10] Some texts (e.g., Cooley & Lohnes, 1971) and early versions of SPSS use the total correlation matrix to find standardized coefficients rather than the within-group matrix.

problems. Interpretation in this example is reasonably straightforward because only one predictor is highly correlated with each discriminant function; interpretation is much more interesting when several predictors correlate with a discriminant function.

Consensus is lacking regarding how high correlations in a loading matrix must be to be interpreted. By convention, correlations in excess of .33 (10% of variance) may be considered eligible while lower ones are not. Guidelines suggested by Comrey and Lee (1992) are included in Section 13.6.5. However, the size of loadings depends both on the value of the correlation in the population and on the homogeneity of scores in the sample taken from it. If the sample is unusually homogeneous with respect to a predictor, the loadings for the predictor are lower and it may be wise to lower the criterion for determining whether or not to interpret the predictor as part of a discriminant function.

Caution is always necessary in interpreting loadings, however, because they are full, not partial or semipartial, correlations. The loading could be substantially lower if correlations with other predictors were partialed out. For a review of this material, read Section 5.6.1. Section 11.6.4 deals with methods for interpreting predictors after variance associated with other predictors is removed, if that is desired.

In some cases, rotation of the loading matrix may facilitate interpretation, as discussed in Chapter 13. SPSS DISCRIMINANT, SYSTAT GLM, and SPSS MANOVA allow rotation of discriminant functions. But rotation of discriminant function loading matrices is still considered problematic and not recommended for the novice.

11.6.4 ■ Evaluating Predictor Variables

Another tool for evaluating contribution of predictors to separation of groups is available through the CONTRAST procedure of BMDP7M and SYSTAT DISCRIM. Means for predictors for each group are contrasted with means for other groups pooled. For instance, if there are three groups, means on predictors for group 1 are contrasted with pooled means from groups 2 and 3; then means for group 2 are contrasted with pooled means from groups 1 and 3; finally means for group 3 are contrasted with pooled means from groups 1 and 2. This procedure is used to determine which predictors are important for isolating one group from the rest.

One BMDP7M or SYSTAT DISCRIM run is required to isolate the means for each group and contrast them with means for other groups. Because there are only two groups (one group versus pooled other groups), there is only one discriminant function per run. F TO ENTER for each predictor at step 0 is univariate F for testing the reliability of the mean difference between the group singled out and the other groups. Therefore, at step 0, F TO ENTER shows how important a predictor is, by itself, in separating the members of a particular group.

At the last step of each contrast run, with all predictors forced into the analysis, F TO REMOVE reflects the reduction in prediction that would result if a predictor were removed from the equation. It is the unique contribution the predictor makes, in this particular set of predictors, to separation of groups. F TO REMOVE is the importance of a predictor after adjustment for all other predictors in the set.

In order to avoid overinterpretation, it is probably best to consider only predictors with F ratios "significant" after adjusting error for the number of predictors in the set. The adjustment is made on the basis of

$$\alpha = 1 - (1 - \alpha_i)^p \tag{11.8}$$

Type I error rate (α) for evaluating contribution of p predictors to between-group contrasts is based on the error rate for evaluating each predictor and the number of predictors, $i = 1, 2, \ldots, p$.

F TO REMOVE for each predictor at the last step, then, is evaluated at α_i. Predictors that meet this criterion contribute reliable unique variance to separation of one group from the others. Prediction of group membership is reliably reduced if the predictor is deleted from the equation.

Even with this adjustment, there is danger of inflation of Type I error rate because multiple nonorthogonal contrasts are performed. If there are numerous groups, further adjustment might be considered such as multiplication of critical F by $k - 1$, where k = number of groups. Or interpretation can proceed very cautiously, de-emphasizing statistical justification.

As an example of use of F TO ENTER and F TO REMOVE for interpretation, partial output from one BMDP7M CONTRAST run for the data of Table 11.1, between the group with MEMORY problems and the other two groups, is shown in Table 11.10.

With 1 and 3 df at the last step, it is not reasonable to evaluate predictors with respect to statistical significance. However, the pattern of F TO ENTER and F TO REMOVE reveals the predictors most likely to separate each group from the other two. The group with MEMORY deficits is characterized by low scores on WISC INFOrmation, as corroborated by the group means in Table 11.10.

A form of squared semipartial correlation is useful as a measure of strength of association between each predictor and dichotomized group membership. The percent of variance contributed at the last step by reliable predictors is

$$sr_i^2 = \frac{F_i}{\text{df}_{\text{res}}} (1 - r_c^2) \tag{11.9}$$

The squared semipartial correlation (sr_i^2) between predictor I and the discriminant function for the difference between two groups (one group vs. all others) is calculated from F TO REMOVE for the ith predictor at the last step with all predictors forced, degrees of freedom for error (df_{res}) at the last step, and r_c^2 (CANONICAL CORRELATIONS, squared).

For example, strength of association between INFO and the group with MEMORY problems vs. the other two groups (assuming statistical reliability) is

$$sr_i^2 = \frac{33.37}{3} (1 - .96357^2) = .7957$$

Almost 80% of the variance in INFO scores overlaps that of the discriminant function that separates the MEMORY group from the PERCEPT and COMMUN groups.

The procedures detailed in this section are most useful when the number of groups is small and the separations among groups are fairly uniform on the discriminant function plot for the first two functions. With numerous groups, some closely clustered, other kinds of contrasts might be suggested by the discriminant function plot (e.g., groups 1 and 2 might be pooled and contrasted with pooled groups 3, 4, and 5). Or, with a very large number of groups, the procedures of Section 11.6.3 may suffice.

If there is logical basis for assigning priorities to predictors, a sequential rather than standard approach to contrasts can be used. Instead of evaluating each predictor after adjustment for all other predictors, it is evaluated after adjustment by only higher priority predictors. This strategy is accomplished through a series of SPSS MANOVA runs, in which Roy-Bargmann stepdown F's (cf. Chapter 9) are evaluated for each contrast.

All the procedures for evaluation of DVs in MANOVA apply to evaluation of predictor variables in DISCRIM. Interpretation of stepdown analysis, univariate F, pooled within-group correlations among predictors, or standardized discriminant function coefficients is as appropriate (or inappropriate) for DISCRIM as for MANOVA. These procedures are summarized in Section 9.5.2.

TABLE 11.10   SETUP AND PARTIAL OUTPUT FOR BMDP7M CONTRAST RUN BETWEEN MEMORY AND OTHER TWO
                   GROUPS FOR SMALL SAMPLE EXAMPLE

```
/INPUT     VARIABLES ARE 6.  FORMAT IS FREE.  FILE = 'TAPE41.DAT'.
/VARIABLE  NAMES ARE SUBJNO,GROUP,PERF,INFO,VERBEXP,AGE.
           LABEL IS SUBJNO.
/GROUP     VAR IS GROUP.
           CODES(GROUP) ARE 1 TO 3.
           NAMES(GROUP) ARE MEMORY, PERCEPT, COMMUN.
/DISCRIM   LEVEL=2*0, 4*1.  FORCE=1.
           CONTRAST = 2, -1, -1.
/PRINT     NO STEP.
/PLOT              NO CANON.   CONTRAST.
/END
```

```
   MEANS
       GROUP =    MEMORY       PERCEPT       COMMUN       ALL GPS.
VARIABLE
   3 PERF       98.66666       87.66666      101.33334     95.88889
   4 INFO        7.00000       11.66667        9.66667      9.44444
   5 VERBEXP    36.33333       38.33333       28.33333     34.33333
   6 AGE         7.30000        6.93333        7.63333      7.28889
COUNTS               3.             3.             3.           9.


****************************************************************************

STEP NUMBER   0
   VARIABLE    F TO   FORCE TOLERNCE *   VARIABLE     F TO   FORCE TOLERNCE
              REMOVE  LEVEL           *                ENTER  LEVEL
       DF =  1     7                  *        DF =  1     6
                                      *   3 PERF        0.16    1   1.00000
                                      *   4 INFO        3.56    1   1.00000
                                      *   5 VERBEXP     0.72    1   1.00000
                                      *   6 AGE         0.00    1   1.00000
****************************************************************************

STEP NUMBER   4
VARIABLE ENTERED   6 AGE
   VARIABLE    F TO   FORCE TOLERNCE *   VARIABLE     F TO   FORCE TOLERNCE
              REMOVE  LEVEL           *                ENTER  LEVEL
       DF =  1     3                  *        DF =  1     2
   3 PERF        2.90    1   0.10696 *
   4 INFO       33.37    1   0.08121 *
   5 VERBEXP     2.72    1   0.19930 *
   6 AGE         0.34    1   0.53278 *


CANONICAL CORRELATIONS
           0.96357
```

11.6.5 ■ Design Complexity: Factorial Designs

The notion of placing cases into groups is easily extended to situations where groups are formed by
differences on more than one dimension. An illustration of factorial arrangement of groups is the
large sample example of Section 9.7, where women are classified by femininity (high or low) and
also by masculinity (high or low) on the basis of scores on the Bem Sex Role Inventory (BSRI).
Dimensions of femininity and masculinity (each with two levels) are factorially combined to form

four groups: high-high, high-low, low-high, low-low. Unless you want to classify cases, factorial designs are best analyzed through MANOVA. If classification is your goal, however, some issues require attention.

As long as sample sizes are equal in all cells, factorial designs are fairly easily analyzed through a factorial DISCRIM program such as BMDP7M or SYSTAT DISCRIM. A separate run is required for each main effect and interaction wherein the main effect or interaction is specified by contrast coding. Discriminant and classification functions are produced for the effect during the run.

When sample sizes are unequal in cells, a two-stage analysis is often best. First, questions about reliability of separation of groups by predictors are answered through MANOVA. Second, if classification is desired after MANOVA, it is found through DISCRIM programs.

Formation of groups for DISCRIM depends on the outcome of MANOVA. If the interaction is statistically significant, groups are formed for the cells of the design. That is, in a two-by-two design, four groups are formed and used as the grouping variable in DISCRIM. Note that main effects as well as interactions influence group means (cell means) in this procedure, but for most purposes classification of cases into cells seems reasonable.

If an interaction(s) is not statistically reliable, classification is based on significant main effects. For example, interaction is not reliable, in the data of Section 9.7, but both main effect of masculinity and main effect of femininity are reliable. One DISCRIM run is used to produce the classification equations for main effect of masculinity and a second run is used to produce the classification equations for main effect of femininity. That is, classification of main effects is based on marginal groups.

SYSTAT GLM makes no distinction between MANOVA and DISCRIM, therefore factorial DISCRIM is produced by the instruction PRINT=LONG with the setup for a factorial design (cf. Table 9.5). In SYSTAT GLM each main effect and interaction is tested separately and classification information is written to a file. Therefore a two-step procedure is still useful, in which the first analysis determines whether classification results are to be saved for the interaction or main effect(s) steps.

11.6.6 ■ Use of Classification Procedures

The basic technique for classifying cases into groups is outlined in Section 11.4.2. Results of classification are presented in tables such as the CLASSIFICATION MATRIX of BMDP7M (Table 11.3) and SYSTAT DISCRIM (Table 11.6), the CLASSIFICATION RESULTS of SPSS (Table 11.4), or NUMBER OF OBSERVATIONS AND PERCENTS CLASSIFIED INTO GROUP of SAS (Table 11.5) where actual group membership is compared to predicted group membership. From these tables, one finds the percent of cases correctly classified and the number and nature of errors of classification.

But how good is the classification? When there are equal numbers of cases in every group, it is easy to determine the percent of cases that should be correctly classified by chance alone to compare to the percent correctly classified by the classification procedure. If there are two equally sized groups, 50% of the cases should be correctly classified by chance alone (cases are randomly assigned into two groups and half of the assignments in each group are correct), while three equally sized groups should produce 33% correct classification by chance, and so forth. However, when there are unequal numbers of cases in the groups, computation of the percent of cases that should be correctly classified by chance alone is a bit more complicated.

The easier way to find it[11] is to first compute the number of cases in each group that should be correct by chance alone and then add across the groups to find the overall expected percent correct. Consider an example where there are 60 cases, 10 in Group 1, 20 in Group 2, and 30 in Group 3. If prior probabilities are specified as .17, .33, and .50, respectively, the programs will assign 10, 20, and 30 cases to the groups. If 10 cases are assigned at random to Group 1, .17 of them (or 1.7) should be correct by chance alone. If 20 cases are randomly assigned to Group 2, .33 (or 6.6) of them should be correct by chance alone, and if 30 cases are assigned to Group 3, .50 of them (or 15) should be correct by chance alone. Adding together 1.7, 6.6, and 15 gives 23.3 cases correct by chance alone, 39% of the total. The percent correct using classification equations has to be substantially larger than the percent expected correct by chance alone if the equations are to be useful.

Some of the computer programs offer sophisticated additional features that are helpful in many classification situations.

11.6.6.1 ■ Cross-Validation and New Cases

Classification is based on classification coefficients derived from samples and they usually work too well for the sample from which they were derived. Because the coefficients are only estimates of population classification coefficients, it is often most desirable to know how well the coefficients generalize to a new sample of cases. Testing the utility of coefficients on a new sample is called cross-validation. One form of cross-validation involves dividing a single large sample randomly in two parts, deriving classification functions on one part, and testing them on the other. A second form of cross-validation involves deriving classification functions from a sample measured at one time, and testing them on a sample measured at a later time. In either case, cross-validation techniques are especially well developed in the BMDP7M and 5M, SAS DISCRIM, SYSTAT DISCRIM, and SPSS DISCRIMINANT programs.

For a large sample randomly divided into parts, you simply omit information about actual group membership for some cases (hide it in the program) as shown in Section 11.8.2. SPSS DISCRIMINANT, SYSTAT DISCRIM, BMDP7M and BMDP5M do not include these cases in the derivation of classification functions, but do include them in the classification phase. In SAS DISCRIM, the withheld cases are put in a separate data file. The accuracy with which the classification functions predict group membership for cases in this data file is then examined. The SAS manual (SAS Institute, 1990, p. 698) provides details for classifying cases in the data file by including "calibration" information for their classification in yet another file.

When the new cases are measured at a later time, classifying them is somewhat more complicated unless you use SAS DISCRIM (in the same way that you would for cross-validation). This is because none of the other computer programs for DISCRIM allows classification of new cases without repeated entry of the original cases to derive the classification functions. You "hide" the new cases, derive the classification functions from the old cases, and test classification on all cases. Or, you can input the classification coefficients along with raw data for the new cases and run the data only through the classification phase. Or, it may be easiest to write your own program based on the classification coefficients to classify cases as shown in Section 11.4.2.

---

[11] The harder way to find it is to expand the multinomial distribution, a procedure that is more technically correct but produces identical results to those of the simpler method presented here.

11.6.6.2 ■ Jackknifed Classification

Bias enters classification if the coefficients used to assign a case to a group are derived, in part, from the case. In jackknifed classification, the data from the case are left out when the coefficients used to assign it to a group are computed. Each case has a set of coefficients that are developed from all other cases. Jackknifed classification gives a more realistic estimate of the ability of predictors to separate groups.

BMDP7M and SYSTAT DISCRIM provide for jackknifed classification. When the procedure is used with all predictors forced into the equation, bias in classification is eliminated. When it is used with stepwise entry of predictors (where they may not all enter), bias is reduced. An application of jackknifed classification is shown in Section 11.8.

11.6.6.3 ■ Evaluating Improvement in Classification

In sequential DISCRIM, it is useful to determine if classification improves as a new set of predictors is added to the analysis. McNemar's repeated-measures chi square provides a simple, straightforward (but tedious) test of improvement. Cases are tabulated one by one, by hand, as to whether they are correctly or incorrectly classified before the step and after the step where the predictors are added.

		Early step classification	
		Correct	Incorrect
Later step	Correct	*(A)*	B
classification	Incorrect	C	*(D)*

Cases that have the same result at both steps (either correctly classified—cell *A*—or incorrectly classified—cell *D*) are ignored because they do not change. Therefore, χ^2 for change is

$$\chi^2 = \frac{(B - C)^2}{B + C} \qquad\qquad \text{df} = 1 \qquad\qquad (11.10)$$

Ordinarily, the researcher is only interested in improvement in χ^2, that is, in situations where $B > C$ because more cases are correctly classified after addition of predictors. When $B > C$ and χ^2 is greater than 3.84 (critical value of χ^2 with 1 df at $\alpha = .05$), the added predictors reliably improve classification.

With very large samples hand tabulation of cases is not reasonable. An alternative, but possibly less desirable, procedure is to test the significance of the difference between two lambdas, as suggested by Frane (personal communication). Wilks' Lambda from the step with the larger number of predictors (Λ_2) is divided by Lambda from the step with fewer predictors (Λ_1) to produce (Λ_D).

$$\Lambda_D = \frac{\Lambda_2}{\Lambda_1} \qquad\qquad (11.11)$$

Wilks' Lambda for testing the significance of the difference between two Lambdas (Λ_D) is calculated by dividing the smaller Lambda (Λ_2) by the larger Lambda (Λ_1).

Λ_D is evaluated with three degree of freedom parameters: p, the number of predictors after addition of predictors; df_{effect}, the number of groups minus 1; and the df_{error} at the step with the added predictors. Approximate F is found according to procedures in Section 11.4.1.

For the sequential example of Table 11.7, one can test whether addition of INFO, PERF, and VERBEXP at the last step ($\Lambda_2 = .0105$) reliably improves classification of cases over that achieved at step 1 with AGE in the equation ($\Lambda_1 = .8819$).

$$\Lambda_D = \frac{.0105}{.8819} = .0119$$

where

$$df_p = p = 4$$

$$df_{effect} = k - 1 = 2$$

$$df_{error} = N - 1 = 6$$

From Section 11.4.1,

$$s = \sqrt{\frac{(4)^2(2)^2 - 4}{(4)^2 + (2)^2 - 5}} = 2$$

$$y = .0119^{\frac{1}{2}} = .1091$$

$$df_1 = (4)(2) = 8$$

$$df_2 = (2)\left[6 - \frac{4 - 2 + 1}{2}\right] - \left[\frac{4(2) - 2}{2}\right] = 6$$

$$\text{Approximate } F(8, 6) = \left(\frac{1 - .1091}{.1091}\right)\left(\frac{6}{8}\right) = 6.124$$

Because critical $F(8, 6)$ is 4.15 at $\alpha = .05$, there is reliable improvement in classification into the three groups when INFO, PERF and VERBEXP scores are added to AGE scores.

11.7 ■ COMPARISON OF PROGRAMS

There are numerous programs for discriminant function analysis in the four packages, some general and some special purpose. Both SPSS and BMDP have a general purpose discriminant function analysis program that performs direct, sequential, or stepwise DISCRIM with classification. In addition, SPSS MANOVA performs DISCRIM, but not classification. BMDP also has a program for quadratic discriminant function analysis. SYSTAT GLM provides discriminant function analysis as part of the general linear model program, but now SYSTAT also has a specialized DISCRIM program, similar to BMDP7M. SAS has three programs, one for inference and discriminant function coefficients, one which adds classification, and a third for stepwise analysis. Finally, if the only question is reliability of predictors to separate groups, any of the MANOVA programs discussed in Chapter 9 is appropriate. Table 11.11 compares features of direct discriminant function programs. Features for stepwise discriminant function are compared in Table 11.12.

TABLE 11.11 COMPARISON OF PROGRAMS FOR DIRECT DISCRIMINANT FUNCTION ANALYSIS

Feature	SAS DISCRIM	SAS CANDISC	SPSS DISCRIMINANT	SPSS MANOVA[a]	SYSTAT DISCRIM	SYSTAT GLM	BMDP7M	BMDP5M
Input								
Optional matrix input	Yes	Yes	Yes	Yes	No	Yes	No	No
Missing data options	No	No	Yes	No	No	No	No	No
Restrict number of discriminant functions	NCAN	NCAN	Yes	No	No	No	No	No
Specify cumulative % of sum of eigenvalues	No	No	Yes	No	No	No	No	No
Specify significance level of functions to retain	No	No	Yes	ALPHA	Yes	No	No	No
Factorial arrangement of groups	No	No	No	Yes	CONTRASTS	CONTRASTS	CONTRASTS	No
Specify tolerance	SINGULAR	SINGULAR	Yes	No	Yes	Yes	Yes	No
Rotation of discriminant functions	No	No	Yes	Yes	No	Yes	No	No
Quadratic discriminant function analysis	POOL=NO	N.A.[b]	No	No	Yes	No	No	No
Optional prior probabilities	Yes	N.A.[b]	Yes	N.A.[b]	No	No	Yes	Yes
Specify separate covariance matrices for classification	POOL=NO	N.A.	Yes	N.A.	No	No	N.A.	Yes
Threshold for classification	Yes	N.A.	No	N.A.	No	No	No	No
Nonparametric classification method	Yes	N.A.	No	N.A.	No	No	No	No
Output								
Wilks' Lambda with approx. F χ^2	Yes	Yes	Yes	Yes	PRINT MEDIUM	Yes	Yes	Yes
Generalized distance between groups (Mahalanobis D^2)	No	No	Yes	No	No	No	No	No
Hotelling's trace criterion	Yes	Yes	No	Yes	PRINT MEDIUM	Yes	No	Yes
Roy's gcr (maximum root)	Yes	Yes	No	Yes	PRINT MEDIUM	THETA	Yes	No
Pillai's criterion	Yes	Yes	No	Yes	PRINT MEDIUM	Yes	Yes	No
Tests of successive dimensions (roots)	Yes	Yes	Yes	Yes	No	RESIDUAL ROOTS	No	No
Univariate F ratios	Yes	Yes	Yes	Yes	No	Yes	Yes[c]	No
Group means	Yes	Yes	Yes	Yes	PRINT MEDIUM	Yes	Yes	Yes

TABLE 11.11 *(CONTINUED)*

Feature	SAS DISCRIM	SAS CANDISC	SPSS DISCRIMINANT	SPSS MANOVA[a]	SYSTAT DISCRIM	SYSTAT GLM	BMDP7M	BMDP5M
Total and within-group standardized group means	Yes	Yes	No	No	No	No	No	No
Group standard deviations	Yes	Yes	Yes	Yes	No	Yes	Yes	Yes
Total, within-group and between-group standard deviations	Yes	Yes	No	No	No	No	No	No
Coefficient of variation	No	No	No	No	No	No	No	No
Standardized discriminant function (canonical) coefficients	Yes	Yes	Yes	Yes	PRINT MEDIUM	Yes	Yes	No
Unstandardized (raw) discriminant function (canonical) coefficients	Yes	Yes	Yes	Yes	PRINT MEDIUM	No	Yes	No
Group centroids	Yes	Yes	Yes	No	Yes	No	Yes	No
Pooled within-groups (residual) SSCP matrix	Yes	Yes	No	Yes	No	Yes	No	No
Between-groups SSCP matrix	Yes	Yes	No	No	No	No	No	No
Hypothesis SSCP matrix	No	No	No	Yes	No	Yes	No	No
Total SSCP matrix	Yes	Yes	No	No	No	Yes	No	No
Group SSCP matrices	Yes	Yes	No	No	No	No	No	No
Pooled within-groups (residual) correlation matrix	Yes	Yes	Yes	Yes	PRINT LONG	Yes	Yes	Yes
Determinant of within-group correlation matrix	No	No	No	Yes	No	No	No	No
Between-groups correlation matrix	Yes	Yes	No	No	No	No	No	Yes
Group correlation matrices	Yes	Yes	No	No	PRINT LONG	No	No	Yes
Total correlation matrix	Yes	Yes	No	No	PRINT LONG	No	No	Yes
Total covariance matrix	Yes	Yes	Yes	No	PRINT LONG	No	No	Yes
Pooled within-groups (residual) covariance matrix	Yes	Yes	Yes	Yes	PRINT LONG	Yes	Yes	Yes
Group covariance matrices	Yes	Yes	Yes	No	No	No	No	Yes
Between-group covariance matrix	Yes	Yes	No	No	No	No	No	Yes

TABLE 11.11 (CONTINUED)

Feature	SAS DISCRIM	SAS CANDISC	SPSS DISCRIMINANT	SPSS MANOVA[a]	SYSTAT DISCRIM	SYSTAT GLM	BMDP7M	BMDP5M
Determinants of group covariance matrices	Yes	No	No	Yes	No	No	No	No
Homogeneity of variance-covariance matrices	Yes	No	Yes	Yes	Yes	No	Yes[d]	Yes
F matrix, pairwise group comparison	No	No	Yes	Yes	Yes	No	Yes	Yes
Canonical correlations	Yes	Yes	Yes	No[e]	Yes	No	Yes	No
Average squared canonical correlation	No	No	No	Yes	No	Yes	Yes	No
Adjusted canonical correlations	Yes	Yes	No	No	No	No	No	No
Standard errors of canonical correlations	Yes	Yes	No	No	No	No	No	No
Eigenvalues	Yes	Yes	Yes	Yes	Yes	No	Yes	No
SMCs for each variable	R-Squared	R-Squared	No	No	No	Yes	No	Yes[f]
SMC divided by tolerance for each variable	RSQ/(1−RSQ)	RSQ/(1−RSQ)	No	No	No	No	No	No
Loading (structure) matrix (pooled within-groups)	Yes	Yes	Yes	No	No	Yes	No	No
Total structure matrix	Yes	Yes	No	No	No	No	No	No
Between structure matrix	Yes	Yes	No	No	No	No	No	No
Individual discriminant (canonical variate) scores	No	No	Yes	No	Yes	Data file	Yes	No
Classification features								
Classification of cases	Yes	N.A[b]	Yes	N.A[b]	Yes	No[g]	Yes	Yes
Classification function coefficients	Yes[h]	N.A.	Yes	N.A.	PRINT MEDIUM	No	Yes	Yes

TABLE 11.11 *(CONTINUED)*

Feature	SAS DISCRIM	SAS CANDISC	SPSS DISCRIMINANT	SPSS MANOVA[a]	SYSTAT DISCRIM	SYSTAT GLM	BMDP7M	BMDP5M
Classification matrix	Yes	N.A.	Yes	N.A.	Yes	No	Yes	Yes
Posterior probabilities for classification	Data file	N.A.	Yes	N.A.	PRINT LONG	Data file	Yes	Yes
Mahalanobis' D^2 for cases (outliers)	No	N.A.	No	N.A.	PRINT LONG	Data file[i]	Yes	No
Jackknifed classification matrix	Cross-validate	N.A.	No	N.A.	Yes	No	Yes	No
Classification with a cross-validation sample	Yes	N.A.	Yes	N.A.	No	Yes	Yes	Yes
Plots								
Plot of group centroids alone	No	N.A.	No	N.A.	No	No	Yes	No
All groups scatterplot	No	N.A.	Yes	N.A.	Yes	No	Yes	No
Separate scatterplots by group	No	N.A.	Yes	N.A.	No	No	Yes	No
Territorial map	No	N.A.	Yes	N.A.	No	No	No	No

[a] Additional features reviewed in Section 8.9
[b] SAS CANDISC and SPSS MANOVA do not classify cases
[c] STEP NUMBER 0, F TO ENTER
[d] Group plots of scores on first two discriminant functions
[e] Can be obtained through CONTRAST procedure
[f] Separately for each group
[g] Classification requires SYSTAT TABLES or SYSTAT DISCRIM
[h] Labeled Linear Discriminant Function

TABLE 11.12 COMPARISON OF PROGRAMS FOR STEPWISE AND SEQUENTIAL DISCRIMINANT FUNCTION ANALYSIS

Feature	SPSS DISCRIMINANT	BMDP7M	SAS STEPDISC	SYSTAT DISCRIM
Input				
Optional matrix input	Yes	No	Yes	Yes
Missing data options	Yes	No	No	No
Specify contrast	No	Yes	No	Yes
Factorial arrangement of groups	No	CONTRASTS	No	CONTRAST
Suppress intermediate steps	No	NO STEP	No	No
Suppress all but summary table	NOSTEP	No	SHORT	No
Optional methods for order of entry/removal	3	3	5	2
Forced entry by level (sequential)	Yes	Yes	No	Yes
Force some variables into model	Yes	Yes	INCLUDE	FORCE
Specify tolerance	Yes	Yes	SINGULAR	Yes
Specify maximum number of steps	Yes	Yes	Yes	No
Specify number of variables in final stepwise model	No	No	STOP=	No
Specify F to enter/remove	FIN/FOUT	ENTER/ REMOVE	No	FEnter/ FRemove
Specify significance of F to enter/remove	PIN/POUT	No	SLE/SLS	Enter/Remove
Specify partial R^2 to enter/remove	No	No	PR2E/PR2S	No
Restrict number of discriminant functions	Yes	No	No	No
Specify cumulative % of sum of eigenvalues	Yes	No	No	No
Specify significance level of functions to retain	Yes	No	No	No
Rotation of discriminant functions	Yes	No	No	No
Prior probabilities optional	Yes	Yes	N.A.[a]	No
Specify separate covariance matrices for classification	Yes	No	N.A.	No
Output				
Wilks' Lambda with approximate F	Yes	Yes	Yes	PRINT MEDIUM
χ^2	Yes	No	No	No
Mahalanobis' D^2 (between groups)	Yes	No	No	No
Rao's V	Yes	No	No	No
Pillai's criterion	No	Yes	Yes	PRINT MEDIUM
Tests of successive dimensions (roots)	Yes	No	No	No
Univariate F ratios	Yes	Yes[b]	STEP 1 F	Yes[b]
Group means	Yes	Yes	Yes	PRINT MEDIUM
Within-group and total standardized group means	No	No	Yes	No
Group standard deviations	Yes	Yes	Yes	No
Total and pooled within-group standard deviations	No	No	Yes	No
Coefficients of variation	No	Yes	No	No
Standardized discriminant function (canonical) coefficients	Yes	Yes	No	PRINT MEDIUM
Unstandardized discriminant function (canonical) coefficients	Yes	Yes	No	PRINT MEDIUM
Group centroids	Yes	Yes	No	Yes[c]

TABLE 11.12 *(CONTINUED)*

Feature	SPSS DISCRIMINANT	BMDP7M	SAS STEPDISC	SYSTAT DISCRIM
Pooled within-group correlation matrix	Yes	Yes	Yes	PRINT LONG
Total correlation matrix	No	No	Yes	PRINT LONG
Total covariance matrix	Yes	No	Yes	PRINT LONG
Total SSCP matrix	No	No	Yes	No
Pooled within-group covariance matrix	Yes	Yes	Yes	PRINT LONG
Pooled within-group SSCP matrix	No	No	Yes	No
Group covariance matrices	Yes	No	Yes	PRINT LONG
Group correlation matrices	No	No	Yes	PRINT LONG
Group SSCP matrices	No	No	Yes	No
Between-group correlation matrix	No	No	Yes	No
Between-group covariance matrix	No	No	Yes	No
Between-group SSCP matrix	No	No	Yes	No
Homogeneity of variance-covariance matrices	Yes	Yes[d]	No	Yes
F matrix, pairwise group comparison	Yes	Yes	No	Yes
Canonical correlations, each discriminant function	Yes	Yes	No	Yes
Canonical correlations, average	No	Yes[e]	Yes	No
Eigenvalues	Yes	Yes	No	Yes
Loading (structure) matrix	Yes	No	No	No
Partial R^2 (or tolerance) to enter/remove, each step	Yes	Yes	Yes	Yes
F to enter/remove, each step	Yes	Yes	Yes	Yes
Classification features				
Classification of cases	Yes	Yes	N.A.[a]	Yes
Classification function coefficients	Yes	Yes	N.A.	PRINT MEDIUM
Classification matrix	Yes	Yes	N.A.	Yes
Individual discriminant (canonical variate) scores	Yes	Yes	N.A.	PRINT LONG
Posterior probabilities for classification	Yes	Yes	N.A.	PRINT LONG
Mahalanobis' D^2 for cases (outliers)	No	Yes	N.A.	PRINT LONG
Jackknifed classification matrix	No	Yes	N.A.	Yes
Classification with a cross-validation sample	Yes	Yes	N.A.	Yes
Classification information at each step	No	Yes	N.A.	No
Plots				
Plot of group centroids alone	No	Yes	N.A.	No
All groups scatterplot	Yes	Yes	N.A.	Yes
Separate scatterplots by group	Yes	Yes	N.A.	No
Territorial map	Yes	No	N.A.	No

[a] SAS STEPDISC does not classify cases (see SAS DISCRIM, Table 11.11)

[b] STEP 0, F TO ENTER

[c] Canonical scores of group means

[d] Group plots of scores on first two discriminant functions

[e] Squared

11.7.1 ■ SPSS Package

SPSS DISCRIMINANT[12], features of which are described in both Tables 11.11 and 11.12, is the basic program in this package for DISCRIM. The program provides direct (standard), sequential, or stepwise entry of predictors with numerous options. Strong points include several types of plots and plenty of information about classification. Territorial maps are handy for classification using discriminant function scores if there are only a few cases to classify. In addition, a test of homogeneity of variance-covariance matrices is provided through plots and, should heterogeneity be found, classification may be based on separate matrices. Other useful features are evaluation of successive discriminant functions and availability of loading matrices.

SPSS MANOVA can also be used for DISCRIM and has some features unobtainable in any of the other DISCRIM programs. SPSS MANOVA is described rather fully in Table 9.11, but some aspects especially pertinent to DISCRIM are featured in Table 11.11. MANOVA offers a variety of statistical criteria for testing the significance of the set of predictors (cf. Section 11.6.1) and routinely prints loading matrices. Many other matrices can be printed out, and these, along with determinants, are useful for the more sophisticated researcher. Successive discriminant functions (roots) are evaluated, as in SPSS DISCRIMINANT.

SPSS MANOVA provides discriminant functions for more complex designs such as factorial arrangements with unequal sample sizes. The program is limited, however, in that it includes no classification phase. Further, only standard DISCRIM is available, with no provision for stepwise or sequential analysis other than Roy-Bargmann stepdown analysis as described in Chapter 9.

11.7.2 ■ BMDP Series

BMDP7M deals well with all varieties of DISCRIM. Although designed as a stepwise program, direct DISCRIM is easily produced by forcing entry of all predictors and suppressing all steps except the last. Several inferential multivariate tests are available, along with automatic printing of pairwise multivariate F ratios between groups.

BMDP7M's main advantage over most other DISCRIM programs is identification of outliers through the case classification procedure. Further, the JACKKNIFED CLASSIFICATION feature in BMDP7M assigns a case to a group without using the case in developing the classification coefficients for it. This eliminates or reduces bias in classification, as discussed in Section 11.6.6.2. However, the program lacks a dimension reduction analysis for successive roots and does not provide a loading matrix.

The BMDP7M provision of stepping by contrasts allows analysis of factorial designs as long as sample sizes are equal in cells (Section 11.6.5). Because this program allows for classification as well as statistical inference, it is the program of choice for equal-n factorial designs. (Unequal-n factorial designs require contrast coefficients that reflect choice of adjustment procedure and some considerable sophistication on the part of the user.)

BMDP5M is designed to do quadratic discriminant function analysis and emphasizes classification and various matrices as a result. The program offers a test for homogeneity of variance-covariance matrices, and can be used for linear (but not stepwise) discriminant function analysis as well. Cross-validation is available, but not jackknifed classification. None of the standard canonical analyses or coefficients are available, nor is there dimension reduction analysis or a loading matrix.

---

[12] DSCRIMINANT in the PC+ version.

11.7.3 ■ SYSTAT System

SYSTAT DISCRIM now is the major discriminant function analysis program, replacing MGLH. The new program deals with all varieties of DISCRIM. Automatic (forward and backward) and interactive stepping are available, as well as a contrast procedure to control entry of variables. Jackknifed classification is produced by default, and the manual shows how to perform cross-validation. Dimension reduction analysis is no longer available, but can be obtained by rephrasing the problem as MANOVA and running it through GLM. Such a strategy also is well suited to factorial arrangements of unequal-n groups. The SYSTAT manual (Wilkinson & Hill, 1994a, pp. 370–373) demonstrates an example of discriminant function analysis through GLM.

The manual shows how to use scatterplot matrices (SYSTAT SPLOM) to evaluate homogeneity of variance-covariance matrices; quadratic discrimination function analysis is available through DISCRIM should the assumption be violated. Several univariate and multivariate inferential tests also are available. SYSTAT DISCRIM can be used to assess outliers through Mahalanobis distance of each case to each group centroid.

11.7.4 ■ SAS System

In SAS, there are three separate programs to deal with different aspects of discriminant analysis, with surprisingly little overlap between the stepwise and direct programs. All of the SAS programs for discriminant function analysis are especially rich in output of SSCP, correlation, and covariance matrices.

The most comprehensive program is DISCRIM. This program now does just about everything that the CANDISC program does, although earlier versions lacked inferential tests. It does not perform stepwise or sequential analysis. This program is especially handy for classifying new cases or performing cross-validation (Section 11.6.6.1). With the expansion of DISCRIM, there seems little advantage to using CANDISC. Both programs offer alternate inferential tests, dimension reduction analysis, and all of the standard matrices of discriminant function results. DISCRIM classifies cases while CANDISC does not, so that some time savings may be gained by using CANDISC for direct DISCRIM when classification is not wanted.

Finally, stepwise (but not sequential) analysis is accomplished through STEPDISC. As seen in Table 11.12, very few additional amenities are available in this program. There is no classification, nor is there information about the discriminant functions. On the other hand, this program offers plenty of options for entry and removal of predictors.

11.8 ■ COMPLETE EXAMPLE OF DISCRIMINANT FUNCTION ANALYSIS

The example of direct discriminant function analysis[13] in this section explores how role-dissatisfied housewives, role-satisfied housewives, and employed women differ in attitudes. The sample of 465 women is described in Appendix B, Section B.1. The grouping variable is role-

---

[13] The example of hierarchical discriminant analysis from earlier editions of this book (e.g., Tabachnick & Fidell, 1989) has been dropped in favor of sequential logistic regression analysis of the same variables (cf. Section 12.8).

dissatisfied housewives (UNHOUSE), role-satisfied housewives (HAPHOUSE), and working women (WORKING).

Predictors are internal vs. external locus of control (CONTROL), satisfaction with current marital status (ATTMAR), attitude toward women's role (ATTROLE), and attitude toward homemaking (ATTHOUSE). A fifth attitudinal variable, attitude toward paid work, was dropped from analysis because data were available only for women who had been employed within the past five years and use of this predictor would have involved nonrandom missing values (cf. Chapter 4). The example of DISCRIM, then, involves prediction of group membership from the four attitudinal variables.

The direct discriminant function analysis allows us to evaluate the distinctions among the three groups on the basis of attitudes. We explore the dimensions on which the groups differ, the predictors contributing to differences among groups on these dimensions, and the degree to which we can accurately classify members into their own groups. We also evaluate efficiency of classification with a cross-validation sample.

11.8.1 ■ Evaluation of Assumptions

The data are first evaluated with respect to practical limitations of DISCRIM.

11.8.1.1 ■ Unequal Sample Sizes and Missing Data

In a screening run through BMDP7D (cf. Chapters 4 or 8), seven cases had missing values among the four attitudinal predictors. Missing data were scattered over predictors and groups in apparently random fashion, so that deletion of the cases was deemed appropriate.[14] The full data set includes 458 cases once cases with missing values are deleted.

During classification, unequal sample sizes are used to modify the probabilities with which cases are classified into groups. Because the sample is randomly drawn from the population of interest, sample sizes in groups are believed to represent some real process in the population that should be reflected in classification. For example, knowledge that over half the women are employed implies that greater weight should be given the WORKING group.

11.8.1.2 ■ Multivariate Normality

After deletion of cases with missing data, there are still over 80 cases per group. Although a BMDP7D run (not shown) reveals skewness in ATTMAR, sample sizes are large enough to suggest normality of sampling distributions of means. Therefore there is no reason to expect distortion of results due to failure of multivariate normality.

11.8.1.3 ■ Linearity

Although ATTMAR and RACE are skewed, there is no expectation of curvilinearity between these two and the remaining predictors. At worst, ATTMAR in conjunction with the remaining continuous, well-behaved predictors may contribute to a mild reduction in association.

---

[14] Alternative strategies for dealing with missing data are discussed in Chapter 4.

11.8.1.4 ■ Outliers

To identify univariate outliers, z scores associated with minimum and maximum values on each of the four predictors are investigated through BMDP7D for each group separately. There are some questionable values on ATTHOUSE, with a few exceptionally positive scores. These values are about 4.5 standard deviations from their group means, making them candidates for deletion or alteration. However, the cases are retained for the search for multivariate outliers.

BMDP7M is used to search for multivariate outliers in each group separately. A portion of BMDP7M output for the WORKING group is shown in Table 11.13. Outliers are identified as cases with too large a Mahalanobis D^2, evaluated as χ^2 with degrees of freedom equal to the number of predictors. Critical χ^2 with 4 df at $\alpha = .001$ is 18.467; any case with $D^2 > 18.467$ is an outlier. In Table 11.13, cases 261 ($D^2 = 25.3$) and 299 ($D^2 = 24.8$) are identified as outliers in the group of WORKING women. No additional outliers were found.

The multivariate outliers are the same cases that have extreme univariate scores on ATTHOUSE. Because transformation is questionable for ATTHOUSE (where it seems unreasonable to transform the predictor for only two cases) it is decided to delete multivariate outliers.

Therefore, of the original 465 cases, 7 are lost due to missing values and 2 are multivariate outliers, leaving a total of 456 cases for analysis.

11.8.1.5 ■ Homogeneity of Variance-Covariance Matrices

A second BMDP7M run, Table 11.14, deletes the outliers in order to evaluate homogeneity of variance-covariance matrices. Examination of sample variances for the 4 predictors reveals no gross discrepancies among the groups. ATTMAR shows the largest ratio of variances, where the ratio is less than 3 ($10.29752^2/6.62350^2 = 2.73$) for the HAPHOUSE vs. UNHOUSE groups. There is somewhat more variance in attitude toward marital status in role-dissatisfied than in role-satisfied housewives. Sample sizes are fairly discrepant, with almost three times as many working women as role-dissatisfied housewives, but with two-tailed tests and reasonable homogeneity of variance, DISCRIM is robust enough to handle the discrepancies.

Homogeneity of variance-covariance matrices is also examined through plots of the first two discriminant functions (called canonical variables) produced by BMDP7M using the setup of Table 11.14. As seen in Figure 11.2, the spread of cases for the three groups is relatively equal. Therefore no further test of homogeneity of variance-covariance matrices is necessary.

11.8.1.6 ■ Multicollinearity and Singularity

Because BMDP7M, used for the major analysis, protects against multicollinearity through checks of tolerance, no formal evaluation is necessary (cf. Chapter 4).

11.8.2 ■ Direct Discriminant Function Analysis

Direct DISCRIM is performed through BMDP7M with the 4 attitudinal predictors all forced into the equation. The program instructions and a segment of the output appear in Table 11.15. Because this is a direct analysis, stepwise output is suppressed through NO STEP.

In Table 11.15 the F TO ENTER values at step 0 are univariate F ratios with 2 and 453 df for the individual predictors. Three of the four predictors, all except CONTROL, show univariate F

TABLE 11.13	IDENTIFICATION OF OUTLIERS. SETUP AND SELECTED OUTPUT FROM BMDP7M

```
/INPUT          VAR=13.    FILE='DISCRIM.DAT'.   FORMAT IS FREE.
/VARIABLE       NAMES ARE CASESEQ, WORKSTAT, MARITAL, CHILDREN,
                    RELIGION, RACE, CONTROL, ATTMAR, ATTROLE, SEL,
                    ATTHOUSE, AGE, EDUC.
                MISSING=2*0, 3*9, 5*0, 1, 2*0.
                LABEL = CASESEQ.
                USE = WORKSTAT, CONTROL, ATTMAR, ATTROLE, ATTHOUSE.
/GROUP          VAR = WORKSTAT.
                CODES(WORKSTAT) ARE 1 TO 3.
                NAMES(WORKSTAT) ARE WORKING, HAPHOUSE, UNHOUSE.
/DISC           LEVEL=6*0, 3*1, 0, 1, 2*0.
                FORCE = 1.
/PRINT          NO STEP.
/END
```

		INCORRECT CLASSIFICATIONS		MAHALANOBIS DSQUARE FROM AND POSTERIOR PROBABILITY FOR GROUP	
GROUP	WORKING		WORKING	HAPHOUSE	UNHOUSE
CASE					
252	337		2.7 0.527	4.0 0.263	4.5 0.210
256	341	HAPHOUSE	1.1 0.325	0.5 0.446	1.8 0.229
259	344	HAPHOUSE	0.7 0.329	0.3 0.405	1.2 0.266
260	345	HAPHOUSE	2.8 0.312	2.2 0.421	3.1 0.267
261	346	HAPHOUSE	25.3 0.328	24.1 0.609	28.6 0.063
262	347	UNHOUSE	1.0 0.365	1.7 0.261	1.0 0.374
263	348		1.6 0.481	3.0 0.236	2.7 0.283
264	349		4.7 0.390	5.6 0.244	4.8 0.366
265	355		3.1 0.427	4.6 0.201	3.4 0.372
266	357	HAPHOUSE	1.1 0.314	0.4 0.447	1.6 0.239
267	358	UNHOUSE	5.6 0.280	7.0 0.141	4.2 0.579
268	359		4.6 0.481	6.6 0.182	5.3 0.337
270	362	UNHOUSE	1.1 0.328	1.3 0.291	0.8 0.380
272	365	HAPHOUSE	3.4 0.353	2.7 0.502	5.1 0.145
274	369	UNHOUSE	6.1 0.359	7.5 0.175	5.6 0.466
276	372	HAPHOUSE	7.1 0.244	5.3 0.586	7.8 0.169
278	378		5.8 0.593	7.8 0.219	8.1 0.188
281	381		4.7 0.486	6.3 0.216	5.7 0.297
283	383	UNHOUSE	3.5 0.229	2.9 0.308	2.1 0.463
284	384	UNHOUSE	2.4 0.246	1.9 0.315	1.3 0.439
286	386		0.7 0.406	1.4 0.290	1.3 0.303
287	387		1.1 0.451	2.0 0.277	2.1 0.272
289	397	HAPHOUSE	5.1 0.386	4.6 0.488	7.3 0.126
290	398		6.5 0.444	8.9 0.130	6.5 0.426
291	399		8.5 0.443	9.3 0.294	9.5 0.264
292	400		3.3 0.475	5.4 0.170	3.9 0.355
293	401		9.2 0.423	10.6 0.209	9.5 0.368
295	403		4.0 0.429	6.2 0.144	4.0 0.427
296	404		1.2 0.436	1.8 0.321	2.4 0.243
298	406		7.0 0.410	8.2 0.228	7.3 0.362
299	407	HAPHOUSE	24.8 0.311	23.4 0.616	27.7 0.073
307	425		1.5 0.417	2.3 0.268	2.0 0.315

ratios that are significant at the .01 level. When all 4 predictors are used, the approximate F of 6.274 (with 8 and 900 df based on Wilks' Lambda) is highly significant. That is, there is reliable separation of the three groups based on all four predictors combined, as discussed in Section 11.6.1.1.

TABLE 11.14   SETUP AND SELECTED OUTPUT FROM BMDP7M RUN TO CHECK HOMOGENEITY OF VARIANCE-
COVARIANCE MATRICES

```
/INPUT          VAR=13.    FILE='DISCRIM.DAT'.    FORMAT IS  FREE.
/VARIABLE       NAMES ARE CASESEQ, WORKSTAT, MARITAL, CHILDREN,
                     RELIGION, RACE, CONTROL, ATTMAR, ATTROLE, SEL,
                     ATTHOUSE, AGE, EDUC.
                MISSING=2*0, 3*9, 5*0, 1, 2*0.
                LABEL = CASESEQ.
                USE = WORKSTAT, CONTROL, ATTMAR, ATTROLE, ATTHOUSE.
/GROUP          VAR = WORKSTAT.
                CODES(WORKSTAT) ARE 1 TO 3.
                NAMES(WORKSTAT) ARE WORKING, HAPHOUSE, UNHOUSE.
/TRANSFORM        DELETE = 261, 299.
/DISC           LEVEL=6*0, 3*1, 0, 1, 2*0.
                FORCE = 1.
/PRINT          NO STEP.
/PLOT           GROUP=1. GROUP=2. GROUP=3.
/END
```

```
    STANDARD DEVIATIONS

        GROUP =    WORKING      HAPHOUSE     UNHOUSE      POOLED
VARIABLE
    7 CONTROL      1.23780      1.30984      1.25401      1.26253
    8 ATTMAR       8.53003      6.62350     10.29753      8.36830
    9 ATTROLE      6.95618      6.45843      5.75977      6.61150
   11 ATTHOUSE     4.45544      3.88348      3.95846      4.20608
```

Table 11.15 also shows the classification functions used to classify cases into the three groups (see Equation 11.3) and the results of that classification, with and without jackknifing (see Section 11.6.6). In this case, classification is made on the basis of a modified equation in which unequal prior probabilities are used to reflect unequal group sizes by the use of the PRIOR convention in the setup.

A total of 54.4% of cases are correctly classified by normal procedures, 53.5% by jackknifed procedures. How do these compare with random assignment? Prior probabilities, specified as .52 (WORKING), .30 (HAPHOUSE), and .18 (UNHOUSE), put 237 cases (.52 × 456) in the WORK-ING group, 137 in the HAPHOUSE group, and 82 in the UNHOUSE group. Of those randomly assigned to the WORKING group, 123 (.52 × 237) should be correct, while 41.1 (.30 × 137) and 14.8 (.18 × 82) should be correct by chance in the HAPHOUSE and UNHOUSE groups, respectively. Over all three groups 178.9 out of the 456 cases or 39% should be correct by chance alone. Both classification procedures correctly classify substantially more than that.

Although BMDP7M does not provide tests of the discriminant functions, it does provide considerable information about them as seen near the end of Table 11.15. Canonical correlations for each discriminant function (.26699 and .18437), although small, are relatively equal for the two discriminant functions. CANONICAL VARIABLES EVALUATED AT GROUP MEANS are centroids on the discriminant functions for the groups, discussed in Sections 11.4.1 and 11.6.3.1.

An additional BMDP7M run for cross-validation is shown in Table 11.16. Approximately 25% of the original cases are randomly selected out in the TRANSFORM paragraph to use for cross-validation. The 111 cases randomly selected out are used to assess the results of classification. New group membership is designated in the GROUP paragraph for these cases: NEWWORK (for WORKING), NEWHAP (for HAPHOUSE), and NEWUN (for UNHOUSE). The other 345 cases

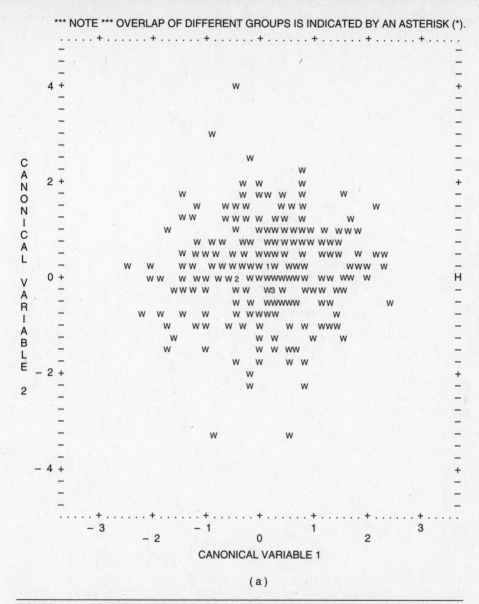

FIGURE 11.2   SCATTERPLOTS OF CASES ON FIRST TWO CANONICAL VARIATES FOR (a)
WORKING WOMEN, (b) ROLE-SATISFIED HOUSEWIVES, AND (c) ROLE-
DISSATISFIED HOUSEWIVES. SETUP APPEARS IN TABLE 11.14.

are used, as specified in the GROUP paragraph, to develop discriminant functions and classification equations. Table 11.16 shows results of classification and jackknifed classification for original
and cross-validation cases. Unexpectedly, classification improves for the cross-validation sample,
although ordinarily it is worse.

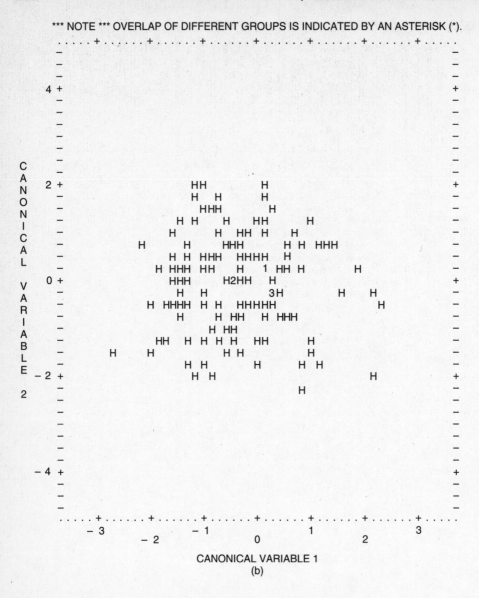

FIGURE 11.2   (CONTINUED)

SPSS DISCRIMINANT is used to test the reliability of the discriminant functions and to provide the loadings for them, as shown in Table 11.17. The total overlap between groups and predictors (AFTER FUNCTION 0) is tested as CHI-SQUARE with 8 df (49.002) and found to be reliable. After the first discriminant function is removed (AFTER FUNCTION 1) CHI-SQUARE with 3 df is 15.614, and still reliable at $p = 0.0014$. This second chi square is a test of the overlap between groups and predictors after the first function is removed. Because there are only two possible discriminant functions with three groups, this is a test of the second discriminant function.

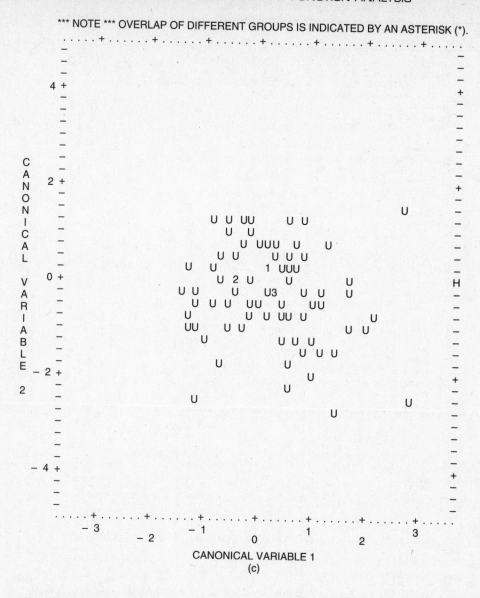

FIGURE 11.2 (CONTINUED)

SPSS DISCRIMINANT, unlike BMDP7M, also provides the loadings for the predictors on the discriminant functions. Labeled Pooled-within groups correlations between discriminating variables, the first column is the correlations between the predictors and the first discriminant function; the second column is the correlations between the predictors and the second discriminant function. The loadings could be calculated by hand according to Equation 11.7, but they are more conveniently found through SPSS, as shown in Table 11.17. For interpretation of this heterogeneous example, loadings less than .50 are not considered.

TABLE 11.15   SETUP AND PARTIAL OUTPUT FROM BMDP7M DISCIRMINANT FUNCTION ANALYSIS OF FOUR
ATTITUDINAL VARIABLES

```
/INPUT          VAR=13.    FILE='DISCRIM.DAT'.    FORMAT IS  FREE.
/VARIABLE       NAMES ARE CASESEQ, WORKSTAT, MARITAL, CHILDREN,
                   RELIGION, RACE, CONTROL, ATTMAR, ATTROLE, SEL,
                   ATTHOUSE, AGE, EDUC.
                MISSING=2*0, 3*9, 5*0, 1, 2*0.
                LABEL = CASESEQ.
                USE = WORKSTAT, CONTROL, ATTMAR, ATTROLE, ATTHOUSE.
/GROUP          VAR = WORKSTAT.
                CODES(WORKSTAT) ARE 1 TO 3.
                NAMES(WORKSTAT) ARE WORKING, HAPHOUSE, UNHOUSE.
                PRIOR = 0.52, 0.30, 0.18.
/TRANSFORM      DELETE = 261, 299.
/DISC           LEVEL=6*0, 3*1, 0, 1, 2*0.
                FORCE = 1.
/PLOT           NO CANON.
/PRINT          NO STEP.  CORR.
/END



PRIOR PROBABILITIES. . . .   0.52000   0.30000   0.18000


WITHIN CORRELATION MATRIX


                  CONTROL     ATTMAR      ATTROLE     ATTHOUSE
                     7           8           9           11


CONTROL     7      1.00000
ATTMAR      8      0.17169     1.00000
ATTROLE     9      0.00912    -0.07010     1.00000
ATTHOUSE   11      0.15500     0.28229    -0.29145     1.00000

*************************************************************************

STEP NUMBER    0

  VARIABLE      F TO   FORCE TOLERNCE  *   VARIABLE      F TO   FORCE TOLERNCE
              REMOVE  LEVEL            *                ENTER  LEVEL
      DF =  2   454                    *        DF =  2   453
                                       *    7 CONTROL     2.96    1   1.00000
                                       *    8 ATTMAR      9.81    1   1.00000
                                       *    9 ATTROLE    11.26    1   1.00000
                                       *   11 ATTHOUSE    8.91    1   1.00000


*************************************************************************

STEP NUMBER    4
VARIABLE ENTERED   7 CONTROL

  VARIABLE      F TO   FORCE TOLERNCE  *   VARIABLE      F TO   FORCE TOLERNCE
              REMOVE  LEVEL            *                ENTER  LEVEL
      DF =  2   450                    *        DF =  2   449
   7 CONTROL     1.08    1   0.95517   *
   8 ATTMAR      4.90    1   0.90351   *
   9 ATTROLE     9.31    1   0.91201   *
  11 ATTHOUSE    3.22    1   0.83315   *

U-STATISTIC(WILKS' LAMBDA) 0.8971503   DEGREES OF FREEDOM    4    2    453
APPROXIMATE FSTATISTIC          6.274   DEGREES OF FREEDOM    8.00   900.00

  F - MATRIX        DEGREES OF FREEDOM =    4   450

            WORKING  HAPHOUSE
HAPHOUSE    7.57
UNHOUSE     4.12      7.30
```

TABLE 11.15   *(CONTINUED)*

CLASSIFICATION FUNCTIONS

GROUP =	WORKING	HAPHOUSE	UNHOUSE
VARIABLE			
7 CONTROL	3.22293	3.21674	3.36966
8 ATTMAR	0.07664	0.04406	0.09771
9 ATTROLE	1.08148	1.15040	1.13735
11 ATTHOUSE	1.64843	1.62485	1.71834
CONSTANT	-50.30873	-52.00294	-56.54160

CLASSIFICATION MATRIX

GROUP	PERCENT CORRECT	NUMBER OF CASES CLASSIFIED INTO GROUP		
		WORKING	HAPHOUSE	UNHOUSE
WORKING	86.2	206	31	2
HAPHOUSE	27.9	96	38	2
UNHOUSE	4.9	66	11	4
TOTAL	54.4	368	80	8

JACKKNIFED CLASSIFICATION

GROUP	PERCENT CORRECT	NUMBER OF CASES CLASSIFIED INTO GROUP		
		WORKING	HAPHOUSE	UNHOUSE
WORKING	84.9	203	33	3
HAPHOUSE	27.2	97	37	2
UNHOUSE	4.9	66	11	4
TOTAL	53.5	366	81	9

EIGENVALUES

              0.07675   0.03519

CUMULATIVE PROPORTION OF TOTAL DISPERSION

              0.68566   1.00000

CANONICAL CORRELATIONS

              0.26699   0.18437

AVERAGE SQUARED CANONICAL CORRELATION   0.05264

GROUP	CANONICAL VARIABLES EVALUATED AT GROUP MEANS	
WORKING	0.14072	0.15053
HAPHOUSE	-0.41601	-0.05393
UNHOUSE	0.28328	-0.35361

```
/INPUT          VAR=13.    FILE='DISCRIM.DAT'.    FORMAT IS  FREE.
/VARIABLE       NAMES ARE CASESEQ, WORKSTAT, MARITAL, CHILDREN,
                   RELIGION, RACE, CONTROL, ATTMAR, ATTROLE, SEL,
                   ATTHOUSE, AGE, EDUC.
                MISSING=2*0, 3*9, 5*0, 1, 2*0.
                LABEL = CASESEQ.
                USE = WORKSTAT, CONTROL, ATTMAR, ATTROLE, ATTHOUSE.
/GROUP          VAR = WORKSTAT.
                CODES(WORKSTAT) ARE 1 TO 6.
                NAMES(WORKSTAT) ARE WORKING, HAPHOUSE, UNHOUSE,
                                    NEWWORK, NEWHAP, NEWUN.
                PRIOR = 0.39, 0.225, 0.135, 0.13, 0.075, 0.045.
                USE = 1 TO 3.
/TRANSFORM      DELETE = 261, 299.
                IF(RNDU(11738) LE .25) THEN WORKSTAT = WORKSTAT + 3.
/DISC           LEVEL=6*0, 3*1, 0, 1, 2*0.
                FORCE = 1.
/PLOT           NO CANON.
/PRINT          NO STEP.  NO POST.  NO POINT.
/END
```

CLASSIFICATION FUNCTIONS

	GROUP =	WORKING	HAPHOUSE	UNHOUSE
VARIABLE				
7 CONTROL		2.84787	2.80040	3.03103
8 ATTMAR		0.09174	0.05442	0.11964
9 ATTROLE		0.98078	1.04430	1.02213
11 ATTHOUSE		1.42151	1.40166	1.47110
CONSTANT		-45.18960	-46.41790	-50.85888

CLASSIFICATION MATRIX

GROUP	PERCENT CORRECT	NUMBER OF CASES CLASSIFIED INTO GROUP		
		WORKING	HAPHOUSE	UNHOUSE
WORKING	86.3	158	24	1
HAPHOUSE	25.0	70	24	2
UNHOUSE	7.6	54	7	5
NEWWORK	0.0	49	7	0
NEWHAP	0.0	28	12	0
NEWUN	0.0	12	3	0
TOTAL	54.2	371	77	8

JACKKNIFED CLASSIFICATION

GROUP	PERCENT CORRECT	NUMBER OF CASES CLASSIFIED INTO GROUP		
		WORKING	HAPHOUSE	UNHOUSE
WORKING	85.8	157	25	1
HAPHOUSE	22.9	72	22	2
UNHOUSE	4.5	56	7	3
NEWWORK	0.0	49	7	0
NEWHAP	0.0	28	12	0
NEWUN	0.0	12	3	0
TOTAL	52.8	374	76	6

TABLE 11.17   SETUP AND PARTIAL OUTPUT FROM SPSS DISCRIMINANT FOR PREDICTION OF   MEMBERSHIP IN THREE GROUPS ON THE BASIS OF FOUR ATTITUDINAL VARIABLES

```
DATA LIST            FILE='DISCRIM.DAT'     FREE
                       /CASESEQ,WORKSTAT,MARITAL,CHILDREN,RELIGION,RACE,
                       CONTROL,ATTMAR,ATTROLE,SEL,ATTHOUSE,AGE,EDUC.
MISSING VALUES       CASESEQ, WORKSTAT, RACE, CONTROL, ATTMAR,
                     ATTROLE, SEL, AGE, EDUC(0)  ATTHOUSE(1)  MARITAL TO
                     RELIGION(9).
SELECT IF            (CASESEQ NE 346).
SELECT IF            (CASESEQ NE 407).
DISCRIMINANT         GROUPS=WORKSTAT(1,3)/
                     VARIABLES=CONTROL ATTMAR ATTROLE ATTHOUSE/
                     ANALYSIS=CONTROL TO ATTHOUSE/
                     PRIORS = SIZE.
```

Prior Probabilities

Group	Prior	Label
1	.52412	
2	.29825	
3	.17763	
Total	1.00000	

Canonical Discriminant Functions

Fcn	Eigenvalue	Pct of Variance	Cum Pct	Canonical Corr		After Fcn	Wilks' Lambda	Chisquare	DF	Sig
					:	0	.8972	49.002	8	.0000
1*	.0768	68.57	68.57	.2670	:	1	.9660	15.614	3	.0014
2*	.0352	31.43	100.00	.1844	:					

* marks the   2 canonical discriminant functions remaining in the analysis.

Structure Matrix:

Pooled-withingroups correlations between discriminating variables
                              and canonical discriminant functions
(Variables ordered by size of correlation within function)

	FUNC 1	FUNC 2
ATTMAR	.71846*	.32299
ATTHOUSE	.67945*	.33332
ATTROLE	-.63925	.72223*
CONTROL	.28168	.44494*

A plot of the placement of the centroids for the three group on the two discriminant functions (canonical variables) as axes appears in Figure 11.3. The points that are plotted are given in Table 11.15 as CANONICAL VARIABLES EVALUATED AT GROUP MEANS.

A summary of information appropriate for publication appears in Table 11.18. In the table are the loadings, univariate F for each predictor, and pooled within-group correlations among predictors (as found through SPSS DISCRIMINANT).

Contrasts run through BMDP7M are shown in Tables 11.19 to 11.21. In Table 11.19, means on predictors for WORKING women are contrasted with the pooled means for HAPHOUSE and

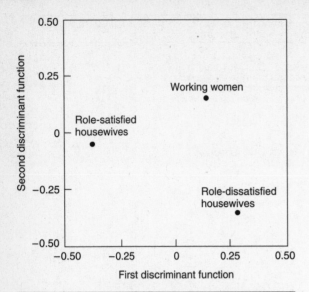

FIGURE 11.3 PLOTS OF THREE GROUPS CENTROIDS ON TWO DISCRIMINANT FUNCTIONS DERIVED FROM FOUR ATTITUDINAL VARIABLES.

TABLE 11.18 RESULTS OF DISCRIMINANT FUNCTION ANALYSIS OF ATTITUDINAL VARIABLES

Predictor variable	Correlations of predictor variables with discriminant functions		Univariate $F(2, 453)$	Pooled within-group correlations among predictors		
	1	2		ATTMAR	ATTROLE	ATTHOUSE
CONTROL	.28	.44	2.96	.17	.01	.16
ATTMAR	.72	.32	9.81		−.07	.28
ATTROLE	−.64	.72	11.26			−.29
ATTHOUSE	.68	.33	8.91			
Canonical R	.27	.18				
Eigenvalue	.08	.04				

UNHOUSE to determine which predictors distinguish WORKING women from others. Table 11.20 has the HAPHOUSE group contrasted with the other two groups; Table 11.21 shows the UNHOUSE group contrasted with the other two groups. F TO ENTER is the univariate F ratio as in ANOVA where each predictor is evaluated separately. F TO REMOVE is the contribution each predictor makes to the contrast after adjustment for all other predictors. Critical F TO REMOVE for the four predictors at $\alpha_i = .01$ (cf. Section 11.6.4) with 1 and 450 df is approximately 6.8.

On the basis of both F TO ENTER and F TO REMOVE, the predictor that most clearly distinguishes the WORKING group from the other two is ATTROLE. The HAPHOUSE group differs from the other two groups on the basis of ATTMAR after adjustment for the remaining predictors; ATTMAR, ATTROLE, and ATTHOUSE reliably separate this group from the other two in ANOVA.

TABLE 11.19	SETUP AND PARTIAL OUTPUT OF BMDP7M CONTRASTING THE WORKING GROUP WITH THE OTHER TWO GROUPS

```
/INPUT          VAR=13.    FILE='DISCRIM.DAT'.   FORMAT IS  FREE.
/VARIABLE       NAMES ARE CASESEQ, WORKSTAT, MARITAL, CHILDREN,
                    RELIGION, RACE, CONTROL, ATTMAR, ATTROLE, SEL,
                    ATTHOUSE, AGE, EDUC.
                MISSING=2*0, 3*9, 5*0, 1, 2*0.
                LABEL = CASESEQ.
                USE = WORKSTAT, CONTROL, ATTMAR, ATTROLE, ATTHOUSE.
/GROUP          VAR = WORKSTAT.
                CODES(WORKSTAT) ARE 1 TO 3.
                NAMES(WORKSTAT) ARE WORKING, HAPHOUSE, UNHOUSE.
/TRANSFORM      DELETE = 261, 299.
/DISC           LEVEL=6*0, 3*1, 0, 1, 2*0.
                FORCE = 1.
                CONTRAST = -2, 1, 1.
/PLOT           NO CANON.
/PRINT          NO STEP.  NO POST.  NO POINT.
/END
```

```
******************************************************************************

STEP NUMBER   0

    VARIABLE    F TO   FORCE TOLERNCE *  VARIABLE     F TO   FORCE TOLERNCE
                REMOVE LEVEL           *              ENTER  LEVEL
         DF =  1  454                  *      DF =  1   453
                                       *   7 CONTROL    1.08    1   1.00000
                                       *   8 ATTMAR     0.13    1   1.00000
                                       *   9 ATTROLE   16.55    1   1.00000
                                       *  11 ATTHOUSE   0.06    1   1.00000

******************************************************************************

STEP NUMBER   4
VARIABLE ENTERED   8 ATTMAR

    VARIABLE    F TO   FORCE TOLERNCE *  VARIABLE     F TO   FORCE TOLERNCE
                REMOVE LEVEL           *              ENTER  LEVEL
         DF =  1  450                  *      DF =  1   449
  7 CONTROL      0.79    1   0.95517  *
  8 ATTMAR       0.22    1   0.90351  *
  9 ATTROLE     16.87    1   0.91201  *
 11 ATTHOUSE     0.83    1   0.83315  *


EIGENVALUES

            0.07675     0.03519


CANONICAL CORRELATIONS

            0.26699     0.18437
```

The UNHOUSE group does not differ from the other two when each predictor is adjusted for all others but does differ on ATTHOUSE and ATTMAR without adjustment.

A checklist for a direct discriminant function analysis appears in Table 11.22. It is followed by an example of a Results section, in journal format, for the analysis just described.

TABLE 11.20	SETUP AND PARTIAL OUTPUT OF BMDP7M CONTRASTING THE HAPHOUSE GROUP WITH THE OTHER TWO GROUPS

```
/INPUT          VAR=13.    FILE='DISCRIM.DAT'.   FORMAT IS FREE.
/VARIABLE       NAMES ARE CASESEQ, WORKSTAT, MARITAL, CHILDREN,
                   RELIGION, RACE, CONTROL, ATTMAR, ATTROLE, SEL,
                   ATTHOUSE, AGE, EDUC.
                MISSING=2*0, 3*9, 5*0, 1, 2*0.
                LABEL = CASESEQ.
                USE = WORKSTAT, CONTROL, ATTMAR, ATTROLE, ATTHOUSE.
/GROUP          VAR = WORKSTAT.
                CODES(WORKSTAT) ARE 1 TO 3.
                NAMES(WORKSTAT) ARE WORKING, HAPHOUSE, UNHOUSE.
/TRANSFORM      DELETE = 261, 299.
/DISC           LEVEL=6*0, 3*1, 0, 1, 2*0.
                FORCE = 1.
                CONTRAST = 1, -2, 1.
/PLOT           NO CANON.
/PRINT          NO STEP.  NO POST.  NO POINT.
/END
```

****

STEP NUMBER   0

VARIABLE	F TO REMOVE	FORCE LEVEL	TOLERNCE	*	VARIABLE	F TO ENTER	FORCE LEVEL	TOLERNCE
DF = 1 454				*	DF = 1 453			
				*	7 CONTROL	3.42	1	1.00000
				*	8 ATTMAR	18.95	1	1.00000
				*	9 ATTROLE	11.73	1	1.00000
				*	11 ATTHOUSE	17.05	1	1.00000

****

STEP NUMBER   4
VARIABLE ENTERED   7 CONTROL

VARIABLE	F TO REMOVE	FORCE LEVEL	TOLERNCE	*	VARIABLE	F TO ENTER	FORCE LEVEL	TOLERNCE
DF = 1 450				*	DF = 1 449			
7 CONTROL	0.78	1	0.95517	*				
8 ATTMAR	9.66	1	0.90351	*				
9 ATTROLE	5.45	1	0.91201	*				
11 ATTHOUSE	4.09	1	0.83315	*				

EIGENVALUES

        0.07675   0.03519

CANONICAL CORRELATIONS

        0.26699   0.18437

TABLE 11.21	SET AND PARTIAL OUTPUT OF BMDP7M CONTRASTING THE UNHOUSE GROUP WITH THE OTHER TWO GROUPS

```
/INPUT          VAR=13.   FILE='DISCRIM.DAT'.   FORMAT IS FREE.
/VARIABLE       NAMES ARE CASESEQ, WORKSTAT, MARITAL, CHILDREN,
                    RELIGION, RACE, CONTROL, ATTMAR, ATTROLE, SEL,
                    ATTHOUSE, AGE, EDUC.
                MISSING=2*0, 3*9, 5*0, 1, 2*0.
                LABEL = CASESEQ.
                USE = WORKSTAT, CONTROL, ATTMAR, ATTROLE, ATTHOUSE.
/GROUP          VAR = WORKSTAT.
                CODES(WORKSTAT) ARE 1 TO 3.
                NAMES(WORKSTAT) ARE WORKING, HAPHOUSE, UNHOUSE.
/TRANSFORM      DELETE = 261, 299.
/DISC           LEVEL=6*0, 3*1, 0, 1, 2*0.
                FORCE = 1.
                CONTRAST = 1, 1, -2.
/PLOT           NO CANON.
/PRINT          NO STEP.  NO POST.  NO POINT.
/END
```

****

STEP NUMBER   0

VARIABLE	F TO REMOVE	FORCE LEVEL	TOLERNCE	*	VARIABLE	F TO ENTER	FORCE LEVEL	TOLERNCE
DF = 1 454				*	DF = 1 453			
				*	7 CONTROL	5.81	1	1.00000
				*	8 ATTMAR	12.27	1	1.00000
				*	9 ATTROLE	0.03	1	1.00000
				*	11 ATTHOUSE	11.58	1	1.00000

****

STEP NUMBER   4
VARIABLE ENTERED   9 ATTROLE

VARIABLE	F TO REMOVE	FORCE LEVEL	TOLERNCE	*	VARIABLE	F TO ENTER	FORCE LEVEL	TOLERNCE
DF = 1 450				*	DF = 1 449			
7 CONTROL	2.13	1	0.95517	*				
8 ATTMAR	5.56	1	0.90351	*				
9 ATTROLE	1.14	1	0.91201	*				
11 ATTHOUSE	6.20	1	0.83315	*				

EIGENVALUES

              0.07675   0.03519

CANONICAL CORRELATIONS

              0.26699   0.18437

TABLE 11.22   CHECKLIST FOR DIRECT DISCRIMINANT FUNCTION ANALYSIS

1. Issues
   a. Unequal sample sizes and missing data
   b. Normality of sampling distributions
   c. Outliers
   d. Linearity
   e. Homogeneity of variance-covariance matrices
   f. Multicollinearity and singularity
2. Major analysis
   a. Significance of discriminant functions. If significant:
      (1) Variance accounted for
      (2) Plot(s) of discriminant functions
      (3) Loading matrix
   b. Variables separating each group
3. Additional analyses
   a. Group means for high-loading variables
   b. Pooled within-group correlations among predictor variables
   c. Classification results
      (1) Jackknifed classification
      (2) Cross-validation
   d. Change in Rao's V (or stepdown F) plus univariate F for predictors

---

Results

A direct discriminant function analysis was performed using four attitudinal variables as predictors of membership in three groups. Predictors were locus of control, attitude toward marital status, attitude toward role of women, and attitude toward homemaking. Groups were working women, role-satisfied housewives, and role-dissatisfied housewives.

Of the original 465 cases, seven were dropped from analysis because of missing data. Missing data appeared to be randomly scattered throughout groups and predictors. Two additional cases were identified as multivariate outliers with $p < .001$, and were also deleted. Both of the outlying cases were in the working group; they were women with extraordinarily favorable attitudes toward housework. For the remaining 456 cases (239 working women, 136 role-satisfied housewives, and 81 role-dissatisfied housewives), evaluation of assumptions of linearity, normality, multicollinearity or singularity, and homogeneity of variance-covariance matrices revealed no threat to multivariate analysis.

Two discriminant functions were calculated, with a combined $\chi^2(8) = 49.00$, $p < .01$. After removal of the first function, there was still strong association between groups and predictors, $\chi^2(3) = 15.61$, $p < .01$. The two discriminant functions accounted for 69% and 31%, respectively, of the between-group variability. [χ^2 values and percent of variance are from Table 11.17; cf. Section 11.6.2.] As shown in

Figure 11.3, the first discriminant function maximally separates role-satisfied housewives from the other two groups. The second discriminant function discriminates role-dissatisfied housewives from working women, with role satisfied housewives falling between these two groups.

The loading matrix of correlations between predictors and discriminant functions, as seen in Table 11.18, suggests that the best predictors for distinguishing between role-satisfied housewives and the other two groups (first function) are attitudes toward current marital status, toward women's role, and toward homemaking. Role-satisfied housewives have more favorable attitudes toward marital status (mean = 20.60) than working women (mean = 23.40) or role-dissatisfied housewives (mean = 25.62), and more conservative attitudes toward women's role (mean = 37.19) than working women (mean = 33.86) or dissatisfied housewives (mean = 35.67). Role-satisfied women are more favorable toward homemaking (mean = 22.51) than either working women (mean = 23.81) or role-dissatisfied housewives (mean = 24.93). [*Group means are shown in Table 11.15.*] Loadings less than .50 are not interpreted.

One predictor, attitudes toward women's role, has a loading in excess of .50 on the second discriminant function, which separates role-dissatisfied housewives from working women. Role-dissatisfied housewives have more conservative attitudes toward the role of women than working women (means have already been cited).

Three contrasts were performed where each group, in turn, was contrasted with the other two groups, pooled, to determine which predictors reliably separate each group from the other two groups. When working women were contrasted with the pooled groups of housewives, after adjustment for all other predictors, and keeping overall $\alpha < .05$ for the four predictors, only attitude toward women's role significantly separates working women from the other two groups, $\underline{F}(1, 450) = 16.87$. The squared semipartial correlation between working women vs. the pooled groups of housewives and attitudes toward women's role is .04. [*Equation 11.9 applied to output in Table 11.19.*] (Attitude toward women's role also separates working women from the other two groups without adjustment for the other predictors.)

Role-satisfied housewives differ from the other two groups on attitudes toward marital status, $\underline{F}(1, 450) = 9.66$, $\underline{p} < .05$. The squared semipartial correlation between this predictor and the role-satisfied group vs. the other two groups pooled is .02. (Attitudes toward marital status, toward the role of women, and toward housework separate role-

satisfied housewives from the other two groups without adjustment for
the other predictors.)

The group of role-dissatisfied housewives does not differ from the
other two groups on any predictor after adjustment for all other predictors. Without adjustment, however, the group differs on attitudes
toward marital status (less favorable) and attitudes toward homemaking
(less favorable).

Pooled within-group correlations among the four predictors are shown
in Table 11.18. [*Produced by run of Table 11.15 in section of output
not shown called* WITHIN CORRELATION MATRIX.] Of the six correlations,
four would show statistical significance at α = .01 if tested individually. There is a small positive relationship between locus of control
and attitude toward marital status, with $r(454)$ = .17, p < .01, indicating that women who are more satisfied with their current marital
status are less likely to attribute control of reinforcements to external sources. Attitude toward homemaking is positively correlated with
locus of control, $r(454)$ = .16, p < .01, and attitude toward marital
status, $r(454)$ = .28, p < .01, and negatively correlated with attitude
toward women's role, $r(454)$ = -.29, p < .01. This indicates that women
with negative attitudes toward homemaking are likely to attribute control to external sources, to be dissatisfied with their current marital
status, and to have more liberal attitudes toward women's role.

With the use of a jackknifed classification procedure for the total
usable sample of 456 women, 244 (53.5%) were classified correctly,
compared to 178.9 (39%) who would be correctly classified by chance
alone. The 53.5% classification rate was achieved by classifying a
disproportionate number of cases as working women. Although 52% of the
women actually were employed, the classification scheme, using sample
proportions as prior probabilities, classified 80.3% of the women as
employed [*366/456 from jackknifed classification matrix in Table
11.15*]. This means that the working women were more likely to be correctly classified (84.9% correct classifications) than either the role-
satisfied housewives (27.2% correct classifications) or the
role-dissatisfied housewives (4.9% correct classifications).

The stability of the classification procedure was checked by a
cross-validation run. Approximately 25% of the cases were withheld from
calculation of the classification functions in this run. For the 75% of
the cases from whom the functions were derived, there was a 52.8% correct classification rate. For the cross-validation cases, classification actually improved to 55%. This indicates a high degree of
consistency in the classification scheme, and an unusual random division of cases into the cross-validation sample.

11.9 ■ SOME EXAMPLES FROM THE LITERATURE

Curtis and Simpson (1977) explored predictors for three types of drug users: daily opioid users, less-than-daily opioid users, and nonusers of opioids. A preliminary discriminant function analysis used 31 predictors related to demographic characteristics, alcohol use, and drug history to predict group membership. Both discriminant functions were statistically reliable and virtually all the predictors showed significant univariate Fs for group differences, not surprising with the sample size greater than 23,000. For the second analysis, only the 13 predictors with univariate $F > 100$ were selected.

The first discriminant function accounted for 20% of the variance between groups, the second for 1% of the variance. The first discriminant function separated daily opioid users from the other two groups, presumably based on comparison of group centroids, although they were not reported. Correlations between predictors and the discriminant function were used to interpret the function. Daily opioid users were likely to be black, to be older, to have used illicit drugs for a longer period of time, to be responsible for slightly more family members, and to have begun daily use of illicit drugs with a drug other than marijuana.

A series of two-group stepwise discriminant functions was run by Strober and Weinberg (1977) on families that did or did not purchase various merchandise. Each item (e.g., dishwasher, hobby and recreational items, college education) was analyzed in a separate DISCRIM. The hypothesis was that predictors based on wife's employment status would be unrelated to purchase of the items considered in the study. Other predictors were family income, net family assets, stage of family life cycle, whether or not the family recently moved, and whether or not the family was in the market for the item being considered. In none of the analyses did predictors related to wife's employment status show significant F to enter the stepwise discriminant function. Different predictors were important for various expenditures. For example, for college education, color TV, and washer, life-cycle stage of the family was the first predictor to enter. For purchases of furniture, net family assets was the first predictor to enter.

For those analyses in which sample size was large enough, part of the sample was reserved for cross-validation. Percent correct classification for original sample and cross-validation sample ranged from 53% to 73%. In all except one analysis in which cross-validation was possible, the difference in correct prediction between the two samples was less than 2%. For hobby and recreational items, however, the classification coefficients did not generalize to the cross-validation sample.

Men involved in midcareer changes were studied by Wiener and Vaitenas (1977). Two personality inventories were used to distinguish midcareer changers from vocationally stable men. Separate discriminant function analyses were run for each inventory. The Edwards Personal Preference Schedule generated 15 predictors. The Gordon Personal Profile and Gordon Personal Inventory generated 8 predictors. For each analysis, Wilks' Lambda showed statistically significant association between groups and predictors, with discriminant functions accounting for 28% and 42% of the variance in the Edwards and Gordon analyses, respectively.

Interpretation was based on standardized discriminant function coefficients (rather than the less biased correlations of predictors with discriminant functions). Midcareer changers scored lower on endurance, dominance, and order scales on the Edwards inventory, and showed less responsibility and ascendancy as measured by the Gordon inventory.

Caffrey and Lile (1976) explored differences between psychology and physics students in attitudes toward science. Predictors were based on a scientific attitude questionnaire, in which the respondent expressed degree of agreement with 15 quotations from various writers on science.

Stepwise discriminant function analysis on upper-division psychology vs. physics majors followed preliminary analyses on a random sample of humanities, social science, and nature science majors.

Three items entered the stepwise analysis. Entering first was a statement suggesting that the results of scientific knowledge about behavior should be used to change behavior. Psychology majors were more likely to agree with this item than physics majors. Second, physics majors were more likely to agree with a statement suggesting that humans are free agents, not amenable to scientific prediction, and third, more likely to agree with a statement that Shakespeare conveys more truth about human nature than results of questionnaires.

Distinction between menopausal and nonmenopausal women in Hawaii was studied by Goodman, Steward, and Gilbert (1977). Predictors included medical, gynecological, and obstetrical history; age and age squared; physical measurements; and blood test results. Separate stepwise discriminant function analyses were run for Caucasian women, Japanese women, and then both groups pooled. Because only two groups were compared at a time, analyses were run through a stepwise multiple regression program. Preliminary runs with 35 predictors allowed elimination of nonsignificant predictors; then the stepwise analyses were rerun. In all analyses, age and age squared were forced into the stepwise series first because of known relationships with the remaining predictors.

Percent of variance accounted for in separating menopausal from nonmenopausal women was 45% and 36% for Caucasian and Japanese women, respectively. After adjustment for effects of age, the only predictor retained in the stepwise analysis for both races was "surgery related to a female disorder." For Caucasian women only, one additional predictor, medication, was selected in the stepwise analysis. For age, age squared, surgery, and medication, discriminant function coefficients and their standard errors were given.[15] Finally, predictors common to both races were selected and heterogeneity of coefficients between Japanese and Caucasian samples were tested.[16] No significant heterogeneity was found—that is, the two races did not produce significantly different discriminant functions. An alternative strategy for this test is a factorial discriminant function analysis, with race as one IV and menopausal vs. nonmenopausal as the other.

---

[15] Because this two-group analysis was run through a multiple regression program, discriminant function coefficients were produced as beta weights (cf. Chapter 5) and standard errors were available.

[16] The test for heterogeneity reported was that of Rao (1952) as applied by Goodman et al. (1974).

CHAPTER 12
Logistic Regression

12.1 ■ GENERAL PURPOSE AND DESCRIPTION

Logistic regression allows one to predict a discrete outcome such as group membership from a set of variables that may be continuous, discrete, dichotomous, or a mix. Because of its popularity in the health sciences, the discrete outcome in logistic regression is often disease/no disease. For example, can presence or absence of hay fever be diagnosed from geographic area, season, degree of nasal stuffiness, and body temperature?

Logistic regression is related to, and answers the same questions as, discriminant function analysis, the logit form of multiway frequency analysis with a discrete DV, and multiple regression analysis with a dichotomous DV. However, logistic regression is more flexible than the other techniques. Unlike discriminant function analysis, logistic regression has no assumptions about the distributions of the predictor variables; in logistic regression, the predictors do not have to be normally distributed, linearly related, or of equal variance within each group. Unlike multiway frequency analysis, the predictors do not need to be discrete; the predictors can be any mix of continuous, discrete and dichotomous variables. Unlike multiple regression analysis, which also has distributional requirements for predictors, logistic regression cannot produce negative predicted probabilities.

There may be two or more outcomes in logistic regression. If there are more than two outcomes, they may or may not have order (e.g., no hay fever, moderate hay fever, severe hay fever). Logistic regression emphasizes the probability of a particular outcome for each case. For example, it evaluates the probability that a given person has hay fever, given that person's pattern of responses to questions about geographic area, season, nasal stuffiness, and temperature.

Logistic regression analysis is especially useful when the distribution of responses on the DV is expected to be nonlinear with one or more of the IVs. For example, the probability of heart disease may be little affected (say 1%) by a 10-point difference among people with low blood pressure (e.g., 110 vs. 120) but may change quite a bit (say 5%) with an equivalent difference among people with high blood pressure (e.g., 180 vs. 190). Thus, the relationship between heart disease and blood pressure is not linear.

Because the model produced by logistic regression is nonlinear, the equations used to describe the outcomes are slightly more complex than those for multiple regression. The outcome variable, $\hat{Y}$, is the probability of having one outcome or another based on a nonlinear function of the best linear combination of predictors; with two outcomes:

$$\hat{Y}_i = \frac{e^u}{1 + e^u} \tag{12.1}$$

where $\hat{Y}_i$ is the estimated probability that the ith case $(i = 1, \cdot \cdot \cdot, n)$ is in one of the categories and u is the usual linear regression equation:

$$u = A + B_1X_1 + B_2X_2 + \cdot \cdot \cdot + B_kX_k \tag{12.2}$$

with constant A, coefficients B_j, and predictors, X_j for k predictors $(j = 1, 2, \cdot \cdot \cdot, k)$.

This linear regression equation creates the *logit* or log of the odds:

$$\ln \left(\frac{\hat{Y}}{1 - \hat{Y}} \right) = A + \Sigma B_j X_{ij} \tag{12.3}$$

That is, the linear regression equation is the (natural log of the) probability of being in one group divided by the probability of being in the other group. The procedure for estimating coefficients is maximum likelihood, and the goal is to find the best linear combination of predictors to maximize the likelihood of obtaining the observed outcome frequencies.

Logistic regression, like multiway frequency analysis, can be used to fit and compare models. The simplest (and worst-fitting) model includes only the constant and none of the predictors. The most complex (and "best"-fitting) model includes the constant, all predictors and, perhaps, interactions among predictors. However, not all predictors (and interactions) may be related to outcome. The researcher uses goodness-of-fit tests to choose the model that does the best job of prediction with the fewest predictors.

12.2 ■ KINDS OF RESEARCH QUESTIONS

The goal of analysis is to correctly predict category of outcome for individual cases. The first step is to establish that there is a relationship between the outcome and the set of predictors. If a relationship is found, one usually tries to simplify the model by eliminating some predictors while still maintaining strong prediction. Once a reduced set of predictors is found, the equation can be used to predict outcomes for new cases on a probabilistic basis.

12.2.1 ■ Prediction of Group Membership or Outcome

Can outcome be predicted from the set of variables? For example, can hay fever be predicted from geographic area, season, degree of nasal stuffiness, and body temperature? Several tests of relationship are available in logistic regression. The most straightforward compares a model with the constant plus predictors with a model that has only the constant. A reliable difference between the models indicates a relationship between the predictors and the outcome. This procedure is demonstrated in Section 12.4.2.

An alternative is to test a model with only some predictors against the model with all predictors (called a full model). The goal is to find a nonsignificant χ^2, indicating no reliable difference between the model with only some predictors and the full model. The use of these and other goodness-of-fit tests is discussed in Section 12.6.1.1.

12.2.2 ■ Importance of Predictors

Which variables predict outcome? How do variables affect the outcome? Does a particular variable increase or decrease the probability of an outcome, or does it have no effect on outcome? Does inclusion of information about geographic area improve prediction of hay fever and is a particular area associated with an increase or decrease in the probability that a case has hay fever? Several methods of answering these questions are available in logistic regression. One may, for instance, ask how much the model is harmed by eliminating a predictor, or one may assess the statistical significance of the coefficients associated with each of the predictors. These procedures are discussed in Sections 12.4 and 12.6.1.2.

12.2.3 ■ Interactions among Predictors

As in multiway frequency (logit) analysis, a model can also include interactions among the predictor variables: two-way interactions and, if there are many predictor variables, higher order interactions. For example, knowledge of geographic area and season, in combination, might be useful in the prediction of hay fever; or knowledge of degree of nasal stuffiness combined with fever. Geographic area may be associated with hay fever only in some seasons; stuffiness might only matter with no fever. Other combinations such as between temperature and geographic areas may, however, not be helpful.

Like individual predictors, interactions may complicate a model without reliably improving prediction. Decisions about including interactions are made in the same way as decisions about including individual predictors. Section 12.6.4 discusses decisions about whether inclusion of interactions also presumes inclusion of their individual components.

12.2.4 ■ Parameter Estimates

The parameter estimates in logistic regression are the coefficients of the predictors included in a model. They are related to the *A* and *B* values of Equation 12.2. Section 12.4.1 discusses methods for calculating parameter estimates. Section 12.6.6 shows how to use parameter estimates to calculate and interpret odds. For example, what are the odds that someone has hay fever in the spring, given residence in the Midwest, nasal stuffiness, and no fever?

12.2.5 ■ Classification of Cases

How good is a reliable model at classifying cases for whom the outcome is known? For example, how many people with hay fever are diagnosed correctly? How many people without hay fever are diagnosed correctly? The researcher establishes a cutpoint (say, .5) and then asks, for instance,

"How many people with hay fever are correctly classified if everyone with a predicted probability of .5 or more is diagnosed as having hay fever?" Classification of cases is discussed in Section 12.6.3.

12.2.6 ■ Significance of Prediction with Covariates

The researcher may consider some of the predictors covariates and others independent variables. For example, the researcher may consider stuffiness and temperature covariates, and geographic area and season independent variables in an analysis that asks if knowledge of geographic area and season added to knowledge of physical symptoms reliably improves prediction over knowledge of physical symptoms alone. Section 12.5.2 discusses sequential logistic regression and the large sample example of Section 12.8.3 demonstrates sequential logistic regression.

12.2.7 ■ Strength of Association

How strong is the relationship between outcome and the set of predictors in the chosen model? What proportion of variance in outcome is associated with the set of predictors? For example, what proportion of the variability in hay fever is accounted for by geographic area, season, stuffiness, and temperature?

The logic of assessing strength of association is different in routine statistical hypothesis testing from situations where models are being evaluated. In routine statistical hypothesis testing, one does not report strength of association for a nonsignificant effect. However, in model testing, the goal is often to find *non*significance, to find a model that is not reliably different from the full model. In that case, it is appropriate to report strength of association for a model that is not reliably different from a full model. However, when samples are large, there may be a reliable deviation from the full model, even when a model does a fine job of prediction. Therefore, strength of association may also sometimes be reported with a reliably deviant model. Computer generated measures of strength of association are mentioned in Sections 12.7.3 and 12.7.4.

12.3 ■ LIMITATIONS TO LOGISTIC REGRESSION ANALYSIS

Logistic regression is relatively free of restrictions and, with the capacity to analyze a mix of all types of predictors (continuous, discrete, and dichotomous) the variety and complexity of data sets that can be analyzed is almost unlimited. The outcome variable does have to be discrete but a continuous variable can be converted to a discrete one when there is reason to do so.

12.3.1 ■ Theoretical Issues

The usual cautions about causal inference apply, as in all analyses in which one variable is an outcome. To say that the probability of correctly diagnosing hay fever is related to geographic area, season, nasal stuffiness, and fever is not to imply that any of those variables cause hay fever.

As a flexible alternative to both discriminant function analysis and the logit form of multiway frequency analysis, the popularity of logistic regression analysis is growing. The technique has the sometimes useful property of producing predicted values that are probabilities between 0 and 1.

However, when assumptions regarding the distributions of predictors are met, discriminant function analysis may be a more powerful and efficient analytic strategy. On the other hand, discriminant function analysis sometimes overestimates the size of the association with dichotomous predictors (Hosmer & Lemeshow, 1989). Multiple regression is likely to be more powerful than logistic regression when the outcome is continuous and assumptions regarding it and the predictors are met.

When all the predictors are discrete, multiway frequency analysis offers some convenient screening procedures that may make it the more desirable option.

12.3.2 ■ Practical Issues

Although assumptions regarding the distributions of predictors are not required for logistic regression, multivariate normality and linearity among the predictors may enhance power, because a linear combination of predictors is used to form the exponent (see Equations 12.1 and 12.2). Other limitations are mentioned below.

12.3.2.1 ■ Ratio of Cases to Variables

A number of problems may occur when there are too few cases relative to the number of predictor variables. Logistic regression may produce extremely large parameter estimates and standard errors, and, possibly, failure of convergence when combinations of discrete variables result in too many cells with no cases. If this occurs, *collapse categories, delete the offending category, or delete the discrete variable if it is not important to the analysis.*

A maximum likelihood solution also is impossible when outcome groups are perfectly separated. Complete separation of groups by a dichotomous predictor occurs when all cases in one outcome group have a particular value of a predictor (e.g., all those with hay fever have sniffles) while all those in another group have another value of a predictor (e.g., no hay fever and no sniffles). This is likely to be a result of too small a sample rather than a fortuitous discovery of the perfect predictor that will generalize to the population. Complete separation of groups also can occur when there are too many variables relative to the few cases in one outcome (Hosmer & Lemeshow, 1989). This is essentially a problem of overfitting, as occurs with a small case-to-variable ratio in multiple regression (cf. Section 5.3.2.1).

Extremely high parameter estimates and standard errors are indications that a problem exists. These estimates also increase with succeeding iterations (or the solution fails to converge). If this occurs, *increase the number of cases or eliminate one or more predictors.*

12.3.2.2 ■ Adequacy of Expected Frequencies

When a goodness-of-fit test is used that compares observed with expected frequencies in cells formed by combinations of discrete variables, the analysis may have little power if expected frequencies are too small. *If you plan to use such a goodness-of-fit test, evaluate expected cell frequencies for all pairs of discrete variables, including the outcome variable.* Recall that expected frequencies = [(row total) × (column total)]/grand total. It is best if all expected frequencies are greater than one, and that no more than 20% are less than five. Should either of these conditions fail, the choices are (1) accept lessened power for the analysis, (2) collapse categories for variables with more than two levels, (3) delete discrete variables to reduce the number of cells, or (4) use a goodness-of-fit criterion that is not based on observed versus expected frequencies of cells formed by categorical variables, as discussed in Section 12.6.1.1. (Consult Section 7.3.2.2 for a further discussion of this issue.)

12.3.2.3 ■ Multicollinearity

Logistic regression, like all varieties of multiple regression, is sensitive to extremely high correlations among predictor variables, signaled by exceedingly high standard errors for parameter estimates and/or failure of a tolerance test in the computer run. To find a source of multicollinearity among the discrete predictors, use multiway frequency analysis (cf. Chapter 7) to find very strong relationships among them. To find a source of multicollinearity among the continuous predictors, replace the discrete predictors with dichotomous dummy variables and then use the procedures of Section 4.1.7. *Delete one or more redundant variables from the model to eliminate multicollinearity.*

12.3.2.4 ■ Outliers in the Solution

One or more of the cases may be very poorly predicted by the solution; a case that actually is in one category of outcome may show a high probability for being in another category. If there are enough cases like this, the model has poor fit. Outlying cases are found by examination of residuals, which can also aid in interpreting the results of the logistic regression analysis. Section 12.4.4 discusses how to examine residuals to evaluate outliers.

12.4 ■ FUNDAMENTAL EQUATIONS FOR LOGISTIC REGRESSION

Table 12.1 shows a hypothetical data set in which falling down (0 = not falling, 1 = falling) on a ski run is tested against the difficulty of the run (on an ordered scale from 1 to 3, treated as if con-

TABLE 12.1 SMALL SAMPLE OF HYPOTHETICAL DATA FOR ILLUSTRATION OF LOGISTIC REGRESSION ANALYSIS

Fall	Difficulty	Season
1	3	1
1	1	1
0	1	3
1	2	3
1	3	2
0	2	2
0	1	2
1	3	1
1	2	3
1	2	1
0	2	2
0	2	3
1	3	2
1	2	2
0	3	1

tinuous) and the season (a categorical variable where 1 = autumn, 2 = winter, and 3 = spring). Data from 15 skiers are presented.

Logistic regression uses procedures similar to both multiple regression and multiway frequency analysis. Like multiple regression, the prediction equation includes a linear combination of the predictor variables. For example, with three predictors and no interactions:

$$\hat{Y}_i = \frac{e^{A+B_1X_1+B_2X_2+B_3X_3}}{1+e^{A+B_1X_1+B_2X_2+B_3X_3}} \tag{12.4}$$

The difference between multiple regression and logistic regression is that the linear portion of the equation $(A+B_1X_1+B_2X_2+B_3X_3)$, the logit, is not the end in itself, but is used to find the odds of being in one of the categories. Similar to multiway frequency analysis, models are evaluated by assessing the (natural log) likelihood for each model. Models are then compared by calculating the difference between their log likelihoods.

In this section the simpler calculations are illustrated in detail, while those involving calculus or matrix inversion are merely described and left to computer software for solution. At the end of the section, the most straightforward program in each package (BMDPLR, SPSS LOGISTIC REGRESSION, SYSTAT LOGIT, and SAS LOGISTIC) is demonstrated for the small data set. Additional programs for analysis of more complex data sets are described in Section 12.7.

Before analysis, discrete variables are recoded into a series of dichotomous (dummy) variables, one fewer than there are categories. Thus two dichotomous variables, called season(1) and season(2), are created to represent the three categories of season. Following the convention of most software, season(1) is coded 1 if the season is winter, and 0 otherwise; season(2) is coded 1 if spring, and 0 otherwise. Autumn is identified by codes of 0 on both dummy variables.

12.4.1 ■ Testing and Interpreting Coefficients

Solving for logistic regression coefficients A and B and their standard errors involves calculus, in which values are found using maximum likelihood methods. These values, in turn, are used to evaluate the fit of one or more models (cf. Section 12.4.2). If an acceptable model is found, the statistical significance of each of the coefficients is evaluated[1] using the Wald test where the coefficient is divided by its standard error:

$$W_j = \frac{B_j}{SE_{Bj}} \tag{12.5}$$

A parameter estimate divided by its standard error is a z statistic. Table 12.2 shows these coefficients, their standard errors, and the Wald test obtained from statistical software. None of the predictors is statistically significant in this tiny data set.

The logistic regression equation is:

$$\text{Prob(fall)} = \hat{Y}_i = \frac{e^{-0.849+(1.011)(DIFF)+(-1.346)(SEAS1)+(-0.928)(SEAS2)}}{1 + e^{-0.849+(1.011)(DIFF)+(-1.346)(SEAS1)+(-0.928)(SEAS2)}}$$

---

[1] It is convenient for didactic purposes to first illustrate coefficients and then show how they are used to develop goodness-of-fit tests for models.

TABLE 12.2   COEFFICIENTS, STANDARD ERRORS AND WALD
                    TEST DERIVED FOR SMALL SAMPLE EXAMPLE

Term	Coefficient	Standard Error	Wald Test (z)
(CONSTANT)	−0.849	2.18	−0.39
DIFFICULTY	1.011	0.90	1.13
SEASON(1)	−1.346	1.48	−0.91
SEASON(2)	−0.928	1.59	−0.58

Because the equation is solved for the outcome "falling", coded 1,[2] the derived probabilities are also for falling. Since none of the coefficients is significant, the equation would normally not be applied to any cases. However, for illustration, the equation is applied below to the first case, a skier who actually did fall on a difficult run in the autumn. The probability is:

$$Prob(fall) = \hat{Y}_1 = \frac{e^{-0.849+(1.011)(3)+(-1.346)(0)+(-0.928)(0)}}{1 + e^{-0.849+(-1.011)(3)+(-1.346)(0)+(0.928)(0)}}$$

$$= \frac{e^{2.184}}{1 + e^{2.184}}$$

$$= \frac{8.882}{9.882} = .899$$

Prediction is quite good for this case, since the probability of falling is .899 (with a residual of 1 − .899 = .101, where 1 represents the actual outcome: falling). Section 12.6.6 discusses further interpretation of coefficients.

12.4.2 ■ Goodness-of-Fit

For a candidate model, a log-likelihood is calculated, based on summing the probabilities associated with the predicted and actual outcomes for each case:

$$\text{log-likelihood} = \sum_{i=1}^{n} [Y_i \ln(\hat{Y}_i) + (1 - Y_i) \ln(1 - \hat{Y}_i)] \qquad (12.6)$$

Table 12.3 shows the actual outcome (Y) and predicted probability of falling for the 15 cases in the small sample example, along with the values needed to calculate log-likelihood.

Two models are compared by computing the difference in their log-likelihoods and using chi square. The bigger model is the one to which predictors have been added to the smaller model. Models must be nested to be compared; all the components of the smaller model must also be in the bigger model.

$$\chi^2 = 2[(\text{log-likelihood for bigger model}) - (\text{log-likelihood for smaller model})] \qquad (12.7)$$

---

[2] Some texts and software solve the equation for the outcome coded 0, in the example above "not falling".

TABLE 12.3 CALCULATION OF LOG-LIKELIHOOD FOR SMALL EXAMPLE

Outcome Y	Predicted Probability $\hat{Y}$	$1 - \hat{Y}$	$Y \ln \hat{Y}$	$(1 - Y) \ln(1 - \hat{Y})$	$\sum[Y \ln \hat{Y} + (1 - Y) \ln(1 - \hat{Y})]$
1	.899	.101	−0.106	0	−0.106
1	.540	.460	−0.616	0	−0.616
0	.317	.683	0	−0.381	−0.381
1	.516	.439	−0.578	0	−0.578
1	.698	.302	−0.360	0	−0.360
0	.457	.543	0	−0.611	−0.611
0	.234	.766	0	−0.267	−0.267
1	.899	.101	−0.106	0	−0.106
1	.451	.439	−0.578	0	−0.578
1	.764	.236	−0.269	0	−0.269
0	.457	.543	0	−0.611	−0.611
0	.561	.439	0	−0.823	−0.823
1	.698	.302	−0.360	0	−0.360
1	.457	.543	−0.783	0	−0.783
0	.899	.101	0	−2.293	−2.293
					SUM = −8.74

For this example, the log-likelihood for the smaller model that contains only the constant is −10.095. When all predictors are in the bigger model, the log-likelihood, as shown in Table 12.3, is −8.740. The difference between log-likelihoods is multiplied by two to create a statistic that is distributed as chi square. In the example, the difference (times two) is:

$$\chi^2 = 2[(-8.740) - (-10.095)] = 2.71$$

Degrees of freedom are the difference between degrees of freedom for the bigger and smaller models. The constant-only model has 1 df (for the constant) and the full model has 4 df (1 df for each individual effect and one for the constant); therefore χ^2 is evaluated with 3 df. Because χ^2 is not reliable at $\alpha = .05$, the model with all predictors is no better than one with no predictors, an expected result because of the failure to find any reliable predictors. Additional goodness-of-fit statistics are described in Section 12.6.1.1.

12.4.3 ■ Comparing Models

This goodness-of-fit χ^2 process is also used to evaluate predictors that are eliminated from the full model, or predictors (and their interactions) that are added to a smaller model. In general, as predictors are added/deleted, log-likelihood decreases/increases. The question in comparing models is, Does the log-likelihood decrease/increase significantly with the addition/deletion of predictor(s)?

For example, the log-likelihood for the small sample example with DIFF removed is −9.433. Compared to the full model, χ^2 is

$$\chi^2 = 2[(-8.740) - (-9.433)] = 1.39$$

with 1 df, indicating no significant enhancement to prediction of falling by knowledge of difficulty of the ski run.

12.4.4 ■ Interpretation and Analysis of Residuals

As shown in Section 12.4.2, the first case has a residual of .101; the predicted probability of falling, .899, was off by .101 from the actual outcome of falling for that case of 1.00. Residuals are calculated for each case and then standardized to assist in the evaluation of the fit of the model to each case.

There are several schemes for standardizing residuals. The one used here is common among software packages, and defines a standardized residual for a case as:

$$\text{std residual}_i = \frac{(Y_i - \hat{Y}_i / \hat{Y}_i(1 - \hat{Y}_i)}{\sqrt{1 - h_i}} \tag{12.8}$$

where

$$h_i = \hat{Y}_i(1 - \hat{Y}_i)x_i'(X'VX)^{-1}x_i \tag{12.9}$$

and where x is the vector of predictors for the case, X is the data matrix for the whole sample including the constant, and V is a diagonal matrix with general element:

$$\hat{Y}_i(1 - \hat{Y}_i)$$

Table 12.4 shows residuals and standardized residuals for each case in the small sample example. There is a very large residual for the last case, a skier who did not fall, but had a predicted probability of falling of about .9. This is the case the model predicts most poorly. With a standardized residual $(z) = 3.326$ in a sample of this size, the case is an outlier in the solution.

TABLE 12.4	RESIDUALS AND STANDARDIZED RESIDUALS FOR SMALL SAMPLE EXAMPLE	
Outcome	Residual	Standardized Residual
1	.101	0.375
1	.460	1.403
0	−.317	−0.833
1	.439	1.034
1	.302	0.770
0	−.457	−1.019
0	−.234	−0.678
1	.101	0.375
1	.439	1.034
1	.236	0.646
0	−.457	−1.019
0	−.561	−1.321
1	.302	0.770
1	.543	1.213
0	−.899	3.326

12.4.5 ■ Computer Analyses of Small Sample Example

Setup and selected output for computer analyses of the data in Table 12.1 appear in Tables 12.5 through 12.8. BMDPLR is in Table 12.5, SYSTAT LOGIT in Table 12.6, SAS LOGISTIC in Table 12.7, and SPSS LOGISTIC REGRESSION in Table 12.8.

BMDPLR is a stepwise program, but with the setup in Table 12.5, all variables are forced into the equation. Continuous variables are indicated by the keyword INTERVAL. The MODEL statement identifies the predictors to be included in the logistic regression analysis; the constant is included by default.

BMDPLR output shows the number of cases at each outcome, and offers descriptive statistics for each continuous variable. The output then shows how the discrete variable, SEASON, is dichotomized into the two DESIGN VARIABLES, SEASON(1) and SEASON(2).

Step Number 0 gives the basic logistic regression statistics: three GOODNESS-OF-FIT CHI-SQ statistics (described in Section 12.6.1.1), and, for each term in the equation, the B regression COEF-FICIENT (for finding the probability of falling), the STANDARD ERROR of B, the Wald test (labeled COEF/SE), EXP(COEF) or e^B, and the 95% confidence interval for e^B. BMDPLR then shows the CORRELATION MATRIX OF COEFFICIENTS (the correlations among the B, the logistic regression coefficients) and, finally, the results of a modeling procedure in which an F test and associated probability are given for removing each variable from the equation. APPROX. F TO REMOVE indicates tests of whether prediction would be significantly degraded by removal of the predictor.

SYSTAT LOGIT (Table 12.6) requires that the number of categories be specified for each categorical variable in the CATEGORY command and that the outcome variable, CONSTANT, and all predictors be specified in the MODEL statement.

SYSTAT LOGIT output shows the number of cases in each outcome group, followed by a history of the iterative process for finding the log-likelihood. Once convergence is achieved, a table shows the B regression coefficient for predicting the probability of being in the group labeled "1" (ESTIMATE), the standard error of the B weight (S.E.), the Wald test (T-RATIO), and the probability associated with the Wald test (P-VALUE) for each term in the equation .

A second table shows the ODDS RATIO for each predictor (see Section 12.6.6) and the 95% confidence interval for each odds ratio. Below the table is the log-likelihood of the model without the predictors (CONSTANTS ONLY MODEL) followed by a CHI-SQUARE test (and its probability) for the difference between the constant-only model and the one specified in the MODEL statement. Finally, MCFADDEN'S RHO-SQUARED is shown as a measure of the strength of association between the outcome and the predictors.

As seen in Table 12.7, SAS LOGISTIC requires user coding for discrete predictor variables with more than two categories. The MODEL statement then includes the continuous predictor and the user-coded dummy variables. The constant does not have to be stated explicitly.

The Response Profile in the output shows the coding and numbers of cases for each outcome group. Descriptive statistics follow and, since all predictor variables are assumed to be continuous, descriptive statistics are given for all of them. Under Criteria for Assessing Model Fit the log-likelihood is multiplied by -2 (shown as -2 LOG L) and then tested against the intercept (constant) only model; the result is reported as Chi-Square for Covariates. Other criteria, AIC and SC, are reported, but without inferential statistics.

The Analysis of Maximum Likelihood Estimates provides B weights (Parameter Estimates) for predicting the probability of *not* falling (code of 0 for FALL), the Standard Error of B, and Wald Chi-Square together with its Pr(obability). Standardized (parameter) Estimates are also provided.

```
/INPUT          VARIABLES ARE 3.  FORMAT IS FREE.  FILE IS 'SSLOGREG.DAT'.
/VARIABLES      NAMES ARE FALL, DIFFCLTY, SEASON.
/GROUP          CODES(FALL) ARE 1, 0.
                NAMES(FALL) ARE YES, NO.
                CODES(SEASON) ARE 1 TO 3.
                NAMES(SEASON) ARE FALL, WINTER, SPRING.
/REGRESS        DEPEND = FALL.
                INTERVAL = DIFFCLTY.
                MODEL = DIFFCLTY, SEASON.
/END
```

NUMBER OF CASES READ.   15

    TOTAL NUMBER OF RESPONSES USED IN THE ANALYSIS   15.
                       NO   . . . . .   6.
                       YES   . . . . .   9.

    NUMBER OF DISTINCT COVARIATE PATTERNS   8

DESCRIPTIVE STATISTICS OF INDEPENDENT VARIABLES

VARIABLE NO. N A M E	MINIMUM	MAXIMUM	MEAN	STANDARD DEVIATION	SKEWNESS	KURTOSIS
2 DIFFCLTY	1.0000	3.0000	2.1333	0.7432	-0.1833	-1.3018

VARIABLE NO. N A M E	GROUP INDEX	FREQ	DESIGN VARIABLES (1)	(2)
3 SEASON	1	5	0	0
	2	6	1	0
	3	4	0	1

STEP NUMBER   0

```
                    LOG LIKELIHOOD =    -8.740
GOODNESS OF FIT CHI-SQ  (2*O*LN(O/E))  =    6.023  D.F.=  4  P-VALUE= 0.197
GOODNESS OF FIT CHI-SQ (HOSMER-LEMESHOW)=   4.892  D.F.=  6  P-VALUE= 0.558
GOODNESS OF FIT CHI-SQ  ( C.C.BROWN )  =    4.723  D.F.=  2  P-VALUE= 0.094
```

	TERM	COEFFICIENT	STANDARD ERROR	COEF/SE	EXP(COEF)	95% C.I. OF EXP(COEF) LOWER-BND	UPPER-BND
DIFFCLTY		1.011	0.896	1.13	2.75	0.382	19.7
SEASON	(1)	-1.346	1.48	-0.911	0.260	0.101E-01	6.73
	(2)	-0.9275	1.59	-0.584	0.396	0.120E-01	13.1
CONSTANT		-0.8493	2.18	-0.390	0.428	0.353E-02	51.8

CORRELATION MATRIX OF COEFFICIENTS

	DIFFCLTY	SEASO(1)	SEASO(2)	CONSTANT
DIFFCLTY	1.000			
SEASO(1)	-0.088	1.000		
SEASO(2)	0.148	0.593	1.000	
CONSTANT	-0.837	-0.369	-0.535	1.000

TABLE 12.5   *(CONTINUED)*

```
STATISTICS TO ENTER OR REMOVE TERMS
------------------------------------
            APPROX.              APPROX.
   TERM     F TO   D.F. D.F.     F TO  D.F. D.F.
            ENTER                REMOVE          P-VALUE

DIFFCLTY                          0.79   1   11   0.3917
DIFFCLTY                 IS IN                MAY NOT BE REMOVED.
SEASON                            0.26   2   11   0.7757
SEASON                   IS IN                MAY NOT BE REMOVED.
CONSTANT                          0.09   1   11   0.7638
CONSTANT                 IS IN                MAY NOT BE REMOVED.

NO TERM PASSES THE REMOVE AND ENTER LIMITS (  0.1500  0.1000 ) .
```

SAS LOGISTIC additionally provides several measures of strength of association between the outcome and the predictors: Somers' D, Gamma, Tau-a, and Tau-c.

SPSS LOGISTIC REGRESSION uses by default a different scheme from other software programs for coding categorical variables with more than two levels. However, as seen in Table 12.8, the CONTRAST instruction can be used to create dummy-variable coding that is consistent with other software. The ENTER instruction assures that all of the predictors enter the logistic regression equation simultaneously on Step Number 1.

The first two tables of output show the coding for the outcome and predictor variables. Then -2 Log Likelihood is given for the constant-only model. After information about the iterative process, several goodness-of-fit statistics are given. First, the fit of the model with constant and all predictors is in the row labeled -2 Log Likelihood (17.481). Model Chi-Square compares the fit of the model with constant and predictors against the constant-only model (2.710). Improvement compares the fit of the model at this step with the fit at the previous step. (Goodness of Fit in the last row of the table is a slight modification of the comparison with the perfect model in the first row.)

The classification table follows, showing the results of classifying all cases with predicted values above .5 as YES and all cases below as NO. Of the skiers who did not fall, 66.67% are correctly classified by the model; of those who did fall, 88.89% are correctly classified. The Variables in the Equation table provides B coefficients, standard errors of B (S.E.), a Wald test (B/SE) for each coefficient, and e^B. The column labeled R is a measure of partial correlation, the correlation of each predictor with the outcome, after adjusting for all other predictors.

12.5 ■ TYPES OF LOGISTIC REGRESSION

As in multiple regression and discriminant function analysis, there are three major types of logistic regression: direct (standard), sequential, and stepwise. Logistic regression programs tend to have more options for controlling equation-building than discriminant function programs, but fewer options than multiple regression programs.

12.5.1 ■ Direct Logistic Regression

In direct logistic regression, all predictors enter the equation simultaneously (as long as tolerance is not violated, cf. Chapter 4). As with multiple regression and discriminant function analyses, this is the

```
USE SSLOGREG
OUTPUT "SSLOGREG.SYT"
CATEGORY SEASON = 3
MODEL FALL = CONSTANT + DIFFCLTY + SEASON
ESTIMATE
```

BINARY LOGIT ANALYSIS

DEPENDENT VARIABLE: FALL

INPUT RECORDS: 15

RECORDS IN RAM FOR ANALYSIS: 15

SAMPLE SPLIT

CATEGORY	CHOICES
RESP	9
REF	6
	15

```
L-L AT ITER    1 IS    -10.397
L-L AT ITER    2 IS     -8.800
L-L AT ITER    3 IS     -8.741
L-L AT ITER    4 IS     -8.740
```

CONVERGENCE ACHIEVED

RESULTS OF ESTIMATION

LOG LIKELIHOOD: -8.740

PARAMETER	ESTIMATE	S.E.	T-RATIO	P-VALUE
1 CONSTANT	-0.849	2.179	-0.390	0.697
2 DIFFCLTY	1.011	0.896	1.128	0.259
3 SEASON_1	-1.346	1.478	-0.911	0.362
4 SEASON_2	-0.928	1.589	-0.584	0.559

PARAMETER	ODDS RATIO	95.0% BOUNDS UPPER	LOWER
2 DIFFCLTY	2.748	15.907	0.475
3 SEASON_1	0.260	4.714	0.014
4 SEASON_2	0.396	8.912	0.018

LOG LIKELIHOOD OF CONSTANTS ONLY MODEL = LL(0) = -10.095
2*[LL(N)-LL(0)] = 2.710 WITH 3 DOF, CHI-SQ P-VALUE = 0.439
MCFADDEN'S RHO-SQUARED = 0.134

```
DATA     SAMPLE;
INFILE   'SSLOGREG.DAT';
INPUT    FALL DIFFCLTY SEASON;
SEASON1=0;
SEASON2=0;
IF SEASON EQ 2 THEN SEASON1=1;
IF SEASON EQ 3 THEN SEASON2=1;
PROC LOGISTIC;
     MODEL FALL=DIFFCLTY SEASON1 SEASON2;
```

The SAS System
The LOGISTIC Procedure

Data Set: WORK.SAMPLE
Response Variable: FALL
Response Levels: 2
Number of Observations: 15
Link Function: Logit

Response Profile

Ordered Value	FALL	Count
1	0	6
2	1	9

Simple Statistics for Explanatory Variables

Variable	Mean	Standard Deviation	Minimum	Maximum
DIFFCLTY	2.133333	0.743223	1.00000	3.00000
SEASON1	0.400000	0.507093	0.00000	1.00000
SEASON2	0.266667	0.457738	0.00000	1.00000

Criteria for Assessing Model Fit

Criterion	Intercept Only	Intercept and Covariates	ChiSquare for Covariates
AIC	22.190	25.481	.
SC	22.898	28.313	.
-2 LOG L	20.190	17.481	2.710 with 3 DF (p=0.4386)
Score	.	.	2.454 with 3 DF (p=0.4837)

Analysis of Maximum Likelihood Estimates

Variable	Parameter Estimate	Standard Error	Wald Chi-Square	Pr > Chi-Square	Standardized Estimate
INTERCPT	0.8493	2.1794	0.1519	0.6968	.
DIFFCLTY	-1.0108	0.8960	1.2726	0.2593	-0.414185
SEASON1	1.3460	1.4781	0.8293	0.3625	0.37633
SEASON2	0.9275	1.5894	0.3406	0.5595	0.234072

Association of Predicted Probabilities and Observed Responses

Concordant = 72.2%	Somers' D	= 0.556
Discordant = 16.7%	Gamma	= 0.625
Tied = 11.1%	Tau-a	= 0.286
(54 pairs)	c	= 0.778

TABLE 12.8   SETUP AND OUTPUT FROM SPSS LOGISTIC REGRESSION ANALYSIS OF SMALL SAMPLE EXAMPLE

```
DATA LIST          FILE='SSLOGREG.DAT'  FREE
                   /FALL, DIFFCLTY, SEASON.
VALUE LABELS       FALL 0 'NO' 1 'YES'
                   /SEASON 1 'AUTUMN' 2 'WINTER' 3 'SPRING'.
LOGISTIC REGRESSION      FALL WITH DIFFCLTY SEASON
                   /CATEGORICAL = SEASON
                   /CONTRAST(SEASON)=INDICATOR(1)
                             /ENTER.
```

```
    Total number of cases:      15 (Unweighted)
    Number of selected cases:   15
    Number of unselected cases: 0

    Number of selected cases:                15
    Number rejected because of missing data:  0
    Number of cases included in the analysis: 15
```

Dependent Variable Encoding:

Original Value	Internal Value
.00	0
1.00	1

	Value	Freq	Parameter Coding	
			(1)	(2)
SEASON				
AUTUMN	1.00	5	.000	.000
WINTER	2.00	6	1.000	.000
SPRING	3.00	4	.000	1.000

Dependent Variable..   FALL

Beginning Block Number 0. Initial Log Likelihood Function

-2 Log Likelihood   20.19035
* Constant is included in the model.

Beginning Block Number 1. Method: Enter

Variable(s) Entered on Step Number
1..   DIFFCLTY
         SEASON

Estimation terminated at iteration number 3 because
Log Likelihood decreased by less than .01 percent.

	Chi-Square	df	Significance
-2 Log Likelihood	17.481	11	.0944
Model Chi-Square	2.710	3	.4386
Improvement	2.710	3	.4386
Goodness of Fit	17.603	11	.0913

TABLE 12.8 *(CONTINUED)*

Classification Table for FALL

```
                     Predicted
                  NO      YES     Percent Correct
                  N  |    Y
Observed        +-------+-------+
    NO      N   |  4    |  2    |   66.67%
                +-------+-------+
    YES     Y   |  1    |  8    |   88.89%
                +-------+-------+
                        Overall   80.00%
```

--------------------- Variables in the Equation ----------------------

Variable	B	S.E.	Wald	df	Sig	R	Exp(B)
DIFFCLTY	1.0105	.8959	1.2720	1	.2594	.0000	2.7469
SEASON			.8316	2	.6598	.0000	
SEASON(1)	-1.3453	1.4778	.8287	1	.3626	.0000	.2605
SEASON(2)	-.9270	1.5892	.3402	1	.5597	.0000	.3957
Constant	-.8493	2.1793	.1519	1	.6968		

method of choice if there are no specific hypotheses about the order or importance of predictor variables. The method allows evaluation of the contribution made by each predictor over and above that of the other predictors. In other words, each predictor is evaluated as if it entered the equation last.

This method has the usual difficulties with interpretation when predictors are correlated. A predictor that is highly correlated with the outcome by itself may show little predictive capability in the presence of the other predictors (cf. Section 5.5.1, Figure 5.3).

SAS LOGISTIC, SYSTAT LOGIT, and SPSS LOGISTIC REGRESSION produce direct logistic regression analysis by default (Tables 12.6 to 12.8). BMDPLR and PR produce direct logistic regression analysis if a MODEL sentence is included in the setup and information about entering and removing variables is omitted (as seen in Table 12.5).

12.5.2 ■ Sequential Logistic Regression

Sequential logistic regression is similar to sequential multiple regression and sequential discriminant function analysis in that the researcher specifies the order of entry of predictors into the model. However, sequential logistic regression is not so easily accomplished within the reviewed computer packages. One option with all programs is simply to do multiple runs, one for each step of the proposed sequence. For example, one might start with a run predicting hay fever from degree of stuffiness and temperature. Then, in a second run, geographic area and season are added to stuffiness and temperature. The difference between the two models is evaluated to determine if geographic area and season significantly add to prediction above that afforded by symptoms alone, using the technique described in Section 12.4.3 and illustrated in the large sample example of Section 12.8.3.

The only procedure available for sequential regression in BMDPPR is to do multiple runs. Interactive execution can be used to specify the order of entry of predictors in a single run of BMD-PLR, but only if each step adds a single predictor. Similarly, interactive stepping can be used in SYSTAT LOGIT to include individual predictors in a user specified sequence. In SAS LOGISTIC you can specify SEQUENTIAL entry of predictors, to enter in the order listed in the MODEL instruction, but also one predictor at a time. Only SPSS LOGISTIC REGRESSION allows sequential entry of one or more predictors by the use of successive ENTER instructions.

Table 12.9 shows sequential logistic regression for the small sample example through SPSS LOGISTIC REGRESSION. DIFFCLTY is given highest priority because it is expected to be the strongest predictor of falling. The sequential process asks if SEASON adds to prediction of FALLing beyond that of difficulty of the ski run. At Step 2, Improvement $\chi^2(2, N = 15) = 0.906, p < .05$, indicating no significant improvement with addition of SEASON as a predictor. Note that this improvement in fit statistic is the difference between -2 Log Likelihood for Step 1 (18.387) and -2 Log Likelihood for Step 2 (17.481). Note also that the logistic regression equation at the end of Step 2, with all predictors in the equation, is the same in direct and sequential logistic regression.

12.5.3 ■ Stepwise (Statistical) Logistic Regression

In stepwise logistic regression, inclusion and removal of predictors from the equation are based solely on statistical criteria. Thus, stepwise logistic regression is best seen as a screening or hypothesis-generating technique, as it suffers from the same problems as statistical multiple regression and discriminant function analysis (Sections 5.5.3 and 11.5.3). When statistical analyses are used, it is very easy to misinterpret the exclusion of a predictor; the predictor may be very highly correlated with the outcome but not included in the equation because it was "bumped" out by another predictor or combination of predictors. Hosmer and Lemeshow (1989) recommend a criterion for inclusion of a variable that is less stringent than .05; they suggest that something in the range of .15 or .20 is more appropriate to ensure entry of variables with coefficients different from zero.

All four computer packages reviewed offer statistical logistic regression, and all allow specification of alternative stepping methods and criteria. BMDPLR and PR permit specification of which predictors (including the constant) are in the equation at the start of modeling and which are out of the equation, how many times each may be moved in or out, the rules for entering and removing predictors and interactions (hierarchically or not), and tail probabilities for entering and removing predictors. BMDPLR also allows specification of either the asymptotic covariance estimate method or the maximum likelihood method for entering or removing terms at each step. BMDPPR uses only the maximum likelihood method.

SPSS LOGISTIC REGRESSION offers forward or backward statistical regression, either of which can be based on either the Wald or maximum likelihood-ratio statistic, with user specified tail probabilities. SYSTAT LOGIT also permits forward or backward statistical regression with user specification of tail probabilities for entry or removal of predictors, predictors to be excluded, predictors to be forced into the model, and the maximum number of steps in the process.

SAS LOGISTIC allows specification of forward, backward or "stepwise" stepping. (In forward selection, a variable once in the equation stays there; if stepwise is chosen, variables once in the equation may leave.) The researcher can specify the maximum number of steps in the process, the significance level for entry or for staying in the model, variables to be included in all models, and maximum number of variables to be included. The researcher can also specify the removal or entry of variables based on the residual chi-square.

Table 12.10 shows stepwise logistic regression for the small sample example using SAS LOGISTIC. Stepwise selection of predictors is specified where predictors enter the equation one at a time, but may be removed if doing so will help prediction. Significance levels of 0.5 were chosen

TABLE 12.9 SETUP AND SELECTED SEQUENTIAL LOGISTIC REGRESSION OUTPUT FROM SPSS LOGISTIC REGRESSION ANALYSIS OF SMALL SAMPLE EXAMPLE

```
DATA LIST         FILE='SSLOGREG.DAT'  FREE
                  /FALL, DIFFCLTY, SEASON.
VALUE LABELS      FALL 0 'NO' 1 'YES'
                  /SEASON 1 'AUTUMN' 2 'WINTER' 3 'SPRING'.
LOGISTIC REGRESSION       FALL WITH DIFFCLTY SEASON
                          /CATEGORICAL = SEASON
                          /CONTRAST(SEASON)=INDICATOR(1)
                                    /ENTER DIFFCLTY /ENTER SEASON.
```

Dependent Variable.. FALL

Beginning Block Number 0. Initial Log Likelihood Function

2 Log Likelihood   20.19035

* Constant is included in the model.

 Beginning Block Number 1. Method: Enter

Variable(s) Entered on Step Number
1..   DIFFCLTY

Estimation terminated at iteration number 3 because
Log Likelihood decreased by less than .01 percent.

	Chi-Square	df	Significance
-2 Log Likelihood	18.387	13	.1434
Model Chi-Square	1.804	1	.1793
Improvement	1.804	1	.1793
Goodness of Fit	14.982	13	.3085

Classification Table for FALL

```
                Predicted
             NO      YES     Percent Correct
              N  |    Y
Observed     +-------+-------+
   NO    N   |   2   |   4   |  33.33%
             +-------+-------+
   YES   Y   |   1   |   8   |  88.89%
             +-------+-------+
                 Overall  66.67%
```

---------------------- Variables in the Equation ----------------------

Variable	B	S.E.	Wald	df	Sig	R	Exp(B)
DIFFCLTY	1.0428	.8262	1.5932	1	.2069	.0000	2.8373
Constant	-1.7709	1.7831	.9864	1	.3206		

Beginning Block Number 2. Method: Enter

Variable(s) Entered on Step Number
1..   SEASON

Estimation terminated at iteration number 3 because
Log Likelihood decreased by less than .01 percent.

	Chi-Square	df	Significance
-2 Log Likelihood	17.481	11	.0944
Model Chi-Square	.906	2	.6358
Improvement	.906	2	.6358
Goodness of Fit	17.603	11	.0913

TABLE 12.9   *(CONTINUED)*

```
Classification Table for FALL
                    Predicted
                 NO      YES      Percent Correct
                 N   |    Y
Observed        +-------+-------+
   NO      N    |   4   |   2   |    66.67%
                +-------+-------+
   YES     Y    |   1   |   8   |    88.89%
                +-------+-------+
                        Overall    80.00%
```

```
--------------------- Variables in the Equation -----------------------

Variable          B        S.E.      Wald      df      Sig       R      Exp(B)

DIFFCLTY        1.0105     .8959    1.2720      1     .2594    .0000    2.7469
SEASON                              .8316       2     .6598    .0000
  SEASON(1)    -1.3453    1.4778     .8287      1     .3626    .0000     .2605
  SEASON(2)     -.9270    1.5892     .3402      1     .5597    .0000     .3957
Constant        -.8493    2.1793     .1519      1     .6968
```

for entering the model and staying in it, so that more than one variable would enter for this tiny data set. With a larger data set, the probabilities are usually smaller.

The output shows that DIFFCLTY enters on the first step, followed by SEASON1, which codes whether the season is winter. SEASON2 never enters the equation. Since entry is stepwise, it is not legitimate to evaluate the model statistically, nor assess whether addition of SEASON1 adds significantly to prediction beyond that afforded by DIFFCLTY.

You may want to consider including interactions as potential predictors in stepwise model building. Hosmer and Lemeshow (1989) discuss issues surrounding use of interactions and appropriate scaling of continuous variables for them (pp. 88–91).

12.6 ■ SOME IMPORTANT ISSUES

12.6.1 ■ Statistical Inference

Logistic regression has two types of inferential tests: tests of models and tests of individual predictors.

12.6.1.1 ■ Assessing Goodness-of-Fit of Models

There are numerous models in logistic regression: a constant- (intercept-)only model that includes no predictors, an incomplete model that includes the constant plus some predictors, a full model that includes the constant plus all predictors (including, possibly, interactions and variables raised to a power), and a perfect (hypothetical) model that would provide exact fit of expected frequencies to observed frequencies if only the right set of predictors were measured.

As a consequence, there are several comparisons possible: between the constant-only model and the full model, between the constant-only model and an incomplete model, between an incomplete model and the full model, between two incomplete models, between a chosen model and the perfect model, between . . . well, you get the picture.

TABLE 12.10. SETUP AND PARTIAL OUTPUT FOR STEPWISE SAS LOGISTIC REGRESSION ANALYSIS

```
DATA    SAMPLE;
INFILE  'SSLOGREG.DAT';
INPUT   FALL DIFFCLTY SEASON;
SEASON1=0;
SEASON2=0;
IF SEASON EQ 2 THEN SEASON1=1;
IF SEASON EQ 3 THEN SEASON2=1;
PROC LOGISTIC;
    MODEL FALL=DIFFCLTY SEASON1 SEASON2
    / SELECTION = STEPWISE
    SLENTRY = 0.5
    SLSTAY = 0.5
```

Stepwise Selection Procedure

Step 0. Intercept entered:

Residual Chi-Square = 2.4539 with 3 DF (p=0.4837)

Step 1. Variable DIFFCLTY entered:

Criteria for Assessing Model Fit

Criterion	Intercept Only	Intercept and Covariates	Chi-Square for Covariates
AIC	22.190	22.387	.
SC	22.898	23.803	.
-2 LOG L	20.190	18.387	1.804 with 1 DF (p=0.1793)
Score	.	.	1.746 with 1 DF (p=0.1864)

Residual Chi-Square = 0.8844 with 2 DF (p=0.6426)

TABLE 12.10 *(CONTINUED)*

Step 2. Variable SEASON1 entered:

Criteria for Assessing Model Fit

Criterion	Intercept Only	Intercept and Covariates	Chi-Square for Covariates
AIC	22.190	23.829	.
SC	22.898	25.953	.
-2 LOG L	20.190	17.829	2.361 with 2 DF (p=0.3071)
Score	.	.	2.230 with 2 DF (p=0.3279)

Residual Chi-Square = 0.3481 with 1 DF (p=0.5552)

NOTE: No (additional) variables met the 0.5 significance level for entry into the model.

Summary of Stepwise Procedure

Step	Variable Entered	Variable Removed	Number In	Score Chi-Square	Wald Chi-Square	Pr > Chi-Square
1	DIFFCLTY		1	1.7457	.	0.1864
2	SEASON1		2	0.5547	.	0.4564

Analysis of Maximum Likelihood Estimates

Variable	Parameter Estimate	Standard Error	Wald Chi-Square	Pr > Chi-Square	Standardized Estimate
INTERCPT	1.5496	1.8202	0.7247	0.3946	.
DIFFCLTY	-1.1108	0.8537	1.6929	0.1932	-0.455155
SEASON1	0.8638	1.1728	0.5425	0.4614	0.241497

Association of Predicted Probabilities and Observed Responses

Concordant = 64.8%		Somers' D = 0.444
Discordant = 20.4%		Gamma = 0.522
Tied = 14.8%		Tau-a = 0.229
(54 pairs)		c = 0.722

Not only are there numerous possible comparisons among models but also there are several tests to evaluate goodness of fit. No single test is universally preferred, so the computer programs use different tests for the models. Further, for some tests, a good fit is indicated by a nonsignificant result (as in loglinear analysis); for others, good fit is indicated by a significant result.

It is also the case that the same statistical result can lead to different conclusions for different analyses; the same result might sometimes indicate an adequate model and at other times indicate an inadequate one. For instance, if sample size is very large, almost any difference between models is likely to be reliable (statistically significant) even if it has no practical importance. Classification of cases may be wonderful, for instance, even though the chosen model deviates significantly from a perfect model. Therefore, a statistically significant difference between a fitted model and the observed frequencies (representing a perfect model) may not indicate a poor model with large samples. On the other hand, a reliable difference between a model and observed frequencies probably does mean that the model is inadequate if sample size is small. Therefore, the analyst needs to keep both the effects of sample size (big = more likely to find significance) and the way the test works (good fit = significant, or good fit = not significant) in mind while interpreting results.

Constant-only vs. full model. A common first step in any analysis is to ask if the predictors, as a group, contribute to prediction of the outcome. In logistic regression, this is the comparison of the constant-only model with a model that has the constant plus all predictors. If no improvement is found when all predictors are added, the predictors are unrelated to outcome.

The log-likelihood technique for comparing the constant-only model with the full model is shown in Section 12.4.2, Equation 12.6. All of the computer programs do a log-likelihood test, but report it differently. For example, BMDPLR and PR provide the log-likelihood directly for each step of a model that is printed out, but provide the information for the constant-only model only when no MODEL instruction is included in the setup. To test the constant-only model against the full model in BMDP, two runs are required, a run with no MODEL specification and another run where all predictors are included in the model. Improvement in fit with inclusion of predictors is computed using Equation 12.7.

SYSTAT LOGIT provides log-likelihood for the constant-only model and the full model, as well as a test for the difference between them, labeled $2*[LL(N)-LL(0)]$. SPSS LOGISTIC REGRESSION reports the difference between the constant-only model and the full model is as Model Chi-Square.

SAS LOGISTIC reports log-likelihood for both the constant-only model and the full model as –2 LOG L in the table labeled Criteria for Assessing Model Fit (cf. Table 12.7). The difference between the models is reported under the column labeled Chi-Square for Covariates, along with degrees of freedom and p level. One hopes for a reliable difference between the full model and the constant- (intercept-)only model at a level of at least $p < .05$. SAS LOGISTIC provides a second statistic labeled Score that is interpreted in the same way as the difference between log-likelihoods.

These same procedures could be used with any of the programs to test the adequacy of an incomplete model (only some predictors) against the constant-only model by inclusion of some but not all predictors in the setup language.

Comparison with a perfect (hypothetical) model. The perfect model contains exactly the right set of predictors to duplicate the observed frequencies. Either the full model (all predictors) or an incomplete model (some predictors) can be tested against the perfect model in several different ways. However, these statistics assume adequate expected cell frequencies between pairs of discrete predictors, as discussed in Section 12.3.2.1, since they are based on differences between observed and expected frequencies. In this context, the set of predictors is sometimes called the covariate pattern where covariate pattern refers to combinations of scores on all predictors, both continuous and discrete.

With these statistics a *nonsignificant* difference is desired. A nonsignificant difference indicates that the full or incomplete model being tested is not reliably different from the perfect model. Put another way, a nonsignificant difference indicates that the full or incomplete model adequately duplicates the observed frequencies at the various levels of outcome.

In the BMDP series, LR provides one statistic that tests for differences in fit between observed and expected frequencies called (2*O*LN(O/E)). A second test in the same program is the C.C. BROWN statistic (Brown, 1982; Prentice, 1976). For this statistic, logistic regression models are compared with a larger family of response models that have two additional parameters. The additional parameters equal 1 for the logistic regression model. The Brown test evaluates whether the additional parameters differ significantly from 1. If they do, the larger ("perfect") model is a better fit to the data than the logistic model. A nonsignificant result indicates that the additional parameters of the larger model provide no better fit than the logistic model (Hosmer & Lemeshow, 1989).

In SYSTAT LOGIT, the log-likelihood test for the difference between observed and expected frequencies is called DEVIANCE, optionally available using the DC instruction.

In SPSS LOGISTIC REGRESSION, the value of -2 Log Likelihood for a model is a comparison of the frequencies produced by the model to observed frequencies (Norušis, 1990). The degrees of freedom and tail probability value for this statistic are also given, as seen in Table 12.8. By this criterion, the small sample model provides adequate fit to a perfect model because $p = .09$.

Deciles of risk. Deciles-of-risk statistics evaluate goodness-of-fit by creating ordered groups of subjects and then comparing the number actually in each group with the number predicted into each group by the logistic regression model.

Subjects are first put in order by their estimated probability on the outcome variable. Then subjects are divided into 10 groups according to their estimated probability; those with estimated probability below .1 (in the lowest decile) form one group, and so on, up to those with estimated probability .9 or higher (in the highest decile).[3] The next step is to further divide the subjects into two groups on the outcome variable (e.g., didn't fall, did fall) to form a 2×10 matrix of observed frequencies. Expected frequencies for each of the 20 cells are obtained from the model. If the logistic regression model is good, then most of the subjects with outcome 1 are in the higher deciles of risk and most with outcome 0 in the lower deciles of risk. If the model is not good, then subjects are roughly evenly spread among the deciles of risk for both outcomes 1 and 0. Goodness-of-fit is formally evaluated using the Hosmer-Lemeshow statistic where a good model produces a nonsignificant chi-square. The Hosmer-Lemeshow statistic is routinely provided by BMDPLR. It is available optionally in SYSTAT LOGIT using the DC instruction. That instruction also produces the observed vs. expected frequencies for each decile of risk, separately for each outcome group.

12.6.1.2 ■ Tests of Individual Variables

Three types of tests are available to evaluate the contribution of an individual predictor to a model: (1) the Wald test, (2) evaluation of the effect of omitting a predictor, and (3) the score (Lagrange multiplier) test. For all these tests, a significant result indicates a predictor that is reliably associated with outcome.

---

[3] Sometimes subjects are divided into 10 groups by putting the first $N/10$ subjects in the first group, the second $N/10$ in the second group, and so on; however, Hosmer and Lemeshow (1989) report that the other procedure is preferable.

The Wald test is the simplest; it is the default option in all reviewed computer programs, but in slightly different forms. For example, the test is called WALD or WALD Chi-Square in SPSS and SAS, while BMDP programs call the statistic COEF/SE, and SYSTAT LOGIT labels the statistic T-RATIO. As seen in Section 12.4.1, all of these tests are (a function of) the logistic regression coefficient divided by its standard error.

However, several sources express doubt about use of the Wald statistic. For instance, Norušis (1990) points out that when the absolute value of the regression coefficient is large, the estimated standard error tends to become too large, resulting in increased Type II error: making the test too conservative. A second disadvantage of the Wald test occurs when predictors have more than two levels. The Wald test as implemented in SPSS, BMDP, and SAS, estimates the reliability of each degree of freedom (each dummy variable) separately, but not a discrete predictor as a whole. In SYSTAT LOGIT, however, the CONSTRAIN procedure can be used to combine single df components of a discrete predictor (or sets of predictors) into a single Wald test.

The test that compares models with and without each predictor (sometimes called the likelihood-ratio or LR test) is considered superior to the Wald test, but is highly computer intensive. Each predictor is evaluated by testing the improvement in model fit when that predictor is added to the model or, conversely, the decrease in model fit when that predictor is removed. The BMDP programs provide these statistics by default (see the final portion of Table 12.5). A predictor with large APPROX. F (or CHI-SQUARE) TO REMOVE and small P-VALUE is one which significantly enhances prediction. The BMDPLR default option is ACE, asymptotic covariance estimate, which produces APPROX. F for entry and removal of predictors. BMDPPR uses only MLR, a maximum likelihood ratio which produces a CHI-SQUARE statistic. MLR is more reliable, but is even more computer-intensive than ACE.

SYSTAT LOGIT requires runs of models with and without each predictor to produce the LR test to assess the reliability of improvement in fit when a predictor is included in the model. SPSS LOGISTIC REGRESSION and SAS LOGISTIC also have no simple, one-run procedure for evaluating differences in direct logistic regression models with and without individual predictors.

The score test for individual PREDICTORS is offered as a simple option only in SYSTAT LOGIT. Its advantage over the LR test is economic; it is not as computer intensive as the tests that compare models with and without individual predictors.

12.6.2 ■ Number and Type of Outcome Categories

Logistic regression analysis can be applied with two or more categories of outcome, and, when there are more that two categories of outcome, they may or may not have order. That is, outcomes with more than two categories can be either nominal (without order) or ordinal (with order). Logistic regression is more appropriate than multiple regression when the distribution of responses over a set of categories seriously departs from normality, making it difficult to justify using an ordered categorical variable as if it were continuous.

When there are more than two categories the analysis is called multinomial or polychotomous logistic regression and there is more than one logistic regression model/equation. In fact, like discriminant function analysis, there are as many models (equations) as there are degrees of freedom for the outcome categories; the number of models is equal to the number of categories minus one.

When the outcome is ordered, the first equation finds the probability that a case is above the first (lowest) category. The second equation finds the probability that the case is above the second category, and so forth, as seen in Equation 12.10,

$$P(Y > j) = \frac{e^u}{1 + e^u} \tag{12.10}$$

where u is the linear regression equation as in Equation 12.1, Y is the outcome, and j indicates the category.

An equation is solved for each category except the last, since there are no cases above the last category. (SAS LOGISTIC bases the equations on the probability that a case is below rather than above a category, so it is the lowest category that is omitted.)

When there are more than two categories of outcome, but they are not ordered, each equation predicts the probability that a case is (or is not) in a particular category. Equations are built for all categories except the last. Logistic regression with this type of outcome is illustrated in the large sample example of Section 12.8.3.

With the exception of SPSS LOGISTIC REGRESSION, which analyzes only two-category outcomes, the programs handle multiple-category outcomes but have different ways of summarizing the results of multiple models.

The BMDP series has LR for two-category outcomes and PR for three- or more-category outcomes. (Although PR can also be used with two-category outcomes, LR is recommended for its richer options.) The default analysis for BMDPPR is for unordered outcome, but ordered outcome can be specified using either ORDINAL or EQUALODDS. EQUALODDS assumes that odds ratios between adjacent categories are the same. Individual predictors are evaluated with respect to the combined equations for all models, regardless of the particular outcome categories with which a predictor is associated. The summed models are tested with and without each predictor.

SYSTAT LOGIT assumes that categories are unordered if more than two are specified; there is no provision for ordered categories. Classification and prediction success tables (cf. Section 12.6.3) are used to evaluate the success of the equations taken together. Predictors are also evaluated for their contribution to prediction across all outcome categories through the Wald tests for "parameters across all choices."

SAS LOGISTIC treats categories in all multinomial models as ordered; there is no provision for unordered categories. The logistic regression coefficients for individual predictors are for the combined set of equations. Classification tables and strength of association measures are used to evaluate the set of equations as a whole.

The following sections demonstrate analyses of ordered response categories with BMDPPR and SAS LOGISTIC. Section 12.8.3 demonstrates analysis of unordered outcome categories.

12.6.2.1 ■ Ordered Response Categories with BMDPPR

Table 12.11 is an example of BMDPPR analysis of ordered categories of the frequency with which psychotherapists reported that they had been sexually attracted to the therapists in their own psychotherapy.[4] Predictor variables are SEX, AGE, and THEORETical orientation (psychodynamic vs. all other orientations).

TYPE = ORDINAL and LEVEL = 5 indicate that scores on the dependent variable, ATTRACT, represent 5 levels of an ordinal outcome variable. The 364 responses are concentrated in the lowest category (NEVER) and fairly equally dispersed over the remaining categories. The GOODNESS-OF-FIT CHI-SQ shows close correspondence between observed and expected

---

[4]Most psychotherapists undergo psychotherapy themselves as part of their training.

```
/INPUT          VARIABLES = 4.  FORMAT IS FREE.  FILE IS 'LOGORDER.DAT'.
/VARIABLES      NAMES ARE AGE, SEX, THEORET, ATTRACT.
                MISSING = 4*99.
/GROUP          CODES(SEX) ARE 0, 1.
                NAMES(SEX) ARE MALE, FEMALE.
                CODES(THEORET) ARE 1 TO 14.
                NAMES(THEORET) ARE PSYCHDYN, OTHER, OTHER, OTHER,
                    OTHER, OTHER, OTHER, OTHER, OTHER, OTHER,
                    OTHER, OTHER, OTHER, OTHER.
                CODES(ATTRACT) ARE 0 TO 4.
                NAMES(ATTRACT) ARE NEVER, ONCE, RARELY,
                    SOMETIME, OFTEN.
/REGRESS        DEPEND = ATTRACT.
                TYPE = ORDINAL.  LEVEL = 5.
                INTERVAL = AGE.
                MODEL = AGE, SEX, THEORET.
/END
```

```
NUMBER OF CASES READ. . . . . . . . . . . . .      476
   CASES WITH DATA MISSING OR BEYOND LIMITS . .    112
       REMAINING NUMBER OF CASES . . . . . . . .   364


   TOTAL NUMBER OF RESPONSES USED IN THE ANALYSIS    364.
                             NEVER    . . . . . .    232.
                             ONCE     . . . . . .     31.
                             RARELY   . . . . . .     52.
                             SOMETIME . . . . . .     26.
                             OFTEN    . . . . . .     23.


   TOTAL NUMBER OF DISTINCT COVARIATE PATTERNS . .    134
```

DESCRIPTIVE STATISTICS OF INDEPENDENT VARIABLES

VARIABLE NO. N A M E	MINIMUM	MAXIMUM	MEAN	STANDARD DEVIATION	SKEWNESS	KURTOSIS
1 AGE	30.0000	82.0000	48.8214	9.5345	0.9135	0.8151

VARIABLE NO. N A M E	GROUP INDEX	FREQ	DESIGN VARIABLES (1)
2 SEX	1	166	0
	2	198	1
3 THEORET	1	140	0
	2	224	1

```
STEP NUMBER    0
--------------

                    LOG LIKELIHOOD =  -395.619
GOODNESS OF FIT CHI-SQ    (2*O*LN(O/E)) =    400.633  D.F.= 529  P-VALUE= 1.000
```

TABLE 12.11 *(CONTINUED)*

	COEFFICIENT	STANDARD ERROR	COEF/SE	EXP(COEF)	95% CI OF EXP(COEF) LOW-END	HIGH-END
ALL OUTCOMES						
1 AGE	0.1110E-02	0.118E-01	0.940E-01	1.0	0.98	1.0
2 SEX	1.127	0.231	4.87	3.1	2.0	4.9
3 THEORET	-0.7554	0.222	-3.40	0.47	0.30	0.73
OUTCOME: NEVER						
4 CONST1	-0.8205	0.638	-1.29	0.44	0.13	1.5
NEVER TO ONCE						
5 CONST2	-1.249	0.640	-1.95	0.29	0.82E-01	1.0
NEVER TO RARELY						
6 CONST3	-2.222	0.648	-3.43	0.11	0.30E-01	0.39
NEVER TO SOMETIME						
7 CONST4	-3.092	0.668	-4.63	0.45E-01	0.12E-01	0.17

STATISTICS TO ENTER OR REMOVE TERMS   MODELS LOG LIKELIHOOD -395.6195

TERM	APPROX. CHI-SQ. D.F. ENTER	APPROX. CHI-SQ. D.F. REMOVE	P-VALUE	LOG LIKELIHOOD
AGE		0.01 1	0.9250	-395.6239
AGE		IS IN		MAY NOT BE REMOVED.
SEX		25.29 1	0.0000	-408.2657
SEX		IS IN		MAY NOT BE REMOVED.
THEORET		11.58 1	0.0007	-401.4093
THEORET		IS IN		MAY NOT BE REMOVED.
CONSTANT		IS IN		MAY NOT BE REMOVED.

frequencies when taking all three predictor variables into account. Looking at variables individually, the WALD (COEF/SE) and STATISTICS TO ENTER AND REMOVE TERM show that SEX and THEORETical orientation are associated with ATTRACTion, but AGE is not.

The signs of the coefficients show the directions of the relationships between response categories and predictors. The sign of the coefficient for SEX is positive, indicating that low coded values for SEX (male = 0) correspond to low response values; that is, men are less frequently sexually attracted to their therapists than are women. The negative sign for THEORETical orientation indicates that low values (psychodynamic orientation) are associated with high response values; psychodynamically-oriented therapists report sexual attraction to their own personal therapists more frequently than therapists of other theoretical orientations. The OUTCOME: statistics are of interest only if the categories are unordered. The assumption is that odds ratios are proportional for all adjacent categories. For example, the odds of moving up one category of attraction (from NEVER to ONCE, or from RARELY to SOMETIME, for example) are about tripled for women.

12.6.2.2 ■ Ordered Response Categories with SAS LOGISTIC

Table 12.12 shows the same analysis through SAS LOGISTIC. There is no need to indicate that the response categories are ordered; that is assumed by the program. The Score Test for the Proportional Odds Assumption shows that the odds ratios between adjacent outcome categories are not significantly different ($p=0.1664$). The value for -2 LOG L in the model fit table shows

TABLE 12.12 LOGISTIC REGRESSION WITH ORDERED CATEGORIES. SETUP AND SAS LOGISTIC OUTPUT

```
DATA    SAMPLE;
INFILE      'LOGORDER.DAT';
INPUT       AGE SEX THEORET ATTRACT;
IF THEORET>1 THEN THEORET=2;
PROC LOGISTIC;
    MODEL ATTRACT = AGE SEX THEORET;
```

The LOGISTIC Procedure

Data Set: WORK.SAMPLE
Response Variable: ATTRACT
Response Levels: 5
Number of Observations: 364
Link Function: Logit

Response Profile

Ordered Value	ATTRACT	Count
1	0	232
2	1	31
3	2	52
4	3	26
5	4	23

WARNING: 112 observation(s) were deleted due to missing values for the response or explanatory variables.

Simple Statistics for Explanatory Variables

Variable	Mean	Standard Deviation	Minimum	Maximum
AGE	48.821429	9.534538	30.0000	82.0000
SEX	0.543956	0.498750	0.0000	1.0000
THEORET	1.615385	0.487174	1.0000	2.0000

Score Test for the Proportional Odds Assumption

Chi-Square = 12.9160 with 9 DF (p=0.1664)

TABLE 12.12 (CONTINUED)

Criteria for Assessing Model Fit

Criterion	Intercept Only	Intercept and Covariates	Chi-Square for Covariates
AIC	836.352	805.239	
SC	851.940	832.519	
-2 LOG L	828.352	791.239	37.113 with 3 DF (p=0.0001)
Score	.	.	35.886 with 3 DF (p=0.0001)

Analysis of Maximum Likelihood Estimates

Variable	Parameter Estimate	Standard Error	Wald Chi-Square	Pr > Chi-Square	Standardized Estimate
INTERCP1	0.0651	0.7299	0.0080	0.9289	.
INTERCP2	0.4938	0.7302	0.4573	0.4989	.
INTERCP3	1.4665	0.7352	3.9790	0.0461	.
INTERCP4	2.3371	0.7497	9.7189	0.0018	.
AGE	-0.00111	0.0116	0.0092	0.9235	-0.005836
SEX	-1.1265	0.2307	23.8531	0.0001	-0.309768
THEORET	0.7554	0.2222	11.5581	0.0007	0.202895

Association of Predicted Probabilities and Observed Responses

Concordant = 62.5%	Somers' D = 0.317
Discordant = 30.8%	Gamma = 0.340
Tied = 6.7%	Tau-a = 0.177
(36901 pairs)	c = 0.659

a significant difference between the constant-only model and the full model, indicating a good model fit with the set of three predictors (covariates).

The following table in the output shows Wald tests for the three predictors, indicating that SEX and THEORETical orientation, but not AGE, significantly predict response category. Parameter Estimates show direction of relationships between predictor and outcome variables (recall that SAS LOGISTIC solves for the probability that a response is below a particular category, opposite from BMDPPR). Thus the negative value for SEX indicates that male therapists are associated with lower numbered categories (less frequently sexually attracted to their own therapists); the positive value for theoretical orientation indicates that psychodynamically-oriented therapists are associated with higher numbered categories (more frequently sexually attracted to their own therapists).

12.6.2.3 ■ Coding Outcome and Predictor Categories

The way that outcome categories are coded determines the direction of the odds ratios as well as the sign of the B coefficient. The interpretation is simplified, therefore, if you pay close attention to coding your categories. Most software programs solve the logistic regression equation for the dichotomous outcome category coded 1 but a few solve for the category coded 0. If the odds ratio is 4 in a problem run through the first type of program, it will be 0.25 in the second type of program.

A convenient way of setting up the coding follows the jargon of SYSTAT LOGIT; for the outcome, the category coded 1 is the "response" category (e.g., illness) and the category coded 0 is the "reference" category (wellness).[5] It is often helpful, then, to think of the response group in comparison to the reference group, e.g., compare people who are ill to those who are well, compare minority persons to white males, etc. The solution tells you the odds of being in the response group given some value on a predictor. If you also give higher codes to the category of a predictor most likely associated with "response," interpretation is facilitated because the parameter estimates are positive. For example, if people over 60 are more likely to be ill, code both "wellness" and "under 60" 0 while both "illness" and "over 60" are coded 1.

This recommendation is extended to predictors with multiple discrete levels where dummy variables are formed for all but one level of the discrete predictor (e.g., Season1 and Season2 for the three seasons in the small sample example). Each dummy variable is coded 1 for one level of a predictor and 0 for the other levels. If possible, code levels likely to be associated with the "reference" group 0 and code levels likely to be associated with the "response" group 1. Odds ratios are calculated in the standard manner, and the usual interpretation is made of each dummy variable (e.g., Season 1).

Other coding schemes may be used, such as orthogonal polynomial coding (trend analysis) but interpretation via odds ratios is far more difficult. However, in some contexts, significance tests for trends may be more interesting than odds ratios.

BMDPLR and PR solve for the category coded 1 unless there is a GROUP paragraph. Then, the programs solve for the group listed first in the CODES sentence. If you follow the practice of listing CODES in numerical order, you may well find yourself solving for the group coded 0. On the other hand, SAS LOGISTIC routinely solves for the group coded 0. You might want to consider recoding your outcome categories to make 1 the reference category if you are using SAS.

There are other methods of coding discrete variables, each with its own impact on interpretation. For example, dichotomous $(1, -1)$ coding, the default in SPSS LOGISTIC REGRESSION, might

---

[5] Hosmer and Lemeshow (1989) recommend coding all dichotomous variables 0 and 1.

be used. Or, discrete categories might be coded for trend analysis, or whatever. Hosmer and Lemeshow (1989) discuss the desirability of various coding schemes and the effects of them on parameter estimates (pp. 44–56). Further discussion of coding schemes available in the computer packages is in Section 12.7.

12.6.3 ■ Classification of Cases

One method of assessing the success of a model is to evaluate its ability to predict correctly the outcome category for cases for whom outcome is known. If a case has hay fever, for instance, we can see if the case is correctly classified as diseased on the basis of degree of stuffiness, temperature, geographic area, and season. With the exception of SYSTAT LOGIT, classification is available for two-category outcomes only.

As in hypothesis testing, there are two types of errors: classifying a truly nondiseased individual as diseased (Type I error) or classifying a truly diseased individual as nondiseased (Type II error). For different research projects, the costs associated with the two types of errors may be different. A Type II error is very costly, for instance, when there is an effective treatment, but the case will not receive it if classified nondiseased. A Type I error is very costly, for instance, when there is considerable risk associated with treatment, particularly for a nondiseased individual. Some of the computer programs allow different cutoffs for asserting "diseased" or "nondiseased." Because extreme cutoffs could result in everyone or no one being classified as diseased, intermediate values for cutoffs are recommended, but these can be chosen to reflect the relative costs of Type I and Type II errors. However, the only way to improve the overall accuracy of classification is to find a better set of predictors.

Because the results of logistic regression analysis are in terms of probability of a particular outcome (e.g., having hay fever), the cutoff chosen for assignment to a category is critical in evaluating the success of the model. BMDPLR, through the COST instruction, provides results for a range of cutoffs, including a plot of the percent of correct classifications of each outcome as a function of cutoff. (Classification is not available in BMDPPR, which deals with models with more than two outcome categories.)

BMDPLR also offers an ROC (receiver operating characteristic) curve, which plots true positives (e.g., correct detection of hay fever) as a function of false positives (e.g., incorrect classification of hay fever). This plot shows how true positives and false positives increase as the cutoff is lowered. For example, a cutoff of 40% (e.g., a .4 probability of having hay fever) for diagnosis could result in a true positive rate of 38% and a false positive rate of 12%. If the cutoff for diagnosis is changed to 15%, the true positive rate is 92% and the false positive rate 70%. Aficionados of the Theory of Signal Detectability will recognize this as $d' = 0.88$ under the most simplifying assumptions. BMD-PLR expresses detectibility as area under the ROC curve. This is interpreted as the probability of a correct diagnosis of a randomly selected pair of cases from each outcome category.

SYSTAT LOGIT has two classification options: classification tables and prediction success tables. For classification tables, a single cutoff strategy is used, If there are only two categories of outcome, a case is assigned to the "disease" outcome if the model predicts an outcome probability of greater than .5. If there are more than two categories of outcome, the model for each outcome category is applied to the scores for a case and the case is assigned to the outcome category for which it has the highest probability, as in discriminant function analysis.

SYSTAT LOGIT also produces prediction success tables, and the manual (Steinberg & Colla, 1991) recommends prediction success tables over classification tables. In these tables, cases are "partially" assigned according to their probability of being in each outcome, so that entries are not necessarily whole numbers. While it is impossible to speak of the number of cases correctly assigned, it is possible to evaluate the percent of correct prediction for each outcome category. An example of a prediction success table is in Section 12.8.3.

Classification is available only for two-category outcomes through SAS LOGISTIC. However, the program does allow specification of the cutoff criterion.

SPSS LOGISTIC REGRESSION, which analyzes data with two-outcome categories only, prints a classification table by default; assignment is based on a cutoff probability criterion of .5. Although the criterion cannot be changed in SPSS, the CLASSPLOT instruction produces a histogram of predicted probabilities, showing whether incorrectly classified cases had probabilities near the criterion. SPSS LOGISTIC REGRESSION also has a SELECT instruction, by which certain cases are selected for use in computing the equations. Classification is then performed on all of the cases in a form of cross validation. This offers a less biased estimate of the classification results.

12.6.4 ■ Hierarchical and Nonhierarchical Analysis

The distinction between hierarchical and nonhierarchical logistic regression is the same as in multiway frequency analysis. When interactions among predictors are included in a model,[6] a restriction may be applied that all main effects and lower order interactions of those predictors are also included in the model. If so, the model is called hierarchical.

In most programs, interactions are specified in the MODEL instruction, using a convention in which interaction components are joined by asterisks (e.g., X1*X2). However, SAS LOGISTIC requires creation of a new variable in the DATA step to form an interaction, which is then used like any other variable in the MODEL instruction.

All of the reviewed programs allow nonhierarchical models, although the SYSTAT LOGIT manual advises against including interactions without their main effects (Steinberg & Colla, 1991). Both BMDPLR and PR default to a hierarchical model if interactions terms are included in the MODEL instruction but allow specification of a nonhierarchical model through the RULE=NONE instruction.

12.6.5 ■ Dosage-Response Relationships

When there is a single predictor, or when one predictor is especially interesting, an informative way of displaying the results of a logistic regression analysis is to plot probability of outcome as a function of the value of the predictor. Logistic (or probit) models form an S-shaped relationship between dosage (predictor) and response (probability of outcome).

SYSTAT LOGIT produces a dosage-response plot using a single predictor in the S-shaped form and also transformed to a linear format. The latter plots the logit (rather than probability) as a function of the value of the predictor. Both of these plots are produced with confidence bands, the 95% confidence interval as default. The plots are produced through the QNTL instruction, as demonstrated in the large sample example of Section 12.8.3.

12.6.6 ■ Interpretation of Coefficients using Odds

Interpretation of odds in logistic regression is similar to that in multiway frequency analysis (Section 7.5.3.2). The odds ratio is the increase (or decrease if the ratio is less than one) in odds of being in one outcome category when the value of the predictor increases by one unit. The coefficients, B, are the natural logs of the odds ratios; odds ratio = e^B. For example, in Table 12.6 the

---

[6] Hosmer and Lemeshow (1989) discuss issues surrounding the use of interactions in a model, particularly the problem of adjusting for confounding variables (covariates) in the presence of interactions (pp. 63–68).

probability of falling on a ski run increases by a multiplicative factor of 2.75 as the difficulty level of the run increases from 1 to 2 (or 2 to 3); a skier is almost three times more likely to fall on a ski run rated 2 as on a ski run rated 1. As in linear regression, coefficients are interpreted in the context of the other predictor variables. That is, the probability of falling as a function of difficulty level is interpreted after adjusting for all other predictors. (Usually only statistically significant coefficients are interpreted; this example is for illustrative purposes only.)

In the health sciences, the odds ratio is often referred to as relative risk when the predictor (symptom) precedes the outcome (disease), that is, directionality has been determined.

The odds ratio has a clear, intuitive meaning for a 2 × 2 table; it is the odds of an outcome for cases in a particular category of a predictor divided by the odds of that outcome for the other category of the predictor. Suppose the outcome is hyperactivity in a child and the predictor is familial history of hyperactivity:

| | | Familial history of hyperactivity | |
		Yes	No
	Yes	15	3
Hyperactivity	No	5	30

$$odds\ ratio = \frac{15/5}{3/30} = 30$$

The relative risk of hyperactivity is 30 given a familial history of hyperactivity; children with a familial history are 30 times more likely to be hyperactive.

Odds ratios are produced directly by SYSTAT LOGIT, SPSS LOGISTIC REGRESSION, SAS LOGISTIC (in most recent versions), and BMDPLR and PR. It is called ODDS RATIO in SYSTAT LOGIT, while it is called Exp(B) by SPSS LOGISTIC REGRESSION, and EXP(COEF) by BMDPLR and PR.

SAS LOGISTIC does not provide an odds ratio directly in some earlier versions, but the value is easily found using a scientific/statistical calculator with an e^x function. The odds ratio is e^B, where B is the logistic regression coefficient estimate for the predictor. Keep in mind that the odds are for the group coded 0 in SAS.

Interpretation of the coefficient for a discrete predictor depends on how the predictor is coded, as indicated in Section 12.6.2.3. The most common coding (0, 1) leads to an interpretation that is consistent with that of a continuous variable: the probability of being in the outcome coded 1 (e.g., hay fever) is multiplied by the odds ratio (e^B) for a case that is in the predictor level coded 1 (e.g., stuffiness), relative to a case that is in the predictor level coded 0 (no stuffiness). However, if some other coding method is used, for example 1 and −1, the odds ratio must be calculated differently, e.g., e^{2B}.

12.6.7 ■ Logistic Regression for Matched Groups

Although usually logistic regression is a between-subjects analysis, there is a form of it called conditional logistic regression for matched subjects or case-control analysis. Cases with disease are matched by cases without disease on variables such as age, gender, socioeconomic status, and so forth. There may be only one matched control subject for each disease subject, or more than one matched control for each disease subject. When there is only one matched control subject, the outcome has two categories (disease and control), while multiple matched control subjects lead to multiple outcome categories (disease, control1, control2, etc.). The model is, as usual, based on the predictors that are included, but there is no constant (or intercept).

BMDPLR uses the CCID instruction to specify a case-control model with a single control for each case. If there is a single matching variable, LR can verify the match for pairs of cases. BMD-PLE, a maximum likelihood program, can be used if you have more than one control for each case. The manual (Dixon, 1992) shows an LE example with four controls.

SYSTAT LOGIT analyzes case-control designs through special language in the MODEL instruction: omitting the CONSTANT and ending the instruction with a # sign. Any number of controls can be used for each case, and some cases may even have different numbers of controls. SAS LOGISTIC uses a conditional logistic regression procedure for case-control studies with a single control by specifying NOINT (no intercept), and requires that each matched pair be transformed into a single observation, where the response variable is the difference in scores between each case and its control.

12.7 ■ COMPARISON OF PROGRAMS

Only the programs in the four reviewed packages that are specifically designed for logistic regression are discussed here. All of the major packages also have programs for nonlinear regression, which perform logistic regression if the researcher specifies the basic logistic regression equation.

The logistic regression programs differ in how they code the outcome: some base probabilities and other statistics on outcome coded "1" (i.e., success, disease) and some base statistics on outcomes coded "0". The manuals are also sometimes inconsistent and confusing in their labeling of predictors where the terms independent variable, predictor, and covariate are used interchangeably. Table 12.13 compares five programs from the four major packages.

12.7.1 ■ SPSS Package

SPSS LOGISTIC REGRESSION[7] handles only dichotomous outcomes and bases statistics on the outcome coded "1". The default coding scheme for categorical predictors, where each category is contrasted with the last category, is unusual in logistic regression. However, the program also has numerous options for alternative coding schemes, including INDICATOR (which most closely mimics the more usual coding strategy), HELMERT, POLYNOMIAL (for trend analysis), and others, including user-specified contrasts. Choice of coding strategy is a nontrivial issue, because both Wald tests and interpretation of coefficients depend on the choice. The program is the only one that provides partial correlations with the outcome, based on the Wald statistic, for each predictor adjusted for other predictors.

SPSS LOGISTIC REGRESSION offers the most flexible procedures for controlling the entry of predictors. All can be entered in one step, or sequential steps can be specified where one or more predictors enter at each step. For stepwise logistic regression both forward and backward stepping are available based on either the Wald or likelihood-ratio statistics at the user's option. The user also has a choice among several criteria for terminating the iterative procedure used to find the optimal solution: change in parameter estimates, maximum number of iterations, percent

---

[7] The PROBIT program can also do logistic regression.

TABLE 12.13 COMPARISON OF PROGRAMS FOR LOGISTIC REGRESSION

Feature	BMDPLR	BMDPPR	SYSTAT LOGIT	SPSS LOGISTIC REGRESSION	SAS LOGISTIC
Input					
Accepts discrete predictors without recoding	Yes	Yes	Yes	Yes	No
Alternative coding schemes for discrete predictors	3	3	2	8	No
Accepts tabulated data	Yes	Yes	Yes	No	Yes
Can specify interactions without recoding	Yes	Yes	Yes	Yes	No
Specify stepping methods and criteria	Yes[a]	No	Yes	Yes	Yes
Specify sequential order of entry and test of predictors	Yes	No	No	Yes	Yes
Specify a case–control design (conditional)	Yes	No	Yes	No	Yes
Can specify size of confidence limits	Yes	No	No	No	No
Specify cutoff probability for classification table	COST	N.A.	Yes	No	Yes
Accepts multiple unordered outcome categories	No	Yes	No	No	No
Deals with multiple ordered outcome categories	No	Yes	No	No	Yes
Can specify equal odds model	N.A.	Yes	No	N.A.	No
Can specify discrete choice models	No	No	No	No	No
Can specify Poisson regression	No	No	SP	No	Yes
Setup to select a subset of cases for classification only	No	No	Yes	Yes	No
Specify quasi-maximum likelihood covariance matrix	No	No	Yes	No	No
Specify case weights	No	No	Yes	No	Yes
Specify start values	No	No	Yes	No	No
Specify link function for response probabilities	No	No	No	No	Yes
Can restrict printing of diagnostics to outliers	No	No	No	Yes	No
Regression output					
Log-likelihood for full model	Yes	Yes	Yes	Yes	Yes
Log-likelihood for constant-only model	Yes[b]	Yes[b]	Yes	Yes	Yes
Hosmer-Lemeshow goodness-of-fit χ^2	Yes	N.A.	Yes	No	No
C. C. Brown goodness-of-fit χ^2	Yes	N.A.	Yes	No	No
Goodness-of-fit χ^2: constant-only vs. full model	Yes[c]	No	Yes	Yes	Yes
Goodness-of-fit χ^2 based on observed vs. expected frequencies	Yes	Yes	Yes	Yes	No
Akaike information index (AIC)	No	No	No	No	Yes
Schwartz criterion	No	No	No	No	Yes
Score statistic	No	No	Yes	No	Yes
Improvement in goodness-of-fit since last step	Yes	Yes	No	Yes	Yes
Goodness-of-fit χ^2 tests for individual predictors in specified model	Yes	Yes	No	Yes(LR)	No

TABLE 12.13 (CONTINUED)

Feature	BMDPLR	BMDPPR	SYSTAT LOGIT	SPSS LOGISTIC REGRESSION	SAS LOGISTIC
Wald tests for predictors combined over multiple categories	No	No	Yes[d]	No	No
Regression coefficient	Yes	Yes	ESTIMATE	B	Parameter estimate
Standard error of the regression coefficient	Yes	Yes	Yes	Yes	Yes
Regression coefficient divided by standard error	COEF/SE	COEF/SE	T-RATIO	Wald	Wald Chi Square
Probability value for coefficient divided by standard error	No	No	Yes	Yes	Yes
Standardized regression coefficient	No	No	No	No	Yes
e^B	EXP(COEF)	EXP(COEF)	No	Yes	No
95% confidence interval of e^B	Yes	Yes	No	Yes	No
Odds ratio	No	No	No	No	No
95% confidence interval for odds ratio	No	No	Yes	No	No
McFadden's rho-squared for model	No	No	Yes	No	No
Association measures between observed responses and predicted probabilities	No	No	No	No	Yes
Partial correlations between outcome and each predictor variable (R)	No	No	No	Yes	No
Correlations among regression coefficients	Yes	Yes	No	Yes	Yes
Covariances among regression coefficients	Yes	Yes	No	No	Yes
Classification table	COST	No	Yes	Yes	Yes
Histograms of predicted probabilities for each group	Yes	Yes	No	CLASSPLOT	No
Scatterplots of observed and predicted proportions	Yes	Yes	No[e]	No	No
Quantile tables and plots	No	No	QNTL	No	No
Derivative tables	No	No	Yes	No[f]	No
Plot of predicted probability as a function of the logit	No	No	No	Yes	No
Diagnostics					
Observed proportion of successes for each pattern of predictors (covariates)	Yes	Yes	No	No	No
Predicted probability of success for each case	Yes	Yes	Yes	Yes	Yes
Standard deviation of predicted probability of success for each case	Yes	No	No	No	No
Residual for each case	No	No	No	Yes	No
Standardized (Pearson) residual for each case	Yes	Yes	Yes	Yes	Yes
Variance of standardized (Pearson) residual for each case	No	No	Yes	No	No
Standardized (normed) residual for each case	Yes	Yes	Yes	No	No
Studentized residual for each case	No	No	No	Yes	No

TABLE 12.13 (CONTINUED)

Feature	BMDPLR	BMDPPR	SYSTAT LOGIT	SPSS LOGISTIC REGRESSION	SAS LOGISTIC
Logit residual for each case	No	No	No	Yes	No
Predicted log odds for each pattern of predictors	Yes	Yes	No	No	No
Deviance for each case	Yes	No	Yes	Yes	Yes
Diagonal of the hat matrix	Yes	No	Yes	No	Yes
Cook's distance for each case	Yes	Yes	No	Yes	Yes
Cumulative residuals	No	Yes	No	No	No
Total χ^2 for pattern of predictors (covariates)	No	Yes	No	No	No
Deviance residual for each case	No	No	Yes	Yes	Yes
Change in Pearson χ^2 for each case	No	No	Yes	Yes	Yes
Change in betas for each case	No	No	Yes	Yes	Yes
Confidence interval displacement diagnostics for each case	No	No	No	No	Yes
Special diagnostics for ordered response variables	No	No	No	No	Yes
Plots of diagnostics	Yes	Yes	No[e]	No[g]	Yes[e]
Saved files of diagnostics	Yes	Yes	Yes	Yes	Yes

[a] Can be accomplished only through interactive model building.

[b] Available only in stepwise logistic regression. Not available if MODEL sentence specified.

[c] Not shown for direct models.

[d] Combines over multiple response categories but not over dummy variables.

[e] Available through plotting programs using saved values.

[f] Available through SPSS PROBIT.

[g] Limited in LOGISTIC procedure, can plot saved values through PLOT.

change in log-likelihood, probability of score statistics for predictor entry, probability of Wald or likelihood ratio statistic to remove a variable, and the epsilon value used for redundancy checking.

SPSS LOGISTIC REGRESSION has a simple SELECT instruction to select a subset of cases on which to compute the logistic regression equation; classification is then performed on all cases as a test of the generalizability of the equation.

This program provides a comprehensive set of residuals statistics which can be displayed or saved, including predicted probability, predicted group, difference between observed and predicted values (residuals), deviance, the logit of the residual, studentized residual, normalized residual, leverage value, Cook's influence, and difference in betas as a result of leaving that case out of the logistic regression equation. The listing can be restricted to outliers with user-specified criteria for determining an outlier.

12.7.2 ■ BMDP Series

BMDP has two programs for logistic regression: LR for dichotomous outcomes and PR for any number of outcome categories, ordered or not. Both are designed as stepwise programs, but do direct analysis as the default if there is a MODEL instruction. Although PR can also analyze dichotomous outcomes, LR is usually the better choice for that design because it offers goodies such as classification analysis and special setups for case-control models. For dichotomous outcomes, both programs base the statistics on the first code listed for the DV in the GROUP paragraph if there is one.[8] If there is no GROUP paragraph and the outcome codes are 0 and 1, BMDP solves for the group coded 1.

For a direct model, BMDPLR and PR are the only programs reviewed that offer goodness-of-fit χ^2 for individual predictors. However, these statistics are highly computer intensive and slow down processing appreciably on slower hardware. Both LR and PR fit any specified direct model; however, only the interactive form of model building available in LR allows user-specified order of entry of predictors. LR also offers special procedures for case-control designs and modification of the size of confidence limits. LR prints a series of cutoff probabilities for the classification table. BMDPLR is the only program that offers an ROC curve on request. PR provides no classification tables but it is the only program in which you can specify an equal odds model when dealing with multiple-category outcomes.

When direct logistic regression is specified, neither LR nor PR print statistics for the constant-only model; Step 0 is the only step in the analysis, and all predictors plus the constant are included. However, when stepwise model building is specified, Step 0 produces statistics for the constant-only model. Both the exponential values for the B coefficients and their confidence intervals are reported by both LR and PR. The correlation matrix of B coefficients is routinely printed in LR and optional in PR; both provide the covariance matrix of B coefficients as an option. In LR, both Hosmer-Lemeshow and C.C. Brown's tests are produced routinely as model goodness-of-fit tests. LR also makes available scatterplots of observed and predicted proportions for each pattern of predictor variables. BMDPLR and PR offer three methods for coding discrete predictor variables; the usual 0, 1 coding is the default.

The many diagnostic measures available (especially through LR) can be printed out, saved, or plotted. They are displayed for each combination of predictor variables (covariate pattern) rather

---

[8] If you follow the practice of listing codes in numerical order, (e.g., 0, 1) in the GROUP CODES instruction, you will end up solving for the group coded "0"!

than for each case. The description of cells is especially handy when all of the predictors are discrete; the predicted probability of outcome for each combination of predictors is an interesting way to describe the study results. It is the average probability of outcome for cases in that cell. The plots can be sorted in various ways.

12.7.3 ■ SAS System

SAS LOGISTIC handles multiple as well as dichotomous response categories, but assumes multiple categories are ordered. There is no default coding of categorical predictors; the coding is user specified before invoking PROC LOGISTIC. Statistics for dichotomous outcomes are based on the category coded "0."

This is the most flexible of the logistic regression programs in that alternative (to logit) link functions can be specified, including normit (inverse standardized normal probability integral function) and complementary log-log functions. The program also does Poisson regression.

SAS LOGISTIC has the basic goodness-of-fit statistics, as well as exclusive ones such as Akaike's Information Criterion and the Schwartz criterion. Strength of association between the set of predictors and the outcome is assessed using Somers' D, Gamma, Tau-a, or Tau-c, a more extensive set of strength of association statistics than the other programs.

A full set of regression diagnostics is available, accessible through saved files or through an "influence plot" to find outliers in the solution. The plot consists of a case-by-case listing of the values of regression statistics along with a plot of deviations from those statistics. Through IPLOT an additional set of plots can be produced in which the value of each statistic is plotted as a function of case numbers.

12.7.4 ■ SYSTAT System

LOGIT is the major program for logistic regression analysis in SYSTAT; however it is an add-on module to the basic program. While logistic regression can also be performed through the NONLIN (nonlinear estimation) module, it is much more painful to do so.

SYSTAT LOGIT is highly flexible in the types of models permitted. For dichotomous outcomes, the statistics are based on the category coded "1." With more than two categories of outcome, the categories are considered unordered. Two types of coding (including the usual dummy variable 0, 1 coding) are permitted for discrete predictor variables. A case-control model can be specified. This is the only program that allows specification of quasi-maximum likelihood covariance, which corrects problems created by misspecified models.

For models with two-category outcomes and a single predictor, a nice-looking "dosage-response" plot can be produced which shows probability of outcome (e.g., disease) as a function of the predictor (e.g., exposure), complete with confidence intervals. This plot is also displayed in standardized (linear) form, with logit scaled on the ordinate.

The basic output is rather sparse, but includes the test of the full model against the constant-only model, with McFadden's *rho* as a measure of strength of association[9], and the odds ratio and its confidence interval for each predictor component. Hosmer-Lemeshow and other χ^2 goodness-of-fit

---

[9] *Rho*-squared tends to be much smaller than a corresponding R^2 for multiple regression; values between .20 and .40 are considered very satisfactory (Steinberg & Colla, 1991).

tests are available through options. Stepping options are plentiful. Although improvement in fit at each step is not given, it can be hand-calculated easily from the log-likelihood ratio that is available at each step. Plenty of diagnostics are available for each case. However they must be saved into a file for viewing; they are not available as part of printout of results.

12.8 ■ COMPLETE EXAMPLES OF LOGISTIC REGRESSION

Data for these analyses are from the data set described in Appendix B. Two complete examples are demonstrated. The first is a simple direct logistic regression analysis, in which work status (employed vs. unemployed) is the two-category outcome that is predicted from four attitudinal variables.

The second analysis is more complex, involving three categories of outcome and sequential entry of predictors. The goal in the second analysis is to predict membership in one of three categories of outcome formed from work status and attitude toward that status. Variables are entered in two sets: demographic and then attitudinal. The major question is whether attitudinal variables significantly enhance prediction of outcome after prediction by demographic variables. Demographic variables include a variety of continuous, discrete and dichotomous variables: marital status (discrete), presence of children (dichotomous), religious affiliation (discrete), race (dichotomous), socioeconomic level (continuous), age (continuous), and attained educational level (continuous). The attitudinal variables used for both analyses are all continuous; they are locus of control, attitude toward current marital status, attitude toward role of women, and attitude toward housework. (Note that prediction of outcome in these three categories on the basis of attitudinal variables alone is addressed in the large sample discriminant function analysis of Chapter 11.)

12.8.1 ■ Evaluation of Limitations

12.8.1.1 ■ Adequacy of Expected Frequencies

Only if a goodness-of-fit criterion is to be used to compare observed and expected frequencies is there a limitation to logistic regression. As discussed in Section 12.3.2.1, the expected frequencies for all pairs of discrete predictors must meet the usual "chi-square" requirements. (This requirement is only for the sequential analysis of Section 12.8.3 because the predictors in the direct analysis are all continuous.)

Table 12.14 shows the results of a BMDP4F run to check the adequacy of expected frequencies for all pairs of discrete predictors. Although only pairs of discrete predictors are evaluated, the cases must match those of the full logistic analysis. Therefore, cases with missing data on continuous variables are deleted by the transform statement, resulting in elimination of 22 cases. An additional 3 cases are deleted because of missing data on one or more of the discrete variables, leaving 440 cases for the sequential analysis. These same 440 cases are also used for the direct analysis.

The first table shows one expected cell frequency just under 5: $(28)(76)/440 = 4.8$ for single women who are unhappy housewives. Another table shows that single women without children are expected to be rare in the sample: $(28)(73)/440 = 4.6$. Even more rare are single, nonwhite women, with an expected frequency of 2.4. However, in no two-way table do more than 20% of the cells have frequencies less than 5, nor are any expected frequencies less than 1. Therefore, there is no restriction on the goodness-of-fit criteria used to evaluate the model.

```
/INPUT      VARIABLES ARE 13.  FORMAT IS FREE.
            FILE='LOGREG.DAT'.
/VARIABLES NAMES ARE SUBNO, WORKSTAT, MARITAL, CHILDREN,
                 RELIGION, RACE, CONTROL, ATTMAR, ATTROLE, SEL,
                 ATTHOUSE, AGE, EDUC.
            LABEL IS SUBNO.
/TRANSFORM    IF (CONTROL EQ 0 OR ATTMAR EQ 0 OR ATTROLE EQ 0
                 OR SEL EQ 0 OR ATTHOUSE EQ 1 OR AGE EQ 0 OR
                 EDUC EQ 0) THEN USE=-1.
/GROUP      CODES(WORKSTAT) ARE 1 TO 3.
            NAMES(WORKSTAT) ARE WORKING,HAPHOUSE,UNHOUSE.
            CODES(MARITAL) ARE 1 TO 3.
            NAMES(MARITAL) ARE SINGLE, MARRIED, BROKEN.
            CODES(CHILDREN) ARE 0, 1.
            NAMES(CHILDREN) ARE NO, YES.
            CODES(RELIGION) ARE 1 TO 4.
            NAMES(RELIGION) ARE NONOTHER, CATHOLIC, PRTSTNT, JEWISH.
            CODES(RACE) ARE 1,2.
            NAMES(RACE) ARE WHITE, NONWHITE.
/TABLE      INDICES ARE WORKSTAT TO RACE.
/PRINT      MARGINALS=2.
/END
```

```
NUMBER OF CASES READ. . . . . . . . . . . . . .      465
   CASES WITH USE SET TO NEGATIVE VALUE . . . .       22
      REMAINING NUMBER OF CASES . . . . . . . .      443
```

***** MARGINAL SUBTABLE -- TABLE 1

MARITAL	WORKSTAT			
	WORKING	HAPHOUSE	UNHOUSE	TOTAL
SINGLE	22	2	4	28
MARRIED	163	125	62	350
BROKEN	50	2	10	62
TOTAL	235	129	76	440

***** MARGINAL SUBTABLE -- TABLE 1

CHILDREN	WORKSTAT			
	WORKING	HAPHOUSE	UNHOUSE	TOTAL
NO	53	10	10	73
YES	182	119	66	367
TOTAL	235	129	76	440

***** MARGINAL SUBTABLE -- TABLE 1

RELIGION	WORKSTAT			
	WORKING	HAPHOUSE	UNHOUSE	TOTAL
NONOTHER	45	21	9	75
CATHOLIC	59	27	26	112
PRTSTNT	89	48	26	163
JEWISH	42	33	15	90
TOTAL	235	129	76	440

TABLE 12.14   *(CONTINUED)*

```
*****  MARGINAL SUBTABLE -- TABLE  1

RACE                    WORKSTAT
------                  ------
          WORKING  HAPHOUSE UNHOUSE    TOTAL
-------------------------------------------------
WHITE       209       126       68  |    403
NONWHITE     26         3        8  |     37
-----------------------------------|---------
TOTAL       235       129       76  |    440


*****  MARGINAL SUBTABLE -- TABLE  1

CHILDREN                MARITAL
------                  ------
          SINGLE   MARRIED   BROKEN    TOTAL
-------------------------------------------------
NO           26        35       12  |     73
YES           2       315       50  |    367
-----------------------------------|---------
TOTAL        28       350       62  |    440

*****  MARGINAL SUBTABLE -- TABLE  1

RELIGION                MARITAL
------                  ------
          SINGLE   MARRIED   BROKEN    TOTAL
-------------------------------------------------
NONOTHER      9        48       18  |     75
CATHOLIC      2        96       14  |    112
PRTSTNT       8       130       25  |    163
JEWISH        9        76        5  |     90
-----------------------------------|---------
TOTAL        28       350       62  |    440


*****  MARGINAL SUBTABLE -- TABLE  1

RACE                    MARITAL
------                  ------
          SINGLE   MARRIED   BROKEN    TOTAL
-------------------------------------------------
WHITE        26       321       56  |    403
NONWHITE      2        29        6  |     37
-----------------------------------|---------
TOTAL        28       350       62  |    440

*****  MARGINAL SUBTABLE -- TABLE  1

RELIGION   CHILDREN
------     ------
            NO      YES     TOTAL
-----------------------------------------
NONOTHER    15       60  |     75
CATHOLIC    10      102  |    112
PRTSTNT     29      134  |    163
JEWISH      19       71  |     90
-------------------------|---------
TOTAL       73      367  |    440
```

TABLE 12.14   *(CONTINUED)*

```
*****  MARGINAL SUBTABLE -- TABLE  1

RACE         CHILDREN
------       ------
             NO     YES     TOTAL
-------------------------------------
WHITE        71     332  |   403
NONWHITE      2      35  |    37
------------------------------|---------
TOTAL        73     367  |   440

*****  MARGINAL SUBTABLE -- TABLE  1

RACE                     RELIGION
------                   ------
        NONOTHER CATHOLIC PRTSTNT  JEWISH   TOTAL
-------------------------------------------------
WHITE       69      90      157      87  |    403
NONWHITE     6      22        6       3  |     37
-------------------------------------------|---------
TOTAL       75     112      163      90  |    440
```

12.8.1.2 ■ Ratio of Cases to Variables

Sections 12.8.2 and 12.8.3 show no inordinately large parameter estimates or standard errors. Therefore, there is no reason to suspect a problem with too many empty cells or with outcome groups perfectly predicted by any variable.

12.8.1.3 ■ Multicollinearity

Analyses in Sections 12.8.2 and 12.8.3 show no problem with convergence after the probability criterion was relaxed for the sequential analysis, nor are the standard errors for parameters exceedingly large. Therefore, no multicollinearity is evident.

12.8.1.4 ■ Outliers in the Solution

Sections 12.8.2 and 12.8.3 show adequate model fits. Therefore, there is no need to search for outliers in the solution.

12.8.2 ■ Direct Logistic Regression with Two-Category Outcome

Table 12.15 shows the results of the logistic regression analysis predicting work status from the set of four attitudinal variables. The first block of instructions, for the SYSTAT DATA run, recodes the outcome, WORKSTAT, into two groups, working and housewives by combining the two housewife groups. This run assigns a code of 0 to the housewives, since SYSTAT LOGIT requires codes of 0 and 1 for a two-category outcome. Code 1 is the "response" group and code 0 is the "reference" group. The instructions for the SYSTAT LOGIT run include a request for tables showing both the success of prediction and the classification of cases.

The sample is split into 235 working women (RESP) and 205 housewives (REF). The comparison of the constant-only model with the full model, as shown in the next to last line of the RESULTS OF ESTIMATION table, shows a highly significant probability value, indicating that the predictors, as a set, reliably predict work status.

TABLE 12.15   SETUP FOR SYSTAT DATA RUN TO CREATE WORKING AND HOUSEWIFE GROUPS; SETUP AND OUTPUT OF SYSTAT LOGIT FOR LOGISTIC REGRESSION ANALYSIS OF WORK STATUS WITH ATTITUDINAL VARIABLES

DATA:

```
USE LOGREGR
SAVE LOGREG1
CODE WORKSTAT/3=0, 2=0
RUN
```

LOGIT:

```
USE LOGREG1
OUTPUT "LOGREG1.SYT"
MODEL WORKSTAT = CONSTANT + CONTROL + ATTMAR + ATTROLE + ATTHOUSE
LOPTION PREDICT / BOTH
ESTIMATE
```

BINARY LOGIT ANALYSIS

DEPENDENT VARIABLE: WORKSTAT

INPUT RECORDS: 440

RECORDS IN RAM FOR ANALYSIS: 440

SAMPLE SPLIT

CATEGORY	CHOICES
RESP	235
REF	205
	440

```
L-L AT ITER   1 IS   -304.985
L-L AT ITER   2 IS   -293.097
L-L AT ITER   3 IS   -293.072
```

CONVERGENCE ACHIEVED

RESULTS OF ESTIMATION

LOG LIKELIHOOD: -293.072

PARAMETER	ESTIMATE	S.E.	T-RATIO	P-VALUE
1 CONSTANT	3.300	0.992	3.327	0.001
2 CONTROL	-0.084	0.080	-1.051	0.293
3 ATTMAR	0.019	0.012	1.504	0.133
4 ATTROLE	-0.066	0.016	-4.170	0.000
5 ATTHOUSE	-0.030	0.024	-1.213	0.225

PARAMETER	ODDS RATIO	95.0% BOUNDS UPPER	LOWER
2 CONTROL	0.919	1.075	0.786
3 ATTMAR	1.019	1.044	0.994
4 ATTROLE	0.936	0.966	0.907
5 ATTHOUSE	0.971	1.018	0.926

TABLE 12.15   *(CONTINUED)*

```
LOG LIKELIHOOD OF CONSTANTS ONLY MODEL = LL(0) = -303.961
2*[LL(N)-LL(0)] = 21.779 WITH 4 DOF, CHI-SQ P-VALUE = 0.000
MCFADDEN'S RHO-SQUARED = 0.036
```

MODEL PREDICTION SUCCESS TABLE

ACTUAL CHOICE	PREDICTED CHOICE RESPONSE	REFERENCE	ACTUAL TOTAL
RESPONSE	130.821	104.179	235.000
REFERENCE	104.179	100.821	205.000
PRED. TOT.	235.000	205.000	440.000
CORRECT	0.557	0.492	
SUCCESS IND.	0.023	0.026	
TOT. CORRECT	0.526		

```
    SENSITIVITY:        0.557    SPECIFICITY:        0.492
FALSE REFERENCE:        0.443    FALSE RESPONSE:     0.508
```

MODEL CLASSIFICATION TABLE

ACTUAL CHOICE	PREDICTED CHOICE RESPONSE	REFERENCE	ACTUAL TOTAL
RESPONSE	164.000	71.000	235.000
REFERENCE	105.000	100.000	205.000
PRED. TOT.	269.000	171.000	440.000
CORRECT	0.698	0.488	
SUCCESS IND.	0.164	0.022	
TOT. CORRECT	0.600		

```
    SENSITIVITY:        0.698    SPECIFICITY:        0.488
FALSE REFERENCE:        0.390    FALSE RESPONSE:     0.415
```

The table of parameters shows that the only successful predictor is attitude toward role of women; working women and housewives differ significantly only in how they view the proper role of women. Nonsignificant coefficients are produced for locus of control, attitude toward marital status, and attitude toward housework. The negative coefficient for attitude toward the role of women means that working women (coded 1) have lower scores on the variable, indicating more liberal attitudes. (Note that these findings are consistent with the results of the contrast of working women vs. the other groups in Table 11.19 of the discriminant function analysis.

The last line of the RESULTS OF ESTIMATION table shows McFadden's *rho*-squared to be .036, indicating less than 4% of shared variance between work status and the set of predictors. Thus, the gain in prediction is minimal, even considering the tendency for *rho* to underestimate strength of association, as discussed in Section 12.7.4.

The MODEL PREDICTION SUCCESS and MODEL CLASSIFICATION tables confirm the finding that, even with attitude toward women's role as a significant predictor, the ability of the model to predict work status is not impressive. The MODEL CLASSIFICATION table shows individual cases fit into actual group membership and predicted membership by assigning each case to the group for which it has the highest probability (similar to discriminant function analysis classification tables). The MODEL PREDICTION SUCCESS table applies actual predicted probabilities and result in fractional numbers of cases. The PREDICTION SUCCESS table shows a success rate of about 53% while the classification table shows 60% correct classification.

Because of difficulties associated with the Wald test (cf. Section 12.6.1.2), an additional run is prudent to evaluate the predictors in the model. The second SYSTAT LOGIT run (Table 12.16) evaluates a model without attitude toward women's role. Applying Equation 12.7, the difference between that model and the model that includes ATTROLE is:

$$\chi^2 = 2[(-293.072) - (-302.261)] = 18.378$$

with df $= 1$, $p < .01$, reinforcing the finding of the Wald test that attitude toward women's role significantly enhances prediction of work status. Note also that Table 12.16 shows no difference between the model with the three remaining predictors and the constant-only model, confirming that these predictors are unrelated to work status.

Because only one predictor reliably predicted work status, a dosage-response curve where attitude toward women's role is the "dose" and work status "response" was requested through SYS-TAT LOGIT. Figure 12.1 shows the setup and resulting relationship. Note that actual ATTROLE scores range only from 18 to 55, representing a fairly narrow range of probabilities. That is, for the lowest score (the most liberal attitude toward women's role) the probability of working outside the home more than half time is about .75. For the women with the highest score on ATT-ROLE, the predicted probability of working is about .25. It would take a score in the 90s (well beyond the scale) for a 0.0 (certain) probability of not working.

Table 12.17 summarizes the statistics for the predictors. Table 12.18 contains a checklist for direct logistic regression with a two-category outcome. A Results section follows, that might be appropriate for submission to a journal.

TABLE 12.16   SETUP AND SELECTED SYSTAT LOGIT OUTPUT FOR MODEL THAT EXCLUDES ATTROLE

```
USE LOGREG1
OUTPUT "LOGREG2.SYT"
MODEL WORKSTAT = CONSTANT + CONTROL + ATTMAR + ATTHOUSE
ESTIMATE


LOG LIKELIHOOD: -302.261
```

PARAMETER	ESTIMATE	S.E.	TRATIO	PVALUE
1 CONSTANT	0.330	0.670	0.492	0.623
2 CONTROL	-0.098	0.078	-1.257	0.209
3 ATTMAR	0.018	0.012	1.466	0.143
4 ATTHOUSE	0.003	0.022	0.119	0.905

PARAMETER	ODDS RATIO	95.0% BOUNDS UPPER	LOWER
2 CONTROL	0.906	1.057	0.777
3 ATTMAR	1.018	1.043	0.994
4 ATTHOUSE	1.003	1.048	0.959

```
LOG LIKELIHOOD OF CONSTANTS ONLY MODEL = LL(0) = -303.961
2*[LL(N)-LL(0)] = 3.401 WITH 3 DOF, CHI-SQ P-VALUE = 0.334
MCFADDEN'S RHO-SQUARED = 0.006
```

TABLE 12.17 LOGISTIC REGRESSION ANALYSIS OF WORK STATUS AS A FUNCTION OF ATTITUDINAL VARIABLES

Variables	B	Wald test (z–ratio)	Odds Ratio	95% Confidence Interval for Odds Ratio	
				Upper	Lower
Locus of control	−0.08	−1.05	0.92	1.08	0.79
Attitude toward marital status	0.02	1.50	1.02	1.04	0.99
Attitude toward role of women	−0.07	−4.17	0.94	0.97	0.91
Attitude toward housework	−0.03	−1.21	0.97	1.02	0.93
(Constant)	3.30	3.33			

TABLE 12.18 CHECKLIST FOR STANDARD LOGISTIC REGRESSION WITH DICHOTOMOUS OUTCOME

1. Issues
   a. Adequacy of expected frequencies (if necessary)
   b. Outliers in the solution (if fit inadequate)
2. Major analyses
   a. Evaluation of overall fit. If adequate:
      (1) Significance tests for each predictor
      (2) Parameter estimates
   b. Evaluation of models without predictors
3. Additional analyses
   a. Odds ratios
   b. Dosage–response plots
   c. Classification and/or prediction success tables
   d. Interpretation in terms of means and/or percentages

```
USE LOGREG1
OUTPUT "LOGREGD.SYT"
MODEL WORKSTAT = CONSTANT + ATTROLE
ESTIMATE
QNTL
```

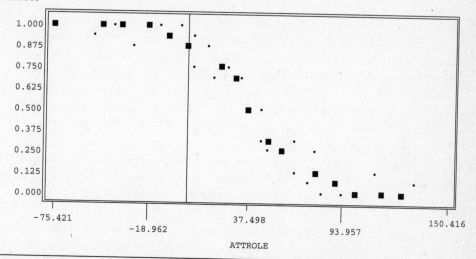

FIGURE 12.1   SETUP AND SELECTED SYSTAT LOGIT OUTPUT FOR PRODUCING
DOSAGE–RESPONSE RELATIONSHIP BETWEEN ATTITUDE TOWARD WOMEN'S ROLE
AND WORK STATUS.

Results

A direct logistic regression analysis was performed on work status
as outcome and four attitudinal predictors: locus of control, attitude
toward current marital status, attitude toward women's rights, and
attitude toward housework. Analysis was performed using SYSTAT LOGIT.
After deletion of 25 cases with missing values, data from 440 women
were available for analysis: 205 housewives and 235 women who work outside the home more than 20 hours a week for pay. Missing data appeared
to be scattered randomly across categories of outcome and predictors.

A test of the full model with all four predictors against a constant-only model was statistically reliable, $\chi^2(4, N = 440) = 21.78$, p
$< .001$, indicating that the predictors, as a set, reliably distin-

guished between working women and housewives. The variance in work status accounted for is small, however, with McFadden's rho = .04.
Prediction success was unimpressive, with 56% of the working women and
49% of the housewives correctly predicted, for an overall success rate
of 53%.

   Table 12.17 shows regression coefficients, Wald statistics, odds
ratios and 95% confidence intervals for odds ratios for each of the
four predictors. According to the Wald criterion, only attitude toward
role of women reliably predicted work status, z = −4.17, p < .001. A
model run with attitude toward role of women omitted was not reliably
different from a constant-only model, however this model was reliably
different from the full model, $\chi^2(1, N = 440) = 18.38$, p < .001. This
confirms the finding that attitude towards women's role is the only
reliable predictor of work status among the four attitudinal variables.
However, the odds ratio of .94 shows little change in the likelihood of
working on the basis of a one unit change in attitude toward women's
role.

   Figure 12.1 shows the probability of working outside the home as a
function of attitude toward women's role. Since higher scores indicate
more conservative attitudes, the more conservative the woman's attitude,
the less likely she is to be working.

12.8.3 ■ Sequential Logistic Regression with Three Categories of Outcome

Table 12.19 shows the results of logistic regression analysis through BMDPPR predicting the three categories of outcome (working, role-satisfied housewives, and role-dissatisfied housewives) from the set of seven demographic and four attitudinal variables. The REGRESS paragraph shows which predictors are continuous (INTERVAL) and sets up the regression model. This first run evaluates individual predictors and the model as a whole, including both demographic and attitudinal predictors. PCONV is set to be different from the default of .0001 to allow the model to converge.

   BMDPPR then shows the number of respondents in each of the three outcomes and how the categorical predictors are coded. The 440 distinct covariate patterns indicate that no two women have exactly the same pattern of responses to the 11 predictors.

   At STEP NUMBER 0, all predictors are in the model and the goodness-of-fit test (comparing observed with expected frequencies) shows an excellent fit with $p = .996$. Two tables of statistics are shown, one for each degree of freedom for outcome. The first table compares working women with the other two groups, while the second compares role-satisfied women with the other groups. Using a criterion $\alpha = .0045$ (to compensate for inflated Type I error rate with 11 predictors), the critical value of z (COEFF/SE, the Wald test) = 2.84 for a two-tailed test. By this criterion, only ATTHOUSE (attitude toward housework) reliably separates role-satisfied housewives from the other two groups. No predictor reliably separates working women from the other two groups.

   Evidence from these two tables is not conclusive regarding the ability of the predictors to distinguish members of the three groups simultaneously. Further, these tables evaluate categorical

```
/INPUT     VARIABLES ARE 13.  FORMAT IS FREE.
           FILE='LOGREG.DAT'.
/VARIABLES NAMES ARE SUBNO, WORKSTAT, MARITAL, CHILDREN,
               RELIGION, RACE, CONTROL, ATTMAR, ATTROLE, SEL,
               ATTHOUSE, AGE, EDUC.
           MISSING = 2*0, 3*9, 5*0, 1, 2*0.
           LABEL IS SUBNO.
/GROUP     CODES(WORKSTAT) ARE 1 TO 3.
           NAMES(WORKSTAT) ARE WORKING,HAPHOUSE,UNHOUSE.
/REGRESS   DEPENDENT = WORKSTAT.  LEVEL = 3.
           INTERVAL = CONTROL, ATTMAR, ATTROLE, SEL, ATTHOUSE, AGE,
               EDUC.
           MODEL = MARITAL, CHILDREN, RELIGION, RACE, SEL, AGE, EDUC,
               CONTROL, ATTMAR, ATTROLE, ATTHOUSE.
           PCONV = .003.
/END
```

```
NUMBER OF CASES READ. . . . . . . . . . . . .    465
    CASES WITH DATA MISSING OR BEYOND LIMITS . .    25
        REMAINING NUMBER OF CASES . . . . . . . .   440

  TOTAL NUMBER OF RESPONSES USED IN THE ANALYSIS    440.

                          WORKING  . . . . . .    235.

                          HAPHOUSE . . . . . .    129.

                          UNHOUSE  . . . . . .     76.

  TOTAL NUMBER OF DISTINCT COVARIATE PATTERNS . .   440
```

VARIABLE NO. N A M E	GROUP INDEX	FREQ	DESIGN VARIABLES (1)	(2)	(3)
3 MARITAL	1	28	0	0	
	2	350	1	0	
	3	62	0	1	
4 CHILDREN	0	73	0		
	1	367	1		
5 RELIGION	1	75	0	0	0
	2	112	1	0	0
	3	163	0	1	0
	4	90	0	0	1
6 RACE	1	403	0		
	2	37	1		

```
STEP NUMBER   0

                     LOG LIKELIHOOD =  -372.176
GOODNESS OF FIT CHI-SQ   (2*O*LN(O/E))  =   744.352  D.F.= 850  P-VALUE= 0.996
```

		COEFFICIENT	STANDARD ERROR	COEF/SE	EXP(COEF)	95% CI OF EXP(COEF) LOW-END	HIGH-END
OUTCOME: WORKING							
1 MARITAL	(1)	-0.3651	0.778	-0.469	0.69	0.15	3.2
2	(2)	0.2897	0.797	0.363	1.3	0.28	6.4
3 CHILDREN		-0.6393	0.502	-1.27	0.53	0.20	1.4
4 RELIGION	(1)	-0.6738	0.487	-1.38	0.51	0.20	1.3
5	(2)	-0.2845	0.464	-0.613	0.75	0.30	1.9
6	(3)	-0.4463	0.507	-0.880	0.64	0.24	1.7
7 RACE		0.8243	0.508	1.62	2.3	0.84	6.2
8 SEL		-0.6797E-02	0.664E-02	-1.02	0.99	0.98	1.0

TABLE 12.19   *(CONTINUED)*

9 AGE		0.1531	0.744E-01	2.06	1.2	1.0	1.3
10 EDUC		0.1800	0.805E-01	2.24	1.2	1.0	1.4
11 CONTROL		-0.9612E-01	0.113	-0.854	0.91	0.73	1.1
12 ATTMAR		-0.2304E-01	0.189E-01	-1.22	0.98	0.94	1.0
13 ATTROLE		-0.5149E-01	0.250E-01	-2.06	0.95	0.90	1.0
14 ATTHOUSE		-0.8470E-01	0.377E-01	-2.25	0.92	0.85	0.99
15 CONST1		4.595	2.12	2.16	99.	1.5	0.64E+04
OUTCOME: HAPHOUSE							
16 MARITAL	(1)	1.337	1.11	1.20	3.8	0.43	34.
17	(2)	-0.8033	1.32	-0.611	0.45	0.34E-01	5.9
18 CHILDREN		0.5542	0.604	0.918	1.7	0.53	5.7
19 RELIGION	(1)	-1.241	0.544	-2.28	0.29	0.99E-01	0.84
20	(2)	-0.7381	0.514	-1.44	0.48	0.17	1.3
21	(3)	-0.3065	0.548	-0.560	0.74	0.25	2.2
22 RACE		-1.406	0.754	-1.86	0.25	0.56E-01	1.1
23 SEL		0.7775E-02	0.748E-02	1.04	1.0	0.99	1.0
24 AGE		0.9277E-03	0.832E-01	0.112E-01	1.0	0.85	1.2
25 EDUC		0.8043E-01	0.921E-01	0.873	1.1	0.90	1.3
26 CONTROL		-0.7127E-01	0.125	-0.570	0.93	0.73	1.2
27 ATTMAR		-0.2767E-01	0.228E-01	-1.21	0.97	0.93	1.0
28 ATTROLE		0.4516E-01	0.281E-01	1.61	1.0	0.99	1.1
29 ATTHOUSE		-0.1266	0.425E-01	-2.98	0.88	0.81	0.96
30 CONST2		0.6310	2.48	0.254	1.9	0.14E-01	0.25E+03

STATISTICS TO ENTER OR REMOVE TERMS   MODELS LOG LIKELIHOOD   -372.1761

TERM	APPROX. CHI-SQ. D.F. ENTER	APPROX. CHI-SQ. D.F. REMOVE	P-VALUE	LOG LIKELIHOOD
MARITAL		24.10 4	0.0001	-384.2245
MARITAL		IS IN	MAY NOT BE REMOVED.	
CHILDREN		7.69 2	0.0214	-376.0213
CHILDREN		IS IN	MAY NOT BE REMOVED.	
RELIGION		8.73 6	0.1895	-376.5401
RELIGION		IS IN	MAY NOT BE REMOVED.	
RACE		15.69 2	0.0004	-380.0235
RACE		IS IN	MAY NOT BE REMOVED.	
SEL		6.01 2	0.0496	-375.1807
SEL		IS IN	MAY NOT BE REMOVED.	
AGE		7.40 2	0.0247	-375.8773
AGE		IS IN	MAY NOT BE REMOVED.	
EDUC		6.07 2	0.0482	-375.2086
EDUC		IS IN	MAY NOT BE REMOVED.	
CONTROL		0.73 2	0.6945	-372.5407
CONTROL		IS IN	MAY NOT BE REMOVED.	
ATTMAR		1.90 2	0.3861	-373.1279
ATTMAR		IS IN	MAY NOT BE REMOVED.	
ATTROLE		19.59 2	0.0001	-381.9723
ATTROLE		IS IN	MAY NOT BE REMOVED.	
ATTHOUSE		9.39 2	0.0092	-376.8696
ATTHOUSE		IS IN	MAY NOT BE REMOVED.	
CONSTANT		7.20 2	0.0274	-375.7751
CONSTANT		IS IN	MAY NOT BE REMOVED.	

predictors with more than two categories by their single degree of freedom dummy variables. Finally, difficulties associated with the Wald test (cf. Section 12.6.1.2) make another type of evaluation desirable.

The final table in the output provides a maximum likelihood test for each predictor, with an indication of the improvement in the model associated with each predictor individually; the P-VALUE

TABLE 12.20   SETUP AND SELECTED OUTPUT OF BMDPPR FOR LOGISTIC REGRESSION ANALYSIS OF WORK STATUS WITH DEMOGRAPHIC VARIABLES ONLY

```
/INPUT      VARIABLES ARE 13.  FORMAT IS FREE.
            FILE='LOGREG.DAT'.
/VARIABLES  NAMES ARE SUBNO, WORKSTAT, MARITAL, CHILDREN,
                     RELIGION, RACE, CONTROL, ATTMAR, ATTROLE, SEL,
                     ATTHOUSE, AGE, EDUC.
            GROUPING IS WORKSTAT.
            MISSING = 2*0, 3*9, 5*0, 1, 2*0.
            LABEL IS SUBNO.
/GROUP      CODES(WORKSTAT) ARE 1 TO 3.
            NAMES(WORKSTAT) ARE WORKING,HAPHOUSE,UNHOUSE.
/REGRESS    DEPENDENT = WORKSTAT.  LEVEL = 3.
            INTERVAL = SEL, AGE, EDUC.
            MODEL = MARITAL, CHILDREN, RELIGION, RACE, SEL, AGE, EDUC.
            PCONV = .003.
/END


        TOTAL NUMBER OF RESPONSES USED IN THE ANALYSIS      440.

                             WORKING . . . . . .      235.

                             HAPHOUSE . . . . . .     129.

                             UNHOUSE  . . . . . .      76.

        TOTAL NUMBER OF DISTINCT COVARIATE PATTERNS . .     422



STEP NUMBER   0


                    LOG LIKELIHOOD =   -393.753
GOODNESS OF FIT CHI-SQ   (2*O*LN(O/E))  =    758.733  D.F.= 822  P-VALUE= 0.944
```

shows if the model is significantly degraded by removal of each predictor. Using $\alpha = .0045$ as a criterion, four predictors reliably distinguish among outcomes: marital status, race, attitude toward role of women, and attitude toward housework.

The second BMDPPR run (Table 12.20) includes only demographic variables, and allows evaluation of improvement in the model by addition of attitudinal predictors. When both demographic and attitudinal predictors are included, log-likelihood = -372.176 with 14 df; when demographic predictors alone are included, log-likelihood = -393.752 with 10 df. Applying Equation 12.7 to evaluate improvement in fit,

$$\chi^2 = 2[(-372.176) - (-393.753)] = 43.15$$

with df = 4, $p < .05$. This indicates reliable improvement in the model with the addition of attitudinal predictors.

Two SYSTAT LOGIT runs (Tables 12.21 and 12.22) produce prediction success tables for demographic predictors alone and demographic plus attitudinal predictors.

The remaining issue is interpretation of the reliable effects, difficult through BMDPPR because of the separation of effects for both outcome and predictors into single degree of freedom dummy variables. For categorical predictors, group differences are observed in proportions of cases in

TABLE 12.21   LARGE SAMPLE EXAMPLE SETUP AND SELECTED OUTPUT FOR SYSTAT LOGIT PREDICTION SUCCESS TABLE: DEMOGRAPHIC VARIABLES ONLY

```
USE LOGREGR
OUTPUT "LOGREGD.SYT"
NCAT = 3
CAT MARITAL=3 RELIGION=4
MODEL WORKSTAT = CONSTANT + MARITAL + CHILDREN + RELIGION + RACE +,
                SEL + AGE + EDUC
LOPTION PREDICT
ESTIMATE
```

MODEL PREDICTION SUCCESS TABLE

ACTUAL CHOICE	PREDICTED CHOICE 1	2	3	ACTUAL TOTAL
1	139.842	57.113	38.045	235.000
2	56.793	49.472	22.735	129.000
3	38.365	22.415	15.220	76.000
PRED. TOT.	235.000	129.000	76.000	440.000
CORRECT	0.595	0.384	0.200	
SUCCESS IND.	0.061	0.090	0.028	
TOT. CORRECT	0.465			

TABLE 12.22   LARGE SAMPLE EXAMPLE SETUP AND SELECTED OUTPUT FOR SYSTAT LOGIT PREDICTION SUCCESS TABLE: ALL VARIABLES

MODEL PREDICTION SUCCESS TABLE

ACTUAL CHOICE	PREDICTED CHOICE 1	2	3	ACTUAL TOTAL
1	144.616	52.745	37.639	235.000
2	52.835	55.794	20.371	129.000
3	37.549	20.461	17.990	76.000
PRED. TOT.	235.000	129.000	76.000	440.000
CORRECT	0.615	0.433	0.237	
SUCCESS IND.	0.081	0.139	0.064	
TOT. CORRECT	0.496			

each category of predictor for each category of outcome. This information is available from the screening run of Table 12.14. For example, we see that 69% (163/235) of the working women are currently married, while 97% and 82% of the role-satisfied and role-dissatisfied housewives, respectively, are currently married. For continuous predictors, interpretation is based on mean differences for significant predictors for each category of outcome. Table 12.23 shows BMDP1D output giving means for each category of outcome on the reliable predictors.

TABLE 12.23 SETUP AND PARTIAL OUTPUT OF BMDP1D SHOWING GROUP MEANS FOR ATTROLE AND ATTHOUSE

```
/INPUT      VARIABLES ARE 13.  FORMAT IS FREE.
            FILE='LOGREG.DAT'.
/VARIABLES NAMES ARE SUBNO, WORKSTAT, MARITAL, CHILDREN,
           RELIGION, RACE, CONTROL, ATTMAR, ATTROLE, SEL,
           ATTHOUSE, AGE, EDUC.
           LABEL IS SUBNO.
           GROUPING IS WORKSTAT.
           USE = WORKSTAT, ATTROLE, ATTHOUSE.
/TRANSFORM  IF (CONTROL EQ 0 OR ATTMAR EQ 0 OR ATTROLE EQ 0
              OR SEL EQ 0 OR ATTHOUSE EQ 1 OR AGE EQ 0 OR
              EDUC EQ 0 OR MARITAL EQ 9 OR CHILDREN EQ 9 OR
              RELIGION EQ 9 OR RACE EQ 0) THEN USE=-1.
/GROUP      CODES(WORKSTAT) ARE 1 TO 3.
            NAMES(WORKSTAT) ARE WORKING,HAPHOUSE,UNHOUSE.
/END
```

VARIABLE NO. NAME	GROUPING VARIABLE LEVEL	TOTAL FREQUENCY	MEAN	STANDARD DEVIATION	ST.ERR OF MEAN	COEFF. OF VARIATION	SMALLEST VALUE	SMALLEST Z-SCORE	LARGEST VALUE	LARGEST Z-SCORE	RANGE
9 ATTROLE		440	35.175	6.7624	.32239	.19225	18.000	-2.54	55.000	2.93	37.000
	WORKSTAT WORKING	235	33.919	6.9748	.45498	.20563	18.000	-2.28	52.000	2.59	34.000
	HAPHOUSE	129	37.202	6.3853	.56219	.17164	24.000	-2.07	55.000	2.79	31.000
	UNHOUSE	76	35.618	5.8400	.66990	.16396	20.000	-2.67	49.000	2.29	29.000
11 ATTHOUSE		440	23.527	4.5066	.21484	.19155	2.0000	-4.78	35.000	2.55	33.000
	WORKSTAT WORKING	235	23.600	4.8589	.31696	.20588	2.0000	-4.45	34.000	2.14	32.000
	HAPHOUSE	129	22.574	3.8705	.34078	.17146	11.000	-2.99	32.000	2.44	21.000
	UNHOUSE	76	24.921	4.0125	.46027	.16101	15.000	-2.47	35.000	2.51	20.000

VARIABLE NO. NAME	CATEGORY NAME	CATEGORY FREQUENCY	TOTAL FREQUENCY	NO. OF VALUES MISSING OR OUTSIDE THE RANGE
2 WORKSTAT			440	0
	WORKING	235		
	HAPHOUSE	129		
	UNHOUSE	76		

TABLE 12.24   LOGISTIC REGRESSION ANALYSIS OF WORK STATUS AS A
FUNCTION OF DEMOGRAPHIC AND ATTITUDINAL VARIABLES

Variables	χ^2 to Remove	df	Log-likelihood
Demographic			
Marital status	24.10*	4	
Presence of children	7.69	2	
Religion	8.73	6	
Race	15.69*	2	
Socioeconomic level	6.01	2	
Age	7.40	2	
Educational level	6.07	2	
All demographic variables			−393.75
Attitudinal			
Locus of control	0.73	2	
Attitude toward marital status	1.90	2	
Attitude toward role of women	19.59*	2	
Attitude toward housework	9.39*	2	
All variables			−372.18

*$p < .0045$

TABLE 12.25   MARITAL STATUS AND RACE AS A FUNCTION OF
WORK STATUS

	Work Status[a]			
	1	2	3	Total
Marital Status				
Single	22	2	4	28
Married	163	125	62	350
Broken	50	2	10	62
Total	235	129	76	440
Race				
White	209	126	68	403
Nonwhite	26	3	8	37
Total	235	129	76	440

[a]1 = Working; 2 = Role-satisfied housewives; 3 = Role-dissatisfied housewives.

Table 12.24 summarizes results of the sequential analysis. Table 12.25 shows contingency tables for reliable discrete predictors. Table 12.26 provides a checklist for sequential logistic regression with more than two outcomes. A Results section in journal format follows.

TABLE 12.26   CHECKLIST FOR SEQUENTIAL LOGISTIC REGRESSION
WITH MULTIPLE OUTCOMES

1. Issues

    a. Adequacy of expected frequencies (if necessary)
    b. Outliers in the solution (if fit inadequate)

2. Major analyses

    a. Evaluation of overall fit at each step. If adequate:
     (1) Significance tests for each predictor
     (2) Parameter estimates
    b. Evaluation of improvement in model at each step

3. Additional analyses

    a. Odds ratios
    b. Dosage−response plots
    c. Classification and/or prediction success tables
    d. Interpretation in terms of means and/or percentages
    e. Evaluation of models without individual predictors

Results

A sequential logistic regression analysis was performed through BMDP-PR to assess prediction of membership in one of three categories of outcome (working women, role-satisfied housewives, and role-dissatisfied housewives), first on the basis of seven demographic predictors and then after addition of four attitudinal predictors. Demographic predictors were children (presence or absence), race (Caucasian or other), socioeconomic level, age, religious affiliation (Protestant, Catholic, Jewish, none/other), and marital status (single, married, broken). Attitudinal predictors were locus of control, attitude toward marital status, attitude toward role of women, and attitude toward homemaking.

Of the original 465 cases, 25 were deleted due to missing data; missing data appeared to be randomly scattered over outcomes and predictors. Evaluation of adequacy of expected frequencies for categorical demographic predictors revealed no need to restrict model goodness-of-fit tests.

There was a good model fit (discrimination among groups) on the basis of the seven demographic predictors alone, $\chi^2(822, \underline{N} = 440) = 758.73$, $\underline{p} = .94$. After addition of the four attitudinal predictors, $\chi^2(850, \underline{N} = 440) = 744.35$, $\underline{p} = .99$. Comparison of log-likelihood ratios (see Table 12.24) for models with and without attitudinal variables showed reliable improvement with the addition of attitudinal predictors, $\chi^2(4, \underline{N} = 440) = 43.15$, $\underline{p} < .05$.

Overall prediction rates were unimpressive. On the basis of 10 demographic variables alone, success rates were 60% for working women, 38% for role-satisfied, and 20% for role-dissatisfied women; the overall correct prediction rate was 47%. The improvement to 50% with the addition of four attitudinal predictors reflected success rates of 62%, 43%, and 24% for the three groups, respectively.

Table 12.24 shows the contribution of the individual predictors to the model by comparing models with and without each predictor. Two predictors from the demographic and two from the attitudinal set reliable enhanced prediction, p < .0045. Outcome was predictable from marital status, race, attitude toward role of women, and attitude toward housework.

Table 12.25 shows the relationship between work status and the two categorical demographic predictors. Working women are less likely to be currently married (69%) than are role-satisfied housewives (97%) or role-dissatisfied housewives (82%). Role-satisfied housewives are more likely to be Caucasian (98%) than are working women or role-dissatisfied housewives (89% for both groups).

Mean group differences in attitudes toward role of women and homemaking were not large. However, role-satisfied housewives had more conservative attitudes (mean = 37.2) than role-dissatisfied housewives (mean = 35.6) or working women (mean = 33.9). Role-satisfied housewives also expressed more positive attitudes toward homemaking (mean = 22.6) than did working women (mean = 23.6) or role-dissatisfied housewives (mean = 24.9).

12.9 ■ SOME EXAMPLES FROM THE LITERATURE

Many logistic regression studies describe themselves as stepwise, despite the fact that a sequential strategy is often employed. Perhaps this is partially due to generic use of the term stepwise to refer to all but standard models, and partially to the self-description of some programs as stepwise.

Thompson (1985) used a stepwise logistic regression strategy to study outcome (success or failure) in a community mental health program. Modeling was based on 519 client admissions with data on 17 client characteristics, consisting of demographic data, dates of admission and discharge, referral data, mental health history, intelligence scores at time of admission and discharge, and follow-up treatment. BMDPLR was used to implement a stepwise strategy in which all 17 potential predictors were initially included in a series of models, with 12 remaining on the basis of a criterion $\alpha = .10$. Of the 12, all but one (age at admission to the program) were significant in the anticipated direction. Since there were four different criteria for evaluating outcome, separate logistic regressions were run for each criterion. Further, separate logistic regressions were run for positive outcome and negative outcome since criteria were not inclusive. For example, a positive Goal Attainment Score (GAS) was defined as ≥ 11.12 (the mean), while a negative GAS was defined as

≤ 0. Cases with scores between 0 and 11.12 were in the "other" category for both the positive and negative outcome analyses.

Overall model goodness-of-fit measures were not presented. Instead, tables were generated in which significant values were given for characteristics associated with positive and negative outcome. For example, diagnosis was significantly associated with positive but not negative outcome on the basis of GAS. Evidently p-values in the tables are based on univariate analysis, since variables which were removed from the models are so annotated. Variables which stayed in the models despite lack of significant prediction on a univariate basis were withheld, pending cross-validation or experimental studies.

A sequential strategy was used to investigate racial bias in psychiatric diagnoses (Pavkov, Lewis, & Lyons, 1989). The outcome was whether a patient was diagnosed as schizophrenic. The major predictor was race, dummy coded into Black vs. White (entered last in the modeling sequence) and Hispanic/other vs. White. Covariates were age, whether professional help had been sought, number of previous admissions, adjustment rating, four dummy variables assessing location of treatment, independent clinical diagnosis, and the interaction between Black and seeking professional help.

A table of results (misleadingly labeled "stepwise" logistic regression) included each β and its standard error, along with χ^2 and p for each predictor (presumably χ^2 to remove from the model), and R. Race/Black significantly improved model fit, and was found to be the strongest predictor of schizophrenic diagnosis after the independent clinical diagnosis.

A two-stage investigation of children who do exceptionally well in first grade was pursued by Pallas, Entwisle, Alexander, and Cadigan (1987). In all of their logistic regression analyses, the outcome was whether or not the child showed exceptional gains in scores on verbal tests. Four clusters of predictors were (1) child's personal characteristics, (2) child's performance in school, (3) family effects, and (4) teacher's characteristics. In the first stage of analysis, separate logistic regressions were run for each cluster. Then, a final logistic regression analysis included predictors from each cluster which were reliable in the first-stage analyses. Presumably all were run as direct logistic regressions. Tables of results include, for each predictor, the logistic regression coefficient with its standard error (Wald test), and the change in probability of outcome associated with a one standard deviation change in the predictor. No mention is made of overall fit of the models.

The combined analysis showed effects of teacher characteristics that were stronger than expected. Four predictors from the teacher characteristic cluster contributed to exceptional gains in scores on verbal tests, along with two characteristics of the students (one each from personal characteristics and performance), and only one related to parents (child's absence).

Bennett, Beaurepaire, Langeluddicke, Kellow, and Tennant, (1991) compared non-ulcer dyspeptic (NUD) patients with controls in an application of logistic regression to a case-control study. Subjects were matched on age, sex, and social status, with one control for each patient. Predictor variables included a variety of life stress, personality, mood state, and coping measures. While univariate analyses showed that patients differed from controls on 17 psychological variables, logistic regression analysis (presumably direct) showed that a single predictor, highly threatening chronic difficulties, alone provided a highly adequate model (significant improvement over, presumably, a constant-only model). The odds ratio was reported to be near infinity, thus incalculable.

While 98% of NUDs were exposed to at least one such stressor, only 2% of controls were so exposed. In its presence, no other predictor could improve the model. A second (presumably stepwise) model was run in which predictors associated with chronic difficulty were omitted. This model showed significant predictability associated with severe long-term threat at one week, with NUDS patients 17 times more likely to experience such threat than controls.

CHAPTER 13

Principal Components and Factor Analysis

13.1 ■ GENERAL PURPOSE AND DESCRIPTION

Principal components analysis (PCA) and factor analysis (FA) are statistical techniques applied to a single set of variables where the researcher is interested in discovering which variables in the set form coherent subsets that are relatively independent of one another. Variables that are correlated with one another but largely independent of other subsets of variables are combined into factors.[1] Factors are thought to reflect underlying processes that have created the correlations among variables.

Suppose, for instance, a researcher is interested in studying characteristics of graduate students. The researcher measures a large sample of graduate students on personality characteristics, motivation, intellectual ability, scholastic history, familial history, health and physical characteristics, etc. Each of these areas is assessed by numerous variables; the variables all enter the analysis individually at one time and correlations among them are studied. The analysis reveals patterns of correlation among the variables that are thought to reflect underlying processes affecting the behavior of graduate students. For instance, several individual variables from the personality measures combine with some variables from the motivation and scholastic history measures to form a factor measuring the degree to which a person prefers to work independently, an independence factor. Several variables from the intellectual ability measures combine with some others from scholastic history to suggest an intelligence factor.

A major use of PCA and FA in psychology is in development of objective tests for measurement of personality and intelligence and the like. The researcher starts out with a very large number of items reflecting a first guess about the items that may eventually prove useful. The items are given to randomly selected subjects and factors are derived. As a result of the first factor analysis, items are added and deleted, a second test is devised, and that test is given to other randomly selected subjects. The process continues until the researcher has a test with numerous items forming several factors that

---

[1] PCA produces components while FA produces factors, but it is less confusing here to call the results of both analyses factors.

represent the area to be measured. The validity of the factors is tested in research where predictions are made regarding differences in behavior of persons who score high or low on a factor.

The specific goals of PCA or FA are to summarize patterns of correlations among observed variables, to reduce a large number of observed variables to a smaller number of factors, to provide an operational definition (a regression equation) for an underlying process by using observed variables, or to test a theory about the nature of underlying processes. Some or all of these goals may be the focus of a particular research project.

PCA and FA have considerable utility in reducing numerous variables down to a few factors. Mathematically, PCA and FA produce several linear combinations of observed variables, each linear combination a factor. The factors summarize the patterns of correlations in the observed correlation matrix and can, with varying degrees of success, be used to reproduce the observed correlation matrix. But since the number of factors is usually far fewer than the number of observed variables, there is considerable parsimony in using the factor analysis. Further, when scores on factors are estimated for each subject, they are often more reliable than scores on individual observed variables.

Steps in PCA or FA include selecting and measuring a set of variables, preparing the correlation matrix (to perform either PCA or FA), extracting a set of factors from the correlation matrix, determining the number of factors, (probably) rotating the factors to increase interpretability, and, finally, interpreting the results. Although there are relevant statistical considerations to most of these steps, an important test of the analysis is its interpretability.

A good PCA or FA "makes sense"; a bad one does not. Interpretation and naming of factors depend on the meaning of the particular combination of observed variables that correlate highly with each factor. A factor is more easily interpreted when several observed variables correlate highly with it and those variables do not correlate with other factors.

Once interpretability is adequate, the last, and very large, step is to verify the factor structure by establishing the construct validity of the factors. The researcher seeks to demonstrate that scores on the latent variables (factors) covary with scores on other variables, or that scores on latent variables change with experimental conditions as predicted by theory.

One of the problems with PCA and FA is that there is no criterion variable against which to test the solution. In regression analysis, for instance, the DV is a criterion and the correlation between observed and predicted DV scores serves as a test of the solution—similarly for the two sets of variables in canonical correlation. In discriminant function analysis, logistic regression, profile analysis, and multivariate analysis of variance, the solution is judged by how well it predicts group membership. But in PCA and FA there is no external criterion such as group membership against which to test the solution.

A second problem with FA or PCA is that, after extraction, there is an infinite number of rotations available, all accounting for the same amount of variance in the original data, but with factors defined slightly differently. The final choice among alternatives depends on the researcher's assessment of its interpretability and scientific utility. In the presence of an infinite number of mathematically identical solutions, researchers are bound to differ regarding which is best. Because the differences cannot be resolved by appeal to objective criteria, arguments over the best solution sometimes become vociferous. However, those who expect a certain amount of ambiguity with respect to choice of the best FA solution will not be surprised when other researchers choose a different one. Nor will they be surprised when results are not replicated exactly if different decisions are made at one, or more, of the steps in performing FA.

A third problem is that FA is frequently used in an attempt to "save" poorly conceived research. If no other statistical procedure is applicable, at least data can usually be factor analyzed. Thus, in the minds of many, the various forms of FA are associated with sloppy research. The very power of PCA and FA to create apparent order from real chaos contributes to their somewhat tarnished reputations as scientific tools.

There are two major types of FA: exploratory and confirmatory. In exploratory FA, one seeks to describe and summarize data by grouping together variables that are correlated. The variables themselves may or may not have been chosen with potential underlying processes in mind. Exploratory FA is usually performed in the early stages of research, when it provides a tool for consolidating variables and for generating hypotheses about underlying processes. Confirmatory FA is a much more sophisticated technique used in the advanced stages of the research process to test a theory about latent processes. Variables are carefully and specifically chosen to reveal underlying processes.

Before we go on, it is helpful to define a few terms. The first terms involve correlation matrices. The correlation matrix produced by the observed variables is called the *observed correlation matrix*. The correlation matrix produced from factors is called the *reproduced correlation matrix*. The difference between observed and reproduced correlation matrices is the *residual correlation matrix*. In a good FA, correlations in the residual matrix are small, indicating a close fit between observed and reproduced matrices.

A second set of terms refers to matrices produced and interpreted as part of the solution. Rotation of factors is a process by which the solution is made more interpretable without changing its underlying mathematical properties. There are two general classes of rotation: orthogonal and oblique. If rotation is *orthogonal* (so that all the factors are uncorrelated with each other), a *loading* matrix is produced. The loading matrix is a matrix of correlations between observed variables and factors. The sizes of the loadings reflect the extent of relationship between each observed variable and each factor. Orthogonal FA is interpreted from the loading matrix by looking at which observed variables correlate with each factor.

If rotation is *oblique* (so that the factors themselves are correlated), several additional matrices are produced. The *factor correlation* matrix contains the correlations among the factors. The loading matrix from orthogonal rotation splits into two matrices for oblique rotation: a *structure* matrix of correlations between factors and variables and a *pattern* matrix of unique relationships (uncontaminated by overlap among factors) between each factor and each observed variable. Following oblique rotation, the meaning of factors is ascertained from the pattern matrix.

Lastly, for both types of rotation, there is a *factor-score* coefficients matrix, a matrix of coefficients used in several regression-like equations to predict scores on factors from scores on observed variables for each individual.

FA produces *factors*, while PCA produces *components*. However, the processes are similar except in preparation of the observed correlation matrix for extraction. The difference between PCA and FA is in the variance that is analyzed. In PCA, all the variance in the observed variables is analyzed. In FA, only shared variance is analyzed; attempts are made to estimate and eliminate variance due to error and variance that is unique to each variable. The term *factor* is used here to refer to both components and factors unless the distinction is critical, in which case the appropriate term is used.

13.2 ■ KINDS OF RESEARCH QUESTIONS

The goal of research using PCA or FA is to reduce a large number of variables to a smaller number of factors, to concisely describe (and perhaps understand) the relationships among observed variables, or to test theory about underlying processes. Some of the specific questions that are frequently asked are presented in Sections 13.2.1 through 13.2.5.

13.2.1 ■ Number of Factors

How many reliable and interpretable factors are there in the data set? How many factors are needed to summarize the pattern of correlations in the correlation matrix? In the graduate student example, two factors are discussed; are these both reliable? Are there any more factors that are reliable? Strategies for choosing an appropriate number of factors and for assessing the correspondence between observed and reproduced correlation matrices are discussed in Section 13.6.2.

13.2.2 ■ Nature of Factors

What is the meaning of the factors? How are the factors to be interpreted? Factors are interpreted by the variables that correlate with them. Rotation to improve interpretability is discussed in Section 13.6.3; interpretation itself is discussed in Section 13.6.5.

13.2.3 ■ Importance of Solutions and Factors

How much variance in a data set is accounted for by the factors? Which factors account for the most variance? In the graduate student example, does the independence or intellectual ability factor account for more of the variance in the measured variables? How much variance does each account for? In a good factor analysis, a high percentage of the variance in the observed variables is accounted for by the first few factors. And, because factors are computed in descending order of magnitude, the first factor accounts for the most variance, with later factors accounting for less and less of the variance until they are no longer reliable. Methods for assessing the importance of solutions and factors are in Section 13.6.4. Confirmatory factor analysis is demonstrated in Section 14.7.

13.2.4 ■ Testing Theory in FA

How well does the obtained factor solution fit an expected factor solution? If the researcher had generated hypotheses regarding both the number and the nature of the factors expected of graduate students, comparisons between the hypothesized factors and the factor solution provide a test of the hypotheses. Tests of theory in FA are addressed, in preliminary form, in Sections 13.6.2 and 13.6.7.

More highly developed techniques are available for testing theory in complex data sets in the form of structural equation modeling, which can also be used to test theory regarding factor structure. These techniques are sometimes known by the names of the most popular programs for doing them, EQS and LISREL. Structural equation modeling is the focus of Chapter 14. Confirmatory factor analysis is demonstrated in Section 14.7.

13.2.5 ■ Estimating Scores on Factors

Had factors been measured directly, what scores would subjects have received on each of them? For instance, if each graduate student were measured directly on independence and intelligence, what scores would each student receive for each of them? Estimation of factor scores is the topic of Section 13.6.6.

13.3 ■ LIMITATIONS

13.3.1 ■ Theoretical Issues

Most applications of PCA or FA are exploratory in nature; FA is used primarily as a tool for reducing the number of variables or examining patterns of correlations among variables. Under these circumstances, both the theoretical and the practical limitations to FA are relaxed in favor of a frank exploration of the data. Decisions about number of factors and rotational scheme are based on pragmatic rather than theoretical criteria.

The research project that is designed specifically to be factor analyzed, however, differs from other projects in several important respects. Among the best detailed discussions of the differences is the one found in Comrey and Lee (1992), from which some of the following discussion is taken.

The first task of the researcher is to generate hypotheses about factors believed to underlie the domain of interest. Statistically, it is important to make the research inquiry broad enough to include five or six hypothesized factors so that the solution is stable. Logically, in order to reveal the processes underlying a research area, all relevant factors have to be included. Failure to measure some important factor may distort the apparent relationships among measured factors. Inclusion of all relevant factors poses a logical, but not statistical, problem to the researcher.

Next, one selects variables to observe. For each hypothesized factor, five or six variables, each thought to be a relatively pure measure of the factor, are included. Pure measures are called marker variables. Marker variables are highly correlated with one and only one factor, and load on it regardless of extractional or rotational technique. Marker variables are useful because they define clearly the nature of a factor; adding potential variables to a factor to round it out is much more meaningful if the factor is unambiguously defined by marker variables to begin with.

The complexity of the variables is also considered. Complexity is indicated by the number of factors with which a variable correlates. A pure variable, which is preferred, is correlated with only one factor, whereas a complex variable is correlated with several. If variables differing in complexity are all included in an analysis, those with similar complexity levels may "catch" each other in factors that have little to do with underlying processes. Variables with similar complexity may correlate with each other because of their complexity and not because they relate to the same factor. Estimating (or avoiding) the complexity of variables is part of generating hypotheses about factors and selecting variables to measure them.

Several other considerations are required of the researcher planning a factor analytic study. It is important, for instance, that the sample chosen exhibit spread in scores with respect to the variables and the factors they measure. If all subjects achieve about the same score on some factor, correlations among the observed variables are low and the factor may not emerge in analysis. Selection of subjects expected to differ on the observed variables and underlying factors is an important design consideration.

One should also be leery about pooling the results of several samples, or the same sample with measures repeated in time, for factor analytic purposes. First, samples that are known to be different with respect to some criterion (e.g., socioeconomic status) may also have different factors. Examination of group differences is often quite revealing. Second, underlying factor structure may shift in time for the same subjects with learning or with experience in an experimental setting and these differences may also be quite revealing. Pooling results from diverse groups in FA may obscure differences rather than illuminate them. On the other hand, if different samples do produce the same factors, pooling them is desirable because of increase in sample size. For example, if men and women produce the same factors, the samples should be combined and the results of the single FA reported.

13.3.2 ■ Practical Issues

Because FA and PCA are exquisitely sensitive to the sizes of correlations, it is critical that honest, reliable correlations be employed. Sensitivity to outlying cases, problems created by missing data, and degradation of correlations between poorly distributed variables all plague FA and PCA. A review of these issues in Chapter 4 is important to FA and PCA. Thoughtful solutions to some of the problems, including variable transformations, may markedly enhance FA, whether performed for exploratory or confirmatory purposes. However, the limitations apply with greater force to confirmatory FA.

13.3.2.1 ■ Sample Size and Missing Data

Correlation coefficients tend to be less reliable when estimated from small samples. Therefore, it is important that sample size be large enough that correlations are reliably estimated. The required sample size also depends on magnitude of population correlations and number of factors: if there are strong, reliable correlations and a few, distinct factors, a smaller sample size is adequate.

Comrey and Lee (1992) give as a guide sample sizes of 50 as very poor, 100 as poor, 200 as fair, 300 as good, 500 as very good, and 1000 as excellent. *As a general rule of thumb, it is comforting to have at least 300 cases for factor analysis.* Solutions that have several high loading marker variables (>.80) do not require such large sample sizes (about 150 cases should be sufficient) as solutions with lower loadings and/or fewer marker variables (Guadagnoli and Velicer, 1988).

If cases have missing data, either the missing values are estimated, the cases deleted, or a missing data (pairwise) correlation matrix is analyzed. Consult Chapter 4 for methods of finding and estimating missing values and cautions about pairwise deletion of cases. Consider the distribution of missing values (is it random?) and remaining sample size when deciding between estimation and deletion. If cases are missing values in a nonrandom pattern or if sample size becomes too small, estimation is in order. However, beware of using estimation procedures (such as regression) that are likely to overfit the data and cause correlations to be too high. These procedures may "create" factors.

13.3.2.2 ■ Normality

As long as PCA and FA are used descriptively as convenient ways to summarize the relationships in a large set of observed variables, assumptions regarding the distributions of variables are not in force. If variables are normally distributed, the solution is enhanced. To the extent that normality fails, the solution is degraded but may still be worthwhile.

However, multivariate normality is assumed when statistical inference is used to determine the number of factors. Multivariate normality is the assumption that all variables, and all linear combinations of variables, are normally distributed. Although tests of multivariate normality are overly sensitive, *normality among single variables is assessed by skewness and kurtosis* (see Chapter 4 and Section 13.8.1.2). If a variable has substantial skewness and kurtosis, variable transformation is considered.

13.3.2.3 ■ Linearity

Multivariate normality also implies that relationships among pairs of variables
sis is degraded when linearity fails because correlation measures linear relat
reflect nonlinear relationship. *Linearity among pairs of variables is assessed t*
scatterplots. Consult Chapter 4 and Section 13.8.1.3 for methods of screening
linearity is found, transformation of variables is considered.

13.3.2.4 ■ Outliers among Cases

As in all multivariate techniques, cases may be outliers either on individual variables (univariate)
or on combinations of variables (multivariate). Such cases have more influence on the factor solution than other cases. Consult Chapter 4 and Section 13.8.1.4 for methods of detecting and reducing the influence of both univariate and multivariate outliers.

13.3.2.5 ■ Multicollinearity and Singularity

In PCA, multicollinearity is not a problem because there is no need to invert a matrix. For most
forms of FA and for estimation of factor scores in any form of FA, singularity or extreme multicollinearity is a problem. For FA, if the determinant of $\mathbf{R}$ and eigenvalues associated with some factors approach 0, multicollinearity or singularity may be present.

To investigate further, look at the SMCs for each variable where it serves as DV with all other
variables as IVs. If any of the SMCs is one, singularity is present; if any of the SMCs is very large
(near one), multicollinearity is present. Delete the variable with multicollinearity or singularity.
Chapter 4 and Section 13.8.1.5 provide examples of screening for and dealing with multicollinearity and singularity.

13.3.2.6 ■ Factorability of R

A matrix that is factorable should include several sizable correlations. The expected size depends,
to some extent, on N (larger sample sizes tend to produce smaller correlations), but if no correlation exceeds .30, use of FA is questionable because there is probably nothing to factor analyze.
Inspect $\mathbf{R}$ for correlations in excess of .30 and, if none is found, reconsider use of FA.

High bivariate correlations, however, are not ironclad proof that the correlation matrix contains
factors. It is possible that the correlations are between only two variables and do not reflect underlying processes that are simultaneously affecting several variables. For this reason, it is helpful to
examine matrices of partial correlations where pairwise correlations are adjusted for effects of all
other variables. If there are factors present, then high bivariate correlations become very low partial correlations. BMDP, SPSS, and SAS produce partial correlation matrices.

Bartlett's (1954) test of sphericity is a notoriously sensitive test of the hypothesis that the correlations in a correlation matrix are zero. The test is available in SPSS FACTOR but because of its
sensitivity and its dependence on N, the test is likely to be significant with samples of substantial
size even if correlations are very low. Therefore, use of the test is recommended only if there are
fewer than, say, five cases per variable.

Several more sophisticated tests of the factorability of $\mathbf{R}$ are available through SPSS and SAS.
Both programs give significance tests of correlations, the anti-image correlation matrix, and

...iser's (1970, 1974) measure of sampling adequacy. Significance tests of correlations in the correlation matrix provide an indication of the reliability of the relationships between pairs of variables. If **R** is factorable, numerous pairs are significant. The anti-image correlation matrix contains the negatives of partial correlations between pairs of variables with effects of other variables removed. If **R** is factorable, there are mostly small values among the off-diagonal elements of the anti-image matrix. Finally, Kaiser's measure of sampling adequacy is a ratio of the sum of squared correlations to the sum of squared correlations plus sum of squared partial correlations. The value approaches 1 if partial correlations are small.[2] Values of .6 and above are required for good FA.

13.3.2.7 ■ Outliers among Variables

After FA, in both exploratory and confirmatory FA, variables that are unrelated to others in the set are identified. These variables are usually not correlated with the first few factors although they often correlate with factors extracted later. These factors are usually unreliable, both because they account for very little variance and because factors that are defined by just one or two variables are not stable. Therefore, one never knows whether these factors are "real." Suggestions for determining reliability of factors defined by one or two variables are in Section 13.6.2.

If the variance accounted for by a factor defined by only one or two variables is high enough, the factor is interpreted with great caution or ignored, as pragmatic considerations dictate. In confirmatory FA, the factor represents either a promising lead for future work or (probably) error variance, but its interpretation awaits clarification by more research.

A variable with a low squared multiple correlation with all other variables and low correlations with all important factors is an outlier among the variables. The variable is usually ignored in the current FA and either deleted or given friends in future research. Screening for outliers among variables is illustrated in Section 13.8.1.7.

13.3.2.8 ■ Outlying Cases among the Factors

Cases in FA and PCA that have unusually large or small scores on the factors as estimated from factor score coefficients are univariate outliers with respect to the solution. The deviant scores are from cases for which the factor solution is inadequate. Examination of these cases for consistency is informative if it reveals the kinds of cases for which the FA is not appropriate.

If BMDP4M is used, *multivariate outliers among the factors are cases with large Mahalanobis distances, estimated as chi-square values, from the location of the case in the space defined by the factors to the centroid of all cases in the same space.* If scatterplots between pairs of factors are requested, these cases appear along the borders. Screening for these outlying cases is in Section 13.8.1.8.

13.4 ■ FUNDAMENTAL EQUATIONS FOR FACTOR ANALYSIS

Because of the variety and complexity of the calculations involved in preparing the correlation matrix, extracting factors, and rotating them, and because, in our judgment, little insight is produced by demonstrations of some of these procedures, this section does not show them all. Instead, the

---

[2] BMDP4M prints partial correlations between pairs of variables with effects of other variables removed through the PARTIAL option so Kaiser's measure of sampling adequacy could (with some pain) be hand-calculated.

relationships between some of the more important matrices are shown, with an assist from SPSS FACTOR for underlying calculations.

Table 13.1 lists many of the important matrices in FA and PCA. Although the list is lengthy, it is composed mostly of *matrices of correlations* (between variables, between factors, and between variables and factors), *matrices of standard scores* (on variables and on factors), *matrices of regression weights* (for producing scores on factors from scores on variables), and the *pattern matrix* of unique relationships between factors and variables after oblique rotation.

Also in the table are the matrix of eigenvalues and the matrix of their corresponding eigenvectors. Eigenvalues and eigenvectors are discussed here and in Appendix A, albeit scantily, because of their importance in factor extraction, the frequency with which one encounters the terminology, and the close association between eigenvalues and variance in statistical applications.

A data set appropriate for FA consists of numerous subjects each measured on several variables. A grossly inadequate data set appropriate for FA is in Table 13.2. Five subjects who were trying on

TABLE 13.1 COMMONLY ENCOUNTERED MATRICES IN FACTOR ANALYSES

Label	Name	Rotation	Size[a]	Description
R	Correlation matrix	Both orthogonal and oblique	$p \times p$	Matrix of correlations between variables
Z	Variable matrix	Both orthogonal and oblique	$N \times p$	Matrix of standardized observed variable scores
F	Factor-score matrix	Both orthogonal and oblique	$N \times m$	Matrix of standardized scores on factors or components
A	Factor loading matrix Pattern matrix	Orthogonal Oblique	$p \times m$	Matrix of regression-like weights used to estimate the unique contribution of each factor to the variance in a variable. If orthogonal, also correlations between variables and factors
B	Factor-score coefficients matrix	Both orthogonal and oblique	$p \times m$	Matrix of regression-like weights used to generate factor scores from variables
C	Structure matrix[b]	Oblique	$p \times m$	Matrix of correlations between variables and (correlated) factors
Φ	Factor correlation matrix	Oblique	$m \times m$	Matrix of correlations among factors
L	Eigenvalue matrix[c]	Both orthogonal and oblique	$m \times m$	Diagonal matrix of eigenvalues, one per factor[e]
V	Eigenvector matrix[d]	Both orthogonal and oblique	$p \times m$	Matrix of eigenvectors, one vector per eigenvalue

[a]Row by column dimensions where

  p = number of variables

  N = number of subjects

  m = number of factors or components

[b]In most textbooks, the structure matrix is labeled **S**. However, we have used **S** to represent the sum-of-squares and cross-products matrix elsewhere and will use **C** for the structure matrix here.

[c]Also called characteristic roots or latent roots.

[d]Also called characteristic vectors.

[e]If the matrix is of full rank, there are actually p rather than m eigenvalues and eigenvectors. Only m are of interest, however, so the remaining $p - m$ are not displayed.

TABLE 13.2 SMALL SAMPLE OF HYPOTHETICAL DATA
FOR ILLUSTRATION OF FACTOR ANALYSIS

Skiers	Variables			
	COST	LIFT	DEPTH	POWDER
S_1	32	64	65	67
S_2	61	37	62	65
S_3	59	40	45	43
S_4	36	62	34	35
S_5	62	46	43	40

Correlation matrix				
	COST	LIFT	DEPTH	POWDER
COST	1.000	−.953	−.055	−.130
LIFT	−.953	1.000	−.091	−.036
DEPTH	−.055	−.091	1.000	.990
POWDER	−.130	−.036	.990	1.000

ski boots late on a Friday night in January were asked about the importance of each of four variables to their selection of a ski resort. The variables were cost of ski ticket (COST), kind of ski lift (LIFT), depth of snow (DEPTH), and kind of snow (POWDER). Larger numbers indicate greater importance. The researcher wanted to investigate the pattern of relationships among the variables in an effort to understand better the dimensions underlying choice of ski area.

Notice the pattern of correlations in the correlation matrix as set off by the vertical and horizontal lines. The strong correlations in the upper left and lower right quadrants show that scores on COST and LIFT are related, as are scores on DEPTH and POWDER. The other two quadrants show that scores on DEPTH and LIFT are unrelated, as are scores on POWDER and LIFT, and so on. With luck, FA will find this pattern of correlations, easy to see in a small correlation matrix but not in a very large one.

13.4.1 ■ Extraction

An important theorem from matrix algebra indicates that, under certain conditions, matrices can be diagonalized. Correlation and covariance matrices are among those that often can be diagonalized. When a matrix is diagonalized, it is transformed into a matrix with numbers in the positive diagonal[3] and zeros everywhere else. In this application, the numbers in the positive diagonal represent variance from the correlation matrix that has been repackaged as follows:

$$L = V'RV \qquad (13.1)$$

Diagonalization of R is accomplished by post- and premultiplying it by the matrix V and its transpose.

The columns in V are called eigenvectors, and the values in the main diagonal of L are called eigenvalues. The first eigenvector corresponds to the first eigenvalue, and so forth.

---

[3] The positive diagonal runs from upper left to lower right in a matrix.

Because there are four variables in the example, there are four eigenvalues with their corresponding eigenvectors. However, because the goal of FA is to summarize a pattern of correlations with as few factors as possible, and because each eigenvalue corresponds to a different potential factor, usually only factors with large eigenvalues are retained. In a good FA, these few factors almost duplicate the correlation matrix.

In this example, when no limit is placed on the number of factors, eigenvalues of 2.02, 1.94, .04, and .00 are computed for each of the four possible factors. Only the first two factors, with values over 1.00, are large enough to be retained in subsequent analyses. FA is rerun specifying extraction of just the first two factors; they have eigenvalues of 2.00 and 1.91, respectively, as indicated in Table 13.3.

Using Equation 13.1 and inserting the values from the example, we obtain

$$\mathbf{L} = \begin{bmatrix} -.283 & .177 & .658 & .675 \\ .651 & -.685 & .252 & .207 \end{bmatrix} \begin{bmatrix} 1.000 & -.953 & -.055 & -.130 \\ -.935 & 1.000 & -.091 & -.036 \\ -.055 & -.091 & 1.000 & .990 \\ -.130 & -.036 & .990 & 1.000 \end{bmatrix} \begin{bmatrix} -.283 & .651 \\ .177 & -.685 \\ .658 & .252 \\ .675 & .207 \end{bmatrix}$$

$$= \begin{bmatrix} 2.00 & .00 \\ .00 & 1.91 \end{bmatrix}$$

(All values agree with computer output. Hand calculation may produce discrepancies due to rounding error.)

The matrix of eigenvectors premultiplied by its transpose produces the identity matrix with ones in the positive diagonal and zeros elsewhere. Therefore, pre- and postmultiplying the correlation matrix by eigenvectors does not change it so much as repackage it.

$$\mathbf{V'V} = \mathbf{I} \tag{13.2}$$

For the example:

$$\begin{bmatrix} -.283 & .177 & .658 & .675 \\ .651 & -.685 & .252 & .207 \end{bmatrix} \begin{bmatrix} -.283 & .651 \\ .177 & -.685 \\ -.658 & .252 \\ .675 & .207 \end{bmatrix} = \begin{bmatrix} 1.000 & .000 \\ .000 & 1.000 \end{bmatrix}$$

TABLE 13.3 EIGENVECTORS AND CORRESPONDING EIGENVALUES FOR THE EXAMPLE

Eigenvector 1	Eigenvector 2
−.283	.651
.177	−.685
.658	.252
.675	.207
Eigenvalue 1	Eigenvalue 2
2.00	1.91

The important point is that because correlation matrices often meet requirements for diagonalizability, it is possible to use on them the matrix algebra of eigenvectors and eigenvalues with FA as the result. When a matrix is diagonalized, the information contained in it is repackaged. In FA, the variance in the correlation matrix is condensed into eigenvalues. The factor with the largest eigenvalue has the most variance and so on, down to factors with small or negative eigenvalues that are usually omitted from solutions.

Calculations for eigenvectors and eigenvalues are extremely laborious and not particularly enlightening (although they are illustrated in Appendix A for a small matrix). They require solving p equations in p unknowns with additional side constraints and are rarely performed by hand. Once the eigenvalues and eigenvectors are known, however, the rest of FA (or PCA) more or less "falls out," as is seen from Equations 13.3 to 13.6.

Equation 13.1 can be reorganized as follows:

$$\mathbf{R} = \mathbf{VLV'} \tag{13.3}$$

The correlation matrix can be considered a product of three matrices—the matrices of eigenvalues and corresponding eigenvectors.

After reorganization, the square root is taken of the matrix of eigenvalues.

or

$$\mathbf{R} = \mathbf{V}\sqrt{\mathbf{L}}\sqrt{\mathbf{L}}\mathbf{V'} \tag{13.4}$$

$$\mathbf{R} = (\mathbf{V}\sqrt{\mathbf{L}})(\sqrt{\mathbf{L}}\mathbf{V'})$$

If $\mathbf{V}\sqrt{\mathbf{L}}$ is called $\mathbf{A}$, and $\sqrt{\mathbf{L}}\mathbf{V'}$ is $\mathbf{A'}$, then

$$\mathbf{R} = \mathbf{AA'} \tag{13.5}$$

The correlation matrix can also be considered a product of two matrices, each a combination of eigenvectors and the square root of eigenvalues.

Equation 13.5 is frequently called the fundamental equation for FA.[4] It represents the assertion that the correlation matrix is a product of the factor loading matrix, $\mathbf{A}$, and its transpose.

Equations 13.4 and 13.5 also reveal that the major work of FA (and PCA) is calculation of eigenvalues and eigenvectors. Once they are known, the (unrotated) factor loading matrix is found by straightforward matrix multiplication, as follows.

$$\mathbf{A} = \mathbf{V}\sqrt{\mathbf{L}} \tag{13.6}$$

For the example:

$$\mathbf{A} = \begin{bmatrix} -.283 & .651 \\ .177 & -.685 \\ .658 & .252 \\ .675 & .207 \end{bmatrix} \begin{bmatrix} \sqrt{2.00} & 0 \\ 0 & \sqrt{1.91} \end{bmatrix} = \begin{bmatrix} -.400 & .900 \\ .251 & -.947 \\ .932 & .348 \\ .956 & .286 \end{bmatrix}$$

---

[4] In order to reproduce the correlation matrix exactly, as indicated in Equations 13.4 and 13.5, all eigenvalues and eigenvectors are necessary, not just the first few of them.

The factor loading matrix is a matrix of correlations between factors and variables. The first column is correlations between the first factor and each variable in turn, COST($-.400$), LIFT ($.251$), DEPTH ($.932$), and POWDER ($.956$). The second column is correlations between the second factor and each variable in turn, COST ($.900$), LIFT ($-.947$), DEPTH ($.348$), and POWDER ($.286$). A factor is interpreted from the variables that are highly correlated with it—that have high loadings on it. Thus, the first factor is primarily a snow conditions factor (DEPTH and POWDER), while the second reflects resort conditions (COST and LIFT). Subjects who score high on the resort conditions factor (Equation 13.11) tend to assign high value to COST and low value to LIFT (the negative correlation); subjects who score low on the resort conditions factor value LIFT more than COST.

Notice, however, that all the variables are correlated with both factors to a considerable extent. Interpretation is fairly clear for this hypothetical example, but most likely would not be for real data. Usually a factor is most interpretable when a few variables are highly correlated with it and the rest are not.

13.4.2 ■ Orthogonal Rotation

Rotation is ordinarily used after extraction to maximize high correlations and minimize low ones. Numerous methods of rotation are available (see Section 13.5.2) but the most commonly used, and the one illustrated here, is varimax. Varimax is a variance maximizing procedure. The goal of varimax rotation is to maximize the variance of factor loadings by making high loadings higher and low ones lower for each factor.

This goal is accomplished by means of a transformation matrix Λ (as defined in Equation 13.8), where

$$\mathbf{A}_{unrotated}\Lambda = \mathbf{A}_{rotated} \tag{13.7}$$

The unrotated factor loading matrix is multiplied by the transformation matrix to produce the rotated loading matrix.

For the example:

$$\mathbf{A}_{rotated} = \begin{bmatrix} -.400 & .900 \\ .251 & -.947 \\ .932 & .348 \\ .956 & .286 \end{bmatrix} \begin{bmatrix} .946 & -.325 \\ .325 & .946 \end{bmatrix} = \begin{bmatrix} -.086 & .981 \\ -.071 & -.977 \\ .994 & .026 \\ .997 & -.040 \end{bmatrix}$$

Compare the rotated and unrotated loading matrices. Notice that in the rotated matrix the low correlations are lower and the high ones higher than in the unrotated loading matrix. Emphasizing differences in loadings facilitates interpretation of a factor by making unambiguous the variables that correlate with it.

The numbers in the transformation matrix have a spatial interpretation.

$$\Lambda = \begin{bmatrix} \cos \Psi & -\sin \Psi \\ \sin \Psi & \cos \Psi \end{bmatrix} \tag{13.8}$$

The transformation matrix is a matrix of sines and cosines of an angle Ψ.

For the example, the angle is approximately 19°. That is, cos 19≈.946 and sin 19≈.325. Geometrically, this corresponds to a 19° swivel of the factor axes about the origin. Greater detail regarding the geometric meaning of rotation is in Section 13.5.2.3.

13.4.3 ■ Communalities, Variance, and Covariance

Once the rotated loading matrix is available, other relationships are found, as in Table 13.4. The communality for a variable is the variance accounted for by the factors. It is the squared multiple correlation of the variable as predicted from the factors. Communality is the sum of squared loadings (SSL) for a variable across factors. In Table 13.4, the communality for COST is $(-.086)^2 + .981^2 = .970$. That is, 97% of the variance in COST is accounted for by Factor 1 plus Factor 2.

The proportion of variance *in the set of variables* accounted for by a factor is the SSL for the factor divided by the number of variables (if rotation is orthogonal).[5] For the first factor, the proportion of variance is $[(-.086)^2 + (-.071)^2 + .994^2 + .997^2]/4 = 1.994/4 = .50$. Fifty percent of the variance in the variables is accounted for by the first factor. The second factor accounts for 48% of the variance in the variables and, because rotation is orthogonal, the two factors together account for 98% of the variance in the variables.

The proportion of variance *in the solution* accounted for by a factor—the proportion of covariance—is the SSL for the factor divided by the sum of communalities (or, equivalently, the sum of the SSLs). The first factor accounts for 51% of the variance in the solution (1.994/3.915) while the second factor accounts for 49% of the variance in the solution (1.919/3.915). The two factors together account for all of the covariance.

The reproduced correlation matrix for the example is generated using Equation 13.5:

$$\bar{\mathbf{R}} = \begin{bmatrix} -.086 & .981 \\ -.071 & -.977 \\ .994 & .026 \\ .997 & -.040 \end{bmatrix} \begin{bmatrix} -.086 & -.071 & .994 & .997 \\ .981 & -.977 & .026 & -.040 \end{bmatrix} = \begin{bmatrix} .970 & -.953 & -.059 & -.125 \\ -.953 & .962 & -.098 & -.033 \\ -.059 & -.098 & .989 & .990 \\ -.125 & -.033 & .990 & .996 \end{bmatrix}$$

TABLE 13.4   RELATIONSHIPS AMONG LOADINGS, COMMUNALITIES, SSLs, VARIANCE, AND COVARIANCE OF ORTHOGONALLY ROTATED FACTORS

	Factor 1	Factor 2	Communalities (h^2)
COST	−.086	.981	$\sum a^2 = .970$
LIFT	−.071	−.977	$\sum a^2 = .960$
DEPTH	.994	.026	$\sum a^2 = .989$
POWDER	.997	−.040	$\sum a^2 = .996$
SSLs	$\sum a^2 = 1.994$	$\sum a^2 = 1.919$	3.915
Proportion of variance	.50	.48	.98
Proportion of covariance	.51	.49	

[5] For unrotated factors only, the sum of the squared loadings for a factor is equal to the eigenvalue. Once loadings are rotated, the sum of squared loadings is called SSL and is no longer equal to the eigenvalue.

Notice that the reproduced correlation matrix differs slightly from the original correlation matrix. The difference between the original and reproduced correlation matrices is the residual correlation matrix:

$$\mathbf{R}_{res} = \mathbf{R} - \overline{\mathbf{R}} \tag{13.9}$$

The residual correlation matrix is the difference between the observed correlation matrix and the reproduced correlation matrix.

For the example, with communalities inserted in the positive diagonal of $\mathbf{R}$:

$$\mathbf{R}_{res} = \begin{bmatrix} .970 & -.953 & -.055 & -.130 \\ -.953 & .960 & -.091 & -.036 \\ -.055 & -.091 & .989 & .990 \\ -.130 & -.036 & .990 & .996 \end{bmatrix} - \begin{bmatrix} .970 & -.953 & -.059 & -.125 \\ -.953 & .962 & -.098 & -.033 \\ -.059 & -.098 & .989 & .990 \\ -.125 & -.033 & .990 & .996 \end{bmatrix}$$

$$= \begin{bmatrix} .000 & .000 & .004 & -.005 \\ .000 & -.002 & .007 & -.003 \\ .004 & .007 & .000 & .000 \\ -.005 & -.003 & .000 & .000 \end{bmatrix}$$

In a "good" FA, the numbers in the residual correlation matrix are small because there is little difference between the original correlation matrix and the correlation matrix generated from factor loadings.

13.4.4 ■ Factor Scores

Scores on factors can be predicted for each case once the loading matrix is available. Regression-like coefficients are computed for weighting variable scores to produce factor scores. Because $\mathbf{R}^{-1}$ is the inverse of the matrix of correlations among variables and $\mathbf{A}$ is the matrix of correlations between factors and variables, Equation 13.10 for factor score coefficients is similar to Equation 5.6 for regression coefficients in multiple regression.

$$\mathbf{B} = \mathbf{R}^{-1}\mathbf{A} \tag{13.10}$$

Factor score coefficients for estimating factor scores from variable scores are a product of the inverse of the correlation matrix and the factor loading matrix.

For the example:[6]

$$\mathbf{B} = \begin{bmatrix} 25.485 & 22.689 & -31.655 & 35.479 \\ 22.689 & 21.386 & -24.831 & 28.312 \\ -31.655 & -24.831 & 99.917 & -103.950 \\ 35.479 & 28.312 & -103.950 & 109.567 \end{bmatrix} \begin{bmatrix} -.087 & .981 \\ -.072 & -.978 \\ .994 & .027 \\ .997 & -.040 \end{bmatrix} = \begin{bmatrix} 0.082 & 0.537 \\ 0.054 & -0.461 \\ 0.190 & 0.087 \\ 0.822 & -0.074 \end{bmatrix}$$

---

[6] The numbers in **B** are different from the factor score coefficients generated by computer for the small data set. The difference is due to rounding error following inversion of a multicollinear correlation matrix. Note also that the **A** matrix contains considerable rounding error.

To estimate a subject's score for the first factor, then, all of the subject's scores on variables are standardized and then the standardized score on COST is weighted by 0.082, LIFT by 0.054, DEPTH by 0.190, and POWDER by 0.822 and the results are added. In matrix form,

$$\mathbf{F} = \mathbf{ZB} \qquad (13.11)$$

Factor scores are a product of standardized scores on variables and factor score coefficients.

For the example:

$$\mathbf{F} = \begin{bmatrix} -1.22 & 1.14 & 1.15 & 1.14 \\ 0.75 & -1.02 & 0.92 & 1.01 \\ 0.61 & -0.78 & -0.36 & -0.47 \\ -0.95 & 0.98 & -1.20 & -1.01 \\ 0.82 & -0.30 & -0.51 & -0.67 \end{bmatrix} \begin{bmatrix} 0.082 & 0.537 \\ 0.054 & -0.461 \\ 0.190 & 0.087 \\ 0.822 & -0.074 \end{bmatrix}$$

The first subject has an estimated standard score of 1.12 on the first factor and -1.16 on the second factor, and so on for the other four subjects. The first subject strongly values both the snow factor and the resort factor, one positive and the other negative (indicating primary value assigned to quality of LIFT). The second subject values both the snow factor and the resort factor (with more value placed on COST than LIFT); the third subject places more value on resort conditions (particularly COST) and less value on snow conditions, and so forth. The sum of standardized factor scores across subjects for a single factor is zero.

Predicting scores on variables from scores on factors is also possible. The equation for doing so is

$$\mathbf{Z} = \mathbf{FA}' \qquad (13.12)$$

Predicted standardized scores on variables are a product of scores on factors weighted by factor loadings.

For example:

$$\mathbf{Z} = \begin{bmatrix} 1.12 & -1.16 \\ 1.01 & 0.88 \\ -0.45 & 0.69 \\ -1.08 & -0.99 \\ -0.60 & 0.58 \end{bmatrix} \begin{bmatrix} -.086 & -.072 & .994 & .997 \\ .981 & -.978 & .027 & -.040 \end{bmatrix}$$

$$= \begin{bmatrix} -1.23 & 1.05 & 1.08 & 1.16 \\ 0.78 & -0.93 & 1.03 & 0.97 \\ 0.72 & -0.64 & -0.43 & -0.48 \\ -0.88 & 1.05 & -1.10 & -1.04 \\ 0.62 & -0.52 & -0.58 & -0.62 \end{bmatrix}$$

That is, the first subject (the first row of $\mathbf{Z}$) is predicted to have a standardized score of -1.23 on COST, 1.05 on LIFT, 1.08 on DEPTH, and 1.16 on POWDER. Like the reproduced correlation matrix, these values are similar to the observed values if the FA captures the relationship among the variables.

It is helpful to see these values written out because they provide an insight into how scores on variables are conceptualized in factor analysis. For example, for the first subject,

$$-1.23 = -.086(1.12) + .981(-1.16)$$

$$1.05 = -.072(1.12) - .978(-1.16)$$

$$1.08 = .994(1.12) + .027(-1.16)$$

$$1.16 = .997(1.12) - .040(-1.16)$$

Or, in algebraic form,

$$z_{COST} = a_{11}F_1 + a_{12}F_2$$

$$z_{LIFT} = a_{21}F_1 + a_{22}F_2$$

$$z_{DEPTH} = a_{31}F_1 + a_{32}F_2$$

$$z_{POWDER} = a_{41}F_1 + a_{42}F_2$$

A score on an observed variable is conceptualized as a properly weighted and summed combination of the scores on factors that underlie it. The researcher believes that each subject has the same latent factor structure, but different scores on the factors themselves. A particular subject's score on an observed variable is produced as a weighted combination of that subject's scores on the underlying factors.

13.4.5 ■ Oblique Rotation

All the relationships mentioned thus far are for orthogonal rotation. Most of the complexities of orthogonal rotation remain and several others are added when oblique (correlated) rotation is used. Consult Table 13.1 for a listing of additional matrices and a hint of the discussion to follow.

SPSS FACTOR is run on the data from Table 13.2 using the default option for oblique rotation (cf. Section 13.5.2.2) to get values for the pattern matrix, **A**, and factor-score coefficients, **B**.

In oblique rotation, the loading matrix becomes the pattern matrix. Values in the pattern matrix, when squared, represent the unique contribution of each factor to the variance of each variable but do not include segments of variance that come from overlap between correlated factors. For the example, the pattern matrix following oblique rotation is

$$\mathbf{A} = \begin{bmatrix} -.079 & .981 \\ -.078 & -.978 \\ .994 & .033 \\ .977 & -.033 \end{bmatrix}$$

The first factor makes a unique contribution of $(-.079)^2$ to the variance in COST, $(-.078)^2$ to LIFT, $.994^2$ to DEPTH, and $.997^2$ to POWDER.

Factor-score coefficients following oblique rotation are also found:

$$\mathbf{B} = \begin{bmatrix} 0.104 & 0.584 \\ 0.081 & -0.421 \\ 0.159 & -0.020 \\ 0.856 & 0.034 \end{bmatrix}$$

Applying Equation 13.11 to produce factor scores results in the following values:

$$\mathbf{F} = \begin{bmatrix} -1.22 & 1.14 & 1.15 & 1.14 \\ 0.75 & -1.02 & 0.92 & 1.01 \\ 0.61 & -0.78 & -0.36 & -0.47 \\ -0.95 & 0.98 & -1.20 & -1.01 \\ 0.82 & -0.30 & -0.51 & -0.67 \end{bmatrix} \begin{bmatrix} 0.104 & 0.584 \\ 0.081 & -0.421 \\ 0.159 & -0.020 \\ 0.856 & 0.034 \end{bmatrix} = \begin{bmatrix} 1.12 & -1.18 \\ 1.01 & 0.88 \\ -0.46 & 0.68 \\ -1.07 & -0.98 \\ -0.59 & 0.59 \end{bmatrix}$$

Once the factor scores are determined, correlations among factors can be obtained. Among the equations used for the purpose is

$$\Phi = \left(\frac{1}{N-1}\right)\mathbf{F'F} \tag{13.13}$$

One way to compute correlations among factors is from cross products of standardized factor scores divided by the number of cases minus one.

The factor correlation matrix is a standard part of computer output following oblique rotation. For the example:

$$\Phi = \frac{1}{4}\begin{bmatrix} 1.12 & 1.01 & -0.46 & -1.07 & -0.59 \\ -1.18 & 0.88 & 0.68 & -0.98 & 0.59 \end{bmatrix} \begin{bmatrix} 1.12 & -1.18 \\ 1.01 & 0.88 \\ -0.46 & 0.68 \\ -1.07 & -0.98 \\ -0.59 & 0.59 \end{bmatrix}$$

$$= \begin{bmatrix} 1.00 & -0.01 \\ -0.01 & 1.00 \end{bmatrix}$$

The correlation between the first and second factor is quite low, $-.01$. For this example, there is almost no relationship between the two factors, although considerable correlation could have been produced had it been warranted. Ordinarily one uses orthogonal rotation in a case like this because complexities introduced by oblique rotation are not warranted by such a low correlation among factors.

However, if oblique rotation is used, the structure matrix, C, is the correlations between variables and factors. These correlations assess the unique relationship between the variable and the factor (in the pattern matrix) plus the relationship between the variable and the overlapping variance among the factors. The equation for the structure matrix is

$$\mathbf{C} = \mathbf{A\Phi} \tag{13.14}$$

The structure matrix is a product of the pattern matrix and the factor correlation matrix.

For example:

$$
\mathbf{C} = \begin{bmatrix} -.079 & .981 \\ -.078 & -.978 \\ .994 & .033 \\ .997 & -.033 \end{bmatrix} \begin{bmatrix} 1.00 & -.01 \\ -.01 & 1.00 \end{bmatrix} = \begin{bmatrix} -.069 & .982 \\ -.088 & -.977 \\ .994 & .023 \\ .997 & -.043 \end{bmatrix}
$$

COST, LIFT, DEPTH, and POWDER correlate $-.069$, $-.088$, $.994$, and $.997$ with the first factor and $.982$, $-.977$, $.023$, and $-.043$ with the second factor, respectively.

There is some debate as to whether one should interpret the pattern matrix or the structure matrix following oblique rotation. The structure matrix is appealing because it is readily understood. However, the correlations between variables and factors are inflated by any overlap between factors. The problem becomes more severe as the correlations among factors increase and it may be hard to determine which variables are related to a factor. On the other hand, the pattern matrix contains values representing the unique contributions of each factor to the variance in the variables. Shared variance is omitted (as it is with standard multiple regression), but the set of variables that composes a factor is usually easier to see. If factors are very highly correlated, it may appear that no variables are related to them because there is almost no unique variance once overlap is omitted.

Most researchers interpret and report the pattern matrix rather than the structure matrix. However, if the researcher reports either the structure or the pattern matrix and also $\mathbf{\Phi}$, then the interested reader can generate the other using Equation 13.14 as desired.

In oblique rotation, $\overline{\mathbf{R}}$ is produced as follows:

$$
\overline{\mathbf{R}} = \mathbf{CA'} \tag{13.15}
$$

The reproduced correlation matrix is a product of the structure matrix and the transpose of the pattern matrix.

Once the reproduced correlation matrix is available, Equation 13.9 is used to generate the residual correlation matrix to diagnose adequacy of fit in FA.

13.4.6 ■ Computer Analyses of Small Sample Example

A two-factor principal factor analysis with varimax rotation using the example is shown for SPSS FACTOR, BMDP4M, SAS FACTOR, and SYSTAT FACTOR in Tables 13.5 through 13.8.

For a principal factor analysis with varimax rotation, SPSS FACTOR requires that you specify EXTRACTION=PAF and ROTATION=VARIMAX[7]. SPSS FACTOR (Table 13.5) begins by printing out SMCs for each variable, labeled Communality in the Initial Statistics portion of the output. In a parallel but unrelated table, Eigenvalues, Pct of Var(iance), and percent of variance cumulated over the four factors (CUM PCT) are printed out for the four initial factors. (Be careful not to confuse factors with variables.) The program then indicates the number of factors extracted with eigenvalues greater than 1 (the default value).

---

[7] The defaults for SPSS FACTOR are principal components analysis with varimax rotation. However, rotation is not automatic if principal factors extraction is requested.

```
DATA LIST   FILE = 'TAPE17.DAT'  FREE
            /SUBJNO, COST, LIFT, DEPTH, POWDER.
FACTOR      VARIABLES = COST TO POWDER/
            EXTRACTION = PAF/
            ROTATION = VARIMAX.
```

            - - - -   F A C T O R   A N A L Y S I S   - - - -

Analysis Number 1 Listwise deletion of cases with missing values

Extraction 1 for Analysis 1, Principal Axis Factoring (PAF)

Initial Statistics:

Variable	Communality	*	Factor	Eigenvalue	Pct of Var	Cum Pct
		*				
COST	.96076	*	1	2.01631	50.4	50.4
LIFT	.95324	*	2	1.94151	48.5	98.9
DEPTH	.98999	*	3	.03781	.9	99.9
POWDER	.99087	*	4	.00437	.1	100.0

  PAF Extracted 2 factors.   4 Iterations required.
            - - - -   F A C T O R   A N A L Y S I S   - - - -

Factor Matrix:

	FACTOR 1	FACTOR 2
COST	-.40027	.89978
LIFT	.25060	-.94706
DEPTH	.93159	.34773
POWDER	.95596	.28615

Final Statistics:

Variable	Communality	*	Factor	Eigenvalue	Pct of Var	Cum Pct
		*				
COST	.96983	*	1	2.00473	50.1	50.1
LIFT	.95973	*	2	1.90933	47.7	97.9
DEPTH	.98877	*				
POWDER	.99574	*				

            - - - -   F A C T O R   A N A L Y S I S   - - - -

Varimax   Rotation 1, Extraction 1, Analysis 1 - Kaiser Normalization.

  Varimax converged in   3 iterations.

Rotated Factor Matrix:

	FACTOR 1	FACTOR 2
COST	-.08591	-.98104
LIFT	-.07100	.97708
DEPTH	.99403	-.02588
POWDER	.99706	.04028

Factor Transformation Matrix:

	FACTOR 1	FACTOR 2
FACTOR 1	.94565	.32519
FACTOR 2	.32519	-.94565

For the two extracted factors, an unrotated Factor (loading) Matrix is then printed. In the output labeled Final Statistics are final Communality values for each variable (h^2 in Table 13.4), and, parallel with them, Eigenvalues for the two retained factors (see Table 13.3), the Pct of Var(iance) for each factor and Cum Pct of variance accounted for by successive factors. The Rotated Factor (loading) Matrix, which matches loadings in Table 13.4, is given along with the Factor Transformation Matrix (Equation 13.8) for orthogonal varimax rotation with Kaiser normalization.

With principal factors extraction (PFA) specified in the control language for BMDP4M (Table 13.6), the CORRELATION MATRIX for the variables is printed out first, followed by SMCs for each variable (initial communalities) and a HISTOGRAM OF EIGENVALUES (scree plot oriented sideways). After some additional pre-extraction information (not shown), BMDP4M prints the COMMUNALITIES (h^2) for each variable, OBTAINED FROM 2 FACTORS with eigenvalues greater than 1. In the next table, eigenvalues are given for all 4 factors along with two types of CUMULATIVE PROPORTION OF VARIANCE, actually cumulative proportion of *co*variance. CARMINES THETA is a measure of the reliability of the first factor that is closely related to Cronbach's alpha.

UNROTATED FACTOR LOADINGS are given for each variable on the two factors, along with the SSLs for each factor, labeled VP. This is followed in Table 13.6 by ROTATED FACTOR LOADINGS (default is varimax rotation), with SSLs again shown as VP. Finally, BMDP4M prints out the FACTOR SCORE COVARIANCE matrix, in which the off-diagonal elements are set to zero by the nature of orthogonal rotation. The diagonal elements are the SMCs of each factor with the variables.

SAS FACTOR (Table 13.7) requires a bit more instruction to produce a principal factor analysis with orthogonal rotation for two factors. You specify the type (METHOD=PRINIT), initial communalities (PRIORS=SMC), number of factors to be extracted (NFACTORS=2) and the type of rotation (ROTATE=V). Prior Communality Estimates—SMCs—are given, followed by Preliminary Eigenvalues for all four factors; also given is the Total of the eigenvalues and their Average. The next row shows Differences between successive eigenvalues. For example, there is a small difference between the first and second eigenvalues (0.099606) and between the third and fourth eigenvalues (0.020622), but a large difference between the second and third eigenvalues (1.897534). Proportion and Cumulative proportion of variance are then printed for each factor. This is followed by corresponding information for the Reduced Correlation Matrix (after factoring). Information on the iterative process is not shown.

The Factor Pattern matrix contains unrotated factor loadings for the first two factors. (Note that the signs of the FACTOR2 loadings are the reverse of those of BMDP and SPSS.) SSLs for each factor are in the table labeled Variance explained by each factor. Both Final Communality Estimates (h^2) and the Total h^2 are then given. The transformation matrix for orthogonal rotation (Equation 13.8) is followed by the rotated factor loadings in the Rotated Factor Pattern matrix. SSLs for rotated factors—Variance explained by each factor—appear below the loadings. Final Communality Estimates are then repeated.

SYSTAT FACTOR (Table 13.8) currently provides principal factor analysis (IPA). Type of rotation must be explicitly requested. The output begins with Initial Communality Estimates (SMCs) and, after information about iteration (not shown), provides Final Communality Estimates (h^2). Latent Roots (eigenvalues) are then given for all four factors, followed by the (unrotated) Factor Pattern (loading) matrix for the two factors with eigenvalues > 1 (which differs slightly from those of the other programs). Variance Explained by Factors (SSLs) follows. Percent of Total Variance Explained by each factor is then shown (matching that of SPSS). Finally, parallel information is

TABLE 13.6 SETUP AND SELECTED BMDP4M OUTPUT FOR FACTOR ANALYSIS ON SAMPLE DATA OF TABLE 13.2

```
/INPUT       VARIABLES = 5.  FORMAT IS FREE.
             FILE = 'TAPE17.DAT'.
/VARIABLE    NAMES ARE SUBJNO, COST, LIFT, DEPTH, POWDER.
             LABEL = SUBJNO.
/FACTOR      METHOD = PFA.
/END
```

CORRELATION MATRIX

```
                 COST      LIFT     DEPTH    POWDER
                   2         3         4         5

COST      2      1.0000
LIFT      3     -0.9530    1.0000
DEPTH     4     -0.0553   -0.0911    1.0000
POWDER    5     -0.1300   -0.0362    0.9902    1.0000
```

SQUARED MULTIPLE CORRELATIONS (SMC) OF
EACH VARIABLE WITH ALL OTHER VARIABLES

```
      2 COST       0.96076
      3 LIFT       0.95324
      4 DEPTH      0.98999
      5 POWDER     0.99087
```

COMMUNALITY ESTIMATES ARE SQUARED MULTIPLE CORRELATIONS (COVARIANCES).

HISTOGRAM OF INITIAL EIGENVALUES

```
      EIGENVALUE    HISTOGRAM

   1   2.00234      ***********************************************************
   2   1.90274      *********************************************************
```

REMAINING EIGENVALUES ARE TOO SMALL TO APPEAR.   THEY ARE--
0.520370E-02-0.154181E-01

TABLE 13.6 *(CONTINUED)*

COMMUNALITIES OBTAINED FROM 2 FACTORS AFTER 4 ITERATIONS.

THE COMMUNALITY OF A VARIABLE IS ITS SQUARED MULTIPLE
CORRELATION WITH THE FACTORS.

2	COST	0.9698
3	LIFT	0.9597
4	DEPTH	0.9888
5	POWDER	0.9957

FACTOR	VARIANCE EXPLAINED	CUMULATIVE PROPORTION OF VARIANCE		CARMINES THETA
		IN DATA SPACE	IN FACTOR SPACE	
1	2.0047	0.5110	0.5122	0.6810
2	1.9093	0.9977	1.0000	
3	0.0090	1.0000		
4	-0.0103			

TOTAL VARIANCE IS DEFINED AS THE SUM OF THE POSITIVE EIGENVALUES OF THE
CORRELATION MATRIX.

NEGATIVE VALUES FOR VARIANCE EXPLAINED INDICATE THE DEGREE
TO WHICH THE COMMUNALITIES HAVE BEEN UNDERESTIMATED.
LARGE NEGATIVE VALUES FOR VARIANCE EXPLAINED INDICATE THAT
THE COVARIANCE OR CORRELATION MATRIX IS POORLY ESTIMATED.

UNROTATED FACTOR LOADINGS (PATTERN)

		FACTOR1	FACTOR2
		1	2
COST	2	-0.400	-0.900
LIFT	3	0.251	0.947
DEPTH	4	0.932	-0.348
POWDER	5	0.956	-0.286
VP		2.005	1.909

THE VP IS THE VARIANCE EXPLAINED BY THE FACTOR.
IT IS COMPUTED AS THE SUM OF SQUARES FOR THE
ELEMENTS OF THE FACTOR'S COLUMN IN THE FACTOR
LOADING MATRIX.

TABLE 13.6 *(CONTINUED)*

ORTHOGONAL ROTATION, GAMMA =   1.0000
ROTATED FACTOR LOADINGS (PATTERN)

	FACTOR1 1	FACTOR2 2
COST 2	-0.086	0.981
LIFT 3	-0.071	-0.977
DEPTH 4	0.994	0.026
POWDER 5	0.997	-0.040
VP	1.995	1.919

THE VP IS THE VARIANCE EXPLAINED BY THE FACTOR.
IT IS COMPUTED AS THE SUM OF SQUARES FOR THE
ELEMENTS OF THE FACTOR'S COLUMN IN THE FACTOR
LOADING MATRIX.

FACTOR SCORE COVARIANCE (COMPUTED FROM FACTOR
STRUCTURE AND FACTOR SCORE COEFFICIENTS)

THE DIAGONAL OF THE MATRIX BELOW CONTAINS THE SQUARED
MULTIPLE CORRELATIONS OF EACH FACTOR WITH THE VARIABLES.

	FACTOR1 1	FACTOR2 2
FACTOR1 1	0.997	
FACTOR2 2	-0.000	0.982

TABLE 13.7   SETUP AND SELECTED SAS FACTOR OUTPUT FOR FACTOR ANALYSIS OF SAMPLE DATA OF
              TABLE 13.2

```
DATA        SSAMPLE;
INFILE      'TAPE17.DAT';
INPUT       SUBJNO COST LIFT DEPTH POWDER;
PROC   FACTOR
       M = PRINIT
       PRIORS = SMC
       NFACTORS = 2
       ROTATE = V;
       VAR COST LIFT DEPTH POWDER
```

Initial Factor Method: Iterated Principal Factor Analysis

Prior Communality Estimates: SMC

COST	LIFT	DEPTH	POWDER
0.960761	0.953241	0.989992	0.990873

Preliminary Eigenvalues: Total = 3.89486621 Average = 0.97371655

	1	2	3	4
Eigenvalue	2.002343	1.902738	0.005204	-0.015418
Difference	0.099606	1.897534	0.020622	
Proportion	0.5141	0.4885	0.0013	-0.0040
Cumulative	0.5141	1.0026	1.0040	1.0000

     2 factors will be retained by the NFACTOR criterion.
WARNING: Too many factors for a unique solution.

Eigenvalues of the Reduced Correlation Matrix:
        Total = 3.91277649 Average = 0.97819412

	1	2	3	4
Eigenvalue	2.004734	1.909335	0.008962	-0.010255
Difference	0.095399	1.900373	0.019217	
Proportion	0.5124	0.4880	0.0023	-0.0026
Cumulative	0.5124	1.0003	1.0026	1.0000

Initial Factor Method: Iterated Principal Factor Analysis

Factor Pattern

	FACTOR1	FACTOR2
COST	-0.40027	0.89978
LIFT	0.25060	-0.94706
DEPTH	0.93159	0.34773
POWDER	0.95596	0.28615

Variance explained by each factor

FACTOR1	FACTOR2
2.004734	1.909335

Initial Factor Method: Iterated Principal Factor Analysis

Final Communality Estimates: Total = 3.914069

COST	LIFT	DEPTH	POWDER
0.969828	0.959725	0.988774	0.995742

TABLE 13.7   *(CONTINUED)*

Rotation Method: Varimax

Orthogonal Transformation Matrix
                              1   2

                    1   0.94565   -0.32519
                    2   0.32519   0.94565

Rotated Factor Pattern

                        FACTOR1   FACTOR2

        COST   -0.08591   0.98104
        LIFT   -0.07100   -0.97708
        DEPTH   0.99403   0.02588
        POWDER   0.99706   -0.04028

Rotation Method: Varimax

Variance explained by each factor

                    FACTOR1   'FACTOR2
                    1.994646   1.919423

Final Communality Estimates: Total = 3.914069

        COST   LIFT   DEPTH   POWDER
    0.969828   0.959725   0.988774   0.995742

shown for the rotated factors, as well as Percent of Common Variance Explained (see Proportion of covariance, Table 13.4).

13.5 ■ MAJOR TYPES OF FACTOR ANALYSIS

Numerous procedures for factor extraction and rotation are available. However, only those procedures available in the BMDP, SPSS, SAS, and SYSTAT packages are summarized here. Other extraction and rotational techniques are described in Mulaik (1972), Harman (1976), Rummel (1970), Comrey and Lee (1992), and Gorsuch (1983).

13.5.1 ■ Factor Extraction Techniques

Among the extraction techniques available in the four packages are principal components (PCA), principal factors, maximum likelihood factoring, Rao's canonical factoring, image factoring, alpha factoring, and unweighted and generalized (weighted) least squares factoring (see Table 13.9). Of these, PCA and principal factors are the most commonly used.

All the extraction techniques calculate a set of orthogonal components or factors that, in combination, reproduce **R.** Criteria used to establish the solution, such as maximizing variance or minimizing residual correlations, differ from technique to technique. But differences in solutions are small for a data set with a large sample, numerous variables and similar communality

TABLE 13.8   SETUP AND SYSTAT FACTOR OUTPUT FOR FACTOR ANALYSIS OF SAMPLE DATA OF TABLE 13.2.

```
USE TAPE17
MODEL COST LIFT DEPTH POWDER
ESTIMATE / METHOD = IPA, ROTATE = VARIMAX
```

Initial Communality Estimates

1	2	3	4
0.961	0.953	0.990	0.991

Final Communality Estimates

1	2	3	4
0.970	0.960	0.989	0.996

Latent Roots (Eigenvalues)

1	2	3	4
2.005	1.909	0.009	-0.010

Factor pattern

	1	2
COST	-0.400	0.900
LIFT	0.251	-0.947
DEPTH	0.932	0.348
POWDER	0.956	0.286

Variance Explained by Factors

1	2
2.005	1.909

Percent of Total Variance Explained

1	2
50.118	47.733

Rotated Pattern Matrix (VARIMAX,Gamma =   1.0000)

	1	2
COST	-0.086	0.981
LIFT	-0.071	-0.977
DEPTH	0.994	0.026
POWDER	0.997	-0.040

"Variance" Explained by Rotated Factors

1	2
1.995	1.919

Percent of Total Variance Explained

1	2
49.866	47.986

Percent of Common Variance Explained

1	2
50.961	49.039

estimates. In fact, one test of the stability of a FA solution is that it appears regardless of which extraction technique is employed. Table 13.10 shows solutions for the same data set after extraction with several different techniques, followed by varimax rotation. Similarities among the solutions are obvious.

None of the extraction techniques routinely provides an interpretable solution without rotation. All types of extraction may be rotated by any of the procedures described in Section 13.5.2, except Kaiser's Second Little Jiffy Extraction, which has its own rotational procedure.

Lastly, when using FA the researcher should hold in abeyance well-learned proscriptions against data snooping. It is quite common to use PCA as a preliminary extraction technique, followed by one or more of the other procedures, perhaps varying number of factors, communality estimates, and rotational methods with each run. Analysis terminates when the researcher decides on the preferred solution.

13.5.1.1 ■ PCA vs. FA

One of the most important decisions is the choice between PCA and FA. Mathematically, the difference involves the contents of the positive diagonal in the correlation matrix (the diagonal that

TABLE 13.9 SUMMARY OF EXTRACTION PROCEDURES

Extraction technique	Program	Goal of analysis	Special features
Principal components	SPSS BMDP4M SAS SYSTAT	Maximize variance extracted by orthogonal components	Mathematically determined, empirical solution with common, unique, and error variance mixed into components
Principal factors	SPSS BMDP4M SAS	Maximize variance extracted by orthogonal factors	Estimates communalities to attempt to eliminate unique and error variance from factors
Image factoring	SPSS BMDP4M (Second Little Jiffy) SAS (Image and Harris)	Provides an empirical factor analysis	Uses variances based on multiple regression of a variable with all other variables as communalities to generate a mathematically determined solution with error variance and unique variance eliminated
Maximum likelihood factoring	BMDP4M SPSS SAS	Estimate factor loadings for population that maximize the likelihood of sampling the observed correlation matrix	Has significance test for factors; especially useful for confirmatory factor analysis
Alpha factoring	SPSS SAS	Maximize the generalizability of orthogonal factors	
Unweighted least squares	SPSS SAS	Minimize squared residual correlations	
Generalized least squares	SPSS SAS	Weights variables by shared variance before minimizing squared residual correlations	

TABLE 13.10   RESULTS OF DIFFERENT EXTRACTION METHODS ON SAME DATA SET

	Factor 1				Factor 2			
Variables	PCA	PFA	Rao	Alpha	PCA	PFA	Rao	Alpha
			Unrotated factor loadings					
1	.58	.63	.70	.54	.68	.68	−.54	.76
2	.51	.48	.56	.42	.66	.53	−.47	.60
3	.40	.38	.48	.29	.71	.55	−.50	.59
4	.69	.63	.55	.69	−.44	−.43	.54	−.33
5	.64	.54	.48	.59	−.37	−.31	.40	−.24
6	.72	.71	.63	.74	−.47	−.49	.59	−.40
7	.63	.51	.50	.53	−.14	−.12	.17	−.07
8	.61	.49	.47	.50	−.09	−.09	.15	−.03
			Rotated factor loadings (varimax)					
1	.15	.15	.15	.16	.89	.91	.87	.92
2	.11	.11	.10	.12	.83	.71	.72	.73
3	−.02	.01	.02	.00	.81	.67	.69	.66
4	.82	.76	.78	.76	−.02	−.01	−.03	.01
5	.74	.62	.62	.63	.01	.04	.03	.04
6	.86	.86	.87	.84	.04	−.02	−.01	−.03
7	.61	.49	.48	.50	.20	.18	.21	.17
8	.57	.46	.45	.46	.23	.20	.20	.19

Note: The largest difference in communality estimates for a single variable between extraction techniques was 0.08.

contains the correlation between a variable and itself). In either PCA or FA, the variance that is analyzed is the sum of the values in the positive diagonal. In PCA ones are in the diagonal and there is as much variance to be analyzed as there are observed variables; each variable contributes a unit of variance by contributing a 1 to the positive diagonal of the correlation matrix. All the variance is distributed to components, including error and unique variance for each observed variable. So if all components are retained, PCA duplicates exactly the observed correlation matrix and the standard scores of the observed variables.

In FA, on the other hand, only the variance that each observed variable shares with other observed variables is available for analysis. Exclusion of error and unique variance from FA is based on the belief that such variance only confuses the picture of underlying processes. Shared variance is estimated by *communalities*, values between 0 and 1 that are inserted in the positive diagonal of the correlation matrix.[8] The solution in FA concentrates on variables with high communality values. The sum of the communalities (sum of the SSLs) is the variance that is distributed among factors and is less than the total variance in the set of observed variables. Because unique and error variances are omitted, a linear combination of factors approximates, but does not duplicate, the observed correlation matrix and scores on observed variables.

PCA analyzes variance; FA analyzes covariance (communality). The goal of PCA is to extract maximum variance from a data set with a few orthogonal components. The goal of FA is to reproduce the correlation matrix with a few orthogonal factors. PCA is a unique mathematical solution, whereas most forms of FA are not unique.

---

[8] Maximum likelihood extraction manipulates off-diagonal elements rather than values in the diagonal.

The choice between PCA and FA depends on your assessment of the fit between the models, the data set, and the goals of the research. If you are interested in a theoretical solution uncontaminated by unique and error variability, FA is your choice. If, on the other hand, you want an empirical summary of the data set, PCA is the better choice.

13.5.1.2 ■ Principal Components

The goal of PCA is to extract maximum variance from the data set with each component. The first principal component is the linear combination of observed variables that maximally separates subjects by maximizing the variance of their component scores. The second component is formed from residual correlations; it is the linear combination of observed variables that extracts maximum variability uncorrelated with the first component. Subsequent components also extract maximum variability from residual correlations and are orthogonal to all previously extracted components.

The principal components are ordered, with the first component extracting the most variance and the last component the least variance. The solution is mathematically unique and, if all components are retained, exactly reproduces the observed correlation matrix. Further, since the components are orthogonal, their use in other analyses (e.g., as DVs in MANOVA) may greatly facilitate interpretation of results.

PCA is the solution of choice for the researcher who is primarily interested in reducing a large number of variables down to a smaller number of components. PCA is also recommended as the first step in FA where it reveals a great deal about probable number and nature of factors. PCA is available through all four computer packages.

13.5.1.3 ■ Principal Factors

Principal factors extraction differs from PCA in that estimates of communality, instead of ones, are in the positive diagonal of the observed correlation matrix. These estimates are derived through an iterative procedure, with SMCs (squared multiple correlations of each variable with all other variables) used as the starting values in the iteration. The goal of analysis, like that for PCA, is to extract maximum orthogonal variance from the data set with each succeeding factor. Advantages to principal factors extraction are that it is widely used (and understood) and that it conforms to the factor analytic model in which common variance is analyzed with unique and error variance removed. Because the goal is to maximize variance extracted, however, principal factors is sometimes not as good as other extraction techniques in reproducing the correlation matrix. Also, communalities must be estimated and the solution is, to some extent, determined by those estimates. Principal factor analysis is available through SPSS, BMDP, SAS, and SYSTAT.

13.5.1.4 ■ Image Factor Extraction

The technique is called image factoring because the analysis distributes among factors the variance of an observed variable that is *reflected* by the other variables, in a manner similar to the SMC. Image factor extraction provides an interesting compromise between PCA and principal factors. Like PCA, image extraction provides a mathematically unique solution because there are fixed values in the positive diagonal of **R**. Like principal factors, the values in the diagonal are communalities with unique and error variability excluded.

Image scores for each variable are produced by multiple regression, with each variable, in turn, serving as a DV. A covariance matrix is calculated from these image (predicted) scores. The variances from the image score covariance matrix are the communalities for factor extraction. Care is necessary in interpreting the results of image analysis, because loadings represent covariances between variables and factors rather than correlations.

Image factoring is available through SPSS FACTOR, BMDP4M (as Kaiser's Second Little Jiffy), and SAS FACTOR (with two types—"image" and Harris component analysis).

13.5.1.5 ■ Maximum Likelihood Factor Extraction

The maximum likelihood method of factor extraction was developed originally by Lawley in the 1940s (see Lawley & Maxwell, 1963). Maximum likelihood extraction estimates population values for factor loadings by calculating loadings that maximize the probability of sampling the observed correlation matrix from a population. Within constraints imposed by the correlations among variables, population estimates for factor loadings are calculated that have the greatest probability of yielding a sample with the observed correlation matrix. This method of extraction also maximizes the canonical correlations between the variables and the factors (see Chapter 6).

Maximum likelihood extraction is available through BMDP4M, SPSS FACTOR, SYSTAT FACTOR, and SAS FACTOR. It is the extraction procedure recommended by BMDP when the common factor model is appropriate, the number of variables is not too many, and the correlation matrix is not singular.

13.5.1.6 ■ Unweighted Least Squares Factoring

The goal of unweighted least squares factor extraction is to minimize squared differences between the observed and reproduced correlation matrices. Only off-diagonal differences are considered; communalities are derived from the solution rather than estimated as part of the solution. Thus, unweighted least squares factoring can be seen as a special case of principal factors analysis in which communalities are estimated after the solution.

The procedure, originally called minimum residual, was developed by Comrey (1962) and later modified by Harman and Jones (1966). The latter procedure is available through SPSS FACTOR and SAS FACTOR.

13.5.1.7 ■ Generalized (Weighted) Least Squares Factoring

Generalized least squares extraction also seeks to minimize (off-diagonal) squared differences between observed and reproduced correlation matrices but in this case weights are applied to the variables. Variables that have substantial shared variance with other variables are weighted more heavily than variables that have substantial unique variance. In other words, variables that are not as strongly related to other variables in the set are not as important to the solution. This relatively new method of extraction is available through SPSS FACTOR and SAS FACTOR.

13.5.1.8 ■ Alpha Factoring

Alpha factor extraction, available through SPSS FACTOR and SAS FACTOR, grew out of psychometric research where the interest is in discovering which common factors are found consistently when repeated samples of variables are taken from a population of *variables*. The problem is the same as identifying mean differences that are found consistently among samples of subjects taken from a population of subjects—a question at the heart of most univariate and multivariate statistics.

In alpha factoring, however, the concern is with the reliability of the common factors rather than with the reliability of group differences. Coefficient alpha is a measure derived in psychometrics for the reliability (also called generalizability) of a score taken in a variety of situations. In alpha factoring, communalities that maximize coefficient alpha for the factors are estimated using iterative procedures (and sometimes exceed 1.0). (BMDP4M does not do alpha factoring, but optionally prints out alpha statistics.)

Probably the greatest advantage to the procedure is that it focuses the researcher's attention squarely on the problem of sampling variables from the domain of variables of interest. Disadvantages stem from the relative unfamiliarity of most researchers with the procedure and the reason for it.

13.5.2 ■ Rotation

The results of factor extraction, unaccompanied by rotation, are likely to be hard to interpret regardless of which method of extraction is used. After extraction, rotation is used to improve the interpretability and scientific utility of the solution. It is *not* used to improve the quality of the mathematical fit between the observed and reproduced correlation matrices because all orthogonally rotated solutions are mathematically equivalent to one another and to the solution before rotation.

Just as different methods of extraction tend to give similar results with a good data set, so also different methods of rotation tend to give similar results if the pattern of correlations in the data is fairly clear. In other words, a stable solution tends to appear regardless of the method of rotation used.

A decision is required between orthogonal and oblique rotation. In orthogonal rotation, the factors are uncorrelated. Orthogonal solutions offer ease of interpreting, describing, and reporting results; yet they strain "reality" unless the researcher is convinced that underlying processes are almost independent. The researcher who believes that underlying processes are correlated uses an oblique rotation. In oblique rotation the factors may be correlated, with conceptual advantages but practical disadvantages in interpreting, describing, and reporting results.

Among the dozens of rotational techniques that have been proposed, only those available in the four reviewed packages are included in this discussion (see Table 13.11). The reader who wishes to know more about these or other techniques is referred to Gorsuch (1983), Harman (1976), or Mulaik (1972). For the industrious, a presentation of rotation by hand is in Comrey and Lee (1992).

13.5.2.1 ■ Orthogonal Rotation

Varimax, quartimax, and equamax—three orthogonal techniques—are available in all four packages. Varimax is easily the most commonly used of all the rotations available.

Just as the extraction procedures have slightly different statistical goals, so also the rotational procedures maximize or minimize different statistics. The goal of varimax rotation is to simplify factors by maximizing the variance of the loadings within factors, across variables. The spread in loadings is maximized—loadings that are high after extraction become higher after rotation and loadings that are low become lower. Interpreting a factor is easier because it is obvious which variables correlate with it. Varimax also tends to reapportion variance among factors so that they

TABLE 13.11 SUMMARY OF ROTATIONAL TECHNIQUES

Rotational technique	Program	Type	Goals of analysis	Comments
Varimax	BMDP4M SAS SPSS SYSTAT	Orthogonal	Minimize complexity of factors (simplify columns of loading matrix) by maximizing variance of loadings on each factor.	Most commonly used rotation; rotation recommended as default option (in BMDP, $\Gamma = 1$)
Quartimax	BMDP4M SAS SPSS SYSTAT	Orthogonal	Minimize complexity of variables (simplify rows of loading matrix) by maximizing variance of loadings on each variable.	First factor tends to be general with others subclusters of variables (in BDMP, $\Gamma = 0$)
Equamax	BMDP4M SAS SPSS SYSTAT	Orthogonal	Simplify both variables and factors (rows and columns); compromise between quartimax and varimax.	May behave erratically ($\Gamma = 1/2$)
Orthogonal with gamma (orthomax)	BMDP4M SAS	Orthogonal	Simplify either factors or variables, depending on the value of gamma (Γ).	Γ continuously variable
Parsimax	SAS	Orthogonal	Simplify both variables and factors: $\Gamma = (p^*(m-1))/p + m - 2$.	
Direct oblimin	BMDP4M SPSS	Oblique	Simplify factors by minimizing cross products of loadings.	Continuous values of Γ (BMDP) or δ (SPSS), available; allows wide range of factor intercorrelations
(Direct) quartimin	BMDP4M SPSS	Oblique	Simplify factors by minimizing sum of cross products of squared loadings in pattern matrix.	Permits fairly high correlations among factors. Recommended oblique rotation by BMDP. In BMDP, $\Gamma = 0$. Achieved in SPSS by setting $\delta = 0$
Orthoblique	BMDP4M SAS (HK)	Both orthogonal and oblique	Rescale factor loadings to yield othogonal solution; nonrescaled loadings may be correlated.	Accompanies Kaiser's Second Little Jiffy (image) extraction in BMDP4M.
Promax	SAS	Oblique	Orthogonal factors rotated to oblique positions.	Fast and inexpensive
Procrustes	SAS	Oblique	Rotate to target matrix.	Useful in confirmatory FA

become relatively equal in importance; variance is taken from the first factors extracted and distributed among the later ones.

Quartimax does for variables what varimax does for factors. It simplifies variables by increasing the dispersion of the loadings within variables, across factors. Varimax operates on the columns of the loading matrix, quartimax operates on the rows. Quartimax is not nearly as popular as varimax because one is usually more interested in simple factors than in simple variables.

Equamax is a hybrid between varimax and quartimax that tries simultaneously to simplify the factors and the variables. Mulaik (1972) reports that equamax tends to behave erratically unless the researcher can specify the number of factors with confidence.

Although varimax rotation simplifies the factors, quartimax the variables, and equamax both, they do so in BMDP4M by setting levels on a simplicity criterion—Γ(gamma)—of 1, 0, and $\frac{1}{2}$, respectively. Gamma can also be continuously varied between 0 (variables simplified) and 1 (factors simplified) by using the orthogonal rotation that allows the user to specify Γ level. In SAS

FACTOR, this is done through orthomax with Γ. Parsimax in SAS uses a formula incorporating numbers of factors and variables to determine Γ (see Table 13.11).

Varimax is the rotation of choice for many applications; it is the default option of all packages that have defaults.

13.5.2.2 ■ Oblique Rotation

An *embarrasse de richesse* awaits the researcher who uses oblique rotation (see Table 13.11). Oblique rotations offer a continuous range of correlations between factors. The amount of correlation permitted between factors is determined by a variable called delta (δ) by SPSS FACTOR and gamma (Γ) by SYSTAT FACTOR and BMDP4M.[9] The values of delta and gamma determine the maximum amount of correlation permitted among factors. When the value is less than zero, solutions are increasingly orthogonal; at about -4 the solution is orthogonal. When the value is zero, solutions can be fairly highly correlated. Values near 1 can produce factors that are very highly correlated. Although there is a relationship between values of delta or gamma and size of correlation, the maximum correlation at a given size of gamma or delta depends on the data set.

It should be stressed that factors do not necessarily correlate when an oblique rotation is used. Often, in fact, they do not correlate and the researcher reports the simpler orthogonal rotation.

The family of procedures used for oblique rotation with varying degrees of correlation in SPSS, SYSTAT, and BMDP is direct oblimin. In the special case where Γ or $\delta = 0$ (the default option for the three programs), the procedure is called direct quartimin. Values of gamma or delta greater than zero permit high correlations among factors, and the researcher should take care that the correct number of factors is chosen. Otherwise, highly correlated factors may be indistinguishable one from the other. Some trial and error, coupled with inspection of the scatterplots of relationships between pairs of factors, may be required to determine the most useful size of gamma or delta. Or, one might simply trust to the default value.

Orthoblique rotation is designed to accompany Kaiser's Second Little Jiffy (Image) Factor extraction, and does so automatically through BMDP4M. Orthoblique rotation uses the quartimax algorithm to produce an orthogonal solution *on rescaled factor loadings*; therefore, the solution may be oblique with respect to the original factor loadings.

Promax and Procrustes are available through SAS. In promax rotation, an orthogonally rotated solution (usually varimax) is rotated again to allow correlations among factors. The orthogonal loadings are raised to powers (usually powers of 2, 4, or 6) to drive small and moderate loadings to zero while larger loadings are reduced, but not to zero. Even though factors correlate, simple structure is maximized by clarifying which variables do and do not correlate with each factor. Promax has the additional advantage of being fast and inexpensive.

In Procrustes rotation, a target matrix of loadings (usually zeros and ones) is specified by the researcher and a transformation matrix is sought to rotate extracted factors to the target, if possible. If the solution can be rotated to the target, then the hypothesized factor structure is said to be confirmed. Unfortunately, as Gorsuch (1983) reports, with Procrustean rotation, factors are often extremely highly correlated and sometimes a correlation matrix generated by random processes is rotated to a target with apparent ease.

---

[9] In BMDP and SYSTAT, gamma is used to indicate the nature of simplicity in orthogonal rotation and the amount of obliqueness in oblique rotation.

13.5.2.3 ■ Geometric Interpretation

A geometric interpretation of rotation is in Figure 13.1 where 13.1(a) is the unrotated and 13.1(b) the rotated solution to the example in Table 13.2. Points are represented in two-dimensional space by listing their coordinates with respect to X and Y axes. With the first two unrotated factors as axes, unrotated loadings are COST $(-.400, .900)$, LIFT $(.251, -.947)$, DEPTH $(.932, .348)$, and POWDER $(.956, .286)$.

The points for these variables are also located with respect to the first two rotated factors as axes in Figure 13.1(b). The position of points does not change, but their coordinates change in the new axis system. COST is now $(-0.86, .981)$, LIFT $(-.071, -.977)$, DEPTH $(.994, .026)$, and POWDER $(.997, -.040)$. Statistically, the effect of rotation is to amplify high loadings and reduce low ones. Spatially, the effect is to rotate the axes so that they "shoot through" the variable clusters more closely.

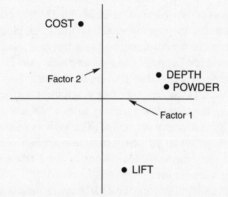

(a) Location of COST, LIFT, DEPTH, and POWDER after extraction, before rotation

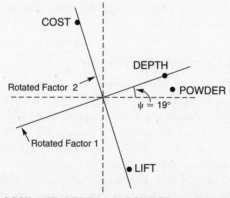

(b) Location of COST, LIFT, DEPTH, and POWDER vis-á-vis rotated axes

---

FIGURE 13.1 ILLUSTRATION OF ROTATION OF AXES TO PROVIDE A BETTER
DEFINITION OF FACTORS VIS Á VIS THE VARIABLES WITH
WHICH THEY CORRELATE.

Factor extraction yields a solution in which observed variables are vectors that run from the origin to the points indicated by the coordinate system. The factors serve as axes for the system. The coordinates of each point are the entries from the loading matrix for the variable. If there are three factors, then the space has three axes and three dimensions, and each observed variable is positioned by three coordinates. The length of the vector for each variable is the communality of the variable.

If the factors are orthogonal, the factor axes are all at right angles to one another and the coordinates of the variable points are correlations between the common factors and the observed variables. Correlations (factor loadings) are read directly from these graphs by projecting perpendicular lines from each point to each of the factor axes.

One of the primary goals of PCA or FA, and the motivation behind extraction, is to discover the minimum number of factor axes needed to reliably position variables. A second major goal, and the motivation behind rotation, is to discover the meaning of the factors that underlie responses to observed variables. This goal is met by interpreting the factor axes that are used to define the space. Factor rotation repositions factor axes so as to make them maximally interpretable. Repositioning the axes changes the coordinates of the variable points but not the positions of the points with respect to each other.

Factors are usually interpretable when some observed variables load highly on them and the rest do not. And, ideally, each variable loads on one, and only one, factor. In graphic terms this means that the point representing each variable lies far out along one axis but near the origin on the other axes, that is, that coordinates of the point are large for one axis and near zero for the other axes.

If you have only one observed variable, it is trivial to position the factor axis—variable point and axis overlap in a space of one dimension. However, with many variables and several factor axes, compromises are required in positioning the axes. The variables form a "swarm" in which variables that are correlated with one another form a cluster of points. The goal is to shoot an axis to the swarm of points. With luck, the swarms are about 90° away from one another so that an orthogonal solution is indicated. And with lots of luck, the variables cluster in just a few swarms with empty spaces between them so that the factor axes are nicely defined.

In oblique rotation the situation is slightly more complicated. Because factors may correlate with one another, factor axes are not necessarily at right angles. And, although it is easier to position each axis near a cluster of points, axes may be very near each other (highly correlated), making the solution harder to interpret. See Section 13.6.3 for practical suggestions of ways to use graphic techniques to judge the adequacy of rotation.

13.5.3 ■ Some Practical Recommendations

Although an almost overwhelmingly large number of combinations of extraction and rotation techniques is available, in practice differences among them are often slight (Velicer & Jackson, 1990; Fava & Velicer, 1992). The results of extraction are similar regardless of which method is used when there is a large number of variables with some strong correlations among them, with the same, well-chosen number of factors, and with similar values for communality. Further, differences that are apparent after extraction tend to disappear after rotation.

Most researchers begin their FA by using principal components extraction and varimax rotation. From the results, one estimates the factorability of the correlation matrix (Sections 13.3.2.6 and 13.8.1.6), the rank of the observed correlation matrix (Sections 13.3.2.5 and 13.8.1.5), the number of factors (Sections 13.6.2), and variables that might be excluded from subsequent analyses (Sections 13.3.2.7 and 13.8.1.7).

During the next few runs, researchers experiment with different numbers of factors, different extraction techniques, and both orthogonal and oblique rotations. Some number of factors with some combination of extraction and rotation produces the solution with the greatest scientific utility, consistency, and meaning; this is the solution that is interpreted.

13.6 ■ SOME IMPORTANT ISSUES

Some of the issues raised in this section can be resolved through several different methods. Usually different methods lead to the same conclusion; occasionally they do not. When they do not, results are judged by the interpretability and scientific utility of the solutions.

13.6.1 ■ Estimates of Communalities

FA differs from PCA in that communality values (numbers between 0 and 1) replace ones in the positive diagonal of **R** before factor extraction. Communality values are used instead of ones to remove the unique and error variance of each observed variable; only the variance a variable shares with the factors is used in the solution. But communality values are estimated, and there is some dispute regarding how that should be done.

The SMC of each variable as DV with the others in the sample as IVs is usually the starting estimate of communality. As the solution develops, communality estimates are adjusted by iterative procedures (which can be directed by the researcher) to fit the reproduced to the observed correlation matrix with the smallest number of factors. Iteration stops when successive communality estimates are very similar.

Final estimates of communality are also SMCs, but now between each variable as DV and the factors as IVs. Final communality values represent the proportion of variance in a variable that is predictable from the factors underlying it. Communality estimates do not change with orthogonal rotation.

Image extraction and maximum likelihood extraction are slightly different. In image extraction, variances from the image covariance matrix are used as the communality values throughout. Image extraction produces a mathematically unique solution because communality values are not changed. In maximum likelihood extraction, number of factors instead of communality values are estimated and off-diagonal correlations are "rigged" to produce the best fit between observed and reproduced matrices.

BMDP, SPSS, and SAS provide several different starting statistics for communality estimation. BMDP4M offers SMCs, user-specified values, or maximum absolute correlation with any other variable as initial communality estimates. SPSS FACTOR permits user supplied values for principal factor extraction only, but otherwise uses SMCs. SAS FACTOR offers, for each variable, a choice of SMC, SMC adjusted so that the sum of the communalities is equal to the sum of the maximum absolute correlations, maximum absolute correlation with any other variable, user-specified values, or random numbers between 0 and 1. SYSTAT FACTOR uses SMCs. Fewer iterations are usually required when starting from SMCs.

The seriousness with which estimates of communality should be regarded depends on the number of observed variables. If the number of variables exceeds, say, 20, sample SMCs probably provide reasonable estimates of communality. Furthermore, with 20 or more variables, the elements in the positive diagonal are few compared with the total number of elements in R, and their sizes do not influence the solution very much. Actually, if the communality values for all variables in FA are of approximately the same magnitude, results of PCA and FA are very similar (Velicer & Jackson, 1990; Fava & Velicer, 1992).

If communality values equal or exceed 1, problems with the solution are indicated. There is too little data, or starting communality values are wrong, or the number of factors extracted is wrong; addition or deletion of factors may reduce the communality below 1. Very low communality values, on the other hand, indicate that the variables with them are unrelated to other variables in the set (Sections 13.3.2.7 and 13.8.1.7). SAS FACTOR has two alternatives for dealing with communalities > 1: HEYWOOD sets them to 1, and ULTRAHEYWOOD allows them to exceed 1, but warns that doing so can cause convergence problems.

13.6.2 ■ Adequacy of Extraction and Number of Factors

Because inclusion of more factors in a solution improves the fit between observed and reproduced correlation matrices, adequacy of extraction is tied to number of factors. The more factors extracted, the better the fit and the greater the percent of variance in the data "explained" by the factor solution. However, the more factors extracted, the less parsimonious the solution. To account for all the variance (PCA) or covariance (FA) in a data set, one would normally have to have as many factors as observed variables. It is clear, then, that a trade-off is required: One wants to retain enough factors for an adequate fit, but not so many that parsimony is lost.

Selection of the number of factors is probably more critical than selection of extraction and rotational techniques or communality values. In confirmatory FA, selection of the number of factors is really selection of the number of theoretical processes underlying a research area. You can partially confirm a hypothesized factor structure by asking if the theoretical number of factors adequately fits the data.

There are several ways to assess adequacy of extraction and number of factors. For a highly readable summary of these methods, not all currently available through the statistical packages, see Gorsuch (1983). Reviewed below are methods available through SPSS, BMDP, SAS, and SYSTAT.

A first quick estimate of the number of factors is obtained from the sizes of the eigenvalues reported as part of an initial run with principal components extraction. Eigenvalues represent variance. Because the variance that each standardized variable contributes to a principal components extraction is 1, a component with an eigenvalue less than 1 is not as important, from a variance perspective, as an observed variable. The number of components with eigenvalues greater than 1 is usually somewhere between the number of variables divided by 3 and the number of variables divided by 5 (e.g., 20 variables should produce between 7 and 4 components with eigenvalues greater than 1). If this is a reasonable number of factors for the data, if the number of variables is 40 or fewer, and if sample size is large, the number of factors indicated by this criterion is probably about right. In other situations, this criterion may either over- or underestimate the number of factors in the data set.

A second criterion is the scree test (Cattell, 1966) of eigenvalues plotted against factors. Factors, in descending order, are arranged along the abscissa with eigenvalue as the ordinate. The plot is appropriately used with principal components or factor analysis at initial and later runs to find the number of factors. The scree plot is available through SPSS, SYSTAT and SAS FACTOR and BMDP4M (where it is called a histogram of eigenvalues).

Usually the scree plot is negatively decreasing—the eigenvalue is highest for the first factor and moderate but decreasing for the next few factors before reaching small values for the last several factors, as illustrated for real data through SPSS in Figure 13.2. You look for the point where a line drawn through the points changes slope. In the example, a single straight line can comfortably fit the first four eigenvalues. After that, another line, with a noticeably different slope, best fits the remaining eight points. Therefore, there appear to be about four factors in the data of Figure 13.2.

Unfortunately, the scree test is not exact; it involves judgment of where the discontinuity in eigenvalues occurs and researchers are not perfectly reliable judges. As Gorsuch (1983) reports, results of the scree test are more obvious (and reliable) when sample size is large, communality values are high, and each factor has several variables with high loadings. Under less than optimal conditions, the scree test is still usually accurate to within one or two factors.

If you are unsure of the number of factors, perform several factor analyses, each time specifying a different number of factors, repeating the scree test, and examining the residual correlation matrix. The residual correlation matrix is available through SPSS, BMDP, and SAS. As discussed in Section 13.4, the residual correlation matrix is obtained by subtracting the reproduced correlation matrix from the observed correlation matrix. The numbers in the residual matrix are actually partial correlations between pairs of variables with effects of factors removed. If the analysis is good, the residuals are small. Several moderate residuals (say, .05 to .10) or a few large residuals (say $>.10$) suggest the presence of another factor.

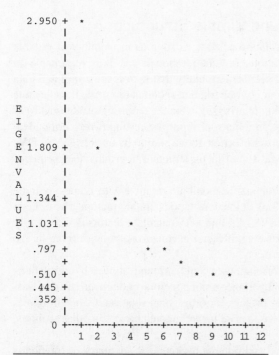

FIGURE 13.2 SCREE OUTPUT FOR SAMPLE DATA PRODUCED BY SPSS FACTOR. NOTE BREAK IN SIZE OF EIGENVALUES BETWEEN THE FOURTH AND FIFTH FACTORS.

Once you have determined the number of factors by these criteria, it is important to look at the rotated loading matrix to determine the number of variables that load on each factor (see Section 13.6.5). If only one variable loads highly on a factor, the factor is poorly defined. If two variables load on a factor, then whether or not it is reliable depends on the pattern of correlations of these two variables with each other and with other variables in **R.** If the two variables are highly correlated with each other (say, $r > .70$) and relatively uncorrelated with other variables, the factor may be reliable. Interpretation of factors defined by only one or two variables is hazardous, however, under even the most exploratory factor analysis.

For principal components extraction and maximum likelihood extraction in confirmatory factor analysis there are significance tests for number of factors. Bartlett's test evaluates all factors together and each factor separately against the hypothesis that there are no factors. However, there is some dispute regarding use of these tests. The interested reader is referred to Gorsuch (1983) or one of the other newer factor analysis texts for discussion of significance testing in FA.

There is debate about whether it is better to retain too many or too few factors if the number is ambiguous. Sometimes a researcher wants to rotate, but not interpret, marginal factors for statistical purposes (e.g., to keep some factors with communality values < 1). Other times the last few factors represent the most interesting and unexpected findings in a research area. These are good reasons for retaining factors of marginal reliability. However, if the researcher is interested in using only demonstrably reliable factors, the fewest possible factors are retained.

13.6.3 ■ Adequacy of Rotation and Simple Structure

The decision between orthogonal and oblique rotation is made as soon as the number of reliable factors is apparent. In many factor analytic situations, oblique rotation seems more reasonable on the face of it than orthogonal rotation because it seems more likely that factors are correlated than that they are not. However, reporting the results of oblique rotation requires reporting the elements of the pattern matrix **(A)** and the factor correlation matrix **(Φ)**, whereas reporting orthogonal rotation requires only the loading matrix **(A)**. Thus, simplicity of reporting results favors orthogonal rotation. Further, if factor scores or factorlike scores (Section 13.6.6) are to be used as IVs or DVs in other analyses, or if a goal of analysis is comparison of factor structure in groups, then orthogonal rotation has distinct advantages.

Perhaps the best way to decide between orthogonal and oblique rotation is to request oblique rotation with the desired number of factors and look at the correlations among factors. The oblique rotations available by default in SPSS, BMDP, SYSTAT, and SAS calculate factors that are fairly highly correlated if necessary to fit the data. However, if factor correlations are not driven by the data, the solution remains nearly orthogonal.

Look at the factor correlation matrix for correlations around .32 and above. If correlations exceed .32, then there is 10% (or more) overlap in variance among factors, enough variance to warrant oblique rotation unless there are compelling reasons for orthogonal rotation. Compelling reasons include a desire to compare structure in groups, a need for orthogonal factors in other analyses, or a theoretical need for orthogonal rotation.

Once the decision is made between orthogonal and oblique rotation, the adequacy of rotation is assessed several ways. Perhaps the simplest way is to compare the pattern of correlations in the correlation matrix with the factors. Are the patterns represented in the rotated solution? Do highly correlated variables tend to load on the same factor? If you included marker variables, do they load on the predicted factors?

Another criterion is simple structure (Thurstone, 1947). If simple structure is present (and factors are not too highly correlated), several variables correlate highly with each factor and only one factor correlates highly with each variable. In other words, the columns of **A,** which define factors vis-à-vis variables, have several high and many low values while the rows of **A,** which define variables vis-à-vis factors, have only one high value. Rows with more than one high correlation correspond to variables that are said to be complex because they reflect the influence of more than one factor. It is usually best to avoid complex variables because they make interpretation of factors more ambiguous.

Adequacy of rotation is also ascertained through the PLOT commands of the four programs. In the figures, factors are considered two at a time with a different pair of factors as axes for each plot. Look at the *distance*, *clustering*, and *direction* of the points representing variables relative to the factor axes in the figures.

The *distance* of a variable point from the origin reflects the size of factor loadings; variables highly correlated with a factor are far out on that factor's axis. Ideally, each variable point is far out on one axis and near the origin on all others. *Clustering* of variable points reveals how clearly defined a factor is. One likes to see a cluster of several points near the end of each axis and all other points near the origin. A smattering of points at various distances along the axis indicates a factor that is not clearly defined, while a cluster of points midway between two axes reflects the presence of another factor or the need for oblique rotation. The *direction* of clusters after orthogonal rotation may also indicate the need for oblique rotation. If clusters of points fall between factor axes after orthogonal rotation, if the angle between clusters with the respect to the origin is not 90°, then a better fit to the clusters is provided by axes that are not orthogonal. Oblique rotation may reveal substantial correlations among factors. Several of these relationships are depicted in Figure 13.3.

13.6.4 ■ Importance and Internal Consistency of Factors

The importance of a factor (or a set of factors) is evaluated by the proportion of variance or covariance accounted for by the factor after rotation. The proportion of variance attributable to individual factors differs before and after rotation because rotation tends to redistribute variance among factors somewhat. Ease of ascertaining proportions of variance for factors depends on whether rotation was orthogonal or oblique.

After orthogonal rotation, the importance of a factor is related to the size of its SSLs (Sum of Squared Loadings from **A** after rotation). SSLs are converted to proportion of variance for a factor by dividing by p, the number of variables. SSLs are converted to proportion of covariance for a factor by dividing its SSL by the sum of SSLs or, equivalently, sum of communalities. These computations are illustrated in Table 13.4 and Section 13.4 for the example.

The proportion of variance accounted for by a factor is the amount of variance in the original variables (where each has contributed one unit of variance) that has been condensed into the factor. Proportion of *variance* is the variance of a factor relative to the variance in the variables. The proportion of covariance accounted for by a factor indicates the relative importance of the factor to the total covariance accounted for by all factors. Proportion of *covariance* is the variance of a factor relative to the variance in the solution. The variance in the solution is likely to account for only a fraction of the variance in the original variables.

In oblique rotation, proportions of variance and covariance can be obtained from **A** *before* rotation by the methods just described, but they are only rough indicators of the proportions of variance and covariance of factors after rotation. Because factors are correlated, they share overlapping variability,

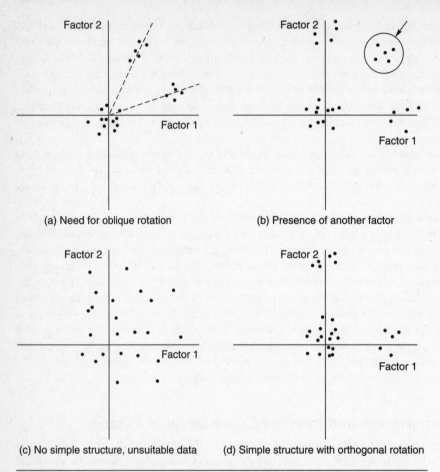

(a) Need for oblique rotation   (b) Presence of another factor

(c) No simple structure, unsuitable data   (d) Simple structure with orthogonal rotation

FIGURE 13.3   PAIRWISE PLOTS OF FACTOR LOADINGS FOLLOWING ORTHOGO-
NAL ROTATION AND INDICATING: (a) NEED FOR OBLIQUE ROTA-
TION; (b) PRESENCE OF ANOTHER FACTOR; (c) UNSUITABLE DATA;
AND (d) SIMPLE STRUCTURE.

and assignment of variance to individual factors is ambiguous. After oblique rotation the size of the SSL associated with a factor is a rough approximation of its importance—factors with bigger SSLs are more important—but proportions of variance and covariance cannot be specified.

An estimate of the internal consistency of the solution—the certainty with which factor axes are fixed in the variable space—is given by the squared multiple correlations of factor scores predicted from scores on observed variables. In a good solution, SMCs range between 0 and 1; the larger the SMCs, the more stable the factors. A high SMC (say, .70 or better) means that the observed variables account for substantial variance in the factor scores. A low SMC means the factors are poorly defined by the observed variables. If an SMC is negative, too many factors have been retained. If an SMC is above 1, the entire solution needs to be reevaluated.

BMDP4M prints these SMCs as the positive diagonal of the factor-score covariance matrix. SPSS FACTOR prints them as the diagonal of the covariance matrix for estimated regression factor scores. In SAS FACTOR, SMCs are printed along with factor score coefficients by the SCORE option.

13.6.5 ■ Interpretation of Factors

To interpret a factor, one tries to understand the underlying dimension that unifies the group of variables loading on it. In both orthogonal and oblique rotations, loadings are obtained from the loading matrix, **A,** but the meaning of the loadings is different for the two rotations.

After orthogonal rotation, the values in the loading matrix are correlations between variables and factors. The researcher decides on a criterion for meaningful correlation (usually .32 or larger), collects together the variables with loadings in excess of the criterion, and searches for a concept that unifies them.

After oblique rotation, the process is the same, but the interpretation of the values in **A,** the pattern matrix, is no longer straightforward. The loading is not a correlation but is a measure of the unique relationship between the factor and the variable. Because factors correlate, the correlations between variables and factors (available in the structure matrix, **C**) are inflated by overlap between factors. A variable may correlate with one factor through its correlation with another factor rather than directly. The elements in the pattern matrix have overlapping variance among factors "partialled out," but at the expense of conceptual simplicity.

Actually, the reason for interpretation of the pattern matrix rather than the structure matrix is pragmatic—it's easier. The difference between high and low loadings is more apparent in the pattern matrix than in the structure matrix.

As a rule of thumb, only variables with loadings of .32 and above are interpreted. The greater the loading, the more the variable is a pure measure of the factor. Comrey and Lee (1992) suggest that loadings in excess of .71 (50% overlapping variance) are considered excellent, .63 (40% overlapping variance) very good, .55 (30% overlapping variance) good, .45 (20% overlapping variance) fair, and .32 (10% overlapping variance) poor. Choice of the cutoff for size of loading to be interpreted is a matter of researcher preference. Sometimes there is a gap in loadings across the factors and, if the cutoff is in the gap, it is easy to specify which variables load and which do not. Other times, the cutoff is selected because one can interpret factors with that cutoff but not with a lower cutoff.

The size of loadings is influenced by the homogeneity of scores in the sample. If homogeneity is suspected, interpretation of lower loadings is warranted. That is, if the sample produces similar scores on observed variables, a lower cutoff is used for interpretation of factors.

At some point, a researcher usually tries to characterize a factor by assigning it a name or a label, a process that involves art as well as science. Rummel (1970) provides numerous helpful hints on interpreting and naming factors. Interpretation of factors is facilitated by output of the matrix of sorted loadings where variables are grouped by their correlations with factors. Sorted loadings are produced routinely by BMDP4M, by REORDER in SAS FACTOR, and SORT in SPSS and SYSTAT FACTOR.

The replicability, utility, and complexity of factors are also considered in interpretation. Is the solution replicable in time and/or with different groups? Is it trivial or is it a useful addition to scientific thinking in a research area? Where do the factors fit in the hierarchy of "explanations" about a phenomenon? Are they complex enough to be intriguing without being so complex that they are uninterpretable?

13.6.6 ■ Factor Scores

Among the potentially more useful outcomes of PCA or FA are factor scores. Factor scores are estimates of the scores subjects would have received on each of the factors had they been measured
directly.

Because there are normally fewer factors than observed variables, and because factor scores are
nearly uncorrelated if factors are orthogonal, use of factor scores in other analyses may be very
helpful. Multicollinear matrices can be reduced to orthogonal components using PCA, for instance.
Or, one could use FA to reduce a large number of DVs to a smaller number of factors for use as
DVs in MANOVA. Alternatively, one could reduce a large number of IVs to a small number of factors for purposes of predicting a DV in multiple regression or group membership in discriminant
analysis or logistic regression. If factors are few in number, stable, and interpretable, their use
enhances subsequent analyses.

Procedures for estimating factor scores range between simple-minded (but frequently adequate)
and sophisticated. Comrey and Lee (1992) describe several rather simple-minded techniques for
estimating factor scores. Perhaps the simplest is to sum scores on variables that load highly on each
factor. Variables with bigger standard deviations contribute more heavily to the factor scores produced by this procedure, a problem that is alleviated if variable scores are standardized first or if
the variables have roughly equal standard deviations to begin with. For many research purposes,
this "quick and dirty" estimate of factor scores is entirely adequate.

There are several sophisticated statistical approaches to estimating factors. All produce factor
scores that are correlated, but not perfectly, with the factors. The correlations between factors and
factor scores are higher when communalities are higher and when the ratio of variables to factors
is higher. But as long as communalities are estimated, factor scores suffer from indeterminacy
because there is an infinite number of possible factor scores that all have the same mathematical
characteristics. As long as factor scores are considered only estimates, however, the researcher is
not overly beguiled by them. It is because of this indeterminancy that SYSTAT FACTOR makes
available component but not factor scores.

The method described in Section 13.4 (especially Equations 13.10 and 13.11) is the regression
approach to estimating factor scores. This approach results in the highest correlations between factors and factor scores. The distribution of each factor's scores has a mean of zero and a standard
deviation of 1 (after PCA) or equal to the SMC between factors and variables (after FA). However,
this regression method, like all others (see Chapter 5), capitalizes on chance relationships among
variables so that factor-score estimates are biased (too close to "true" factor scores). Further, there
are often correlations among scores for factors even if factors are orthogonal and factor scores
sometimes correlate with other factors (in addition to the one they are estimating).

The regression approach to estimating factor scores is available through SYSTAT (component
scores only) as well as BMDP, SAS, and SPSS. All four packages write component/factor scores to
files for use in other analyses and all four print standardized component/factor score coefficients—
routinely through SYSTAT and BMDP, and through FSCORE in SPSS and SCORE in SAS. BMDP
also routinely prints out factor scores.

SPSS FACTOR provides two additional methods of estimating factor scores. In the Bartlett method,
factor scores correlate only with their own factors and the factor scores are unbiased (that is, neither
systematically too close nor too far away from "true" factor scores). The factor scores correlate with
the factors almost as well as in the regression approach and have the same mean and standard deviation as in the regression approach. However, factor scores may still be correlated with each other.

The Anderson-Rubin approach (discussed by Gorsuch, 1983) produces factor scores that are uncorrelated with each other even if factors are correlated. Factor scores have mean zero, standard deviation 1. Factor scores correlate with their own factors almost as well as in the regression approach, but they sometimes also correlate with other factors (in addition to the one they are estimating) and they are somewhat biased. If you need uncorrelated scores, the Anderson-Rubin approach is best; otherwise the regression approach is probably best simply because it is best understood and most widely available.

13.6.7 ■ Comparisons among Solutions and Groups

Frequently, a researcher is interested in deciding whether or not two groups that differ in experience or characteristics have the same factors. Comparisons among factor solutions involve the *pattern* of the correlations between variables and factors, or both the *pattern and magnitude* of the correlations between them. Rummel (1970), Levine (1977), and Gorsuch (1983) have excellent summaries of several comparisons that might be of interest. Some of the simpler of these techniques are described in an earlier version of this book (Tabachnick & Fidell, 1989).

Tests of theory (in which theoretical factor loadings are compared with those derived from a sample) and comparisons among groups are currently the province of structural equation modeling. These techniques are discussed in Chapter 14.

13.7 ■ COMPARISON OF PROGRAMS

BMDP, SPSS, SAS, and SYSTAT each have a single program to handle both FA and PCA. The first three programs have numerous options for extraction and rotation and give the user considerable latitude in directing the progress of the analysis. Features of all four programs are described in Table 13.12.

13.7.1 ■ SPSS Package

The program in the SPSS package developed for PCA and FA is FACTOR. A number of features have been added since the original SPSS FACTOR program. SPSS FACTOR is a much more appealing program than the original. In fact, this program is now one of the most attractive of those available for FA.

SPSS FACTOR does a PCA or FA on a correlation matrix or a factor loading matrix, helpful to the researcher who is interested in higher-order factoring (extracting factors from previous FAs). Several extraction methods and a variety of orthogonal rotation methods are available. Oblique rotation is done using direct oblimin, one of the best methods currently available (see Section 13.5.2.2).

Univariate output is limited to means, standard deviations, and number of cases per variable, so that the search for univariate outliers must be conducted through other programs. Similarly, there is no provision for screening for multivariate outliers among cases. But the program is very helpful in assessing factorability of **R,** as discussed in Section 13.3.2.6.

Output of extraction and rotation information is extensive. The residual and reproduced correlation matrices are provided as an aid to diagnosing adequacy of extraction and rotation. SPSS

TABLE 13.12 COMPARISON OF FACTOR ANALYSIS PROGRAMS

Feature	SPSS FACTOR	BMDP4M	SAS FACTOR	SYSTAT FACTOR
Input				
Correlation matrix	Yes	Yes	Yes	Yes
About origin	No	Yes	Yes	No
Covariance matrix	No	Yes	Yes	Yes
About origin	No	Yes	Yes	No
SSCP matrix	No	No	No	Yes
Factor loadings (unrotated pattern)	Yes	Yes	Yes	Yes
Factor-score coefficients	No	Yes	Yes	Data file
Factor loadings (rotated pattern) and factor correlations	No	Yes	Yes	Yes
Options for missing data	Yes	No	No	Yes
Analyze partial correlation or covariance matrix	No	No	Yes	No
Specify maximum number of factors	Yes	Yes	Yes	Yes
Extraction method (see Table 13.9)				
PCA	PC	PCA	PRIN	PCA
PFA	PAF	PFA	PRINIT	IPA
Image (Little Jiffy, Harris)	IMAGE	LJIFFY	Yes[a]	No
Maximum likelihood (Rao's canonical)	ML	MLFA	ML	MLA
Alpha	Yes	No	Yes	No
Unweighted least squares	ULS	No	ULS	No
Generalized least squares	GLS	No	Yes	No
Specify communalities	Yes	Yes	Yes	No
Specify minimum eigenvalues	Yes	Yes	Yes	Yes
Specify proportion of variance to be accounted for	No	No	Yes	No
Specify maximum number of iterations	Yes	Yes	Yes	Yes
Option to allow communalities > 1	No	No	Yes	No
Specify tolerance	No	Yes	SING	Yes
Specify convergence criterion for extraction	ECONVERGE	EPS	CONV	CONV
Specify convergence criterion for rotation	RCONVERGE	No	No	No
Rotation method (see Table 13.11)				
Varimax	Yes	Yes	Yes	Yes
Quartimax	Yes	Yes	Yes	Yes
Equamax	Yes	Yes	Yes	Yes
Orthogonal with gamma	No	Yes	ORTHOMAX	ORTHOMAX
Parsimax	No	No	Yes	No
Direct oblimin	Yes	Yes	No	Yes
Direct quartimin	DELTA = 0	Yes	No	No
Orthoblique	No	Yes	HK	No
Promax	No	No	Yes	No
Procrustes	No	No	Yes	No
Prerotation criteria	No	No	Yes	No
Optional Kaiser's normalization	Yes	Yes	Yes	Normalized only
Optional weighting by Cureton-Mulaik technique	No	No	Yes	No

TABLE 13.12 *(CONTINUED)*

Feature	SPSS FACTOR	BMDP4M	SAS FACTOR	SYSTAT FACTOR
Optional rescaling of pattern matrix to covariances	No	No	Yes	No
Differential case weighting	No	Yes	No	No
Differential variable weighting	No	No	Yes	No
Alternate methods for computing factor scores	Yes	No	No	No
Output				
Means and standard deviations	Yes	Yes	Yes	No
Number of cases per variable (missing data)	Yes	Yes	No	No
Coefficients of variation	No	Yes	No	No
Skewness and kurtosis	No	Yes	No	No
Minimums and maximums	No	Yes	No	No
z for minimums and maximums	No	Yes	No	No
First case for minimums and maximums	No	Yes	No	No
Correlation matrix	Yes	Yes	Yes	Yes
Significance of correlations	Yes	No	No	No
Covariance matrix	No	Yes	Yes	Yes
Initial communalities	Yes	Yes	Yes	Yes
Final communalities	Yes	Yes	Yes	Yes
Eigenvalues	Yes	Yes	Yes	Yes
Difference between successive eigenvalues	No	No	Yes	No
Standard error for each eigenvector element	No	No	No	Yes
Cronbach's alpha	No	CRON	ALPHA	No
Second order factors	No	CRON	No	No
Carmine's theta	No	Yes	No	No
Percent of variance explained by factors	Yes	No	No	Yes
Cumulative percent of variance	Yes	No	No	No
Percent of covariance	No	No	Yes	Yes
Cumulative percent of covariance	No	Yes[b]	Yes	No
Unrotated factor loadings	Yes	Yes	Yes	Yes
Variance explained by factors for all loading matrices	No	Yes	Yes	Yes
Simplicity criterion, each rotation iteration	δ[c]	Γ	No	No
Rotated factor loadings (pattern)	Yes	Yes	Yes	Yes
Rotated factor loadings (structure)	Yes	Yes	Yes	Yes
Eigenvectors	No	No	Yes	No
Transformation matrix	Yes	No	Yes	No
Factor-score coefficients	Yes	Yes	Yes	Data file[d]
Standardized factor scores	Data file	Yes	Data file	Data file[d]
Unstandardized factor scores	No	No	No	Data file[d]
Residual component scores	No	No	No	Data file[d]
Sum of squared residuals (Q)	No	No	No	Data file[d]
Probability for Q	No	No	No	Data file[d]

TABLE 13.12 *(CONTINUED)*

Feature	SPSS FACTOR	BMDP4M	SAS FACTOR	SYSTAT FACTOR
Scree plot	Yes	Yes	Yes	Yes
Plots of unrotated factor loadings	No	Yes	Yes	No
Plots of rotated factors loadings	Yes	Yes	Yes	Yes
Plots of factor scores	No	Yes	No	No
Sorted rotated factor loadings	Yes	Yes	Yes	Yes
χ^2 test for number of factors (with maximum likelihood estimation)	No	Yes	No	Yes
χ^2 test that all eigenvalues are equal	No	No	No	Yes[d]
χ^2 test that last n eigenvalues are equal	No	No	No	Yes[d]
Inverse of correlation matrix	Yes	Yes	Yes	No
Determinant of correlation matrix	Yes	No	No	No
Factor-score covariance matrices	No	Yes	No	No
Mahalanobis distance of cases	No	Yes	No	No
Standard scores—each variable for each case	No	Yes	No	No
Partial correlations (anti-image matrix)	AIC	PART	MSA	No
Measure of sampling adequacy	AIC, KMO	No	MSA	No
Anti-image covariance matrix	AIC	No	No	No
Bartlett's test of sphericity	KMO	No	No	No
Residual correlation matrix	Yes	Yes	Yes	Yes
Reproduced correlation matrix	Yes	No	No	No
Shaded correlations	No	Yes	No	No
Correlations among factors	Yes	Yes	Yes	Yes
Print matrix entries $\times$ 100 and rounded	No	No	ROUND	No
Direct and indirect contribution of factors to variance	No	No	No	Yes
Print low matrix values as missing values	No	No	FUZZ	No

[a] Two types.

[b] Labeled proportion of variance.

[c] Oblique only.

[d] PCA only.

FACTOR is the only program reviewed that, under conditions requiring matrix inversion, prints out the determinant of the correlation matrix, helpful in signaling the need to check for multi-collinearity and singularity (Sections 13.3.2.5 and 4.1.7). Determination of number of factors is aided by an optional printout of a scree plot (Section 13.6.2).

Several estimation procedures for factor scores (Section 13.6.6) are available as output to a file; they are not given as part of regular printed output. Information about cases that are poorly fit by the factors is not available.

13.7.2 ■ BMDP Series

BMDP4M is the program for FA and PCA in the BMDP package. BMDP4M is flexible in its input, performing PCA and FA on raw data or either correlation or covariance matrices, with an option for analysis to proceed around the origin (setting all variable means to zero).

Several varieties of factor extraction and both orthogonal and oblique rotation are available. Information is given for all varieties of outlier detection—variables, univariate and multivariate outliers among cases, and outliers in the solution. About the only features missing are measures of the factorability of **R,** the determinant and the transformation matrix, generally of limited interest.

A large variety of plots can be obtained. Both rotated and unrotated factor loadings can be plotted, as well as factor scores. A helpful table contains rotated factor loadings that are sorted so variables that load highly on each factor can be easily identified visually. (Loadings below a particular cutoff, specified by the user, are printed out as 0.0.) The residual correlation matrix is printed (see Section 13.6.2) as an aid to evaluating adequacy of extraction and rotation. The factor-score covariance matrix is useful for assessing internal consistency (Section 13.6.5). Scale evaluation is facilitated by specifying PRINT=CRON, which provides second-order factors as well as Cronbach's alpha for each variable, all variables combined, and factors.

13.7.3 ■ SAS System

SAS FACTOR is another highly flexible, full-featured program for FA and PCA. About the only weakness is in screening for outliers. Like BMDP4M, SAS FACTOR accepts rotated loading matrices, as long as factor correlations are provided, and, unlike BMDP4M, can analyze a partial correlation or covariance matrix (with specification of variables to partial out). There are several options for extraction, as well as orthogonal and oblique rotation. And a target pattern matrix can be specified as a criterion for oblique rotation in confirmatory FA. Additional options include specification of proportion of variance to be accounted for in determining the number of factors to retain, and the option to allow communalities to be greater than 1.0. Variables in the data set can be weighted differentially to allow the generalized least squares method of extraction.

Factor scores can be written to a data file, but information based on those scores, such as Mahalanobis distance for detection of outliers in the factor solution, is not available. A factor score covariance matrix is not available but SMCs of factors as DVs with variables as IVs are given. Two options unavailable in any of the other programs are the ability to print whole-number matrices (in which entry is multiplied by 100 and rounded to the nearest whole number) and to substitute low matrix values as missing values.

13.7.4 ■ SYSTAT System

The current SYSTAT FACTOR program is less limited than earlier versions. Wilkinson (1990) advocated the use of PCA rather than FA because of the indeterminacy problem (Section 13.6.6). However, the program now does PFA (called IPA) as well as PCA and maximum likelihood (MLA) extraction. Four common methods of orthogonal rotation are provided, as well as provision for oblique rotation. SYSTAT FACTOR can accept correlation or covariance matrices as well as raw data.

The SYSTAT FACTOR program provides scree plots and plots of factor loadings and will optionally sort the loading matrix by size of loading to aid interpretation. Additional information is available by requesting that standardized and unstandardized component scores, their coefficients, and loadings be sent to a data file. Factor scores (from PFA or MLA) cannot be saved. Residual scores (actual minus predicted z scores) also can be saved, as well as the sum of the squared residuals and a probability value for it.

13.8 ■ COMPLETE EXAMPLE OF FA

During the second year of the panel study described in Appendix B Section B.1, participants completed the Bem Sex Role Inventory (BSRI; Bem, 1974). The sample included 369 middle-class, English-speaking women between the ages of 21 and 60 who were interviewed in person.

Forty five items from the BSRI were selected for this research, where 20 items measure femininity, 20 masculinity,[10] and 5 social desirability. Respondents attribute traits (e.g., "gentle," "shy," "dominant") to themselves by assigning numbers between 1 ("never or almost never true of me") and 7 ("always or almost always true of me") to each of the items. Responses are summed to produce separate masculine and feminine scores. Masculinity and femininity are conceived as orthogonal dimensions of personality with both, one, or neither descriptive of any given individual.

Previous factor analytic work had indicated the presence of between three and five factors underlying the items of the BSRI. Investigation of the factor structure for this sample of women is a goal of this analysis.

13.8.1 ■ Evaluation of Limitations

Because the BSRI was neither developed through nor designed for factor analytic work, it meets only marginally the requirements listed in Section 13.3.1. For instance, marker variables are not included and variables from the feminine scale differ in social desirability as well as in meaning (e.g., "tender" and "gullible"), so some of these variables are likely to be complex.

Nonetheless, the widespread use of the instrument and its convergent validity with some behavioral measures make factor structure an important research question.

13.8.1.1 ■ Sample Size and Missing Data

Missing and out-of-range values are replaced with the mean response for the variable in question. That is, if a woman is missing a score on TENDER, the mean on TENDER is inserted.

Data are available initially from 369 women. With outlying cases deleted (see below), the FA is conducted on responses of 344 women. Using the guidelines of Section 13.3.2.1, over 300 cases provide a good sample size for factor analysis.

13.8.1.2 ■ Normality

Distributions of the 44 variables are examined for skewness through BMDP2D (cf. Chapter 4). Many of the variables are negatively skewed and a few are positively skewed. However, because the BSRI is already published and in use, no deletion of variables or transformations of them is performed.

---

[10] Due to clerical error, one of the masculine items, "aggression," was omitted from the questionnaires.

Because the variables fail in normality, significance tests are inappropriate. And because the direction of skewness is different for different variables, we also anticipate a weakened analysis due to lowering of correlations in **R**.

13.8.1.3 ■ Linearity

The differences in skewness for variables suggest the possibility of curvilinearity for some pairs of variables. With 44 variables, however, examination of all pairwise scatterplots (about 1000 plots) is impractical. Therefore, a spot check on a few plots is run through BMDP6D. Figure 13.4 shows the plot expected to be among the worst—between LOYAL (with strong negative skewness) and MASCULIN (with strong positive skewness). Although the plot is far from pleasing, and shows departure from linearity as well as the possibility of outliers, there is no evidence of true curvilinearity (Section 4.1.5.2). And again, transformations are viewed with disfavor considering the variable set and the goals of analysis.

13.8.1.4 ■ Outliers Among Cases

Multivariate outliers are identified using BMDPAM (cf. Chapter 4) and deleted from subsequent analysis. Using a criterion of $\alpha = .001$, 25 women are deleted.

Because of the large number of outliers and variables, a case-by-case analysis (cf. Chapter 4) is not feasible. Instead, a discriminant function analysis (BMDP7M) is used to identify variables that significantly discriminate between outliers and nonoutliers. On the last step of the discriminant analysis, four variables (RELIANT, FLATTER, TRUTHFUL, and LEADACT) discriminate outliers as a group with $p < .001$. Table 13.13 shows a portion of the BMDP7M run. Included in the table are means for outliers and nonoutliers (COMPLETE cases) on the four relevant variables. Also included is an edited part of the last step in the analysis, showing F TO REMOVE for the two groups (outliers vs. nonoutliers) on those four variables.

13.8.1.5 ■ Multicollinearity and Singularity

The original nonrotated PCA runs through BMDP4M reveal that the smallest eigenvalue (the one associated with the 44th factor) is 0.126, not dangerously close to zero. Simultaneously, it is observed that SMCs between variables where each, in turn, serves as DV for the others, do not approach 1. The largest SMC among the variables is .76 (Table 13.14). Multicollinearity and singularity are not a threat in this data set.

13.8.1.6 ■ Factorability of **R**

Correlation matrices among the 44 items produced by BMDP4M reveal numerous correlations in excess of .30 and some considerably higher. Patterns in responses to variables are therefore anticipated. (More sophisticated measures of factorability are available through SPSS or SAS FACTOR.)

13.8.1.7 ■ Outliers among Variables

SMCs among variables (Table 13.14) are also used to screen for outliers among variables, as discussed in Section 13.3.2.7. The lowest SMC among variables is .11. It is decided to retain all 44

```
/INPUT       VARIABLES ARE 45.  FORMAT IS ' (A3, 44F1.0) '.  FILE = 'WFACTOR.DAT'.
/VARIABLE    NAMES ARE SUBNO, HELPFUL, RELIANT, DEFBEL, YIELDING,
             CHEERFUL, INDPT, ATHLET, SHY, ASSERT, STRPRS,
             FORCEFUL, AFFECT, FLATTER, LOYAL, ANALYT, FEMININE,
             SYMPATHY, MOODY, SENSITIV, UNDSTAND, COMPASS,
             LEADERAB, SOOTHE, RISK, DECIDE, SELFSUFF, CONSCIEN,
             DOMINANT, MASCULINE, STAND, HAPPY, SOFTSPOK, WARM,
             TRUTHFUL, TENDER, GULLIBLE, LEADACT, CHILDLIK,
             INDIVID, FOULLANG, LOVECHIL, COMPETE, AMBITIOU,
             GENTLE.
             LABEL = SUBNO.
             USE = 2 TO 45.
/PLOT        XVAR = LOYAL.   YVAR = MASCULIN.
             SIZE = 40 , 25.

/END
```

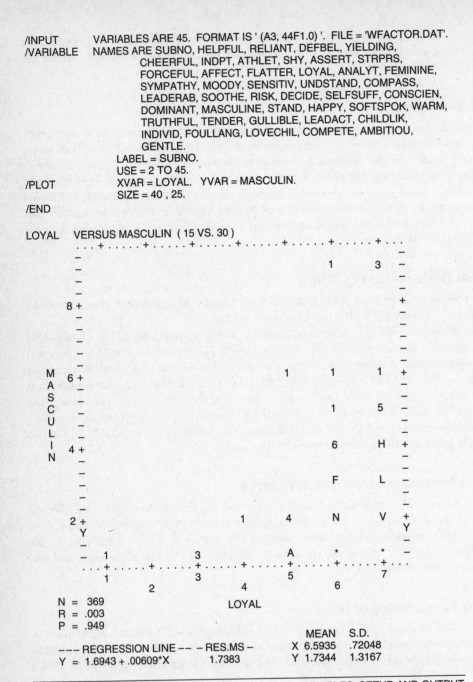

```
LOYAL   VERSUS MASCULIN  ( 15 VS. 30 )
        . . . + . . . . . + . . . . . + . . . . . + . . . . . + . . . . . + . . . . . + . . .
        -                                                             -
        -                                          1        3         -
        -                                                             -
        -                                                             -
     8 +                                                              +
        -                                                             -
        -                                                             -
        -                                                             -
        -                                                             -
        -                                                             -
   M    -                                                             -
   A 6 +                              1        1        1             +
   S    -                                                             -
   C    -                                          1        5         -
   U    -                                                             -
   L    -                                                             -
   I 4 +                              6        H             +
   N    -                                                             -
        -                                          F        L         -
        -                                                             -
        -                                                             -
     2 +                    1        4        N        V         +
     Y                                                                 Y
        -                                                             -
        -     1              3              A        *        *         -
        . . . + . . . . . + . . . . . + . . . . . + . . . . . + . . . . . + . . .
        1              3              5              7
             2              4              6
```

N = 369
R = .003   LOYAL
P = .949

                                    MEAN   S.D.
                              X 6.5935   .72048
– – – REGRESSION LINE – – – RES.MS –   Y 1.7344   1.3167
Y = 1.6943 + .00609*X   1.7383

FIGURE 13.4   SPOT CHECK FOR LINEARITY AMONG VARIABLES. SETUP AND OUTPUT
                  FROM BMDP6D.

TABLE 13.13   DESCRIPTION OF OUTLIERS USING BMDP7M.

```
/INPUT      VARIABLES ARE 45. FORMAT IS '(A3,44F1.0)'.  FILE='WFACTOR.DAT'.
/VARIABLE   NAMES ARE SUBNO, HELPFUL, RELIANT, DEFBEL, YIELDING,
              CHEERFUL, INDPT, ATHLET, SHY, ASSERT, STRPERS,
              FORCEFUL, AFFECT, FLATTER, LOYAL, ANALYT, FEMININE,
              SYMPATHY, MOODY, SENSITIV, UNDSTAND, COMPASS,
              LEADERAB, SOOTHE, RISK, DECIDE, SELFSUFF, CONSCIEN,
              DOMINANT, MASCULINE, STAND, HAPPY, SOFTSPOK, WARM,
              TRUTHFUL, TENDER, GULLIBLE, LEADACT, CHILDLIK,
              INDIVID, FOULLANG, LOVECHIL, COMPETE, AMBITIOU,
              GENTLE.
            LABEL = SUBNO.
            USE = 2 TO 46.
/TRANSFORM  DUMMY = 0.
              IF (KASE EQ 15 OR KASE EQ 20 OR KASE EQ 42 OR KASE EQ
              58 OR KASE EQ 64 OR KASE EQ 88 OR KASE EQ 89 OR KASE
              EQ 109 OR KASE EQ 126 OR KASE EQ 140 OR KASE EQ 147 OR
              KASE EQ 153 OR KASE EQ 169 OR KASE EQ 199 OR KASE EQ
              201 OR KASE EQ 203 OR KASE EQ 214 OR KASE
              EQ 220 OR KASE EQ 225 OR KASE EQ 284 OR KASE EQ 310 OR
              KASE EQ 312 OR KASE EQ 341 OR KASE EQ 343 OR KASE EQ
              359) THEN DUMMY = 1.0.
/GROUP      VAR = DUMMY.
            CODES(DUMMY) ARE 0, 1.
            NAMES(DUMMY) ARE COMPLETE, OUTLIERS.
/PRINT      NO STEP.  NO POST.  NO POINT.
/PLOT       NO CANON.
/END
```

MEANS

GROUP = VARIABLE	COMPLETE	OUTLIERS	ALL GPS.
3 RELIANT	6.01163	4.80000	5.92954
14 FLATTER	4.51453	3.28000	4.43089
17 FEMININE	5.57849	5.68000	5.58537
23 LEADERAB	4.65407	3.84000	4.59892
30 MASCULIN	1.70930	2.08000	1.73442
35 TRUTHFUL	6.42151	5.80000	6.37940
38 LEADACT	4.32558	3.00000	4.23577
41 FOULLANG	4.29651	3.36000	4.23306
.			
.			

COUNTS	344.	25.	369.

STEP NUMBER   8
VARIABLE ENTERED   41 FOULLANG

VARIABLE	F TO REMOVE	FORCE LEVEL	TOLERNCE
DF =	1	360	
3 RELIANT	20.92	1	0.86454
14 FLATTER	13.62	1	0.95690
17 FEMININE	8.69	1	0.85305
23 LEADERAB	6.79	1	0.36130
30 MASCULIN	6.18	1	0.85108
35 TRUTHFUL	12.95	1	0.97139
38 LEADACT	16.07	1	0.36292
41 FOULLANG	4.04	1	0.99155

TABLE 13.14   SETUP AND SELECTED BMDP4M OUTPUT TO ASSESS MULTICOLLINEARITY

```
/INPUT        VARIABLES ARE 45. FORMAT IS '(A3,44F1.0)'.  FILE='WFACTOR.DAT'.
/VARIABLE     NAMES ARE SUBNO, HELPFUL, RELIANT, DEFBEL, YIELDING,
                 CHEERFUL, INDPT, ATHLET, SHY, ASSERT, STRPERS,
                 FORCEFUL, AFFECT, FLATTER, LOYAL, ANALYT, FEMININE,
                 SYMPATHY, MOODY, SENSITIV, UNDSTAND, COMPASS,
                 LEADERAB, SOOTHE, RISK, DECIDE, SELFSUFF, CONSCIEN,
                 DOMINANT, MASCULINE, STAND, HAPPY, SOFTSPOK, WARM,
                 TRUTHFUL, TENDER, GULLIBLE, LEADACT, CHILDLIK,
                 INDIVID, FOULLANG, LOVECHIL, COMPETE, AMBITIOU,
                 GENTLE.
              LABEL = SUBNO.
              USE = 2 TO 45.
/TRANSFORM    DELETE = 15,20,42,58,64,88,89,109,126,140,147,153,169,
                 199,201,203,214,220,225,284,310,312,341,343,359.
/PRINT        FSCORE = 0.
/PLOT         FINAL=0.  FSCORE=0.
/END
```

```
SQUARED MULTIPLE CORRELATIONS (SMC) OF
EACH VARIABLE WITH ALL OTHER VARIABLES
--------------------------------------

    2 HELPFUL    0.37427
    3 RELIANT    0.46116
    4 DEFBEL     0.41691
    5 YIELDING   0.22995
    6 CHEERFUL   0.49202
    7 INDPT      0.53847
    8 ATHLET     0.25761
    9 SHY        0.32498
   10 ASSERT     0.53767
   11 STRPERS    0.59340
   12 FORCEFUL   0.56565
   13 AFFECT     0.55264
   14 FLATTER    0.29586
   15 LOYAL      0.39073
   16 ANALYT     0.24184
   17 FEMININE   0.35791
   18 SYMPATHY   0.45290
   19 MOODY      0.38091
   20 SENSITIV   0.48604
   21 UNDSTAND   0.61663
   22 COMPASS    0.64934
   23 LEADERAB   0.76269
   24 SOOTHE     0.43514
   25 RISK       0.42237
   26 DECIDE     0.48930
   27 SELFSUFF   0.63268
   28 CONSCIEN   0.39916
   29 DOMINANT   0.56213
   30 MASCULIN   0.31595
   31 STAND      0.57284
   32 HAPPY      0.53577
   33 SOFTSPOK   0.40179
   34 WARM       0.61523
   35 TRUTHFUL   0.35627
   36 TENDER     0.60454
   37 GULLIBLE   0.29683
   38 LEADACT    0.76136
   39 CHILDLIK   0.29604
   40 INDIVID    0.37905
   41 FOULLANG   0.11346
   42 LOVECHIL   0.28419
   43 COMPETE    0.46468
   44 AMBITIOU   0.45870
   45 GENTLE     0.57954
```

variables although many are largely unrelated to others in the set. (In fact, 45% of the 44 variables in the analysis have loadings too low on all the factors to assist interpretation in the final solution.)

13.8.1.8 ■ Outlying Cases among the Factors

The ESTIMATED FACTOR SCORES table produced by the final PFA run is used to find cases that are not well fit by the solution. The second column of CHISQ/DF (see Table 13.15) gives Mahalanobis distance of each case from the centroid of the factor scores evaluated as χ^2/df. Cases with large values are deviant cases in the space of the factor solution. Critical χ^2 (cf. Table C.4) at the chosen probability level is divided by number of factors as degrees of freedom to provide the cutoff value.

At $\alpha = .001$, the cutoff for four factors is 4.62 (18.47/4). Cases with scores that produced values larger than this in the column are outliers. No cases produced scores that made them outliers in the space of the solution.

13.8.2 ■ Principal Factors Extraction with Varimax Rotation

Principal components extraction with varimax rotation through BMDP4M is used in an initial run, to evaluate assumptions and limitations as discussed in Section 13.8.1 and to estimate the number of factors from eigenvalues. The first 13 eigenvalues are shown in Table 13.16. The maximum number of factors (eigenvalues larger than 1) is 11. However, retention of 11 factors seems unreasonable so sharp breaks in size of eigenvalues are sought using the scree test (Section 13.6.2).

Eigenvalues (VARIANCE EXPLAINED) for the first four factors are all larger than two and, after the sixth factor, changes in successive eigenvalues are small. This is taken as evidence that

---

TABLE 13.15 SELECTED BMDP4M PFA OUTPUT USED TO EVALUATE OUTLYING CASES IN SOLUTION. SETUP SHOWN IN TABLE 13.17.

ESTIMATED FACTOR SCORES AND MAHALANOBIS DISTANCES (CHISQUARE S) FROM
EACH CASE TO THE CENTROID OF ALL CASES FOR THE ORIGINAL DATA
(44 D.F.) FACTOR SCORES (4 D.F.) AND THEIR DIFFERENCE (40 D.F.).
EACH CHI-SQUARE HAS BEEN DIVIDED BY ITS DEGREES OF FREEDOM.
A MINUS SIGN AFTER THE CASE NUMBER DENOTES A CASE WITH NONPOSITIVE WEIGHT.

CASE LABEL	NO.	CHISQ/DF 44	CHISQ/DF 4	CHISQ/DF 40	FACTOR1	FACTOR2	FACTOR3	FACTOR4
1	1	1.715	1.004	1.786	0.248	0.774	1.534	0.704
2	2	1.317	1.101	1.338	-1.066	-1.227	-1.019	-0.636
3	3	0.758	0.689	0.765	-1.096	-0.201	-0.033	0.987
4	4	0.449	0.224	0.472	-0.602	-0.074	0.043	0.588
5	5	0.613	1.348	0.540	1.724	0.856	0.444	-0.866
7	6	0.980	1.254	0.953	-1.194	-0.218	0.119	1.563
8	7	1.091	1.800	1.020	-1.893	0.411	-0.660	-1.459
9	8	0.464	0.186	0.492	0.575	0.391	0.085	0.428
10	9	1.550	1.411	1.564	0.922	-0.207	-1.897	-0.401
11	10	1.137	1.037	1.147	0.156	0.005	0.880	-1.545
13	11	1.247	0.873	1.284	-0.423	1.481	-0.523	-0.326
14	12	1.853	4.445	1.594	-1.253	1.576	0.074	-3.250
15	13	0.596	1.668	0.489	1.800	0.896	1.323	0.591
23	14	1.050	0.770	1.078	0.380	-0.024	-0.952	-1.171
26	16	0.777	0.174	0.837	-0.287	-0.385	0.183	0.532
27	17	0.652	0.726	0.644	1.021	0.157	0.489	1.115

TABLE 13.16   EIGENVALUES AND PROPORTIONS OF VARIANCE FOR FIRST 13 COMPONENTS.   BMDP4M PCA
                   SETUP AND SELECTED OUTPUT

```
/INPUT       VARIABLES ARE 45. FORMAT IS '(A3,44F1.0)'. FILE='WFACTOR.DAT'.
/VARIABLE    NAMES ARE SUBNO, HELPFUL, RELIANT, DEFBEL, YIELDING,
                CHEERFUL, INDPT, ATHLET, SHY, ASSERT, STRPERS,
                FORCEFUL, AFFECT, FLATTER, LOYAL, ANALYT, FEMININE,
                SYMPATHY, MOODY, SENSITIV, UNDSTAND, COMPASS,
                LEADERAB, SOOTHE, RISK, DECIDE, SELFSUFF, CONSCIEN,
                DOMINANT, MASCULINE, STAND, HAPPY, SOFTSPOK, WARM,
                TRUTHFUL, TENDER, GULLIBLE, LEADACT, CHILDLIK,
                INDIVID, FOULLANG, LOVECHIL, COMPETE, AMBITIOU,
                GENTLE.
             LABEL = SUBNO.
             USE = 2 TO 45.
/TRANSFORM   DELETE = 15,20,42,58,64,88,89,109,126,140,147,153,169,
                199,201,203,214,220,225,284,310,312,341,343,359.
/PRINT       FSCORE = 0.
/PLOT        FINAL=0.  FSCORE=0.
/END
```

FACTOR	VARIANCE EXPLAINED	CUMULATIVE PROPORTION OF VARIANCE IN DATA SPACE	IN FACTOR SPACE	CARMINES THETA
1	8.1940	0.1862	0.3057	0.8984
2	5.1535	0.3034	0.4980	
3	2.5905	0.3622	0.5946	
4	2.0730	0.4093	0.6719	
5	1.6476	0.4468	0.7334	
6	1.4152	0.4789	0.7862	
7	1.2907	0.5083	0.8344	
8	1.2212	0.5360	0.8799	
9	1.1095	0.5613	0.9213	
10	1.0776	0.5857	0.9615	
11	1.0317	0.6092	1.0000	
12	0.9513	0.6308		
13	0.9417	0.6522		

there are probably between 4 and 6 factors. These results are consistent with earlier research suggesting 3 to 5 factors on the BSRI.

A common factor extraction model that removes unique and error variability from each variable is used for the next several runs and the final solution. Principal factors is chosen from among methods for common factor extraction. Several PFA runs specifying 4 to 6 factors are planned to find the optimal number of factors.

The trial PFA run with 5 factors has 5 eigenvalues larger than 1 among unrotated factors. But after rotation, the eigenvalue for the fifth factor is below 1 and it has no loadings larger than .45, the criterion for interpretation chosen for this research. The solution with four factors, on the other hand, meets the goals of interpretability, so four factors are chosen for follow-up runs. The first six eigenvalues from the four-factor solution are shown in Table 13.17.

As another test of adequacy of extraction and number of factors, it is noted (but not shown) that most values in the residual correlation matrix for the four-factor orthogonal solution are near zero. This is further confirmation that a reasonable number of factors is 4.

The decision between oblique and orthogonal rotation is made by requesting principal factor extraction with oblique rotation of four factors. Direct quartimin (DQUART) is the oblique method employed. The highest correlation (.333) is between factors 2 and 3 (see Table 13.18). The generally oblong shape of the scatterplot of factor scores between these two factors (see Figure 13.5) confirms the correlation. This level of correlation can be considered borderline between accepting an

TABLE 13.17	EIGENVALUES AND PROPORTIONS OF VARIANCE FOR FIRST SIX FACTORS. PRINCIPAL FACTORS EXTRACTION AND VARIMAX ROTATION. BMDP4M SETUP AND SELECTED OUTPUT

```
/INPUT       VARIABLES ARE 45. FORMAT IS '(A3,44F1.0)'.  FILE='WFACTOR.DAT'.
/VARIABLE    NAMES ARE SUBNO, HELPFUL, RELIANT, DEFBEL, YIELDING,
                CHEERFUL, INDPT, ATHLET, SHY, ASSERT, STRPERS,
                FORCEFUL, AFFECT, FLATTER, LOYAL, ANALYT, FEMININE,
                SYMPATHY, MOODY, SENSITIV, UNDSTAND, COMPASS,
                LEADERAB, SOOTHE, RISK, DECIDE, SELFSUFF, CONSCIEN,
                DOMINANT, MASCULINE, STAND, HAPPY, SOFTSPOK, WARM,
                TRUTHFUL, TENDER, GULLIBLE, LEADACT, CHILDLIK,
                INDIVID, FOULLANG, LOVECHIL, COMPETE, AMBITIOU,
                GENTLE.
             LABEL = SUBNO.
             USE = 2 TO 45.
/TRANSFORM   DELETE = 15,20,42,58,64,88,89,109,126,140,147,153,169,
                199,201,203,214,220,225,284,310,312,341,343,359.
/FACTOR      METHOD = PFA.  NUMBER = 4.
/PRINT       NO CORR.  LOLEV = .45.  RESI.
/END
```

FACTOR	VARIANCE EXPLAINED	CUMULATIVE PROPORTION OF VARIANCE		CARMINES THETA
		IN DATA SPACE	IN FACTOR SPACE	
1	7.6193	0.3669	0.4871	0.9599
2	4.6011	0.5885	0.7812	
3	1.9576	0.6828	0.9064	
4	1.4646	0.7533	1.0000	
5	0.9360	0.7984		
6	0.7715	0.8355		

TABLE 13.18	SETUP AND SELECTED BMDP4M PFA OUTPUT OF CORRELATIONS AMONG FACTORS FOLLOWING DIRECT QUARTIMIN ROTATION.

```
/INPUT       VARIABLES ARE 45. FORMAT IS '(A3,44F1.0)'.  FILE='WFACTOR.DAT'.
/VARIABLE    NAMES ARE SUBNO, HELPFUL, RELIANT, DEFBEL, YIELDING,
                CHEERFUL, INDPT, ATHLET, SHY, ASSERT, STRPERS,
                FORCEFUL, AFFECT, FLATTER, LOYAL, ANALYT, FEMININE,
                SYMPATHY, MOODY, SENSITIV, UNDSTAND, COMPASS,
                LEADERAB, SOOTHE, RISK, DECIDE, SELFSUFF, CONSCIEN,
                DOMINANT, MASCULINE, STAND, HAPPY, SOFTSPOK, WARM,
                TRUTHFUL, TENDER, GULLIBLE, LEADACT, CHILDLIK,
                INDIVID, FOULLANG, LOVECHIL, COMPETE, AMBITIOU,
                GENTLE.
             LABEL = SUBNO.
             USE = 2 TO 45.
/TRANSFORM   DELETE = 15,20,42,58,64,88,89,109,126,140,147,153,169,
                199,201,203,214,220,225,284,310,312,341,343,359.
/FACTOR      METHOD = PFA.  NUMBER = 4.
/ROTATE      METHOD = DQUART.
/PRINT       NO CORR.  LOLEV = .45.  RESI.
/PLOT        FINAL = 0.
/END
```

FACTOR CORRELATIONS FOR ROTATED FACTORS

		FACTOR1 1	FACTOR2 2	FACTOR3 3	FACTOR4 4
FACTOR1	1	1.000			
FACTOR2	2	0.173	1.000		
FACTOR3	3	0.089	0.333	1.000	
FACTOR4	4	0.113	0.013	0.034	1.000

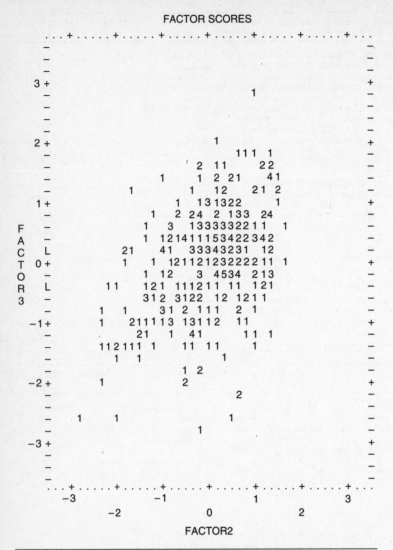

FIGURE 13.5 SCATTERPLOT OF FACTOR SCORES WITH PAIRS OF FAC-
TORS (2 AND 3) AS AXES FOLLOWING OBLIQUE ROTA-
TION. (SETUP IN TABLE 13.18).

orthogonal solution versus dealing with the complexities of interpreting an oblique solution. The
simpler, orthogonal, solution is chosen.

The solution that is evaluated, interpreted, and reported is the run with principal factors extraction, varimax rotation, and 4 factors. In other words, after "trying out" oblique rotation, the decision is made to interpret the earlier run with orthogonal rotation. Setup for this run is in Table 13.17.

Communalities are inspected to see if the variables are well defined by the solution.
Communalities indicate the percent of variance in a variable that overlaps variance in the factors.

TABLE 13.19   COMMUNALITY VALUES (FOUR FACTORS).   SELECTED OUTPUT FROM BMDP4M PFA.   SEE TABLE 13.17 FOR SETUP.

```
COMMUNALITIES OBTAINED FROM  4 FACTORS AFTER    6 ITERATIONS.
-------------------------------------------------------------

THE COMMUNALITY OF A VARIABLE IS ITS SQUARED MULTIPLE
CORRELATION WITH THE FACTORS.

     2 HELPFUL     0.2825
     3 RELIANT     0.3979
     4 DEFBEL      0.2490
     5 YIELDING    0.1511
     6 CHEERFUL    0.3598
     7 INDPT       0.4541
     8 ATHLET      0.1837
     9 SHY         0.1568
    10 ASSERT      0.4403
    11 STRPERS     0.5074
    12 FORCEFUL    0.4635
    13 AFFECT      0.4796
    14 FLATTER     0.2002
    15 LOYAL       0.2938
    16 ANALYT      0.1514
    17 FEMININE    0.1562
    18 SYMPATHY    0.4405
    19 MOODY       0.2713
    20 SENSITIV    0.4440
    21 UNDSTAND    0.5813
    22 COMPASS     0.6846
    23 LEADERAB    0.5771
    24 SOOTHE      0.3877
    25 RISK        0.2763
    26 DECIDE      0.3765
    27 SELFSUFF    0.6365
    28 CONSCIEN    0.3502
    29 DOMINANT    0.5400
    30 MASCULIN    0.1894
    31 STAND       0.4385
    32 HAPPY       0.4421
    33 SOFTSPOK    0.2775
    34 WARM        0.6316
    35 TRUTHFUL    0.1683
    36 TENDER      0.5346
    37 GULLIBLE    0.2214
    38 LEADACT     0.5407
    39 CHILDLIK    0.1920
    40 INDIVID     0.2362
    41 FOULLANG    0.0248
    42 LOVECHIL    0.1366
    43 COMPETE     0.3376
    44 AMBITIOU    0.2645
    45 GENTLE      0.5139
```

As seen in Table 13.19, communality values for a number of variables are quite low (e.g., FOUL-LANG). Ten of the variables have communality values lower than .2 indicating considerable heterogeneity among the variables. It should be recalled, however, that factorial purity was not a consideration in development of the BSRI.

Adequacy of rotation (Section 13.6.3) is assessed, in part, by scatterplots with pairs of rotated factors as axes and variables as points, as shown in Figure 13.6. Ideally, variable points are at the

ROTATED FACTOR LOADINGS

VARIABLES ARE DENOTED BY 1,..., 9, A,..., Z
OVERLAPS ARE DENOTED BY AN ASTERISK.

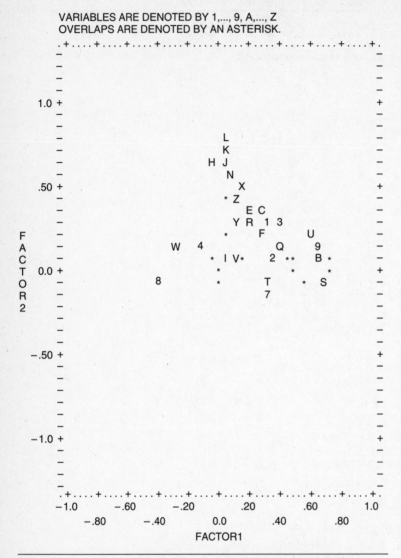

FIGURE 13.6   SELECTED BMDP4M PFA OUTPUT SHOWING SCATTER-
              PLOTS OF VARIABLE LOADINGS WITH FACTORS AS AXES
              FOR ALL FOUR FACTORS. (SETUP IN TABLE 12.19).

origin (the unmarked middle of figures) or in clusters at the ends of factor axes. Scatterplots between factor 1 and factor 2, between factor 2 and factor 4, and between factor 3 and factor 4 seem reasonably clear. The scatterplots between other pairs of factors show evidence of correlation among factors as found during oblique rotation. Otherwise, the scatterplots are disappointing but consistent with the other evidence of heterogeneity among the variables in the BSRI.

ROTATED FACTOR LOADINGS

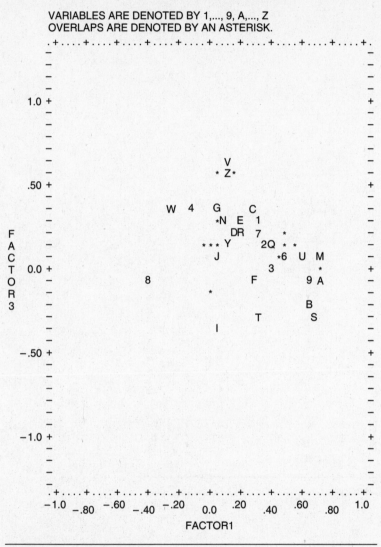

FIGURE 13.6 *(CONTINUED)*

Simplicity of structure (Section 13.6.3) in factor loadings following orthogonal rotation is assessed from the table of ROTATED FACTOR LOADINGS (see Table 13.20). In each column there are a few high and many low correlations between variables and factors. There are also numerous moderate loadings so several variables will be complex (load on more than one factor) unless a fairly high cutoff for interpreting loadings is established. Complexity of variables (Section 13.6.5) is assessed by examining loadings for a variable across factors. With a loading cut of .45 only two variables, WARM and INDPT load on more than one factor.

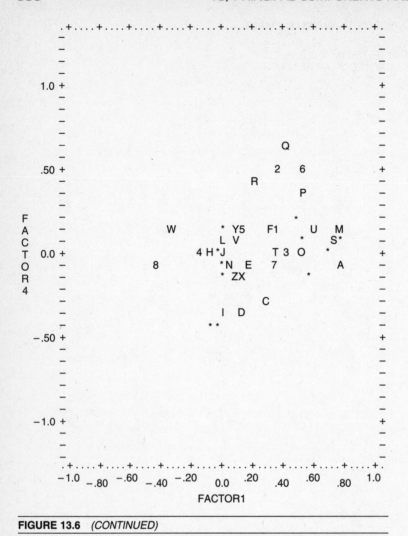

FIGURE 13.6   *(CONTINUED)*

The importance of each factor (Sections 13.4 and 13.6.4) is assessed by the percent of variance and covariance it represents. SSLs, called VP at the bottom of each column of loadings shown in Table 13.20, are used in the calculations. It is important to use SSLs from rotated factors, because the variance is redistributed during rotation. Proportion of variance for a factor is SSL for the factor divided by number of variables. Proportion of covariance is SSL divided by sum of SSLs. Results, converted to percent, are shown in Table 13.21. Each of the factors accounts for between 4 and 16% of the variance in the set of variables, not an outstanding performance. Only the first factor accounts for substantial covariance.

Internal consistency of the factors (Section 13.6.4) is assessed through SMCs in the diagonal of the FACTOR SCORE COVARIANCE matrix where factors serve as DVs with variables as IVs.

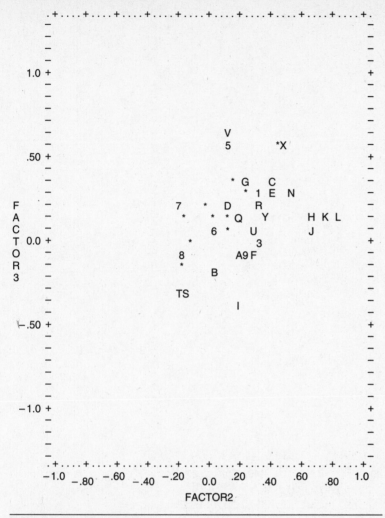

FIGURE 13.6 *(CONTINUED)*

Factors that are well defined by the variables have high SMCs, whereas poorly defined factors have low SMCs. As can be seen in Table 13.22, all factors are internally consistent. (The off-diagonal elements in these matrices are correlations among factor scores. Although uniformly low, the values are not zero. As discussed in Section 13.6.6, low correlations among scores on factors are often obtained even with orthogonal rotation.)

Factors are interpreted through their factor loadings (Section 13.6.5). The process is facilitated with BMDP4M by requesting output of SORTED ROTATED FACTOR LOADINGS through the PRINT paragraph (see setup, Table 13.17). In the sorted, rotated loading matrix, variables are grouped by factors and reordered by size of loading. An example is shown in Table 13.23.

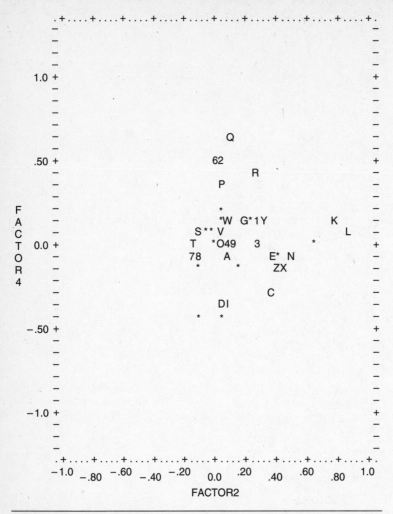

FIGURE 13.6 *(CONTINUED)*

Although a default cut of .25 is used for inclusion of variables in factors by BMDP4M, it is decided to use a loading of .45 (20% variance overlap between variable and factor). With the use of the .45 cut, Table 13.24 is generated to further assist interpretation. In more informal presentations of factor analytic results, this table might be reported instead of Table 13.25. Factors are put in columns and variables with the largest loadings are put on top. In interpreting a factor, items near the top of the columns are given somewhat greater weight. Variable names are written out in full

ROTATED FACTOR LOADINGS

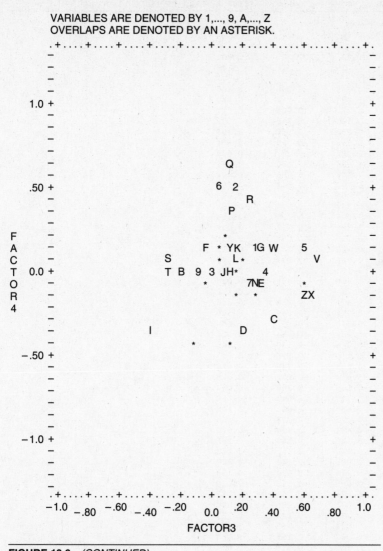

FIGURE 13.6   *(CONTINUED)*

detail and labels for the factors (e.g., Dominance) are suggested at the top of each column. Table 13.25 shows a more formal summary table of factor loadings, including communalities as well as percents of variance and covariance.

Table 13.26 provides a checklist for FA. A Results section in journal format follows for the data analyzed in this section.

TABLE 13.20   FACTOR LOADINGS FOR PRINCIPAL FACTORS EXTRACTION AND VARIMAX ROTATION OF FOUR
                    FACTORS. SELECTED BMDP4M OUTPUT. SETUP APPEARS IN TABLE 13.17

ROTATED FACTOR LOADINGS (PATTERN)

		FACTOR1 1	FACTOR2 2	FACTOR3 3	FACTOR4 4
HELPFUL	2	0.311	0.269	0.296	0.160
RELIANT	3	0.365	0.083	0.167	0.480
DEFBEL	4	0.413	0.280	0.001	0.017
YIELDING	5	-0.139	0.113	0.345	-0.009
CHEERFUL	6	0.167	0.088	0.559	0.109
INDPT	7	0.466	0.043	0.036	0.484
ATHLET	8	0.324	-0.122	0.247	-0.054
SHY	9	-0.383	-0.074	-0.050	-0.042
ASSERT	10	0.643	0.142	-0.084	0.014
STRPERS	11	0.701	0.102	-0.052	-0.054
FORCEFUL	12	0.645	0.055	-0.210	-0.013
AFFECT	13	0.300	0.392	0.391	-0.289
FLATTER	14	0.165	0.095	0.215	-0.343
LOYAL	15	0.200	0.388	0.319	-0.038
ANALYT	16	0.277	0.233	-0.055	0.131
FEMININE	17	0.057	0.189	0.324	0.111
SYMPATHY	18	-0.042	0.649	0.129	-0.024
MOODY	19	0.030	0.103	-0.374	-0.346
SENSITIV	20	0.056	0.660	0.069	0.028
UNDSTAND	21	0.024	0.731	0.177	0.124
COMPASS	22	0.052	0.811	0.151	0.036
LEADERAB	23	0.739	0.086	0.046	0.146
SOOTHE	24	0.070	0.540	0.297	-0.060
RISK	25	0.496	0.082	0.154	0.016
DECIDE	26	0.483	0.085	0.130	0.345
SELFSUFF	27	0.418	0.110	0.136	0.657
CONSCIEN	28	0.203	0.285	0.235	0.416
DOMINANT	29	0.675	-0.064	-0.281	0.036
MASCULIN	30	0.308	-0.105	-0.287	-0.032
STAND	31	0.593	0.244	0.043	0.162
HAPPY	32	0.122	0.069	0.641	0.106
SOFTSPOK	33	-0.290	0.129	0.388	0.160
WARM	34	0.149	0.483	0.594	-0.150
TRUTHFUL	35	0.139	0.320	0.139	0.166
TENDER	36	0.107	0.446	0.551	-0.142
GULLIBLE	37	-0.041	0.085	0.110	-0.448
LEADACT	38	0.727	-0.024	0.020	0.106
CHILDLIK	39	0.005	-0.068	-0.111	-0.418
INDIVID	40	0.435	0.094	0.069	0.182
FOULLANG	41	-0.017	0.032	0.150	0.030
LOVECHIL	42	0.024	0.201	0.282	-0.127
COMPETE	43	0.541	-0.083	0.163	-0.109
AMBITIOU	44	0.466	0.000	0.199	0.087
GENTLE	45	0.023	0.447	0.554	-0.084
	VP	6.014	4.002	3.400	2.227

THE VP IS THE VARIANCE EXPLAINED BY THE FACTOR.
IT IS COMPUTED AS THE SUM OF SQUARES FOR THE
ELEMENTS OF THE FACTOR'S COLUMN IN THE FACTOR
LOADING MATRIX.

TABLE 13.21 PERCENTS OF VARIANCE AND COVARIANCE
EXPLAINED BY EACH OF THE ROTATED
ORTHOGONAL FACTORS

	Factors			
	1	2	3	4
SSL	6.01	4.00	3.40	2.23
Percent of variance	13.66	9.09	7.73	5.07
Percent of covariance	38.42	25.57	21.74	14.26

TABLE 13.22   SMCs FOR FACTORS WITH VARIABLES AS IVs, AND CORRELATIONS AMONG FACTOR SCORES.
SELECTED OUTPUT FROM BMDP4M PFA WITH ORTHOGONAL (VARIMAX) ROTATION. SETUP IN
TABLE 13.17

```
FACTOR SCORE COVARIANCE (COMPUTED FROM FACTOR
STRUCTURE AND FACTOR SCORE COEFFICIENTS)
---------------------------------------------

THE DIAGONAL OF THE MATRIX BELOW CONTAINS THE SQUARED
MULTIPLE CORRELATIONS OF EACH FACTOR WITH THE VARIABLES.


              FACTOR1     FACTOR2     FACTOR3     FACTOR4
                 1           2           3           4

FACTOR1     1    0.903
FACTOR2     2    0.021       0.849
FACTOR3     3    0.001       0.087       0.805
FACTOR4     4    0.044      -0.006       0.010       0.779
```

SORTED ROTATED FACTOR LOADINGS (PATTERN)

		FACTOR1	FACTOR2	FACTOR3	FACTOR4
		1	2	3	4
LEADERAB	23	0.739	0.000	0.000	0.000
LEADACT	38	0.727	0.000	0.000	0.000
STRPERS	11	0.701	0.000	0.000	0.000
DOMINANT	29	0.675	0.000	0.000	0.000
FORCEFUL	12	0.645	0.000	0.000	0.000
ASSERT	10	0.643	0.000	0.000	0.000
STAND	31	0.593	0.000	0.000	0.000
COMPETE	43	0.541	0.000	0.000	0.000
COMPASS	22	0.000	0.811	0.000	0.000
UNDSTAND	21	0.000	0.731	0.000	0.000
SENSITIV	20	0.000	0.660	0.000	0.000
SYMPATHY	18	0.000	0.649	0.000	0.000
SOOTHE	24	0.000	0.540	0.000	0.000
HAPPY	32	0.000	0.000	0.641	0.000
WARM	34	0.000	0.483	0.594	0.000
CHEERFUL	6	0.000	0.000	0.559	0.000
GENTLE	45	0.000	0.000	0.554	0.000
TENDER	36	0.000	0.000	0.551	0.000
SELFSUFF	27	0.000	0.000	0.000	0.657
INDPT	7	0.466	0.000	0.000	0.484
RELIANT	3	0.000	0.000	0.000	0.480
GULLIBLE	37	0.000	0.000	0.000	0.000
CHILDLIK	39	0.000	0.000	0.000	0.000
CONSCIEN	28	0.000	0.000	0.000	0.000
MOODY	19	0.000	0.000	0.000	0.000
DECIDE	26	0.483	0.000	0.000	0.000
FLATTER	14	0.000	0.000	0.000	0.000
AFFECT	13	0.000	0.000	0.000	0.000
INDIVID	40	0.000	0.000	0.000	0.000
TRUTHFUL	35	0.000	0.000	0.000	0.000
HELPFUL	2	0.000	0.000	0.000	0.000
SOFTSPOK	33	0.000	0.000	0.000	0.000
ANALYT	16	0.000	0.000	0.000	0.000
LOVECHIL	42	0.000	0.000	0.000	0.000
FEMININE	17	0.000	0.000	0.000	0.000
AMBITIOU	44	0.466	0.000	0.000	0.000
ATHLET	8	0.000	0.000	0.000	0.000
SHY	9	0.000	0.000	0.000	0.000
LOYAL	15	0.000	0.000	0.000	0.000
MASCULIN	30	0.000	0.000	0.000	0.000
FOULLANG	41	0.000	0.000	0.000	0.000
DEFBEL	4	0.000	0.000	0.000	0.000
RISK	25	0.496	0.000	0.000	0.000
YIELDING	5	0.000	0.000	0.000	0.000
	VP	6.014	4.002	3.400	2.227

TABLE 13.24 ORDER (BY SIZE OF LOADINGS) IN WHICH VARIABLES CONTRIBUTE TO FACTORS

Factor 1: *Dominance*	Factor 2: *Empathy*	Factor 3: *Positive Affect*	Factor 4: *Independence*
Has leadership abilities	Compassionate	Happy	Self-sufficient
Acts as a leader	Understanding	Warm	Independent
Strong personality	Sensitive to needs of others	Cheerful	Self-reliant
Dominant	Sympathetic	Gentle	
Forceful	Eager to soothe hurt feelings	Tender	
Assertive	Warm		
Willing to take a stand			
Competitive			
Willing to take risks			
Makes decisions easily			
Independent			
Ambitious			

Note: Variables with higher loadings on the factor are nearer the top of the columns. Proposed labels are in italics.

TABLE 13.25 FACTOR LOADINGS, COMMUNALITIES (h^2), PERCENTS OF VARIANCE AND COVARIANCE FOR PRINCIPAL FACTORS EXTRACTION AND VARIMAX ROTATION ON BSRI ITEMS.

Item	$F_1{}^a$	F_2	F_3	F_4	h^2
Leadership ability	.74	.00	.00	.00	.58
Acts as leader	.73	.00	.00	.00	.54
Strong personality	.70	.00	.00	.00	.51
Dominant	.68	.00	.00	.00	.54
Forceful	.65	.00	.00	.00	.46
Assertive	.64	.00	.00	.00	.44
Takes stand	.59	.00	.00	.00	.44
Competitive	.54	.00	.00	.00	.34
Takes risks	.50	.00	.00	.00	.28
Makes decisions	.48	.00	.00	.00	.38
Independent	.47	.00	.00	.48	.25
Ambitious	.47	.00	.00	.00	.26
Compassionate	.00	.81	.00	.00	.68
Understanding	.00	.73	.00	.00	.58
Sensitive	.00	.66	.00	.00	.44
Sympathetic	.00	.65	.00	.00	.44
Eager to soothe hurt feelings	.00	.54	.00	.00	.39
Warm	.00	.48	.72	.00	.63
Happy	.00	.00	.64	.00	.44
Cheerful	.00	.00	.56	.00	.36
Gentle	.00	.00	.55	.00	.51
Tender	.00	.00	.55	.00	.53
Self-sufficient	.00	.00	.00	.66	.64
Self-reliant	.00	.00	.00	.48	.40
Affectionate	.00	.00	.00	.00	.48
Conscientious	.00	.00	.00	.00	.35
Defends beliefs	.00	.00	.00	.00	.25
Masculine	.00	.00	.00	.00	.19
Truthful	.00	.00	.00	.00	.17
Feminine	.00	.00	.00	.00	.16
Helpful	.00	.00	.00	.00	.28
Individualistic	.00	.00	.00	.00	.24
Shy	.00	.00	.00	.00	.16
Moody	.00	.00	.00	.00	.27
Percent of variance	13.66	9.09	7.73	5.06	
Percent of covariance	38.42	25.57	21.74	14.26	

[a] Factor labels:

F_1 Dominance

F_2 Empathy

F_3 Positive Affect

F_4 Independence

TABLE 13.26 CHECKLIST FOR FACTOR ANALYSIS

1. Limitations
   a. Outliers among cases
   b. Sample size
   c. Factorability of **R**
   d. Normality and linearity of variables
   e. Multicollinearity and singularity
   f. Outliers among variables
2. Major analyses
   a. Number of factors
   b. Nature of factors
   c. Type of rotation
   d. Importance of factors
3. Additional analyses
   a. Factor scores
   b. Distinguishability and simplicity of factors
   c. Complexity of variables
   d. Internal consistency of factors
   e. Outlying cases among the factors

Results

Principal factors extraction with varimax rotation was performed through BMDP4M on 44 items from the BSRI for a sample of 344 women. Principal components extraction was used prior to principal factors extraction to estimate number of factors, presence of outliers, absence of multicollinearity, and factorability of the correlation matrices. With an α =.001 cutoff level, 25 of 369 women produced scores that identified them as outliers; these cases were deleted from principal factors extraction.[11]

Four factors were extracted. As indicated by SMCs, all factors were internally consistent and well defined by the variables; the lowest of the SMCs for factors from variables was .78. [*Information on SMCs is from Table 13.22.*] The reverse was not true, however; variables were, by and large, not well-defined by this factor solution. Communality values, as seen in Table 13.25, tended to be low. With a cut of .45 for inclusion of a variable in interpretation of a factor, 20 of 44 variables did not load on any factor. Failure of numerous variables to load on a factor reflects heterogeneity of items on the BSRI. However, only two of the variables in the solution, "warm" and "independent," were complex.

When oblique rotation was requested, factors interpreted as Empathy and Positive Affect correlated .33. However, because the correlation

---

[11] Outliers were compared as a group to nonoutliers through discriminant function analysis. As a group, at p < .01, the 25 women were more feminine, less self-reliant, more easily flattered, less truthful, and less likely to act as leaders than women who were not outliers.

was modest and limited to one pair of factors, and because remaining
correlations were low, orthogonal rotation was chosen.

    Loadings of variables on factors, communalities, and percents of
variance and covariance are shown in Table 13.25. Variables are ordered
and grouped by size of loading to facilitate interpretation. Loadings
under .45 (20% of variance) are replaced by zeros. Interpretive labels
are suggested for each factor in a footnote.

13.9 ■ SOME EXAMPLES FROM THE LITERATURE

McLeod, Brown, and Becker (1977) used principal components extraction with varimax rotation to
collapse responses to 17 items into 4 factors for use in a study of the relationship between perceived
locus of blame for the Watergate scandal and several different indicators of political orientation and
behaviors. Perceived locus of blame for Watergate among the 617 respondents was composed of 4
factors accounting for a total of 48.7% of the variance and labeled the Media, the Regime
(Republican and Democratic Parties, Congress), Nixon and Entourage, and the System (party politics, the political system, the economic system). Factor scores were generated for 181 respondents
who were part of a panel study; scores on each factor were correlated with such indicators of political behavior as political trust, party affiliation, campaign participation, and voting direction.
Results were somewhat different for those who had or had not supported Nixon in 1972 and indicated that deleterious effects of Watergate were short-term and should be considered in the context
of a more general and longer-term decline in political trust and participation.

    Menozzi, Piazza, and Cavalli-Sforza (1978) studied the relationship between distributions of
genetic characteristics and the expansion of early farming in Europe. The overall goal of analysis
was to determine, if possible, if early farmers themselves had migrated or if farming techniques had
been adopted by widening circles of local hunters and gatherers. PCA was used to collapse data
from 10 genetic loci and 38 alleles (for blood group substances and the like) for 400 map locations
into three components. Gradients (or clines) were then formed of the densities of each component
over the geographical area of Europe and the Mid-East. The first component showed greatest density in the southeast (the Mid-East) and least in the northwest (Northern Europe and Scandinavia).
This map was in remarkable agreement with a map showing archaeological evidence of spread of
early farming; and both argue, but not exclusively, for a spread of farmers rather than a spread of
only their technology. The second and third components show changes in genetic densities corresponding to east-west gradients and northeast to southwest gradients, respectively.

    In a study of willingness to donate human body parts (Pessemier, Bemmoar and Hanssens,
1977), FA was used to collapse 10 items measuring willingness to donate into three factors: donation of blood, skin, marrow; donations upon death; and donation of kidney. FA was performed
after a Guttman scale analysis revealed that willingness to donate was not unidimensional.
Responses to items composing factors were summed to yield three factor scores for each respondent that served as DVs in separate analyses of variance.[12] Middle-aged female respondents with
above-average incomes were most likely to donate regenerative tissue or parts after death.

---

[12] The authors might well have used MANOVA instead of ANOVA.

To investigate the relationship between certain attitudinal constructs and willingness to donate, an attempt was made to reduce 36 attitudinal questions to a smaller set of underlying variables using FA. When the determinant of the covariance matrix was near-zero (indicating extreme multicollinearity), FA was abandoned in favor of inspection of partial correlations between the 36 variables, guided by face validity of the items. Of the 36 items, 25 were retained, composing 9 attitudinal constructs for which responses were summed. A high-medium-low split was used on each attitudinal construct to generate groups for IVs in ANOVA, with the willingness-to-donate variables used as DVs in three separate analyses of each IV. Significant differences were found between most attitudinal variables (e.g., liberalism, interest in physical attractiveness) and one or more of the DVs. However, because of the number of analyses performed among both IVs and DVs that are likely to be correlated, it is difficult to have much faith in any one of the significant differences.

A study by Wong and Allen (1976) combined categories of drugs with three major cognitive dimensions in FAs of semantic differentials to study the relationships between 18 drugs and various attitudes. The three major dimensions were "pleasantness," "strength," and "dangerousness." Four hundred and fifty respondents evaluated each of 18 drugs on each dimension using a Likert-type scale with 10 gradations (from "very safe" to "very dangerous," for instance). For the set of 18 drugs, principal factors extraction of 4 factors followed by rotation were performed separately for each attitudinal dimension plus usage. Only drugs loading .60 or higher with the factors were retained. Usage and "pleasantness" tended to have similar factor structures, as did "dangerousness" and "strength." Other analyses revealed that usage is inversely related to both perceived dangerousness and strength, but directly related to pleasantness. Alcoholic beverages and marijuana tended to show deviant patterns from the rest.

The meaning of money was investigated in a study by Wernimont and Fitzpatrick (1972). Principal components extraction and varimax rotation were used to identify 7 components underlying patterns of responses of 533 subjects to 40 adjective pairs, each rated on a 7-point scale. The 7 components were labeled "shameful failure," "social acceptability," "pooh-pooh attitude," "moral evil," "comfortable security," "social unacceptability," and "conservative business attitudes." Following PCA for the group as a whole, subjects were subdivided into 11 groups on the basis of occupation, which served as IVs in one-way ANOVAs, with component scores as DVs. Separate ANOVAs were performed for each component. The groups differed significantly in their component scores, with a break usually found between the meaning of money to the unemployed (hardcore trainees, college persons) versus the employed (scientists, managers, salespeople, and the like). In general, it was concluded that money means very different things to different segments of the population.

CHAPTER 14
Structural Equation Modeling[1]

JODIE B. ULLMAN

University of California, Los Angeles

14.1 ■ GENERAL PURPOSE AND DESCRIPTION

Structural equation modeling (SEM) is a collection of statistical techniques that allow examination of a set of relationships between one or more IVs, either continuous or discrete, and one or more DVs, either continuous or discrete. The IVs and DVs can be either factors or measured variables. Structural equation modeling is known by several names; it is sometimes referred to as causal modeling, causal analysis, simultaneous equation modeling, or analysis of covariance structures. SEM also is sometimes referred to as path analysis or confirmatory factor analysis, actually, both of these are simply special types of SEM.

SEM allows questions to be answered that involve multiple regression analyses of factors. Combine exploratory factor analysis (EFA, Chapter 13) with multiple regression (Chapter 3) and you discover SEM. At the simplest level, a researcher posits a relationship between a single measured variable (say, success in graduate school) and other measured variables (say, undergraduate gpa, gender, and average daily caffeine consumption). This simple model is multiple regression presented in diagram form in Figure 14.1. All four of the measured variables appear in boxes connected by lines with arrows indicating that gpa, gender, and caffeine (the IVs) predict graduate school success (the DV). A line with two arrows indicates a correlation among the IVs. The presence of a residual indicates imperfect prediction.

A more complicated model of success in graduate school appears in Figure 14.2. In this model, Graduate School Success is a latent variable (a factor) that is not directly measured but rather assessed indirectly using number of publications, grades, and faculty evaluations: three measured variables. Graduate School Success in turn, is predicted by gender (a measured variable) and by Undergraduate Success, a second factor that is assessed through undergraduate gpa, faculty

[1] I would like to thank Barbara Tabachnick and Linda Fidell for the opportunity to write this chapter and also for their helpful comments on an earlier draft. I also would like to thank Peter Bentler, who not only performed a detailed review of this chapter, but who also has been responsible for shaping my thinking on SEM. Support for this chapter was funded in part by Grant DA01070 from the National Institute on Drug Abuse.

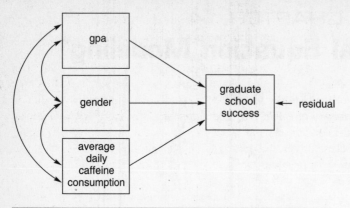

FIGURE 14.1 PATH DIAGRAM OF MULTIPLE REGRESSION.

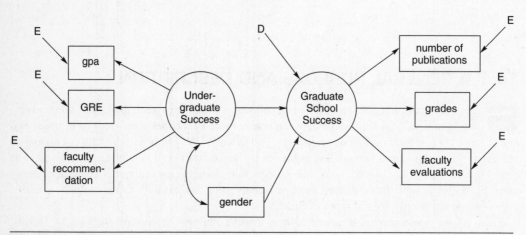

FIGURE 14.2 PATH DIAGRAM OF A STRUCTURAL MODEL.

recommendations, and GRE scores (three additional measured variables). For clarity in the text, initial capitals are used for names of factors and lower case letters for names of measured variables.

Figures 14.1 and 14.2 are examples of *path diagrams*. These diagrams are fundamental to SEM because they allow the researcher to diagram the hypothesized set of relationships—the model. The diagrams are helpful in clarifying a researcher's ideas about the relationships among variables and they can be directly translated into the equations needed for the analysis.

Several conventions are used in developing SEM diagrams. Measured variables, also called *observed variables, indicators*, or *manifest variables,* are represented by squares or rectangles. Factors have two or more indicators and are also called *latent variables*, *constructs*, or *unobserved variables*. Factors are represented by circles or ovals in path diagrams. Relationships between variables are indicated by lines; lack of a line connecting variables implies that no direct relationship has been hypothesized. Lines have either one or two arrows. A line with one arrow represents a hypothesized direct relationship between two variables, and the variable with the arrow pointing to it is the DV. A line with arrows at both ends indicates an unanalyzed relationship; it is simply a covariance between the two variables with no implied direction of effect.

In the model of Figure 14.2, Graduate School Success is a latent variable (factor) that is predicted by gender (a measured variable), and Undergraduate Success (a factor). Notice the line with the arrow at either end connecting Undergraduate Success and gender. This line with arrows at both ends implies that there is a relationship between the variables but the model makes no prediction regarding the direction of effect. Also notice the direction of the arrows connecting the Graduate School Success construct to its indicators; the construct *predicts* the measured variables. The implication is that Graduate School Success drives, or creates, the number of publications, grades, and faculty evaluations of graduate students. It is impossible to measure this construct directly, so we do the next best thing and measure several indicators of success. We hope that we are able to tap into graduate students' true level of success by measuring a lot of observable indicators. This is the same logic as in factor analysis (Chapter 13)[2].

In Figure 14.2, gpa, GRE, faculty recommendation, Graduate School Success, number of publications, grades, and faculty evaluations are all DVs. They all have one-way arrows pointing to them. Gender and Undergraduate Success are IVs in the model. They have no one-way arrows pointing to them. Notice that all the DVs, both observed and unobserved, also have arrows labeled "E" or "D" pointing toward them. Es (errors) point to measured variables; Ds (disturbances) point to latent variables. As in multiple regression, nothing is predicted perfectly; there is always residual or error. In SEM, the residual not predicted by the IV(s) is included in the diagram with these paths.

The part of the model that relates the measured variables to the factors is sometimes called the *measurement model*. In this example, the two constructs (factors) Undergraduate Success and Graduate School Success and the indicators of these constructs form the *measurement model*. The hypothesized relationship among the constructs, in this example the one path between Undergraduate Success and Graduate School Success, is called the *structural model*.

Note that both models presented so far include hypotheses about relationships among variables (covariances) but not about means or mean differences. Mean differences associated with group membership can also be tested within the SEM framework.

When experiments are analyzed, with proper data collection, the adequacy of the manipulation can also be accounted for within the analysis (Blalock, 1985). Experiments with or without a mean structure can be analyzed using SEM. Even in a simple experiment, researchers are often interested in processes that are more complex than a standard analysis suggests. Consider the diagram in Figure 14.3.

Students are randomly assigned at the start of a semester to one of two treatment conditions, study skills training or waiting-list control. X_1 is a dummy-coded variable (cf. Section 1.2.1) that indicates the assigned treatment group, e.g., 0 = control, 1 = treatment. Final exam scores are recorded at the end of the semester. An ANOVA essentially tests path *a*. But is it reasonable to suggest that merely assignment to a treatment group creates the change? Perhaps not. Maybe, instead, the treatment manipulation increases a student's motivation level and this higher motivation leads to a higher grade. The relationship between the treatment and the exam score is *mediated* by motivation level. This is a different question than is posed in ANOVA or even ANCOVA or sequential regression. ANOVA asks simply "is there a difference between the treatment and control group on

---

[2] Arrows pointing in the opposite direction, to the Graduate School Success construct from the measured indicators, would imply a principal component or linear combination.

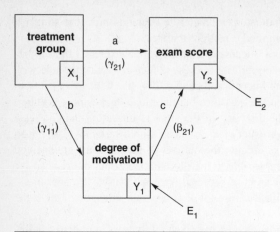

FIGURE 14.3   PATH DIAGRAM OF AN EXPERIMENT.

exam score?" ANCOVA and sequential regression ask "is there a difference between groups after the DV has been adjusted by a covariate (e.g., degree of motivation)?" These questions are distinct from the hypotheses illustrated in Figure 14.3 that involve a process or an *indirect effect*. The indirect effect can be tested by testing the product of paths *b* and *c*. This example uses only measured variables and is called *path analysis*; however, mediational hypotheses can be tested using both latent and observed variables.

The first step in a SEM analysis is specification of a model, so this is a *confirmatory* rather than an exploratory technique. The model is estimated, evaluated, and perhaps modified. The goal of the analysis might be to test a model, to test specific hypotheses about a model, to modify an existing model, or to test a set of related models.

There are a number of advantages to SEM. When relationships among factors are examined, the relationships are free of measurement error because the error has been estimated and removed, leaving only common variance. Reliability of measurement can be accounted for explicitly within the analysis by estimating and removing the measurement error. Additionally, as was seen in Figure 14.2, complex relationships can be examined. When the phenomena of interest are complex and multidimensional, SEM is the only analysis that allows complete and simultaneous tests of all the relationships.

Unfortunately, there is a small price to pay for the flexibility that SEM offers. With the ability to analyze complex relationships among combinations of discrete and continuous variables, both observed and latent, comes more complexity and more ambiguity. Indeed, there is quite a bit of jargon and many choices of analytic techniques. But if you love to wallow in data, you'll adore SEM!

14.2 ■ KINDS OF RESEARCH QUESTIONS

The data set is an empirical covariance matrix and the model produces an estimated population covariance matrix. The major question asked by SEM is, "Does the model produce an estimated population covariance matrix that is consistent with the sample (observed) covariance matrix?"

After the adequacy of the model is assessed, various other questions about specific aspects of the model are addressed.

14.2.1 ■ Adequacy of the Model

Parameters (regression coefficients, variances, and covariances) are estimated to create an estimated population covariance matrix. If the model is good the parameter estimates produce an estimated matrix that is close to the sample covariance matrix. "Closeness" is evaluated primarily with chi-square tests and fit indices. For the Graduate School Success model of Figure 14.2, Is the estimated population covariance matrix generated by the model consistent with the covariance matrix of the data (the sample covariance matrix)? This is discussed in Sections 14.5.3, 14.7.3, and 14.7.7.

14.2.2 ■ Testing Theory

Each theory (model) generates its own covariance matrix. Which theory produces an estimated population covariance matrix that is most consistent with the sample covariance matrix? Models representing competing theories in a specific research area are estimated, pitted against each other, and evaluated, as demonstrated in Sections 14.5.4.1 and 14.7.4.

14.2.3 ■ Amount of Variance in a Variable Accounted for by a Factor

How much of the variance in the DVs, both latent and observed, is accounted for by the IVs? For example, How much variance in Graduate School Success is accounted for by gender and Undergraduate Success? Which of the variables included in the analysis account for the most variance? This question is answered through R^2-type statistics discussed in Section 14.5.5.

14.2.4 ■ Reliability of the Indicators

How reliable are each of the measured variables? For the example, Is the measure of faculty evaluations reliable? Reliability of measured variables derived from SEM analyses is discussed in Section 14.5.5.

14.2.5 Parameter Estimates

Estimates of parameters are fundamental to SEM analyses because they are used to generate the estimated population covariance matrix for the model. What is the regression coefficient for a specific path? For example, What is the coefficient for predicting Graduate School Success from Undergraduate Success? Does the coefficient differ significantly from 0? Within the model, what is the relative importance of various paths? For instance, Is the path from Undergraduate Success more or less important to prediction of Graduate School Success than the path from gender? Parameter estimates also can be compared across SEM models. When a single path is tested, it is called a test of *direct effects*. Assessment of parameters is demonstrated in Sections 14.7.3 , 14.7.4, and 14.7.8.

14.2.6 ■ Mediation

Does an IV directly affect a specific DV or does the IV affect the DV through an intermediary, or mediating variable? In the example of Figure 14.3, Is the relationship between treatment group and exam score mediated by degree of motivation? Because motivation is a mediational variable, this is a test of *indirect effects*. Tests of indirect effects are demonstrated in Section 14.7.8.

14.2.7 ■ Group Differences

Do two or more groups differ in their covariance matrices, regression coefficients, or means? For example, if the experiment described by Figure 14.3 is performed for both grade school and high school youngsters, does the same model fit both age groups? This analysis could be performed with or without means (c.f., Section 14.5.8). Multiple group modeling is discussed briefly in Section 14.5.7.

14.2.8 ■ Longitudinal Differences (Repeated Measures)

Differences within people across time also can be examined. The time interval can be years, days, or microseconds. For the experimental example, How, if at all, does treatment change performance and motivation at several different points in the semester? Longitudinal modeling is not illustrated in this chapter, but differs primarily in type of data. An example of a longitudinal model of the D.A.R.E program is Dukes, Ullman, and Stein (1995). An excellent reference for longitudinal structural equation models is Collins and Horn (1991).

14.3 ■ LIMITATIONS TO STRUCTURAL EQUATION MODELING

14.3.1 ■ Theoretical Issues

SEM *is a confirmatory technique* in contrast to exploratory factor analysis. It is used most often to test a theory: maybe just a personal theory, but a theory nonetheless. Indeed, one cannot do SEM without prior knowledge of, or hypotheses about, potential relationships among variables. This is perhaps the largest difference between SEM and other techniques in this book and one of its greatest strengths. Planning, driven by theory, is essential to any SEM analysis. The guidelines for planning an exploratory factor analysis as outlined in Section 13.3.1, are also applicable to SEM analyses.

Although SEM is a confirmatory technique, there are many ways in which a variety of different models can be examined after a model has been estimated: models that either test specific hypotheses, or perhaps develop a better fitting model. If numerous modifications of a model are tested in hopes of finding the best-fitting model, the researcher has moved to exploratory data analysis and appropriate steps need to be taken to protect against inflated Type I error levels. Searching for the best model is appropriate provided significance levels are viewed cautiously and cross-validation with another sample is performed whenever possible.

SEM has developed a bad reputation in some circles, in part because of the use of SEM for exploratory work without the necessary controls. It may also be due, in part, to the use of the term "causal modeling" to refer to structural equation modeling. There is nothing causal, in the sense of inferring causality, about use of SEM. Attributing causality is a design issue, not a statistical issue.

Unfortunately, SEM often is thought of as a technique strictly for non-experimental or correlational designs. This is overly limiting. SEM, like regression, can be applied to both experimental and non-experimental designs. In fact there are some advantages to using SEM in the analysis of experiments; mediational processes can be tested and information regarding the adequacy of the manipulations can be included in the analysis (Blalock, 1985).

The same caveats regarding generalizing results apply to SEM as to the other techniques in this book. Results can only be generalized to the type of samples that were used to estimate and test the SEM model.

14.3.2 ■ Practical Issues

14.3.2.1 ■ Sample Size and Missing Data

Covariances, like correlations, are less stable when estimated from small samples. SEM is based on covariances. Parameter estimates and chi-square tests of fit also are very sensitive to sample size. SEM, then, like factor analysis, is a large sample technique. The rules of thumb for sample size presented in Chapter 13 apply to SEM as well. However, instead of thinking about number of subjects per measured variable, it is probably more helpful to think about number of subjects per *estimated parameter* (in SEM there is no linear relationship between number of variables and number of parameters) as noted by Bentler (1995). In many cases a sample size of about 200 is adequate for small to medium size models (Boomsma, 1983). As with multiple regression (Chapter 5), expected effect size and the distributions of the measured variables also influence the power of the analysis, so that fewer than 10 subjects per estimated parameter may be adequate if the estimated size of effect is large and the measured variables are normally distributed.

Chapter 4 guidelines for the treatment of missing data apply to SEM analyses. However, as discussed in Chapter 4, problems are associated with either deleting or estimating missing data. An advantage of structural modeling is that the missing data mechanism can be included in the model. Treatment of missing data patterns through SEM is not demonstrated in this chapter but the interested reader is referred to Allison (1987), Muthén, Kaplan, and Hollis (1987), and Bentler (1995).

14.3.2.2 ■ Multivariate Normality and Outliers

Most of the estimation techniques used in SEM assume multivariate normality. To determine the extent and shape of non-normally distributed data, *screen the measured variables for outliers, both univariate and multivariate, and examine the skewness and kurtosis of the measured variables in the manner described in Chapter 4.* All measured variables, regardless of their status as DVs or IVs, are screened together for outliers. (Some SEM programs test for the presence of multivariate outliers, skewness, and kurtosis.) If significant skewness is found, transformations can be attempted; however, often variables are still highly skewed or highly kurtotic even after transformation. Some variables, such as drug use variables, are not expected to be normally distributed in the population, anyway. If transformations do not restore normality, or a variable is not expected to be normally distributed in the population, an estimation method can be selected that addresses the non-normality (Sections 14.5.2, 14.6.1 and 14.6.2).

14.3.2.3 ■ Linearity

SEM techniques examine only linear relationships among variables. Linearity among latent variables is difficult to assess; however, *linear relationships among pairs of measured variables can be assessed through inspection of scatterplots*. If nonlinear relationships among measured variables are hypothesized, these relationships are included by raising the measured variables to powers, as in multiple regression. For example if the relationship between graduate school success and average daily caffeine consumption is quadratic (a little caffeine is not enough, a few cups is good, but more than a few is detrimental), the square of average daily caffeine consumption is used.

14.3.2.4 ■ Multicollinearity and Singularity

As with most other techniques discussed in the book, matrices need to be inverted in SEM. Therefore if variables are perfect linear combinations of one another or are *extremely* highly correlated, the necessary matrices cannot be inverted. If possible, *inspect the determinant of the covariance matrix. An extremely small determinant may indicate a problem with multicollinearity or singularity*. Generally, SEM programs abort and provide warning messages if the covariance matrix is singular. If you get such a message, check your data set. It often is the case that linear combinations of variables have been inadvertently included. Simply delete the variable causing the singularity. If true singularity or multicollinearity is found, create composite variables and use them in the analysis.

14.3.2.5 ■ Analyzability of Covariances

The factorability of the correlation matrix **R** was discussed in Chapter 13. The concept holds for SEM except that the covariance matrix is the one of concern. SEM models may include variables thought to relate to one another and other variables thought to be independent of one another; these relationships should be roughly reflected in the sample covariance matrix. Unfortunately, the sample covariance matrix is not standardized, making it sometimes difficult to determine the magnitude of the relationships among variables. As a very rough guide, the correlation matrix can be inspected and should contain the predicted pattern of correlations. The problem with inspecting the correlation matrix is that it contains correlations among variables *with measurement error included*. When measurement error is removed, the relationships among the variables may change; typically, they increase in magnitude. Nonetheless, a look at the correlation matrix may still be helpful.

14.3.2.6 ■ Residuals

After model estimation, *the residuals should be small and centered around zero. The frequency distribution of the residual covariances should be symmetric*. Residuals in the context of SEM are residual *covariances,* not residual *scores* as discussed in other chapters. SEM programs provide diagnostics of residuals. Non-symmetrically distributed residuals in the frequency distribution may signal a poorly fitting model; the model is estimating some of the covariances well and others poorly. It sometimes happens that one or two residuals remain quite large, although the model fits reasonably well and the residuals appear to be symmetrically distributed and centered around zero. When large residuals are found it is often helpful to examine the Lagrange Multiplier (LM) test, discussed in Section 14.5.4.2 and consider adding paths to the model.

14.4 ■ FUNDAMENTAL EQUATIONS FOR STRUCTURAL EQUATIONS MODELING

14.4.1 ■ Covariance Algebra

The idea behind SEM is that the hypothesized model has a set of underlying parameters that correspond to (1) the regression coefficients, and (2) the variances and covariances of the independent variables in the model (Bentler, 1995). These parameters are estimated from the sample data as a "best guess" about population values. The estimated parameters are then combined by means of covariance algebra to produce an estimated population covariance matrix. This estimated population covariance matrix is compared with the sample covariance matrix and, ideally, the difference is very small and not statistically significant.

Covariance algebra is a helpful tool in calculating estimated population variances and covariances in SEM models; however, matrix methods are generally employed because covariance algebra becomes extremely tedious as models become increasingly complex. Covariance algebra is useful to demonstrate how parameters estimates are combined to produce an estimated population covariance matrix for a small example.

The three basic rules in covariance algebra, follow where c is a constant and X_i is a random variable:

$$1.\ COV(c, X_1) = 0$$

$$2.\ COV(cX_1, X_2) = cCOV(X_1, X_2) \tag{14.1}$$

$$3.\ COV(X_1 + X_2, X_3) = COV(X_1, X_3) + COV(X_2, X_3)$$

By the first rule, the covariance between a variable and a constant is zero. By the second rule, the covariance between two variables where one is multiplied by a constant is the same as the constant multiplied by the covariance between the two variables. By the third rule, the covariance between the sum of two variables and a third variable is the sum of the covariance of the first variable and the third and the covariance of the second variable and the third.

Figure 14.3 is used to illustrate some of the principles of covariance algebra. (Ignore for now the difference between γ and β; the difference is explained in Section 14.4.3). In SEM, as in multiple regression, we assume that the residuals do not correlate with each other or with other variables in the models. In this model, both degree of motivation (Y_1) and exam score (Y_2) are DVs. Recall that a DV in SEM is any variable with a single headed arrow pointing toward it. Treatment group (X_1) with no single headed arrows pointing to it is an IV. To specify the model, a separate equation is written for each DV. For motivation, Y_1:

$$Y_1 = \gamma_{11}X_1 + \epsilon_1 \tag{14.2}$$

Degree of motivation is a weighted function of treatment group plus error.

Note that ϵ_1 in the equation corresponds to E1 in Figure 14.3.
and for exam score, Y_2,

$$Y_2 = \beta_{21}Y_1 + \gamma_{21}X_1 + \epsilon_2 \tag{14.3}$$

Exam score is a weighted function of treatment group plus a weighted function of degree of motivation plus error.

To calculate the covariance between X_1 (treatment group) and Y_1 (degree of motivation) the first step is substituting in the equation for Y_1:

$$COV(X_1,Y_1) = COV(X_1,\gamma_{11}X_1 + \epsilon_1) \tag{14.4}$$

The second step is distributing the first term, in this case X_1:

$$COV(X_1,Y_1) = COV(X_1\gamma_{11}X_1 + X_1\epsilon_1) \tag{14.5}$$

The last term in this equation , $COV(X_1\epsilon_1)$, is equal to zero by assumption because it is assumed that there are no covariances between errors and other variables. Now,

$$COV(X_1Y_1) = \gamma_{11}COV(X_1X_1) \tag{14.6}$$

by rule 2, and because the covariance of a variable with itself is just a variance,

$$COV(X_1Y_1) = \gamma_{11}\sigma_{x_1x_1} \tag{14.7}$$

The estimated population covariance between X_1 and Y_1 is equal to the regression coefficient times the estimated variance of X_1.

This is the population covariance between X_1 and Y_1 as estimated from the model. If the model is good, the product of $\gamma_{11}\sigma_{x1x1}$ produces a covariance that is very close to the sample covariance.
Following the same procedures, the covariance between Y_1 and Y_2 is:

$$COV(Y_1,Y_2) = COV(\gamma_{11}X_1 + \epsilon_1, \beta_{21}Y_1 + \gamma_{21}X_1 + \epsilon_2)$$

$$= COV(\gamma_{11}X_1\beta_{21}Y_1 + \gamma_{11}X_1\gamma_{21}X_1 + \gamma_{11}X_1\epsilon_2 + \epsilon_1\beta_{21}Y_1 + \epsilon_1\gamma_{21}X_1 + \epsilon_1\epsilon_2) \tag{14.8}$$

$$= \gamma_{11}\beta_{21}\sigma_{x_1y_1} + \gamma_{11}\gamma_{21}\sigma_{x_1x_1}$$

because, as can be seen in the diagram, the error terms E_1 and E_2 do not correlate with any other variables or each other.

All of the estimated covariances in the model could be derived in the same manner; but as is apparent even in this small example, covariance algebra rapidly becomes somewhat tedious. The "take home point" of this example is that covariance algebra can be used to estimate parameters and then estimate a population covariance matrix from them. Estimated parameters give us the estimated population covariance matrix.

14.4.2 ■ Model Hypotheses

A truncated raw data set and corresponding covariance matrix appropriate for SEM analysis are presented in Table 14.1. This very small data set contains five continuous measured variables: (1) NUMYRS, the number of years a participant has skied, (2) DAYSKI, the total number of days a person has skied, (3) SNOWSAT, a Likert scale measure of overall satisfaction with the snow conditions, (4) FOODSAT, a Likert scale measure of overall satisfaction with the quality of the food at the resort, and (5) SENSEEK, a Likert scale measure of degree of sensation seeking. Note that hypothetical data are included for only 5 skiers although the analysis is performed with hypothetical data from 100 skiers. Matrix computations in SEM by hand are tedious, at best. Therefore, MATLAB, a matrix manipulation program, is used to perform the calculations. Grab MATLAB or SYSTAT MATRIX or SAS IML to perform matrix manipulations yourself as the example develops. Note also that the calculations presented here are rounded to two decimal places.

The hypothesized model for these data is diagrammed in Figure 14.4. Latent variables are represented by circles and measured variables are represented by squares. A line with an arrow indicates a hypothesized direct relationship between the variables. Absence of a line implies no hypothesized direct relationship. The asterisks indicate parameters to be estimated. Shading indicates that the variable is an IV. The variances of IVs are parameters of the model and are estimated or fixed to a particular value. The number 1 indicates that a parameter, either a regression coefficient or a variance, has been set, (fixed) to the value of 1.

This example contains two hypothesized latent variables (factors): Love of Skiing (LOVESKI), and Ski Trip Satisfaction (SKISAT). The Love of Skiing (LOVESKI) factor is hypothesized to have

TABLE 14.1 SMALL SAMPLE OF HYPOTHETICAL DATA FOR STRUCTURAL EQUATION MODELING

	Variables				
Vacationers	NUMYRS	DAYSKI	SNOWSAT	FOODSAT	SENSEEK
S_1	5	50	3	8	16
S_2	7	100	6	7	43
S_3	14	200	10	4	23
S_4	5	25	8	6	18
.					
.					
.					
S_{100}	10	300	10	3	30

	Covariance Matrix				
	NUMYRS	DAYSKI	SNOWSAT	FOODSAT	SKISAT
NUMYRS	2.74	0.80	0.68	0.65	2.02
DAYSKI	0.80	3.25	0.28	0.35	2.12
SNOWSAT	0.68	0.28	1.23	0.72	2.02
FOODSAT	0.65	0.35	0.72	1.87	2.12
SKISAT	2.02	2.12	2.02	2.12	27.00

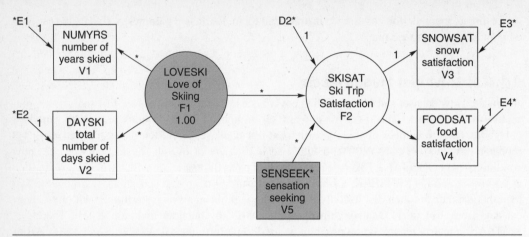

FIGURE 14.4 HYPOTHESIZED MODEL FOR SMALL SAMPLE EXAMPLE.

two indicators, number of years skied (NUMYRS) and number of days skied (DAYSKI). Greater Love of Skiing predicts more numerous years skied and days skied. Note that the direction of the prediction matches the direction of the arrows. The Ski Trip Satisfaction (SKISAT) factor also has two indicators; snow satisfaction (SNOWSAT) and food satisfaction (FOODSAT). Higher Ski Trip Satisfaction predicts a higher degree of satisfaction with both the snow and the food. This model also hypothesizes that both Love of Skiing and degree of sensation seeking (SENSEEK) predict level of Ski Trip Satisfaction; greater levels of Love of Skiing and sensation seeking predict higher levels of Ski Trip Satisfaction. Also notice that no arrow directly connects Love of Skiing with degree of sensation seeking. There is no hypothesized relationship, either predictive or correlational, between these variables.

As in the discussion of covariance algebra, these relationships are directly translated into equations and the model then estimated. The analysis proceeds by specifying the model in the diagram and then translating the model into a series of equations or matrices. Population parameters that imply a covariance matrix are then estimated. This estimated population covariance matrix is compared with the sample covariance matrix. The goal, as you might have guessed, is to estimate parameters that produce an estimated population covariance matrix that is not significantly different from the sample covariance matrix. This is similar to factor analysis (Chapter 13) where the reproduced correlation matrix is compared with the observed correlation matrix. One distinction between SEM and EFA is that in SEM the difference between the sample covariance matrix and the estimated population covariance typically is evaluated with a chi-square test statistic.[3]

14.4.3 ■ Model Specification

One method of model specification is the Bentler-Weeks method (Bentler & Weeks, 1980). In this method every variable in the model, latent or measured, is either an IV or a DV. The parameters to be estimated are (1) the regression coefficients, and (2) the variances and the covariances of the independent variables in the model (Bentler, 1995).

---

[3] A chi-square test statistic can be used in EFA when maximum likelihood factor extraction is employed.

In the example, SKISAT, SNOWSAT, FOODSAT, NUMYRS, and DAYSKI are all DVs because they all have at least one line with a single-headed arrow pointing to them. Notice that SKISAT is a latent variable and also a dependent variable. Whether or not a variable is observed makes no difference as to its status as a DV or IV. Although SKISAT is a factor, it is also a DV because it has arrows from both LOVESKI and SENSEEK. The seven IVs in this example are SENSEEK, LOVESKI, D2, E1, E2, E3, E4.

Residual variables (errors) of measured variables are labeled E and errors of latent variables (called disturbances) are labeled D. It may seem odd that a residual variable is considered an IV but remember the familiar regression equation:

$$Y = X\beta + \epsilon \tag{14.9}$$

where Y is the DV and X and ϵ are both IVs.

In fact the Bentler-Weeks model *is* a regression model, expressed in matrix algebra:

$$\boldsymbol{\eta} = \mathbf{B}\boldsymbol{\eta} + \boldsymbol{\gamma}\boldsymbol{\xi} \tag{14.10}$$

where, if q is the number of DVs and r is the number of IVs, then $\boldsymbol{\eta}$ (eta) is a $q \times 1$ vector of DVs, $\mathbf{B}$ (beta) is a $q \times q$ matrix of regression coefficients among DVs, $\boldsymbol{\gamma}$ (gamma) is a $q \times r$ matrix of regression coefficients among DVs and IVs, and $\boldsymbol{\xi}$ (xi) is a $r \times 1$ vector of IVs.

In the Bentler-Weeks model only independent variables have variances and covariances as parameters of the model and these variances and covariances are in $\boldsymbol{\Phi}$ (phi), an $r \times r$ matrix. Therefore, the parameter matrices of the model are $\mathbf{B}$, $\boldsymbol{\gamma}$, and $\boldsymbol{\Phi}$. Unknown parameters in these matrices need to be estimated. The vectors of dependent variables, $\boldsymbol{\eta}$, and independent variables, $\boldsymbol{\xi}$ are not estimated. The diagram for the example is translated into the Bentler-Weeks model, with $r = 7$ and $q = 5$:

$$
\begin{matrix}
\boldsymbol{\eta} & = & \mathbf{B} & \boldsymbol{\eta} & + & \boldsymbol{\gamma} & & \boldsymbol{\xi}
\end{matrix}
$$

$$
\begin{bmatrix} V1 \text{ or } \eta_1 \\ V2 \text{ or } \eta_2 \\ V3 \text{ or } \eta_3 \\ V4 \text{ or } \eta_4 \\ F2 \text{ or } \eta_5 \end{bmatrix}
=
\begin{bmatrix} 0&0&0&0&0 \\ 0&0&0&0&0 \\ 0&0&0&0&1 \\ 0&0&0&0&* \\ 0&0&0&0&0 \end{bmatrix}
\begin{bmatrix} V1 \text{ or } \eta_1 \\ V2 \text{ or } \eta_2 \\ V3 \text{ or } \eta_3 \\ V4 \text{ or } \eta_4 \\ F2 \text{ or } \eta_5 \end{bmatrix}
+
\begin{bmatrix} 0&*&1&0&0&0&0 \\ 0&*&0&1&0&0&0 \\ 0&0&0&0&1&0&0 \\ 0&0&0&0&0&1&0 \\ *&*&0&0&0&0&1 \end{bmatrix}
\begin{bmatrix} V5 \text{ or } \xi_1 \\ F1 \text{ or } \xi_2 \\ E1 \text{ or } \xi_3 \\ E2 \text{ or } \xi_4 \\ E3 \text{ or } \xi_5 \\ E4 \text{ or } \xi_6 \\ D2 \text{ or } \xi_7 \end{bmatrix}
$$

Notice that $\boldsymbol{\eta}$ is on both sides of the equation. This is because DVs can predict one another in SEM. The diagram and matrix equations are identical. Notice that the asterisks in Figure 14.4 directly correspond to the asterisks in the matrices and these matrix equations directly correspond to simple regression equations. In the matrix equations, the number 1 indicates that we have "fixed" the parameter, either a variance or a regression coefficient, to the specific value of 1. Parameters are generally fixed for identification purposes. Identification is discussed in more detail in Section 14.5.1. Parameters can be fixed to any number; most often, however, parameters are fixed to 1 or zero. The parameters that are fixed to zero are also included in the path diagram but are easily over-looked. The zero parameters are represented by the *absence* of a line in the diagram!

Carefully compare the model in Figure 14.4 with this matrix equation. The 5×1 vector of values to the left of the equal sign, the eta (η) vector, is a vector of DVs listed in the order indicated, NUMYRS (V1), DAYSKI (V2), SNOWSAT (V3), FOODSAT (V4), and SKISAT (F2). The next matrix, just to the right of the equal sign, is a 5×5 matrix of regression coefficients among the DVs. The DVs are in the same order. The matrix contains 23 zeros, one 1, and one *. Remember that matrix multiplication involves cross multiplying and then summing the elements in the first row of the beta (**B**) matrix with the first column in the eta (η) matrix , and so forth (consult Appendix A as necessary). The zeros in the first, second, and fifth rows of the beta matrix indicate that no regression coefficients are to be estimated between DVs for V1, V2, and F2. The 1 at the end of the third row is the regression coefficient between F2 and SNOWSAT that was fixed to 1. The * at the end of the fourth row is the regression coefficient between F2 and V4 that is to be estimated.

Now look to the right of the plus sign. The 5×7 gamma matrix contains the regression coefficients that are used to predict the DVs from the IVs. The five DVs that are associated with the rows of this matrix are in the same order as above. The seven IVs that identify the columns are, in the order indicated, SENSEEK (V5), LOVESKI (F1), the four (errors) for V1 to V4, and the (Disturbance) of F2. The 7×1 vector of IVs is in the same order. The first row of the γ (gamma) matrix times the ξ (xi) vector produces the equation for NUMYRS. The * is the regression coefficient for predicting NUMYRS from LOVESKI (F1) and the 1 is the fixed regression coefficient for the relationship between NUMYRS and its E1. For example, consider the equation for NUMYRS (V1) reading from the first row in the matrices:

$$\eta_1 = 0 \cdot \eta_1 + 0 \cdot \eta_2 + 0 \cdot \eta_3 + 0 \cdot \eta_4 + 0 \cdot \eta_5 + 0 \cdot \xi_1 + * \cdot \xi_2$$
$$+ 1 \cdot \xi_3 + 0 \cdot \xi_4 + 0 \cdot \xi_5 + 0 \cdot \xi_6 + 0 \cdot \xi_7$$

or, by dropping the zero-weighted products, and using the diagram's notation,

$$V1 = *F1 + E1$$

Continue in this fashion for the next four rows to be sure you understand their relationship to the diagrammed model.

In the Bentler-Weeks model only IVs have variances and covariances as parameters of the model and these are in Φ (phi), an $r \times r$ matrix. For the example, with seven IVs:

	V5 or ξ_1	F1 or ξ_2	E1 or ξ_3	E2 or ξ_4	E3 or ξ_5	E4 or ξ_6	D2 or ξ_7
V5 or ξ_1	*	0	0	0	0	0	0
F1 or ξ_2	0	1	0	0	0	0	0
E1 or ξ_3	0	0	*	0	0	0	0
E2 or ξ_4	0	0	0	*	0	0	0
E3 or ξ_5	0	0	0	0	*	0	0
E4 or ξ_6	0	0	0	0	0	*	0
D2 or ξ_7	0	0	0	0	0	0	*

$\Phi =$ (heading for the matrix above)

This 7×7 Phi matrix contains the variances and covariances that are to be estimated for the IVs. The *s on the diagonal indicate the variances to be estimated for SENSEEK (V5), LOVESKI (F1), E1, E2, E3, E4, and D2. The 1 in the second row corresponds to the variance of LOVESKI (F1) that was fixed to 1. There are no covariances among IVs to be estimated, as indicated by the zeros in all the off-diagonal positions.

14.4.4 ■ Model Estimation

Initial guesses (start values) for the parameters are needed to begin the modeling process. The more similar the guess and the final value, the fewer iterations needed to find a solution. There are many options available for start values (Bollen, 1989b). Often it is perfectly reasonable to allow the SEM computer program to supply initial start values.

When user-specified start values are supplied, the first decision about start values is whether the valence of the population coefficient is likely to be positive or negative. These will always be positive for variances; however, regression coefficients and covariances could be either positive or negative, so the start values should reflect the valences.

There are several options for valence and size of start values. The sample variances and covariances can be used as the start values for the estimated variances and covariances. Prior research may suggest a value for regression coefficients or the reliability of a measured variable can be used, if known. Or, one can use the ratio of the standard deviation of the DV to the standard deviation of the IV multiplied by an arbitrary fraction (less than one) that reflects the strength of the estimated effect: the stronger the estimated effect, the higher the fraction. Or, start values for regression coefficients can be selected from prior multiple regression analyses. In this example, the start values supplied by the EQS program were employed.

The process begins with start values inserted into each of the three parameter matrices in the Bentler-Weeks model, $\hat{\mathbf{B}}, \hat{\gamma}$, and $\hat{\Theta}$. The ^ (hat) over the matrices indicates that these are matrices of *estimated* parameters. The $\hat{\mathbf{B}}$ (beta hat) matrix is the matrix of regression coefficients between DVs where start values have been substituted for * (the parameters to be estimated). For the example:

$$\hat{\mathbf{B}} = \begin{bmatrix} 0.00 & 0.00 & 0.00 & 0.00 & 0.00 \\ 0.00 & 0.00 & 0.00 & 0.00 & 0.00 \\ 0.00 & 0.00 & 0.00 & 0.00 & 1.00 \\ 0.00 & 0.00 & 0.00 & 0.00 & 1.00 \\ 0.00 & 0.00 & 0.00 & 0.00 & 0.00 \end{bmatrix}$$

The matrix containing the start values for regression coefficients between DVs and IVs is (gamma hat). For the example:

$$\hat{\gamma} = \begin{bmatrix} 0.00 & 1.13 & 1.00 & 0.00 & 0.00 & 0.00 & 0.00 \\ 0.00 & 0.67 & 0.00 & 1.00 & 0.00 & 0.00 & 0.00 \\ 0.00 & 0.00 & 0.00 & 0.00 & 1.00 & 0.00 & 0.00 \\ 0.00 & 0.00 & 0.00 & 0.00 & 0.00 & 1.00 & 0.00 \\ 0.04 & 1.48 & 0.00 & 0.00 & 0.00 & 0.00 & 1.00 \end{bmatrix}$$

Finally, the matrix containing the start values for the variances and covariances of the IVs is $\hat{\Phi}$ (phi hat). For the example:

$$\hat{\Phi} = \begin{bmatrix} 27.00 & 0.00 & 0.00 & 0.00 & 0.00 & 0.00 & 0.00 \\ 0.00 & 1.00 & 0.00 & 0.00 & 0.00 & 0.00 & 0.00 \\ 0.00 & 0.00 & 1.48 & 0.00 & 0.00 & 0.00 & 0.00 \\ 0.00 & 0.00 & 0.00 & 2.91 & 0.00 & 0.00 & 0.00 \\ 0.00 & 0.00 & 0.00 & 0.00 & 0.62 & 0.00 & 0.00 \\ 0.00 & 0.00 & 0.00 & 0.00 & 0.00 & 1.04 & 0.00 \\ 0.00 & 0.00 & 0.00 & 0.00 & 0.00 & 0.00 & 27.54 \end{bmatrix}$$

To calculate the estimated population covariance matrix implied by the parameter estimates, selection matrices are first used to pull the measured variables out of the full parameter matrices. (Remember, the parameter matrices have both measured *and* latent variables as components). The selection matrix is simply labeled $\mathbf{G}$ and has elements that are either 1's or 0's. The selection matrix to pull the measured DVs out of the matrix of all DVs is $\mathbf{G}_y$. For the example:

$$\mathbf{G}_y = \begin{bmatrix} 1.00 & 0.00 & 0.00 & 0.00 & 0.00 \\ 0.00 & 1.00 & 0.00 & 0.00 & 0.00 \\ 0.00 & 0.00 & 1.00 & 0.00 & 0.00 \\ 0.00 & 0.00 & 0.00 & 1.00 & 0.00 \end{bmatrix}$$

Take a minute and multiply $\mathbf{G}_y * \boldsymbol{\eta}$ to see for yourself that the resulting matrix contains only measured variables,

$$\mathbf{Y} = \mathbf{G}_y * \boldsymbol{\eta} = \begin{bmatrix} V1 \\ V2 \\ V3 \\ V4 \end{bmatrix} \tag{14.11}$$

where $\mathbf{Y}$ is our name for those measured variables that are dependent.

The selection matrix to pull the measured IVs out of all the IVs is a vector, $\mathbf{G}_x$: For this example:

$$\mathbf{G}_x = [1.00\ 0.00\ 0.00\ 0.00\ 0.00\ 0.00\ 0.00]$$

so that,

$$\mathbf{X} = \mathbf{G}_x * \boldsymbol{\xi} = V5 \tag{14.12}$$

where $\mathbf{X}$ is our name for the measured independent variables.

Computation of the estimated population covariance matrix proceeds by rewriting the basic structural modeling equation (14.10) as[4]:

$$\boldsymbol{\eta} = (\mathbf{I} - \mathbf{B})^{-1}\ \boldsymbol{\gamma}\boldsymbol{\xi} \tag{14.13}$$

---

[4] This rewritten equation is often called the "reduced form".

where $\mathbf{I}$ is simply an identity matrix the same size as $\mathbf{B}$. This equation expresses the DVs as a linear combination of the IVs[5]

At this point the estimated population covariance matrix for the DVs, $\hat{\mathbf{\Sigma}}_{yy}$ is estimated using:

$$\hat{\mathbf{\Sigma}}_{yy} = \mathbf{G}_y(\mathbf{I} - \hat{\mathbf{B}})^{-1}\hat{\boldsymbol{\gamma}}\hat{\boldsymbol{\Phi}}\hat{\boldsymbol{\gamma}}'(\mathbf{I} - \hat{\mathbf{B}})^{-1'}\mathbf{G}_y' \tag{14.14}$$

For the example,

$$\hat{\mathbf{\Sigma}}_{yy} = \begin{bmatrix} 2.76 & 0.76 & -1.68 & -1.68 \\ 0.76 & 3.36 & -0.99 & -1.00 \\ -1.68 & -0.99 & 30.41 & 29.91 \\ -1.68 & -1.00 & 29.91 & 31.06 \end{bmatrix}$$

The estimated population covariance matrix between IVs and DVs is similarly obtained by:

$$\hat{\mathbf{\Sigma}}_{yx} = \mathbf{G}_y(\mathbf{I} - \hat{\mathbf{B}})^{-1}\hat{\boldsymbol{\gamma}}\hat{\boldsymbol{\Phi}}\mathbf{G}_x' \tag{14.15}$$

For the example:

$$\hat{\mathbf{\Sigma}}_{yx} = \begin{bmatrix} 0.00 \\ 0.00 \\ 0.19 \\ 0.19 \end{bmatrix}$$

Finally, (phew!) the estimated population covariance matrix between IVs is estimated:

$$\hat{\mathbf{\Sigma}}_{xx} = \mathbf{G}_x\hat{\boldsymbol{\Phi}}\mathbf{G}_x' \tag{14.16}$$

for the example,

$$\hat{\mathbf{\Sigma}}_{xx} = 27$$

In practice, a "super $\mathbf{G}$" matrix is used so that all the covariances are estimated in one step. The components of $\hat{\mathbf{\Sigma}}$ are then combined to produce the estimated population covariance matrix after one iteration.

---

[5] To move from the estimated population variances and covariances to the population variances and covariances,
$\mathbf{y} = \mathbf{G}_y\boldsymbol{\eta}$
$\quad = \mathbf{G}_y(\mathbf{I} - \mathbf{B})^{-1}\boldsymbol{\gamma}\boldsymbol{\xi}$
so if we obtain the product of $\mathbf{yy}'$
$\mathbf{yy}' = \mathbf{G}_y(\mathbf{I} - \mathbf{B})^{-1}\boldsymbol{\gamma}\boldsymbol{\xi}\boldsymbol{\xi}'\boldsymbol{\gamma}'(\mathbf{I} - \mathbf{B})^{-1'}\mathbf{G}_y'$
and take expectations or covariances we end up with Eq. 14.14
$\mathbf{\Sigma}_{yy} = \mathbf{G}_y(\mathbf{I} - \mathbf{B})^{-1}\boldsymbol{\gamma}\boldsymbol{\Phi}\boldsymbol{\gamma}'(\mathbf{I} - \mathbf{B})^{-1'}\mathbf{G}_y'$ with no hats!

For the example, using initial start values supplied by EQS:

$$\hat{\Sigma} = \begin{bmatrix} 2.76 & 0.76 & -1.68 & -1.68 & 0.00 \\ 0.77 & 3.36 & -0.99 & -1.00 & 0.00 \\ -1.68 & -0.99 & 30.41 & 29.91 & 1.19 \\ -1.68 & -1.00 & 29.91 & 31.06 & 1.19 \\ 0.00 & 0.00 & 1.19 & 1.19 & 27.00 \end{bmatrix}$$

The sample covariance matrix is **S**. For the sample, the covariance matrix is:

$$\mathbf{S} = \begin{bmatrix} 2.74 & 0.80 & 0.68 & 0.65 & 2.02 \\ 0.80 & 3.25 & 0.28 & 0.35 & 2.12 \\ 0.68 & 0.28 & 1.23 & 0.72 & 2.02 \\ 0.65 & 0.35 & 0.72 & 1.87 & 2.12 \\ 2.02 & 2.12 & 2.02 & 2.12 & 27.00 \end{bmatrix}$$

The residual covariance matrix is found by subtracting the estimated population covariance matrix from the sample covariance matrix, $\mathbf{S} - \hat{\Sigma}$. For the example,

$$\text{Residual Covariance Matrix} = \begin{bmatrix} -0.02 & 0.04 & 2.36 & 2.33 & 2.02 \\ 0.04 & -0.11 & 1.27 & 1.35 & 2.12 \\ 2.36 & 1.27 & -29.18 & -29.19 & 0.83 \\ 2.33 & 1.35 & -29.19 & -29.19 & 0.93 \\ 2.02 & 2.12 & 0.83 & 0.93 & 0.00 \end{bmatrix}$$

These are the values in the residual covariance matrix after the first iteration. Parameter estimates are changed incrementally (iterations continue) until the pre-specified (in this case maximum likelihood) function (Section 14.5.2) is minimized (converges). This example took nine iterations to converge. Although results of all nine iterations are not particularly enlightening, it might be interesting to look at the residual matrix after 5 iterations:

$$\text{Residual Covariance Matrix} = \begin{bmatrix} 0.00 & 0.00 & 0.14 & 0.12 & 2.02 \\ 0.00 & 0.00 & 0.12 & 0.19 & 2.10 \\ 0.14 & 0.12 & 0.05 & 0.05 & 0.39 \\ 0.12 & 0.19 & 0.05 & 0.05 & 0.49 \\ 2.02 & 2.12 & 0.39 & 0.49 & 0.00 \end{bmatrix}$$

At this step, it looks as if the model is estimating the variances (the values in the main diagonal) quite well, but some of the covariances are still rather large i.e., between NUMYRS and SENSEEK (2.02). This residual seems to indicate a possible relationship between these two variables, although there was no hypothesized relationship between them in the original model, i.e., no paths directly or indirectly connect these variables.

After nine iterations, the maximum likelihood function is at a minimum and the solution converges. The final estimated parameters are presented for comparison purposes in the **B**, γ, and Φ matrices; these unstandardized parameters are also presented (in parentheses) in Figure 14.5.

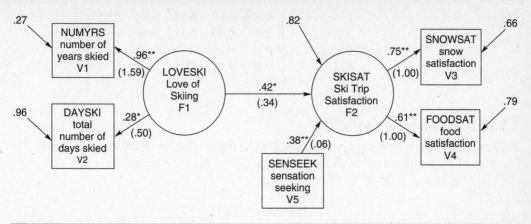

FIGURE 14.5   FINAL MODEL FOR SMALL SAMPLE EXAMPLE WITH STANDARDIZED AND (UNSTANDARDIZED) COEFFICIENTS.

$$\hat{\mathbf{B}} = \begin{bmatrix} 0.00 & 0.00 & 0.00 & 0.00 & 0.00 \\ 0.00 & 0.00 & 0.00 & 0.00 & 0.00 \\ 0.00 & 0.00 & 0.00 & 0.00 & 1.00 \\ 0.00 & 0.00 & 0.00 & 0.00 & 1.00 \\ 0.00 & 0.00 & 0.00 & 0.00 & 0.00 \end{bmatrix}$$

$$\hat{\boldsymbol{\gamma}} = \begin{bmatrix} 0.00 & 1.59 & 1.00 & 0.00 & 0.00 & 0.00 & 0.00 \\ 0.00 & 0.50 & 0.00 & 1.00 & 0.00 & 0.00 & 0.00 \\ 0.00 & 0.00 & 0.00 & 0.00 & 1.00 & 0.00 & 0.00 \\ 0.00 & 0.00 & 0.00 & 0.00 & 0.00 & 1.00 & 0.00 \\ 0.06 & 0.34 & 0.00 & 0.00 & 0.00 & 0.00 & 1.00 \end{bmatrix}$$

$$\hat{\boldsymbol{\Phi}} = \begin{bmatrix} 27.00 & 0.00 & 0.00 & 0.00 & 0.00 & 0.00 & 0.00 \\ 0.00 & 1.00 & 0.00 & 0.00 & 0.00 & 0.00 & 0.00 \\ 0.00 & 0.00 & 0.20 & 0.00 & 0.00 & 0.00 & 0.00 \\ 0.00 & 0.00 & 0.00 & 3.00 & 0.00 & 0.00 & 0.00 \\ 0.00 & 0.00 & 0.00 & 0.00 & 0.51 & 0.00 & 0.00 \\ 0.00 & 0.00 & 0.00 & 0.00 & 0.00 & 1.15 & 0.00 \\ 0.00 & 0.00 & 0.00 & 0.00 & 0.00 & 0.00 & 0.45 \end{bmatrix}$$

The final estimated population covariance matrix is given by $\hat{\boldsymbol{\Sigma}}$. For the example:

$$\hat{\boldsymbol{\Sigma}} = \begin{bmatrix} 2.74 & 0.80 & 0.55 & 0.55 & 0.00 \\ 0.80 & 3.25 & 0.17 & 0.17 & 0.00 \\ 0.55 & 0.17 & 1.18 & 0.67 & 1.64 \\ 0.55 & 0.17 & 0.67 & 1.82 & 1.64 \\ 0.00 & 0.00 & 1.64 & 1.64 & 27.00 \end{bmatrix}$$

$$\text{The final residual matrix} = \begin{bmatrix} 0.00 & 0.00 & 0.13 & 0.10 & 2.02 \\ 0.00 & 0.00 & 0.11 & 0.18 & 2.12 \\ 0.13 & 0.11 & 0.05 & 0.05 & 0.38 \\ 0.10 & 0.18 & 0.05 & 0.05 & 0.48 \\ 2.02 & 2.12 & 0.38 & 0.48 & 0.00 \end{bmatrix}$$

14.4.5 ■ Model Evaluation

A χ^2 statistic is computed based upon the function minimum when the solution has converged. The minimum of the function was .08924 in this example. This value is multiplied by $N - 1$, ($N =$ number of participants) to yield the χ^2 value:

$$(.08924)(99) = 8.835$$

This χ^2 is evaluated with degrees of freedom equal to the difference between the total number of degrees of freedom and the number of parameters estimated. The total number of degrees of freedom equals the number of data points but *the data points in SEM are the variances and covariances of the observed variables*. In a model with a few variables it is easy to count the number of variances and covariances; however, the number of data points in larger models is calculated as,

$$\text{number of data points} = \frac{p(p + 1)}{2} \tag{14.17}$$

where p equals the number of observed variables.

In this example with 5 measured variables, there are $(5)(6)/2 = 15$ data points (5 variances and 10 covariances). The estimated model includes 11 parameters (5 regression coefficients and 6 variances) so χ^2 is evaluated with 4 df, $\chi^2 (4, N = 100) = 8.835, p = .065$.

Because the goal is to develop a model that fits the data, a *nonsignificant* chi square is desired. This χ^2 is not statistically significant at $p = .05$ so we conclude that the model fits the data. However, chi-square values depend on sample sizes; in models with large samples, trivial differences often cause the χ^2 to be significant solely because of sample size. For this reason, many fit indices have been developed that assess model fit while eliminating or minimizing the effect of sample size. All fit indices for this model indicate an adequate, but not spectacular, fit. Fit indices are discussed fully in Section 14.5.3.

The model fits, but what does it mean? The hypothesis is that the observed covariances among the measured variables occur because of the linkages among variables specified in the model; because the chi–square is not significant, we conclude that we should not reject our hypothesized model.

Next, researchers usually examine the statistically significant relationships within the model. If the unstandardized coefficients in the three parameter matrices are divided by their respective standard errors, a z score is obtained for each parameter that is evaluated in the usual manner,

$$z = \frac{\text{parameter estimate}}{\text{std. error for estimate}} \tag{14.18}$$

For NUMYRS predicted from LOVESKI: $\dfrac{1.59}{.74} = 2.16, p < .05$

DAYSKI predicted from LOVESKI: $\dfrac{.50}{.29} = 1.73, p < .05$

FOODSAT predicted from SKISAT: $\dfrac{1.00}{.28} = 3.53, p < .01$

SKISAT predicted from SENSEEK: $\dfrac{.06}{.02} = 3.33, p < .01$

SKISAT predicted from LOVESKI: $\dfrac{.34}{.18} = 1.86, p < .05$

Because of differences in scales, it is sometimes difficult to interpret unstandardized regression coefficients; therefore, researchers often examine standardized coefficients. Both the standardized and unstandardized regression coefficients for the final model are in Figure 14.5. The unstandardized coefficients are in parentheses. The paths from the factors to the variables are just standardized factor loadings. It could be concluded that number of years skied (NUMYRS) is a significant indicator of Love of Skiing (LOVESKI): the greater the Love of Skiing, the higher the number of years skied. Number of total days skied (DAYSKI) is a significant indicator of Love of Skiing, (i.e., greater Love of Skiing predicts more total days skied), because there was an a priori hypothesis that stated a positive relationship. Degree of food satisfaction (FOODSAT) is a significant indicator of ski trip satisfaction (SKISAT), higher Ski Trip Satisfaction predicts greater satisfaction with the food. Because the path from SKISAT to SNOWSAT is fixed to 1 for identification, a standard error is not calculated. If this standard error is desired, a second run is performed with the FOODSAT path fixed instead. Higher SENSEEK predicts higher SKISAT. Lastly, greater Love of Skiing (LOVESKI) significantly predicts higher Ski Trip Satisfaction (SKISAT) because this relationship is also tested as an a priori, unidirectional hypothesis.

14.4.6 ■ Computer Analysis of Small Sample Example

Tables 14.2, 14.3, 14.5, 14.6, and 14.7 show setup and minimal selected output for computer analyses of the data in Table 14.1 using EQS, SAS CALIS, LISREL, SIMPLIS LISREL, and RAMONA SYSTAT respectively. The setup and output for the programs are all quite different.

As seen in Table 14.2 and described in Section 14.4.3, the model is specified in EQS using a series of regression equations. In the /EQUATIONS section, as in ordinary regression, the DV appears on the left side of the equation, the IVs on the right side. Measured variables are referred to by the letter V and the number corresponding to the variable given in the /LABELS section. Errors associated with measured variables are indicated by the letter E and the number of the variable. Factors are referred to with the letter F and a number given in the /LABELS section. The errors, or disturbances, associated with factors are referred to by the letter D and the number corresponding to the factor. An asterisk indicates a parameter to be estimated. Variables included in the equation without asterisks are considered fixed to the value 1. In this example start values are not specified and are estimated automatically by the program, hence the asterisks. The variances of IVs

TABLE 14.2 STRUCTURAL EQUATION MODEL OF SMALL SAMPLE EXAMPLE THROUGH EQS (SETUP AND SELECTED OUTPUT)

```
   /TITLE
    SMALL SAMPLE STRUCTURAL EQUATION MODEL
   /SPECIFICATIONS
    CAS = 100; VAR = 5; ME = ML;
   /LABELS
    V1 = NUMYRSS; V2 = DAYSKI; V3 = SNOWSAT;
    V4 = FOODSAT; V5 = SENSEEK;
    F1 = LOVESKI; F2= SKISAT;
   /EQUATIONS
    V1 = *F1 + E1;
    V2 = *F1 + E2;

    V3 = F2 + E3;
    V4 =*F2 + E4;

    F2 =  *V5 + *F1 + D2;
   /VAR
    F1 = 1;
    E1 to E4 = *;
    V5 = *;
    D2 = *;
   /PRINT
    FIT = ALL;
   /MATRIX
    2.74
     .80 3.25
     .68  .28 1.23
     .65  .35  .72 1.87
    2.02 2.12 2.02 2.12  27.00
   /END
```

GOODNESS OF FIT SUMMARY

INDEPENDENCE MODEL CHI-SQUARE =   68.865 ON   10 DEGREES OF FREEDOM

INDEPENDENCE AIC =   48.86550   INDEPENDENCE CAIC =   12.81380
      MODEL AIC =   0.83463   MODEL CAIC =   -13.58605

CHI-SQUARE =   8.835 BASED ON   4 DEGREES OF FREEDOM
PROBABILITY VALUE FOR THE CHI-SQUARE STATISTIC IS   0.06537
THE NORMAL THEORY RLS CHI-SQUARE FOR THIS ML SOLUTION IS   8.471.

BENTLER-BONETT NORMED   FIT INDEX=   0.872
BENTLER-BONETT NONNORMED FIT INDEX=   0.795
COMPARATIVE FIT INDEX   =   0.918
BOLLEN (IFI)   FIT INDEX=   0.925
McDonald (MFI)   FIT INDEX=   0.976
LISREL GFI   FIT INDEX=   0.967
LISREL AGFI   FIT INDEX=   0.876
ROOT MEAN SQUARED RESIDUAL   =   0.776
STANDARDIZED RMR   =   0.021

MEASUREMENT EQUATIONS WITH STANDARD ERRORS AND TEST STATISTICS

NUMYRS =V1 =   1.5934*F1   + 1.0000 E1
                 .7374
                2.1609

DAYSKI=V2 =   .5021*F1   + 1.0000 E2
                 .2899
                1.7317
```
```

TABLE 14.2   *(CONTINUED)*

```
SNOWSAT =V3   =    1.0000 F2      + 1.0000 E3



FOODSAT =V4   =    1.0017*F2      + 1.0000 E4
                    .2834
                   3.5350


CONSTRUCT EQUATIONS WITH STANDARD ERRORS AND TEST STATISTICS


SKISAT  =F2   =     .0607*V5   +   .3437*F1    + 1.0000 D2
                    .0182          .1850
                   3.3306         1.8576

   STANDARDIZED SOLUTION:


NUMYRS  =V1   =     .9626*F1   +   .2710 E1
DAYSKI  =V2   =     .2785*F1   +   .9604 E2
SNOWSAT =V3   =     .7529 F2   +   .6582 E3
FOODSAT =V4   =     .6073*F2   +   .7945 E4
SKISAT  =F2   =     .3858*V5   +   .4202*F1   +   .8213 D2
```

are parameters of the model and are indicated in the /VAR paragraph. The data appear as a covariance matrix in the paragraph labeled /MATRIX. In the /PRINT paragraph, FIT=ALL requests all available goodness-of-fit indices.

The output is heavily edited. After much diagnostic information (not included here), goodness-of-fit indices are given in the section labeled GOODNESS OF FIT SUMMARY. The independence model chi square is labeled INDEPENDENCE CHI-SQUARE. The independence model chi square tests the hypothesis that there is no relationship among the variables. This chi square should always be significant, indicating that there is *some* relationship among the variables. CHI-SQUARE is the model chi square that ideally should be nonsignificant. Several different goodness-of-fit indices are given (cf. Section 14.5.3) beginning with BENTLER-BONETT NORMED FIT INDEX. Significance tests for each parameter of the measurement portion of the model are found in the section labeled MEASUREMENT EQUATIONS WITH STANDARD ERRORS AND TEST STATISTICS. The unstandardized coefficient appears on the first line; immediately below it is the standard error for that parameter. The z score associated with the parameter (the unstandardized coefficient divided by its standard error) is given on the third line. The section labeled CONSTRUCT EQUATIONS WITH STANDARD ERRORS AND TEST STATISTICS contains the unstandardized regression coefficients, standard errors, and z score significance tests for predicting factors from other factors and measured variables. The standardized parameter estimates appear in the section labeled STANDARDIZED SOLUTION.

SAS CALIS allows models to be specified in three different ways: two based on regression equations and one based on matrices. The LINEQS option, a regression equation format, is illustrated in Table 14.3. The method of specification, LINEQS, is given, followed by a set of equations. As in EQS, Vs are measured variables, Es are errors associated with Vs (measured variables), Fs are factors and Ds are disturbances (or errors) associated with factors. In contrast to the asterisks in EQS, each parameter to be estimated is given a label. In this example the parameters are labeled X1 to

```
   TITLE "SMALL SAMPLE EXAMPLE IN CALIS";
   DATA CMAT(TYPE=COV);
   _TYPE_='COV';
   INPUT _NAME_ $ V1-V5;
   LABEL V1='NUMYRS' V2='DAYSKI' V3='SNOWSAT'
      V4='FOODSAT' V5='SENSEEK';
   CARDS;
V1   2.74     .       .       .       .
V2    .80    3.25     .       .       .
V3    .68     .28    1.23     .       .
V4    .65     .35     .72    1.87     .
V5   2.02    2.12    2.02    2.12    27.00
;

PROC CALIS COV DATA=CMAT OMETHOD=NR EDF=99 ALL1;
   LINEQS
      V1=X1 F1+E1,
      V2=X2 F1+E2,
      V3=1 F2+E3,
      V4=X4 F2+E4,
      F2=X5 F1+ X6 V5+D2;
   STD
      E1-E4=X7-X10,
      V5=X11,
      D2=X12,
      F1=1;
RUN;
```

              Covariance Structure Analysis: Maximum Likelihood Estimation

              Fit criterion 0.0892
              Goodness of Fit Index (GFI) 0.9669
              GFI Adjusted for Degrees of Freedom (AGFI) . . .   0.8759
              Root Mean Square Residual (RMR) 0.7757
              Chi-square = 8.8346   df = 4   Prob>chi**2 = 0.0654
              Null Model Chi-square: df = 10   68.8655
              Bentler's Comparative Fit Index 0.9179
              Normal Theory Reweighted LS Chi-square 8.4705
              Akaike's Information Criterion 0.8346
              Consistent Information Criterion . . . .   -13.5861
              Schwarz's Bayesian Criterion . . . .   -9.5861
              McDonald's (1989) Centrality. 0.9761
              Bentler & Bonett's (1980) Non-normed Index. . . .   0.7947
              Bentler & Bonett's (1980) Normed Index. 0.8717
              James, Mulaik, & Brett (1982) Parsimonious Index.   0.3487
              Z-Test of Wilson & Hilferty (1931) 1.5182
              Bollen (1986) Normed Index Rho1 . . . .   0.6793
              Bollen (1988) Non-normed Index Delta2 . . . . . .   0.9255
              Hoelter's (1983) Critical N   107

                          Manifest Variable Equations

                    V1   =   1.5927*F1 + 1.0000 E1
                    Std Err   0.7326 X1
                    t Value   2.1740

                    V2   =   0.5023*F1 + 1.0000 E2
                    Std Err   0.2884 X2
                    t Value   1.7418

                    V3   =   1.0000 F2 + 1.0000 E3

                    V4   =   1.0015*F2 + 1.0000 E4
                    Std Err   0.2819 X4
                    t Value   3.5533

TABLE 14.3   *(CONTINUED)*

```
              Latent Variable Equations

    F2       =    0.0607*V5 + 0.3439*F1 + 1.0000 D2
    Std Err       0.0181 X6   0.1841 X5
    t Value       3.3477      1.8688


              Equations with Standardized Coefficients

    V1       =    0.9622*F1 + 0.2723 E1
                       X1

    V2       =    0.2786*F1 + 0.9604 E2
                       X2

    V3       =    0.7529 F2 + 0.6581 E3

    V4       =    0.6072*F2 + 0.7945 E4
                       X4

    F2       =    0.3858*V5 + 0.4205*F1 + 0.8212 D2
                       X6           X5
```

X12. Parameters included in the model without a label are fixed at 1. The variances are specified in the section labeled STD.

Again, much output has been deleted. The goodness-of-fit information begins with the Fit criterion. The model chi square is labeled Chi-square. The independence chi-square is labeled Null Model Chi-Square. Many fit indices are also included. The unstandardized regression coefficients among the measured variables appear in the section labeled Manifest Variable Equations. The standard errors appear on the line labeled Std Err, the *t* values on the line labeled t Value. These *t* values are the same as the z tests in EQS and LISREL. The unstandardized regression coefficients for the factors are given in the section labeled Latent Variable Equations. The standardized solution is given in the section labeled Equations with Standardized Coefficients.

LISREL offers two very different methods of specifying models. SIMPLIS uses equations and LISREL employs matrices. Neither program allows the exact model specified in Figure 14.4 to be tested. Underlying both these programs is the LISREL model which, although similar to the Bentler-Weeks models, employs eight matrices instead of three. The matrices of the LISREL model that correspond to the Bentler-Weeks model are given in Table 14.4. There is no matrix of regression coefficients for predicting latent DVs from measured IVs in the LISREL model. Figure 14.6 illustrates a little trick to estimate these parameters. A dummy "latent" variable with one indicator[6] is specified, in this example SENSEEK. The dummy latent variable then predicts SKISAT. The regression coefficient from the dummy "latent" variable to SENSEEK is fixed to one and the error variance of SENSEEK is fixed to zero. With this modification, the solutions are identical, because SENSEEK = (Dummy latent variable) + 0.

The setup and edited output for SIMPLIS appear in Table 14.5. After the covariance matrix and sample size are specified, the latent variables are named. The latent DVs must proceed the latent IVs. The regression equations are specified in the section labeled RELATIONSHIPS. Note that, unlike EQS, asterisks mean that *the parameter is fixed at 1, not estimated.* The error variance of

---

[6] Note this dummy variable is not a true latent variable. A single indicator latent variable is simply a measured variable.

TABLE 14.4 EQUIVALENCE OF MATRICES IN BENTLER-WEEKS AND LISREL MODEL SPECIFICATIONS

Bentler-Weeks Model			LISREL Model			
Symbol	Name	Contents	Symbol	Name	LISREL Two Letter Specification	Contents
B	Beta	matrix of regression coefficients of DVs predicting other DVs	1. **B**	1. Beta	1. BE	1. matrix of regression coefficients of latent DVs predicting other latent DVs
			2. Λ_y	2. Lambda y	2. LY	2. matrix of regression coefficients of measured DVs predicted by latent DVs
γ	Gamma	matrix of regression coefficients of DVs predicted by IVs	1. Γ	1. Gamma	1. GA	1. matrix of regression coefficient of latent DVs predicted by latent IVs
			2. Λ_x	2. Lambda x	2. LX	2. matrix of regression coefficients of measured DVs predicted by latent IVs
Φ	Phi	matrix of covariances among the IVs	1. Φ	1. Phi	1. PI	1. matrix of covariances among the latent IVs
			2. Ψ	2. Psi	2. PS	2. matrix of covariances of errors associated with latent DVs
			3. Θ_δ	3. Theta-Delta	3. TD	3. matrix of covariances among errors associated with measured DVs predicted from latent IVs
			4. Θ_ϵ	4. Theta-Epsilon	4. TE	4. matrix of covariances among errors associated with measured DVs predicted from latent DVs.

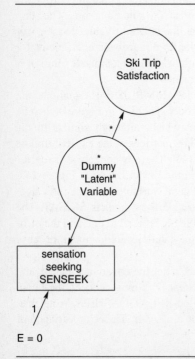

FIGURE 14.6 LISREL ADAPTATION FOR SMALL SAMPLE EXAMPLE.

TABLE 14.5 STRUCTURAL EQUATION MODEL OF SMALL SAMPLE EXAMPLE THROUGH LISREL SIMPLIS LANGUAGE (SETUP AND SELECTED OUTPUT)

```
SMALL SAMPLE EXAMPLE - SIMPLIS
OBSERVED VARIABLES
NUMYRS DAYSKI SNOWSAT FOODSAT SENSEEK
COVARIANCE MATRIX
2.74
 .80   3.25
 .68    .28   1.23
 .65    .35    .72   1.87
2.02   2.12   2.02   2.12   27.00
SAMPLE SIZE 100
LATENT VARIABLES SKISAT LOVESKI DUMMY
RELATIONSHIPS
NUMYRS =  1*LOVESKI
DAYSKI = LOVESKI
SNOWSAT  = 1*SKISAT
FOODSAT = SKISAT
SENSEEK = 1*DUMMY
SKISAT = DUMMY LOVESKI
SET THE ERROR VARIANCE OF SENSEEK TO 0
SET THE COVARIANCES OF LOVESKI - DUMMY TO 0
NUMBER OF DECIMALS= 3
END OF PROBLEM



LISREL ESTIMATES (MAXIMUM LIKELIHOOD)

 SNOWSAT = 1.000*SKISAT, Errorvar.= 0.511 , R2 = 0.567
                                    (0.189)
                                     2.701


 FOODSAT = 1.002*SKISAT, Errorvar.= 1.149 , R2 = 0.369
           (0.283)                  (0.240)
            3.535                    4.796



 NUMYRS = 1.000*LOVESKI, Errorvar.= 0.203 , R2 = 0.926
                                    (2.315)
                                     0.0875


 DAYSKI = 0.315*LOVESKI, Errorvar.= 2.998 , R2 = 0.0776
          (0.306)                  (0.484)
           1.030                    6.191


 SENSEEK = 1.000*DUMMY,, R2 = 1.000


 SKISAT = 0.216*LOVESKI + 0.0607*DUMMY, Errorvar.= 0.451 , R2 = 0.326
          (0.205)          (0.0182)               (0.214)
           1.052            3.331                   2.109
```

GOODNESS OF FIT STATISTICS

CHI-SQUARE WITH 4 DEGREES OF FREEDOM = 8.835 (P = 0.0654)
ESTIMATED NON-CENTRALITY PARAMETER (NCP) = 4.835
90 PERCENT CONFIDENCE INTERVAL FOR NCP = (0.0 ; 17.537)

MINIMUM FIT FUNCTION VALUE = 0.0892
POPULATION DISCREPANCY FUNCTION VALUE (F0) = 0.0488
90 PERCENT CONFIDENCE INTERVAL FOR F0 = (0.0 ; 0.177)
ROOT MEAN SQUARE ERROR OF APPROXIMATION (RMSEA) = 0.110
90 PERCENT CONFIDENCE INTERVAL FOR RMSEA = (0.0 ; 0.210)
P-VALUE FOR TEST OF CLOSE FIT (RMSEA < 0.05) = 0.130

TABLE 14.5   *(CONTINUED)*

```
             EXPECTED CROSS-VALIDATION INDEX (ECVI) = 0.311
      90 PERCENT CONFIDENCE INTERVAL FOR ECVI = (0.263 ; 0.440)
                 ECVI FOR SATURATED MODEL = 0.303
              ECVI FOR INDEPENDENCE MODEL = 0.797

CHI-SQUARE FOR INDEPENDENCE MODEL WITH 10 DEGREES OF FREEDOM = 68.865
                    INDEPENDENCE AIC = 78.865
                          MODEL AIC = 30.835
                      SATURATED AIC = 30.000
                   INDEPENDENCE CAIC = 96.891
                         MODEL CAIC = 70.491
                     SATURATED CAIC = 84.078
           ROOT MEAN SQUARE RESIDUAL (RMR) = 0.776
                   STANDARDIZED RMR = 0.0949
             GOODNESS OF FIT INDEX (GFI) = 0.967
    ADJUSTED GOODNESS OF FIT INDEX (AGFI) = 0.876
  PARSIMONY GOODNESS OF FIT INDEX (PGFI) = 0.258
                 NORMED FIT INDEX (NFI) = 0.872
             NON-NORMED FIT INDEX (NNFI) = 0.795
        PARSIMONY NORMED FIT INDEX (PNFI) = 0.349
           COMPARATIVE FIT INDEX (CFI) = 0.918
           INCREMENTAL FIT INDEX (IFI) = 0.925
             RELATIVE FIT INDEX (RFI) = 0.679
                        CRITICAL N (CN) = 149.779
```

SENSEEK is set to zero. SIMPLIS automatically correlates IVs, so the instruction, SET THE COVARIANCES OF LOVESKI – DUMMY TO 0 fixes the covariance to zero. The edited output contains the unstandardized regression coefficients, standard errors and z tests for the parameters estimated. An R2 for each equation is given; this is the SMC for the variable, indicating the amount of variance in the variable accounted for by the factor.

After printing out the covariance matrix, the goodness-of-fit information appears on the section labeled GOODNESS OF FIT STATISTICS. The model chi square is given first, labeled CHI-SQUARE WITH 4 DEGREES OF FREEDOM. The independence chi square is labeled CHI-SQUARE FOR INDEPENDENCE MODEL WITH 10 DEGREES OF FREEDOM. Again, many additional fit indices are given. SIMPLIS does not offer a standardized (or partially standardized) solution unless LISREL output is requested.

LISREL uses matrices, rather than equations to specify the model. Setup and edited output are presented in Table 14.6. Matrices and commands are given with the two-letter specifications defined in Table 14.4. CM with an asterisk indicates analysis of a covariance matrix. Following LA (for label), the measured variable names are given in the same order as the data. LISREL requires that the DVs appear before the IVs, so the specification SE (select) reorders the variables. The model specification begins with MO. The number of measured DVs is indicated after the key letters NY (number of *Y*s). The number of measured IVs is specified after the key letters NX (number of *X*s). The latent DVs are specified after NE and the latent IVs are specified after NK. Labels are optional, but helpful. The labels for the latent DVs follow the key letters LE and labels for the latent IVs follow the key letters LK.

By default, elements of the matrices are either fixed at zero or are freely estimated. Each matrix is one of four possible shapes: full nonsymmetrical, symmetrical, diagonal, or zero. Matrices are referred to by their two-letter designations. For example, LX (LAMBDA-X) is a full, non-symmetrical, and fixed matrix of regression coefficients for predicting measured DVs from latent IVs.

```
SMALL SAMPLE EXAMPLE LISREL8
DA NI = 5 NO = 100
CM
*
2.74
.80   3.25
.68   .28   1.23
.65   .35   .72   1.87
2.02  2.12  2.02  2.12  27.00
LA
NUMYRS DAYSKI SNOWSAT FOODSAT SENSEEK
SE
SNOWSAT FOODSAT NUMYRS DAYSKI SENSEEK
MO NY=2 NX=3 NE=1 NK=2
LE
SKISAT
LK
LOVESKI DUMMY
FR LX(1,1) LX(2,1) LY(2,1)
FI PH(2,1) TD(3,3)
VA 1 LX(3,2) LY(1,1) PH(1,1)
OU SC SE TV RS SS ND=3
```

LISREL ESTIMATES (MAXIMUM LIKELIHOOD)

LAMBDA-Y

	SKISAT
SNOWSAT	1.000
FOODSAT	1.002
	(.283)
	3.535

LAMBDA-X

	LOVESKI	DUMMY
NUMYRS	1.593	- -
	(.737)	
	2.162	
DAYSKI	.502	- -
	(.290)	
	1.733	
SENSEEK	- -	1.000

GAMMA

	LOVESKI	DUMMY
SKISAT	.344	.061
	(.185)	(.018)
	1.859	3.331

COVARIANCE MATRIX OF ETA AND KSI

	SKISAT	LOVESKI	DUMMY
SKISAT	.669		
LOVESKI	.344	1.000	
DUMMY	1.640	- -	27.000

TABLE 14.6 *(CONTINUED)*

```
PHI

    LOVESKI      DUMMY
    --------    --------
     1.000      27.000
                (3.838)
                 7.036


PSI

    SKISAT
   --------
     .451
    (.214)
     2.109


THETA-EPS

    SNOWSAT    FOODSAT
   --------   --------
     .511       1.149
    (.189)     (.240)
     2.701      4.796
```

```
                GOODNESS OF FIT STATISTICS

     CHI-SQUARE WITH 4 DEGREES OF FREEDOM = 8.835 (P = 0.0654)
          ESTIMATED NON-CENTRALITY PARAMETER (NCP) = 4.835
        90 PERCENT CONFIDENCE INTERVAL FOR NCP = (0.0 ; 17.537)


              MINIMUM FIT FUNCTION VALUE = 0.0892
       POPULATION DISCREPANCY FUNCTION VALUE (F0) = 0.0488
        90 PERCENT CONFIDENCE INTERVAL FOR F0 = (0.0 ; 0.177)
      ROOT MEAN SQUARE ERROR OF APPROXIMATION (RMSEA) = 0.110
      90 PERCENT CONFIDENCE INTERVAL FOR RMSEA = (0.0 ; 0.210)
        P-VALUE FOR TEST OF CLOSE FIT (RMSEA < 0.05) = 0.130


          EXPECTED CROSS-VALIDATION INDEX (ECVI) = 0.311
      90 PERCENT CONFIDENCE INTERVAL FOR ECVI = (0.263 ; 0.440)
             ECVI FOR SATURATED MODEL = 0.303
            ECVI FOR INDEPENDENCE MODEL = 0.797


  CHI-SQUARE FOR INDEPENDENCE MODEL WITH 10 DEGREES OF FREEDOM = 68.865
                  INDEPENDENCE AIC = 78.865
                     MODEL AIC = 30.835
                    SATURATED AIC = 30.000
                  INDEPENDENCE CAIC = 96.891
                     MODEL CAIC = 70.491
                   SATURATED CAIC = 84.078


           ROOT MEAN SQUARE RESIDUAL (RMR) = 0.776
                  STANDARDIZED RMR = 0.0949
            GOODNESS OF FIT INDEX (GFI) = 0.967
        ADJUSTED GOODNESS OF FIT INDEX (AGFI) = 0.876
       PARSIMONY GOODNESS OF FIT INDEX (PGFI) = 0.258


              NORMED FIT INDEX (NFI) = 0.872
            NON-NORMED FIT INDEX (NNFI) = 0.795
        PARSIMONY NORMED FIT INDEX (PNFI) = 0.349
          COMPARATIVE FIT INDEX (CFI) = 0.918
          INCREMENTAL FIT INDEX (IFI) = 0.925
            RELATIVE FIT INDEX (RFI) = 0.679
                 CRITICAL N (CN) = 149.779
```

TABLE 14.6 *(CONTINUED)*

```
COMPLETELY STANDARDIZED SOLUTION

        LAMBDA-Y

              SKISAT
              --------
SNOWSAT        .753
FOODSAT        .607

        LAMBDA-X

              LOVESKI      DUMMY
              --------    --------
NUMYRS         .962         - -
DAYSKI         .279         - -
SENSEEK        - -         1.000

        GAMMA

              LOVESKI      DUMMY
              --------    --------
SKISAT         .420         .386

        CORRELATION MATRIX OF ETA AND KSI

              SKISAT      LOVESKI      DUMMY
              --------    --------    --------
SKISAT        1.000
LOVESKI        .420       1.000
DUMMY          .386        - -       1.000

        PSI

              SKISAT
              --------
               .674

        THETA-EPS

              SNOWSAT     FOODSAT
              --------    --------
               .433        .631

        THETA-DELTA

              NUMYRS      DAYSKI      SENSEEK
              --------    --------    --------
               .074        .922        - -

        REGRESSION MATRIX ETA ON KSI (STANDARDIZED)

              LOVESKI      DUMMY
              --------    --------
SKISAT         .420         .386
```

The model is specified by a combination of freeing (FR) or fixing (FI) elements of the relevant matrices. "Freeing" a parameter means estimating the parameter. When an element of a matrix is fixed with the key letter FI it is fixed at zero. A command line begins with either FI (for fix) or FR (for free). Following this FI or FR specification, the particular matrix and specific element (row, column) that is to be freed or fixed is indicated. For example, from Table 14.6, FR LX(1,1) means free (FR) the element of the LAMBDA-X matrix (LX) that is in the first row and the first column (1,1), i.e., the factor loading of NUMYRS on LOVESKI. Similarly FI PH(2,1) indicates that the covariance that is in the 2nd row, 1st column (2,1) of the phi matrix (PH) is fixed to zero (FI) (i.e., there is no relationship between LOVESKI and DUMMY).

In this example, LX (LAMBDA–X) is a 3 × 2 full and fixed matrix of regression coefficients of measured variables predicted by latent IVs. The rows are the three measured variables that are the indicators of latent IVs: NUMYRS, DAYSKI, SENSEEK. The columns are the latent IVs: LOVE-SKI and DUMMY. LY (LAMBDA–Y) is a full and fixed matrix of the regression coefficients predicting measured DVs from the latent DV. In this example LY is a 2 × 1 vector. The rows are the measured variables SNOWSAT and FOODSAT and the column is SKISAT. The PH (PHI matrix) of variances and covariances among latent IVs is by default symmetrical and free. In this example phi is a 2 × 2 matrix. No covariance is specified between the dummy latent variable and LOVE-SKI therefore PH(2,1) is fixed, FI. To estimate this model the error variance associated with SENSEEK must be fixed to zero. This is done by specifying FI TD(3,3). TD refers to the THETA DELTA matrix (errors associated with measured variables serving as indicators of latent IVs); by default this matrix is diagonal and free. A diagonal matrix has zeros everywhere but the main diagonal. In the small sample example it is a 3 × 3 matrix.

Only four of the eight LISREL matrices (LX, LY, PH, and TD) are included on the model (MO) line. LISREL matrices have particular shapes and elements specified by default. If these defaults are appropriate for the model there is no need to mention the unmodified matrices on the MO line. In this example the default specifications for TE, GA, PS, and BE are all appropriate. TE (theta epsilon) is diagonal and free by default. TE contains the covariances associated with the measured DVs associated with the latent DVs. In this example, it is a 2 × 2 matrix. Gamma (GA) contains the regression coefficients of latent IVs predicting latent DVs. By default this matrix is full and free. In this example GA is a 1 × 2 vector. PS contains the covariances among errors associated with latent DVs; by default it is diagonal and free. In the small sample example there is only 1 latent DV, therefore, PS is simply a scalar (a number). BE contains the regression coefficients among the latent DVs; by default a matrix of zeros. The small sample example contains no relationships among latent DVs; there is only 1 latent DV so there is no need to mention BE.

Finally, for identification, a path is fixed to 1 on each factor and the variance of LOVESKI is fixed at the value 1. (See Section 14.5.1 for a discussion of identification). This is accomplished with the key letters VA 1 and the relevant matrices and corresponding elements. The OU line specifies output options (SC completely standardized solution, SE standard errors, TV *t* values, RS residual information, SS standardized solution, ND number of decimal places), not all of which are included in the edited output.

The highly edited output provides the unstandardized regression coefficients, standard errors for the regression coefficients, and *t* tests (unstandardized regression coefficient divided by standard error) by matrix in the section labeled LISREL ESTIMATES (MAXIMUM LIKELIHOOD). The statistical significance of parameters estimates is determined with a table of *t* distributions (Table C.2). A *t* statistic greater than 1.96 is needed for significance at $p < .05$ and 2.56 for significance at $p < .01$. These are two tailed tests. If the direction of the effect has been hypothesized a priori a one-tailed test is appropriate, $t = 1.65$, $p < .05$, one-tailed. The goodness-of-fit summary is labeled GOODNESS OF FIT STATISTICS and is identical in format and content to the SIMPLIS OUTPUT. A partially standardized solution appears, by matrix, in the section labeled COMPLETELY STANDARDIZED SOLUTION. The regression coefficients, and variances and covariances, are completely standardized (latent variable mean of 0, sd = 1, observed variable mean = 0, sd = 1) and are identical to the standardized solution in EQS and CALIS.

The error variances given in COMPLETELY STANDARDIZED SOLUTION for both measured variables and latent variables actually are not completely standardized and differ in terms of the residuals from both EQS and CALIS (Chou & Bentler, 1993). An option in LISREL (not shown) is the STANDARDIZED SOLUTION, a second type of partially standardized solution in which the latent variables are standardized to a mean = 1 and sd = 0 but the observed variables remain in their original scale.

SYSTAT RAMONA employs an equation method to specify models. Setup and edited output are presented in Table 14.7. After the SYSTAT subprogram name, RAMONA, is specified and the data file named, (SMSAMDAT), the observed variables are listed in the order of the data set following the MANIFEST keyword. The latent variables including all residual variables are listed following LATENT. In RAMONA, residuals associated with DVs need labels, but these labels can be anything you like. In this example, residuals are labeled to be consistent with the small sample example diagram. Residuals of measured variables are labeled E with the corresponding measured variable number; residuals of latent variables are labeled D with the corresponding latent variable number.

The model is specified after the MODEL command. The DV appears on the left-hand side of the equation and is referred to by name. The symbol $<-$ (a single-headed arrow) means the DV is predicted by the variables on the right-hand side of the equation. The information about the estimation status of each parameter is given in parentheses following the parameter. For example, consider the equation: NUMYRS $<-$ LOVESKI (*,*) E1 (0, 1.0). Within the parentheses the first column numbers the parameter. If the parameter is estimated, either a parameter number or an asterisk can be given. If the parameter is fixed, a zero must be placed in the first column. The second column indicates the start value for the parameter. Either a start value or an asterisk can be provided. An asterisk indicates that the program will estimate start values. If a parameter is fixed, i.e., a 0 is in the first column, a value for the parameter must be given in the second column. In the NUMYRS example the path predicting NUMYRS from LOVESKI is estimated (first column asterisk) and RAMONA will provide the start value (second column asterisk). NUMYRS is also predicted by a residual, E1. The path predicting NUMYRS from E1 is fixed (zero first column) to one (one in the second column). If parameters are constrained to be equal they are given the same parameter number. Variances and covariances are indicated with a double-headed arrow, $<->$, with the same format within the parentheses to indicate the estimation status of the parameter.

After the model has been specified, the output file is named (SMSAM.DOC). The estimation method information is indicated after the ESTIMATE keyword. The default estimation method is maximum likelihood. The type of matrix analyzed is given after DISP =. The number of cases is indicated after NCASES=.

The edited output contains the unstandardized coefficients labeled Point Estimate, the 90% confidence interval for each parameter estimate, the standard error associated with the estimate, and the t value associated with each parameter. In the section labeled Maximum Likelihood Discrepancy Function the model chi-square value (8.835) is labeled Test statistic. The degrees of freedom associated with the chi-square value appear on the next to last line of Table 14.7. The degrees of freedom associated with the chi-square value are labeled Degrees of Freedom. The independence model chi-square is not given[7]. Three fit indices are included with their associated confidence intervals.

---

[7] If the independence chi-square value is needed a model can be estimated with only the variances of the measured variables as parameters.

```
RAMONA
USE SMSAMDAT
MANIFEST  NUMDAYS NUMMILES SNOWSAT FOODSAT SENSEEK
LATENT   LOVESKI SKISAT E1 E2 E3 E4 D1
MODEL NUMDAYS <- LOVESKI(*,*) E1(0,1.0),
     NUMMILES <- LOVESKI(*,*) E2(0,1.0),
     SNOWSAT <- SKISAT(0,1.0) E3(0,1.0),
     FOODSAT <- SKISAT(*,*) E4(0,1.0),
     SKISAT <- LOVESKI(*,*) SENSEEK(*,*) D1(0,1.0),
      LOVESKI <-> LOVESKI(0,1.0),
     SENSEEK <-> SENSEEK(*,*),
     E1 <-> E1(*,*),
     E2 <-> E2(*,*),
     E3 <-> E3(*,*),
     E4 <-> E4(*,*),
     D1 <-> D1(*,*)
OUTPUT SMSAMD.DOC
ESTIMATE / DISP=COVA NCASES = 100
```

ML Estimates of Free Parameters in Dependence Relationships

	Path	Param #	Point Estimate	90.00% Conf. Int. Lower	Upper	Standard Error	T Value
NUMDAYS	<- LOVESKI	1	1.593	0.381	2.805	0.737	2.16
NUMMILES	<- LOVESKI	2	0.502	0.025	0.979	0.290	1.73
FOODSAT	<- SKISAT	3	1.002	0.536	1.468	0.283	3.54
SKISAT	<- LOVESKI	4	0.344	0.040	0.648	0.185	1.86
SKISAT	<- SENSEEK	5	0.061	0.031	0.091	0.018	3.33

Maximum Likelihood Discrepancy Function

Measures of fit of the model

Sample Discrepancy Function Value   : 0.089   (8.923864E-02)

Population discrepancy function value, Fo
Bias adjusted point estimate   : 0.049
90.000 percent confidence interval   :(0.0,0.177)

Root mean square error of approximation
Steiger-Lind : RMSEA = SQRT(Fo/DF)
Point estimate   : 0.110
90.000 percent confidence interval   :(0.0,0.210)

Expected cross-validation index
Point estimate (modified aic)   : 0.311
90.000 percent confidence interval   :(0.263,0.440)
CVI (modified AIC) for the saturated model   : 0.303

Test statistic: : 8.835
Exceedance probabilities:-
Ho: perfect fit (RMSEA = 0.0)   : 0.065
Ho: close fit   (RMSEA <=   0.050)   : 0.130

Multiplier for obtaining test statistic =   99.000
Degrees of freedom   = 4
Effective number of parameters   = 11

14.5 ■ SOME IMPORTANT ISSUES

14.5.1 ■ Model Identification

In SEM a model is specified, parameters for the model are estimated using sample data, and the parameters are combined to produce the estimated population covariance matrix. But only models that are identified can be estimated. A model is said to be identified if there is a unique numerical solution for each of the parameters in the model. For example, say that both the variance of $Y = 10$ and the variance of $Y = \alpha + \beta$. Any two values can be substituted for α and β as long as they sum to 10. There is no unique numerical solution for either α or β, i.e., there are an infinite number of combinations of two numbers that would sum to 10. Therefore, this single equation model is not identified. However, if we fix α to 0 then there is a unique solution for β, 10, and the equation is identified. It is possible to use covariance algebra to calculate equations and assess identification in very simple models; however, in large models this procedure quickly becomes unwieldy. For a detailed, technical discussion of identification, see Bollen (1989b). The following guidelines are rough, but may suffice for many models.

The first step is to count the number of data points and the number of parameters that are to be estimated. *The data in SEM are the variances and covariances in the sample covariance matrix.* The number of data points is the number of sample variances and covariances (found through Equation. 14.17). The number of parameters is found by adding together the number of regression coefficients, variances, and covariances that are to be estimated (i.e., the number of asterisks in a diagram).

If there are more data points than parameters to be estimated, the model is said to be over-identified, a necessary condition for proceeding with the analysis. If there are the same number of data points as parameters to be estimated, the model is said to be just-identified. In this case, the estimated parameters perfectly reproduce the sample covariance matrix, chi-square and degrees of freedom are equal to zero, and the analysis is uninteresting because hypotheses cannot be tested. If there are fewer data points than parameters to be estimated, the model is said to be under-identified and parameters cannot be estimated. The number of parameters needs to be reduced by fixing, constraining, or deleting some of them. A parameter may be fixed by setting it to a specific value or constrained by setting the parameter equal to another parameter.

In the small sample example of Figure 14.4, there are 5 measured variables, so there are 15 data points: $5(5 + 1)/2 = 15$ (5 variances and 10 covariances). There are 11 parameters to be estimated in the hypothesized model: 5 regression coefficients and 6 variances. The hypothesized model has 4 fewer parameters than data points, so the model may be identified.

The second step in determining model identifiability is to examine the measurement portion of the model. The measurement part of the model deals with the relationship between the measured variables and the factors. It is necessary both to establish the scale of each factor and to assess the identifiability of this portion of the model.

To establish the scale of a factor, you either fix the variance of the factor to 1, or fix to 1 the regression coefficient from the factor to one of the measured variables (perhaps a marker variable cf, Section 13.3.1). Fixing the regression coefficient to 1 gives the factor the same variance as the measured variable. If the factor is an IV, either alternative is acceptable. If the factor is a DV, most researchers fix the regression coefficient to 1. In the small sample example, the variance of the Love of Skiing factor was set to 1 (normalized), while the scale of the Ski Trip Satisfaction factor was set equal to the scale of the snow satisfaction variable.

To establish the identifiability of the measurement portion of the model, look at the number of factors and the number of measured variables (indicators) loading on each factor. If there is only one factor, the model may be identified if the factor has at least three indicators with nonzero loading and the errors (residuals) are uncorrelated with one another. If there are two or more factors, again consider the number of indicators for each factor. If each factor has three or more indicators, the model may be identified if errors associated with the indicators are not correlated, each indicator loads on only one factor and the factors are allowed to covary. If there are only two indicators for a factor, the model may be identified if there are no correlated errors, each indicator loads on only one factor, and none of the variances or covariances among factors is equal to zero.

In the small sample example, there are two indicators for each factor. The errors are uncorrelated and each indicator loads on only one factor. Additionally, the covariance between the factors is not zero. Therefore, this part of the model may be identified. Please note that identification may still be possible if errors are correlated or variables load on more than one factor, but determining it is more complicated.

The third step in establishing model identifiability is to examine the structural portion of the model, looking only at the relationships among the latent variables (factors). Ignore the measured variables for the moment; consider only the structural portion of the model that deals with the regression coefficients relating latent variables to one another. If none of the latent DVs predict each other (the beta matrix is all zeros) the structural part of the model may be identified. The small sample example has only one latent DV so this part of the model may be identified. If the latent DVs do predict one another, look at the latent DVs in the model and ask if they are recursive or nonrecursive. If the latent DVs are recursive there are no feedback loops among them, and there are no correlated disturbances (errors) among them. (In a feedback loop, DV1 predicts DV2 and DV2 predicts DV1. That is, there are two lines linking the factors, one with an arrow in one direction and the other line with an arrow in the other direction. Correlated disturbances are linked by single curved lines with double-headed arrows.) If the structural part of the model is recursive, it may be identified. These rules also apply to path analysis models with only measured variables. The small sample example is a recursive model and therefore may be identified.

If a model is nonrecursive either there are feedback loops among the DVs or there are correlated disturbances among the DVs, or both. Two additional conditions are necessary for identification of nonrecursive models, each condition applying to each equation in the model separately. The first condition is that each equation has at least the number of latent DVs minus 1 excluded from it. The second condition is that the *information matrix* (a matrix necessary for calculating standard errors) is full rank and can be inverted. The inverted information matrix can be examined in the output from most SEM programs. If after examining the model, the number of data points exceeds the number of parameters estimated and both the measurement and structural parts of the model are identified there is good evidence that the whole model is identified.

Identification is often difficult to establish and frequently, despite the best laid plans, problems emerge. One extremely common error that leads to identification problems is failure to set the scale of a factor. In the small sample example, if we had forgotten to set the scale of the Ski Trip Satisfaction factor, each of the programs would have indicated a problem, as follows.

EQS signals this type of identification problem with the following message:

```
PARAMETER        CONDITION CODE
V3,F2            LINEARLY DEPENDENT ON OTHER PARAMETERS
```

This message usually indicates an identification problem either with the particular variables mentioned, as in this case, or in the general neighborhood of the variables mentioned.

LISREL prints the following message, given the same identification problem,

W_A_R_N_I_N_G: TE 2,2 may not be identified.
Standard Errors, T-Values, Modification Indices,
and Standardized Residuals cannot be computed.

TE 2, 2 refers to an element in the theta epsilon matrix (cf. Table 14.4). This indicates that the error variance for SNOWSAT may not be identified.

SIMPLIS automatically fixes this error and does not print any warning about identification or about its solution to the problem. Potentially, this could lead to some confusion.

SAS CALIS prints out the following message,

NOTE: Hessian matrix is not full rank. Not all parameters are identified.
Some parameter estimates are linearly related to other parameter estimates as shown in the following equations:
X12 = 1.5727 X3 + 1.5751 X4 − 0.0756 X6 − 0.4283 X5 −3.0478

Note that SAS CALIS also includes the equation in which the identification problem may have occurred.

SYSTAT RAMONA automatically fixes the problem, in the sense that the χ^2 is correct, and indicates the problem in the table of maximum likelihood estimates (Table 14.8)

When these messages occur, and despite the best of intentions they will, often it is helpful to compare the diagram of the model with the program input to be absolutely certain that everything on the diagram matches the input, and that every factor has a scale. Application of the few basic principles in this section will solve many identification problems.

Another common error is to fix both the factor variance to 1 and a path from the factor to an indicator to 1. This does not lead to an identification problem but does imply a very restricted model that almost certainly will not fit your data.

TABLE 14.8   EDITED OUTPUT FROM SYSTAT RAMONA SHOWING IDENTIFICATION PROBLEM MESSAGE WHEN SCALE OF SKISAT NOT SPECIFIED

ML Estimates of Free Parameters in Dependence Relationships

	Path	Param #	Point Estimate	90.00% Conf. Int. Lower	Upper	Standard Error	T Value
NUMYRS	<- LOVESKI	1	1.593	0.381	2.805	0.737	2.16
DAYSKI	<- LOVESKI	2	0.502	0.025	0.979	0.290	1.73
SNOWSAT	<- SKISAT	3 **	0.438				0.0
FOODSAT	<- SKISAT	4	0.438	0.234	0.642	0.124	3.54
SKISAT	<- LOVESKI	5	0.786	0.091	1.481	0.423	1.86
SKISAT	<- SENSEEK	6	0.139	0.070	0.207	0.042	3.33

** Denotes a redundant parameter

14.5.2 ■ Estimation Techniques

After a model is specified, population parameters are estimated with the goal of minimizing the difference between the observed and estimated population covariance matrices. To accomplish this goal a function, Q, is minimized where

$$Q = (s - \sigma(\Theta))'W(s - \sigma(\Theta)) \tag{14.19}$$

> s is the vector of data (the observed sample covariance matrix stacked into a vector); σ is the vector of the estimated population covariance matrix (again, stacked into a vector) and (Θ) indicates that σ is derived from the parameters (the regression coefficients, variances and covariances) of the model. W is the matrix that weights the squared differences between the sample and estimated population covariance matrix.

Recall that in factor analysis (Chapter 13) the observed and reproduced correlation matrices are compared. This notion is extended in SEM to include a statistical test of the difference. If the weight matrix, W, is chosen correctly to minimize Q, Q multiplied by $(N-1)$ yields a chi-square test statistic.

The trick is to select W to minimize the squared differences between observed and estimated population covariance matrices. In an ordinary chi square (Chapter 3), the weights are the set of expected frequencies in the denominators of the cells. If we use some other numbers instead of the expected frequencies, the result might be some sort of test statistic, but it would not be a χ^2 statistic; the weight matrix would be wrong.

In SEM, estimation techniques vary by the choice of W. A summary of the most popular estimation techniques and the corresponding functions minimized is presented in Table 14.9[8]. Unweighted least squares estimation (ULS) does not standardly yield a χ^2 statistic or standard errors. Because researchers are usually interested in the test statistic, ULS estimation is not discussed further (see Bollen, 1989b, for further discussion of ULS).

Other estimation methods are GLS (Generalized Least Squares), ML (Maximum Likelihood), EDT (Elliptical Distribution Theory), and ADF (Asymptotically Distribution Free). Satorra and Bentler (1988) have also developed an adjustment for nonnormality that can be applied to the chi-square test statistic following any of the estimation procedures. Briefly, the Satorra-Bentler Scaled χ^2 is a correction to the χ^2 test statistic. EQS also corrects the standard errors to adjust for the extent of nonnormality (Bentler & Dijkstra, 1985). This adjustment to the standard errors and the Satorra-Bentler scaled chi-square have so far been implemented only in the ML estimation procedure in EQS.

The performance of the χ^2 test statistic derived from these different estimation procedures is affected by several factors, among them (1) sample size, (2) nonnormality of the distribution of errors, of factors, and of errors and factors, and (3) violation of the assumption of independence of factors and errors. A goal is to select an estimation procedure that, in Monte Carlo studies, produces a test statistic that neither rejects nor accepts the true model too many times as defined by a pre-specified alpha level, commonly $p < .05$. The following sections summarize the performance of estimation procedures examined in Monte Carlo studies by Hu, Bentler, and Kano (1992). They varied sample size from 150 to 5000 and examined the performance of test statistics

---

[8] Really, it's not *that* technical! See Appendix A for additional guidance in deciphering the equations.

TABLE 14.9 SUMMARY OF ESTIMATION TECHNIQUES AND CORRESPONDING FUNCTION MINIMIZED

Estimation Method	Function minimized	Interpretation of $\mathbf{W}$, the weight matrix
Unweighted Least Squares[a] (ULS)	$F_{ULS} = \frac{1}{2}\text{tr}\Big[(\mathbf{S} - \sum(\Theta))^2\Big]$	$\mathbf{W} = \mathbf{I}$, the identity matrix
Generalized Least Squares (GLS)	$F_{GLS} = \frac{1}{2}\text{tr}\left\{\left[(\mathbf{S} - \sum(\Theta))\mathbf{W}^{-1}\right]^2\right\}$	$\mathbf{W} = \mathbf{S}$. $\mathbf{W}$ is any consistent estimator of $\sum$. Often the sample covariance matrix, $\mathbf{S}$, is used
Maximum Likelihood (ML)	$F_{ML} = \log|\sum| - \log|\mathbf{S}| + \text{tr}\left(\mathbf{S}\sum^{-1}\right) - p$	$\mathbf{W} = \sum^{-1}$, the inverse of the estimated population covariance matrix. The number of measured variables is p.
Elliptical Distribution Theory (EDT)	$F_{EDT} = \frac{1}{2}(\kappa + 1)^{-1}\text{tr}\left\{\left[\mathbf{S} - \sum(\Theta)\right]\mathbf{W}^{-1}\right\}^2 -$ $\delta\left\{\text{tr}\left[\mathbf{S} - \sum(\Theta)\right]\mathbf{W}^{-1}\right\}^2$	$\mathbf{W} =$ any consistent estimator of $\sum$. κ and δ are measures of kurtosis
Asymptotically Distribution Free (ADF)	$F_{ADF} = \left[\mathbf{s} - \sigma(\Theta)\right]'\mathbf{W}^{-1}\left[\mathbf{s} - \sigma(\Theta)\right]$	$\mathbf{W}$ has elements, $w_{ijkl} = \sigma_{ijkl} - \sigma_{ij}\sigma_{kl}$ (σ_{ijkl} is the kurtosis, σ_{ij} is the covariance)

[a] No χ^2 statistics or standard errors are available by the usual formulae, but some programs give these using more general computations.

derived from several estimation methods when the assumptions of normality and independence of factors were violated.

14.5.2.1 ■ Estimation Methods and Sample Size

Hu and colleagues found that when the normality assumption was reasonable, both the ML and the Scaled ML performed well with sample sizes over 500. GLS performed slightly better when the sample size was less than 500. Interestingly, the EDT test statistic performed a little better than ML with small sample sizes. It should be noted that the elliptical distribution theory estimator (EDT) takes into account the kurtosis of the variables and assumes that all variables have the same kurtosis although the variables need not be normally distributed. (If the distribution is normal there is no excess kurtosis.) Finally, The ADF estimator was poor with sample sizes under 2500.

14.5.2.2 ■ Estimation Methods and Non-normality

When the normality assumption was violated, Hu, et al. found that the ML and GLS estimators worked well with sample sizes of 2500 and greater. The GLS estimator was a little better with smaller sample sizes but led to acceptance of too many models. The EDT estimator accepted far too many models. The ADF estimator was poor with sample sizes under 2500. Finally, the Scaled ML performed about the same as the ML and GLS estimators and better than the ADF estimator at all but the largest sample sizes.[9]

---

[9] This is interesting in that the ADF estimator has no distribution assumptions and, theoretically, should perform quite well under conditions of nonnormality.

14.5.2.3 ■ Estimation Methods and Dependence

The assumption that errors are independent underlies SEM and other multivariate techniques. Hu et al. also investigated estimation methods and test statistic performance when the errors and factors were dependent but uncorrelated.[10] ML and GLS performed poorly, always rejecting the true model. ADF was poor unless the sample size was greater than 2500. EDT was better than ML, GLS, and ADF, but still rejected too many true models. The Scaled ML was better than the ADF at all but the largest sample sizes. The Scaled ML performed best overall.

14.5.2.4 ■ Some Recommendations for Choice of Estimation Method

Sample size and plausibility of the normality and independence assumptions need to be considered in selection of the appropriate estimation technique. ML, the Scaled ML, or GLS estimators may be good choices with small samples and evidence of the plausibility of the normality and independence assumptions. The Scaled ML is fairly computer intensive. Therefore, if time or cost is an issue, ML and GLS are better choices when the assumptions seem plausible. ML estimation is currently the most frequently used estimation method in SEM. The Scaled ML test statistic is a good choice with nonnormality or suspected dependence among factors and errors. Because scaled ML is computer intensive and many model estimations may be required, it is often reasonable to use ML during initial model estimation and then scaled ML for the final estimation. The ADF estimator seems to be a poor choice under all conditions unless sample size is very large (> 2500).

14.5.3 ■ Assessing the Fit of the Model

After the model has been specified and then estimated, the major question is, "Is it a good model?" One component of a "good" model is the fit between the sample covariance matrix and the estimated population covariance matrix. Like multiway frequency analysis (Chapter 7) and logistic regression (Chapter 12), a good fit is sometimes indicated by a nonsignificant χ^2. Unfortunately, assessment of fit is not always as straightforward as assessment of χ^2. With large samples, trivial differences between sample and estimated population covariance matrices are often significant because the minimum of the function is multiplied by $N - 1$. With small samples, the computed χ^2, actually may not be distributed as χ^2, leading to inaccurate probability levels. Finally, when assumptions underlying the χ^2 test statistic are violated, the probability levels are inaccurate (Bentler, 1995).

Because of these problems, numerous measures of model fit have been proposed. In fact, this is a lively area of research with new indices seemingly developed daily. One very rough rule of thumb, though unfortunately directly related to the χ^2 value, is that a good fitting model may be indicated when the ratio of the χ^2 to the degrees of freedom is less than 2. The following discussion presents only some examples of frequently encountered fit indices. The interested reader is referred to Tanaka (1993), Browne and Cudeck (1993), and Williams and Holahan (1994) for excellent discussions of fit indices.

---

[10] Factors were made dependent but uncorrelated by creating a curvilinear relationship between the factors and the errors. Pearson Product Moment Correlation coefficients examine only linear relationships; therefore, although the correlation is zero between factors and errors, they are dependent.

14.5.3.1 ■ Comparative Fit Indices

One way of conceptualizing goodness of fit is by thinking of a series of models all nested within one another. Nested models are like the hierarchical models in loglinear modeling discussed in Chapter 7. Nested models are models that are subsets of one another. At one end of the continuum is the independence model: the model that corresponds to completely unrelated variables. This model has degrees of freedom equal to the number of data points minus the variances that are estimated. At the other end of the continuum is the saturated (full or perfect) model with zero degrees of freedom. Fit indices that employ a comparative fit approach place the estimated model somewhere along this continuum. The Bentler-Bonett (1980) normed fit index (NFI) evaluates the estimated model by comparing the χ^2 value of the model to the χ^2 value of the independence model:

$$\text{NFI} = \frac{\chi^2_{\text{indep}} - \chi^2_{\text{model}}}{\chi^2_{\text{indep}}} \tag{14.20}$$

This yields a descriptive fit index that lies in the 0 to 1 range. For the small sample example,

$$\text{NFI} = \frac{68.865 - 8.835}{68.865} = .872$$

High values (greater than .90) are indicative of a good-fitting model. Therefore, the NFI for the small sample example indicates only a marginal fit as compared with a model with completely uncorrelated variables. Unfortunately, the NFI may underestimate the fit of the model in good-fitting models with small samples (Bearden, Sharma, & Teel, 1982). An adjustment to the NFI that incorporates the degrees of freedom in the model yields the nonnormed fit index (NNFI),

$$\text{NNFI} = \frac{\chi^2_{\text{indep}} - \dfrac{\text{df}_{\text{indep}}}{\text{df}_{\text{model}}} \chi^2_{\text{model}}}{\chi^2_{\text{indep}} - \text{df}_{\text{indep}}} \tag{14.21}$$

The adjustment reduces the problem of underestimating the fit in extremely good fitting models but sometimes yields numbers outside of the 0–1 range. The NNFI also can be much too small in small samples, indicating a poor fit when other indices indicate an adequate fit (Anderson & Gerbing, 1984).

The problem of the large variability in the NNFI is addressed by the incremental fit index (IFI) (Bollen, 1989b),

$$\text{IFI} = \frac{\chi^2_{\text{indep}} - \chi^2_{\text{model}}}{\chi^2_{\text{indep}} - \text{df}_{\text{model}}} \tag{14.22}$$

The comparative fit index (CFI; Bentler, 1988) also assesses fit relative to other models as the name implies, but uses a different approach. The CFI employs the noncentral χ^2 distribution with noncentrality parameters, τ_i. The larger the value of τ_i, the greater the model misspecification: i.e., if the estimated model is perfect $\tau_i = 0$. The CFI is defined as,

$$\text{CFI} = 1 - \frac{\tau_{\text{est. model}}}{\tau_{\text{indep. model}}} \tag{14.23}$$

So, clearly, the smaller the noncentrality parameter, τ_i, for the estimated model relative to the τ_i for the independence model, the larger the CFI and the better the fit. The τ value for a model can be estimated by,

$$\tau_{\text{indep. model}} = \chi^2_{\text{indep. model}} - \text{df}_{\text{indep. model}} \qquad (14.24)$$

$$\tau_{\text{est. model}} = \chi^2_{\text{est. model}} - \text{df}_{\text{est. model}}$$

$$\text{CFI} = 1 - \frac{4.835}{58.865} = .918$$

For the small sample example,

$$\tau_{\text{indep. model}} = 68.865 - 10 = 58.865$$

$$\tau_{\text{est. model}} = 8.835 - 4 = 4.835$$

CFI values greater than .90 are indicative of good fitting models. The CFI is normed to the 0 -1 range, and does a good job of estimating model fit even in small samples (Bentler, 1995). It should be noted that the values of all of these indices depend on the estimation method used.

14.5.3.2 ■ Absolute Fit Index

McDonald and Marsh (1990) proposed an index, MFI, that is absolute in that it does not depend on a comparison with another model such as the independence or saturated models (CFI) or the observed data (GFI). This index is illustrated with the small sample example,

$$\text{MFI} = \exp\left[-.5 \frac{(\chi^2_{\text{model}} - \text{df}_{\text{model}})}{N} \right] \qquad (14.25)$$

$$\text{MFI} = \exp\left[-.5 \frac{(8.835 - 4)}{100} \right] = .976$$

14.5.3.3 ■ Indices of Proportion of Variance Accounted For

Two widely available fit indices calculate a weighted proportion of variance in the sample covariance matrix accounted for by the estimated population covariance matrix (Bentler, 1983; Tanaka & Huba, 1989). The goodness of fit index, GFI, can be defined by,

$$\text{GFI} = \frac{\text{tr}(\hat{\sigma}'\mathbf{W}\hat{\sigma})}{\text{tr}(\hat{s}'\mathbf{W}\hat{s})} \qquad (14.26)$$

where the numerator is the sum of the squared weighted variances from the estimated model covariance matrix and the denominator is the sum of the squared weighted variances from the sample covariance matrix. $\mathbf{W}$ is the weight matrix that is selected by the choice of estimation method (Table 14.9).

Tanaka and Huba (1989) suggest that GFI is analogous to R^2 in multiple regression. This fit index also can be adjusted for the number of parameters estimated in the model. The adjusted fit index, labeled AGFI, is estimated by:

$$\text{AGFI} = 1 - \frac{1 - \text{GFI}}{1 - \dfrac{\text{Numbers of est. parameters}}{\text{Number of data points}}} \qquad (14.27)$$

For the small sample example,

$$AGFI = 1 - \frac{1 - .967}{1 - \frac{11}{15}} = .876$$

The fewer the number of estimated parameters relative to the number of data points, the closer the AGFI is to the GFI. In this way the AGFI adjusts the GFI for the number of parameters estimated. The fit improves by estimating lots of parameters in SEM. However, a second goal of modeling is to develop a parsimonious model with as few parameters as possible.

14.5.3.4 ■ Degree of Parsimony Fit Indices

Several indices have been developed that take into account the degree of parsimony in the model. Most simply, an adjustment is made to the GFI (Mulaik, et al. 1989), to produce PGFI:

$$PGFI = \left[1 - \left(\frac{\text{Number of est. parameters}}{\text{Number of data points}} \right) \right] (GFI) \tag{14.28}$$

For the small sample example,

$$PGFI = \left[1 - \left(\frac{11}{15} \right) \right] (.967) = .258$$

The larger the fit index (values closer to 1.00) the better. Clearly, there is a heavy penalty for estimating a lot of parameters with this index. This index will always be substantially smaller than other indices unless the number of parameters estimated is much smaller than the number of data points.

Completely different methods of assessing fit that include a parsimony adjustment are the Akaike Information Criterion (AIC) and the Consistent Akaike Information Criterion (CAIC) (Akaike, 1987; Bozdogan, 1987). These indices are also functions of χ^2 and df:

$$\text{Model AIC} = \chi^2_{\text{model}} - 2\text{df}_{\text{model}} \tag{14.29}$$

$$\text{Model CAIC} = \chi^2_{\text{model}} - (\ln N + 1)\text{df}_{\text{model}} \tag{14.30}$$

For the small sample example,

$$\text{Model AIC} = 8.835 - 2(4) = .835$$

$$\text{Model CAIC} = 8.835 - (\ln 100 + 1)4 = -13.586$$

Small values indicate good fitting, parsimonious models. How small is small enough? There is no clear answer because these indices are not normed to a 0–1 scale. "Small enough" is small as compared with other competing models. This index is applicable to models estimated with maximum likelihood methods. It is useful for cross-validation because it is not dependent on sample data (Tanaka, 1993). EQS and SAS CALIS use Equations 14.29 and 14.30 to calculate the AIC and CAIC. LISREL, however, uses

$$AIC = \chi^2_{\text{model}} + 2(\text{df}_{\text{number of est. parameters}}) \tag{14.31}$$

$$CAIC = \chi^2_{\text{model}} + (1 + \ln N)\text{df}_{\text{number of parameters}} \tag{14.32}$$

Both sets of equations are correct. LISREL computes the AIC and CAIC with a constant included; EQS and SAS CALIS compute the AIC and CAIC without the constant. Therefore, although both sets of computations are correct, the AIC and CAIC computed in EQS and SAS CALIS are always smaller than the same values in LISREL. SYSTAT RAMONA does not provide the AIC or CAIC.

14.5.3.5 ■ Residual Based Fit Indices

Finally, there are indices based on the residuals. The root mean square residual (RMR) and the standardized root mean square residual (SRMR) are the average differences between the sample variances and covariances and the estimated population variances and covariances. The root mean square residual is given by

$$
\text{RMR} = \left[2 \sum_{i=1}^{q} \sum_{j=1}^{i} \frac{(s_{ij} - \hat{\sigma}_{ij})^2}{q(q+1)} \right]^{1/2}
\tag{14.33}
$$

Therefore, the RMR is the square root (indicated by the power of 1/2) of two times the sum, over all of the variables in the covariance matrix, of the average squared differences between each of the sample covariances (or variances) and the estimated covariances (or variances).

Good-fitting models have small RMR. It is sometimes difficult to interpret an unstandardized residual because the scale of the variables affects the size of the residual; therefore, a standardized root mean square residual (SRMR) is also available. The SRMR has a range of 0 to 1, values of .05 or less are desired (Sörbom & Jöreskog, 1982).

14.5.3.6 ■ Choosing Among Fit Indices

Good-fitting models produce consistent results on many different indices in many, if not most, cases. If all the indices lead to similar conclusions, the issue of which indices to report is a matter of personal preference, and perhaps, the preference of the journal editor. Often multiple indices are reported. If the results of the fit indices are inconsistent, the model should probably be re-examined; if the inconsistency cannot be resolved, consider reporting multiple indices.

14.5.4 ■ Model Modification

There are at least two reasons for modifying a SEM model: to improve fit (especially in exploratory work) and to test hypotheses (in theoretical work). The three basic methods of model modification are chi-square difference tests, Lagrange multiplier tests (LM), and Wald tests. All are asymptotically equivalent under the null hypothesis (i.e., act the same as the sample size approaches infinity) but approach model modification differently.

14.5.4.1 ■ Chi-Square Difference Test

If models are nested (models are subsets of each other), the χ^2 value for the larger model is subtracted from the χ^2 value for the smaller nested model and the difference, also a χ^2, is evaluated with degrees of freedom equal to the difference between the degrees of freedom in the two models.

Recall that the residual covariance between LOVESKI and SENSEEK is very high in the small sample example. We might allow these IVs to correlate and ask, Does adding (estimating) this covariance improve the fit of the model? Although our "theory" is that these variables are uncorrelated, is this aspect of theory supported by the data? To examine these questions, a second model is estimated in which LOVESKI and SENSEEK are allowed to correlate. The resulting model produces $\chi^2 = 2.034$, df = 3. In the small sample example solution in Section 14.4.5, $\chi^2 = 8.835$, df = 4. The χ^2 difference test (or likelihood ratio for maximum likelihood) is $8.835 - 2.034 = 6.801$, df = $4 - 3 = 1, p < .05$. The model is significantly improved with the addition of this covariance; in fact, one of the fit indices (CFI) increases to 1. Although the theory specifies independence between sensation seeking and Love of Skiing, the data support the notion that, indeed, these variables are correlated.

There are some disadvantages to the χ^2 difference test. Two models need to be estimated to get the χ^2 difference value and estimating two models for each parameter is time consuming with very large models and/or a slow computer. A second problem relates to χ^2 itself. Because of the relationship between sample size and χ^2, it is hard to detect a difference between models when sample sizes are small.

14.5.4.2 ■ Lagrange Multiplier Test (LM)

The LM test also compares nested models but requires estimation of only one model. The LM test asks if the model is improved if one or more of the parameters in the model that are currently fixed are estimated. Or, equivalently, What parameters should be added to the model to improve the fit? This method of model modification is analogous to forward stepwise regression.

The LM test applied to the small sample example indicates that if we add a covariance between LOVESKI and SENSEEK, the approximate drop in χ^2 value is 5.633. This is one path, so the χ^2 value of 5.633 is evaluated with 1 df. The p level of this difference is .018. The model is re-estimated if the decision is made to add the path. When the path is added, the drop is slightly larger, 6.801, but yields the same result.

The LM test can be examined either univariately or multivariately. There is a danger in examining only the results of univariate LM tests because overlapping variance between parameter estimates may make several parameters appear as if their addition would significantly improve the model. All of these parameters are candidates for inclusion by the results of univariate LM tests but the multivariate LM test identifies the single parameter that would lead to the largest drop in model χ^2 and calculates the expected change in χ^2. After this variance is removed, the parameter that accounts for the next largest drop in model χ^2 is assessed in a manner analogous to Roy-Bargmann stepdown analysis in MANOVA (Chapter 9).

EQS provides both univariate and multivariate LM tests. Several options are also available for LM tests on specific sets of matrices and in specific orders of testing. The default LM test was requested for the small sample example. Portions of the LM test output are presented in Table 14.10.

LM univariate output is presented first. The parameter that the LM test suggests adding is listed under the column labeled PARAMETER. The convention used in EQS is DV, IV or IV, IV. Because both F1 and V5 are IVs this refers to a covariance between LOVESKI and SENSEEK. The CHI-SQUARE column indicates the approximate chi square associated with this path: 5.63. The probability level of this χ^2 is in the PROBABILITY column, $p = .018$. Significant χ^2 values are sought if improvement of the model is the goal. If hypotheses testing guides the use of the LM test, then the desired significance (or lack of significance) depends on the specific hypothesis. The

```
    /TITLE
     Small Sample Structural Equation Model
    /SPECIFICATIONS
      CAS = 100; VAR = 5; ME = ML;
    /LABELS
      V1 = NUMYRS; V2 = DAYSKI; V3 = SNOWSAT;
      V4 = FOODSAT; V5 = SENSEEK;
      F1 = LOVESKI; F2= SKISAT;
    /EQUATIONS
      V1 = *F1 + E1;
      V2 = *F1 + E2;

      V3 = F2 + E3;
      V4 =*F2 + E4;

      F2 =  *V5 + *F1 + D2;
    /VAR
      F1 = 1;
      E1 to E4 = *;
      V5 = *;
      D2 = *;
    /WTEST
    /LMTEST
    /PRINT
      FIT = ALL;
    /MATRIX
      2.74
       .80 3.25
       .68  .28 1.23
       .65  .35  .72 1.87
      2.02 2.12 2.02 2.12  27.00
    /END
```

LAGRANGIAN MULTIPLIER TEST REQUIRES   2161 WORDS OF MEMORY.
PROGRAM ALLOCATES   50000 WORDS.

LAGRANGE MULTIPLIER TEST (FOR ADDING PARAMETERS)

          ORDERED UNIVARIATE TEST STATISTICS:

NO	CODE		PARAMETER	CHI-SQUARE	PROBABILITY	PARAMETER CHANGE
1	2	2	F1,V5	5.633	0.018	1.287
2	2	11	V1,V5	3.236	0.072	0.055
3	2	20	V1,F2	3.236	0.072	0.914
4	2	11	V2,V5	2.846	0.092	0.057
5	2	20	V2,F2	2.846	0.092	0.933
6	2	11	V4,V5	0.039	0.842	0.007
7	2	12	V4,F1	0.039	0.842	-0.039
8	2	12	V3,F1	0.039	0.842	0.039
9	2	11	V3,V5	0.039	0.842	-0.007
10	2	0	F1,F1	0.000	1.000	0.000
11	2	0	F2,D2	0.000	1.000	0.000
12	2	0	V3,F2	0.000	1.000	0.000

MULTIVARIATE LAGRANGE MULTIPLIER TEST BY SIMULTANEOUS PROCESS IN STAGE 1

PARAMETER SETS (SUBMATRICES) ACTIVE AT THIS STAGE ARE:

   PVV PFV PFF PDD GVV GVF GFV GFF BVF BFF

TABLE 14.10   *(CONTINUED)*

		CUMULATIVE MULTIVARIATE STATISTICS				UNIVARIATE INCREMENT	
STEP	PARAMETER	CHI-SQUARE	D.F.	PROBABILITY	CHI-SQUARE	PROBABILITY
----	---------	----------	----	-----------	----------	-----------
1	F1,V5	5.633	1	0.018	5.633	0.018

PARAMETER CHANGE column indicates the approximate coefficient for the added parameter. But this series of univariate tests may include overlapping variance between parameters, so the multivariate LM is examined.

Before presenting the multivariate test, EQS presents the list of default parameter matrices active at this stage of analysis. The matrices beginning with the letter **P** are phi matrices of covariances between IVs; the last two letters indicate what type of variables are included. For instance, VV indicates covariances between measured variables and FV indicates covariances between factors and measured variables. The matrices beginning with the letter **G** refer to gamma matrices of regression coefficients between DVs and IVs. The matrices beginning with the letter **B** refer to beta matrices of regression coefficients of DVs with other DVs. At times only very particular parameters will be of interest for potential inclusion in the model. For example, maybe the only parameters of interest are regression paths among the latent dependent variables; in these cases it is helpful to check that only the appropriate matrix is active (in this example BFF).

The multivariate test also suggests adding the F1,V5 parameter. Indeed, after adding this one parameter, none of the other parameters significantly improves the model χ^2, therefore, no other multivariate LM tests are shown.

LISREL presents only univariate LM tests, called MODIFICATION INDICES. Edited LISREL printout for the small sample example is given in Table 14.11. Modification indices are presented matrix by matrix. For each matrix in the LISREL model, four matrices of model modifications are included. The first, labeled MODIFICATION INDICES FOR . . . , contains χ^2 values for the specific parameter. The second matrix, labeled EXPECTED CHANGE FOR . . . , contains unstandardized changes in parameter values, corresponding to the parameter change value given in EQS. The third matrix, labeled STANDARDIZED EXPECTED CHANGE FOR . . . , contains the parameter changes where the latent variable has been standardized with a standard deviation of 1, but the original scale of the measured variables has been retained. The last matrix, labeled COMPLETELY STANDARDIZED EXPECTED CHANGE FOR . . . , contains the approximate parameter change when both the measured and latent variables (with the exception of the latent and observed errors) have been standardized with a standard deviation of 1. After all the LISREL matrices have been given, the parameter with the largest chi-square value is reported under MAXIMUM MODIFICATION INDEX . . . .

SIMPLIS presents model modification indices only if the path diagram option is selected. By default, only large ($p > .005$), univariate model modification indices are included in SIMPLIS output. Modification indices are given on the path diagram and in LISREL notation as in Table 14.11. If LISREL output is selected within the SIMPLIS procedure file, the LISREL model modification indices of Table 14.11 are presented.

SAS CALIS presents the univariate LM test in a series of matrices: PHI (covariances between IVs), GAMMA (regression coefficients predicting DVs from IVs), and BETA (regression coefficients

TABLE 14.11 SETUP AND EDITED OUTPUT FROM LISREL FOR MODIFICATION INDICES

```
SMALL SAMPLE EXAMPLE LISREL8
DA NI = 5 NO = 100
CM
*
    2.74
     .80  3.25
     .68   .28  1.23
     .65   .35   .72  1.87
    2.02  2.12  2.02  2.12  27.00
LA
NUMYRS DAYSKI SNOWSAT FOODSAT SENSEEK
SE
SNOWSAT FOODSAT NUMYRS DAYSKI SENSEEK
MO NY=2 NX=3 NE=1 NK=2
LE
SKISAT
LK
LOVESKI DUMMY
FR LX(1,1) LX(2,1) LY(2,1)
FI PH(2,1) TD(3,3)
VA 1 LX(3,2) LY(1,1) PH(1,1)
OU SC SE TV RS SS ND=3
```

```
 SMALL SAMPLE EXAMPLE LISREL8
 MODIFICATION INDICES AND EXPECTED CHANGE

NO NON-ZERO MODIFICATION INDICES FOR LAMBDA-Y

         MODIFICATION INDICES FOR LAMBDA-X

              LOVESKI      DUMMY
              --------    --------
NUMYRS          - -        3.236
DAYSKI          - -        2.846
SENSEEK        5.634        - -

         EXPECTED CHANGE FOR LAMBDA-X

              LOVESKI      DUMMY
              --------    --------
NUMYRS          - -         .055
DAYSKI          - -         .057
SENSEEK        1.288        - -

         STANDARDIZED EXPECTED CHANGE FOR LAMBDA-X

              LOVESKI      DUMMY
              --------    --------
NUMYRS          - -         .288
DAYSKI          - -         .294
SENSEEK        1.288        - -

         COMPLETELY STANDARDIZED EXPECTED CHANGE FOR LAMBDA-X

              LOVESKI      DUMMY
              --------    --------
NUMYRS          - -         .174
DAYSKI          - -         .163
SENSEEK         .248        - -
```

TABLE 14.11 *(CONTINUED)*

NO NON-ZERO MODIFICATION INDICES FOR GAMMA

     MODIFICATION INDICES FOR PHI

	LOVESKI	DUMMY
	--------	-------- LOVESKI - -
DUMMY	5.634	- -

     EXPECTED CHANGE FOR PHI

	LOVESKI	DUMMY
	--------	-------
LOVESKI	- -	
DUMMY	1.288	- -

     STANDARDIZED EXPECTED CHANGE FOR PHI

	LOVESKI	DUMMY
	--------	-------
LOVESKI	- -	
DUMMY	.248	- -

NO NON-ZERO MODIFICATION INDICES FOR PSI

     MODIFICATION INDICES FOR THETA-DELTA-EPS

	SNOWSAT	FOODSAT
	---------	--------
NUMYRS	.105	.105
DAYSKI	.122	.122
SENSEEK	.039	.039

     EXPECTED CHANGE FOR THETA-DELTA-EPS

	SNOWSAT	FOODSAT
	---------	--------
NUMYRS	.093	-.093
DAYSKI	-.079	.079
SENSEEK	-.185	.185

     COMPLETELY STANDARDIZED EXPECTED CHANGE FOR THETA-DELTA-EPS

	SNOWSAT	FOODSAT
	---------	--------
NUMYRS	.051	-.042
DAYSKI	-.040	.032
SENSEEK	-.033	.026

     MODIFICATION INDICES FOR THETA-DELTA

	NUMYRS	DAYSKI	SENSEEK
	--------	--------	--------
NUMYRS	- -		
DAYSKI	- -	- -	
SENSEEK	3.236	2.846	- -

     EXPECTED CHANGE FOR THETA-DELTA

	NUMYRS	DAYSKI	SENSEEK
	--------	--------	--------
NUMYRS	- -		
DAYSKI	- -	- -	
SENSEEK	1.498	1.530	- -

     COMPLETELY STANDARDIZED EXPECTED CHANGE FOR THETA-DELTA

	NUMYRS	DAYSKI	SENSEEK
	--------	--------	---------
NUMYRS	- -		
DAYSKI	- -	- -	
SENSEEK	.174	.163	- -

MAXIMUM MODIFICATION INDEX IS   5.63 FOR ELEMENT (3, 1) OF LAMBDA-X

between DVs). For each matrix, CALIS presents the χ^2 value, the probability associated with the χ^2, and the approximate parameter value for each potential parameter. These are the same as the parameter change in EQS and the expected change in LISREL. After each matrix is printed, the 10 largest chi-square values are listed in rank order. CALIS does not provide multivariate LM tests. Edited CALIS output for the LM (and WALD) test is presented in Table 14.12

SYSTAT RAMONA does not provide the LM test.

14.5.4.3 ■ Wald Test

While the LM test asks which parameters, if any, should be added to a model, the Wald test asks which, if any, could be deleted. Are there any parameters that are currently being estimated that could, instead, be fixed to zero? Or, equivalently, Which parameters are not necessary in the model? The Wald test is analogous to backward deletion of variables in stepwise regression where one seeks a nonsignificant change in the equation when variables are left out.

When the Wald test is applied to the small sample example, the only candidate for deletion is the error variance associated with NUMYRS. If this parameter is dropped, the χ^2 value increases by .008, a nonsignificant change ($p = .931$). The model is not significantly degraded by deletion of this parameter. However, because it is generally not reasonable to drop an error variance from a model, the decision is made to retain the error variance associated with NUMYRS. Notice that, unlike the LM test, *nonsignificance* is desired when using the Wald test.

This illustrates an important point. Both the LM and Wald tests are based on statistical, not substantive, criteria. If there is conflict between these two criteria, substantive criteria are more important.

Table 14.13 presents edited output of both the univariate and multivariate Wald test from EQS. The specific candidate for deletion is indicated and the approximate multivariate χ^2 and its probability value are provided. The univariate tests are shown in the last two columns. In this example, only the one parameter is suggested for deletion. By default, parameters are considered for deletion only if deletion does not cause the multivariate χ^2 associated with the Wald test to become significant. Remember, the goal of the Wald test is to drop parameters that do not contribute significantly to the model.

SAS CALIS also computes multivariate Wald tests and includes the results in the same matrices as the LM test (Table 14.12). In the matrices, the Wald test is identified by the parameter label in brackets, the same label that was given to the parameter in the procedure file. After all the matrices are printed, SAS CALIS gives the Multivariate Wald test, with associated χ^2 and probability value. As with EQS, only one parameter is included in the multivariate Wald test; the value of the Wald test is the approximate amount by which the model χ^2 increases if the parameter is fixed to zero, i.e., not estimated. LISREL and SYSTAT CALIS do not provide the Wald test.

14.5.4.4 ■ Some Caveats and Hints on Model Modification

Because the LM test and Wald test are stepwise procedures, Type I error rates are inflated. However, there are, as yet, no adjustments like the type available for ANOVA. A simple approach is to use a conservative probability value (say, $p < .01$) for adding parameters with the LM test. Cross validation with another sample is also highly recommended if modifications are made. If numerous modifications are made and new data are not available for cross validation, compute the correlation between the estimated parameters from the original, hypothesized, model and the estimated parameters from the final model, using only parameters common to both models. If this correlation is high ($> .90$), relationships within the model have been retained despite the modifications.

```
TITLE "SMALL SAMPLE EXAMPLE IN CALIS";
DATA CMAT(TYPE=COV);
   _TYPE_='COV';
   INPUT _NAME_ $ V1-V5;
   LABEL V1='NUMYRS' V2='DAYSKI' V3='SNOWSAT'
      V4='FOODSAT' V5='SENSEEK';
   CARDS;
V1   2.74    .      .      .      .
V2    .80   3.25    .      .      .
V3    .68    .28   1.23    .      .
V4    .65    .35    .72   1.87    .
V5   2.02   2.12   2.02   2.12   27.00
;

PROC CALIS COV DATA=CMAT OMETHOD=NR EDF=99 ALL1;
   LINEQS
    V1=X1 F1+E1,
    V2=X2 F1+E2,
    V3=1 F2+E3,
    V4=X4 F2+E4,
    F2=X5 F1+ X6 V5+D2;
   STD
    E1-E4=X7-X10,
      V5=X11,
    D2=X12,
    F1=1;
RUN;
```

Lagrange Multiplier and Wald Test Indices _PHI_[7:7]
Diagonal Matrix
Univariate Tests for Constant Constraints
```
-------------------------------------------
| Lagrange Multiplier  or  Wald Index  |
-------------------------------------------
| Probability | Approx Change of Value |
-------------------------------------------
```

		V5		F1		E1		E2
V5	50.000	[X11]	5.692		3.269		2.875	
	0.017		1.288		0.071	1.498	0.090	1.530
F1	5.692		sing		sing		sing	
	0.017	1.288	.	.	.	.	.	.
E1	3.269		sing		0.008	[X7]	sing	
	0.071	1.498	.	.	.		.	.
E2	2.875		sing		sing		38.714	[X8]
	0.090	1.530	.	.	.	.	.	
E3	0.040		0.040		0.105		0.124	
	0.841	-0.184	0.842	0.039	0.745	0.093	0.725	-0.079
E4	0.040		0.040		0.106		0.123	
	0.842	0.185	0.842	-0.039	0.745	-0.093	0.725	0.079
D2	sing		sing		sing		sing	
	.	.	.	.	.	.	.	.

		E3		E4		D2	
V5	0.040		0.040		sing		
	0.841	-0.184	0.842	0.185	.	.	

759

TABLE 14.12   *(CONTINUED)*

F1	0.040		0.040		sing
	0.842	0.039	0.842	-0.039	.	.
E1	0.105		0.106		sing
	0.745	0.093	0.745	-0.093	.	.
E2	0.124		0.123		sing
	0.725	-0.079	0.725	0.079	.	.
E3	7.368	[X9]	sing		sing
			.		.	.
D2	sing		sing		4.494	[X12]
	.	.	.	.	

Rank order of 10 largest Lagrange multipliers in _PHI_

F1 : V5	E1 : V5	E2 : V5
5.6916 : 0.017	3.2691 : 0.071	2.8751 : 0.090
E3 : E2	E4 : E2	E4 : E1
0.1236 : 0.725	0.1233 : 0.725	0.1057 : 0.745
E3 : E1	E3 : V5	E4 : V5
0.1054 : 0.745	0.0401 : 0.841	0.0399 : 0.842
	E4 : F1	
	0.0399 : 0.842	

Lagrange Multiplier and Wald Test Indices _GAMMA_[5:2]
General Matrix
Univariate Tests for Constant Constraints

	V5		F1	
V1	3.269		4.726	[X1]
	0.071	0.055		
V2	2.875		3.034	[X2]
	0.090	0.057		
V3	0.040		0.040	
	0.841	-0.007	0.842	0.039
V4	0.040		0.040	
	0.842	0.007	0.842	-0.039
F2	11.207	[X6]	3.492	[X5]

Rank order of 5 largest Lagrange multipliers in _GAMMA_

V1 : V5	V2 : V5	V3 : V5
3.2691 : 0.071	2.8751 : 0.090	0.0401 : 0.841
	V4 : V5	V4 : F1
	0.0399 : 0.842	0.0399 : 0.842

TABLE 14.12   *(CONTINUED)*

```
                Lagrange Multiplier and Wald Test Indices _BETA_[5:5]
                                General Matrix
                        Identity-Minus-Inverse Model Matrix
                     Univariate Tests for Constant Constraints

                        V1                    V2                    V3
        V1        sing                  sing                  2.424
                   .          .          .          .          0.120     0.583

        V2        sing                  sing                  0.607
                   .          .          .          .          0.436     0.267

        V3        0.044                 0.094                 sing
                   0.833      0.024      0.759     -0.022       .          .

        V4        0.045                 0.094                 sing
                   0.833     -0.024      0.760      0.022       .          .

        F2        sing                  sing                  sing
                   .          .          .          .          .          .


                            V4                    F2
        V1        0.256                 3.268
                   0.613      0.113      0.071      0.914

        V2        0.806                 2.873
                   0.369      0.165      0.090      0.933

        V3        sing                  sing

        V4        sing                  12.626    [X4]
                   .          .

        F2        sing                  sing
                   .          .          .          .


              Rank order of 10 largest Lagrange multipliers in _BETA_

              V1 : F2               V2 : F2               V1 : V3
        3.2682 : 0.071        2.8734 : 0.090        2.4239 : 0.120

              V2 : V4               V2 : V3               V1 : V4
        0.8055 : 0.369        0.6074 : 0.436        0.2560 : 0.613

              V3 : V2               V4 : V2               V4 : V1
        0.0940 : 0.759        0.0937 : 0.760        0.0445 : 0.833

                                   V3 : V1
                             0.0442 : 0.833
```

Stepwise Multivariate Wald Test

	Cumulative Statistics			Univariate Increment	
Parameter	Chi-Square	D.F.	Prob	Chi-Square	Prob
X7	0.007790	1	0.9297	0.007790	0.9297

MODIFICATION INDICES FOR PHI

	LOVESKI	DUMMY
LOVESKI	- -	
DUMMY	5.634	- -

EXPECTED CHANGE FOR PHI

	LOVESKI	DUMMY
LOVESKI	- -	
DUMMY	1.288	- -

TABLE 14.12 *(CONTINUED)*

STANDARDIZED EXPECTED CHANGE FOR PHI

	LOVESKI	DUMMY
LOVESKI	- -	
DUMMY	.248	- -

NO NON-ZERO MODIFICATION INDICES FOR PSI

MODIFICATION INDICES FOR THETA-DELTA-EPS

	SNOWSAT	FOODSAT
NUMYRS	.105	.105
DAYSKI	.122	.122
SENSEEK	.039	.039

EXPECTED CHANGE FOR THETA-DELTA-EPS

	SNOWSAT	FOODSAT
NUMYRS	.093	-.093
DAYSKI	-.079	.079
SENSEEK	-.185	.185

COMPLETELY STANDARDIZED EXPECTED CHANGE FOR THETA-DELTA-EPS

	SNOWSAT	FOODSAT
NUMYRS	.051	-.042
DAYSKI	-.040	.032
SENSEEK	-.033	.026

MODIFICATION INDICES FOR THETA-DELTA

	NUMYRS	DAYSKI	SENSEEK
NUMYRS	- -		
DAYSKI	- -	- -	
SENSEEK	3.236	2.846	- -

EXPECTED CHANGE FOR THETA-DELTA

	NUMYRS	DAYSKI	SENSEEK
NUMYRS	- -		
DAYSKI	- -	- -	
SENSEEK	1.498	1.530	- -

Unfortunately, the order that parameters are freed or estimated can affect the significance of the remaining parameters. MacCallum (1986) suggests adding all necessary parameters before deleting unnecessary parameters. In other words, do the LM test before the Wald test. Because model modification easily gets very confusing, it is often wise to add, or delete, parameters one at a time.

A more subtle limitation is that tests leading to model modification examine overall changes in χ^2, not changes in individual parameter estimates. Large changes in χ^2 are sometimes associated with very small changes in parameter estimates. A missing parameter may be statistically needed but the estimated coefficient may have an uninterpretable sign. If this happens it is best not to add the parameter. Finally, if the hypothesized model is wrong, tests of model modification, by themselves, may be insufficient to reveal the true model. In fact, the "trueness" of any model is never tested directly, although cross validation does add evidence that the model is correct.

TABLE 14.13 SETUP AND EDITED OUTPUT FROM EQS FOR WALD TEST

```
/TITLE
 Small Sample Structural Equation Model
/SPECIFICATIONS
 CAS = 100; VAR = 5; ME = ML;
/LABELS
 V1 = NUMYRS; V2 = DAYSKI; V3 = SNOWSAT;
 V4 = FOODSAT; V5 = SENSEEK;
 F1 = LOVESKI ; F2= SKISAT;
/EQUATIONS
 V1 = *F1 + E1;
 V2 = *F1 + E2;

 V3 = F2 + E3;
 V4 =*F2 + E4;

 F2 =  *V5 + *F1 + D2;
/VAR
 F1 = 1;
 E1 to E4 = *;
 V5 = *;
 D2 = *;
/WTEST
/LMTEST
/PRINT
 FIT = ALL;
/MATRIX
 2.74
  .80 3.25
  .68  .28 1.23
  .65  .35  .72 1.87
 2.02 2.12 2.02 2.12  27.00
/END
```

WALD TEST (FOR DROPPING PARAMETERS)
MULTIVARIATE WALD TEST BY SIMULTANEOUS PROCESS

		CUMULATIVE MULTIVARIATE STATISTICS			UNIVARIATE INCREMENT	
STEP	PARAMETER	CHI-SQUARE	D.F.	PROBABILITY	CHI-SQUARE	PROBABILITY
1	E1,E1	0.008	1	0.931	0.008	0.931

If model modifications are done in hopes of developing a good-fitting model, the fewer modifications the better, especially if a cross-validation sample is not available. If the LM test, Wald tests, and chi-square difference tests are used to test specific hypotheses, the hypotheses will dictate the number of necessary tests.

14.5.5 ■ Reliability and Proportion of Variance

Reliability is defined in the classic sense as the proportion of true variance in total variance (true plus error variance). Both the reliability and the proportion of variance of a measured variable are assessed thorough squared multiple correlation (SMC) where the measured variable is the DV and the factor is the IV. Each SMC is interpreted as the reliability of the measured variable in the analysis and as the proportion of variance in the variable that is accounted for by the factor, conceptually the same as a communality estimate in factor analysis. To calculate a SMC:

$$\text{SMC}_{var\ i} = \frac{\lambda^2_i}{\lambda^2_i + \Theta_i} \qquad (14.34)$$

The factor loading for variable i is squared and divided by that value plus the residual variance associated with the variable i.

This formula is applicable only when there are no complex factor loadings or correlated errors.[11] The proportion of variance in the DV factor accounted for by the IVs is assessed as:

$$R^2_j = 1 - D^2_j \qquad (14.35)$$

The disturbance (residual) for the DV factor j is squared and subtracted from 1.

14.5.6 ■ Discrete and Ordinal Data

SEM assumes that measured variables are continuous and measured on an interval scale. Often however, a researcher desires to include discrete and/or ordinally measured categorical variables in an analysis. Because the data points in SEM are variances and covariances, the goal is to produce reasonable values for these types of variables for analysis.

Discrete (nominal level) measured variables such as favorite baseball team are included as IVs in a model by either dummy coding the variable e.g., Dodger fan or other, or using a multiple group model where a model is tested for each team preference, as discussed in the next section.

Ordered (ordinal level) categorical variables require special handling in SEM. Imagine that there is a normally distributed, continuous variable underlying each ordinal variable. To convert an ordinal variable to a continuous variable, the categories of the ordinal variables are converted to thresholds of the underlying (latent), normally distributed, continuous variable.

For example, say we ask people lingering outside a candy store if they: (1) hate, (2) like, or (3) love chocolate. As in Figure 14.7, underlying the ordinal variable is a normally distributed, latent, continuous construct representing love of chocolate. It is assumed that people who hate or absolutely loathe chocolate fall at, or below, the first threshold. Those who like chocolate fall between the two thresholds, and people who love or who are enraptured by chocolate, fall at or above the second threshold . The proportion of people falling into each category is calculated and this proportion is used to calculate a z score from a standardized normal table. The z score is the threshold.

SEM proceeds by using polychoric correlations (between two ordinal variables) or polyserial correlations (between an ordinal and an interval variable), rather than covariances, as the basis of the analysis.

Both EQS and LISREL (in PRELIS) compute thresholds and appropriate correlations. To incorporate categorical dependent variables in EQS, the statement CATEGORY = in the SPECIFICATIONS section is followed by the discrete variable labels, e.g., V1, V3. *All* measured variables must be DVs when models with categorical variables are estimated in EQS (Lee, Poon, & Bentler,1994).

---

[11] Note, factor loadings are denoted by λ_i in this formula, to stay consistent with general SEM terminology. In Chapter 13 factor loadings are denoted by a_{ij}.

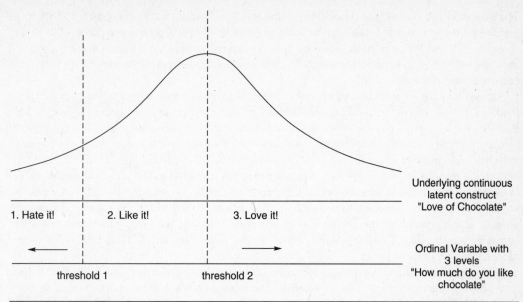

1. Hate it!   2. Like it!   3. Love it!

Ordinal Variable with
3 levels
"How much do you like
chocolate"

threshold 1   threshold 2

FIGURE 14.7   REPRESENTATION OF THRESHOLDS UNDERLYING ORDINAL DATA CATEGORIES.

If a model contains measured IVs, these IVs are first converted to factors in a method similar to LISREL. For example if V1 was a measured IV, it is converted to a measured DV in the \EQUA-TIONS section with V1 = F1. F1 is then used in equations in place of V1.

Using PRELIS, the procedure file specifies both the continuous and ordinal variables and requests matrix output of polyserial and polychoric correlations with the instruction OUTPUT-MATRIX = PMATRIX. This matrix is then used in LISREL as the sample correlation matrix: PM is substituted for CM in the LISREL procedure file. SAS CALIS and SYSTAT RAMONA do not accommodate categorical data.

14.5.7 ■ Multiple Group Models

Although each of the models estimated in this chapter uses data from a single sample, it is also possible to estimate and compare models that come from two or more samples, called multiple group models. The general null hypothesis tested in multiple group models is that the data from each group are from the same population. For example, if data from a sample of men and a sample of women are drawn for the small sample example, the general null hypothesis tested is that the two groups are drawn from the same population.

The analysis begins by developing good-fitting models in separate runs for each group. The models are then tested simultaneously in one run with none of the parameters across models constrained to be equal. This unconstrained multiple group model serves as the baseline against which to judge more restricted models. Following baseline model estimation, progressively more stringent constraints are specified by constraining various parameters across groups. When parameters are constrained they are forced to be equal to one another. After each set of constraints is added, a chi-square difference test is performed between the less restrictive and more restrictive model.

The goal is to not degrade the models by constraining parameters across the groups; therefore, a *nonsignificant* χ^2 is desired. If a significant difference in χ^2 is found between the models at any stage, the LM test is examined to locate the specific parameters that are different in the groups and these parameters are estimated separately in each group, i.e., the specific *across group* parameter constraints are released.

Various hypotheses are tested in a specific order. The first step is usually to constrain the factor loadings (regression coefficients) between factors and their indices to equality across groups. This step tests the hypothesis that the factor structure is the same in the different groups. Using the small sample example with one model for men and another for women, we ask if men and women have the same underlying structure for the Love of Skiing and Ski Trip Satisfaction factors. If these constraints are reasonable, the χ^2 difference test between the restricted model and the baseline model is nonsignificant. If the difference between the restricted and non-restricted models is significant, we need not throw in the towel immediately; rather, results of the LM test are examined and some equality constraints across the groups can be released. Naturally, the more parameters that differ across groups, the less alike the groups. Consult Byrne, Shavelson, and Muthén (1989) for a technical discussion of these issues.

If the equality of the factor structure is established, the second step is to ask if the factor variances and covariances are equal. In the small sample example this is equivalent to asking if the variance of Ski Trip Satisfaction is equal for men and women. (Recall that the variance of Love of Skiing was set to one for identification). If these constraints are feasible, the third step examines equality of the regression coefficients. This is equivalent to testing the equality of the regression coefficient predicting Ski Trip Satisfaction from Love of Skiing in the small sample example. We could also test the equality of the regression path predicting Ski Trip Satisfaction from sensation seeking for men and women. If all of these constraints are reasonable, the last step is to examine the equality of residual variances across groups, an extremely stringent hypothesis not often tested. If all the regression coefficients, variances, and covariances are the same across groups, it is concluded that men and women represent samples from the same population.

The groups are often similar in some respects but not others. For example, men and women may have identical factor structures except for one indicator on one factor. In this case, that loading is estimated separately for the two groups before further, more stringent, constraints are considered. Or men and women may have the same factor structure for Love of Skiing and Ski Trip Satisfaction but differ in the size of the regression coefficient predicting Ski Trip Satisfaction from Love of Skiing.

Some cautions about multiple group modeling are in order before you dive head-first into this type of analysis. Multiple group models are often quite difficult to estimate. Critical to estimating a good-fitting multiple group model are single group models that fit well. It is extremely unlikely that the multiple group model will fit better than the individual group models. Good user-specified start values also seem to be more critical when estimating multiple group models.

A demonstration of multiple group modeling is outside of the scope of this chapter, but see Bentler (1995) and Byrne *et al.* (1989) for examples and detailed discussion of the process.

14.5.8 ■ Mean and Covariance Structure Models

The discussion so far has centered around modeling regression coefficients, variances and covariances; however, means, both latent and observed, can also be modeled (Sörbom, 1974; 1982). Means of latent variables can be particularly interesting. Means are estimated in SEM models by adding a special intercept variable to the model.

Typically, latent means are estimated in the context of a multiple group model. To ask, in the small sample example, if men and women have the same mean Love of Skiing or if gender makes a difference in the mean of Ski Trip Satisfaction, the data are estimated as a two-group model, constraints on the factor structure are made, the measured variable means estimated, and the latent means for both Love of Skiing and Ski Trip Satisfaction estimated. The most interpretable latent mean models are those in which the factor structure is identical or highly similar in both groups. For identification, the latent mean for one group is fixed at zero and the other estimated. The difference between the means is then estimated and evaluated with a z test like any other parameter, where the estimated parameter is divided by standard error. Demonstration of latent mean models is outside the scope of this chapter. However, Bentler (1995), and Byrne *et al.* (1989) also show examples and detailed discussion of these models.

14.6 ■ COMPARISON OF PROGRAMS

The four SEM programs discussed, EQS, LISREL, SAS CALIS, and SYSTAT RAMONA are full service, multi-option programs. A list of options included in each package appears in Table 14.14

14.6.1 ■ EQS

EQS is by far the most user-friendly of the programs. The equation method of specifying the model is clear and easy to use and the output is well organized. EQS offers numerous diagnostics for evaluation of assumptions and handles deletion of cases very simply. Evaluation of multivariate outliers and normality can be performed within this program. Several methods of estimation are offered. EQS is the only program that offers the correct adjusted standard errors and Satorra-Bentler scaled χ^2 for model evaluation. This is the program of choice when data are nonnormal. A specific estimation technique is available if the measured variables are nonnormal with common kurtosis. A second method for treatment of nonnormal data, through estimation of polychoric or polyserial correlations, is also available.

EQS is also the program of choice if model modifications are to be performed. EQS is very flexible, offering both multivariate and univariate Lagrange Multiplier tests as well as the multivariate Wald test. EQS offers several options for matrices to be considered for modification, and allows specification of order of consideration of these matrices. Ordinal categorical variables can be included within the model without preprocessing. EQS allows multiple group models to be specified and tested easily. Diagrams are available in EQS.

14.6.2 ■ LISREL

LISREL is a set of three programs: PRELIS, SIMPLIS, and LISREL. PRELIS preprocesses data, e.g., ordinal categorical or nonnormal data, for SEM analyses through LISREL. SIMPLIS is a program that allows models to be specified with equations. SIMPLIS is very simple to use but is somewhat limited in options, and some output, i.e., a standardized solution, must be requested in LISREL output form. LISREL specifies SEM models with matrices and some models become quite complicated to specify when using this method.

LISREL offers residual diagnostics, several estimation methods, and many fit indices. LISREL includes two types of partially standardized solutions. The univariate Lagrange multiplier test is also

TABLE 14.14 COMPARISON OF PROGRAMS FOR STRUCTURAL EQUATIONS MODELING

Feature	EQS	LISREL	SYSTAT RAMONA	SAS CALIS
Input				
Covariance matrix				
Lower triangular	Yes	Yes	No	Yes
Full symmetric	Yes	Yes	Yes	Yes
Input stream	Yes	Yes	No	No
Asymptotic covariance matrix	Yes	Yes	No	No
Multiple covariance matrices				
Correlations matrix				
Lower triangular	Yes	Yes	No	Yes
Full symmetric	Yes	Yes	Yes	Yes
Input stream	Yes	Yes	No	No
Matrix of polychoric, polyserial correlations	Yes	Yes	No	No
Correlation matrix based on optimal scores	No	Yes	No	No
Multiple correlation matrices	Yes	Yes	No	Yes
Moment matrices	Yes	Yes	No	Yes
Sum of squares and cross products matrix	No	No	No	Yes
Raw data	Yes	Yes	Yes	Yes
User specified weight matrix	Yes	Yes	No	No
Categorical (ordinal) data	Yes	Yes[a]	No	No
Means and standard deviations	Yes	Yes	No	Yes
Delete cases	Yes	Yes[a]	Yes	Yes
Estimation methods				
Maximum likelihood (ML)	Yes	Yes	Yes	Yes
Unweighted least squares (ULS)	Yes	Yes	Yes	Yes
Generalized least squares (GLS)	Yes	Yes	Yes	Yes
Two-stage least squares	No	Yes	No	No
Diagonally weighted least squares	No	Yes	No	No
Elliptical least squares (ELS)	Yes	No	No	No
Elliptical generalized least squares (EGLS)	Yes	No	No	No
Elliptical reweighted least squares (ERLS)	Yes	No	No	No
Arbitrary distribution generalized least squares (AGLS)	Yes	Yes	Yes	No
Satorra-Bentler scaled chi-square	Yes	No	No	No
Robust standard errors	Yes	No	No	No
Elliptical cor. chi-square	No	No	No	Yes
Instrumental variables	No	Ye	No	No
Specify number of groups	Yes	Yes	No	Yes
Specify elliptical kurtosis parameter–kappa	Yes	No	No	No
Specify model with equations	Yes	Yes[b]	Yes	Yes
Specify model with matrix elements	No	Yes	No	Yes
Specify models with intercepts	Yes	Yes	No	Yes
Start values				
Automatic	Yes	Yes	Yes	Yes[c]
User specified	Yes	Yes	Yes	Yes
Specify confidence interval range	No	No	Yes	No
Automatically scale latent variables	No	Yes	Yes	No
Specify covariances	Yes	Yes	Yes	Yes
Specify general linear constraints	Yes	Yes	Yes	Yes
Nonlinear constraints	No	Yes	No	No
Specify cross-group constraints	Yes	Yes	No	No
Specify inequalities	Yes	Yes	No	Yes

TABLE 14.14 *(CONTINUED)*

Feature	EQS	LISREL	SYSTAT RAMONA	SAS CALIS
Lagrange multiplier test—Univariate	Yes	Yes	No	Yes
Lagrange multiplier test—Multivariate	Yes	No	No	No
Lagrange multiplier options				
Indicate parameters to be considered first for addition	Yes	No	No	No
Indicate specific order for entry consideration	Yes	No	No	No
Specify LM testing process	Yes	No	No	No
Specify specific matrices only	Yes	Yes	No	No
Set probability value for criterion for inclusion	Yes	Yes	No	No
Specify parameters not to be included in LM test	Yes	Yes	No	No
Wald test - Univariate	Yes	No	No	Yes
Wald test- Multivariate	Yes	No	No	Yes
Wald test options				
Indicate parameters to be considered first for dropping	Yes	No	No	No
Indicate specific order for dropping consideration	Yes	No	No	No
Set probability value	Yes	No	No	Yes
Specify parameters not to be included in Wald test	Yes	No	No	No
Specify number of iterations	Yes	Yes	Yes	Yes
Specify maximum CPU time used	No	Yes	No	No
Specify optimization method	No	No	No	Yes
Specific convergence criterion	Yes	Yes	Yes	Yes
Specify tolerance	Yes	No	No	No
Specify a ridge factor	No	Yes	No	Yes
Diagram	Yes	Yes	No	No
Effect decomposition	Yes	Yes	No	Yes
Simulation	Yes	Yes	No	Yes[d]
Output				
Means	Yes	No	No	Yes
Skewness and kurtosis	Yes	No	Yes[f]	Yes
Mardia's coefficient	Yes	No	No	Yes
Normalized estimate	Yes	No	No	Yes
Mardia based kappa	Yes	No	No	Yes
Mean scaled univariate kurtosis	Yes[e]	No	Yes	Yes
Multivariate least squares kappa	Yes[e]	No	No	No
Multivariate mean kappa	Yes	No	No	No
Adjusted mean scaled univariate kurtosis	Yes	No	No	Yes
Relative multivariate kurtosis coefficient	No	No	Yes	Yes
Case numbers with largest contribution to				
normalized multivariate kurtosis	Yes	No	No	Yes
Sample covariance matrix	Yes	Yes	Yes	Yes
Sample correlation matrix	Yes	Yes	Yes	Yes
Estimated model covariance matrix	Yes	Yes	Yes	Yes
Correlations among parameter estimates	Yes	Yes	Yes	Yes
Asymptotic covariance matrix of parameters	No	Yes	No	No
Iteration summary	Yes	No	Yes	Yes
Determinant of input matrix	Yes	No	No	Yes
Residual covariance matrix	Yes	Yes	Yes	Yes
Largest raw residuals	Yes	Yes	Yes	Yes
Completely standardized residual matrix	Yes	No	No	Yes
Largest completely standardized residuals	Yes	No	No	Yes
Frequency distribution of standardized residuals	Yes	No	No	Yes

TABLE 14.14 *(CONTINUED)*

Feature	EQS	LISREL	SYSTAT RAMONA	SAS CALIS
Largest partially standardized residual	No	Yes	No	No
Partially standardized residual matrix	No	Yes	No	No
Q Plot of partially standardized residuals	No	Yes	No	No
Frequency distribution of partially standardized residuals	No	Yes	No	No
Estimated covariance matrix	Yes	Yes	Yes	Yes
Estimated correlation matrix	Yes	Yes	Yes	No
Largest eigenvalue of B*B'	No	Yes	No	No
Goodness-of-fit indices				
Normed Fit Index (NFI) (Bentler & Bonett, 1980)	Yes	Yes	No	Yes
Non-Normed Fit Index (NNFI) (Bentler & Bonett, 1980)	Yes	Yes	No	Yes
Comparative Fit Index (CFI) (Bentler, 1995)	Yes	Yes	No	Yes
Minimum of the fit function	Yes	Yes	Yes	Yes
Non-Centrality Parameter (NCP)	No	Yes	No	No
Confidence Interval of NCP	No	Yes	No	No
Goodness-of-Fit index (GFI)	Yes	Yes	No	Yes
Adjusted Goodness-of-Fit index	Yes	Yes	No	Yes
Root Mean Square residual	Yes	Yes	No	Yes
Standardized root mean square residual	Yes	Yes	No	No
Population discrepancy function (PDF)	No	Yes	Yes	No
Confidence interval for PDF	No	Yes	Yes	No
Root mean square error of approximation (RMSEA)	No	Yes	Yes	No
Confidence Interval for RMSEA	No	No	Yes	No
Akaike Information Criterion Model	Yes	Yes	No	Yes
Akaike Information Criterion—Independence Model	Yes	Yes	No	No
Akaike Information Criterion—Saturated Model	No	Yes	No	No
Consistent Information Criterion Model	Yes	Yes	No	Yes
Consistent Information Criterion—Independence Model	Yes	Yes	No	No
Consistent Information Criterion—Saturated Model	No	Yes	No	No
Schwartz Baysian Criterion	No	No	No	Yes
McDonald's Centrality (1989)	Yes	No	No	Yes
James, Mulaik & Brett (1982) Parsimony Index	No	Yes	No	Yes
Z test (Wilson & Hilferty, 1931)	No	No	No	Yes
Normed Index Rho 1 (Bollen, 1986)	No	Yes	No	Yes
Non-Normed Index Delta 2 (Bollen, 1989a)	Yes	Yes	No	Yes
Expected cross validation index (ECVI)	No	Yes	Yes	No
Confidence interval for ECVI	No	Yes	Yes	No
ECVI for saturated model	No	Yes	No	No
ECVI for independence model	No	Yes	No	No
Hoelter's Critical N	No	Yes	No	Yes
R^2 square for dependent variables	No	Yes	No	Yes
SMC for structural equations	No	Yes	No	Yes
Coefficient of determination for structural equations	No	Yes	No	Yes
Latent variable score regression coefficients	No	Yes	No	Yes
Unstandardized parameter estimates	Yes	Yes	Yes	Yes
Completely standardized parameter estimates	Yes	No	No	Yes
Partially standardized solution	No	Yes	No	No
Standard errors for parameter estimates	Yes	Yes	Yes	Yes
Variances of independent variables	Yes	Yes	Yes	Yes
Covariances of independent variables	Yes	Yes	Yes	Yes
Test statistics for parameter estimates	Yes	Yes	Yes	Yes

TABLE 14.14 *(CONTINUED)*

Feature	EQS	LISREL	SYSTAT RAMONA	SAS CALIS
Save output to file				
Condition code flag	Yes	No	No	Yes
Convergence flag	Yes	No	No	No
Function minimum	No	No	No	Yes
Independence model χ^2	Yes	No	No	Yes
Model χ^2 value	Yes	No	No	Yes
Model degress of freedom	Yes	No	No	Yes
Probability level	Yes	No	No	Yes
Bentler-Bonett normed fit index	Yes	Yes	No	No
Bentler-Bonett nonnormed fit index	Yes	Yes	No	No
Comparative fit index	Yes	Yes	No	Yes
GFI	No	Yes	No	Yes
AGFI	No	Yes	No	Yes
Root mean square residual	No	Yes	No	Yes
Number of parameters in model	No	No	No	Yes
AIC	No	Yes	No	Yes
CAIC	No	Yes	No	Yes
Scwartz's Baysesian Criterion	No	No	No	Yes
James, Mulaik, & Hilferty parsimony index	No	Yes	No	Yes
z-test of Wilson & Hilferty	No	No	No	Yes
Hoelter's critical N	No	Yes	No	Yes
Generated data	Yes	Yes[a]	No	Yes[d]
Derivatives	Yes	No	No	Yes
Gradients	Yes	No	No	Yes
Matrix analyzed	Yes	Yes	No	Yes
Means	No	No	No	Yes
Standard deviations	No	No	No	Yes
Sample size	No	No	No	Yes
Univariate skewness	No	No	No	Yes
Univariate kurtosis	No	No	No	Yes
Information matrix	No	No	No	Yes
Inverted information matrix	Yes	No	No	Yes
Weight matrix	Yes	No	No	No
Estimated population covariance matrix	No	Yes	No	Yes
Asymptotic covariance matrix	No	Yes	No	No
Asymptotic covariance matrix of parameter estimates	No	Yes	No	No
Parameter estimates	Yes	No	No	Yes
Residual matrix	Yes	No	No	No
Standard errors	Yes	No	No	Yes
LM test results	Yes	No	No	No
Wald test results	Yes	No	No	No
Updated Start Values	Yes	No	Yes	Yes
Automatic model modification	Yes	Yes	No	No

[a]. In PRELIS

[b]. In SIMPLIS

[c]. Not available with COSAN model specification

[d]. with SAS IML

[e]. with AGLS estimation only

[f]. kurtosis only

available. Nonnormal and ordinal categorical data can be included in LISREL by first preprocessing the data in PRELIS to calculate polyserial and polychoric correlations. LISREL also calculates the SMC for each variable in the equations. Coefficients of determination are calculated for the latent DVs in the model. Diagrams are available in SIMPLIS.

14.6.3 ■ SAS

SAS CALIS offers a choice of model specification methods: LINEQS, RAM, (both equation specification methods) and COSAN (a form of matrix specification). Diagnostics are available for evaluation of assumptions; for instance, evaluations of multivariate outliers and multivariate normality are available within CALIS. If data are nonnormal, but have homogenous kurtosis, the chi-square test statistics can be adjusted within the program. Several estimation techniques are available and lots of information about the estimation process is given. Ordinal categorical data are not treated in CALIS.

14.6.4 ■ SYSTAT

SYSTAT RAMONA employs an equation method of specifying models (McArdle & McDonald, 1984). RAMONA is simple to use but somewhat limited. For example, no standardized or partially standardized solution is available. Only three fit indices are included and no model modification information, such as LM or Wald tests, is available. RAMONA does offer several choices of estimation techniques. Much information on the iteration history is provided, and RAMONA is the only program to provide confidence intervals for parameter estimates. RAMONA does not accommodate ordinal categorical data, models with a mean structure, or multiple group models.

14.7 ■ COMPLETE EXAMPLES OF STRUCTURAL EQUATION MODELING ANALYSIS

The first example is a confirmatory factor analysis (CFA) model performed through LISREL. The data used in this example are from the learning disabled data bank, described in Appendix B. The factor structure underlying the subscales of the WISC in a sample of learning disabled children is examined. The model assesses the relationship between the indicators of IQ and two potential underlying constructs representing IQ. This type of model is sometimes referred to as a *measurement model*.

The second example is performed through EQS and has both measurement and structural components. In this example, mediators of the relationship between age, a life change measure, and latent variables representing Poor Sense of Self, Perceived Ill Health, and Health Care Utilization are examined. Data for the second example are from the women's health and drug study, described in Appendix B.1.

14.7.1 ■ Model Specification for CFA

The hypothesized model is presented in Figure 14.8. A two-factor model is hypothesized: a Verbal factor (with the information, comprehension, arithmetic, similarities, vocabulary, and digit span subscales of the WISC as indicators) and a Performance factor (with the picture completion, picture arrangement, block design, object assembly, and coding subscales of the WISC serving as indicators). For clarity within the text, the labels of latent variables have intial capital letters and the measured variables do not. Two main hypotheses are of interest: (1) Does a two-factor model with simple structure (each variable loading only on one factor) fit the data? and (2) Is there a significant covariance between the Verbal and Performance factors?

After the hypotheses are formulated and the model diagrammed, the first step of the modeling process is complete. At this point it is a good idea to do a preliminary check of the identifiability of the model. Count the number of data points and the number of parameters to be estimated in the model. With 11 observed variables there are $[11(11 + 1)]/2 = 66$ data points. The hypothesized model

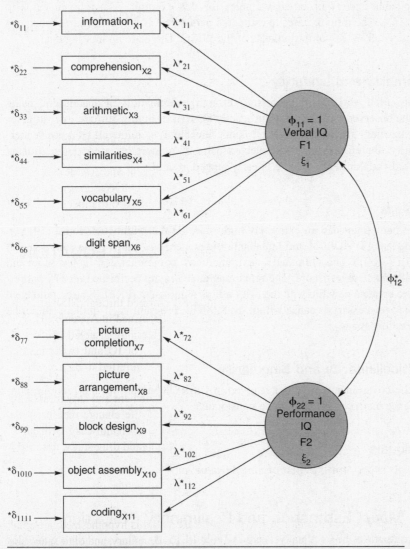

FIGURE 14.8 HYPOTHESIZED CFA MODEL.

indicates that 23 parameters are to be estimated (11 regression coefficients, 1 covariance, and 11 variances have asterisks); therefore, the model is over-identified and is tested with 43 df ($66 - 23$).

14.7.2 ■ Evaluation of Assumptions for CFA

Computer evaluation of assumptions is shown only when the procedure or output differs from that of previous chapters in this book.

14.7.2.1 ■ Sample Size and Missing Data

There are 177 participants and 11 observed variables for this example. The ratio of cases to observed variables is 16:1. The ratio of cases to estimated parameters is 8:1. This ratio is adequate given the high reliability (all $> .85$) of the subtests of the WISC. There are no missing data.

14.7.2.2 ■ Normality and Linearity

Normality of the observed variables is assessed through examination of histograms using BMDP2D. None of the observed variables is significantly skewed or highly kurtotic. No variables have a standardized skewness greater than 3.75. It is not feasible to examine all pairwise scatterplots to assess linearity; therefore, randomly selected pairs of scatterplots are examined using BMDP6D. All observed variables appear to be linearly related, if at all.

14.7.2.3 ■ Outliers

Using BMDP2D, one participant had an extremely high score on the arithmetic subtest (19, $z = 4.11$) and was deleted. Using BMDPAM and Mahalanobis distance, one multivariate outlier is also detected and deleted ($p < .001$). This child had an extremely low comprehension subtest score and an extremely high arithmetic subtest score. The remaining analyses are performed on 175 participants. BMDP is used to create a new file with the two outliers eliminated. (PRELIS also could have been used.) A new file is necessary because within the LISREL program itself outliers cannot be deleted (nor transformations made).

14.7.2.4 ■ Multicollinearity and Singularity

The determinant of the covariance matrix is not given in LISREL output, but the program converged, so the covariance matrix is assumed to be nonsingular.

14.7.2.5 ■ Residuals

Analysis of the residuals is performed as part of model evaluation.

14.7.3 ■ CFA Model Estimation and Preliminary Evaluation

The setup and edited output for the CFA analysis are in Table 14.15. As a first step it is a good idea to check that the covariance matrix matches the covariance matrix from preliminary analyses. It is

TABLE 14.15 SETUP AND PARAMETER SPECIFICATIONS FOR CFA USING LISREL

```
CONFIRMATORY FACTOR ANALYSIS OF THE WISC
DA NI = 13 NO = 175
RA FI = WISCSEM.DAT
LA
CLIENT, AGEMATE, INFO, COMP, ARITH, SIMIL, VOCAB,
DIGIT, PICTCOMP, PARANG, BLOCK, OBJECT, CODING
SE
INFO, COMP, ARITH, SIMIL, VOCAB,
DIGIT, PICTCOMP, PARANG, BLOCK, OBJECT, CODING/
MO NX=11 NK=2
LK
VERBAL PERFOR
FR LX(1,1) LX(2,1) LX(3,1) LX(4,1) LX(5,1) LX(6,1)
FR LX(7,2) LX(8,2) LX(9,2) LX(10,2) LX(11,2)
VA 1 PH(1,1) PH(2,2)
OU SC SE TV RS SS ND=3
```

CONFIRMATORY FACTOR ANALYSIS OF THE WISC-R

NUMBER OF INPUT VARIABLES	13
NUMBER OF Y - VARIABLES	0
NUMBER OF X - VARIABLES	11
NUMBER OF ETA - VARIABLES	0
NUMBER OF KSI - VARIABLES	2
NUMBER OF OBSERVATIONS	175

COVARIANCE MATRIX TO BE ANALYZED

	INFO	COMP	ARITH	SIMIL	VOCAB	DIGIT
INFO	8.481					
COMP	4.034	8.793				
ARITH	3.322	2.684	5.322			
SIMIL	4.758	4.816	2.713	10.136		
VOCAB	5.338	4.621	2.621	5.022	8.601	
DIGIT	2.720	1.891	1.678	2.234	2.334	7.313
PICTCOMP	1.965	3.540	1.052	3.450	2.456	.597
PARANG	1.561	1.471	1.391	2.524	1.031	1.066
BLOCK	1.808	2.966	1.701	2.255	2.364	.533
OBJECT	1.531	2.718	.282	2.433	1.546	.267
CODING	.059	.517	.598	-.372	.842	1.344

COVARIANCE MATRIX TO BE ANALYZED

	PICTCOMP	PARANG	BLOCK	OBJECT	CODING
PICTCOMP	8.610				
PARANG	1.941	7.074			
BLOCK	3.038	2.532	7.343		
OBJECT	3.032	1.916	3.077	8.088	
CODING	-.605	.289	.832	.433	8.249

PARAMETER SPECIFICATIONS
    LAMBDA-X

	VERBAL	PERFOR
INFO	1	0
COMP	2	0
ARITH	3	0
SIMIL	4	0
VOCAB	5	0
DIGIT	6	0
PICTCOMP	0	7
PARANG	0	8
BLOCK	0	9
OBJECT	0	10
CODING	0	11

TABLE 14.15   *(CONTINUED)*

```
        PHI

               VERBAL     PERFOR
               --------   --------
    VERBAL          0
    PERFOR         12          0


        THETA-DELTA

               INFO       COMP       ARITH      SIMIL      VOCAB      DIGIT
               --------   --------   --------   --------   --------   --------
                 13         14         15         16         17         18


        THETA-DELTA

               PICTCOMP   PARANG     BLOCK      OBJECT     CODING
               --------   --------   --------   --------   --------
                 19         20         21         22         23
```

also helpful to assure that the parameters that are indicated as free are those that were really intended to be estimated. The output labeled PARAMETER SPECIFICATIONS lists each matrix specified in the model section and numbers each free parameter. LAMBDA-X is the matrix of regression coefficients to be estimated between indicators and factors. PHI is the matrix of covariances among factors. THETA-DELTA is the diagonal matrix of errors to be estimated for each measured variable. Only the diagonal of the THETA-DELTA matrix is shown as all other entries in this matrix are zero (no correlated errors). In the other matrices, zeros indicate parameters that are fixed (not estimated). After checking the parameter specifications we confirm that they match the path diagram.

Next, it is helpful to assess the overall fit of the model by looking at the χ^2 and fit indices that appear in the section labeled GOODNESS OF FIT STATISTICS (Table 14.16). The CHI-SQUARE FOR INDEPENDENCE MODEL WITH 55 DEGREES OF FREEDOM is χ^2_{indep} (55, N = 175) = 516.237, $p < .01$. This χ^2 tests the hypothesis that the variables are unrelated; it should always be significant. If it is not, as is possible with very small samples, modeling should be reconsidered.

The model chi-square is significant, χ^2 (43, N = 175) = 70.24, $p = .005$. Ideally, a nonsignificant chi-square is desired. (If the model were perfect, the chi square would equal zero and the probability level of a zero chi square would equal 1.) The model χ^2 in this case is significant, but it is also less than two times the model degrees of freedom. This ratio gives a *very* rough indication that the model may fit the data. LISREL output includes many other fit indices, including the CFI = .94, GFI = .93, and the standardized RMR = .06. These indices all seem to indicate a reasonably good-fitting model.

Residuals are examined after evaluation of fit. Residual diagnostics are requested with RS on the OU (output line) of the procedure file in Table 14.15. LISREL gives numerous residual diagnostics (Table 14.17). Residuals are printed in both the original scale of the variables, labeled FITTED RESIDUALS (not shown), and partially standardized, labeled STANDARDIZED RESIDUALS. For both types of residuals, the full residual covariance matrix, summary statistics, and a stemleaf plot are given. Although the model fits the data well, there is a sizable residual (standardized residual 3.06) between picture completion (PICTCOMP) and comprehension (COMP). This indicates

TABLE 14.16 GOODNESS OF FIT STATISTICS FOR CFA MODEL USING LISREL. SETUP APPEARS IN TABLE 14.15

```
                    GOODNESS OF FIT STATISTICS

    CHI-SQUARE WITH 43 DEGREES OF FREEDOM = 70.236 (P = 0.00545)
            ESTIMATED NON-CENTRALITY PARAMETER (NCP) = 27.236
       90 PERCENT CONFIDENCE INTERVAL FOR NCP = (8.130 ; 54.240)

                    MINIMUM FIT FUNCTION VALUE = 0.404
           POPULATION DISCREPANCY FUNCTION VALUE (F0) = 0.157
        90 PERCENT CONFIDENCE INTERVAL FOR F0 = (0.0467 ; 0.312)
       ROOT MEAN SQUARE ERROR OF APPROXIMATION (RMSEA) = 0.0603
    90 PERCENT CONFIDENCE INTERVAL FOR RMSEA = (0.0330 ; 0.0851)
        P-VALUE FOR TEST OF CLOSE FIT (RMSEA < 0.05) = 0.239

            EXPECTED CROSS-VALIDATION INDEX (ECVI) = 0.668
        90 PERCENT CONFIDENCE INTERVAL FOR ECVI = (0.558 ; 0.823)
                    ECVI FOR SATURATED MODEL = 0.759
                   ECVI FOR INDEPENDENCE MODEL = 3.093

  CHI-SQUARE FOR INDEPENDENCE MODEL WITH 55 DEGREES OF FREEDOM = 516.237
                    INDEPENDENCE AIC = 538.237
                        MODEL AIC = 116.236
                       SATURATED AIC = 132.000
                    INDEPENDENCE CAIC = 584.050
                        MODEL CAIC = 212.026
                      SATURATED CAIC = 406.876

              ROOT MEAN SQUARE RESIDUAL (RMR) = 0.468
                     STANDARDIZED RMR = 0.0585
                 GOODNESS OF FIT INDEX (GFI) = 0.931
            ADJUSTED GOODNESS OF FIT INDEX (AGFI) = 0.894
           PARSIMONY GOODNESS OF FIT INDEX (PGFI) = 0.606

                  NORMED FIT INDEX (NFI) = 0.864
                NON-NORMED FIT INDEX (NNFI) = 0.924
             PARSIMONY NORMED FIT INDEX (PNFI) = 0.675
               COMPARATIVE FIT INDEX (CFI) = 0.941
               INCREMENTAL FIT INDEX (IFI) = 0.942
                RELATIVE FIT INDEX (RFI) = 0.826

                    CRITICAL N (CN) = 168.123
```

that the model does not adequately estimate the relationship between these two variables. The large residual between PICTCOMP and COMP is also clearly indicated in the stemleaf plot. The stemleaf plot also shows that the residuals are centered around zero, and symmetrically distributed. The median residual is zero. LISREL also provides a Q PLOT of partially standardized residuals as shown in Figure 14.9. If the residuals are normally distributed the xs hover around the diagonal. As in multiple regression, large deviations from the diagonal indicate nonnormality. Once again the large residual between PICTCOMP and COMP is clearly evident in the upper right-hand corner of the plot.

Finally, estimates of the parameters are examined (Table 14.18). In the section labeled LISREL ESTIMATE (MAXIMUM LIKELIHOOD) are, by row, the unstandardized regression coefficients, standard errors in parentheses, and z scores (coefficient/standard error) for each indicator.[12] All of

---

[12] Pop Quiz! With your knowledge of both SEM and exploratory factor analysis (Chapter 13), What are the regression coefficients in this CFA equivalent to in exploratory factor analysis? Answer: Elements in the pattern matrix.

TABLE 14.17   PARTIALLY STANDARIZED REISUDALS FOR CFA USING LISREL. SETUP APPEARS IN
TABLE 14.15

STANDARDIZED RESIDUALS

	INFO	COMP	ARITH	SIMIL	VOCAB	DIGIT
INFO	.000					
COMP	-2.279	.000				
ARITH	2.027	.054	.000			
SIMIL	-.862	.817	-.748	.000		
VOCAB	2.092	-.008	-1.527	-.138	.000	
DIGIT	1.280	-.753	.897	-.342	-.168	.000
PICTCOMP	-.734	3.065	-.716	2.314	.319	-.942
PARANG	-.176	-.096	1.092	1.759	-1.488	.579
BLOCK	-1.716	1.844	.795	-.441	-.276	-1.341
OBJECT	-1.331	1.673	-2.394	.633	-1.403	-1.443
CODING	-.393	.472	.950	-1.069	1.051	2.146

STANDARDIZED RESIDUALS

	PICTCOMP	PARANG	BLOCK	OBJECT	CODING
PICTCOMP	.000				
PARANG	-.779	.000			
BLOCK	-1.004	.861	.000		
OBJECT	.751	-.319	.473	.000	
CODING	-2.133	.059	1.284	.215	.000

SUMMARY STATISTICS FOR STANDARDIZED RESIDUALS
SMALLEST STANDARDIZED RESIDUAL =   -2.394
  MEDIAN STANDARDIZED RESIDUAL =   .000
 LARGEST STANDARDIZED RESIDUAL =   3.065

STEMLEAF PLOT
- 2|431
- 1|755443310
- 0|9988777443332211000000000000
  0|1123556688899
  1|01133788
  2|0113
  3|1
LARGEST POSITIVE STANDARDIZED RESIDUALS
RESIDUAL FOR PICTCOMP AND   COMP   3.065

the indicators are significant ($p < .01$) with the exception of coding. Sometimes the different scales of the measured variables make the unstandardized coefficients difficult to interpret and the scales of the measured variables often lack inherent meaning. The partially standardized solution, labeled COMPLETELY STANDARDIZED SOLUTION in Table 14.18, is often easier to interpret in such cases. Completely standardized output is requested with SC [on the OU (output line)] in the procedure file of Table 14.15. Note this output is not completely standardized with respect to error variances.

The first hypothesis, that the model fits the data, has been evaluated and supported, although there is a large residual between PICTCOMP and COMP. The final model, with significant parameter estimates presented in standardized form, appears in Figure 14.9. Other questions of interest are now examined. Is there a significant correlation between the Verbal and Performance factors? Looking at the PHI matrix in the completely standardized solution (or Figure 14.10), the Verbal and

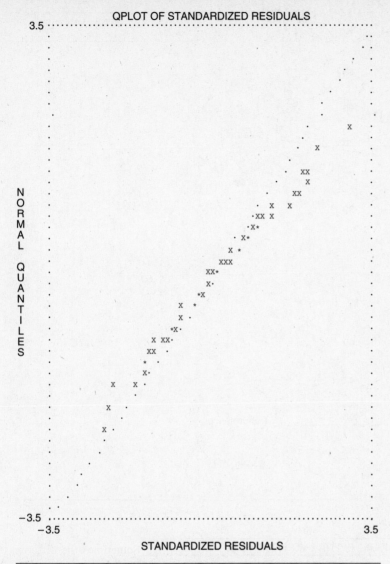

FIGURE 14.9 Q PLOT OUTPUT OF PARTIALLY STANDARDIZED RESIDU-
ALS FOR CFA MODEL IN LISREL. SET-UP APPEARS IN
TABLE 14.15.

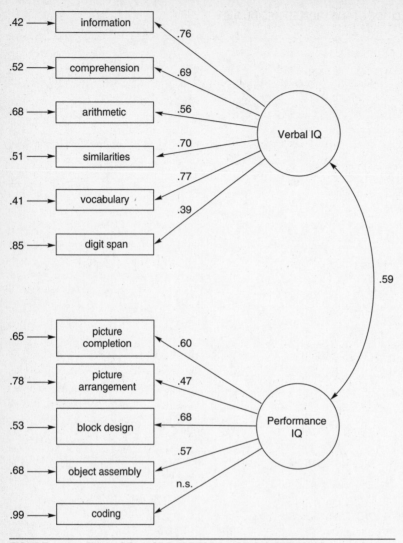

FIGURE 14.10 FINAL CFA MODEL BEFORE MODIFICATIONS.

Performance factors are significantly correlated, $r = .589$, supporting the hypothesis of a relationship between the factors.

Only LISREL provides estimates of the squared multiple correlations of the variables with the factors, in the section labeled SQUARED MULTIPLE CORRELATIONS FOR X − VARIABLES in Table 14.19. It is clear upon examining these SMCs that coding, with an SMC of .005, is not related to the Performance factor.

LISREL ESTIMATES (MAXIMUM LIKELIHOOD)

LAMBDA-X

	VERBAL	PERFOR
INFO	2.212	- -
	(.201)	
	10.997	
COMP	2.048	- -
	(.212)	
	9.682	
ARITH	1.304	- -
	(.173)	
	7.534	
SIMIL	2.238	- -
	(.226)	
	9.911	
VOCAB	2.257	- -
	(.202)	
	11.193	
DIGIT	1.056	- -
	(.213)	
	4.952	
PICTCOMP	- -	1.747
		(.244)
		7.166
PARANG	- -	1.257
		(.226)
		5.566
BLOCK	- -	1.851
		(.223)
		8.287
OBJECT	- -	1.609
		(.237)
		6.781
CODING	- -	.208
		(.256)
		.811

PHI

	VERBAL	PERFOR
VERBAL	1.000	
PERFOR	.589	1.000
	(.076)	
	7.792	

TABLE 14.18   *(CONTINUED)*

THETA-DELTA

	INFO	COMP	ARITH	SIMIL	VOCAB	DIGIT
	3.586	4.599	3.623	5.125	3.507	6.198
	(.511)	(.590)	(.424)	(.667)	(.511)	(.686)
	7.014	7.793	8.547	7.680	6.866	9.030

THETA-DELTA

PICTCOMP	PARANG	BLOCK	OBJECT	CODING
5.558	5.494	3.916	5.499	8.206
(.764)	(.664)	(.646)	(.726)	(.882)
7.276	8.275	6.066	7.578	9.309

COMPLETELY STANDARDIZED SOLUTION

LAMBDA-X

	VERBAL	PERFOR
INFO	.760	- -
COMP	.691	- -
ARITH	.565	- -
SIMIL	.703	- -
VOCAB	.770	- -
DIGIT	.390	- -
PICTCOMP	- -	.595
PARANG	- -	.473
BLOCK	- -	.683
OBJECT	- -	.566
CODING	- -	.072

PHI

	VERBAL	PERFOR
VERBAL	1.000	
PERFOR	.589	1.000

THETA-DELTA

INFO	COMP	ARITH	SIMIL	VOCAB	DIGIT
.423	.523	.681	.506	.408	.848

THETA-DELTA

PICTCOMP	PARANG	BLOCK	OBJECT	CODING
.646	.777	.533	.680	.995

TABLE 14.19   OUTPUT OF SQUARED MULTIPLE CORRELATIONS FOR INDICATORS OF VERBAL AND PERFOMANCE
FACTOR   FROM LISREL.   SETUP APPEARS IN TABLE 14.15

```
SQUARED MULTIPLE CORRELATIONS FOR X - VARIABLES

      INFO       COMP       ARITH      SIMIL      VOCAB      DIGIT
    --------   --------   --------   --------   --------   --------
      .577       .477       .319       .494       .592       .152

SQUARED MULTIPLE CORRELATIONS FOR X - VARIABLES

    PICTCOMP    PARANG     BLOCK      OBJECT     CODING
    --------   --------   --------   --------   --------
      .354       .223       .467       .320       .005
```

14.7.4 ▪ Model Modification

At this point in the analysis there are several choices. The model fits the data, and we have confirmed that there is a significant correlation between the factors. Therefore, we could stop here and report the results. Generally, however, several additional models are examined that test further hypotheses (either a priori or post hoc) and/or attempt to improve the fit of the model. At least two post hoc hypotheses are of interest in this model: (1) Could the residual between comprehension and picture completion be reduced by adding additional paths to the model? and (2) Could a good fitting, more parsimonious model be estimated without the data from coding subtest?

Before demonstrating model modification, be warned that when adding and deleting parameters, there too often comes a point of almost total confusion: What have I added?, What have I deleted? and What *am* I doing anyway? One hint for avoiding this sort of confusion is to diagram the estimated model, prior to any modifications, and make a few copies of it. Then, as parameters are added and deleted, draw the modifications on the copies. In this way the model can be viewed at each stage without having to redraw the diagram each time. When one copy gets too messy, move to next, and make more copies as necessary.

With copies of diagram firmly in hand, modification indices are examined. Completely standardized modification indices are presented in Table 14.20. The largest univariate modification index is for the regression path predicting comprehension from the Performance factor, $\chi^2 = 9.767$, with an approximate standardized parameter value of .317. Because this path may also reduce the residual between the comprehension and picture completion subtests; a model is run with this path estimated, χ^2 (42, $N = 172$) $= 60.29$, $p = .03$, CFI $= .96$. The estimated (first) model and the newly modified model are nested within one another; the estimated model is a subset of this modified model. Therefore, a chi-square difference test is performed to see if the addition of this path significantly improves the model. The estimated (first) model has 43 df, the modified model had 42 df; therefore, this is a 1 df test. The χ^2 for the first model was 70.236, for the second, 60.295. The difference between the two chi-square values is a χ^2 equal to 9.941, χ^2_{diff} (1, $N = 172$) $= 9.941$, $p < .01$, about the value of 9.767 anticipated from the modification index (LM test). We conclude that the addition of a path predicting comprehension from the Performance factor significantly improves the model. The largest standardized residual is now 2.614 and the plot of residuals is improved.

Additional paths could be added to the model but the decision is made to test a third model with the coding subtest removed. The Wald test is unavailable in LISREL so we delete coding and estimate

TABLE 14.20   OUTPUT FROM LISREL OF MODIFICATION INDICES FOR CFA MODEL.   SETUP APPEARS IN TABLE 14.15

MODIFICATION INDICES AND EXPECTED CHANGE

MODIFICATION INDICES FOR LAMBDA-X

	VERBAL	PERFOR
INFO	- -	4.451
COMP	- -	9.767
ARITH	- -	.177
SIMIL	- -	2.556
VOCAB	- -	1.364
DIGIT	- -	1.852
PICTCOMP	1.763	- -
PARANG	.020	- -
BLOCK	.181	- -
OBJECT	1.174	- -
CODING	.199	- -

COMPLETELY STANDARDIZED EXPECTED CHANGE FOR LAMBDA-X

	VERBAL	PERFOR
INFO	- -	-.205
COMP	- -	.317
ARITH	- -	-.046
SIMIL	- -	.161
VOCAB	- -	-.113
DIGIT	- -	-.161
PICTCOMP	.156	- -
PARANG	.016	- -
BLOCK	-.054	- -
OBJECT	-.126	- -
CODING	.052	- -

NO NON-ZERO MODIFICATION INDICES FOR PHI

MODIFICATION INDICES FOR THETA-DELTA

	INFO	COMP	ARITH	SIMIL	VOCAB	DIGIT
INFO	- -					
COMP	5.192	- -				
ARITH	4.110	.003	- -			
SIMIL	.744	.668	.559	- -		
VOCAB	4.378	.000	2.332	.019	- -	
DIGIT	1.637	.567	.804	.117	.028	- -
PICTCOMP	1.318	4.659	1.672	3.251	.000	.767
PARANG	.087	1.543	2.081	3.354	4.122	1.208
BLOCK	1.415	1.205	2.561	2.145	.325	.923
OBJECT	.101	2.798	6.326	.803	.686	.873
CODING	.762	.035	.832	3.252	1.509	4.899

MODIFICATION INDICES FOR THETA-DELTA

	PICTCOMP	PARANG	BLOCK	OBJECT	CODING
PICTCOMP	- -				
PARANG	.607	- -			
BLOCK	1.008	.742	- -		
OBJECT	.564	.102	.223	- -	
CODING	4.549	.004	1.648	.046	- -

TABLE 14.20 *(CONTINUED)*

COMPLETELY STANDARDIZED EXPECTED CHANGE FOR THETA-DELTA

	INFO	COMP	ARITH	SIMIL	VOCAB	DIGIT
INFO	- -					
COMP	-.115	- -				
ARITH	.104	.003	- -			
SIMIL	-.043	.041	-.040	- -		
VOCAB	.107	.000	-.078	-.007	- -	
DIGIT	.069	-.043	.055	-.019	-.009	- -
PICTCOMP	-.056	.112	-.074	.093	.000	-.054
PARANG	.015	-.068	.086	.099	-.103	.072
BLOCK	-.056	.055	.088	-.073	.027	-.057
OBJECT	-.016	.088	-.145	.047	-.041	-.059
CODING	-.048	.011	.059	-.105	.067	.156

COMPLETELY STANDARDIZED EXPECTED CHANGE FOR THETA-DELTA

	PICTCOMP	PARANG	BLOCK	OBJECT	CODING
PICTCOMP	- -				
PARANG	-.054	- -			
BLOCK	-.081	.062	- -		
OBJECT	.054	-.022	.037	- -	
CODING	-.146	.004	.087	.015	- -

MAXIMUM MODIFICATION INDEX IS   9.77 FOR ELEMENT (2, 2) OF LAMBDA-X

the third model, $x^2(33, N = 172) = 45.018$, $p = .08$, CFI = .974. By dropping the coding subtest completely we have changed *the data* and the parameters so the model is no longer nested and the chi-square difference test is no longer appropriate. Although a statistical test of improvement is not available, other fit indices can be examined. The model AIC and CAIC can be compared between the models with small values indicating good fitting, parsimonious models. The AIC for the model with the coding subtest is 108.30; without the coding subtest the AIC drops to 89.018. The CAIC also drops after the coding subtest is deleted: with coding CAIC = 208.25, and without coding CAIC = 180.64. It is unclear if this drop is large enough, as the AIC and CAIC are not normed; however, there does seem to be a sizable change in fit and parsimony when the coding subtest is removed. The third model, with significant coefficients included in standardized form, is presented in Figure 14.11

The model modifications in this example were post hoc and may have capitalized on chance. Ideally, these results would be cross-validated with a new sample. However, in the absence of a new sample, a helpful measure of the extent to which the parameters changed in the course of modifications is the bivariate correlation between the parameter estimates of the first and third models. This correlation, as calculated by BMDPAM, is $r(18) = .947$, $p < .01$, which indicates that, although model modifications were made, the relative size of the parameters hardly changed.

Table 14.31, at the end of Section 14.7, contains a checklist for SEM. A results section for the CFA analysis follows, in journal format.

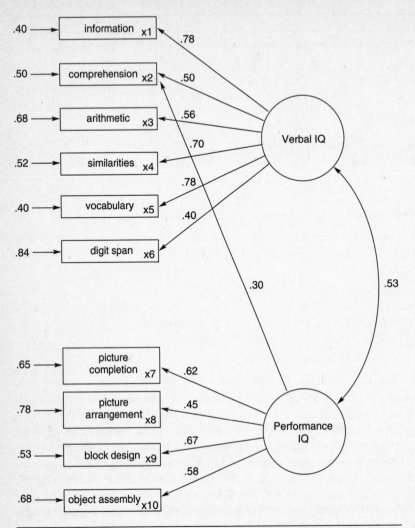

FIGURE 14.11   FINAL MODIFIED CFA MODEL WITH SIGNIFICANT COEFFI-CIENTS PRESENTED IN STANDARDIZED FORM.

TABLE 14.31 CHECKLIST FOR STRUCTURAL
EQUATIONS MODELING

1. Issues
   a. Sample size and missing data
   b. Normality of sampling distributions
   c. Outliers
   d. Linearity
   e. Adequacy of covariances
   f. Identification
   g. Path diagram–hypothesized model
   h. Estimation method
2. Major Analyses
   a. Assessment of fit
      (1) Residuals
      (2) Model chi square
      (3) Fit indices
   b. Significance of specific parameters
   c. Variance in a variable accounted for by a factor
3. Additional Analyses
   a. Lagrange Multiplier Test
      (1) Tests of specific parameters
      (2) Addition of parameters to improve fit
   b. Wald Test for dropping parameters
   c. Correlation between hypothesized and final model
   d. Diagram–final model

Results

The Hypothesized Model

   A confirmatory factor analysis, based on data from learning-disabled children, was performed through LISREL on eleven subtests of the WISC (the mazes subtest was not available for this sample). The hypothesized model is presented in Figure 14.8 where circles represent latent variables, and rectangles represent measured variables. Absence of a line connecting variables implies no hypothesized direct effect. A two factor model of IQ, Verbal and Performance, is hypothesized. The information, comprehension, arithmetic, similarities, vocabulary, and digit span subtests serve as indicators of the Verbal IQ factor. The picture completion, picture arrangement, block design, object assembly, and coding subtests serve as indicators of the Performance IQ factor. The two factors are hypothesized to covary with one another.

Assumptions

The assumptions of multivariate normality and linearity were evaluated through BMDP. One child had an extremely high score on the arithmetic subtest (19, z = 4.11, p < .01) and his data were deleted from the analysis. Using Mahalanobis distance, another child was a multivariate outlier, p < .001, and the data from this child were also deleted. This child had an extremely low comprehension subtest score and an extremely high arithmetic score. Structural equation modeling (SEM) analyses were performed, using data from 175 children. There were no missing data.

Model Estimation

Maximum likelihood estimation was employed to estimate all models. The independence model that tests the hypothesis that all variables are uncorrelated was easily rejectable, χ^2(55, N = 175) = 516.24, p < .01. The hypothesized model was tested next and support for it was found, χ^2(43, N = 175) = 70.24, p = .005, comparative fix index (CFI) = .94. A chi-square difference test indicated a significant improvement in fit between the independence model and the hypothesized model.

Post hoc model modifications were performed in an attempt to develop a better fitting and possibly more parsimonious model. On the basis of the Lagrange multiplier test, a path predicting the comprehension subtest from the Performance factor was added, χ^2(42, N = 172) = 60.29, p = .03, CFI = .96, CAIC = 208.25, AIC = 108.30. A chi-square difference test indicated that the model was significantly improved by addition of this path, χ^2_{diff}(1, N = 172) = 9.94, p < .01. Second, because the coefficient predicting the coding subscale from the Performance factor (.072) was not significant, SMC = .005, this variable was dropped and the model re-estimated, χ^2(33, N = 172) = 45.02, p = .08, CFI = .97, CAIC = 180.64, AIC = 89.02. Both the CAIC and AIC indicated a better fitting, more parsimonious model after the coding subtest is dropped

Because post hoc model modifications were performed, a correlation was calculated between the hypothesized model parameter estimates and the parameter estimates from the finalmodel, r(18) = .95, p < .01; this indicates that parameter estimates were hardly changed despite modification of the model. The final model, including significant coefficients in standardized form, is illustrated in Figure 14.11.

14.7.5 ■ SEM Model Specification

EQS is used to assess the fit of the hypothesized model in Figure 14.12 to the data in Appendix B.1. The model includes three hypothesized factors: Poor Sense of Self (with attitudes toward women's roles—ATTROLE, self-esteem—ESTEEM, satisfaction with marital status—ATTMAR, and locus of control—CONTROL, as indicators), Perceived Ill Health (with number of mental health problems—MENHEAL, and number of physical health problems—PHYHEAL as indicators), and Health Care Utilization (with number of visits to health professionals—TIMEDRS, and extent of drug use—DRUGUSE, as indicators). It is hypothesized that age, a measured variable, predicts number of life change units (STRESS), which, in turn, predicts Perceived Ill Health. Poor Sense of Self also predicts Perceived Ill Health. Additionally, age and Poor Sense of Self covary (hence the two headed arrow). Perceived Ill Health predicts Health Care Utilization.

Several questions are of interest: (1) Does this model fit the data? (2) Does the number of life change units mediate the relationship between age and Perceived Ill Health? (3) Does a Poor Sense of Self predict Perceived Ill Health? (4) Do number of life change units and Perceived Ill Health mediate the relationship between age and Health Care Utilization?

As a preliminary check of the identifiability of the model, the number of data points and parameters to be estimated are counted. With 10 variables there are $[10(10 + 1)]/2 = 55$ data points. The hypothesized model contains 23 parameters to be estimated (10 regression coefficients, 1 covariance, 12 variances); therefore, the model is over-identified and is tested with 32 df. To set the scales of the factors, the path predicting number of mental health problems from Perceived Ill Health and the path predicting number of visits to health professionals from Health Care Utilization are fixed to 1, while the variance of the Poor Sense of Self factor is fixed to 1. EQS setup and summary statistics appear in Table 14.21.

14.7.6 ■ Evaluation of Assumptions for SEM

Output from the evaluation of assumptions is shown only when the procedure or output is different from that in other chapters or the CFA example.

14.7.6.1 ■ Sample Size and Missing Data

There are 459 participants and 10 observed variables for this example. The ratio of cases to observed variables is 46:1; the ratio of cases to parameters is 20:1; both excellent ratios. There are no missing data.

14.7.6.2 ■ Normality and Linearity

Normality of the observed variables is assessed through examination of histograms using BMDP2D[13] and summary descriptive statistics in EQS. Eight of the ten observed variables are significantly skewed:[14]

(1) TIMEDRS z $= 15.66$ (2) PHYHEAL z $= 9.43$;

(3) MENHEAL z $= 5.41$ (4) DRUGUSE z $= 11.32$;

---

[13] Scatterplots also could have been examined through the EQS Windows program.

[14] Skewness z values are from BMDP2D (not shown). EQS shows skewness values, not z, in Table 14.21.

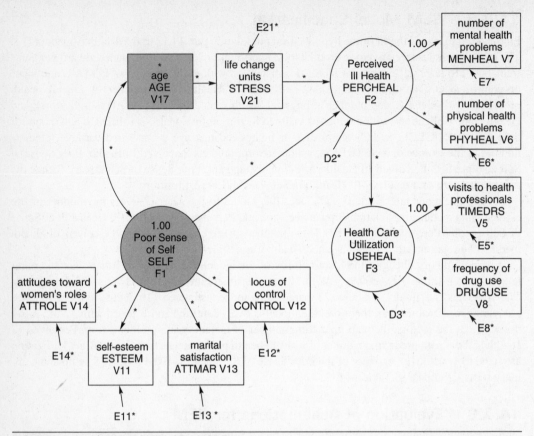

FIGURE 14.12 HYPOTHESIZED SEM MODEL.

(5) ESTEEM z = 3.97   (6) CONTROL z = 4.08;

(7) STRESS z = 6.59   (8) ATTMAR z = 5.69.

EQS also provides information on multivariate normality (see Table 14.21). In the section labeled, MULTIVARIATE KURTOSIS Mardia's coefficient and a normalized estimate of the coefficient are given; the normalized estimate can be interpreted as a z score. In this example, after deletion of all outliers, NORMALIZED ESTIMATE = 15.57, suggesting that the measured variables are not distributed normally.

It is not feasible to examine all pairwise scatterplots to assess linearity; therefore, randomly selected pairs of scatterplots are examined using BMDP6D. All observed pairs appear to be linearly related, if at all. Transformations are not made to these variables because it is reasonable to expect these variables to be skewed in the population (most women use few drugs, are not ill, and do not go to the doctor often). Instead, the decision is made to use provisions in the EQS program to take the nonnormality into account, when estimating χ^2 statistics and standard errors, by use of maximum likelihood estimation with the Satorra-Bentler scaled chi-square and adjustment to the standard errors to the extent of the nonnormality. This analysis is requested from EQS by the ME=ML,ROBUST instruction (see Table 14.21).

TABLE 14.21   SETUP AND PARTIAL OUTPUT FROM EQS SEM LARGE SAMPLE EXAMPLE 2

```
 1  /TITLE
 2    LARGE SAMPLE EXAMPLE 2 HYPOTHESIZED MODEL
 3  /SPECIFICATIONS
 4    CASES = 459;VAR = 21; ME=ML,ROBUST;MA=RAW;
 5    DATA = 'HLTHSEM.DAT';
 6    FO = '(10F8.3/10F8.3/F8.3)';
 7    DEL = 140,163,168,275,279,193,206,249,368,67,
 8          213,286,363,389;
 9   /LABELS
10    V1= SUBNO;V2= EDCODE;V3=INCODE;V4=EMPLMNY; V5=TIMEDRS;
11    V6=PHYHEAL; V7=MENHEAL; V8=DRUGUSE; V9=STRESS;
12    V10=ATTMED; V11=ESTEEM; V12=CONTROL;V13= ATTMAR;
13    V14=ATTROLE; V15=ATTHOUSE; V16=ATTWORK;V17= AGE; V18=SEL;
14    V19= LTIMEDRS; V20= LPHYHEAL; V21= SCSTRESS;
15     F1 = SELF; F2 = PERCHEAL; F3=USEHEAL;
16  /EQUATIONS
17
18  !SENSE OF SELF FACTOR
19    V14 = *F1 + E14;
20    V11 = *F1 + E11;
21    V12 = *F1 + E12;
22    V13 = *F1 + E13;
23
24  !PERCEIVED POOR HEALTH
25    V7 =  F2 + E7;
26    V6 = *F2 + E6;
27
28  !UTILIZATION OF HEALTH CARE
29    V5 =  F3 + E5;
30    V8 = *F3 + E8;
31
32
33    V21 = *V17 + E21;
34    F2 =  *F1 + *V21 + D2;
35    F3 =  *F2 + D3;
36
37  /VARIANCES
38    E11,E12,E13,E14 = *;
39    E6,E7 =*;
40    E5,E8 = *;
41    E21,D2,D3 = *;
42    V17 = *;
43    F1 = 1;
44  /COVARIANCES
45     F1,V17 = *;
46
47  /PRINT
48    EFFECT = YES;
49    FIT = ALL;
50  /WTEST
51  /LMTEST
52  /END
```

CASE NUMBERS DELETED FROM RAW DATA ARE: 67 140 163 168 193 206 213 249 275 279 286 363 368
389

   SAMPLE STATISTICS

                          UNIVARIATE STATISTICS

   VARIABLE   TIMEDRS   PHYHEAL   MENHEAL   DRUGUSE   ESTEEM

   MEAN   7.6270   4.9438   6.1506   8.6562   15.7910

TABLE 14.21 *(CONTINUED)*

SKEWNESS (G1)	2.8748	0.9632	0.6001	1.2613	0.5057
KURTOSIS (G2)	9.7186	0.7879	-0.2965	1.0330	0.3041

VARIABLE	CONTROL	ATTMAR	ATTROLE	AGE	SCSTRESS
MEAN	6.7371	22.8270	35.1978	4.3865	2.0094
SKEWNESS (G1)	0.5089	0.7760	0.0445	0.0323	0.7672
KURTOSIS (G2)	-0.3770	0.8350	-0.4095	-1.1761	0.2241

MULTIVARIATE KURTOSIS

MARDIA'S COEFFICIENT (G2,P) =   22.8685
NORMALIZED ESTIMATE =   15.5697

ELLIPTICAL THEORY KURTOSIS ESTIMATES

MARDIA-BASED KAPPA =   0.1906 MEAN SCALED UNIVARIATE KURTOSIS =   0.3548

MARDIA-BASED KAPPA IS USED IN COMPUTATION. KAPPA=   0.1906

CASE NUMBERS WITH LARGEST CONTRIBUTION TO NORMALIZED MULTIVARIATE KURTOSIS:

CASE NUMBER	39	167	205	273	277
ESTIMATE	971.7036	1213.1013	950.8789	886.6394	1091.0170

COVARIANCE MATRIX TO BE ANALYZED: 10 VARIABLES (SELECTED FROM 21 VARIABLES)
BASED ON   445 CASES.

		TIMEDRS V 5	PHYHEAL V 6	MENHEAL V 7	DRUGUSE V 8	ESTEEM V 11
TIMEDRS	V 5	102.045				
PHYHEAL	V 6	10.884	5.522			
MENHEAL	V 7	10.818	4.932	17.709		
DRUGUSE	V 8	26.795	8.900	14.117	83.591	
ESTEEM	V 11	0.649	0.736	3.712	-1.804	15.747
CONTROL	V 12	0.303	0.341	1.524	0.810	1.735
ATTMAR	V 13	4.253	1.668	9.040	6.866	10.549
ATTROLE	V 14	-5.570	-0.795	-2.100	-6.047	5.280
AGE	V 17	-0.493	0.080	-0.853	-0.716	0.146
SCSTRESS	V 21	3.407	0.922	2.118	3.660	-0.526

		CONTROL V 12	ATTMAR V 13	ATTROLE V 14	AGE V 17	SCSTRESS V 21
CONTROL	V 12	1.609				
ATTMAR	V 13	2.288	79.598			
ATTROLE	V 14	0.034	-4.155	45.992		
AGE	V 17	-0.385	-1.955	3.279	4.954	
SCSTRESS	V 21	0.107	1.289	-2.125	-0.844	1.705

TABLE 14.21 *(CONTINUED)*

```
BENTLER-WEEKS STRUCTURAL REPRESENTATION:

     NUMBER OF DEPENDENT VARIABLES = 11
          DEPENDENT V'S :    5    6    7    8   11   12   13   14   21
          DEPENDENT F'S :    2    3

     NUMBER OF INDEPENDENT VARIABLES = 13
          INDEPENDENT V'S :   17
          INDEPENDENT F'S :    1
          INDEPENDENT E'S :    5    6    7    8   11   12   13   14   21
          INDEPENDENT D'S :    2    3


 3RD STAGE OF COMPUTATION REQUIRED    6415 WORDS OF MEMORY.
 PROGRAM ALLOCATE   50000 WORDS

 DETERMINANT OF INPUT MATRIX IS    0.12213E+12

 MAXIMUM LIKELIHOOD SOLUTION (NORMAL DISTRIBUTION THEORY)

 PARAMETER ESTIMATES APPEAR IN ORDER,
 NO SPECIAL PROBLEMS WERE ENCOUNTERED DURING OPTIMIZATION
```

14.7.6.3 ■ Outliers

Using BMDP2D, seven univariate outliers are detected and deleted: two women with extremely high scores on TIMEDRS, ($z = 6.68$ and 6.13 respectively), three with extremely high DRUGUSE scores, ($z = 5.64, 5.04,$ and 4.55), one with a very high STRESS score ($z = 5.27$), and one with an extremely low CONTROL score ($z = -5.17$). Using Mahalanobis distance (through BMDPAM) and cases with the largest contributions to Mardia's coefficient (through EQS) at $p < .001$, 14 multivariate outliers are detected and deleted. SEM analysis is performed on 445 participants.

The multivariate outliers are described by creating a dummy variable: 1 if multivariate outlier and 0 if not. This variable is then used as the grouping variable in a discriminant function analysis through BMDP7M as illustrated in Section 13.8.1.4. The predictors are the 10 observed variables. The multivariate outliers are women who made more numerous visits to the health professional and had more physical symptoms. Deletion of these outliers, seemingly sicker women, may bias the sample. A better approach might be to explicitly model the differences between the group of outliers and the group of nonoutliers.

14.7.6.4 ■ Multicollinearity and Singularity

The determinant of the matrix, given in EQS (Table 14.21), is $0.12213E+12$. This is much larger than 0, so there is no singularity.

14.7.6.5 ■ Adequacy of Covariances

SEM programs have difficulty with computations if the scale of the variables, and therefore the covariances, are of vastly different sizes. In this example the largest variance is 17049.99 for STRESS, the smallest variance is 1.61 for CONTROL. This difference is large and a preliminary run of the model does not converge after 200 iterations. Therefore, the STRESS variable is multiplied by .01. After this rescaling, the new variable SCSTRESS has a variance of 1.70 and there are no further convergence problems.

TABLE 14.22   STANDARDIZED RESIDUALS AND GOODNESS OF FIT INFORMATION FROM EQS LARGE SAMPLE
EXAMPLE 2.   SETUP APPEARS IN TABLE 14.21

STANDARDIZED RESIDUAL MATRIX:

		TIMEDRS V 5	PHYHEAL V 6	MENHEAL V 7	DRUGUSE V 8	ESTEEM V 11
TIMEDRS	V 5	-0.004				
PHYHEAL	V 6	0.081	-0.008			
MENHEAL	V 7	-0.099	-0.016	-0.007		
DRUGUSE	V 8	-0.004	0.007	-0.014	-0.004	
ESTEEM	V 11	-0.091	-0.078	0.076	-0.166	0.000
CONTROL	V 12	-0.040	0.022	0.199	0.001	-0.009
ATTMAR	V 13	-0.006	0.002	0.168	0.027	0.000
ATTROLE	V 14	-0.106	-0.085	-0.107	-0.124	0.061
AGE	V 17	0.054	0.126	0.012	0.047	0.027
SCSTRESS	V 21	0.003	-0.072	0.037	0.031	-0.104

		CONTROL V 12	ATTMAR V 13	ATTROLE V 14	AGE V 17	SCSTRESS V 21
CONTROL	V 12	0.000				
ATTMAR	V 13	0.026	0.000			
ATTROLE	V 14	-0.076	-0.136	0.000		
AGE	V 17	-0.130	-0.093	0.220	0.000	
SCSTRESS	V 21	0.063	0.109	-0.241	0.000	0.000

AVERAGE ABSOLUTE STANDARDIZED RESIDUALS	=	0.0585
AVERAGE OFF-DIAGONAL ABSOLUTE STANDARDIZED RESIDUALS	=	0.0710

LARGEST STANDARDIZED RESIDUALS:

V 21,V 14	V 17,V 14	V 12,V 7	V 13,V 7	V 11,V 8
-0.241	0.220	0.199	0.168	-0.166

V 14,V 13	V 17,V 12	V 17,V 6	V 14,V 8	V 21,V 13
-0.136	-0.130	0.126	-0.124	0.109

V 14,V 7	V 14,V 5	V 21,V 11	V 7,V 5	V 17,V 13
-0.107	-0.106	-0.104	-0.099	-0.093

V 11,V 5	V 14,V 6	V 6,V 5	V 11,V 6	V 11,V 7
-0.091	-0.085	0.081	-0.078	0.076

14.7.6.6 ■ Residuals

Residuals analysis is performed as part of evaluating the model.

14.7.7 ■ SEM Model Estimation and Preliminary Evaluation

The setup and edited output for SEM analysis are presented in Table 14.21. As a first step, look in
the printout for the statement, PARAMETER ESTIMATES APPEAR IN ORDER, shown at the
end of Table 14.21. If there are identification problems or other problems with estimation, this
statement does not appear and, instead, a message about CONDITION CODES is printed, as discussed in section 14.5.1. This message is used to diagnose and deal with problems that come up
during the analysis.

TABLE 14.22 *(CONTINUED)*

DISTRIBUTION OF STANDARDIZED RESIDUALS

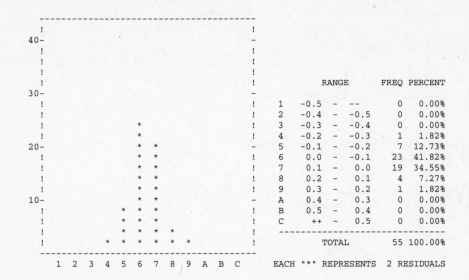

```
      --------------------------------------
      !                                    !
   40-!                                    -
      !                                    !
      !                                    !
      !                                    !
      !                                    !     RANGE       FREQ  PERCENT
   30-!                                    -
      !                                    !  1  -0.5  -  --      0   0.00%
      !                                    !  2  -0.4  - -0.5     0   0.00%
      !              *                     !  3  -0.3  - -0.4     0   0.00%
      !              *                     !  4  -0.2  - -0.3     1   1.82%
   20-!              *   *                 -  5  -0.1  - -0.2     7  12.73%
      !              *   *                 !  6   0.0  - -0.1    23  41.82%
      !              *   *                 !  7   0.1  -  0.0    19  34.55%
      !              *   *                 !  8   0.2  -  0.1     4   7.27%
      !              *   *                 !  9   0.3  -  0.2     1   1.82%
   10-!              *   *                 -  A   0.4  -  0.3     0   0.00%
      !          *   *   *                 !  B   0.5  -  0.4     0   0.00%
      !          *   *   *                 !  C   ++   -  0.5     0   0.00%
      !          *   *   *   *             !     -------------------------------
      !      *   *   *   *   *   *         !          TOTAL      55 100.00%
      --------------------------------------
        1   2   3   4   5   6   7   8   9  A  B  C     EACH "*" REPRESENTS  2 RESIDUALS
```

GOODNESS OF FIT SUMMARY

INDEPENDENCE MODEL CHI-SQUARE =   744.591 ON   45 DEGREES OF FREEDOM

INDEPENDENCE AIC =   654.59068   INDEPENDENCE CAIC =   425.17733
        MODEL AIC =   98.70300   MODEL CAIC =   -64.43538

CHI-SQUARE =   162.703 BASED ON   32 DEGREES OF FREEDOM
PROBABILITY VALUE FOR THE CHI-SQUARE STATISTIC IS LESS THAN 0.001
THE NORMAL THEORY RLS CHI-SQUARE FOR THIS ML SOLUTION IS   173.488.

SATORRA-BENTLER SCALED CHI-SQUARE =   146.6412
PROBABILITY VALUE FOR THE CHI-SQUARE STATISTIC IS   0.00000

BENTLER-BONETT NORMED   FIT INDEX=   0.781
BENTLER-BONETT NONNORMED FIT INDEX=   0.737
COMPARATIVE FIT INDEX   =   0.813
BOLLEN (IFI)   FIT INDEX=   0.817
McDonald (MFI)   FIT INDEX=   0.863
LISREL GFI   FIT INDEX=   0.928
LISREL AGFI   FIT INDEX=   0.875
ROOT MEAN SQUARED RESIDUAL   =   2.475
STANDARDIZED RMR   =   0.012

After checking the covariance matrix for reasonable relationships, examine the portion of the printout labeled BENTLER-WEEKS STRUCTURAL REPRESENTATION that lists the IVs and DVs as specified by the model. These should, and do, match the path diagram.

Table 14.22 presents residuals and goodness-of-fit information for the estimated model in the section labeled GOODNESS OF FIT SUMMARY. The model is a significant improvement over the independence model; the difference in χ^2 is $(744.59 - 162.70 =) 581.89$ with $(45-32) = 13$ degrees of freedom, χ^2_{diff} $(13, N = 445) = 581.89, p < .001$.

TABLE 14.23   Comparison of Nested Models for SEM Large Sample Example

Model	Scaled $\chi 2$	df	CFI	$\chi 2$ difference test
Model 1 Hypothesized Model	146.64	32	.81	
Model 2 Path added–mental health (V7) predicted by Perceived Poor Health (F1)	112.73	31	.86	M1 − M2 = 33.91˙
Model 3 Path added–attitudes toward women's roles (V14) predicted by age (V17)	90.70	30	.89	M2 − M3 = 22.03˙
Model 4 Path added–Residual covariance between attitudes toward women's roles (E14) and self-esteem (E17)	75.18	29	.92	M3 − M4=15.52˙
Model 5 Path added–Residual covariance between physical health symptoms (E6) and visits to health professionals (E7)	63.20	28	.94	M4 − M5 = 11.98˙
Model 6 Final Model – Path dropped – attitudes toward women's roles (V14) predicted from Poor Sense of Self (F1)	63.33	29	.94	M5 − M6 = 0.13

* $p < .01$

However, the Satorra-Bentler scaled chi-square test for ML estimation is also significant, χ^2 (32, $N = 445$) = 162.70, $p < .001$, indicating a difference between the estimated and observed covariance matrices. In a large sample like this, trivial differences can produce a statistically significant χ^2 so fit indices often provide a better gauge of fit when sample size is large. However, with the exception of LISREL GFI and STANDARDIZED RMR, the fit indices do not show a good fitting model. The residuals are symmetrically distributed around zero, but large. The largest standardized residual is −.242, and several are greater than .10. The standardized residuals can be intrepreted like correlations: therefore −.242 is a large residual. The smaller the residuals, generally the better the model fit. Because the hypothesized model does not fit the data, further inspection of parameters is deferred, and instead the Lagrange multiplier test is examined.

14.7.8 ■ Model Modification

Several models are tested and summarized in Table 14.23, where the Satorra-Bentler Scaled χ^2, and χ^2 difference tests for each model are presented. Following MacCallum's (1986) suggestions for model modification, the LM test is requested (by /LMTEST, as seen in Table 14.21) and parameters are added before unnecessary parameters are deleted. Results of the univariate and multivariate LM tests are presented in Table 14.24. Note that these tests are based on ML statistics because EQS does not yet print out Satorra-Bentler LM tests.

The multivariate LM test suggests that adding a path predicting V7 from F1 (predicting number of mental health problems from Poor Sense of Self), would significantly improve the model and lead to an approximate drop in model χ^2 of 29.71. This is a reasonable parameter to add. It may be that women who have low self-esteem, external locus of control and dissatisfaction with their marital status also report more mental health problems. The path is added and the model re-estimated (Table 14.25).

TABLE 14.24   EDITED EQS OUTPUT OF UNIVARIATE AND MULTIVARIATE LAGRANGE MULTIPLIER TEST. SETUP
APPEARS IN TABLE 14.21.

```
LAGRANGE MULTIPLIER TEST (FOR ADDING PARAMETERS)

            ORDERED UNIVARIATE TEST STATISTICS:

  NO    CODE   PARAMETER    CHI-SQUARE   PROBABILITY   PARAMETER CHANGE
  --    ----   ---------    ----------   -----------   ----------------

   1    2  12    V7,F1        29.710       0.000            1.250
   2    2  20    V6,F3        28.868       0.000            3.882
   3    2  11    V14,V17      22.425       0.000            0.680
   4    2  20    V7,F3        19.354       0.000           -5.225
   5    2  16    F3,F1        17.268       0.000           -1.769
   6    2  20    V11,F3       13.349       0.000           -0.192
   7    2  20    V14,F2       13.145       0.000           -0.483
   8    2  20    V14,F3       13.144       0.000           -0.270
   9    2  20    V11,F2       12.198       0.000           -0.327
  10    2  12    V8,F1        11.134       0.001           -1.708
  11    2  11    V12,V17      10.974       0.001           -0.085
  12    2  11    V6,V17        8.253       0.004            0.118

            --- OUTPUT EDITED ------

MULTIVARIATE LAGRANGE MULTIPLIER TEST BY SIMULTANEOUS PROCESS IN STAGE  1

PARAMETER SETS (SUBMATRICES) ACTIVE AT THIS STAGE ARE:

  PVV PFV PFF PDD GVV GVF GFV GFF BVF BFF


        CUMULATIVE MULTIVARIATE STATISTICS          UNIVARIATE INCREMENT
        ----------------------------------          --------------------

  STEP  PARAMETER   CHI-SQUARE  D.F.  PROBABILITY   CHI-SQUARE  PROBABILITY
  ----  ---------   ----------  ----  -----------   ----------  -----------

   1    V7,F1         29.710     1     0.000          29.710      0.000
   2    V14,V17       52.135     2     0.000          22.425      0.000
   3    V6,F3         69.897     3     0.000          17.762      0.000
   4    V11,F3        83.570     4     0.000          13.673      0.000
   5    V14,F3        95.344     5     0.000          11.773      0.001
   6    V6,V17       102.852     6     0.000           7.508      0.006
   7    V12,V17      109.434     7     0.000           6.582      0.010
   8    V13,V17      113.394     8     0.000           3.960      0.047
```

Instead of the message, PARAMETER ESTIMATES APPEAR IN ORDER, however, the following message is found:

PARAMETER   CONDITION CODE
  D3,D3   CONSTRAINED AT LOWER BOUND

This indicates that during estimation, EQS held the disturbance (residual variance) of the third factor, the Health Care Utilization factor, at zero rather than permit it to become negative. Although this message may well indicate a potential problem with interpretation (negative error variance?), the decision is made to continue evaluation of the model in the hopes that subsequent modifications solve this problem. The goodness-of-fit statistics for this second model are presented in Table

TABLE 14.25 GOODNESS OF FIT INFORMATION FROM EQS AFTER THE ADDITION OF MENTAL HEALTH (V7) PREDICTED BY POOR SENSE OF SELF (F1). FULL SETUP IN TABLE 14.21. ADDITION TO SETUP OF TABLE 14.21 AND PARTIAL OUTPUT

```
!PERCEIVED POOR HEALTH
 V7 =  F2 + *F1 + E7;
 V6 = *F2 + E6;



GOODNESS OF FIT SUMMARY

 INDEPENDENCE MODEL CHI-SQUARE =        744.591 ON     45 DEGREES OF FREEDOM

 INDEPENDENCE AIC =    654.59068    INDEPENDENCE CAIC =    425.17733
         MODEL AIC =     65.61850         MODEL CAIC =    -92.42180

 CHI-SQUARE =      127.619 BASED ON   31 DEGREES OF FREEDOM
 PROBABILITY VALUE FOR THE CHI-SQUARE STATISTIC IS LESS THAN 0.001
 THE NORMAL THEORY RLS CHI-SQUARE FOR THIS ML SOLUTION IS         135.801.

 SATORRA-BENTLER SCALED CHI-SQUARE =    112.7296
 PROBABILITY VALUE FOR THE CHI-SQUARE STATISTIC IS      0.00000

 BENTLER-BONETT NORMED    FIT INDEX=      0.829
 BENTLER-BONETT NONNORMED FIT INDEX=      0.800
 COMPARATIVE FIT INDEX            =      0.862
 BOLLEN (IFI)             FIT INDEX=      0.865
 McDonald (MFI)           FIT INDEX=      0.897
 LISREL GFI               FIT INDEX=      0.942
 LISREL AGFI              FIT INDEX=      0.898
 ROOT MEAN SQUARED RESIDUAL       =      2.044
 STANDARDIZED RMR                 =      0.010
```

14.25. Addition of the V7,F1 path significantly improves the model; however, the fit indices are still rather low so the LM test results are again evaluated.

As seen in Table 14.26, the multivariate LM test suggests that a path predicting ATTROLE (V14) from AGE (V17) would significantly improve the model and lead to an approximate drop in χ^2 of 23.266. The hypothesized model postulated a relationship between age and Poor Sense of Self, but, perhaps, age directly affects attitudes toward the role of women rather than the whole Poor Sense of Self factor. As seen in Table 14.23, the model is significantly improved with the addition of this path; however, the condition code remains.

On the next run, the LM tests suggest that addition of a path predicting attitudes towards women's roles from Perceived Ill Health (V14,F2) would significantly improve the model. This path makes very little sense from a substantive view and is not added. Instead the LM test is employed to examine the usefulness of adding correlated errors to the model.[15] Correlated errors are requested in EQS by the inclusion of

/LMTEST

SET = PEE;

Correlated errors are generally cautiously considered, if still necessary, as a final step when adding parameters. Table 14.27 shows that correlating the residuals associated with attitudes

---

[15] Adding post hoc paths is a little like eating salted peanuts—one is never enough. Extreme caution should be used when adding paths as they are generally post hoc and therefore potentially capitalize on chance. Conservative p levels ($p < .001$) are appropriate for adding post hoc paths to the model.

TABLE 14.26   EDITED EQS OUTPUT OF MULTIVARIATE LM TEST FOR ADDING A PATH PREDICITNG
ATTROLE(V14) PREDICTED FROM AGE (V17)

```
/TITLE
  LARGE SAMPLE EXAMPLE 2 HYPOTHESIZED MODEL
/SPECIFICATIONS
  CASES = 459;VAR = 21; ME=ML,ROBUST;MA=RAW;
  DATA = 'HLTHSEM.DAT';
  FO = '(10F8.3/10F8.3/F8.3)';
  DEL = 140,163,168,275,279,193,206,249,368,67,
         213,286,363,389;
/LABELS
  V1= SUBNO;V2= EDCODE;V3=INCODE;V4=EMPLMNY; V5=TIMEDRS;
  V6=PHYHEAL; V7=MENHEAL; V8=DRUGUSE; V9=STRESS;
  V10=ATTMED; V11=ESTEEM; V12=CONTROL;V13= ATTMAR;
  V14=ATTROLE; V15=ATTHOUSE; V16=ATTWORK;V17= AGE; V18=SEL;
  V19= LTIMEDRS; V20= LPHYHEAL; V21= SCSTRESS;
  F1 = SELF; F2 = PERCHEAL; F3=USEHEAL;
/EQUATIONS
 !SENSE OF SELF FACTOR
  V14 = *F1 + E14;
  V11 = *F1 + E11;
  V12 = *F1 + E12;
  V13 = *F1 + E13;
 !PERCEIVED POOR HEALTH
  V7 =  F2 + *F1 + E7;
  V6 = *F2 + E6;
 !UTILIZATION OF HEALTH CARE
  V5 =  F3 + E5;
  V8 = *F3 + E8;
  V21 = *V17 + E21;

  F2 =  *F1 + *V21 + D2;
  F3 =  *F2 + D3;
/VARIANCES
 E11,E12,E13,E14 = *;
 E6,E7 =*;
 E5,E8 = *;
 E21,D2,D3 = *;
 V17 = *;
 F1 = 1;
/COVARIANCES
 F1,V17 = *;
/PRINT
 EFFECT = YES;
 FIT = ALL;
/WTEST
/LMTEST
/END
```

MULTIVARIATE LAGRANGE MULTIPLIER TEST BY SIMULTANEOUS PROCESS IN STAGE 1

PARAMETER SETS (SUBMATRICES) ACTIVE AT THIS STAGE ARE:

   PVV PFV PFF PDD GVV GVF GFV GFF BVF BFF

		CUMULATIVE MULTIVARIATE STATISTICS			UNIVARIATE INCREMENT	
STEP	PARAMETER	CHI-SQUARE	D.F.	PROBABILITY	CHI-SQUARE	PROBABILITY
1	V14,V17	23.266	1	0.000	23.266	0.000
2	V11,V17	32.327	2	0.000	9.061	0.003
3	F2,V17	42.982	3	0.000	10.654	0.001
4	V14,F3	49.728	4	0.000	6.746	0.009
5	V11,F3	54.827	5	0.000	5.099	0.024
6	F3,F1	59.187	6	0.000	4.360	0.037

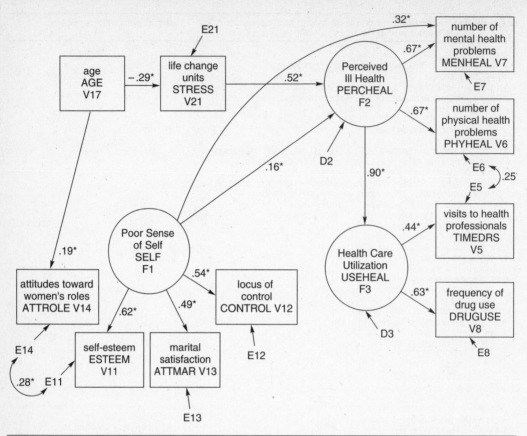

FIGURE 14.13   FINAL SEM MODEL.

towards women's roles and self-esteem will reduce the χ^2 by approximately 16.868. Correlating these errors essentially covaries the parts of attitudes towards women's' roles and self-esteem that are not common to the Poor Sense of Self factor. It seems plausible that women with low self-esteem endorse more traditional ideas about the proper role of women in society, so this covariance is added. The model is significantly improved (Table 14.23), but the condition code remains.

Because the model fits well after addition of these three paths it would be quite reasonable to stop adding paths at this point. However, the condition code is bothersome and the next correlated error that is suggested for estimation (see Table 14.27) is E6,E5. Perhaps estimation of this covariance will address the problem indicated by the condition code. The model is estimated with this covariance added and the negative variance problem signaled by the condition code is eliminated. This correlated error may indicate that number of physical problems is directly related to number of visits to health professionals. With addition of this path the model fits very well and no further parameters are considered for estimation.

It would be entirely reasonable to stop model modification at this point and move to interpretation. However, a goal in modeling is often the development of a parsimonious, good-fitting model with unimportant parameters deleted. The results of the Wald test, presented in Table 14.28, are use-

TABLE 14.27   SETUP MODIFICATIONS AND EDITED EQS MULTIVARIATE LM TEST FOR ADDING CORRELATED ERRORS. (SETUP WITHOUT PATH BETWEEN ATTROLE AND AGE APPEARS IN TABLE 14.26.)

```
/LMTEST
 SET = PEE;


MAXIMUM LIKELIHOOD SOLUTION (NORMAL DISTRIBUTION THEORY)

  MULTIVARIATE LAGRANGE MULTIPLIER TEST BY SIMULTANEOUS PROCESS IN STAGE  1

  PARAMETER SETS (SUBMATRICES) ACTIVE AT THIS STAGE ARE:

    PEE
```

		CUMULATIVE MULTIVARIATE STATISTICS			UNIVARIATE INCREMENT	
STEP	PARAMETER	CHI-SQUARE	D.F.	PROBABILITY	CHI-SQUARE	PROBABILITY
1	E14,E11	16.868	1	0.000	16.868	0.000
2	E6,E5	33.606	2	0.000	16.739	0.000
3	E21,E14	49.628	3	0.000	16.022	0.000
4	E21,E11	60.021	4	0.000	10.393	0.001
5	E7,E5	64.798	5	0.000	4.778	0.029
6	E11,E6	69.262	6	0.000	4.464	0.035

TABLE 14.28   EDITED EQS OUTPUT FOR EXAMINATION OF MULTIVARIATE WALD TEST. (FULL SETUP APPEARS IN TABLE 14.30.)

```
WALD TEST (FOR DROPPING PARAMETERS)
ROBUST INFORMATION MATRIX USED IN THIS WALD TEST
MULTIVARIATE WALD TEST BY SIMULTANEOUS PROCESS
```

		CUMULATIVE MULTIVARIATE STATISTICS			UNIVARIATE INCREMENT	
STEP	PARAMETER	CHI-SQUARE	D.F.	PROBABILITY	CHI-SQUARE	PROBABILITY
1	V14,F1	0.000	1	0.989	0.000	0.989
2	D3,D3	1.699	2	0.428	1.698	0.192
3	F1,V17	4.084	3	0.252	2.386	0.122

ful in evaluating unnecessary parameters. If the path predicting attitudes toward women's roles from the Poor Sense of Self factor (V14,F1) is deleted, there is virtually no change in χ^2. This path is deleted and the model re-estimated. The final model goodness-of-fit information is presented in Table 14.29. The final model with significant parameter estimates presented in standardized form is diagrammed in Figure 14.13

Specific parameter estimates are now examined. The setup for the final model, portions of the printout related to parameter estimates, and the standardized solution are shown in Table 14.30. The section labeled MEASUREMENT EQUATIONS WITH STANDARD ERRORS AND TEST

TABLE 14.29   EDITED EQS OUTPUT FOR FINAL MODEL GOODNESS OF FIT SUMMARY

GOODNESS OF FIT SUMMARY

  INDEPENDENCE MODEL CHI-SQUARE =   744.591 ON   45 DEGREES OF FREEDOM

  INDEPENDENCE AIC =   654.59068   INDEPENDENCE CAIC =   425.17733
          MODEL AIC =   11.06446   MODEL CAIC =   -136.77969

  CHI-SQUARE =   69.064 BASED ON   29 DEGREES OF FREEDOM
  PROBABILITY VALUE FOR THE CHI-SQUARE STATISTIC IS LESS THAN 0.001
  THE NORMAL THEORY RLS CHI-SQUARE FOR THIS ML SOLUTION IS   67.763.

  SATORRA-BENTLER SCALED CHI-SQUARE =   63.3314
  PROBABILITY VALUE FOR THE CHI-SQUARE STATISTIC IS   0.00023

  BENTLER-BONETT NORMED   FIT INDEX=   0.907
  BENTLER-BONETT NONNORMED FIT INDEX=   0.911
  COMPARATIVE FIT INDEX   =   0.943
  BOLLEN (IFI)   FIT INDEX=   0.944
  McDonald (MFI)   FIT INDEX=   0.956
  LISREL GFI   FIT INDEX=   0.970
  LISREL AGFI   FIT INDEX=   0.944
  ROOT MEAN SQUARED RESIDUAL   =   1.421
  STANDARDIZED RMR   =   0.009

STATISTICS contains the parameter estimates, standard errors, and, because robust estimation was employed, the robust statistics in parentheses. These are the standard errors and the z tests to interpret. All of the path coefficients between measured variables and factors in the model are significant, $p < .001$. The section labeled CONSTRUCT EQUATIONS WITH STANDARD ERRORS contains the same information for the equations that relate one factor to another. All of the coefficients are significant.

   Indirect (mediating) effects are evaluated from the section labeled DECOMPOSITION OF EFFECTS WITH NON STANDARDIZED VALUES PARAMETER INDIRECT EFFECTS. Of primary interest to this model, the indirect effect of V17, age, on F2, Perceived Ill Health, is significant. This means that the relationship between age and Perceived Ill Health is mediated (goes through), number of life change units, (unstandardized coefficient for indirect effect $= -.19$, $z = -2.529$). Surprisingly, it is older women who have lower Perceived Ill Health (fewer physical and mental health symptoms). Age, number of life change units, and Poor Sense of Self all indirectly affect Health Care Utilization. The relationship between age and Health Care Utilization is mediated by both life change units and Perceived Ill Health. Older women, in this sample, use health care services less (unstandardized coefficient $= -.27$, $z = -4.268$). Perceived Ill Health mediates the relationship between both number of life change units and Health Care Utilization, and Poor Sense of Self and Health Care Utilization. More life change units lead to greater Perceived Ill Health and greater Health Care Utilization (unstandardized coefficient $= 1.597$, $z = 5.915$). Poor Sense of Self leads to greater Perceived Ill Health which in turn leads to greater Health Care Utilization (unstandardized coefficient $= .596$, $z = 1.869$). The standardized solution for the indirect effects appears in the section labeled DECOMPOSITION OF EFFECTS WITH STANDARDIZED VALUES. The standardized direct effects included in the model are shown in the section labeled, STANDARDIZED SOLUTION.

TABLE 14.30 EDITED OUTPUT FROM EQS FINAL MODEL SETUP, PARAMETER ESTIMATES, AND STANDARDIZED SOLUTION

```
 1  /TITLE
 2    LARGE SAMPLE EXAMPLE 2 FINAL MODEL
 3  /SPECIFICATIONS
 4    CASES = 459;VAR = 21; ME=ML,ROBUST;MA=RAW;
 5    DATA = 'HLTHSEM.DAT';
 6    FO = '(10F8.3/10F8.3/F8.3)';
 7    DEL = 140,163,168,275,279,193,206,249,368,67,
 8         213,286,363,389;
 9   /LABELS
10    V1= SUBNO;V2= EDCODE;V3=INCODE;V4=EMPLMNY; V5=TIMEDRS;
11    V6=PHYHEAL; V7=MENHEAL; V8=DRUGUSE; V9=STRESS;
12    V10=ATTMED; V11=ESTEEM; V12=CONTROL;V13= ATTMAR;
13    V14=ATTROLE; V15=ATTHOUSE; V16=ATTWORK;V17= AGE; V18=SEL;
14    V19= LTIMEDRS; V20= LPHYHEAL; V21= SCSTRESS;
15    F1 = SELF; F2 = PERCHEAL; F3=USEHEAL;
16  /EQUATIONS
17
18  !SENSE OF SELF FACTOR
19    V14 = *V17 + E14;
20    V11 = *F1 + E11;
21    V12 = *F1 + E12;
22    V13 = *F1 + E13;
23
24  !PERCEIVED POOR HEALTH
25    V7 =   F2 + *F1 + E7;
26    V6 = *F2 + E6;
27
28  !UTILIZATION OF HEALTH CARE
29    V5 =   F3 + E5;
30    V8 = *F3 + E8;
31
32
33    V21 = *V17 + E21;
34    F2 =   *F1 + *V21 + D2;
35    F3 =   *F2 + D3;
36
37  /VARIANCES
38    E11,E12,E13,E14 = *;
39    E6,E7 =*;
40    E5,E8 = *;
41    E21,D2,D3 = *;
42    V17 = *;
43    F1 = 1;
44  /COVARINACES
45    V17,F1 = *;
46    E14,E11= *;
47    E5,E6=*;
48  /PRINT
49    EFFECT = YES;
50    FIT = ALL;
51  /WTEST
52  /LMTEST
53    SET = PEE;
54  /END
```

Using Equation 14.35 and the disturbances from the STANDARDIZED SOLUTION, $[1 - (.440)^2] = .8037$ or 80% of the variance in the Health Care Utilization construct is accounted for by its predictors. The predictors of the Perceived Ill Health construct account for $[1 - (.841)^2)] = .2927$ or 29% of its variance. The checklist for the SEM analysis is in Table 14.31. A results section for the SEM analysis follows, in journal format.

TABLE 14.30 *(CONTINUED)*

MEASUREMENT EQUATIONS WITH STANDARD ERRORS AND TEST STATISTICS
(ROBUST STATISTICS IN PARENTHESES)

```
TIMEDRS =V5  =    1.000 F3    + 1.000 E5



PHYHEAL =V6  =     .572*F2    + 1.000 E6
                   .060
                  9.593
              (   .061)
              (  9.343)

MENHEAL =V7  =    1.000 F2    + 1.396*F1    + 1.000 E7
                                  .235
                                 5.950
                             (   .247)
                             (  5.656)

DRUGUSE =V8  =    1.275*F3    + 1.000 E8
                   .205
                  6.231
              (   .206)
              (  6.200)

ESTEEM  =V11 =    2.459*F1    + 1.000 E11
                   .243
                 10.109
              (   .257)
              (  9.550)

CONTROL =V12 =     .690*F1    + 1.000 E12
                   .077
                  9.015
              (   .081)
              (  8.505)

ATTMAR  =V13 =    4.081*F1    + 1.000 E13
                   .529
                  7.710
              (   .569)
              (  7.166)

ATTROLE =V14 =     .588*V17   + 1.000 E14
                   .139
                  4.242
              (   .140)
              (  4.207)

SCSTRESS=V21 =    -.170*V17   + 1.000 E21
                   .027
                 -6.398
              (   .027)
              ( -6.279)
```

CONSTRUCT EQUATIONS WITH STANDARD ERRORS AND TEST STATISTICS
(ROBUST STATISTICS IN PARENTHESES)

```
PERCHEAL=F2  =    1.100*V21   +  .411*F1    + 1.000 D2
                   .125          .210
                  8.798         1.954
              (   .125)     (   .210)
              (  8.781)     (  1.954)

USEHEAL =F3  =    1.452*F2    + 1.000 D3
                   .236
                  6.140
              (   .236)
              (  6.154)
```

TABLE 14.30   (CONTINUED)

COVARIANCES AMONG INDEPENDENT VARIABLES

```
                    V                          F
                    ---                        ---
F1  -  SELF              -.220*I                        I
V17 -  AGE                .135 I                        I
                       -1.631 I                         I
                       (  .136)I                        I
                       ( -1.612)I                       I
                             I                          I


                    E                          D
                    ---                        ---
E6  -PHYHEAL           4.396*I                          I
E5  -TIMEDRS          1.057 I                           I
                      4.157 I                           I
                    (  1.370)I                          I
                    (  3.209)I                          I
                            I                           I
E14 -ATTROLE          5.201*I                           I
E11 -ESTEEM           1.149 I                           I
                      4.526 I                           I
                    (  1.143)I                          I
                    (  4.552)I                          I
```

DECOMPOSITION OF EFFECTS WITH NONSTANDARDIZED VALUES

PARAMETER INDIRECT EFFECTS

```
TIMEDRS =V5  =   1.597 V21   + 1.452 F2    + -.000 F3    + -.272 V17
                  .270          .236          .000          .043
                 5.915         6.140         -6.341        -6.398

                  .596 F1    + 1.597 E21   + 1.452 D2    + 1.000 D3
                  .305
                 1.954


PHYHEAL =V6  =    .630 V21   + -.000 F3    + -.107 V17   +  .235 F1
                  .072          .000          .020          .120
                 8.713         -3.806        -5.323        1.956

                  .630 E21   +  .572 D2    + -.000 D3
                  .066          .060          .000
                 9.593         9.593         -3.512


MENHEAL =V7  =   1.100 V21   + -.000 F2    + -.000 F3    + -.187 V17
                  .125          .000          .000          .029
                 8.798         -3.147        -43.661       -6.398

                  .411*F1    + 1.100 E21   + 1.000 D2    + -.000 D3
                  .210
                 1.954


DRUGUSE =V8  =   2.036 V21   + 1.851 F2    + -.000*F3    + -.347 V17
                  .258          .210          .000          .078
                 7.884         8.832         -7.344        -4.464

                  .760 F1    + 2.036 E21   + 1.851 D2    + 1.275 D3
                  .410          .327          .297          .205
                 1.852         6.231         6.231         6.231


SCSTRESS=V21 =   -.000 F2    + -.000 F3    + -.000 F1    + -.000 D2
                  .000          .000          .170          .413
                 -4.385        -:0.000       -.000         -.000

                 -.000 D3
                  .417
                 -.000
```

TABLE 14.30 *(CONTINUED)*

```
PERCHEAL=F2  =    -.000*V21    + -.000 F2     + -.000 F3     + -.187 V17
                   .000           .000           .000           .074
                 -5.828         -3.147        -43.661         -2.529

                  -.000*F1     + 1.100 E21    + -.000 D2     + -.000 D3
                   .154           .400           .376           .379
                  -.000          2.753          -.000          -.000


USEHEAL =F3  =    1.597 V21    + -.000*F2     + -.000 F3     + -.272 V17
                   .270           .000           .000           .064
                  5.915         -2.870         -6.341         -4.268

                   .596 F1      + 1.597 E21    + 1.452 D2     + -.000 D3
                   .319           .279           .253           .261
                  1.869          5.728          5.728          -.000
```

DECOMPOSITION OF EFFECTS WITH STANDARDIZED VALUES

  PARAMETER INDIRECT EFFECTS

```
TIMEDRS =V5  =     .207 V21    +  .400 F2     + -.000 F3     + -.060 V17
                   .059 F1     +  .198 E21    +  .336 D2     +  .196 D3

PHYHEAL =V6  =     .350 V21    + -.000 F3     + -.102 V17    +  .100 F1
                   .335 E21    +  .569 D2     + -.000 D3

MENHEAL =V7  =     .342 V21    + -.000 F2     + -.000 F3     + -.099 V17
                   .098*F1     +  .327 E21    +  .556 D2     + -.000 D3

DRUGUSE =V8  =     .291 V21    +  .563 F2     + -.000*F3     + -.085 V17
                   .083 F1     +  .278 E21    +  .473 D2     +  .276 D3

SCSTRESS=V21 =    -.000 F2     + -.000 F3     + -.000 F1     + -.000 D2
                  -.000 D3

PERCHEAL=F2  =    -.000*V21    + -.000 F2     + -.000 F3     + -.150 V17
                  -.000*F1     +  .495 E21    + -.000 D2     + -.000 D3

USEHEAL =F3  =     .464 V21    + -.000*F2     + -.000 F3     + -.135 V17
                   .133 F1     +  .444 E21    +  .755 D2     + -.000 D3
```

STANDARDIZED SOLUTION:

```
TIMEDRS =V5  =     .445 F3     +  .896 E5
PHYHEAL =V6  =     .677*F2     +  .736 E6
MENHEAL =V7  =     .661 F2     +  .332*F1    +  .617 E7
DRUGUSE =V8  =     .627*F3     +  .779 E8
ESTEEM  =V11 =     .620*F1     +  .785 E11
CONTROL =V12 =     .544*F1     +  .839 E12
ATTMAR  =V13 =     .457*F1     +  .889 E13
ATTROLE =V14 =     .194*V17    +  .981 E14
SCSTRESS=V21 =    -.291*V17    +  .957 E21
PERCHEAL=F2  =     .517*V21    +  .148*F1    +  .841 D2
USEHEAL =F3  =     .898*F2     +  .440 D3
```

TABLE 14.30 (CONTINUED)

```
CORRELATIONS AMONG INDEPENDENT VARIABLES
----------------------------------------


                    V                             F
                    ---                           ---
F1  -  SELF               -.099*I                    I
V17 -  AGE                     I                      I
                               I                      I
    CORRELATIONS AMONG INDEPENDENT VARIABLES
    ----------------------------------------


                    E                             D
                    ---                           ---
E6  -PHYHEAL              .281*I                     I
E5  -TIMEDRS                  I                      I
                             I                      I
E14 -ATTROLE             .252*I                     I
E11 -ESTEEM                  I                      I
                             I                      I
```

TABLE 14.31 CHECKLIST FOR STRUCTURAL EQUATIONS MODELING

1. Issues
   a. Sample size and missing data
   b. Normality of sampling distributions
   c. Outliers
   d. Linearity
   e. Adequacy of covariances
   f. Identification
   g. Path diagram–hypothesized model
   h. Estimation method
2. Major Analyses
   a. Assessment of fit
      (1) Residuals
      (2) Model chi square
      (3) Fit indices
   b. Significance of specific parameters
   c. Variance in a variable accounted for by a factor
3. Additional Analyses
   a. Lagrange Multiplier Test
      (1) Tests of specific parameters
      (2) Addition of parameters to improve fit
   b. Wald Test for dropping parameters
   c. Correlation between hypothesized and final model
   d. Diagram–final model

Results

The Hypothesized Model

Using EQS, relationships were examined between Poor Sense of Self, a
latent variable with four indicators (locus of control, satisfaction
with marital status, self-esteem, and attitudes toward women's roles),
Perceived Ill Health, a latent variable with two indicators (number of
mental health problems and number of physical health problems), and
Health Care Utilization, a latent variable with two indicators (number
of visits to health professionals, and extent of drug use). Also
included in the analysis were measured indicators of age and number of
life change units. The hypothesized model is presented in Figure
14.12. Circles represent latent variables, rectangles represent measured variables. Absence of a line connecting variables implies lack
of a hypothesized direct effect. Within the text latent variables are
referred to with initial capital letters, measured variables are fully
lower case

Figure 14.12 illustrates the hypotheses that Perceived Ill Health
directly affects Health Care Utilization. A Poor Sense of Self predicts
higher levels of Perceived Ill Health. The relationship between Poor
Sense of Self and Health Care Utilization is mediated by Perceived Ill
Health. Age predicts number of life change units and more life change
units predicts greater Perceived Ill Health. The relationship between
age, life change units and Health Care Utilization is mediated by
Perceived Ill Health.

Assumptions

The assumptions of multivariate normality and linearity were evaluated through BMDP and EQS. There were 7 univariate and 14 multivariate
outliers among the women in the sample. Two women were found to have
extremely high scores on visits to health professionals (z = 6.68 and
6.13), three participants had extremely high scores on drug use, (z =
5.64, 5.04, and 4.55), one woman had a very high number of life change
units (z = 5.27), and one woman had an extremely low locus of control
score (z = −5.17). Using Mahalanobis distance and cases with largest
contribution to Mardia's coefficient, 14 multivariate outliers were
detected and deleted (p < .001). The analysis was performed on 445
participants. There were no missing data.

A discriminant function analysis using BMDP7M was performed to examine the nature of the multivariate outliers. The predictors were the 10
observed variables in this analysis. The multivariate outliers were

found to be women who made more numerous visits to health care professionals and had more physical symptoms.

After deletion of outliers, eight measured variables (TIMEDRS, PHY-
HEAL, MENHEAL, DRUGUSE, ESTEEM, CONTROL, STRESS, and ATTMAR) were still
significantly skewed, $p < .001$. Therefore, maximum likelihood estimation with the Satorra-Bentler scaled chi-square was employed. The standard errors were also adjusted for the extent of the non-normality
(Bentler & Dijkstra, 1985).

Model Estimation

The independence model that tests the hypothesis that the variables
are uncorrelated with one another was easily rejected, $\chi^2(45, \underline{N} = 445)$
$= 744.59$, $p < .01$. The hypothesized model was tested next. A chi-square
difference test indicated a significant improvement in fit between the
independence model and the hypothesized model but only marginal support
was found for the hypothesized model in terms of the Satorra-Bentler
scaled χ^2 test statistic and comparative fit (CFI) index, $\chi^2(32, \underline{N} =$
$445) = 146.64$, $p < .001$, CFI $= .81$.

Post hoc model modifications were performed in an attempt to develop
a better fitting, more parsimonious model. On the basis of the
Lagrange multiplier test, the Wald test and theoretical relevance,
four paths were added and one deleted. Table 14.23 presents the models
tested, scaled chi-square value, CFI, and chi-square difference tests.
The final model fit the data well, $\chi^2(29, \underline{N} = 445) = 63.33$, $p < .05$,
CFI $= .94$. Eighty percent of the variance in the Health Care
Utilization construct is accounted for by it predictors. Twenty-nine
percent of the variance in the Perceived Ill Health construct is
accounted for by its predictors. Because post hoc model modifications
were performed, a correlation was calculated between the hypothesized
model estimates and the estimates from the final model, $\underline{r}(20) = .96$, p
$< .01$. This high correlation indicates the relationships among the
parameters hardly changed as a result of the model modifications. The
final model with significant coefficients presented in standardized
form is in Figure 14.13.

Direct Effects

Perceived Ill Health was strongly predictive of greater Health Care
Utilization (standardized coefficient $= .90$). Perceived Ill Health
increased as number of life change units increased (standardized coefficient $= .53$) and as Poor Sense of Self increased (standardized coefficient $= .16$). Poor Sense of Self also directly increased the number
of reported mental health problems (standardized coefficient $= .67$).

```
Increasing age led to fewer life change units (standardized coefficient
= −.29) and more conservative attitudes toward women's roles (stan-
dardized coefficient = .19).

Indirect Effects
   The relationship between Perceived Ill Health and age was mediated
by number of life change units (standardized coefficient for indirect
effect = −.15 p < .05); older women had better perceived health.
Perceived Ill Health mediated the relationship between Health Care
Utilization and number of life change units (standardized coefficient
for indirect effect = .46, p < .05). More life change units predicted
greater Health Care Utilization. Both the number of life change units
and the degree of Perceived Ill Health mediated the relationship
between age and Health Care Utilization, (standardized coefficient for
indirect effect = −.14 p < .05). Older women used health care less
frequently than younger women. Finally, the relationship between a Poor
Sense of Self and Health Care Utilization was mediated by Perceived Ill
Health (standardized coefficient for indirect effect = .13, p < .05,
one tailed). A Poorer Sense of Self led to greater Health Care
Utilization.
```

14.8 ■ SOME EXAMPLES FROM THE LITERATURE

Stein, Newcomb, and Bentler (1993) examined the effects of grandparent and parent drug use on behavior problems in boys and girls aged 2–8. Separate structural equations models were developed for boys and girls. Six measured child behavior variables served as DVs: (1) developmental problems, (2) fearfulness, (3) hyperactivity, (4) acting out, (5) psychosomatic complaints and (6) social problems. The IVs of interest were four latent variables; (1) grandparent drug use, (2) mother's drug problem, (3) mate's drug use, and (4) dyadic adjustment.

Stein et al. found that boys were more likely than girls to have developmental and social problems. Drug use among grandparents and mothers was associated with more problems in boys than in girls. Grandparent drug use directly predicted greater hyperactivity, more acting out, more psychosomatic complaints, and greater social problems in boys. In girls, grandparent drug use was associated with more acting out. Maternal drug use predicted more developmental problems for both boys and girls. Maternal drug use also predicted more fearfulness, hyperactivity, and social problems for boys.

Harlow and Newcomb (1990) tested several confirmatory factor analytic models to describe the relationships among 25 measured variables related to meaning and satisfaction in life. The best model consisted of 9 first-order specific factors, 3 second-order primary factors, and one third-order general factor. Beginning at the top, the third-order factor representing Meaning and Satisfaction in Life predicted the three second-order factors; (1) Relationship Satisfaction, (2) Purposeful Living, (3) Work and Health Satisfaction. The Relationship Satisfaction predicted three specific first-order factors; (1) Peer Relationships, (2) Intimate Relationships, (3)Family Relationships. The Purposeful Living factor also predicted three specific first-order factors; (1) Purpose in Life, (2) Meaninglessness, and (3) Powerlessness. Finally, the Work and Health Satisfaction factor predicted two specific first-order factors; (1) Work Satisfactions and (2) Health Satisfaction.

Fullagar, McCoy, and Shull (1992) examined the effects of union socialization programs on union attitudes and loyalty in a path analysis model. The participants in this study were 71 apprentices undergoing a union-management training program. Measures of (1) subjective norms, (2) charismatic leadership, (3) individual consideration, (4) intellectual stimulation and (5) training satisfaction were employed as IVs to predict (1) union attitudes, (2) union socialization and (3) union loyalty. Union socialization also served to predict union attitudes.

Fullagar et al. found the most important predictors of union loyalty were positive union attitudes, and training satisfaction. Charismatic leadership and individual consideration both predicted union socialization. Union socialization predicted positive union attitudes. Subjective norms, individual consideration, and intellectual stimulation predicted positive union attitudes.

Duffy, Martin, Battistutta, Hopper, and Mathews (1990) studied the occurrence of asthma and hay fever in Australian twins. This genetics model included both monozygotic and dizygotic female-female, male-male, and female-male twins. A total of 3,808 pairs of twins participated in the study. Monozygotic twins were more likely than dizygotic twins to report hay fever or asthma. Male monozygotic twins were more likely than female monozygotic twins to report either disease. The estimated inheritability of either hay fever or asthma was 60 to 70%. Environmental causes also correlated with disease: .53 for men and .33 for women.

Era, Jokela, Qvarnverg, and Heikkinen (1986) studied pure-tone thresholds and speech understanding in three different age groups of men: (1) 31–35, (2) 51–55, (3) 71–75. Cognitive capacity, occupational status, number of symptoms, and noise at work were used to predict hearing threshold. The strongest predictors of hearing threshold at 4,000 Hz in the 31–35 year old group were occupational status and number of symptoms. The equation accounted for 22% of the variance in hearing threshold. Overall, the strongest predictors of auditory functioning within each age group were occupational status, noise exposure at work, cognitive capacity, education, and number of disease symptoms.

Dukes, Ullman, and Stein (1995) examined the short-term effects of the D.A.R.E. (Drug Abuse Resistance Education) program using classroom level data and a Solomon 4-group design. In series of 2 group models with latent means, pretest sensitization, maturation, and treatment effects were examined for the core concepts (latent variables) targeted by D.A.R.E.: (1) self-esteem, (2) bonds to family, police, and teachers, (3) endorsement of risky behaviors, and (4) resistance to peer pressure.

The maturation analysis of the non-treated classrooms revealed that self-esteem and bonds to family, police, and teacher significantly decreased over time. This is opposite to the effects of the D.A.R.E. program. There was evidence of pre-test sensitization for resistance to peer pressure. Participating in the pre-test alone led to greater resistance to peer pressure. Regardless of participation in the pre-test, students who participated in the D.A.R.E. program reported significantly higher self-esteem, and stronger bonds to family, teachers, and police. They also endorsed fewer risky behaviors than students who did not participate in D.A.R.E.

CHAPTER 15

An Overview of the General Linear Model

15.1 ■ LINEARITY AND THE GENERAL LINEAR MODEL

To facilitate choice of the most useful technique to answer your research question, the emphasis has been on differences among statistical methods. We have repeatedly hinted, however, that most of these techniques are special applications of the general linear model (GLM). The goal of this chapter is to introduce the GLM and to fit the various techniques into the model. In addition to the aesthetic pleasure provided by insight into the GLM, an understanding of it provides a great deal of flexibility in data analysis by promoting use of more sophisticated statistical techniques and computer programs. Most data sets are fruitfully analyzed by one or more of several techniques. Section 15.3 presents an example of the use of alternative research strategies.

Linearity and additivity are important to the GLM. Pairs of variables are assumed to have a linear relationship with each other; that is, it is assumed that relationships between pairs of variables are adequately represented by a straight line. Additivity is also relevant, because if one set of variables is to be predicted by a set of other variables, the effects of the variables within the set are additive in the prediction equation. The second variable in the set adds predictability to the first one, the third adds to the first two, and so on. In all multivariate solutions, the equation relating sets of variables is composed of a series of weighted terms added together.

These assumptions, however, do not prevent inclusion of variables with curvilinear or multiplicative relationships. As is discussed throughout this book, variables can be multiplied together, raised to powers, dichotomized, transformed, or recoded so that even complex relationships are evaluated within the GLM.

15.2 ■ BIVARIATE TO MULTIVARIATE STATISTICS AND OVERVIEW OF TECHNIQUES

15.2.1 ■ Bivariate Form

The GLM is based on prediction or, in jargon, regression. A regression equation represents the value of a DV, Y, as a combination of one or more IVs, Xs, plus error. The simplest case of the GLM, then, is the familiar bivariate regression:

$$A + BX + e = Y \qquad (15.1)$$

where B is the change in Y associated with a one-unit change in X; A is a constant representing the expected value of Y when X is zero; and e is a random variable representing error of prediction.

If X and Y are converted to standard z scores, z_x and z_y, they are now measured on the same scale and cross at the point where both z scores equal zero. The constant A automatically becomes zero because z_y is zero when z_x is zero. Further, after standardization of variances to 1, slope is measured in equal units (rather than the possibly unequal units of X and Y raw scores) and now represents strength of the relationship between X and Y; in bivariate regression with standardized variables, β is equal to the Pearson product-moment correlation coefficient. The closer β is to 1.00 or -1.00, the better the prediction of Y from X (or X from Y). Equation 15.1 then simplifies to

$$\beta z_x + e = z_y \qquad (15.2)$$

As discussed in Chapters 1 and 2, one distinction that is sometimes important in statistics is whether data are continuous or discrete.[1] There are, then, three forms of bivariate regression for situations where X and Y are (1) both continuous, analyzed by Pearson product-moment correlation, (2) mixed, with X dichotomous and Y continuous, analyzed by point biserial correlation, and (3) both dichotomous, analyzed by phi coefficient. In fact, these three forms of correlation are identical. Even with dichotomous variables, all the correlations can be calculated using the equation for Pearson product-moment correlation. Table 15.1 compares the three bivariate forms of the GLM.

15.2.2 ■ Simple Multivariate Form

The first generalization of the simple bivariate form of the GLM is to increase the number of IVs, Xs, used to predict Y. It is here that the additivity of the model first becomes apparent. In standardized form:

$$\sum_{i=1}^{k} \beta_i z_{x_i} + e = z_y \qquad (15.3)$$

---

[1] When discrete variables have more than two levels, they are dummy variable coded into $k - 1$ (df) dichotomous variables to eliminate the possibility of nonlinear relationships. In this section, when we speak of statistical techniques using discrete variables, we imply that recoding is unnecessary or is handled internally in computer programs designed for the particular analysis.

TABLE 15.1 OVERVIEW OF TECHNIQUES IN THE GENERAL LINEAR MODEL

A. Bivariate form (Equation 15.2)
1. Pearson product-moment correlation: X continuous, Y continuous
2. Point biserial correlation: X dichotomous, Y continuous
3. Phi coefficient: X dichotomous, Y dichotomous

B. Simple multivariate form (Equation 15.3)
1. Multiple regression: all Xs continuous, Y continuous
2. ANOVA: all Xs discrete, Y continuous
3. ANCOVA: some Xs continuous and some discrete, Y continuous
4. Two-group discriminant function analysis: all Xs continuous, Y dichotomous
5. Multiway frequency analysis (logit): all Xs discrete, Y discrete
6. Two-group logistic regression analysis: Xs continuous and/or discrete, Y dichotomous

C. Full multivariate form (Equation 15.4)
1. Canonical correlation: all Xs continuous, all Ys continuous
2. MANOVA: all Xs discrete, all Ys continuous
3. MANCOVA: some Xs continuous and some discrete, all Ys continuous
4. Profile analysis: all Xs discrete, all Ys continuous and commensurate
5. Discriminant function analysis: all Xs continuous, all Ys discrete
6. Factor analysis/principal component analysis: all Ys continuous, all Xs latent
7. Structural equations modeling: Xs continuous and/or latent, Ys continuous and/or latent
8. Multiway frequency analysis: all Xs discrete, Y is category frequency
9. Polychotomous logistic regression analysis: Xs continuous and/or discrete, Y discrete

That is, Y is predicted by a weighted sum of Xs. The weights, β_i, no longer reflect the correlation between Y and each X because they are also affected by correlations among the Xs. Here, again, as seen in Table 15.1, there are special statistical techniques associated with whether all Xs are continuous; here also, with appropriate coding, the most general form of the equation can be used to solve all the special cases.

If Y and all Xs are continuous, the special statistical technique is multiple regression. Indeed, as seen in Chapter 5, Equation 15.3 is used to describe the multiple regression problem. But if Y is continuous while all Xs are discrete, we have the special case of regression known as analysis of variance. The values of X represent groups and the emphasis is on finding mean differences in Y between groups rather than on predicting Y, but the basic equation is the same. A significant difference between groups implies that knowledge of X can be used to predict performance on Y.

Analysis of variance problems can be solved through multiple regression computer programs. There are as many Xs, for example, as there are degrees of freedom for the effects. For example, in a one-way design, three groups are recoded into two dichotomous Xs, one representing the first group vs. the other two, and the second X representing the second group vs. the other two. The third group includes those who are not in either of the other two groups. Inclusion of a third X would produce singularity because it is perfectly predictable from the combination of the other two.

If IVs are factorially combined, main effects and interactions are still coded into a series of dichotomous X variables. Consider an example of one IV, anxiety level, divided into three groups, and a second IV, task difficulty, divided into two groups. There are two X components for the 2 df associated with anxiety level and one X component for the 1 df associated with task difficulty. An additional two X components are needed for the 2 df associated with the interaction of anxiety level and task difficulty. The five X components are combined to test each of the two main effects and the interaction, or are tested individually if the comparisons coded into each component are of interest. Because a number of excellent books (e.g., Cohen & Cohen, 1983; Keppel & Zedeck, 1989) describe analysis of variance through multiple regression, that fascinating topic has not been pursued in this book.

If some Xs are continuous and others are discrete, with Y continuous, we have analysis of covariance. The continuous Xs are the covariates and the discrete ones are the IVs. The effects of IVs on Y are assessed after adjustments are made for the effects of the covariates on Y. Actually, the GLM can deal with combinations of continuous and discrete Ys in much more general ways than traditional analysis of covariance, as alluded to in Chapters 5 and 8.

If Y is dichotomous (two groups), with Xs continuous, we have the simple multivariate form of discriminant function analysis. The aim is to predict group membership on the basis of the Xs. There is a reversal in terminology between ANOVA (or MANOVA) and discriminant function analysis; in ANOVA the groups are represented by X, while in discriminant function analysis the groups are represented by Y. The distinction, although confusing, is trivial within the GLM. As seen in forthcoming sections, all the special techniques are simply special cases of the full GLM.

If Y and all Xs are discrete we have multiway frequency analysis. The loglinear, rather than simple linear, model is required to evaluate relationships among variables. Logarithmic transforms are applied to cell frequencies and the weighted sum of these cell frequencies is used to predict group membership. Because the equation eventually boils down to a weighted sum of terms, it is considered here part of the GLM.

Finally, if Y is dichotomous and Xs are continuous and/or discrete, we have logistic regression analysis. Again a nonlinear model, in this case the logistic model, is required to evaluate relationships among variables. Y is expressed in terms of the probability of being in one or the other level. The linear regression equation is the (natural log of the) probability of being in one group divided by the probability of being in the other group. Because the linear regression equation does appear in the model, it can be considered part of the GLM.

15.2.3 ■ Full Multivariate Form

The GLM takes a major leap when the Y side of the equation is expanded. More than one equation may be required to relate the Xs to the Ys:

Equation number

1:
$$\sum_{i=1}^{k} \beta_{i1} z_{xi1} = \sum_{i=1}^{p} \gamma_{j1} z_{yj1}$$

2:
$$\sum_{i=1}^{k} \beta_{i2} z_{xi2} = \sum_{i=1}^{p} \gamma_{j2} z_{yj2} \qquad (15.4)$$

m:
$$\sum_{i=1}^{k} \beta_{im} z_{xim} + e = \sum_{i=m}^{p} \gamma_{jm} z_{yjm}$$

where m equals k or p, whichever is smaller, and γ are regression weights for the standardized Y variables.

In general, there are as many equations as the number of X or Y variables, whichever is smaller. When there is only one Y, Xs are combined to produce one straight-line relationship with Y. Once there is more than one Y, however, combined Ys and combined Xs may fit together in several different ways. Consider first the simplest example where Y is a dichotomous variable representing two groups. The Xs are combined to form a single variable that best separates the mean of Y for the first group from the mean of Y for the second group, as illustrated in Figure 15.1(a). A line parallel to the imaginary line that connects the two means represents the linear combination of Xs, or the first root of X. Once a third group is added, however, it may not fall along that line. To maximally separate the three groups, it may be necessary to add a second linear combination of Xs, or second root. In the example in Figure 15.1(b), the first root separates the means of the first group from the means of the other two groups but does not distinguish the means for the second and third groups. The second root separates the mean of the third group from the other two but does not distinguish between the first two groups.

Roots are called by other names in the special statistical technique in which they are developed: discriminant functions, principal components, canonical variates, and so forth. Full multivariate techniques need multidimensional space to describe relationships among variables. With 2 df, two dimensions might be needed. With 3 df, up to three dimensions might be needed, and so on.

With three groups, however, means for all groups might fall along a single straight line [i.e. $\bar{Y}_3$, of Figure 15.1(b) could also fall along the line representing the first root of X]. Then only the first root is needed to describe the relationship. The number of roots necessary to describe the relationship between two sets of variables may be smaller than the number of roots maximally available. For this reason, the error term in Equation 15.4 is not necessarily associated with the mth root. It is associated with the last necessary root, with "necessary" statistically or psychometrically defined.

As with simpler forms of the GLM, specialized statistical techniques are associated with whether variables are continuous, as summarized in Table 15.1. Canonical correlation is the most general form and the noble ancestor of the GLM where all Xs and Ys are continuous. With appropriate recoding, all bivariate and multivariate problems (with the exception of PCA, FA, MFA, and logistic regression) could be solved through canonical correlation. Practically, however, the programs for canonical correlation tend not to give the kinds of information usually desired when one or more of

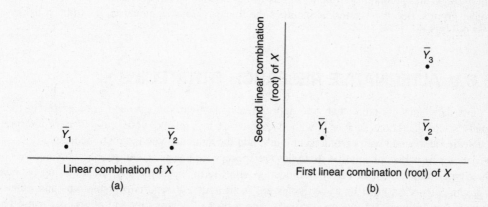

FIGURE 15.1 (a) PLOT OF TWO GROUP MEANS, $\bar{Y}_1$ AND $\bar{Y}_2$, ON A SCALE REPRESENTING THE LINEAR COMBINATION OF X. (b) PLOT OF TWO LINEAR COMBINATIONS OF X REQUIRED TO DISTINGUISH AMONG THREE GROUP MEANS: $\bar{Y}_1$, $\bar{Y}_2$, AND $\bar{Y}_3$.

the X or Y variables is discrete. Programs for the "multivariate general linear model" tend to be rich, but much more difficult to use.

With all Xs discrete and all Ys continuous, we have multivariate analysis of variance. The discrete X variables represent groups, and combinations of Y variables are examined to see how their centroids differ as a function of group membership. If some Xs are continuous, they can be analyzed as covariates, just as in ANCOVA; MANCOVA is used to discover how groups differ on Ys after adjustment for the effects of covariates.

If the Ys are all measured on the same scale and/or represent levels of a within-subjects IV, profile analysis is available—a form of MANOVA that is especially informative for these kinds of data. And if there are multiple DVs at each level of a within-subjects IV, doubly multivariate analysis of variance is used to discover the effects of the IVs on the Ys.

When Y is discrete (more than two groups) and Xs are continuous, the full multivariate form of discriminant function analysis is used to predict membership in Y.

There is a family of procedures—FA and PCA—where the continuous Ys are measured empirically but the Xs are latent. It is assumed that a set of roots underlies the Ys; the purpose of analysis is to uncover the set of roots, or factors, or Xs.

In structural equations modeling, continuous and latent variables are acceptable on both sides of the equations, the X side as well as the Y side. For each Y, whether continuous (an observed indicator variable) or latent (a factor composed of multiple observed indicator variables) there is an equation involving continuous and/or latent Xs. Ys for some equations may serve as Xs for other equations, and vice versa. It is these equations that render structural equations modeling part of the GLM.

MFA is also appropriate when there is a set of discrete Xs among which relationships are sought and none of the variables is designated Y. Application of a loglinear form of the GLM allows prediction of frequencies in the cells from factorially combined Xs, so the implicit DV is cell frequency.

Finally, if Y is discrete and Xs are continuous and/or discrete, we have logistic regression analysis. As for MFA, a nonlinear model, the logistic model, is required to evaluate relationships among variables. Y is expressed in terms of the probability of being in one vs. any of the other levels, with a separate equation for each level of Y but one. For each level, the linear regression equation is the (natural log of the) probability of being in one group divided by the probability of being in any of the other groups. Because the model includes the linear regression equation, it can be considered part of the GLM.

15.3 ■ ALTERNATIVE RESEARCH STRATEGIES

For most data, there is more than one appropriate analytical strategy, and choice among them depends on considerations such as how the variables are interrelated, your preference for interpreting statistics associated with certain techniques, and the audience you intend to address.

A data set for which alternative strategies are appropriate has groups of people who receive one of three types of treatment, behavior modification, short-term psychotherapy, and a waiting-list control group. Suppose a great many variables are measured: self-reports of symptoms and moods, reports of family members, therapist reports, and a host of personality and attitudinal tests. The major goal of analysis is probably to find out if, and on which variable(s), the groups differ after treatment.

The obvious strategy is MANOVA, but a likely problem is that the number of variables exceeds the number of clients in some group, leading to singularity. Further, with so many variables, some are likely to be highly related to combinations of others. You could choose among them or combine them on some rational basis, or you might choose first to look at empirical relationships among them.

A first step in reducing the number of variables might be examination of squared multiple correlations of each variable with all the others through regression or factor analysis programs. But the SMCs might or might not provide sufficient information for a judicious decision about which variables to delete and/or combine. If not, the next likely step is a principal component analysis on the pooled within-cells correlation matrix.

The usual procedures for deciding number of components and type of rotation are followed. Out of this analysis come scores for each client on each component and some idea of the meaning of each component. Depending on the outcome, subsequent strategies might differ. If the principal components are orthogonal, the component scores can serve as DVs in a series of univariate ANOVAs, with adjustment for experimentwise Type I error. If the components are correlated, then MANOVA is used with component scores as DVs. The stepdown sequence might well correspond to the order of components (the scores on the first component enter first, and so on).

Or, you might want to analyze the component scores through a discriminant function analysis to learn, for instance, that differences between behavior modification and short-term psychotherapy are most notable on components loaded heavily with attitudes and self-reports, while differences between the treated groups and the control group are associated with components loaded with therapist reports and personality measures.

You could, in fact, solve the entire problem through discriminant function analysis or logistic regression. Both types of analysis protect against multicollinearity and singularity by setting a tolerance level so that variables that are highly predicted by the other variables do not participate in the solution. Logistic regression is especially handy when the predictors are a mix of many different types of variables.

These strategies are all "legitimate" and simply represent different ways of getting to the same goal. In the immortal words spoken one Tuesday night in the Jacuzzi by Sanford A. Fidell, "You mean you only know one thing, but you have a dozen different names for it?"

APPENDIX A

A Skimpy Introduction to Matrix Algebra

The purpose of this appendix is to provide readers with sufficient background to follow, and duplicate as desired, calculations illustrated in the fourth sections of Chapters 5 through 14. The purpose is not to provide a thorough review of matrix algebra or even to facilitate an in-depth understanding of it. The reader who is interested in more than calculational rules has several excellent discussions available, particularly those in Tatsuoka (1971) and Rummel (1970).

Most of the algebraic manipulations with which the reader is familiar—addition, subtraction, multiplication, and division—have counterparts in matrix algebra. In fact, the algebra that most of us learned is a special case of matrix algebra involving only a single number, a scalar, instead of an ordered array of numbers, a matrix. Some generalizations from scalar algebra to matrix algebra seem "natural" (i.e., matrix addition and subtraction) while others (multiplication and division) are convoluted. Nonetheless, matrix algebra provides an extremely powerful and compact method for manipulating sets of numbers to arrive at desirable statistical products.

The matrix calculations illustrated here are calculations performed on square matrices. Square matrices have the same number of rows as columns. Sums-of-squares and cross-products matrices, variance-covariance matrices, and correlation matrices are all square. In addition, these three very commonly encountered matrices are symmetrical, having the same value in row 1, column 2, as in column 1, row 2, and so forth. Symmetrical matrices are mirror images of themselves about the main diagonal (the diagonal going from top left to bottom right in the matrix).

There is a more complete matrix algebra that includes nonsquare matrices as well. However, once one proceeds from the data matrix, which has as many rows as research units (subjects) and as many columns as variables, to the sum-of-squares and cross-products matrix, as illustrated in Section 1.5, most calculations illustrated in this book involve square, symmetrical matrices. A further restriction on this appendix is to limit the discussion to only those manipulations used in the fourth sections of Chapters 5 through 14. For purposes of numerical illustration, two very simple matrices, square, but not symmetrical (to eliminate any uncertainty regarding which elements are involved in calculations), will be defined as follows:

$$\mathbf{A} = \begin{bmatrix} a & b & c \\ d & e & f \\ g & h & i \end{bmatrix} = \begin{bmatrix} 3 & 2 & 4 \\ 7 & 5 & 0 \\ 1 & 0 & 8 \end{bmatrix}$$

$$\mathbf{B} = \begin{bmatrix} r & s & t \\ u & v & w \\ x & y & z \end{bmatrix} = \begin{bmatrix} 6 & 1 & 0 \\ 2 & 8 & 7 \\ 3 & 4 & 5 \end{bmatrix}$$

A.1 ■ THE TRACE OF A MATRIX

The trace of a matrix is the sum of the numbers on the diagonal that runs from the upper left to lower right. For matrix $\mathbf{A}$, the trace is 16 (3 + 5 + 8); for matrix $\mathbf{B}$ it is 19. If the matrix is a sum-of-squares and cross-products matrix, then the trace is the sum of squares. If it is a variance-covariance matrix, the trace is the sum of variances. If it is a correlation matrix, the trace is the number of variables (each having contributed a value of 1 to the trace).

A.2 ■ ADDITION OR SUBTRACTION OF A CONSTANT TO A MATRIX

If one has a matrix, $\mathbf{A}$, and wants to add or subtract a constant, k, to the elements of the matrix, one simply adds (or subtracts) the constant to every element in the matrix.

$$\mathbf{A} + k = \begin{bmatrix} a & b & c \\ d & e & f \\ g & h & i \end{bmatrix} + k$$

$$= \begin{bmatrix} a+k & b+k & c+k \\ d+k & e+k & f+k \\ g+k & h+k & i+k \end{bmatrix} \tag{A.1}$$

If $k = -3$, then

$$\mathbf{A} + k = \begin{bmatrix} 0 & -1 & 1 \\ 4 & 2 & -3 \\ -2 & -3 & 5 \end{bmatrix}$$

A.3 ■ MULTIPLICATION OR DIVISION OF A MATRIX BY A CONSTANT

Multiplication or division of a matrix by a constant is a straightforward process.

$$k\mathbf{A} = k \begin{bmatrix} a & b & c \\ d & e & f \\ g & h & i \end{bmatrix}$$

$$k\mathbf{A} = \begin{bmatrix} ka & kb & kc \\ kd & ke & kf \\ kg & kh & ki \end{bmatrix} \tag{A.2}$$

and

$$\frac{1}{k}\mathbf{A} = \begin{bmatrix} \dfrac{a}{k} & \dfrac{b}{k} & \dfrac{c}{k} \\[2ex] \dfrac{d}{k} & \dfrac{e}{k} & \dfrac{f}{k} \\[2ex] \dfrac{g}{k} & \dfrac{h}{k} & \dfrac{i}{k} \end{bmatrix} \tag{A.3}$$

Numerically, if $k = 2$, then

$$k\mathbf{A} = \begin{bmatrix} 6 & 4 & 8 \\ 14 & 10 & 0 \\ 2 & 0 & 16 \end{bmatrix}$$

A.4 ■ ADDITION AND SUBTRACTION OF TWO MATRICES

These procedures are also straightforward. If matrices **A** and **B** are as defined at the beginning of this appendix, one simply performs the addition or subtraction of corresponding elements.

$$\mathbf{A} + \mathbf{B} = \begin{bmatrix} a & b & c \\ d & e & f \\ g & h & i \end{bmatrix} + \begin{bmatrix} r & s & t \\ u & v & w \\ x & y & z \end{bmatrix}$$

$$= \begin{bmatrix} a+r & b+s & c+t \\ d+u & e+v & f+w \\ g+x & h+y & i+z \end{bmatrix} \tag{A.4}$$

and

$$\mathbf{A} - \mathbf{B} = \begin{bmatrix} a-r & b-s & c-t \\ d-u & e-v & f-w \\ g-x & h-y & i-z \end{bmatrix} \tag{A.5}$$

For the numerical example:

$$\mathbf{A} + \mathbf{B} = \begin{bmatrix} 3 & 2 & 4 \\ 7 & 5 & 0 \\ 1 & 0 & 8 \end{bmatrix} + \begin{bmatrix} 6 & 1 & 0 \\ 2 & 8 & 7 \\ 3 & 4 & 5 \end{bmatrix}$$

$$\mathbf{A} + \mathbf{B} = \begin{bmatrix} 9 & 3 & 4 \\ 9 & 13 & 7 \\ 4 & 4 & 13 \end{bmatrix}$$

Calculation of a difference between two matrices is required when, for instance, one desires a residuals matrix, the matrix obtained by subtracting a reproduced matrix from an obtained matrix (as in factor analysis, Chapter 13). Or, if the matrix that is subtracted happens to consist of columns with appropriate means of variables inserted in every slot, then the difference between it and a matrix of raw scores produces a deviation matrix.

A.5 ■ MULTIPLICATION, TRANSPOSES, AND SQUARE ROOTS OF MATRICES

Matrix multiplication is both unreasonably complicated and undeniably useful. Note that the ijth element of the resulting matrix is a function of row i of the first matrix and column j of the second.

$$\mathbf{AB} = \begin{bmatrix} a & b & c \\ d & e & f \\ g & h & i \end{bmatrix} \begin{bmatrix} r & s & t \\ u & v & w \\ x & y & z \end{bmatrix}$$

$$= \begin{bmatrix} ar+bu+cx & as+by+cy & at+bw+cz \\ dr+eu+fx & ds+ev+fy & dt+ew+fz \\ gr+hu+ix & gs+hv+iy & gt+hw+iz \end{bmatrix} \tag{A.6}$$

Numerically,

$$\mathbf{AB} = \begin{bmatrix} 3 & 2 & 4 \\ 7 & 5 & 0 \\ 1 & 0 & 8 \end{bmatrix} \begin{bmatrix} 6 & 1 & 0 \\ 2 & 8 & 7 \\ 3 & 4 & 5 \end{bmatrix}$$

$$= \begin{bmatrix} 3 \cdot 6 + 2 \cdot 2 + 4 \cdot 3 & 3 \cdot 1 + 2 \cdot 8 + 4 \cdot 4 & 3 \cdot 0 + 2 \cdot 7 + 4 \cdot 5 \\ 7 \cdot 6 + 5 \cdot 2 + 0 \cdot 3 & 7 \cdot 1 + 5 \cdot 8 + 0 \cdot 4 & 7 \cdot 0 + 5 \cdot 7 + 0 \cdot 5 \\ 1 \cdot 6 + 0 \cdot 2 + 8 \cdot 3 & 1 \cdot 1 + 0 \cdot 8 + 8 \cdot 4 & 1 \cdot 0 + 0 \cdot 7 + 8 \cdot 5 \end{bmatrix}$$

$$= \begin{bmatrix} 34 & 35 & 34 \\ 52 & 47 & 35 \\ 30 & 33 & 40 \end{bmatrix}$$

Regrettably, $\mathbf{AB} \neq \mathbf{BA}$ in matrix algebra. Thus

$$\mathbf{BA} = \begin{bmatrix} 6 & 1 & 0 \\ 2 & 8 & 7 \\ 3 & 4 & 5 \end{bmatrix} \begin{bmatrix} 3 & 2 & 4 \\ 7 & 5 & 0 \\ 1 & 0 & 8 \end{bmatrix}$$

$$= \begin{bmatrix} 25 & 17 & 24 \\ 69 & 44 & 64 \\ 42 & 26 & 52 \end{bmatrix}$$

If another concept of matrix algebra is introduced, some useful statistical properties of matrix algebra can be shown. The transpose of a matrix is indicated by a prime (′) and stands for a rearrangement of the elements of the matrix such that the first row becomes the first column, the second row the second column, and so forth. Thus

$$\mathbf{A}' = \begin{bmatrix} a & d & g \\ b & e & h \\ c & f & i \end{bmatrix} \tag{A.7}$$

$$= \begin{bmatrix} 3 & 7 & 1 \\ 2 & 5 & 0 \\ 4 & 0 & 8 \end{bmatrix}$$

When transposition is used in conjunction with multiplication, then some advantages of matrix multiplication become clear, namely,

$$\mathbf{AA}' = \begin{bmatrix} a & b & c \\ d & e & f \\ g & h & i \end{bmatrix} \begin{bmatrix} a & d & g \\ b & e & h \\ c & f & i \end{bmatrix}$$

$$= \begin{bmatrix} a^2+b^2+c^2 & ad+be+cf & ag+bh+ci \\ ad+be+cf & d^2+e^2+f^2 & dg+eh+fi \\ ag+bh+ci & dg+eh+fi & g^2+h^2+i^2 \end{bmatrix} \tag{A.8}$$

The elements in the main diagonal are the sums of squares while those off the diagonal are cross products.

Had **A** been multiplied by itself, rather than by a transpose of itself, a different result would have been achieved.

$$\mathbf{AA} = \begin{bmatrix} a^2+bd+cg & ab+be+ch & ac+bf+ci \\ da+ed+fg & db+e^2+fh & dc+ef+fi \\ ga+hd+ig & gb+he+ih & gc+hf+i^2 \end{bmatrix}$$

If $\mathbf{AA} = \mathbf{C}$, then $\mathbf{C}^{1/2} = \mathbf{A}$. That is, there is a parallel in matrix algebra to squaring and taking the square root of a scalar, but it is a complicated business because of the complexity of matrix multiplication. If, however, one has a matrix **C** from which a square root is desired (as in canonical correlation, Chapter 6), one searches for a matrix, **A**, which, when multiplied by itself, produces **C**. If, for example,

$$\mathbf{C} = \begin{bmatrix} 27 & 16 & 44 \\ 56 & 39 & 28 \\ 11 & 2 & 68 \end{bmatrix}$$

then

$$\mathbf{C}^{1/2} = \begin{bmatrix} 3 & 2 & 4 \\ 7 & 5 & 0 \\ 1 & 0 & 8 \end{bmatrix}$$

A.6 ■ MATRIX "DIVISION" (INVERSES AND DETERMINANTS)

If you liked matrix multiplication, you'll love matrix inversion. Logically, the process is analogous to performing division for single numbers by finding the reciprocal of the number and multiplying by the reciprocal: if $a^{-1} = 1/a$, then $(a)(a^{-1}) = a/a = 1$. That is, the reciprocal of a scalar is a number that, when multiplied by the number itself, equals 1. Both the concepts and the notation are similar in matrix algebra, but they are complicated by the fact that a matrix is an array of numbers.

To determine if the reciprocal of a matrix has been found, one needs the matrix equivalent of the 1 as employed in the preceding paragraph. The identity matrix, **I**, a matrix with 1s in the main diagonal and zeros elsewhere, is such a matrix. Thus

$$\mathbf{I} = \begin{bmatrix} 1 & 0 & 0 \\ 0 & 1 & 0 \\ 0 & 0 & 1 \end{bmatrix} \tag{A.9}$$

Matrix division, then, becomes a process of finding $\mathbf{A}^{-1}$ such that

$$\mathbf{A}^{-1}\mathbf{A} = \mathbf{AA}^{-1} = \mathbf{I} \tag{A.10}$$

One way of finding A^{-1} requires a two-stage process, the first of which consists of finding the determinant of A, noted $|A|$. The determinant of a matrix is sometimes said to represent the generalized variance of the matrix, as most readily seen in a 2×2 matrix. Thus we define a new matrix as follows:

$$D = \begin{bmatrix} a & b \\ c & d \end{bmatrix}$$

where

$$|D| = ad - bc \qquad (A.11)$$

If D is a variance-covariance matrix where a and d are variances while b and c are covariances, then $ad - bc$ represents variance minus covariance. It is this property of determinants that makes them useful for hypothesis testing (see, for example, Chapter 9, Section 9.4, where Wilks' Lambda is used in MANOVA).

Calculation of determinants becomes rapidly more complicated as the matrix gets larger. For example, in our 3 by 3 matrix,

$$|A| = a(ei - fh) + b(fg - di) + c(dh - eg) \qquad (A.12)$$

Should the determinant of A equal zero, then the matrix cannot be inverted because the next operation in inversion would involve division by zero. Multicollinear or singular matrices (those with variables that are linear combinations of one another, as discussed in Chapter 4) have zero determinants that prohibit inversion.

A full inversion of A is

$$AA^{-1} = \begin{bmatrix} a & b & c \\ d & e & f \\ g & h & i \end{bmatrix}^{-1}$$

$$= \frac{1}{|A|} \begin{bmatrix} ei-fh & ch-bi & bf-cd \\ fg-di & ai-cg & cd-af \\ dh-eg & bg-ah & ae-bd \end{bmatrix} \qquad (A.13)$$

Please recall that because in this example A is not a variance-covariance matrix, a negative determinant is possible, even somewhat likely. Thus, in the numerical example,

$$|A| = 3(5 \cdot 8 - 0 \cdot 0) + 2(0 \cdot 1 - 7 \cdot 8) + 4(7 \cdot 0 - 5 \cdot 1)$$

$$= -12$$

and

$$A^{-1} = \frac{1}{-12} \begin{bmatrix} 5 \cdot 8 - 0 \cdot 0 & 4 \cdot 0 - 2 \cdot 8 & 2 \cdot 0 - 4 \cdot 5 \\ 0 \cdot 1 - 7 \cdot 8 & 3 \cdot 8 - 4 \cdot 1 & 4 \cdot 7 - 3 \cdot 0 \\ 7 \cdot 0 - 5 \cdot 1 & 2 \cdot 1 - 3 \cdot 0 & 3 \cdot 5 - 2 \cdot 7 \end{bmatrix}$$

$$= \begin{bmatrix} \dfrac{4}{-12} & \dfrac{-16}{-12} & \dfrac{-20}{-12} \\[2mm] \dfrac{-56}{-12} & \dfrac{20}{-12} & \dfrac{28}{-12} \\[2mm] \dfrac{-5}{-12} & \dfrac{2}{-12} & \dfrac{1}{-12} \end{bmatrix}$$

$$= \begin{bmatrix} -3.33 & 1.33 & 1.67 \\ 4.67 & -1.67 & -2.33 \\ 0.42 & -0.17 & -0.08 \end{bmatrix}$$

Confirm that, within rounding error, Equation A.10 is true. Once the inverse of **A** is found, "division" by it is accomplished whenever required by using the inverse and performing matrix multiplication.

A.7 ■ EIGENVALUES AND EIGENVECTORS: PROCEDURES FOR CONSOLIDATING VARIANCE FROM A MATRIX

We promised you a demonstration of computation of eigenvalues and eigenvectors for a matrix, so here it is. If you find that this discussion only partially satisfies your appetite, round up Tatsuoka (1971), get the cat off your favorite chair, and prepare for an intelligible, if somewhat lengthy, description of the same subject.

Most of the multivariate procedures rely on eigenvalues and their corresponding eigenvectors (also called characteristic roots and vectors) in one way or another because they consolidate the variance in a matrix (the eigenvalue) while providing the linear combination of variables (the eigenvector) to do it. The coefficients applied to variables to form linear combinations of variables in all the multivariate procedures are rescaled elements from eigenvectors. The variance that the solution "accounts for" is associated with the eigenvalue, and is sometimes called so directly.

Calculation of eigenvalues and eigenvectors is best left up to a computer with any realistically sized matrix. For illustrative purposes, a 2×2 matrix will be used here. The logic of the process is also somewhat difficult, involving several of the more abstract notions and relations in matrix algebra, including the equivalence between matrices, systems of linear equations with several unknowns, and roots of polynomial equations.

Solutions of an eigenproblem involves solution of the following equation:

$$(\mathbf{D} - \lambda \mathbf{I})V = 0 \tag{A.14}$$

where λ is the eigenvalue and V the eigenvector to be sought. Expanded, this equation becomes

$$\left[\begin{bmatrix} a & b \\ c & d \end{bmatrix} - \lambda \begin{bmatrix} 1 & 0 \\ 0 & 1 \end{bmatrix} \right] \begin{bmatrix} v_1 \\ v_2 \end{bmatrix} = 0$$

or

$$\left[\begin{bmatrix} a & b \\ c & d \end{bmatrix} - \begin{bmatrix} \lambda & 0 \\ 0 & \lambda \end{bmatrix}\right]\begin{bmatrix} v_1 \\ v_2 \end{bmatrix} = 0$$

or, by applying Equation A.5,

$$\begin{bmatrix} a-\lambda & b \\ c & d-\lambda \end{bmatrix}\begin{bmatrix} v_1 \\ v_2 \end{bmatrix} = 0 \tag{A.15}$$

If one considers the matrix **D**, whose eigenvalues are sought, a variance-covariance matrix, one can see that a solution is desired to "capture" the variance in **D** while rescaling the elements in **D** by v_1 and v_2 to do so.

It is obvious from Equation A.15 that a solution is always available when v_1 and v_2 are zero. A nontrivial solution may also be available when the determinant of the leftmost matrix in Equation A.15 is zero.[1] That is, if (following Equation A.11)

$$(a - \lambda)(d - \lambda) - bc = 0 \tag{A.16}$$

then there may exist values of λ and values of v_1 and v_2 that satisfy the equation and are not zero. However, expansion of Equation A.16 gives a polynomial equation, in λ, of degree 2:

$$\lambda^2 - (a + d)\lambda + ad - bc = 0 \tag{A.17}$$

Solving for the eigenvalues, λ, requires solving for the roots of this polynomial. If the matrix has certain properties (see footnote 1), there will be as many positive roots to the equation as there are rows (or columns) in the matrix.

If Equation A.17 is rewritten as $x\lambda^2 + y\lambda + z = 0$, the roots may be found by applying the following equation:

$$\lambda = \frac{-y \pm \sqrt{y^2 - 4xz}}{2x} \tag{A.18}$$

For a numerical example, consider the following matrix.

$$\mathbf{D} = \begin{bmatrix} 5 & 1 \\ 4 & 2 \end{bmatrix}$$

Applying Equation A.17, we obtain

$$\lambda^2 - (5 + 2)\lambda + 5\cdot2 - 1\cdot4 = 0$$

---

[1] Read Tatsuoka (1971); a matrix is said to be positive definite when all $\lambda > 0$, positive semidefinite when all $\lambda \geq 0$, and ill-conditioned when some $\lambda < 0$.

or

$$\lambda^2 - 7\lambda + 6 = 0$$

The roots to this polynomial may be found by Equation A.18 as follows:

$$\lambda_1 = \frac{-(-7) + \sqrt{(-7)^2 - 4 \cdot 1 \cdot 6}}{2 \cdot 1} = 6$$

and

$$\lambda_2 = \frac{-(-7) - \sqrt{(-7)^2 - 4 \cdot 1 \cdot 6}}{2 \cdot 1} = 1$$

(The roots could also be found by factoring to get $[\lambda - 6][\lambda - 1]$.)

Once the roots are found, they may be used in Equation A.15 to find v_1 and v_2, the eigenvector. There will be one set of eigenvectors for the first root and a second set for the second root. Both solutions require solving sets of two simultaneous equations in two unknowns, to wit, for the first root, 6, and applying Equation A.15.

$$\begin{bmatrix} 5-6 & 1 \\ 4 & 2-6 \end{bmatrix}\begin{bmatrix} v_1 \\ v_2 \end{bmatrix} = 0$$

or

$$\begin{bmatrix} -1 & 1 \\ 4 & -4 \end{bmatrix}\begin{bmatrix} v_1 \\ v_2 \end{bmatrix} = 0$$

so that

$$-1v_1 + 1v_2 = 0$$

and

$$4v_1 - 4v_2 = 0$$

When $v_1 = 1$ and $v_2 = 1$, a solution is found.

For the second root, 1, the equations become

$$\begin{bmatrix} 5-1 & 1 \\ 4 & 2-1 \end{bmatrix}\begin{bmatrix} v_1 \\ v_2 \end{bmatrix} = 0$$

or

$$\begin{bmatrix} 4 & 1 \\ 4 & 1 \end{bmatrix}\begin{bmatrix} v_1 \\ v_2 \end{bmatrix} = 0$$

so that

$$4v_1 + 1v_2 = 0$$

and

$$4v_1 + 1v_2 = 0$$

When $v_1 = -1$ and $v_2 = 4$, a solution is found. Thus the first eigenvalue is 6, with [1, 1] as a corresponding eigenvector, while the second eigenvalue is 1, with [−1, 4] as a corresponding eigenvector.

Because the matrix was 2×2, the polynomial for eigenvalues was quadratic and there were two equations in two unknowns to solve for eigenvectors. Imagine the joys of a matrix 15×15, a polynomial with terms to the 15th power for the first half of the solution and 15 equations in 15 unknowns for the second half. A little more appreciation for the computer, please, next time you use it.

APPENDIX B
Research Designs for Complete Examples

B.1 ■ WOMEN'S HEALTH AND DRUG STUDY

Data used in most of the large sample examples were collected with the aid of a grant from the National Institute on Drug Abuse (#DA 00847) to L. S. Fidell and J. E. Prather in 1974–1976. Methods of collecting the data and references to the measures included in the study are described here approximately as in Hoffman and Fidell, 1979.

Method

A structured interview, containing a variety of health, demographic, and attitudinal measures, was given to a randomly selected group of 465 female, 20– to 59–year-old, English-speaking residents of the San Fernando Valley, a suburb of Los Angeles, in February 1975. A second interview, focusing primarily on health variables but also containing the Bem Sex Role Inventory (BSRI; Bem, 1974) and the Eysenck Personality Inventory (EPI; Eysenck, 1963), was conducted with 369 (79.4%) of the original respondents in February 1976.

The 1975 target sample of 703 names was approximately a .003 probability sample of appropriately aged female residents of the San Fernando Valley, and was randomly drawn from lists prepared by listers during the weeks immediately preceding the sample selection. Lists were prepared for census blocks that had been randomly drawn (proportional to population) from 217 census tracks, which were themselves randomly drawn after they were stratified by income and assigned probabilities proportional to their populations. Respondents were contacted after first receiving a letter soliciting their cooperation. Substitutions were not allowed. A minimum of four callbacks was required before the attempt to obtain an interview was terminated. The completion rate for the target sample was 66.1%, with a 26% refusal rate and a 7.9% "unobtainable" rate.

The demographic characteristics of the 465 respondents who cooperated in 1975 confirmed the essentially white, middle- and working-class composition of the San Fernando Valley, and agreed, for the most part, with the profile of characteristics of women in the valley that was calculated from

1970 Census Bureau data. The final sample was 91.2% white, with a median family income (before taxes) of $17,000 per year and an average Duncan scale (Featherman, 1973) socioeconomic level (SEL) rating of 51. Respondents were also well educated (13.2 years of school completed, on average), and predominantly Protestant (38%), with 26% Catholic, 20% Jewish, and the remainder "None" or "Other." A total of 52.9% worked (either full-time—33.5%, or part-time—19.4%). Seventy-eight percent were living with husbands at the time of the first interview, with 9% divorced, 6% single, 3% separated, 3% widowed, and fewer than 1% "living together." Altogether, 82.4% of the women had children; the average number of children was 2.7, with 2.1 children, on the average, still living in the same house as the respondent.

Of the original 465 respondents, 369 (79.4%) were reinterviewed a year later. Of the 96 respondents who were not reinterviewed, 51 refused, 36 had moved and could not be relocated, 8 were known to be in the Los Angeles area but were not contacted after a minimum of 5 attempts, and 1 was deceased. Those who were and were not reinterviewed were similar (by analyses of variance) on health and attitudinal variables. They differed, however, on some demographic measures. Those who were reinterviewed tended to be higher-SEL, higher-income white women who were better-educated, were older, and had experienced significantly fewer life change units (Rahe, 1974) in 1975.

The 1975 interview schedule was composed of items assessing a number of demographic, health, and attitudinal characteristics (see Table B.1). Insofar as possible, previously tested and validated items and measures were used, although time constraints prohibited including all items from some measures. Coding on most items was prearranged so that responses given large numbers reflected increasingly unfavorable attitudes, dissatisfaction, poorer health, lower income, increasing stress, increasing use of drugs, and so forth.

The 1976 interview schedule repeated many of the health items, with a shorter set of items assessing changes in marital status and satisfaction, changes in work status and satisfaction, and so forth. The BSRI and EPI were also included, as previously mentioned. The interview schedules for both 1975 and 1976 took 75 minutes on average to administer and were conducted in respondent's homes by experienced and trained interviewers.

To obtain median values for the masculine and feminine scores of the BSRI for a comparable sample of men, the BSRI was mailed to the 369 respondents who cooperated in 1976, with instructions to ask a man near to them (husband, friend, brother, etc.) to fill out and return it. The completed BSRI was received from 162 (46%) men, of whom 82% were husbands, 8.6% friends, 3.7% fiancees, 1.9% brothers, 1.2% sons, 1.2% ex-husbands, 0.6% brothers-in-law, and 0.6% fathers. Analyses of variance were used to compare the demographic characteristics of the men who returned the BSRI with those who did not (insofar as such characteristics could be determined by responses of the women to questions in the 1975 interview). The two groups differed in that, as with the reinterviewed women, the men who responded presented an advantaged socioeconomic picture relative to those who did not. Respondents had higher SEL2 ratings, were better educated, and enjoyed higher income. The unweighted averages of the men's and women's median masculine scores and median feminine scores were used to split the sample of women into those who were feminine, masculine, androgynous, and undifferentiated.

B.2 ■ LEARNING DISABILITIES DATA BANK

Data for large sample examples in Chapter 10, Profile Analysis of Repeated Measures and Chapter 14, Section 14.7, confirmatory factor analysis, were taken from a data bank developed at the California Center for Educational Therapy (CCET) in the San Fernando Valley.

TABLE B.1 DESCRIPTION OF SOME OF THE VARIABLES AVAILABLE FROM 1975–1976 INTERVIEWS

Variable	Abbreviation	Brief Description	Source
Demographic variables			
Socioeconomic level	SEL, SEL2	Measure of deference accorded employment categories (SEL2 from second interview)	Featherman (1973), update of Duncan Scale
Education	EDUC	Number of years completed	
Income	INCOME	Categorical variable assessing family income	
Age	AGE	Chronological age in 5-year categories	
Marital status	MARITAL	A categorical variable assessing current marital status	
	MSTATUS	A dichotomous variable assessing whether or not currently married.	
Parenthood	CHILDREN	A categorical variable assessing whether or not one has children	
Ethnic group membership	RACE	A categorical variable assessing ethnic affiliation	
Employment status	EMPLMNT	A categorical variable assessing whether or not one is currently employed	
	WORKSTAT	A categorical variable assessing current employment status and, if not, attitude toward unemployed status	
Religious affiliation	RELIGION	A categorical variable assessing religious affiliation	
Attitudinal variables			
Attitudes toward housework	ATTHOUSE	Frequency of experiencing various favorable and unfavorable attitudes toward homemaking	Derived from Johnson (1955)
Attitudes toward paid work	ATTWORK	Frequency of experiencing various favorable and unfavorable attitudes toward paid work	Johnson (1955)
Attitudes toward role of women	ATTROLE	Measure of conservative or liberal attitudes toward role of women	Spence & Helmreich (1972)
Locus of control	CONTROL	Measure of control ideology: internal or external	Rotter (1966)
Attitudes toward marital status	ATTMAR	Satisfaction with current marital status	From Burgess & Locke (1960); Locke & Wallace (1959); and Rollins & Feldman (1970)

TABLE B.1 *(CONTINUED)*

Variable	Abbreviation	Brief Description	Source
Personality variables			
Self-esteem	ESTEEM	Measures of self-esteem and confidence in various situations	Rosenberg (1965)
Neuroticism-stability index	NEUROTIC	A scale derived from factor analysis to measure neuroticism vs. stability	Eysenck & Eysenck (1963)
Introversion-extroversion index	INTEXT	A scale derived from factor analysis to measure introversion vs. extraversion	Eysenck & Eysenck (1963)
Androgyny measure	ANDRM	A categorical variable based on femininity and masculinity	Derived from Bem (1974)
Health variables			
Mental health	MENHEAL	Frequency count of mental health problems (feeling somewhat apart, can't get along, etc.)	Langner (1965)
Physical health	PHYHEAL	Frequency count of problems with various body systems (circulation, digestion, etc.) general description of health	
Number of visits	TIMEDRS	Frequency count of visits to physical and mental health professionals	
Use of psychotropic drugs	DRUGUSE	A frequency, recency measure of involvement with prescription and nonprescription major and minor tranquilizers, sedatives-hypnotics, antidepressants, and stimulants	Balter & Levine (1971)
Use of psychotropic and over-the-counter drugs	PSYDRUG	DRUGUSE plus a frequency, recency measure of over-the-counter mood modifying drugs	
Attitudes toward medication	ATTDRUG	Items concerning attitudes toward use of medication	
Life change units	STRESS	Weighted items reflecting number and importance of change in life situation	Rahe (1974)

Method

All children who were referred to the CCET were given an extensive battery of psychodiagnostic tests to measure current intellectual functioning, perceptual development, psycholinguistic abilities, visual and auditory functioning, and achievement in a number of academic subjects. In addition, an extensive Parent Information Outline queried parents about demographic variables, family health history, as well as child's developmental history, strengths, weaknesses, preferences, and the like. The entire data bank consists of 112 variables from the testing battery plus 155 variables from the Parent Information Outline.

Data collection began in July 1972 and continued to 1993. The Chapters 10 and 14 sample includes children tested before February 1984 who were administered the Wechsler Intelligence Scale for Children, who were diagnosed as learning-disabled, whose parents agreed to be included in the data bank, and whose parents answered a question about the child's preference for playmates' age. Available answers to this question were: (1) older, (2) younger, (3) same age, and (4) no preference. The latter two categories were combined into a single one for the Chapter 10 analysis since either category by itself would have been too small. Of the 261 children tested between 1972 and 1984, 177 were eligible for inclusion in the Chapter 10 and 14 sample.

For the entire sample of 261 cases, average age is 10.58 years with a range of 5 to 61 years. (The Chapters 10 and 14 sample consists of school-age children only.) About 75% of the entire sample is male. At the time of testing, 63% of the entire sample attended public school; 33% were enrolled in various types of private schools. Of the 94% of parents who revealed their educational level, mothers had completed an average of 13.6 years of schooling and fathers had completed an average of 14.9 years.

B.3 ■ SEXUAL ATTRACTION STUDY

Data used in the large sample Multiway Frequency Analysis example (Chapter 7) were collected in 1984 as part of a survey assessing issues surrounding the nature of sexual attraction to clients among clinical psychologists. Data-collection methods and demographic characteristics that follow are approximately as they appear in an *American Psychologist* paper (Pope, Keith-Speigel, & Tabachnick, 1986).

Method

A cover letter, a brief 17-item questionnaire (15 structured questions and 2 open-ended questions), and a return envelope were sent to 1000 psychologists (500 men and 500 women) randomly selected from the 4356 members of Division 42 (Psychologists in Private Practice) as listed in the 1983 Membership Register of the American Psychological Association.

The questionnaire requested respondents to provide information about their gender, age group, and years of experience in the field. Information was elicited about the respondent's incidence of sexual attraction to male and female clients; clients' reactions to this experience of attraction; beliefs about the clients' awareness of and reciprocation of the attraction; the impact of the attraction on the therapy process; how such feelings were managed; the incidence of sexual fantasies about clients; why, if relevant, respondents chose to refrain from acting out their attraction through actual sexual intimacies with clients; what features determined which clients would be perceived as sexually attractive; incidence of actual sexual activity with clients; and the extent to which the respondents' graduate training and internship experiences had dealt with issues related to sexual attraction to clients.

Questionnaires were returned by 585 respondents. Of these, 59.7% were men. Sixty-eight percent of the male respondents returned their questionnaires as compared with 49% of the female respondents. Approximately half of the respondents were 45 years of age and under. The sample's median age was approximately 46 years as compared with the median age of 40 years reported in a 1983 survey of mental health service providers (VandenBos & Stapp, 1983).

Respondents averaged 16.99 (SD = 8.43) years of professional experience with no significant differences between male and female psychologists. Younger therapists averaged 11.36 (SD = 8.43) years of experience and older therapists averaged 21.79 (SD = 8.13) years of experience. Only 77 of the 585 therapists reported never being attracted to any client.

APPENDIX C
Statistical Tables

TABLE C.1   NORMAL CURVE AREAS

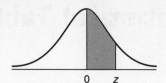

z	.00	.01	.02	.03	.04	.05	.06	.07	.08	.09
0.0	.0000	.0040	.0080	.0120	.0160	.0199	.0239	.0279	.0319	.0359
0.1	.0398	.0438	.0478	.0517	.0557	.0596	.0636	.0675	.0714	.0753
0.2	.0793	.0832	.0871	.0910	.0948	.0987	.1026	.1064	.1103	.1141
0.3	.1179	.1217	.1255	.1293	.1331	.1368	.1406	.1443	.1480	.1517
0.4	.1554	.1591	.1628	.1664	.1700	.1736	.1772	.1808	.1844	.1879
0.5	.1915	.1950	.1985	.2019	.2054	.2088	.2123	.2157	.2190	.2224
0.6	.2257	.2291	.2324	.2357	.2389	.2422	.2454	.2486	.2517	.2549
0.7	.2580	.2611	.2642	.2673	.2704	.2734	.2764	.2794	.2823	.2852
0.8	.2881	.2910	.2939	.2967	.2995	.3023	.3051	.3078	.3106	.3133
0.9	.3159	.3186	.3212	.3238	.3264	.3289	.3315	.3340	.3365	.3389
1.0	.3413	.3438	.3461	.3485	.3508	.3531	.3554	.3577	.3599	.3621
1.1	.3643	.3665	.3686	.3708	.3729	.3749	.3770	.3790	.3810	.3830
1.2	.3849	.3869	.3888	.3907	.3925	.3944	.3962	.3980	.3997	.4015
1.3	.4032	.4049	.4066	.4082	.4099	.4115	.4131	.4147	.4162	.4177
1.4	.4192	.4207	.4222	.4236	.4251	.4265	.4279	.4292	.4306	.4319
1.5	.4332	.4345	.4357	.4370	.4382	.4394	.4406	.4418	.4429	.4441
1.6	.4452	.4463	.4474	.4484	.4495	.4505	.4515	.4525	.4535	.4545
1.7	.4554	.4564	.4573	.4582	.4591	.4599	.4608	4616	.4625	.4633
1.8	.4641	.4649	.4656	.4664	.4671	.4678	.4686	.4693	.4699	.4706
1.9	.4713	.4719	.4726	.4732	.4738	.4744	.4750	.4756	.4761	.4767
2.0	.4772	.4778	.4783	.4788	.4793	.4798	.4803	.4808	.4812	.4817
2.1	.4821	.4826	.4830	.4834	.4838	.4842	.4846	.4850	.4854	.4857
2.2	.4861	.4864	.4868	.4871	.4875	.4878	.4881	.4884	.4887	.4890
2.3	.4893	.4896	.4898	.4901	.4904	.4906	.4909	.4911	.4913	.4916
2.4	.4918	.4920	.4922	.4925	.4927	.4929	.4931	.4932	.4934	.4936
2.5	.4938	.4940	.4941	.4943	.4945	.4946	.4948	.4949	.4951	.4952
2.6	.4953	.4955	.4956	.4957	.4959	.4960	.4961	.4962	.4963	.4964
2.7	.4965	.4966	.4967	.4968	.4969	.4970	.4971	.4972	.4973	.4974
2.8	.4974	.4975	.4976	.4977	.4977	.4978	.4979	.4979	.4980	.4981
2.9	.4981	.4982	.4982	.4983	.4984	.4984	.4985	.4985	.4986	.4986
3.0	.4987	.4987	.4987	.4988	.4988	.4989	.4989	.4989	.4990	.4990

Source: Abridged from Table 1 of *Statistical Tables and Formulas,* by A. Hald. Copyright © 1952, John Wiley & Sons, Inc. Reprinted by permission of A. Hald.

TABLE C.2	CRITICAL VALUES OF THE T DISTRIBUTION FOR $\alpha = .05$ AND .01, TWO-TAILED TEST	
Degrees of freedom	.05	.01
1	12.706	63.657
2	4.303	9.925
3	3.182	5.841
4	2.776	4.604
5	2.571	4.032
6	2.447	3.707
7	2.365	3.499
8	2.306	3.355
9	2.262	3.250
10	2.228	3.169
11	2.201	3.106
12	2.179	3.055
13	2.160	3.012
14	2.145	2.977
15	2.131	2.947
16	2.120	2.921
17	2.110	2.898
18	2.101	2.878
19	2.093	2.861
20	2.086	2.845
21	2.080	2.831
22	2.074	2.819
23	2.069	2.807
24	2.064	2.797
25	2.060	2.787
26	2.056	2.779
27	2.052	2.771
28	2.048	2.763
29	2.045	2.756
30	2.042	2.750
40	2.021	2.704
60	2.000	2.660
120	1.980	2.617
∞	1.960	2.576

Source: Adapted from Table 9 in *Biometrika Tables for Statisticians,* vol. 1, 2d ed., edited by E. S. Pearson and H. O. Hartley (New York: Cambridge University Press, 1958). Reproduced with the permission of the *Biometrika* trustees.

TABLE C.3 CRITICAL VALUES OF THE F DISTRIBUTION

df_2	df_1	1	2	3	4	5	6	8	12	24	∞
1	0.1%	405284	500000	540379	562500	576405	585937	598144	610667	623497	636619
	0.5%	16211	20000	21615	22500	23056	23437	23925	24426	24940	25465
	1 %	4052	4999	5403	5625	5764	5859	5981	6106	6234	6366
	2.5%	647.79	799.50	864.16	899.58	921.85	937.11	956.66	976.71	997.25	1018.30
	5 %	161.45	199.50	215.71	224.58	230.16	233.99	238.88	243.91	249.05	254.32
	10 %	39.86	49.50	53.59	55.83	57.24	58.20	59.44	60.70	62.00	63.33
2	0.1	998.5	999.0	999.2	999.2	999.3	999.3	999.4	999.4	999.5	999.5
	0.5	198.50	199.00	199.17	199.25	199.30	199.33	199.37	199.42	199.46	199.51
	1	98.49	99.00	99.17	99.25	99.30	99.33	99.36	99.42	99.46	99.50
	2.5	38.51	39.00	39.17	39.25	39.30	39.33	39.37	39.42	39.46	39.50
	5	18.51	19.00	19.16	19.25	19.30	19.33	19.37	19.41	19.45	19.50
	10	8.53	9.00	9.16	9.24	9.29	9.33	9.37	9.41	9.45	9.49
3	0.1	167.5	148.5	141.1	137.1	134.6	132.8	130.6	128.3	125.9	123.5
	0.5	55.55	49.80	47.47	46.20	45.39	44.84	44.13	43.39	42.62	41.83
	1	34.12	30.81	29.46	28.71	28.24	27.91	27.49	27.05	26.60	26.12
	2.5	17.44	16.04	15.44	15.10	14.89	14.74	14.54	14.34	14.12	13.90
	5	10.13	9.55	9.28	9.12	9.01	8.94	8.84	8.74	.8.64	8.53
	10	5.54	5.46	5.39	5.34	5.31	5.28	5.25	5.22	5.18	5.13
4	0.1	74.14	61.25	56.18	53.44	51.71	50.53	49.00	47.41	45.77	44.05
	0.5	31.33	26.28	24.26	23.16	22.46	21.98	21.35	20.71	20.03	19.33
	1	21.20	18.00	16.69	15.98	15.52	15.21	14.80	14.37	13.93	13.46
	2.5	12.22	10.65	9.98	9.60	9.36	9.20	8.98	8.75	8.51	8.26
	5	7.71	6.94	6.59	6.39	6.26	6.16	6.04	5.91	5.77	5.63
	10	4.54	4.32	4.19	4.11	4.05	4.01	3.95	3.90	3.83	3.76
5	0.1	47.04	36.61	33.20	31.09	29.75	28.84	27.64	26.42	25.14	23.78
	0.5	22.79	18.31	16.53	15.56	14.94	14.51	13.96	13.38	12.78	12.14
	1	16.26	13.27	12.06	11.39	10.97	10.67	10.29	9.89	9.47	9.02
	2.5	10.01	8.43	7.76	7.39	7.15	6.98	6.76	6.52	6.28	6.02
	5	6.61	5.79	5.41	5.19	5.05	4.95	4.82	4.68	4.53	4.36
	10	4.06	3.78	6.62	3.52	3.45	3.40	3.34	3.27	3.19	3.10
6	0.1	35.51	27.00	23.70	21.90	20.81	20.03	19.03	17.99	16.89	15.75
	0.5	18.64	14.54	12.92	12.03	11.46	11.07	10.57	10.03	9.47	8.88
	1	13.74	10.92	9.78	9.15	8.75	8.47	8.10	7.72	7.31	6.88
	2.5	8.81	7.26	6.60	6.23	5.99	5.82	5.60	5.37	5.12	4.85
	5	5.99	5.14	4.76	4.53	4.39	4.28	4.15	4.00	3.84	3.67
	10	3.78	3.46	3.29	3.18	3.11	3.05	2.98	2.90	2.82	2.72
7	0.1	29.22	21.69	18.77	17.19	16.21	15.52	14.63	13.71	12.73	11.69
	0.5	16.24	12.40	10.88	10.05	9.52	9.16	8.68	8.18	7.65	7.08
	1	12.25	9.55	8.45	7.85	7.46	7.19	6.84	6.47	6.07	5.65
	2.5	8.07	6.54	5.89	5.52	5.29	5.12	4.90	4.67	4.42	4.14
	5	5.59	4.74	4.35	4.12	3.97	3.87	3.73	3.57	3.41	3.23
	10	3.59	3.26	3.07	2.96	2.88	2.83	2.75	2.67	2.58	2.47
8	0.1	25.42	18.49	15.83	14.39	13.49	12.86	12.04	11.19	10.30	9.34
	0.5	14.69	11.04	9.60	8.81	8.30	7.95	7.50	701	6.50	5.95
	1	11.26	8.65	7.59	7.01	6.63	6.37	6.03	5.67	5.28	4.86
	2.5	7.57	6.06	5.42	5.05	4.82	4.65	4.43	4.20	3.95	3.67
	5	5.32	4.46	4.07	3.84	3.69	3.58	3.44	3.28	3.12	2.93
	10	3.46	3.11	2.92	2.81	2.73	2.67	2.59	2.50	2.40	2.29
9	0.1	22.86	16.39	13.90	12.56	11.71	11.13	10.37	9.57	8.72	7.81
	0.5	13.61	10.11	8.72	7.96	7.47	7.13	6.69	6.23	5.73	5.19
	1	10.56	8.02	6.99	6.42	6.06	5.80	5.47	5.11	4.73	4.31

TABLE C.3 *(CONTINUED)*

df$_1$ df$_2$		1	2	3	4	5	6	8	12	24	∞
	2.5%	7.21	5.71	5.08	4.72	4.48	4.32	4.10	3.87	3.61	3.33
	5%	5.12	4.26	3.86	3.63	3.48	3.37	3.23	3.07	2.90	2.71
	10%	3.36	3.01	2.81	2.69	2.61	2.55	2.47	2.38	2.28	2.16
10	0.1	21.04	14.91	12.55	11.28	10.48	9.92	9.20	8.45	7.64	6.76
	0.5	12.83	8.08	7.34	6.87	6.54	6.12	5.66	5.17	4.64	
	1	10.04	7.56	6.55	5.99	5.64	5.39	5.06	4.71	4.33	3.91
	2.5	6.94	5.46	4.83	4.47	4.24	4.07	3.85	3.62	3.37	3.08
	5	4.96	4.10	3.71	3.48	3.33	3.22	3.07	2.91	2.74	2.54
	10	3.28	2.92	2.73	2.61	2.52	2.46	2.38	2.28	2.18	2.06
11	0.1	19.69	13.81	11.56	10.35	9.58	9.05	8.35	7.63	6.85	6.00
	0.5	12.23	8.91	7.60	6.88	6.42	6.10	5.68	5.24	4.76	4.23
	1	9.65	7.20	6.22	5.67	5.32	5.07	4.74	4.40	4.02	3.60
	2.5	6.72	5.26	4.63	4.28	4.04	4.88	3.66	3.43	3.17	2.88
	5	4.84	3.98	3.59	3.36	3.20	3.09	2.95	2.79	2.61	2.40
	10	3.23	2.86	2.66	2.54	2.45	2.39	2.30	2.21	2.10	1.97
12	0.1	18.64	12.97	10.80	9.63	8.89	8.38	7.71	7.00	6.25	5.42
	0.5	11.75	8.51	7.23	6.52	6.07	5.76	5.35	4.91	4.43	3.90
	1	9.33	6.93	5.95	5.41	5.06	4.82	4.50	4.16	3.78	3.36
	2.5	6.55	5.10	4.47	4.12	3.89	3.73	3.51	3.28	3.02	2.72
	5	4.75	3.88	3.49	3.26	3.11	3.00	2.85	2.69	2.50	2.30
	10	3.18	2.81	2.61	2.48	2.39	2.33	2.24	2.15	2.04	1.90
13	0.1	17.81	12.31	10.21	9.07	8.35	7.86	7.21	6.52	5.78	4.97
	0.5	11.37	8.19	6.93	6.23	5.79	5.48	5.08	4.64	4.17	3.65
	1	9.07	6.70	5.74	5.20	4.86	4.62	4.30	3.96	3.59	3.16
	2.5	6.41	4.97	4.35	4.00	3.77	3.60	3.39	3.15	2.89	2.60
	5	4.67	3.80	3.41	3.18	3.02	2.92	2.77	2.60	2.42	2.21
	10	3.14	2.76	2.56	2.43	2.35	2.28	2.20	2.10	1.98	1.85
14	0.1	17.14	11.78	9.73	8.62	7.92	7.43	6.80	6.13	5.41	4.60
	0.5	11.06	7.92	6.68	6.00	5.56	5.26	4.86	4.43	3.96	3.44
	1	8.86	6.51	5.56	5.03	4.69	4.46	4.14	3.80	3.43	3.00
	2.5	6.30	4.86	4.24	3.89	3.66	3.50	3.27	3.05	2.79	2.49
	5	4.60	3.74	3.34	3.11	2.96	2.85	2.70	2.53	2.35	2.13
	10	3.10	2.73	2.52	2.39	2.31	2.24	2.15	2.05	1.94	1.80
15	0.1	16.59	11.34	9.3	8.25	7.57	7.09	6.47	5.81	5.10	4.31
	0.5	10.80	7.70	6.48	5.80	5.37	5.07	4.67	4.25	3.79	3.26
	1	8.68	6.36	5.42	4.89	4.56	4.32	4.00	3.67	3.29	2.87
	2.5	6.20	4.77	4.15	3.80	3.58	3.41	3.20	2.96	2.70	2.40
	5	4.54	3.8	3.29	3.06	2.90	2.79	2.64	2.48	2.29	2.07
	10	3.07	2.70	2.49	2.36	2.27	2.21	2.12	2.02	1.90	1.76
16	0.1	16.12	10.97	9.00	7.94	7.27	6.81	6.19	5.55	4.85	4.06
	0.5	10.58	7.51	6.30	5.64	5.21	4.91	4.52	4.10	3.64	3.11
	1	8.53	6.23	5.29	4.77	4.44	4.20	3.89	3.55	3.18	2.75
	2.5	6.12	4.69	4.08	3.73	3.50	3.34	3.12	2.89	2.63	2.32
	5	4.49	3.63	3.24	3.01	2.85	2.74	2.59	2.42	2.24	2.01
	10	3.05	2.67	2.46	2.33	2.24	2.18	2.09	1.99	1.87	1.72
17	0.1	15.72	10.66	8.73	7.68	7.02	6.56	5.96	5.32	4.63	3.85
	0.5	10.38	7.35	6.16	5.50	5.07	4.78	4.39	3.97	3.51	2.98
	1	8.40	6.11	5.18	4.67	4.34	4.10	3.79	3.45	3.08	2.65
	2.5	6.04	4.62	4.01	3.66	3.44	3.28	3.06	2.82	2.56	2.25
	5	4.45	3.59	3.20	2.96	2.81	2.70	2.55	2.38	2.19	1.96
	10	3.03	2.64	2.44	2.31	2.22	2.15	2.06	1.96	1.84	1.69

TABLE C.3 *(CONTINUED)*

df_1 df_2		1	2	3	4	5	6	8	12	24	∞
18	0.1%	15.38	10.39	8.49	7.46	6.81	6.35	5.76	5.13	4.45	3.67
	0.5%	10.22	7.21	6.03	5.37	4.96	4.66	4.28	3.86	3.40	2.87
	1%	8.28	6.01	5.09	4.58	4.25	4.01	3.71	3.37	3.00	2.57
	2.5%	5.98	4.56	3.95	3.61	3.38	3.22	3.01	2.77	2.50	2.19
	5%	4.41	3.55	3.16	2.93	2.77	2.66	2.51	2.34	2.15	1.92
	10%	3.01	2.62	2.42	2.29	2.20	2.13	2.04	1.93	1.81	1.66
19	0.1	15.08	10.16	8.28	7.26	6.61	6.18	5.59	4.97	4.29	3.52
	0.5	10.07	7.09	5.92	5.27	4.85	4.56	4.18	3.76	3.31	2.78
	1	8.18	5.93	5.01	4.50	4.17	3.94	3.63	3.30	2.92	2.49
	2.5	5.92	4.51	3.90	3.56	3.33	3.17	2.96	2.72	2.45	2.13
	5	4.38	3.52	3.13	2.90	2.74	2.63	2.48	2.31	2.11	1.88
	10	2.99	2.61	2.40	2.27	2.18	2.11	2.02	1.91	1.79	1.63
20	0.1	14.82	9.95	8.10	7.10	6.46	6.02	5.44	4.82	4.15	3.38
	0.5	9.94	6.99	5.82	5.17	4.76	4.47	4.09	3.68	3.22	2.69
	1	8.10	5.85	4.94	4.43	4.10	3.87	3.56	3.23	2.86	2.42
	2.5	5.87	4.46	3.86	3.51	3.29	3.13	2.91	2.68	2.41	2.09
	5	4.35	3.49	3.10	2.87	2.71	2.60	2.45	2.28	2.08	1.84
	10	2.97	2.59	2.38	2.25	2.16	2.09	2.00	1.89	1.77	1.61
21	0.1	14.59	9.77	7.94	6.95	6.32	5.88	5.31	4.70	4.03	3.26
	0.5	9.83	6.89	5.73	5.09	4.68	4.39	4.01	3.60	3.15	2.61
	1	8.02	5.78	4.87	4.37	4.04	3.81	3.51	3.17	2.80	2.36
	2.5	5.83	4.42	3.82	3.48	3.25	3.09	2.87	2.64	2.37	2.04
	5	4.32	3.47	3.07	2.84	2.68	2.57	2.42	2.25	2.05	1.81
	10	2.96	2.57	2.36	2.23	2.14	2.08	1.98	1.88	1.75	1.59
22	0.1	14.38	9.61	7.80	6.81	6.19	5.76	5.19	4.58	3.92	3.15
	0.5	9.73	6.81	5.65	5.02	4.61	4.32	3.94	3.54	3.08	2.55
	1	7.94	5.72	4.82	4.31	3.99	3.76	3.45	3.12	2.75	2.31
	2.5	5.79	4.38	3.78	3.44	3.22	3.05	2.84	2.60	2.33	2.00
	5	4.30	3.44	3.05	2.82	2.66	2.55	2.40	2.23	2.03	1.78
	10	2.95	2.56	2.35	2.22	2.13	2.06	1.97	1.86	1.73	1.57
23	0.1	14.19	9.47	7.67	6.69	6.08	5.65	5.09	4.48	3.82	3.05
	0.5	9.63	6.73	5.58	4.95	4.54	4.26	3.88	3.47	3.02	2.48
	1	7.88	5.66	4.76	4.26	3.94	3.71	3.41	3.07	2.70	2.26
	2.5	5.75	4.35	3.75	3.41	3.18	3.02	2.81	2.57	2.30	1.97
	5	4.28	3.42	3.03	2.80	2.64	2.53	2.38	2.20	2.00	1.76
	10	2.94	2.55	2.34	2.21	2.11	2.05	1.95	1.84	1.72	1.55
24	0.1	14.03	9.34	7.55	6.59	5.98	5.55	4.99	4.39	3.74	2.97
	0.5	9.55	6.66	5.52	4.89	4.49	4.20	3.83	3.42	2.97	2.43
	1	7.82	5.61	4.72	4.22	3.90	3.67	3.36	3.03	2.66	2.21
	2.5	5.72	4.32	3.72	3.38	3.15	2.99	2.78	2.54	2.27	1.94
	5	4.26	3.40	3.01	2.78	2.62	2.51	2.36	2.18	1.98	1.73
	10	2.93	2.54	2.33	2.19	2.10	2.04	1.94	1.83	1.70	1.53
25	0.1	13.88	9.22	7.45	6.49	5.88	5.46	4.91	4.31	3.66	2.89
	0.5	9.48	6.60	5.46	4.84	4.43	4.15	3.78	3.37	2.92	2.38
	1	7.77	5.57	4.68	4.18	3.86	3.63	3.32	2.99	2.62	2.17
	2.5	5.69	4.29	3.69	3.35	3.13	2.97	2.75	2.51	2.24	1.91
	5	4.24	3.38	2.99	2.76	2.60	2.49	2.34	2.16	1.96	1.71
	10	2.92	2.53	2.32	2.18	2.09	2.02	1.93	1.82	1.69	1.52
26	0.1	13.74	9.12	7.36	6.41	5.80	5.38	4.83	4.24	3.59	2.82
	0.5	9.41	6.54	5.41	4.79	4.38	4.10	3.73	3.33	2.87	2.33
	1	7.72	5.53	4.64	4.14	3.82	3.59	3.29	2.96	2.58	2.13
	2.5	5.66	4.27	3.67	3.33	3.10	2.94	2.73	2.49	2.22	1.88

TABLE C.3 *(CONTINUED)*

df₁ df₂		1	2	3	4	5	6	8	12	24	∞
	5%	4.22	3.37	2.98	2.74	2.59	2.47	2.32	2.15	1.95	1.69
	10%	2.91	2.52	2.31	2.17	2.08	2.01	1.92	1.81	1.68	1.50
27	0.1	13.61	9.02	7.27	6.33	5.73	5.31	4.76	4.17	3.52	2.75
	0.5	9.34	6.49	5.36	4.74	4.34	4.06	3.69	3.28	2.83	2.29
	1	7.68	5.49	4.60	4.11	3.78	3.56	3.26	2.93	2.55	2.10
	2.5	5.63	4.24	3.65	3.31	3.08	2.92	2.71	2.47	2.19	1.85
	5	4.21	3.35	2.96	2.73	2.57	2.46	2.30	2.13	1.93	1.67
	10	2.90	2.51	2.30	2.17	2.07	2.00	1.91	1.80	1.67	1.49
28	0.1	13.50	8.93	7.19	6.25	5.66	5.24	4.69	4.11	3.46	2.70
	0.5	9.28	6.44	5.32	4.70	4.30	4.02	3.65	3.25	2.79	2.25
	1	7.64	5.45	4.57	4.07	3.75	3.53	3.23	2.90	2.52	2.06
	2.5	5.61	4.22	3.63	3.29	2.06	2.90	2.69	2.45	2.17	1.83
	5	4.20	3.34	2.95	2.71	2.56	2.44	2.29	2.12	1.91	1.65
	10	2.89	2.50	2.29	2.16	2.06	2.00	1.90	1.79	1.66	1.48
29	0.1	13.39	8.85	7.12	6.19	5.59	5.18	4.64	4.05	3.41	2.64
	0.5	9.23	6.40	5.28	4.66	4.26	3.98	3.61	3.21	2.76	2.21
	1	7.60	5.42	4.54	4.04	3.73	3.50	3.20	2.87	2.49	2.03
	2.5	5.59	4.20	3.61	3.27	3.04	2.88	2.67	2.43	2.15	1.81
	5	4.18	3.33	2.93	2.70	2.54	2.43	2.28	2.10	1.90	1.64
	10	2.89	2.50	2.28	2.15	2.06	1.99	1.89	1.78	1.65	1.47
30	0.1	13.29	8.77	7.05	6.12	5.53	5.12	4.58	4.00	3.36	2.59
	0.5	9.18	6.35	5.24	4.62	4.23	3.95	3.58	3.18	2.73	2.18
	1	7.56	5.39	4.51	4.02	3.70	3.47	3.17	2.84	2.47	2.01
	2.5	5.57	4.18	3.59	3.25	3.03	2.87	2.65	2.41	2.14	1.79
	5	4.17	3.32	2.92	2.69	2.53	2.42	2.27	2.09	1.89	1.62
	10	2.88	2.49	2.28	2.14	2.05	1.98	1.88	1.77	1.64	1.46
40	0.1	12.61	8.25	6.60	5.70	5.13	4.73	4.21	3.64	3.01	2.23
	0.5	8.83	6.07	4.98	4.37	3.99	3.71	3.35	2.95	2.50	1.93
	1	7.31	5.18	4.31	3.83	3.51	3.29	2.99	2.66	2.29	1.80
	2.5	5.42	4.05	3.46	3.13	2.90	2.74	2.53	2.29	2.01	1.64
	5	4.08	3.23	2.84	2.61	2.45	2.34	2.18	2.00	1.79	1.51
	10	2.84	2.44	2.23	2.09	2.00	1.93	1.83	1.71	1.57	1.38
60	0.1	11.97	7.76	6.17	5.31	4.76	4.37	3.87	3.31	2.69	1.90
	0.5	8.49	5.80	4.73	4.14	3.76	3.49	3.13	2.74	2.29	1.69
	1	7.08	4.98	4.13	3.65	3.34	3.12	2.82	2.50	2.12	1.60
	2.5	5.29	3.93	3.34	3.01	2.79	2.63	2.41	2.17	1.88	1.48
	5	4.00	3.15	2.76	2.52	2.37	2.25	2.10	1.92	1.70	1.39
	10	2.79	2.39	2.18	2.04	1.95	1.87	1.77	1.66	1.51	1.29
120	0.1	11.38	7.31	5.79	4.95	4.42	4.04	3.55	3.02	2.40	1.56
	0.5	8.18	5.54	4.50	3.92	3.55	3.28	2.93	2.54	2.09	1.43
	1	6.85	4.79	3.95	3.48	3.17	2.96	2.66	2.34	1.95	1.38
	2.5	5.15	3.80	3.23	2.89	2.67	2.52	2.30	2.05	1.76	1.31
	5	3.92	3.07	2.68	2.45	2.29	2.17	2.02	1.83	1.61	1.25
	10	2.75	2.35	2.13	1.99	1.90	1.82	1.72	1.60	1.45	1.19
∞	0.1	10.83	6.91	5.42	4.62	4.10	3.74	3.27	2.74	2.13	1.00
	0.5	7.88	5.30	4.28	3.72	3.35	3.09	2.74	2.36	1.90	1.00
	1	6.64	4.60	3.78	3.32	3.02	2.80	2.51	2.18	1.79	1.00
	2.5	5.02	3.69	3.12	2.79	2.57	2.41	2.19	1.94	1.64	1.00
	5	3.84	2.99	2.60	2.37	2.21	2.09	1.94	1.75	1.52	1.00
	10	2.71	2.30	2.08	1.94	1.85	1.77	1.67	1.55	1.35	1.00

TABLE C.4 CRITICAL VALUES OF CHI SQUARE (χ^2)

df	0.250	0.100	0.050	0.025	0.010	0.005	0.001
1	1.32330	2.70554	3.84146	5.02389	6.63490	7.87944	10.828
2	2.77259	4.60517	5.99147	7.37776	9.21034	10.5966	13.816
3	4.10835	6.25139	7.81473	9.34840	11.3449	12.8381	16.266
4	5.38527	7.77944	9.48773	1.1433	13.2767	14.8602	18.467
5	6.62568	9.23635	11.0705	12.8325	15.0863	16.7496	20.515
6	7.84080	10.6446	12.5916	14.4494	16.8119	18.5476	22.458
7	9.03715	12.0170	14.0671	16.0128	18.4753	20.2777	24.322
8	10.2188	13.3616	15.5073	17.5346	20.0902	21.9550	26.125
9	11.3887	14.6837	16.9190	19.0228	21.6660	23.5893	27.877
10	12.5489	15.9871	18.3070	20.4831	23.2093	25.1882	29.588
11	13.7007	17.2750	19.6751	21.9200	24.7250	26.7569	31.264
12	14.8454	18.5494	21.0261	23.3367	26.2170	28.2995	32.909
13	15.9839	19.8119	22.3621	24.7356	27.6883	29.8194	34.528
14	17.1770	21.0642	23.6848	26.1190	29.1413	31.3193	36.123
15	18.2451	22.3072	24.9958	27.4884	30.5779	32.8013	37.697
16	19.3688	23.5418	26.2962	28.8454	31.9999	34.2672	39.252
17	20.4887	24.7690	27.5871	30.1910	33.4087	35.7185	40.790
18	21.6049	25.9894	28.8693	31.5264	34.8053	37.1564	42.312
19	22.7178	27.2036	30.1435	32.8523	36.1908	38.5822	43.820
20	23.8277	28.4120	31.4104	34.1696	37.5662	39.9968	45.315
21	24.9348	29.6151	32.6705	35.4789	38.9321	41.4010	46.797
22	26.0393	30.8133	33.9244	36.7807	40.2894	42.7956	48.268
23	27.1413	32.0069	35.1725	38.0757	41.6384	44.1813	49.728
24	28.2412	33.1963	36.4151	39.3641	42.9798	45.5585	51.179
25	29.3389	34.3816	37.6525	40.6465	44.3141	46.9278	52.620
26	30.4345	35.5631	38.8852	41.9232	45.6417	48.2899	54.052
27	31.5284	36.7412	40.1133	43.1944	46.9630	49.6449	55.476
28	32.6205	37.9159	41.3372	44.4607	48.2782	50.9933	56.892
29	33.7109	39.0875	42.5569	45.7222	49.5879	52.3356	58.302
30	34.7998	40.2560	43.7729	46.9792	50.8922	53.6720	59.703
40	45.6160	51.8050	55.7585	59.3417	63.6907	66.7659	73.402
50	56.3336	63.1671	67.5048	71.4202	76.1539	79.4900	86.661
60	66.9814	74.3970	79.0819	83.2976	88.3794	91.9517	99.607
70	77.5766	85.5271	90.5312	95.0231	100.425	104.215	112.317
80	88.1303	96.5782	101.879	106.629	112.329	116.321	124.839
90	98.6499	107.565	113.145	118.136	124.116	128.299	137.208
100	109.141	118.498	124.342	129.561	135.807	140.169	149.449

Source: Adapted from Table 8 in *Biometrika Tables for Statisticians,* vol. 1, 2d ed., edited by E. S. Pearson and H. O. Hartley (New York: Cambridge University Press, 1958). Reproduced with the permission of the *Biometrika* trustees.

TABLE C.5 CRITICAL VALUES FOR SQUARED MULTIPLE CORRELATION (R^2) IN FORWARD STEPWISE SELECTION

$\alpha = .05$

		$N - k - 1$														
k	F	10	12	14	16	18	20	25	30	35	40	50	60	80	100	150	200
2	2	43	38	33	30	27	24	20	16	14	13	10	8	6	5	3	2
2	3	40	36	31	27	24	22	18	15	13	11	9	7	5	4	2	2
2	4	38	33	29	26	23	21	17	14	12	10	8	7	5	4	3	2
3	2	49	43	39	35	32	29	24	21	18	16	12	10	8	7	4	2
3	3	45	40	36	32	29	26	22	19	17	15	11	9	7	6	4	3
3	4	42	36	33	29	27	25	20	17	15	13	11	9	7	5	4	3
4	2	54	48	44	39	35	33	27	23	20	18	15	12	10	8	5	4
4	3	49	43	39	36	33	30	25	22	19	17	14	11	8	7	5	4
4	4	45	39	35	32	29	27	22	19	17	15	12	10	8	6	5	3
5	2	58	52	47	43	39	36	31	26	23	21	17	14	11	9	6	5
5	3	52	46	42	38	35	32	27	24	21	19	16	13	9	8	5	4
5	4	46	41	38	35	52	29	24	21	18	16	13	11	9	7	5	4
6	2	60	54	50	46	41	39	33	29	25	23	19	16	12	10	7	5
6	3	54	48	44	40	37	34	29	25	22	20	17	14	10	8	6	5
6	4	48	43	39	36	33	30	26	23	20	17	14	12	9	7	5	4
7	2	61	56	51	48	44	41	35	30	27	24	20	17	13	11	7	5
7	3	59	50	46	42	39	36	31	26	23	21	18	15	11	9	7	5
7	4	50	45	41	38	35	32	27	24	21	18	15	13	10	8	6	4
8	2	62	58	53	49	46	43	37	31	28	26	21	18	14	11	8	6
8	3	57	52	47	43	40	37	32	28	24	22	19	16	12	10	7	5
8	4	51	46	42	39	36	33	28	25	22	19	16	14	11	9	7	5
9	2	63	59	54	51	47	44	38	33	30	27	22	19	15	12	9	6
9	3	58	53	49	44	41	38	33	29	25	23	20	16	12	10	7	6
9	4	52	46	43	40	37	34	29	25	23	20	17	14	11	10	7	6
10	2	64	60	55	52	49	46	39	34	31	28	23	20	16	13	10	7
10	3	59	54	50	45	42	39	34	30	26	24	20	17	13	11	8	6
10	4	52	47	44	41	38	35	30	26	24	21	18	15	12	10	8	6
12	2	66	62	57	54	51	48	42	37	33	30	25	22	17	14	10	8
12	3	60	55	52	47	44	41	36	31	28	25	22	19	14	12	9	7
12	4	53	48	45	41	39	36	31	27	25	22	19	16	13	11	9	7
14	2	68	64	60	56	53	50	44	39	35	32	27	24	18	15	11	8
14	3	61	57	53	49	46	43	37	32	29	27	23	20	15	13	10	8
14	4	43	49	46	42	40	37	32	29	26	23	20	17	13	11	9	7
16	2	69	65	61	58	55	53	46	41	37	34	29	25	20	17	12	9
16	3	61	58	54	50	47	44	38	34	31	28	24	21	17	14	11	8
16	4	53	50	46	43	40	38	33	30	27	24	21	18	14	12	10	8
18	2	70	67	63	60	57	55	49	44	40	36	31	27	21	18	13	9
18	3	62	59	55	51	49	46	40	35	32	30	26	23	18	15	12	9
18	4	54	50	46	44	41	38	34	31	28	25	22	19	15	13	11	8
20	2	72	68	64	62	59	56	50	46	42	38	33	28	22	19	14	10
20	3	62	60	56	52	50	47	42	37	34	31	27	24	19	16	12	9
20	4	54	50	46	44	41	37	35	32	29	26	23	20	16	14	11	8

TABLE C.5 *(CONTINUED)*
$\alpha = .01$

								$N - k - 1$								
k	*F*	10	12	14	16	18	20	25	30	35	40	50	60	80	100	150	200
2	2	59	53	48	43	40	36	30	26	23	20	17	14	11	9	7	5
2	3	58	52	46	42	38	35	30	25	22	19	16	13	10	8	6	4
2	4	57	49	44	39	36	32	26	22	19	16	13	11	8	7	5	4
3	2	67	60	55	50	46	42	35	30	27	24	20	17	13	11	7	5
3	3	63	58	52	47	43	40	34	29	25	22	19	16	12	10	7	5
3	4	61	54	48	44	40	37	31	26	23	20	16	14	11	9	6	5
4	2	70	64	58	53	49	46	39	34	30	27	23	19	15	12	8	6
4	3	67	62	56	51	47	44	37	32	28	25	21	18	14	11	8	6
4	4	64	58	52	47	43	40	34	29	26	23	19	16	13	11	7	6
5	2	73	67	61	57	52	49	42	37	32	29	25	21	16	13	9	7
5	3	70	65	59	54	50	46	39	34	30	27	23	19	15	12	9	7
5	4	65	60	55	50	46	43	36	31	28	25	20	17	14	12	8	6
6	2	74	69	63	59	55	51	44	39	34	31	26	23	18	14	10	8
6	3	72	67	61	56	51	48	41	36	32	28	24	20	16	13	10	7
6	4	66	61	56	52	48	45	38	33	29	26	22	19	15	13	9	7
7	2	76	70	65	60	56	53	46	40	36	33	28	25	19	15	11	9
7	3	73	68	62	57	53	50	42	37	33	30	25	21	17	14	10	8
7	4	67	62	58	54	49	46	40	35	31	28	23	20	16	14	10	8
8	2	77	72	66	62	58	55	48	42	38	34	29	26	20	16	12	9
8	3	74	69	63	58	54	51	44	39	34	31	26	22	18	15	11	9
8	4	67	63	59	55	50	47	41	36	32	29	24	21	17	15	11	9
9	2	78	73	67	63	60	56	49	43	39	36	31	27	21	17	12	10
9	3	74	69	64	59	56	52	45	40	35	32	27	23	19	16	12	9
9	4	68	63	60	56	51	48	42	37	33	30	25	22	18	16	12	9
10	2	79	74	68	65	61	58	51	45	40	37	32	28	22	18	13	10
10	3	74	69	65	50	57	53	47	41	37	33	28	24	20	17	13	10
10	4	68	64	61	56	52	49	43	38	34	31	26	23	19	17	13	9
12	2	80	75	70	66	63	60	53	48	43	39	34	30	24	20	14	11
12	3	74	70	66	62	58	55	48	43	39	35	30	26	21	18	14	10
12	4	69	65	61	57	53	50	44	40	35	32	27	24	20	18	13	10
14	2	81	76	71	68	65	62	55	50	45	41	36	32	25	21	15	11
14	3	74	70	67	63	60	56	50	45	41	37	31	27	22	19	15	11
14	4	69	65	61	57	54	52	45	41	36	33	28	25	21	19	14	10
16	2	82	77	72	69	66	63	57	52	47	43	38	34	27	22	16	12
16	3	74	70	67	64	61	58	52	47	42	39	33	29	23	20	15	11
16	4	70	66	62	58	55	52	46	42	37	34	29	26	22	20	14	11
18	2	82	78	73	70	67	65	59	54	49	45	39	35	28	23	17	12
18	3	74	70	67	65	62	59	53	48	44	41	35	30	24	21	16	12
18	4	70	65	62	58	55	53	47	43	38	35	30	27	23	20	15	11
20	2	82	78	74	71	68	66	60	55	50	46	41	36	29	24	18	13
20	3	74	70	67	65	62	60	55	60	46	42	36	32	26	22	17	12
20	4	70	66	62	58	55	53	47	43	39	36	31	28	24	21	16	11

Note: Decimals are omitted; *k* = number of candidate predictors; *n* = sample size; *F* = criterion F-to-enter.

Source: Adapted from Tables 1 and 2 in "Tests of significance in forward selection regression," by L. Wilkinson and G. E. Dallal, *Technometrics*, 1981, *23*(4), 377–380. Reprinted with permission from *Technometrics*. Copyright 1981 by the American Statistical Association and the American Society for Quality Control. All rights reserved.

TABLE C.6 CRITICAL VALUES FOR *F*MAX (S^2_{MAX}/S^2_{MIN}) DISTRIBUTION FOR $\alpha = .05$ AND .01

$\alpha = .05$

k df	2	3	4	5	6	7	8	9	10	11	12
4	9.60	15.5	20.6	25.2	29.5	33.6	37.5	41.1	44.6	48.0	51.4
5	7.15	10.8	13.7	16.3	18.7	20.8	22.9	24.7	26.5	28.2	29.9
6	5.82	8.38	10.4	12.1	13.7	15.0	16.3	17.5	18.6	19.7	20.7
7	4.99	6.94	8.44	9.70	10.8	11.8	12.7	13.5	14.3	15.1	15.8
8	4.43	6.00	7.18	8.12	9.03	9.78	10.5	11.1	11.7	12.2	12.7
9	4.03	5.34	6.31	7.11	7.80	8.41	8.95	9.45	9.91	10.3	10.7
10	3.72	4.85	5.67	6.34	6.92	7.42	7.87	8.28	8.66	9.01	9.34
12	3.28	4.16	4.79	5.30	5.72	6.09	6.42	6.72	7.00	7.25	7.48
15	2.86	3.54	4.01	4.37	4.68	4.95	5.19	5.40	5.59	5.77	5.93
20	2.46	2.95	3.29	3.54	3.76	3.94	4.10	4.24	4.37	4.49	4.59
30	2.07	2.40	2.61	2.78	2.91	3.02	3.12	3.21	3.29	3.36	3.39
60	1.67	1.85	1.96	2.04	2.11	2.17	2.22	2.26	2.30	2.33	2.36
∞	1.00	1.00	1.00	1.00	1.00	1.00	1.00	1.00	1.00	1.00	1.00

$\alpha = .01$

k df	2	3	4	5	6	7	8	9	10	11	12
4	23.2	37	49	59	69	79	89	97	106	113	120
5	14.9	22	28	33	38	42	46	50	54	57	60
6	11.1	15.5	19.1	22	25	27	30	32	34	36	37
7	8.89	12.1	14.5	16.5	18.4	20	22	23	24	26	27
8	7.50	9.9	11.7	13.2	14.5	15.8	16.9	17.9	18.9	19.8	21
9	6.54	8.5	9.9	11.1	12.1	13.1	13.9	14.7	15.3	16.0	16.6
10	5.85	7.4	8.6	9.6	10.4	11.1	11.8	12.4	12.9	13.4	13.9
12	4.91	6.1	6.9	7.6	8.2	8.7	9.1	9.5	9.9	10.2	10.6
15	4.07	4.9	5.5	6.0	6.4	6.7	7.1	7.3	7.5	7.8	8.0
20	3.32	3.8	4.3	4.6	4.9	5.1	5.3	5.5	5.6	5.8	5.9
30	2.63	3.0	3.3	3.4	3.6	3.7	3.8	3.9	4.0	4.1	4.2
60	1.96	2.2	2.3	2.4	2.4	2.5	2.5	2.6	2.6	2.7	2.7
∞	1.00	1.0	1.0	1.0	1.0	1.0	1.0	1.0	1.0	1.0	1.0

Note: S^2_{max} is the largest and S^2_{min} the smallest in a set of k independent mean squares, each based on degrees of freedom (df).

Source: Adapted from Table 31 in *Biometrika Tables for Statisticians*, vol. 1, 2d ed., edited by E. S. Pearson and H. O. Hartley (New York: Cambridge University Press, 1958). Reproduced with the permission of the *Biometrika* trustees.

References

Aiken, L. S., & West, S. G. (1991). *Multiple regression: testing and interpreting interactions.* Newbury Park, CA: Sage Publications.

Akaike, H. (1987). Factor analysis and AIC. *Psychometrika, 52,* 317–332.

Algina, J., & Swaminathan, H. (1979). Alternatives to Simonton's analyses of the interrupted and multiple-group time-series design. *Psychological Bulletin, 86,* 919–926.

Allison, P. D. (1987). Estimation of linear models with incomplete data. In C. Cogg (ed.)., *Sociological Methodology 1987* (pp. 71–103). San Francisco: Jossey Bass.

Anderson, J. C., & Gerbing, D. W. (1984). The effect of sampling error on convergence, improper solutions, and goodness-of-fit indices for maximum likelihood confirmatory factor analysis. *Psychometrika, 49,* 155–173..

Asher, H. B. (1976). *Causal Modeling.* Beverly Hills, CA: Sage.

Balter, M. D., & Levine, J. (1971). Character and extent of psychotropic drug usage in the United States. Paper presented at the Fifth World Congress on Psychiatry, Mexico City.

Bartlett, M. S. (1941). The statistical significance of canonical correlations. *Biometrika, 32,* 29–38.

Bartlett, M. S. (1954). A note on the multiplying factors for various chi square approximations. *Journal of the Royal Statistical Society, 16* (Series B), 296–298.

Bearden, W. O., Sharma, S., & Teel, J. E. (1982). Sample size effects on chi-square and other statistics used in evaluating causal models. *Journal of Marketing Research, 19,* 425–430.

Beatty, J. R. (1977). Identifying decision-making policies in the diagnosis of learning disabilities. *Journal of Learning Disabilities, 10*(4), 201–209.

Belsley, D. A., Kuh, E., & Welsch, R. E. (1980). *Regression Diagnostics: Identifying Influential Data and Sources of Collinearity.* New York: Wiley.

Bem, S. L. (1974). The measurement of psychological androgyny. *Journal of Consulting and Clinical Psychology, 42,* 155–162.

Bendel, R. B., & Afifi, A. A. (1977). Comparison of stopping rules in forward regression. *Journal of the American Statistical Association, 72,* 46–53.

Benedetti, J. K., & Brown, M. B. (1978). Strategies for the selection of log-linear models. *Biometrics, 34*, 680–686.

Bennett, E., Beaurepaire, J., Langeluddicke, P. Kellow, J., & Tennant, C. (1991). Life stress and non-ulcer dyspepsia: A case-control study. *Journal of Psychosomatic Research, 35*, 579–590.

Bentler, P. M. (1980). Multivariate analysis with latent variables: Causal modeling. *Annual Review of Psychology, 31*, 419–456.

———. (1983). Some contributions to efficient statistics in structural models: Specifications and estimation of moment structures. *Psychometrika, 48*, 493–517.

———. (1988). Comparative fit indexes in structural models. *Psychological Bulletin, 107*, 238–246.

———. (1995). *EQS: Structural Equations Program Manual*. Encino, CA. Multivariate Software, Inc.

Bentler, P. M., & Bonett, D. G. (1980). Significance tests and goodness of fit in the analysis of covariance structures. *Psychological Bulletin, 88*, 588–606.

Bentler, P. M., & Dijkstra, T. (1985). Efficient estimation via linearization in structural models. In P. R. Krishnaiah (Ed.), *Multivariate analysis VI* (pp. 9–42). Amsterdam: North-Holland.

Bentler, P. M., & Weeks, D. G. (1980). Linear structural equation with latent variables. *Psychometrika, 45*, 289–308.

Berry, W. D. (1993). *Understanding Regression Assumptions*. Newbury Park, CA: Sage.

Blalock, H. M. (Ed.) (1985). *Causal Models in Panel and Experimental Designs*. New York: Adline.

BMDP Statistical Software, Inc. (1992). *BMDP User's Digest*. Berkeley, CA: University of California Press.

Bock, R. D. (1966). Contributions of multivariate experimental designs to educational research. In Cattell, R. B., ed., *Handbook of Multivariate Experimental Psychology*. Chicago: Rand McNally.

———. (1975). *Multivariate Statistical Methods in Behavioral Research*. New York: McGraw-Hill.

Bock, R. D., & Haggard, E. A. (1968). The use of multivariate analysis of variance in behavioral research. In D. K. Whitla, ed., *Handbook of Measurement and Assessment in Behavioral Sciences*. Reading, Mass: Addison-Wesley.

Bollen, K. A. (1986). Sample size and Bentler and Bonett's nonnormed fit index. *Psychometrika, 51*, 375–377.

———. (1989a). A new incremental fit index for general structural equation models. *Sociological Methods & Research, 17*, 303–316.

———. (1989b). *Structural Equations with Latent Variables*. New York: Wiley.

Bonett, D. G., & Bentler, P. M. (1983). Goodness-of-fit procedures for evaluation and selection of log-linear models, *Psychological Bulletin, 93*(1), 149–166.

Boomsma, A. (1983). On the robustness of LISREL (maximum likelihood estimation) against small sample size and nonnormality. Ph.D. Thesis, University of Groningen, The Netherlands.

Box, G. E. P., & Cox, D. R. (1964). An analysis of transformations. *Journal of the Royal Statistical Society, 26*(Series B), 211–243.

Bozdogan, H. (1987). Model selection and Akaike's information criteria (AIC): The general theory and its analytical extensions. *Psychometrika, 52*, 345–370.

Bradley, J. V. (1982). The insidious L-shaped distribution. *Bulletin of the Psychonomic Society, 20*(2), 85–88.

————. (1984). The complexity of nonrobustness effects. *Bulletin of the Psychonomic Society, 22*(3), 250–253.

Bradley, R. H., & Gaa, J.P. (1977). Domain specific aspects of locus of control: Implications for modifying locus of control orientation. *Journal of School Psychology, 15*(1), 18–24.

Brown, C. C. (1982). On a goodness-of-fit test for the logistic model based on score statistics. *Communications in Statistics, 11*, 1087–1105.

Brown, M. B. (1976). Screening effects in multidimensional contingency tables. *Applied Statistics, 25*, 37–46.

Brown, R. D., Braskamp, L. A., & Newman, D. L. (1978). Evaluator credibility as a function of report style. *Evaluation Quarterly, 2*(2), 331–334.

Browne, M. W. (1975). Predictive validity of a linear regression equation. *British Journal of Mathematical and Statistical Psychology, 28*, 79–87.

Browne, M. W., & Cudeck R. (1993). Alternative ways of assessing model fit. In K. A. Bollen and J. S. Long (Eds.), *Testing Structural Models*. Newbury Park: Sage.

Burgess, E., & Locke, H. (1960). *The Family*. 2nd ed. New York: American Book.

Byrne, B. M., Shavelson, R. J., & Muthén, B. (1989). Testing for the equivalence of factor covariance and mean structures: The issue of partial measurement invariance. *Psychological Bulletin, 105*, 456–466.

Caffrey, B., & Lile, S. (1976). Similarity of attitudes toward science on the part of psychology and physics students. *Teaching of Psychology, 3*(1), 24–26.

Campbell, D. R., & Stanley, J. C. (1966). *Experimental and Quasi-experimental Designs for Research*. New York: Rand McNally.

Cattell, R. B. (1957). *Personality and Motivation Structures and Measurement*. Yonkers-on-Hudson, N.Y.: World Book.

————. (1966). The scree test for the number of factors. *Multivariate Behavioral Research, 1*, 245–276.

Cattell, R. B., & Baggaley, A. R. (1960). The salient variable similarity index for factor matching. *British Journal of Statistical Psychology, 13*, 33–46.

Cattell, R. B., Balcar, K. R., Horn, J. L., & Nesselroade, J. R. (1969). Factor matching procedures: An improvement of the *s* index; with tables. *Educational and Psychological Measurement, 29*, 781–792.

Cattin, P. (1980). Note on the estimation of the squared cross-validated multiple correlation of a regression model. *Psychological Bulletin, 87*(1), 63–65.

Chou, C. P., & Bentler, P. M. (1993). Invariant standardized estimated parameter change for model modification in covariance structure analysis. *Multivariate Behavioral Research, 28*, 97–110.

Cohen, J. (1994). The earth is round (p < .05). *American Psychologist, 49*, 997–1003.

Cohen, J., & Cohen, P. (1983). *Applied Multiple Regression/Correlation Analysis for the Behavioral Sciences. 2nd Ed.* New York: Erlbaum.

Cohen, P., Gaughran, E., & Cohen, J. (1979). Age patterns of childbearing: A canonical analysis. *Multivariate Behavioral Research, 14*, 75–89.

Collins, L. M. & Hoen, J. L. (1991). *Best methods for the analysis of change: Recent advances, unanswered questions, future directions*. American Psychological Association: Washington, D.C.

Comrey, A. L. (1962). The minimum residual method of factor analysis. *Psychological Reports, 11*, 15–18.

Comrey, A. L. & Lee, H. B. (1992). *A first course in factor analysis.* (2nd. Ed.) Hillsdale, NJ: Erlbaum.

Cooke, T. D., & Campbell, D. T. (1979). *Quasi-experimentation: Design and Analysis Issues for Field Settings.* Chicago: Rand McNally College Publishing Co.; Boston: Houghton Mifflin.

Cooley, W. W., & Lohnes, P. R. (1971). *Multivariate Data Analysis.* New York: John Wiley & Sons.

Cornbleth, T. (1977). Effects of a protected hospital ward area on wandering and nonwandering geriatric patients. *Journal of Gerontology, 32*(5), 573–577.

Costanza, M. C., & Afifi, A. A. (1979). Comparison of stopping rules in forward stepwise discriminant analysis. *Journal of the American Statistical Association, 74,* 777–785.

Curtis, B., & Simpson, D. D. (1977). Differences in background and drug use history among three types of drug users entering drug therapy programs. *Journal of Drug Education, 7*(4), 369–379.

Dillon, W. R., & Goldstein, M. (1984). *Multivariate Analysis: Methods and Applications.* New York: John Wiley & Sons.

Dixon, W. J., ed. (1992). *BMDP Statistical Software: Version 7.0.* Berkeley: University of California Press.

Duffy, D. L., Martin, N. G., Battistutta, D., Hopper, J. L., & Mathews, J.D. (1990). Genetics of asthma and hay fever in Australian twins. *American Review of Respiratory Disease, 142,* 1351–1358.

Dukes, R. L., Ullman, J. B., & Stein, J. A. (in press). An evaluation of D.A.R.E. (Drug Abuse Resistance Education) using a Solomon Four-Group Design with latent variables. *Evaluation Review.*

Edwards, A. L. (1976). *An Introduction to Linear Regression and Correlation.* San Francisco: Freeman.

Era, P., Jokela, J., Qvarnverg, Y., & Heikkinen E. (1986). Pure-tone threshold, speech understanding, and their correlates in samples of men of different ages. *Audiology, 25,* 338–352.

Eysenck, H. J., & Eysenck, S. B. G. (1963). *The Eysenck Personality Inventory.* San Diego, Calif.: Educational and Industrial Testing Service; London: University of London Press.

Fava, J. L., & Velicer, W. F. (1992). An empirical comparison of factor, image, component, and scale scores. *Multivariate Behavioral Research, 27,* 301-322.

Featherman, D. (1973). Metrics of occupational status reconciled to the 1970 Bureau of Census Classification of Detailed Occupational Titles (based on Census Technical Paper No. 26, "1970 Occupation and Industry Classification Systems in Terms of their 1960 Occupational and Industry Elements"). Washington, D. C.: Government Printing Office. (Update of Duncan's socioeconomic status metric described in Reiss, J., et al., *Occupational and Social Status.* New York: Free Press of Glencoe).

Fidell, S. (1978). Nationwide urban noise survey. *Journal of the Acoustical Society of America, 16*(1), 198–206.

Fillenbaum, G. G., & Wallman, L. M. (1984). Change in household composition of the elderly: a preliminary investigation. *Journal of Gerontology, 39,* 342–349.

Fornell, C. (1979). External single-set components analysis of multiple criterion/multiple predictor variables. *Multivariate Behavioral Research, 14,* 323–338.

Fox, J. (1991). *Regression Diagnostics.* Newbury Park, CA: Sage.

Frane, J. W. (August 3, 1977). Personal communication. Health Sciences Computing Facility, University of California, Los Angeles.

———. (1980). The univariate approach to repeated measures-foundation, advantages, and caveats. BMD Technical Report No. 69. Health Sciences Computing Facility, University of California, Los Angeles.

Fullagar, C., McCoy, D., & Shull, C. (1992). The socialization of union loyalty. *Journal of Organization Behavior, 13*, 13–26.

Giarrusso, R. (1977). The effects of attitude similarity and physical attractiveness on romantic attraction and time perception. Masters thesis, California State University, Northridge.

Gini, C. (1912) *Variabilité Mutabilitá: Contributo allo Studio delle Distribuzioni e delle Relazioni Statistiche.* Bologna: Cuppini.

Glock, C., Ringer, B., & Babbie, E. (1967). *To Comfort and to Challenge.* Berkeley: University of California Press.

Goodman, L. A. (1978). *Analyzing Qualitative/Categorical Data.* Cambridge, MT: Abt Books.

Goodman, M. J., Chung, C. S., & Gilbert, F. (1974). Racial variation in diabetes mellitus in Japanese and Caucasians living in Hawaii. *Journal of Medical Genetics, 11,* 328–338.

Goodman, M. J., Steward, C. J., & Gilbert, F. Jr. (1977). Patterns of menopause: A study of certain medical physiological variables among Caucasian and Japanese women living in Hawaii. *Journal of Gerontology, 32*(3), 291–298.

Gorsuch, R. L. (1983). *Factor Analysis.* Hillsdale, N. J.: Erlbaum.

Gray-Toft, P. (1980). Effectiveness of a counseling support program for hospice nurses. *Journal of Counseling Psychology, 27*, 346–354.

Green, S. B. (1991). How many subjects does it take to do a regression analysis? *Multivariate Behavioral Research, 26,* 499-510.

Greenhouse, S. W., & Geisser, S. (1959). On the methods in the analysis of profile data. *Psychometrika, 24*, 95–112.

Guadagnoli, E., & Velicer, W. F. (1988). Relation of sample size to the stability of component patterns. *Psychological Bulletin, 103,* 265-275.

Haberman, S. J. (1982). Analysis of dispersion of multinomial responses. *Journal of the American Statistical Association, 77*, 568–580.

Hakstian, A. R., Roed, J. C., & Lind, J. C. (1979). Two-sample T^2 procedure and the assumption of homogeneous covariance matrices. *Psychological Bulletin, 86,* 1255–1263.

Hall, S. M., Hall, R. G., DeBoer, G., & O'Kulitch, P. (1977). Self and external management compared with psychotherapy in the control of obesity. *Behavior Research and Therapy, 15,* 89–95.

Harlow, L. L., & Newcomb, M. D. (1990). Towards a general hierarchical model of meaning and satisfaction in life. *Multivariate Behavioral Research, 25*, 387–405.

Harman, H. H. (1967). *Modern Factor Analysis* (2nd ed.). Chicago: University of Chicago Press.

———. (1976). Modern Factor Analysis (3rd edition). Chicago: University of Chicago Press.

Harman, H. H., & Jones, W. H. (1966). Factor analysis by minimizing residuals (Minres). *Psychometrika, 31,* 351–368.

Harris, R. J. (1975). *Primer of Multivariate Statistics.* New York: Academic Press.

Heise, D. R. (1975). *Causal Analysis.* New York: John Wiley & Sons.

Hoffman, D., & Fidell, L. S. (1979). Characteristics of androgynous, undifferentiated, masculine and feminine middle class women. *Sex Roles, 5*(6), 765–781.

Hosmer, D. W., & Lemeshow, S. (1989). *Applied Logistic Regression.* New York: John Wiley & Sons.

Hu, L-T., Bentler, P. M., & Kano Y. (1992). Can test statistics in covariance structure analysis be trusted? *Psychological Bulletin, 112*, 351–362.

Hull, C. H., & Nie, N. H. (1981). SPSS Update 7–9. New York: McGraw-Hill.

Huynh, H., & Feldt, L. S. (1976). Estimation of the Box correction for degrees of freedom from sample data in the randomized block and split-plot designs. *Journal of Educational Statistics, 1,* 69–82.

Jakubczak, L. F. (1977). Age differences in the effects of palatability of diet on regulation of caloric intake and body weight of rats. *Journal of Gerontology, 32*(1), 49–57.

James, L. R., Mulaik, S. A., & Brett, J. M. (1982). *Causal Analysis: Assumptions, Models, and Data.* Beverly Hills: Sage.

Johnson, G. (1955). An instrument for the assessment of job satisfaction. *Personnel Psychology, 8,* 27–37.

Jöreskog, K. G., & Sörbom, D. (1988). *LISREL 7: A Guide to the Program and Applications.* Chicago: SPSS Inc.

Kaiser, H. F. (1970). A second-generation Little Jiffy. *Psychometrika, 35,* 401–415.

———. (1974). An index of factorial simplicity. *Psychometrika, 39,* 31–36.

Keppel, G. (1991). *Design and Analysis: A Researcher's Handbook* (3rd .ed). Englewood Cliffs, N. J.: Prentice-Hall.

Keppel, G., & Zedeck, S. (1989). *Data Analysis for Research Designs: Analysis of Variance and Multiple Regression/Correlation Approaches.* New York: Freeman.

Kish, L. (1965). *Survey Sampling.* New York: John Wiley & Sons.

Knoke, D., & Burke, P. J. (1980). *Log-linear Models.* Beverly Hills, CA: Sage.

Langner, T. (1962). A 22-item screening score of psychiatric symptoms indicating impairment. *Journal of Health and Human Behavior, 3,* 269–276.

Larzelere, R. E., & Mulaik, S. A. (1977). Single-sample test for many correlations. *Psychological Bulletin, 84*(3), 557–569.

Lawley, D. N., & Maxwell, A. E. (1963). *Factor Analysis as a Statistical Method.* London: Butterworth.

Lee, S. Y., Poon, W.Y., & Bentler, P. M. (1994). Covariance and correlation structure analyses with continuous and polyomous variables. *Multivariate analysis and its applications, 24,* 347–356.

Lee, W. (1975). *Experimental Design and Analysis.* San Francisco: Freeman.

Levine, M. S. (1977). *Canonical Analysis and Factor Comparison.* Beverly Hills: Sage.

Locke, H., & Wallace, K. (1959). Short marital-adjustment and prediction tests: Their reliability and validity. *Marriage and Family Living, 21,* 251–255.

Lunneborg, C. E. (1994). *Modeling Experimental and Observational Data.* Belmont, CA: Duxbury Press.

MacCallum, R. (1986). Specification searches in covariance structure modeling. *Psychological Bulletin, 100,* 107–120.

McArdle, J. J., & McDonald, R. P. (1984). Some algebraic properties of the Reticular Action Model for moment structures. British Journal of Mathematical and Statistical Psychology, *37,* 234–251.

McDonald, R. P. (1989). An index of goodness-of-fit based on non-centrality. *Journal of classification, 6,* 97–103.

McDonald, R. P., & Marsh, H. W. (1990) Choosing a multivariate model: Noncentrality and goodness of fit. *Psychological Bulletin, 107,* 247–255.

McLeod, J. M., Brown, J. D., & Becker, L. B. (1977). Watergate and the 1974 congressional elections. *Public Opinion Quarterly, 41,* 181–195.

McNeil, K. S., Kelly, F. J., & McNeil, J. T. (1975). *Testing Research Hypotheses Using Multiple Linear Regression.* Carbondale: Southern Illinois University Press.

Maki, J. E., Hoffman, D. M. & Berk, R. A. (1978). A time series analysis of the impact of a water conservation campaign. *Evaluation Quarterly, 2*(1), 107–118.

Marascuilo, L. A., & Levin, J. R. (1983). *Multivariate Statistics in the Social Sciences: A Researcher's Guide.* Monterey, Ca.: Brooks/Cole.

Mardia, K. V. (1971). The effect of nonnormality on some multivariate tests and robustness to non-normality in the linear model. *Biometrika, 58*(1), 105–121.

Marshall, S. P. (1983). Sex differences in mathematical errors: an analysis of distractor choices. *Journal for Research in Mathematics Education, 14*, 325–336.

Menozzi, P., Piazza, A., & Cavalli-Sforza, L. (1978). Synthetic maps of human gene frequencies in Europeans. *Science, 201,* 786–792.

Merrill, P. F., & Towle, J. J. (1976). The availability of objectives and performance in a computer-managed graduate course. *Journal of Experimental Education, 45*(1), 12–29.

Miller, J. K., & Farr, S. D. (1971). Bimultivariate redundancy: A comprehensive measure of inter-battery relationship. *Multivariate Behavioral Research, 6*, 313–324.

Milligan, G. W. (1980). Factors that affect Type I and Type II error rates in the analysis of multidimensional contingency tables. *Psychological Bulletin, 87,* 238–244.

Milligan, G. W., Wong, D. S., & Thompson, P. A. (1987). Robustness properties of nonorthogonal analysis of variance. *Psychological Bulletin, 101*(3), 464–470.

Mitchell, L. K., & Krumboltz, J. D. (1984, April). The effect of training in cognitive restructuring on the inability to make career decisions. Paper presented at the meeting of the Western Psychological Association, Los Angeles, CA.

Moser, C. A., & Kalton, G. (1972). *Survey Methods in Social Investigation.* New York: Basic Books.

Mosteller, F., & Tukey, J. W. (1977). *Data analysis and Regression.* Reading, Mass.: Addison-Wesley.

Mulaik, S. A. (1972). *The Foundation of Factor Analysis.* New York: McGraw-Hill.

Mulaik, S. A., James, L. R., Van Alstine, J., Bennett, N., Lind, S., & Stillwell, C. D. (1989). An evaluation of goodness of fit indices for structural equation models. *Psychological Bulletin, 105,* 430–445.

Muthén, B., Kaplan, D., & Hollis, M. (1987). On structural equation modeling with data that are not missing completely at random. *Psychometrika, 52,* 431–462.

Myers, J. L., & Well, A. D. (1991). *Research Design and Statistical Analysis.* New York: HarperCollins.

Nicholson, W., & Wright, S. R. (1977). Participants' understanding of the treatment in policy experimentation. *Evaluation Quarterly, 1*(2), 245–268.

Nie, N. H., Hull, C. H., Jenkins, J. G., Steinbrenner, K., & Bent, D. H. (1975). *Statistical Package for the Social Sciences.* 2nd ed. New York: McGraw-Hill.

Norušis, M. J. (1990). *SPSS Advanced Statistics User's Guide.* Chicago: SPSS Inc.

O'Kane, J. M., Barenblatt, L., Jensen, P. K., & Cochran, L. T. (1977). Anticipatory socialization and male Catholic adolescent sociopolitical attitude. *Sociometry, 40*(1), 67–77.

Olson, C. L. (1976). On choosing a test statistic in multivariate analysis of variance. *Psychological Bulletin, 83*(4), 579–586.

———. (1979). Practical considerations in choosing a MANOVA test statistic: A rejoinder to Stevens. *Psychological Bulletin, 86,* 1350–1352.

Overall, J. E., & Spiegel, D. K. (1969). Concerning least squares analysis of experimental data. *Psychological Bulletin, 72*(5), 311–322.

Overall, J. E., & Woodward, J. A. (1977). Nonrandom assignment and the analysis of covariance. *Psychological Bulletin, 84*(3), 588–594.

Pallas, A. M., Entwisle, D. R., Alexander, K. L., & Cadigan, D. (1987). Children who do exceptionally well in first grade. *Sociology of Education, 60,* 257–271.

Pavkov, T.W., Lewis, D. A., & Lyons, J. S. (1989). Psychiatric diagnoses and racial bias: An empirical investigation. *Professional Psychology: Research and Practice, 20,* 364–368.

Pessemier, E. A., Bemmoar, A. C., & Hanssens, D. M. (1977). Willingness to supply human body parts: Some empirical results. *Journal of Consumer Research, 4*(3), 131–140.

Pope, K. S., Keith-Spiegel, P., & Tabachnick, B. (1986). Sexual attraction to clients: The human therapist and the (sometimes) inhuman training system. *American Psychologist, 41,* 147–158.

Porter, J. R., & Albert, A. A. (1977). Subculture or assimilation? A cross-cultural analysis of religion and women's role. *Journal for the Scientific Study of Religion, 16,* 234–359.

Prentice, R. L. (1976). A generalization of the probit and logit methods for dose response curves. *Biometrika, 32,* 761–768.

Price, B. (1977). Ridge regression: Application to nonexperimental data. *Psychological Bulletin, 84*(4), 759–766.

Rahe, R. H. (1974). The pathway between subjects' recent life changes and their near-future illness reports: Representative results and methodological issues. In B. S. Dohrenwend & B. P. Dohrenwend, eds., *Stressful Life Events: Their Nature and Effects.* New York: John Wiley & Sons.

Rao, C. R. (1952). *Advanced Statistical Methods in Biometric Research.* New York: John Wiley & Sons.

Rock, D. L. (1982). Appointments for intake interviews: an analysis of appointment-keeping behavior at an urban community mental health center. *Psychological Reports, 50,* 863–868.

Rollins, B., & Feldman, H. (1970). Marital satisfaction over the family life cycle. *Journal of Marriage and Family, 32,* 29–38.

Rosenberg, M. (1965). *Society and the Adolescent Self-Image.* N.J.: Princeton University Press.

Rosenthal, R., & Rubin, D. B. (1986). Meta-analytic procedures for combining studies with multiple effect sizes. *Psychological Bulletin, 99*(3), 400–406.

Rotter, J. B. (1966). Generalized expectancies for internal versus external control of reinforcement. *Psychological Monographs, 80*(1, Whole No. 609).

Rozeboom, W. W. (1979). Ridge regression: Bonanza or beguilement? *Psychological Bulletin, 82*(6), 242–249.

Rummel, R. J. (1970). *Applied Factor Analysis.* Evanston: Northwestern University Press.

SAS Institute Inc. (1990). SAS/STAT User's Guide, Version 6, 4th Edition. Cary, North Carolina: SAS Institute Inc.

St. Pierre, R. G. (1978). Correcting covariables for unreliability. *Evaluation Quarterly, 2*(3), 401–420.

Satorra, A. & Bentler, P. M. (1988). Scaling corrections for chi-square statistics in covariance structure analysis. *Proceedings of the American Statistical Association,* 308–313.

Schall, J. J., & Pianka, E. R. (1978). Geographical trends in numbers of species. *Science, 201,* 679–686.

Scheffé, H. A. (1953). A method of judging all contrasts in the analysis of variance. *Biometika, 40,* 87–104.

Shannon, C. E. (1948). A mathematical theory of communication. *Bell System Technical Journal, 50,* 379–423 and 623–656.

Singh, B., Greer, P. R., & Hammond, R. (1977). An evaluation of the use of the law in a free society materials on "responsibility." *Evaluation Quarterly, 1*(4), 621–628.

Sörbom, D. (1974). A general method for studying differences in factor means and factor structures between groups. British Journal of Mathematical and Statistical Psychology, 27, 229–239.

————. (1982). Structural equation models with structured means. In K. G. Jöreskog & H. Wold (Eds.), *Systems under indirect observation: Causality, structure, prediction I* (pp. 183–195). Amsterdam: North-Holland.

Sörbom, D., & Jöreskog, K. G. (1982). The use of structural equation models in evaluation research. In C. Fornell (Ed.), *A Second Generation of Mulitvariate Analysis: Vol 2. Measurement and Evaluation* (pp. 381–418). New York: Praeger.

Spence, J., & Helmreich, R. (1972). The attitudes toward women scale: An objective instrument to measure attitude towards rights and roles of women in contemporary society. *Journal Supplementary Abstract Service (Catalogue of selected Documents in Psychology), 2,* 66.

SPSS Inc. (1991). SPSS^x Statistical Algorithms, 2nd. ed.. Chicago: SPSS Inc.

————. (1986). SPSS^x User's Guide, Edition 2. New York: McGraw-Hill.

————. (1990). *SPSS Reference Guide.* Chicago: SPSS Inc.

————. (1993). *SPSS Base System Syntax Reference Guide, Release 6.0.* Chicago: SPSS Inc.

Steiger, J. H. (1980). Tests for comparing elements of a correlation matrix. *Psychological Bulletin, 87*(2), 245–251.

Steiger, J. H., & Browne, M. S. (1984). The comparison of interdependent correlations between optimal linear composites. *Psychometrika, 49,* 11-24.

Stein, J. A., Newcomb, M. D., & Bentler, P. M. (1993) Differential effects of parent and grandparent drug use on behavior problems of male and female children. *Developmental Psychology, 29,* 31–43.

Steinberg, D., & Colla, P. (1991). *LOGIT: A Supplementary Module for SYSTAT.* Evanston, IL: SYSTAT Inc.

Stewart, D., & Love, W. (1968). A general canonical index. *Psychological Bulletin, 70,* 160–163.

Strober, M. H., & Weinberg, C. B. (1977). Working wives and major family expenditures. *Journal of Consumer Research, 4*(3), 141–147.

Tabachnick, B. G., & Fidell, L. S. (1983). *Using Multivariate Statistics.* New York: Harper and Row.

————. (1989). Using Multivariate Statistics (2nd ed.). New York: HarperCollins.

Tanaka J. S. (1993) Multifaceted conceptions of fit. In K. A. Bollen and J. S. Long (Eds.), *Testing Structural Models.* Newbury Park: Sage.

Tanaka J. S., & Huba G. J. (1989). A general coefficient of determination for covariance structure models under arbitrary GLS estimation. *British Journal of Mathematical and Statistical Psychology, 42,* 233–239.

Tatsuoka, M. M. (1971). *Multivariate Analysis: Techniques for Educational and Psychological Research.* New York: John Wiley & Sons.

————. (1975). Classification procedures. In D. J. Amick and H. J. Walberg, eds., *Introductory Multivariate Analysis.* Berkeley: McCutchan.

Theil, H. (1970). On the estimation of relationships involving qualitative variables. *American Journal of Sociology, 76,* 103–154.

Thompson, C. M. (1985). Characteristics associated with outcome in a community mental health partial hospitalization program. *Community Mental Health Journal, 21,* 179–188.

Thurstone, L. L. (1947). *Multiple Factor Analysis.* Chicago: University of Chicago Press.

Timm, N. H. (1975). *Multivariate Statistics with Applications in Education and Psychology.* Belmont, CA.: Brooks/Cole.

Vandenbos, G. R., & Stapp, J. (1983). Service providers in psychology: Results of the 1982 APA human resources survey. *American Psychologist, 38,* 1330–1352.

Vaughn, G. M., & Corballis, M. C. (1969). Beyond tests of significance: Estimating strength of effects in selected ANOVA designs. *Psychological Bulletin, 72*(3), 204–213.

Velicer, W. F., & Jackson, D. N. (1990). Component analysis vs. common factor analysis: Some issues in selecting an appropriate procedure. *Multivariate Behavioral Research, 25*, 1-28.

Wade, T. C., & Baker, T. B. (1977). Opinions and use of psychological tests: A survey of clinical psychologists. *American Psychologist, 32*(10), 874–882.

Waternaux, C. M. (1976). Asymptotic distribution of the sample roots for a nonnormal population. *Biometrika, 63*(3), 639–645.

Wernimont, P. F., & Fitzpatrick, S. (1972). The meaning of money. *Journal of Applied Psychology, 56*(3), 218–226.

Wesolowsky, G. O. (1976). *Multiple Regression and Analysis of Variance*. New York: Wiley-Interscience.

Wherry, R. J., Sr. (1931). A new formula for predicting the shrinkage of the coefficient of multiple correlation. *Annals of Mathematical Statistics, 2*, 440–457.

Wiener, Y., & Vaitenas, R. (1977). Personality correlates of voluntary midcareer change in enterprising occupation. *Journal of Applied Psychology, 62*(6), 706–712.

Wilkinson, L. (1979). Tests of significance in stepwise regression. *Psychological Bulletin, 86*(1), 168–174.

———. (1990). *SYSTAT: The System for Statistics*. Evanston, IL: SYSTAT, Inc.

Wilkinson, L., & Dallal, G. E. (1981). Tests of significance in forward selection regression with an *F*-to-enter stopping rule. *Technometrics, 23*(4), 377–380.

Wilkinson, L., & Hill, M. (1994a). *SYSTAT for DOS: Advanced Applications, Version 6 Edition*. Evanston, IL: SYSTAT, Inc.

———. (1994b). *SYSTAT for DOS: Using SYSTAT, Version 6 Edition*. Evanston, IL: SYSTAT, Inc.

———. (1994c). *SYSTAT for DOS: Quick Reference, Version 6 Edition*. Evanston, IL: SYSTAT, Inc.

Williams, L. J., & Holahan, P. J. (1994). Parsimony-based fit indices for multiple-indicator models: Do they work? *Structural Equation Modeling, 1*, 161–189.

Willis, J. W., & Wortman, C. B. (1976). Some determinants of public acceptance of randomized control group experimental designs. *Sociometry, 39*(2), 91–96.

Wilson, E. B. & Hiferty, M. M. (1931). The distribution of chi-square. *Proceedings of the National Academy of Science, 17*, 694.

Winer, B. J. (1971). *Statistical Principles in Experimental Design*. 2nd ed. New York: McGraw-Hill.

Wingard, J. A., Huba, G. J., & Bentler, P. M. (1979). The relationship of personality structure to patterns of adolescent substance use. *Multivariate Behavioral Research, 14*, 131–143.

Wong, M. R., & Allen, T. A. (1976). A three dimensional structure of drug attitudes. *Journal of Drug Education, 6*(2), 181–191.

Woodward, J. A., & Overall, J. E. (1975). Multivariate analysis of variance by multiple regression methods. *Psychological Bulletin, 82*(1), 21–32.

Young, R. K., & Veldman D. J. (1981). *Introductory Statistics for the Behavioral Sciences*. 4th ed. New York: Holt, Rinehart and Winston.

INDEX